FOUNDATIONS
OF NEW WORLD CULTURAL ASTRONOMY

FOUNDATIONS of NEW WORLD CULTURAL ASTRONOMY

A READER WITH COMMENTARY

Edited by
ANTHONY AVENI

UNIVERSITY PRESS OF COLORADO

© 2008 by Anthony Aveni

Published by the University Press of Colorado
5589 Arapahoe Avenue, Suite 206C
Boulder, Colorado 80303

All rights reserved

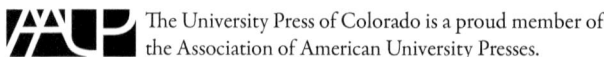 The University Press of Colorado is a proud member of
the Association of American University Presses.

The University Press of Colorado is a cooperative publishing enterprise supported, in part, by Adams State College, Colorado State University, Fort Lewis College, Mesa State College, Metropolitan State College of Denver, University of Colorado, University of Northern Colorado, and Western State College of Colorado.

Library of Congress Cataloging-in-Publication Data

Foundations of new world cultural astronomy : a reader with commentary / edited by Anthony Aveni.
   p. cm.
Includes index.
ISBN 978-0-87081-900-1 (pbk. : alk. paper) 1. Astronomy, Ancient. 2. Archaeoastronomy. 3. Astronomy—Social aspects. I. Aveni, Anthony F.
QB16.F69 2008
520—dc22

                                                              2008020144

Design by Daniel Pratt

Every reasonable effort has been made to trace the ownership of all copyrighted material included in this anthology. Any errors that may have occurred are inadvertent and will be corrected in subsequent editions, provided notification is sent to the publisher.

To Diane Janney, my extraordinary assistant
without whom I could never keep my alignments straight.

CONTENTS

Preface and Acknowledgments—*Anthony Aveni*   xi
Introduction: The Unwritten Record—*Anthony Aveni*   1

Part I: Archaeoastronomy: Establishing a Method and Applying a Paradigm—*Anthony Aveni*   13

  1. Astronomical Alignment of the Big Horn Medicine Wheel—*John A. Eddy*   19
  2. Geometry and Astronomy in Prehistoric Ohio—*Ray Hively and Robert Horn*   39
  3. Archaeoastronomy at Machu Picchu—*D. S. Dearborn and R. E. White*   61

Part II: Acquiring Cultural Context—*Anthony Aveni*   71

  4. The Inca Calendar—*R. T. Zuidema*   77
  5. Horizon Astronomy in Incaic Cuzco—*Anthony Aveni*   109

CONTENTS

6. Here Comes the Sun: The Cuzco–Machu Picchu
   Connection—*David S.P. Dearborn and Katharina J. Schreiber* — 125
      On Seeing the Light: A Reply to *Here Comes the Sun* by Dearborn
   and Schreiber—*Anthony Aveni* — 158
      Blinded by the Light: A Response to *On Seeing the Light* by
   Anthony Aveni—*David S. P. Dearborn* — 162
      Here Comes the Sun: the Cuzco–Machu Picchu Connection: A
   Reply to Dearborn and Schreiber—*R. T. Zuidema* — 168
      Stand and Deliver: A Comment on Zuidema's Reply to Dearborn
   and Schreiber—*David S. P. Dearborn* — 171
      On Shadows of Doubt: A Reply to Dearborn's Review of *Inkawasi:
   The New Cuzco*—*John Hyslop* — 173
      Lengthening Shadows: A Response to Hyslop's Comments—*David
   S.P. Dearborn* — 177
7. Chankillo: A 2,300-Year-Old Solar Observatory in Coastal
   Peru—*Ivan Ghezzi and Clive Ruggles* — 181
8. Keeping the Sacred and Planting Calendar: Archaeoastronomy in the
   Pueblo Southwest—*Michael Zeilik* — 199
9. Native Astronomy in Mesoamerica—*Michael D. Coe* — 231
10. The Role of Astronomical Orientation in the Delineation of World
    View: A Center and Periphery Model—*Anthony Aveni* — 253
11. Astronomical Alignments at the Templo Mayor of Tenochtitlan,
    Mexico—*Ivan Šprajc* — 273
12. Astronomical Observations from the Temple of the Sun—*Alonso
    Mendez, Edwin L. Barnhart, Christopher Powell, and Carol Karasik* — 305
13. Astronomy, Ritual, and the Interpretation of Maya "E-Group"
    Architectural Assemblages—*James J. Aimers and Prudence M. Rice* — 347
14. A Reflection of the Ancient Mesoamerican Ethos—*Miguel León-
    Portilla* — 383

Part III: The Role of Ethnoastronomy—*Anthony Aveni* — **395**

15. Current Astronomical Practices among the Maya—*Judith A.
    Remington* — 399
16. Quichean Time Philosophy—*Barbara Tedlock* — 413
17. Astronomical Models of Social Behavior among Some Indians of
    Colombia—*G. Reichel-Dolmatoff* — 425
18. The *Hooghan* and the Stars—*Trudy Griffin-Pierce* — 439
19. Animals and Astronomy in the Quechua Universe—*Gary Urton* — 457
20. Culture Confronts Nature in the Dialectical World of the
    Tropics—*Billie Jean Isbell* — 485

21. Ethnoastronomy and the Problem of Interpretation: A Zuni Example—*M. Jane Young* — 495

Part IV: The Classic Maya: A Testing Ground for Precise Astronomy in the Written Record—*Anthony Aveni* — 503

22. Ancient Maya Ethnoastronomy: An Overview of Hieroglyphic Sources—*John S. Justeson* — 509
23. Astronomical Knowledge and Its Uses at Bonampak, Mexico—*Floyd G. Lounsbury* — 565
24. A Palenque King and the Planet Jupiter—*Floyd G. Lounsbury* — 587
25. Archaeoastronomical Implications of an Agricultural Almanac in the Dresden Codex—*Victoria R. Bricker and Harvey M. Bricker* — 607

Part V: Cultural Astronomy's Greatest Mysteries—*Anthony Aveni* — 617

26. Between the Lines: Reading the Nazca Markings as Rituals Writ Large—*Anthony Aveni and Helaine Silverman* — 621
27. Possible Rock Art Records of the Crab Nebula Supernova in the Western United States—*John C. Brandt, Stephen P. Maran, Ray Williamson, Robert S. Harrington, Clarion Cochran, Muriel Kennedy, William J. Kennedy, and Von Del Chamberlain* — 635
28. A Thousand Years of the Pueblo Sun-Moon-Star Calendar—*Florence Hawley Ellis* — 647
29. Astronomical Markings at Three Sites on Fajada Butte—*Anna P. Sofaer and Rolf M. Sinclair* — 669
30. Romancing the Stone, or Moonshine on the Sun Dagger—*John B. Carlson* — 693
31. Lunar Standstills at Chimney Rock—*J. McKim Malville, Frank W. Eddy, and Carol Ambruster* — 711

Part VI: The Present and Future of Cultural Astronomy—*Anthony Aveni* — 721

32. The Study of Cultural Astronomy—*Clive L.N. Ruggles and Nicholas J. Saunders* — 725
33. Cosmograms and Maya City Planning: Selected Articles—*Michael E. Smith, Wendy Ashmore, Jeremy A. Sabloff, and Ivan Šprajc* — 751
34. Archaeology and Astronomy: A View from the Southwest—*W. James Judge* — 793

35. I Wasn't Going to Say Anything, but Since You Asked:
    Archaeoastronomy and Archaeology—*Keith W. Kintigh*   803
       Nobody Asked, but I Couldn't Resist: A Response to Keith Kintigh
    on Archaeoastronomy and Archaeology—*Anthony Aveni*   806

Index   **809**

PREFACE

In 2006–2007 the tiny visitor's center near the normally desolate Anasazi outpost known as Chimney Rock, located in the San Juan National Forest in southern Colorado, burgeoned with cosmic curiosity seekers. They had flown into Albuquerque and Durango, driven and bussed to the site to witness a major lunar standstill known as a "lunistice"—a stretching to the limit of the moon's position on the horizon. It happens every 18.6 years and was due to peak during that interval.

What made the moon's appearance so spectacular, and the reason why sellout crowds paid $50 apiece to hike up to the place where they could document one of the downbeats in heaven's eternal rhythm, was that when the moon did rise, it would be framed perfectly between the lofty sandstone spires of Chimney Rock. The viewpoint was an accommodating wide platform atop the thousand-year-old Great House, an edifice made of 6 million stones. Can it be that the Anasazi, who built the great city of Chaco Canyon a thousand years ago, were so preoccupied with astronomy that they saw fit to travel 90 miles to the top of a 7,500-foot-high mesa, far from access to water and firewood, just to build an enormous observatory high in the Rockies?

Lunistice was also celebrated by visitors to the Hopewell Mounds in Newark, Ohio. There the moon makes its appearance along the main axis between the Circle and Octagon of the Newark Earthworks, appearing through an opening in the octagon as viewed from the Observatory Mound at the opposite end of the complex. The same at Chaco Canyon, New Mexico, where another lunar standstill was witnessed over the Chacoan building known as Pueblo del Arroyo. And at the Maya ruins of Chichén Itzá, more than 50,000 celebrants visit the Castillo pyramid on the afternoon of the spring equinox to watch the shadow of a descending serpent cast by one corner of the structure on its northwest balustrade—a spectacular hierophany, or manifestation of sacred phenomena in the landscape. Were the Chimney Rock, Newark, Chaco, and Chichén astronomical events guaranteed by willful design on the part of their ancient builders? Or are they fabrications of the supercharged imaginations of contemporary skywatchers, now exploited by a hungry tourist industry? How can we decide? What is the evidence? All of these questions are part of the business of cultural astronomy.

Actually, the search for ancient observatories is a relatively minor item on the agenda of today's cultural astronomer, but it is the one people hear most about, thanks to the popular media. *Foundations of New World Cultural Astronomy* tells, in the words of the investigators themselves, the story of how and why so much attention has been drawn to the many ways ancient cultures of the world understood the heavens around them, and how we have come to know, with varying degrees of certainty, the motives of the people who expressed their awareness of what happens in the sky. I hope that reading the foundational articles I have selected and assembled will offer insight into the diverse ways ancient and contemporary non-Western cultures expressed and interpreted what they saw when they looked up.

ACKNOWLEDGMENTS

To Kristin Landau, Diane Janney, Lorraine Aveni: thank you for helping me put this work together, and to Darrin Pratt, Laura Furney, Dan Pratt, the rest of the staff, and several anonymous readers: thank you for all the helpful suggestions and implementation.

ANTHONY AVENI
AUGUST 2008
HAMILTON, NEW YORK

FOUNDATIONS
OF NEW WORLD
CULTURAL ASTRONOMY

INTRODUCTION

The Unwritten Record

Anthony Aveni

What area of study can be more exciting than one that combines looking up at the stars and digging down into the earth for ancient remains? Archaeoastronomy, the study of the practice of astronomy in ancient civilizations gleaned from both the written and the unwritten record, had its beginnings in the 1960s, a time of intellectual ferment and questioning of established foundations. This was also the heyday of processual archaeology, which drew heavily on the role played by the environment's effect on culture change and adaptation. Quantitative methods and scientific rigor were other staples of processualism.

Conceived in the aftermath of the great Stonehenge controversy (was Great Britain's most famous ancient monument a scientific observatory, a sacred temple—or both?), the interdiscipline of archaeoastronomy (then called "astroarchaeology") took root. Since then, archaeoastronomy has had more than its share of wild claims and crackpot ideas. Always a part of the diet of fantasy pop literature, it has frequently (and in many cases justifiably) been attacked by critics.

After nearly half a century of tumultuous activity, I believe archaeoastronomy has emerged in a more rational, controlled form as one of the hallmarks of true interdisciplinary research, thanks to the integration of the attitudes and methodologies of the many disciplines that have contributed to it. With an increased emphasis on the role of symbolism and belief systems, it has adapted to the hallmarks of the newer postprocessual archaeology. Along with ethnoastronomy, its sister interdiscipline, which deals with aspects of astronomical practice in contemporary cultures, archaeoastronomy offers novel ways of examining the pursuit of knowledge about the natural world in other societies. Indeed, my own students have told me that studying the astronomies of cultures other than our own is like looking into a mirror at our own scientific approach to the study of nature.

With an ever-increased focus on the development of astronomy in the context of the culture in which it is embedded, archaeo- and ethnoastronomy have now merged under the more aptly descriptive new name of "cultural astronomy," reflected in the title of this text. Especially since the mid-1990s, there has been a resurgence of interest in the astronomies and the role of cosmology in native North, Meso-, and South American cultures. (Mesoamerica covers the distinct culture area that ranges roughly from the U.S.-Mexico border through Central America.) But accessing the research corpus that lies at the foundation of these interdisciplinary fields has proven difficult for at least two reasons. First, the results of investigations in diverse culture areas appear in far-flung specialized journals; and second, many of the proceedings from conferences and seminars held across the Americas and Europe that contain seminal articles are published in costly volumes that are not readily available, even to teachers and their students, not to mention lay readers.

Based on a thorough review of the literature, *Foundations of New World Cultural Astronomy* represents my attempt to bring together some of the most important contributions in these areas. I have tried to balance my choices by including approximately equal numbers of articles from the three basic culture areas (North, Meso-, and South America) and by incorporating a fair share of contemporary selections, even though much of the basic groundwork was laid in the late decades of the twentieth century. Also, I have woven the selections together with helpful commentary designed both to offer some background for uninitiated readers and to direct them to updated resource materials on topics of special interest. Not all scholars will agree with my choices. For want of space, the omission of some important articles will be duly noted. In my introduction and brief interpretive essays to each part of the reader, I cite some of the excluded material and provide a context for it.

Having taught courses in archaeoastronomy for four decades, I am especially interested to provoke my student audience with "things to think about." Consequently, brief segments bearing that title follow each thematic section of the work. There I pose questions for discussion, essay topics, even a few quantitative problems and activities. I hope students and their teachers will devise more things worth thinking about.

I thought a great deal about how to organize my selections. Had I taken the easy route and arranged them by broad culture area, I would have passed up the opportunity to emphasize the common ground sought by the interdiscipline of cultural astronomy. Moreover, I felt that these studies have matured beyond the point of mere categorization by region. Instead I chose to focus the organizational plan for *Foundations* more on *process* than on *information*. Accordingly, I designed the volume not so much to give an up-to-date catalog of the latest results but rather to address the larger issue of how new areas of study originate and how they develop. Another design element of *Foundations* lies in my choice to follow many of the articles with "counterpoint pieces" that challenge the methods and/or conclusions in a given article. This emphasis on skepticism, colored by controversy, reflects a very real part of the story of how cultural astronomy (and indeed most interdisciplines) has progressed. I believe it is as important—not to mention as intellectually exciting—to expose the pitfalls as to demonstrate the progress in the study of cultural astronomy. I hope that by reading these original contributions by scholars who emanate from a wide variety of backgrounds and training, and consequently who pose different questions, readers will discover for themselves that cultural astronomy as we know it today has emerged out of encounters—indeed, intellectual struggles—among the diverse disciplines that have contributed to it.

To guide the reader through this historically based, process-oriented approach to cultural astronomy, I open with Part I, "Archaeoastronomy: Establishing a Method and Applying a Paradigm." The clash of the disciplines of astronomy and archaeology, which ultimately gave rise to the birth of archaeoastronomy as "the practice of astronomy in ancient civilizations using both the written and *unwritten* record" (Aveni 2001:2), actually began with the study of Stonehenge.

Like Sir Norman Lockyer (1894), the "outsider" who studied the orientations of Stonehenge and the Egyptian pyramids three generations before, astronomer Gerald Hawkins wrote a controversial piece in 1963 for the international scientific journal *Nature* titled "Stonehenge Decoded." He followed that with a popular book bearing the same title (Hawkins and White 1965). Based on an analysis of alignments between standing stones at the site, Hawkins concluded that the ancient structure was not so much a temple and a center of worship but rather a 5,000-year old astronomical observatory that precisely marked out the horizon limits of the sun and moon, as well as a computer devised to predict eclipses (Hawkins 1964).

Eschewing archaeological and historical evidence that posited scientific astronomy as a westward-moving phenomenon born in the Fertile Crescent, Hawkins's findings implied the existence of a sophisticated astronomy already in place on a remote island in the northwest of the Eurasian continent.

Archaeologists and historians responded both harshly and swiftly to Hawkins's invasion of their disciplinary turf. In a caustic critique, cleverly titled "Moonshine on Stonehenge" (which appeared in *Antiquity*, the premier international journal of archaeology and history, in 1966), archaeologist Richard Atkinson demonstrated how competence in one area of expertise is not necessarily transferable to another. Atkinson attacked Hawkins's claims to accuracy by noting the astronomer's use of a tourist map to acquire his alignments as well as misuse of the laws of probability to claim his million-to-one-odds certainty of conclusion. He also pointed out that Hawkins disregarded the archaeological record by connecting stones put in place in different epochs to create some of his alignments.

Historian Jacquetta Hawkes followed suit with her 1967 article cleverly titled "God in the Machine." She attacked Hawkins's emphasis on truth derived from computers as well as his lack of knowledge of historical context to back up his claims. Both critics offered a perspective on what kinds of questions the nonconverging disciplines of astronomy and archaeology normally addressed and how each highly valued its own epistemology. (The general subject of conflicting viewpoints among the organized disciplines is further dealt with in my book *Uncommon Sense* [Aveni 2006].)

Part I of *Foundations* consists of three articles that demonstrate how scholars trained largely in the sciences attempted to apply Hawkins's technique for discovering astronomical alignments in ancient architecture in the Americas. Their degrees of success should be judged by the reader. Two of these articles deal with North, and one with South, American sites. All follow the "Thom paradigm," a term I coined (Aveni 1988a) after Alexander Thom, who investigated a large number of alignments on megalithic structures in the British Isles that stressed precision but suffered from a lack of cultural context (e.g., Thom 1967, 1971).

Together these pieces raise interesting questions concerning what we can determine about the function of oddly shaped and/or oddly oriented structures based on standing architecture alone. I also chose them because they demonstrate, again to varying degrees, a deficiency of questions being asked by those who study culture and cultural remains, the anthropologist and the archaeologist, respectively. Following the old comparative tradition, the articles in this section emerge almost as a challenge to Old World archaeoastronomers. "Our" astronomers were indeed as skilled as yours! They too climbed a ladder of progress toward great intellectual heights. Technology, precision, and maybe even scientific theorizing about the nature of the universe fit in on this side of the Atlantic as well—the very points raised in the Stonehenge controversy.

Acquiring cultural context is the next major chronological theme in *Foundations*. It is addressed by the eleven pieces that make up Part II of *Foundations*. Articles by Mesoamerican archaeologist Michael Coe and Andean anthropologist R. Tom Zuidema show how the disciplines that derive from the humanities and the social sciences can serve as the basis for formulating hypotheses concerning the uses of astronomy in Meso- and South America, respectively. I also have included point-counterpoint pieces on the still-controversial status of ancient Andean astronomy (Dearborn and Schrieber; Aveni; Zuidema; and Hyslop 1986–1988). My piece on Cuzco tests a number of Zuidema's hypotheses in the field. Representing the U.S. Southwest, astronomer Michael Zeilik's article draws on a knowledge of Zuni cultural concepts to frame astronomical hypotheses. In "The Role of Astronomical Orientation in the Delineation of World View," I attempt to advance earlier work I did (Aveni 1975) by offering a social model for celestial alignments. The 2007 discovery of Chankillo as the oldest astronomically aligned site in South America is also presented, as are recent articles by Mendez and colleagues, Šprajc, and Aimers and Rice, which offer fresh arguments and new evidence about well-known astronomically related Mesoamerican structures in Maya Palenque, the ancient Aztec capital of Tenochtitlan, and the specialized Maya E Groups, respectively. Finally, Mexican humanist Miguel León-Portilla's short paper on the Mesoamerican ethos, which caps this section, offers a broad cultural perspective in which to frame such studies.

Especially in the past decade, those of us who work in cultural astronomy have come to realize the importance of the study of the cosmos in contemporary cultures other than our own. Part III comprises a collection of seven articles on the role of ethnoastronomy. Full of surprises, these pieces reveal how much richer the study of ancient astronomy might be if one could assume that the habits and folkways of ancient cultures that developed prior to Western influence had continued more or less unchanged up to the present. For example, we learn in Urton's piece, on Quechua animals and cosmology, that Andean skywatchers imagine "dark cloud" constellations, whereas Griffin-Pierce and Reichel-Dolmatoff demonstrate how Navajo and Amazonian people, respectively, pattern their houses after the celestial dome. Presentations by Remington and Tedlock reflect the progress that has been made in contemporary highland Maya ethnoastronomy in the decade that separates them, whereas folklorist Mary Jane Young, using Zuni data, addresses the ever more seminal issue of cultural continuity by responding to the problem of how to validate evidence acquired from contemporary indigenous people regarding their astronomical past. Finally, B. J. Isbell's piece underlines why ethnoastronomies studied in tropical latitudes may have peculiar aspects of their own. For a more detailed perspective on this issue, see Aveni (1981a). Other articles on ethnoastronomy, many of which have appeared in regional journals, are cited in the suggested reading list in Part III.

Whereas the archaeological record and studies of iconography are both a part of the *unwritten* record, and have contributed substantially to our understanding of the role of astronomy in culture, the database for astronomical studies in the Classical (Old) World has always resided in the *written* record. Writing undeniably offers exquisitely detailed information that can carry us much farther along the road toward understanding what ancient astronomers were up to, precisely who they were, exactly what they did, and what fundamental ends were served by their knowledge of the sky. Part IV focuses on the written record. Here I have brought together four articles on the Classic Maya, the only culture on the American continents demonstrated to have developed a writing system of coordinate rank with those of the Old World. Think of this section as a test case for the extent and practice of precise scientific astronomy in cultures other than our own. These articles are admittedly a bit more complex than those in the other parts, but then Maya astronomy is a complex affair and anyone who wishes to penetrate it must persist.

As J.E.S. Thompson (1974:97) candidly stated: "Maya astronomy is too important to be left to the astronomers." As you read the findings of John Justeson, Floyd Lounsbury, and Victoria Bricker and Harvey Bricker, all anthropologists who have worked on the frontiers of Maya epigraphy, note that each tries to address the important question, How were Maya astronomers like our own—and how were they different?

As I suggested earlier, one of the setbacks of archaeoastronomy as a professional interdiscipline had been its continued association, via the media, with unsolved mysteries. This is the topic addressed in Part V, titled "Cultural Astronomy's Greatest Mysteries." One notch down from runways for alien astronauts on the ladder of fantasy explanations for the Nasca lines is the idea that the giant geoglyphs etched on the pampas of southern Peru were part of the largest astronomy book in the world—a vast compendium of astronomical alignments. Astronomy is still the most well-recognized explanation, and has even been proffered as the sole interpretation in a number of textbooks. The short piece by Aveni and Silverman, centered on ground-based research, suggests that this hypothesis is largely untrue. Cultural astronomy, then, is not only about finding astronomically significant artifacts but also about revealing when artifacts studied are not astronomically significant.

North American petroglyphs in the U.S. Southwest that display a crescent moon alongside a star are still widely accepted by lay readers as ancient renditions of the 1054 Crab Nebula supernova explosion. The article by Brandt and colleagues explores this theme; however, anthropologist Florence Hawley Ellis's less well-known counterpoint piece to the "cataclysmic event hypothesis" offers an explanation that may have a better fit with what we actually know about the people who made the petroglyphs. The case of the Fajada Butte sun dagger is debated pro and con in a pair of pieces by Sofaer and Sinclair on the one hand, and Carlson on the other. Similar case studies have been picked up by the popular media; among them are the alignments at Chimney Rock (Colorado), dealt with in the article by Malville, Eddy, and Ambruster. Other such articles are cited in the bibliography.

Finally, Part VI offers a selection of articles that debate the future of cultural astronomy. Important among them is the keynote address by Ruggles and Saunders presented at the Oxford III Archaeoastronomy Conference, held in Scotland in 1990. It proposes for the first time renaming archaeo- and ethnoastronomy "cultural astronomy," a label more recently attached to one of the two international journals on these subjects. Now this label can be misleading, for it may lead some readers to think that our own scientific astronomy, by contrast, is not embedded in Western cultural traditions. Still, I believe this name is better suited to placing the study of astronomy in other cultures, past and present, where it belongs: in the domain of cultural anthropology. There follow more point-counterpoint pieces on the role of cosmology in Maya urban planning (Smith; Šprajc, Ashmore, and Sabloff). Archaeologist Jim Judge adds his perspective to the overemphasis on the archaeoastronomer's search for precision. As historian of astronomy Olaf Pedersen (1982:273) once wrote: "It is one thing to verify that a 'line' can be related to a certain azimuth with a definite degree of accuracy. It seems to be a very different thing to assume that this degree of accuracy was intended by the architect who first laid out the line." Finally, an agenda for dialogue among the disciplines that impinge upon cultural astronomy is offered in the form of a brief discussion between an archaeologist (Keith Kintigh) and an astronomer (myself) in a pair of short pieces written in 1992 that discuss the natural methodological mismatch between disciplines based in science versus social science.

The habit of seeking out ancestor-scientists who behaved like us has been the source of a number of critiques of those who practice cultural astronomy. These articles, along with others I cite in the introduction to this section and its bibliography, reveal the complexity of attempting to understand the role of astronomy in culture. Research needs to be directed toward broader questions capable of yielding a deeper knowledge of the place of astronomy in the ancient world.

Being as dedicated to teaching as I am to conducting research in the interdisciplines of archaeo- and ethnoastronomy, I offer, at the end of the overview essay that commences each of the six parts that make up *Foundations*, a list of "Things to Think About." Many of these topics have arisen out of questions posed by students I have encountered in my forty years of teaching archaeoastronomy both in the field and in the classroom. My broad goal is to help teachers and students gear their thoughts toward ways of pondering problems that arise in interdisciplinary learning and to understand how to balance two or more normally divergent approaches to acquiring human knowledge.

In order to enable readers to access relevant resources and to probe deeper into literature that could not be included in the reader for want of space, I also have provided each section with some Suggestions for Further Reading. The introductory section also contains a general bibliography. Included is my 2003 piece "Archaeoastronomy in the Ancient Americas," which itself contains a bibliography consisting of several hundred references for those who would like to probe further into particular areas. Finally, at the end of this introduction I offer "A Brief History of Archaeoastronomy," which fills in some of the background on the early history of what we today call "cultural astronomy."

Whether parts of *Foundations of New World Cultural Astronomy* are used in the classroom as supplementary reading in courses on astronomy, history of science, cultural anthropology, ethnohistory, and art history; in specialized courses in archaeo- and ethnoastronomy; or simply read cover to cover or in bits and pieces by the lay reader, I hope that the collection and commentary brought together here will help to convey the idea that knowledge of the world around us is as rich and varied as the communities that once sought—and still seek—it, and that the process of advancing knowledge in any healthy discipline or interdiscipline necessitates critical questioning. May cultural astronomy remain dynamic and ever changing!

A BRIEF HISTORY OF ARCHAEOASTRONOMY

Following conferences and edited collected works on archaeoastronomy in the Americas in the 1970s (cf., e.g., Aveni 1975, 1977; Williamson 1981), the first of what would become several Oxford International Conferences on Archaeoastronomy (1981) brought together for the first time scholars conducting research in both the New and Old Worlds. Perhaps reflecting differences in methodology, *two* edited volumes of conference proceedings were published (Aveni 1982; Heggie 1982). In the Old World (green-colored) volume, all sixteen contributed articles dealt with the possible astronomical orientations of alignments at prehistoric sites consisting of standing stones in the British Isles and Central Europe. But in the New World (brown-colored) volume, only three of the nine papers dealt substantially with alignments, two of them referring directly to historical and/or ethnohistorical material as a means of establishing explanations. Two-thirds of the latter papers discussed time reckoning in the context of some sort of written or pictographic record. In fact, one-third of the New World papers made either no reference at all or only an indirect reference to alignments. Yet, both these books carried the word "archaeoastronomy" in their titles.

When I introduced the proceedings of the second Oxford conference in 1986 (Aveni 1989; succeeding conferences adopted the place-name of the first conference), I characterized this development as the green-brown dichotomy: "Judging from the content of the European (megalithic) work (let us call it "green archaeoastronomy"), the principal inquiry would appear to consist of determining whether alignments among standing stones, collected carefully, precisely, and (usually) with due regard to the archaeological record, might bear any relationship to astronomical events at the horizon" (Aveni 1989:3). Furthermore,

> Much effort has been expended in green archaeoastronomy exploring the evolutionary question of defining the level of absolute intelligence attained by ancient people. For preliterate, low-technology civilizations, the definition of the level of astronomical intelligence is likened to a step on a pyramid. The setting of that level is based upon two false assumptions: (a) that there exists a *universal* scale of degrees of difficulty of perception of astronomical phenomena. Thus, recognition of the lunar standstills ranks as a higher achievement than that of the solar extrema, and precession of the equinoxes still higher; and (b) that levels of intelligence can be related to the degree of precision with respect to which a given phenomenon is observed. (1989:3–4)

On the other hand, brown (New World) archaeoastronomy drew much of its research corpus from laying down hypotheses based upon anthropological inquiry. It attempted to address broader questions, such as, What is the nature of the relationship between astronomical phenomena and cultural behavior? What did astronomical phenomenon X mean to the people who practiced it? Why were they interested in phenomenon X instead of Y? How did they conceive of that phenomenon in their ritual, myth, calendar, religion, architecture, and historical chronology? What role did it play in shaping their ideology? What did it mean to them? How did people employ phenomenon X as a way of creating order in their cultural reference frame? Historian of astronomy Olaf Pedersen had remarked that New World studies seemed to operate much like Old World history of astronomy, using data from written and picture manuscripts to extract information about astronomical practice (Pedersen 1982:266).

"We find ourselves in the archaeoastronomy of the mid-1980s in a curious quandary," I commented. "The questions asked by those trained in the quantitative sciences do not seem to elicit all that much interest in anthropological circles, and those astronomical matters that impress people engaged in the study of culture are given too little attention by the astronomers" (Aveni 1989:10). Each discipline pursues what it finds interesting. But was this important to the people whose astronomies they study?

Clive Ruggles (1984), although acknowledging perhaps too heavy an emphasis in green archaeoastronomy on seeking out astronomical alignments without paying much attention to whether other factors might have played a role in determining site plans and orientation (there are, after all, few other data to draw upon in megalithic astronomy), argued that brown archaeoastronomy should not employ other kinds of data as a substitute for scientific rigor.

To date, the number of Oxford conferences has run to eight in twenty-five years. Along with regional meetings in the Americas and abroad, and sessions on archaeoastronomy incorporated within conventions in the standard disciplines, the quality of work in all of archaeoastronomy has, in my view, improved greatly, thanks to collaborative projects devised by experts from diverse fields, along with more critical assessments of work offered for publication via the process of peer review. In the New World especially, a higher percentage of work in archaeo- and ethnoastronomy now appears in the archaeological and anthropological journals, and increasing numbers of graduate students in the standard disciplines have chosen various facets of cultural astronomy as dissertation topics (cf. McCluskey 2004 for details).

These developments prove that practicing the interdiscipline of cultural astronomy requires demonstrable knowledge of the culture about which one theorizes. The emphasis has switched from disclosing "firsts" to providing converging missing pieces that help give shape to the puzzle of cultural belief systems. Still, one need only read the latest keynote at the most recent Oxford conference, and my retort to it (Schaefer 2006; Aveni 2006a, both cited in Part VII), to appreciate the ongoing controversy about what cultural astronomy *is* and how one ought to *do* it.

BIBLIOGRAPHY

Atkinson, R. 1966. "Moonshine on Stonehenge." *Antiquity* 40:212–216.
Aveni, A., ed. 1975. *Archaeoastronomy in Pre-Columbian America*. Austin, University of Texas Press.
Aveni, A., ed. 1977. *Native American Astronomy*. Austin, University of Texas Press.
Aveni, A. 1981a. "Tropical Archaeoastronomy." *Science* 213:161–171.
Aveni, A. 1981b. "Archaeoastronomy." *Advances in Archaeological Method Theory* 4:1–81.
Aveni, A., ed. 1982. *Archaeoastronomy in the New World*. Cambridge, Cambridge University Press.
Aveni, A. 1988a. "The Thom Paradigm in the Americas: The Case of the Cross-Circle Designs." In *Records in Stone: Essays in Memory of Alexander Thom*, ed. C.L.N. Ruggles, 442–472. Cambridge, Cambridge University Press.
Aveni, A., ed. 1988b. *New Directions in American Archaeoastronomy*. Proceedings of the 46th International Congress of Americanists, Amsterdam, British Archaeological Reports International Series 454.

Aveni. A., ed. 1989. "Whither Archaeoastronomy?" In *Archaeoastronomy in the New World*, ed. A. Aveni, 3–12. Cambridge, Cambridge University Press.

Aveni, A. 2001. *Skywatchers: A Revised and Updated Version of Skywatchers of Ancient Mexico*. Austin, University of Texas Press.

Aveni, A. 2003. "Archaeoastronomy in the Ancient Americas." *Journal of Archaeological Research* 11(2):149–191.

Aveni, A. 2006. *Uncommon Sense: Understanding Nature's Truths across Time and Culture*. Boulder, University Press of Colorado.

Aveni, A. 2008. *People and the Sky: Our Ancestors and the Cosmos*. London, Thames and Hudson.

Aveni, A., and G. Brotherston, eds. 1983. *Calendars in Mesoamerica and Peru: Native Computations of Time*. British Archaeological Reports International Series S174.

Aveni, A., and G. Urton, eds. 1982. *Ethnoastronomy and Archaeoastronomy in the American Tropics*. New York Academy of Sciences, vol. 385.

Baity, E. C. 1973. "Archaeoastronomy and Ethnoastronomy so Far." *Cultural Anthropology* 14(4): 389–449.

Bostwick, T., and B. Bates, eds. 2006. *Viewing the Sky through Past and Present Cultures*. (Oxford VII) Pueblo Grande Anthropological Papers No. 15. Phoenix, Parks and Recreation Department.

Brecher, K., ed. 1979. *Astronomy of the Ancients*. Cambridge, MIT Press.

Broda, J., S. Iwaniszewski, and L. Maupomé, eds. 1991. *Arqueoastronomía y Etnoastronomía en Mesoamerica*. Mexico City, INAH, UNAM.

Chamberlain, V. del, J. Carlson, and M. J. Young, eds. 2005. *Songs from the Sky: Indigenous Astronomical and Cosmological Traditions of the World*. Austin, University of Texas Press.

Coyne, G., and R. Sinclair, eds. 1996. *The Inspiration of Astronomical Phenomena (INSAP) Vistas in Astronomy*. Special issue. Oxford, Elsevier.

Hawkes, J. 1967. "God in the Machine." *Antiquity* 41:174–180.

Hawkins, G. S. 1963. "Stonehenge Decoded." *Nature* 200:306–308.

Hawkins, G. S. 1964. "Stonehenge: A Neolithic Computer." *Nature* 202:1258–1261.

Hawkins, G. S., and J. B. White. 1965. *Stonehenge Decoded*. New York, Delta Dell.

Heggie, D., ed. 1982. *Archaeoastronomy in the Old World*. Cambridge, Cambridge University Press.

Iwaniszewski, S. 1989. "Exploring Some Anthropological and Theoretical Foundations for Archaeoastronomy." In *World Archaeoastronomy*, ed. A. Aveni, 27–37. Cambridge, Cambridge University Press.

Iwaniszewski, S. 2001. "Time and Space in Social Systems: Further Issues for Theoretical Archaeoastronomy." Proceedings of the SEAC 1998 Meeting, Dublin, Ireland, ed. C. Ruggles, F. Pendergast, and T. Ray, 1–7. UK, Ocarina.

Krupp, E. 1983. *Echoes of the Ancient Skies: The Astronomy of Lost Civilizations*. New York: Harper.

Krupp, E. 1997. *Skywatchers, Shamans and Kings*. New York: Wiley.

Krupp, E., ed. 1984. *Archaeoastronomy and the Roots of Science*. Boulder, Westview.

Lockyer, N. [1894] 1964. *The Dawn of Astronomy*. Cambridge, MA, MIT Press.

Malmstrom, V. 1997. *Cycles of the Sun, Mysteries of the Moon*. Austin, University of Texas Press.

McCluskey, S. 2004. "The Study of Astronomies in Cultures as Reflected in Dissertations and Theses." *Archaeoastronomy Journal of Astronomy in Culture* 18:20–27.

Milbrath, S. 1999. *Star Gods of the Maya: Astronomy in Art, Folklore and Calendars.* Austin, University of Texas Press.

Pedersen, O. 1982. "The Present Position of Archaeoastronomy." In *Archaeoastronomy in the Old World*, ed. D. Heggie, 265–274. Cambridge, Cambridge University Press.

Ruggles, C. 1984. "Megalithic Astronomy: The Last Five Years." *Vistas in Astronomy* 27:231–289.

Ruggles, C.L.N., ed. 1993. Archaeoastronomy in the 1990s. Loughborough, Group D.

Ruggles, C.L.N., and N. J. Saunders, eds. 1993. *Astronomies and Cultures.* Niwot, University Press of Colorado.

Selin, H. 2001. *Astronomy across Cultures: The History of Non-Western Astronomy.* Dordrecht, Kluwer.

Thom, A. 1967. *Megalithic Sites in Britain.* Oxford, Clarendon Press.

Thom, A. 1971. *Megalithic Lunar Observatories.* Oxford, Clarendon Press.

Thompson, J.E.S. 1974. "Maya Astronomy." In *The Place of Astronomy in the Ancient World,* ed. F. R. Hodson, 83–98. *Philosophical Transactions of the Royal Society, London* A 276.

Walker, C., ed. 1996. *Astronomy before the Telescope.* London, British Museum.

Williamson, R., ed. 1981. *Archaeoastronomy in the Americas.* Los Altos, CA, Ballena and College Park, Center for Archaeoastronomy.

Ziolkowski, M., and R. Sadowski, eds. 1989. *Time and Calendars in the Inca Empire.* Oxford, British Archaeological Reports International Series 479.

PART I

Archaeoastronomy
Establishing a Method and Applying a Paradigm

"If I can see any alignment, general relationship or use for the various parts of Stonehenge, then these facts were also known to the ancients" (Hawkins and White 1965: vii [see introduction]). Thus, brash young astronomer Gerald Hawkins unleashed his hypothesis that Stonehenge, Britain's famous wonder of the ancient world, was originally intended by its builders to be an astronomical observatory. He made that bold assertion in his popular book, *Stonehenge Decoded*, a spinoff best seller of the short, controversial piece of the same title that appeared in *Nature*, one of science's flagship journals.

In the youthful age of the electronic computer, the public, perhaps awed by the power of "The Machine" (the title of a chapter in Hawkins's book), enthusiastically embraced his notion of a cult of prehistoric scientists pursuing precise lunar alignments and integrating them into a gigantic Bronze Age computational device. We were discovering that our ancestors were just like us!

Setting history on its ear by placing the ancient Britons intellectually ahead of the cultures of the Middle East was difficult enough for some scholars to swallow (e.g., see historian Jacquetta Hawkes's 1967 critique of Hawkins's piece [see introduction], which addresses precisely this perspective), but stepping on the turf of an archaeologist and claiming to decode an artifact that one archaeologist in particular had spent his life studying is quite another matter. Richard Atkinson, then dean of the British megalithic archaeologists, leveled a scathing critique of Hawkins's work (Atkinson 1966 [see introduction]). He attacked not only the assumption that one can comprehend Stonehenge the way its builders did, but also the claimed level of accuracy. But was the attack too personal? Or was it simply a case of one reactionary practitioner being threatened by another from a different discipline? Were the criticisms from the archaeological quarter valid? Even though *Foundations* is about New World cultural astronomy, both Hawkins's and Atkinson's articles are important. They laid the foundation for a debate between archaeologists and astronomers that lasted for decades, and although it is not reproduced here, it should be consulted by those who have an interest in the general history of cultural astronomy.

Hawkins may have reignited the flame of controversy over whether ancient structures are astronomically aligned, but it was engineer Alexander Thom who exploited the method of alignment hunting in prehistoric structures. He did it in a series of books (Thom 1967, 1971 [see introduction]) and articles published in the British *Journal for the History of Astronomy* and its supplement, *Archaeoastronomy*. (The recent absorption of the supplement into the main journal may be a sign of the disciplinary mainstreaming of archaeoastronomy.) These works gave rise to what I have called, following Thomas Kuhn's *Structure of Scientific Revolutions* on how sciences develop, the "Thom paradigm"—a method of investigation that can be applied to the unwritten record elsewhere as a test for the astronomical orientation hypothesis (Aveni 1988). This method consists of searching out alignments related to astronomical phenomena presumed to have been observed universally, beginning with the solstice and equinox sunrises and sunsets followed by the lunar standstills.

Like Stonehenge, Big Horn Medicine Wheel, located in the mountains of Wyoming, is a stone ring. Moreover, there are some hints that parts of it (in this case "spokes" and cairns made out of stones rather than an avenue and slotted trilithons) might have functioned to point to astronomically significant horizon positions. Astronomer John Eddy's study of Big Horn, originally published in 1973 and reproduced in this section, shows how the search for alignments in unusual-looking structures came to be applied in pre-contact North American archaeological sites. For further details, readers should consult Williamson (1984:ch. 9). Although critiques of this piece (e.g., Ovenden and Rodger 1981; Vogt 1993) have placed some of Eddy's conclusions in doubt, his work still stands as a classic example of the application (or misapplication?) of the Stonehenge paradigm.

My second choice representing early North American archaeoastronomy was published in 1982 by physicist Ray Hively and philosopher Robert Horn. They conducted a search for solar and lunar alignments in the Hopewell Circle and Octagon earthworks in Newark, Ohio. To bolster their argument, a later paper (Hively and Horn 1984) revealed some of the same alignments in another Ohio earthwork located at High Bank. Readers who would like to follow the course of investigations of Hopewellian geometry and astronomy, including pieces by some of its critics, should consult Hively and Horn (2006) and Connolly and Lepper's (2004) edited volume (see especially the articles by Romain and Aveni).

South America is represented by "Archaeoastronomy at Machu Picchu," written by astronomers David Dearborn and Ray White. They report on an investigation of the alignment and positioning of a window in a curious D-shaped structure known as the "Torreón," located at the famous Inca ruins of Machu Picchu (see the discussion by Dearborn and Schreiber and others reprinted in Part II of this volume). For more on Andean archaeoastronomy, readers should consult Bauer and Dearborn's 1995 text.

Although not entirely devoid of cultural context, these pieces, written in the early stages of the development of archaeoastronomy, are largely concerned with showing that astronomy exists. Because most of these investigations were undertaken by individuals trained strictly in the sciences who, unfortunately, took minimal advantage of outside consultation and collaboration with professionals who study indigenous cultures and their remains, these selections exhibit a rather tentative and uncertain quality. Readers will understand why archaeologist Keith Kintigh (see Part VI) termed archaeoastronomy so much "butterfly collecting."

THINGS TO THINK ABOUT

1. What motives can you suggest for the Anasazi and Hopewell to construct observatories such as Chimney Rock and the Newark Circle-Octagon complex? Think beyond the basic wonder and awe that draw *us* to the sky—for example, think about politics, religion, kinship, social relations, and so forth.

2. Compare and contrast Hawkins's and Eddy's studies. Do you think Eddy heeded any of Atkinson's criticisms regarding Hawkins's analysis of Stonehenge?

3. Research the literature for some of the case studies mentioned in *Foundations*. For example, compare the content, the arguments, and the kinds of evidence presented in the earlier reports on archaeoastronomy at Chimney Rock (e.g., Malville 1991; Malville, Eddy, and Ambruster 1991) and later (e.g., Fairchild, Malville, and Malville 2006; Sutcliffe 1996). Do you find the case more convincing now than then? Why or why not? Trace the development of the investigations in these reports. What, if anything, has changed?

4. Exactly what are the criticisms leveled at Eddy for his work on the Medicine Wheel by Ovenden and Rodger (1981) and by Vogt (1993)? What alternative explanations do they offer? Do these explanations make more sense to you? Why or why not?

SUGGESTIONS FOR FURTHER READING

Aveni, A. 1975. "Possible Astronomical Orientations in Ancient Mesoamerica." In *Archaeoastronomy in Pre-Columbian America,* ed. A. Aveni, 163–190. Austin, University of Texas Press.

Aveni, A. 1988. "The Thom Paradigm in the Americas: The Case of the Cross-Circle Designs." In *Records on Stone: Papers in Memory of Alexander Thom,* ed. C. Ruggles, 442–472. Cambridge, Cambridge University Press.

Aveni, A. 2004. "Zapotec Astronomy: Reconsideration of an Earlier Study." *Archaeoastronomy: The Journal of Astronomy in Culture* 18:26–31.

Aveni, A. 2006a. "Evidence and Intentionality: On Method in Archaeoastronomy." In *Viewing the Sky through Past and Present Cultures,* ed. T. Bostwick and B. Bates, 57–70. Pueblo Grande Anthropological Papers No. 15, Phoenix.

Aveni, A. 2006b. "Schaefer's Rigid Ethnocentric Criteria." In *Viewing the Sky through Past and Present Cultures,* ed. T. Bostwick and B. Bates, 79–83. Pueblo Grande Anthropological Papers No. 15, Phoenix.

Aveni, A., S. Milbrath, and C. Peraza L. 2004. "Chichén Itzá's Legacy in the Astronomically Oriented Architecture of Mayapán." *RES* 45:123–143.

Bauer, B., and D. Dearborn. 1995. *Astronomy and Empire in the Ancient Andes.* Austin, University of Texas Press.

Bricker, H., and V. Bricker. 1996. "Astronomical References in the Throne Inscription of the Palace of the Governor at Uxmal." *Cambridge Archaeological Journal* 6(2):191–229.

Connolly, R., and B. Lepper, eds. 2004. *The Fort Ancient Earthworks: Prehistoric Lifeways of the Hopewell Culture in Southeastern Ohio.* Columbus, Ohio Historical Society.

Fairchild, G., J. M. Malville, and N. Malville. 2006. "Chimney Rock as a Ceremonial Center and Port-of-Trade within the Chaco System." In *Viewing the Sky through Past and Present Cultures,* ed. T. Bostwick and B. Bates, 275–286. Pueblo Grande Museum Anthropology Papers No. 15, Phoenix.

Hawkins, G. 1964. "Stonehenge: A Neolithic Computer." *Nature* 202:1258–1261.

Hively, R., and R. Horn. 1982. "Geometry and Astronomy in Prehistoric Ohio." *Archaeoastronomy* (supplement to *Journal for the History of Astronomy* 13) 4:S1–S20.

Hively, R., and R. Horn. 1984. "Hopewellian Geometry and Astronomy at High Bank." *Archaeoastronomy* (supplement to *Journal for the History of Astronomy* 15) 7:S856–S100.

Hively, R., and R. Horn. 2006. "A Statistical Study of Lunar Alignments at the Newark Earthworks." *Midcontinental Journal of Archaeology* 32(1):281–322.

Hoyle, F. 1979. *On Stonehenge.* London, Heinemann.

Krupp, E. C., ed. *In Search of Ancient Astronomies.* New York, McGraw-Hill.

Kuhn, T. S. 1970. *The Structure of Scientific Revolutions.* Chicago, University of Chicago Press.

Malville, J. M. 1991. "Prehistoric Astronomy in the American Southwest." *Astronomy Quarterly* 8:1–20.

Malville, J. M., F. Eddy, and C. Ambruster. 1991. "Lunar Standstills at Chimney Rock." *Archaeoastronomy* (supplement to *Journal for the History of Astronomy*) 16:543–550.

Ovenden, M., and D. Rodger. 1981. "Megaliths and Medicine Wheels." In *Proceedings of the 11th Chacmool Conference,* University of Calgary, ed. M. Wilson, 371–386. Calgary, Archaeological Association of the University of Calgary.

Ruggles, C. 1999. *Astronomy in Prehistoric Britain and Ireland.* New Haven, CT, Yale University Press.

Schaefer, B. 2006a. "Case Studies of Three of the Most Famous Claimed Archaeoastronomical Alignments in America." In *Viewing the Sky through Past and Present Cultures,* ed. T. Bostwick and B. Bates, 27–56. Pueblo Grande Anthropological Papers No. 15, Phoenix.

Schaefer, B. 2006b. "No Astronomical Alignments at the Caracol." In *Viewing the Sky through Past and Present Cultures,* ed. T. Bostwick and B. Bates, 71–78. Pueblo Grande Anthropological Papers No. 15, Phoenix.

Šprajc, I. 1993a. "The Venus-Rain-Maize Complex in the Mesoamerican World View: Part I." *Journal for the History of Astronomy* 24:17–70.

Šprajc, I. 1993b. "The Venus-Rain-Maize Complex in the Mesoamerican World View: Part II." *Archaeoastronomy* 18 (Supplement to *Journal for the History of Astronomy* 24):S27–S53.

Sutcliffe, R. 2006. "Evaluating the Chimney Rock Pueblo with Respect to Observing the Major Lunar Standstill Moonrises: Potential Architectural Encoding of Astronomical Knowledge." In *Viewing the Sky through Past and Present Cultures,* ed. T. Bostwick and B. Bates, 275–286. Pueblo Grande Museum Anthropology Papers No. 15, Phoenix.

Taube, K. 2001. "The Breath of Life: The Symbolism of Wind in Mesoamerica and the Southwest." In *The Road to Aztlan: Art from a Mythic Homeland*, ed. V. Fields and V. Zamudio-Taylor, 168–177. Los Angeles, Los Angeles County Art Museum.

Vogt, D. 1993. "Medicine Wheel Astronomy." In *Astronomies and Cultures,* ed. C.L.N. Ruggles and N. J. Saunders, 163–201. Niwot, University Press of Colorado.

Williamson, R. 1984. *Living the Sky: The Cosmos of the American Indian*. Boston, Houghton Mifflin.

CHAPTER ONE

Astronomical Alignment of the Big Horn Medicine Wheel

John A. Eddy

The stone Medicine Wheel in the Big Horn Mountains of northern Wyoming is a well-known archeological structure whose origin and purpose have long remained unexplained. It lies on an exposed shoulder of Medicine Mountain (44°49.6'N, 107°55.3'W) at an altitude of about 9640 feet (2940 meters), just above the timberline in the Big Horn National Forest. The "wheel" is a pattern on the surface of the ground, made up of an imperfect circle of stones, about 25 meters in diameter, with a central cairn about 4 meters in diameter. From this inner cairn or hub radiate 28 unevenly spaced spokes which connect to the rim. Five smaller cairns, each an open circle 1 to 1.5 meters in diameter and several courses high, are placed at irregular intervals along the periphery of the wheel. A sixth cairn, of similar construction, lies about 4 meters beyond the rim on an extended, southwestern spoke. Figure 1.1 is an early

From *Science* 184 (1974):1034–1043.

1.1 Photograph of the Medicine Wheel, about 1926, by Jessamine Spear Johnson, Sheridan, Wyoming, made when a stone wall protected the site. Source: Western History Research Center, University of Wyoming.

photograph (about 1926) of the site; Figure 1.2 is a plan view made as part of a recent archeological survey.[1]

BACKGROUND

The Medicine Wheel is generally believed to be of considerable antiquity and is commonly attributed to early Plains Indians, which might include Crow, Sioux, Arapahoe, Shoshone, or Cheyenne, all of whom lived nomadically in the region and for whom the Big Horn Mountains had especial significance. This association is based principally on the circumstances of its discovery, on its location—near a well-worn travois trail—and its yield of a few Amerindian artifacts. At the time of its discovery by whites in the late 19th century and in subsequent archeological investigation,[2] local Indians were found to be aware of the wheel but none interviewed knew its precise location or its purpose. They reported that it was there when they came, or that it had been made "by people who had no iron."

Less is known about the date of construction, although several factors indicate that it has been there for possibly 200 years. These include depositions taken at the time of its discovery, the weathered and partially sunken appearance of the stones in the pattern, and a tree-ring analysis (earliest date, AD 1760) of a piece of wood found in one of the cairns in 1958 by Grey.[3] In the 1958 study the interiors of all the cairns and the sectors between about one-third of the spokes were excavated. A few potsherds, beads, and points were found which were generally consistent with the

dendrochronological date, although these could represent a later use of the wheel. Beneath the central cairn was found a filled, conical hole in the bedrock about 1 meter deep, of the sort that could have served to step a vertical pole in the center of the cairn.

Perhaps the most interesting feature of the Big Horn Medicine Wheel is its unique nature, for the early inhabitants of the northern Great Plains were not known as builders or monument makers. The most common stone artifacts of the region are tipi rings, but these apparently utilitarian structures are simpler and much smaller than the spoked Medicine Wheel and are most probably unrelated to it.[4] Other stone structures are rare. They include surface effigy monuments,[5] linear stone alignments,[6] and a few other spoked wheels,[7] all found along the Rocky Mountains from northern Alberta to southern Wyoming. Of the spoked wheels found thus far, the Big Horn Wheel is clearly the largest and most elaborately constructed, and it is the only one with a rim and outlying cairns.

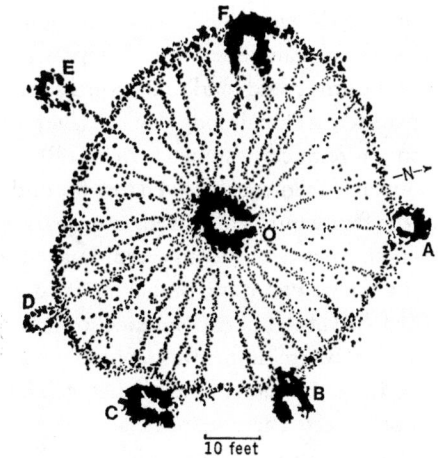

1.2 Plan of the Medicine Wheel with letters added, from a 1958 survey by D. Grey.

1.3 The Sun Dance Lodge (Medicine Lodge). Illustration from *The Sacred Pipe: Black Elk's Account of the Seven Rites of the Oglala Sioux,* Joseph Epes Brown, Ed. © 1953 by the University of Oklahoma Press.

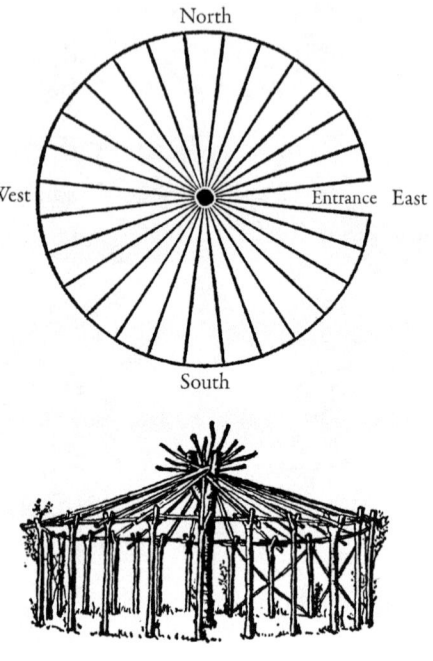

The word "medicine" was used by Indians to mean "magic" or "supernatural," and the Medicine Wheel is associated in most accounts with religious use. Its general similarity to the floor plan of a ceremonial Medicine Lodge[8] led Grinnell in 1922[9] to suggest that this may have been its pattern or original use (see Fig. 1.3). In this analogy, the central cairn (labeled 0 in Fig. 1.2) represents the center lodge pole and the 28 spokes represent the 28 rafters. Cairn F, which opens to the center on the west side, portrays an altar which was by tradition located there; cairn B, on the east and opening east, represents the entry which opened to the rising sun. Other cairns could mark traditional stations of worship; in particular, outlying cairn E marked the "lonely lodge" of the Cheyenne Medicine Lodge ceremony. This association was restated in similar terms by Robert Yellowtail, chairman of the Crow Tribal Council.[10] Yellowtail believed that the outlying cairns marked the (cardinal) directions of the four winds which were customary stations of fasting or worship. Based on his own knowledge of Indian ceremonial customs, he felt that the Medicine Wheel was most likely a two-dimensional replica of the Medicine Lodge (or Sun Dance Lodge) which had been built at one time to allow the observance of the Sun Dance ceremony at a place where timber was not easily available.

The association of the Medicine Wheel with the floor plan of a ceremonial lodge is not perfect. There are too many cairns, and they do not lie at or even very near the cardinal directions. The crucial cairns B and F are distinctly not diametrically opposite as are their counterparts in the Cheyenne Lodge. The ceremonial use of the site by large groups is also subject to doubt. There is little evidence that the site was ever occupied by any number of people for any length of time,[11] and indeed it marks a most inhospitable location—wind- and snow-swept and far from water and plentiful wood. The site seems better suited as a place of personal spiritual quest, in the custom of the Plains Indian, or, by virtue of its altitude and clear horizons, as a primitive astronomical observatory.

ASTRONOMICAL USE

Stone alignments such as the Medicine Wheel could have been used as horizon markers, to identify the directions of rise or set of selected celestial bodies. A pole stepped vertically in the central cairn could serve as a gnomon or foresight, which, in conjunction with a backsight point at a peripheral cairn, would define the azimuth of rising or setting of some important object.[12] We might first suspect the sun, because of its religious importance in Amerindian culture and its frequent association with the Big Horn Wheel in form, name, and legend. The spoked pattern resembles a common sun symbol. A Crow name for the Big Horn Wheel was "the Sun's Tipi," and in one Crow legend "the Sun built it to show us how to build a teepee."[13] "It was built before the light came," according to another reported Crow explanation[14] which was interpreted, perhaps, mistakenly, as indicating that the wheel was of ancient age. A more illuminating translation could well have been "before the light came," or "to mark the sunrise."

The horizon points of sunrise and sunset shift daily; their paths go through reversals of directions at the times of summer and winter solstice when the sun rises and sets at its northern and most southern positions, respectively. These singularities were commonly noted by many primitive peoples for calendar, ritual, or agricultural purposes. The solstitial alignments of Stonehenge and other European megalithic monuments, of Egyptian pyramids and temples, and of Mayan temples are by now generally recognized.[15] Recently, solstitial alignments have been proposed to explain architectural structures of the Anasazi in New Mexico and southern Colorado,[16] post-hole patterns at the Cahokia mounds in Illinois,[17] and the alignment of ceremonial circles on the central Kansas plain.[18] Contemporary Pueblo ritual has required the identification of the solstices by specific tribal officers who have used horizon markers for this purpose.[19]

Of the two solstices only during the summer solstice is there practical access to the Big Horn Wheel. Even then the trail up Medicine Mountain leads through frequent deep snow, and, if the experience of the last two summers is typical, much of the wheel itself can be covered at the time of the solstice. On 19 June 1972, a boot-deep snow fell on the mountain; on 18 June 1973 a blowing snowstorm halted traffic on the nearby highway. In both cases, although drifts were deep at lower points, the barren shoulder on which the wheel is built was swept clean by wind soon after the storm. For at least 3 days around the 1972 solstice almost the entire wheel was clear of snow; in 1973 several inches of snow covered the spokes on the eastern half although the cairns were clear. Frequent snow cover may explain the unique use of built-up cairns at the Big Horn Wheel, as opposed to more common surface alignments, and the selection of this windswept shoulder, since these would seem to ensure the visibility of the reference marks in summer snowfalls. Thus the fundaments of the Big Horn Wheel are probably the cairns; the spokes and rim, which for so long have captured the attention of observers and interfered with serious explanation, are likely of secondary importance and may well have been added later as a day counter or simply as a decoration.

How useful, meteorologically, is the Big Horn site for observing the sun at solstice? The chances of a clear horizon at sunrise or sunset are probably worse for a mountain location at high latitude than for known sites of solstice markers in the American Southwest, Mexico, or Egypt, but probably as good as Stonehenge, for example, which lies at higher northern latitude. At the time of summer solstice in 1972, one morning out of three at the Big Horn Wheel was clear at sunrise; in 1973, three mornings out of four were clear. The solar declination changes very slowly through the solstice—no more than 2' arc per day in the week before and after—and to the accuracy of stone-and-post alignments the solstice is not a sharply defined reversal but a pause of several days in the place of sunrise or sunset. One clear sunrise per week would probably suffice, particularly if the trend of shift in preceding and subsequent days were noted. These considerations would seem to establish the site as a practical one for watching for the summer solstice. But the choice of a cold and arduously reached mountaintop in preference to the equally usable nearby plains must be justified on other grounds—possibly mystical or purely aesthetic.

SOLSTICE CAIRNS

To test for possible astronomical alignment it is necessary first to establish the positions of the cairns and then to check for coincidences with important azimuths of rise and set. This test had apparently not been made at this site in the past; most earlier workers have concentrated on establishing the general pattern and history of the wheel, and on searching for surface or buried artifacts. The early maps by Simms[20] and Grinnell[21] were only approximate sketches, with no accurate compass reference. A transit survey of spoke angles was made in October 1917 by Stockwell, although never published.[22] The first comprehensive map was published in 1963 by Grey,[23] as a result of a 1958 survey of the site by the (amateur) Wyoming Archaeological Society. This map (Fig. 1.2) faithfully portrays the general plan and shows accurate relative bearings of spokes and cairns, although it contains an unfortunate error in scale (the distances shown are 15 percent too small) and the direction reference is approximate.

In the summer of 1972, using a surveyor's transit and steel tape, I made a new survey of the cairns at the Big Horn Wheel to establish their positions and directional alignments. The true compass reference was determined by observation of solar azimuths and checked by triangulation with landmarks from the U.S. Geological Survey chart. In addition, the solstice sunrise was observed on the morning of 20 June in line with cairns E and O, for which solstitial sunrise alignment was indicated. In the summer of 1973 I rechecked the true compass reference on the site, again with solar azimuths, and observed sunrise on 20, 21, and 22 June and sunset on 20 June.

The 1972–1973 survey is summarized in the first six lines of Table 1.1, which lists the position of the center of each peripheral cairn with reference to the center of the hub cairn O. Table 1.1 also gives the celestial declination circle which corresponds to the measured azimuth alignment, taking into account the effects of measured horizon dip angle and calculated refraction. In this procedure I have taken cairn O as a com-

John A. Eddy

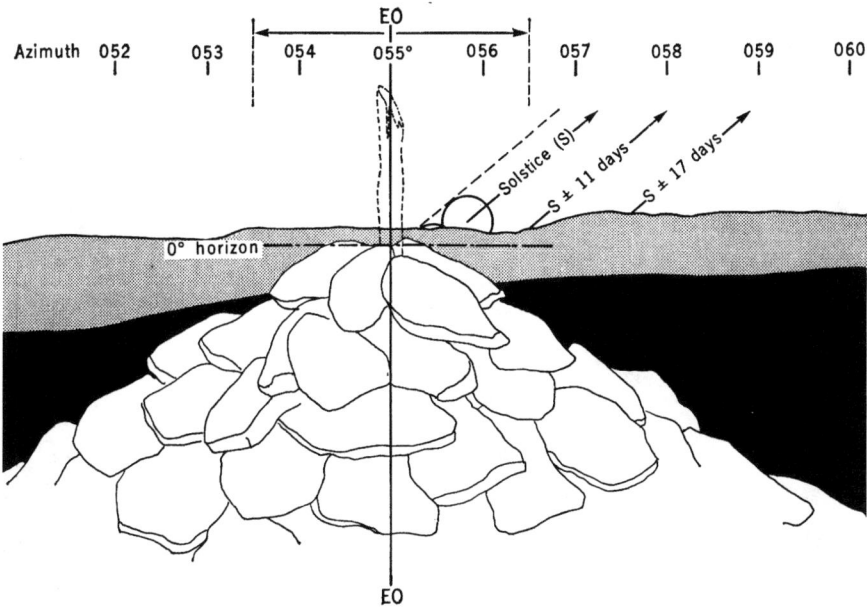

1.4 View of the Medicine Wheel horizon in the EO direction, showing calculated lines of rise for the center (solid line) and upper limb (dashed line) of the sun. The present position of the EO alignment is shown, with an hypothesized foresight pole in cairn O. Limits of certainty of the survey are ±1.5° (vertical dashed bars).

mon foresight for each peripheral backsight cairn. The survey indicates that the line EO intersects the horizon where an object of declination +23.6° rises, as compared to +23.4° for the solstice sun. This is illustrated in Figure 1.4. The other likely solstice sun alignment, line CO, corresponds to the setting point of declination +25.5°. This is about 2° from the expected value; however, the solstice sunset line from foresight O passes easily through the central part of the widely spread cairn C. In Table 1.1, I have given for the solstice lines EO and CO the difference between the measured alignment azimuth and that calculated for a declination of 23°26′, which measures the fit of these lines to the present position of the solstice. These alignments are convincingly confirmed by observation of the rising and setting solstice sun (Figs. 1.5 and 1.6). From backsight E one sees the flash of the dawn sun at the center of foresight O; from cairn C one sees the last of the setting sun very nearly in line with O. One should keep in mind that the original reference point in each cairn is not precisely known and has here been presumed to be the center of the present clear area in the middle of each cairn. For these rough, unstabilized cairns this reference point may not be precisely valid and in any case cannot be determined to better than ±15 centimeters, or about ± 1.5° in alignment. Moreover, we are ignorant of the diameter of post used for a gnomon and its manner of use, if indeed one was stepped in the central cairn as was suggested by Grey's excavation.

1.5 Solstice sunrise at Big Horn Medicine Wheel, 22 June 1973, with cairn E (backsight) in the foreground and cairn O (with tall vertical survey pole) a distant foresight. Curved steel fenceposts enclosed the wheel. Photograph by author.

The centers (defined above) of the distinctive cairn E and the central cairn are aligned to the solstice sunrise to a tolerance which is much less than my measurement uncertainty. The statistical probability of a chance coincidence of one of six arbitrary lines to a predetermined direction to within a small angular tolerance $\Delta°$ is 1: $(360/6\Delta_1)$ or, with $\Delta=0.3°$, 1 in 200. The center of the more amorphous cairn C is aligned to the central cairn in agreement with the solstice sunset to within an error that is about twice my estimated accuracy of fixing the cairn reference points. If the lines of sunrise and sunset at summer solstice are the most likely candidates for possible astronomical alignment of the Medicine Wheel, then the probability that by chance alone two or six peripheral cairns would be aligned to these directions to within the measured tolerances $\Delta_1 = 0.3°$ and $\Delta_2 = 3.4°$ is 1: $(360/6\Delta_1)(360/5\Delta_2)$, or less than 1 in 4000. Moreover, as will be demonstrated below, the coincidences of the other cairn alignments with other logical celestial risings further decrease the probability of chance placement of the cairns. When we allow for the unstabilized nature of the cairns and the present uncertainty in their original reference centers, the statistical argument for solstitial placement seems even more compelling. The case seems strong that these cairns of the Big Horn Medicine Wheel were built for the specific purpose of marking the summer solstice. Cairns E, O, and C, with but little additional specification of their real reference points, would have permitted their builders to identify the time of the solstice with a precision of several days. Such refinement would have followed quite naturally from repeated annual observation and use.

Table 1.1 Summary of 1972 survey of cairn positions.

Cairns		1972 survey			Celestial declination of rise (R) or set (S) (deg)	Indicated alignment (epoch 1972)			Azimuth difference (deg)
Backsight	Foresight	Separation (m)	Horizontal dip angle (deg)	Azimuth (deg)		Declination (deg)	Object	Declination difference (deg)	
E	O	13.4	+0.18	055.0	+23.6 (R)	+23.4	Solstice sunrise	0.2	0.3
C	O	12.5	−0.25	309.6	+25.5 (S)	+23.4	Solstice sunset	2.1	3.4
A	O	12.4	−0.5	196.5	−43.8 (S)		?		
B	O	12.3	−1.0	263.6	−5.7 (S)		?		
D	O	12.9	−0.2	349.5	+43.4 (S)		?		
F	O	12.1	+1.9	114.5	−15.7 (R)	−16.7	Sirius rise	1.0	1.6
F	A		−0.6	065	+16.6 (R)	+16.5	Aldebaran rise	0.1	0.1
F	B		0	099	6.7 (R)	−8.2	Rigel rise	1.5	2.1
E	B		−0.4	069	+14.1 (R)	+16.5	Aldebaran rise	2.4	3.8

1.6 Solstice sunset at the Big Horn Medicine Wheel, 20 June 1973, with cairn C in the foreground and cairn O (with tall survey pole) a distant foresight. The U.S. Forest Service signpost to the left of the survey pole stands outside the enclosure. Photograph by author.

OTHER CAIRNS

This leaves unexplained the four remaining peripheral cairns, of which cairns A, B, and F appear to equal the size, and presumably the significance, of the solstice pair, cairns E and C. We might suspect that the central cairn served as a foresight for each of these other cairns, to mark directions of rise or set of other prominent celestial features, such as the nodal points of the paths of the moon or planets, or the fixed points of rise or set of bright stars or asterisms. When the declinations derived in this way (lines 3 through 6 in Table 1.1) are compared with the declinations of known prominent celestial features, we find only one likely coincidence: backsight F with cairn O points to the rising point of the brightest star, Sirius. Cairns A, B, and D show no convincing alignments with the center cairn.

It is conceivable that cairn O, the center of the spoked, sun-symbol wheel, was chiefly used for solar marking, and that for other objects combinations of the peripheral cairns were used. In testing for astronomical coincidences in this way some logic or restraint should be used, since a large number of lines can be defined by pairing all possible cairns. A convincing case might be that in which one cairn is a common fore- or backsight for more than one other cairn, to mark a set of significant celestial objects. Of all combinations, only when cairn F is taken as a backsight is this criterion met. An observer at cairn F, the largest peripheral cairn, sees Aldebaran (magnitude

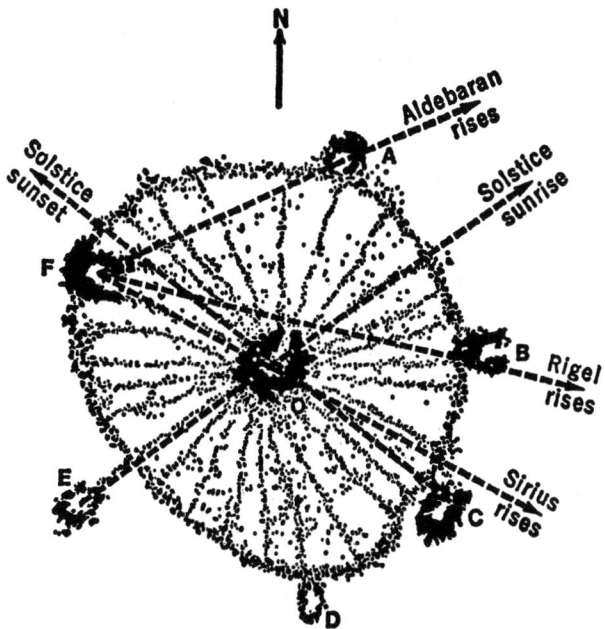

1.7 Summary of adopted cairn alignments, superposed as direction vectors on the diagram by D. Grey. The direction of north has been rectified about 2° to agree with the present survey. Indicated direction vectors accurately depict the true differences between presumed alignments and measured cairn positions.

0.8) rise over cairn A, Rigel (magnitude 0.11) rise near cairn B, and Sirius (magnitude −1.44) rise over cairn O (see Fig. 1.7 and lines 6, 7, and 8 of Table 1.1).

These three stars are the brightest in a compact region of the sky which is near the path of the sun in summer. Thus they rise near dawn at the only season of practical occupancy of the Big Horn Wheel (Fig. 1.8). The nearest stars of comparable brightness are Betelgeuse (variable magnitude 0.1 to 1.2, labeled point B in Fig. 1.8) and Capella (magnitude 0.2, labeled point C in Fig. 1.8). Although Capella is brighter than Aldebaran, it is circumpolar at the latitude of the Big Horn Mountains and hence cannot be marked at the horizon. Atmospheric extinction will limit the number of stars which can be observed near the horizon and will restrict the altitude angle at which even the brightest can be seen. This effect, which has been discussed by Thom,[24] will, for the elevation of the Big Horn Mountains, shift azimuths of apparent rise southward, about 0.5° for Aldebaran and negligibly for brighter Rigel and Sirius. For crude cairns these are insignificant changes. It is more to the point to recognize the effect as probably limiting likely summer dawn-marked stars to the five here, thus strengthening the significance of the apparent marking of Aldebaran, Rigel, and Sirius.

Aldebaran is the brightest star in the well-known Hyades asterism in Taurus. On 21 June it now rises, unseen in the dawn light, about 1 hour before the sun. At the

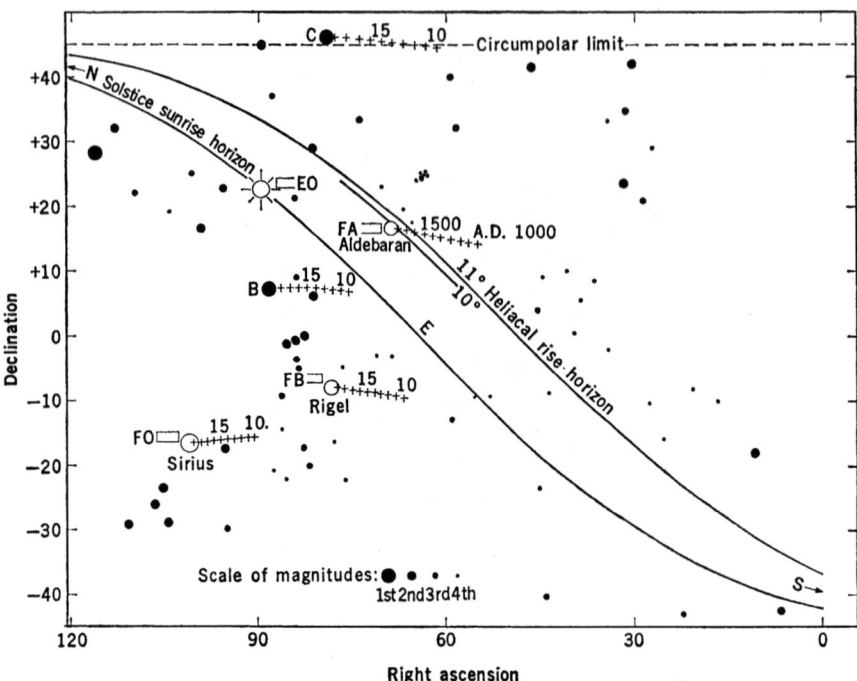

1.8 Diagram of the sky in the vicinity of the solstice sun, with horizon lines calculated for the latitude of the Big Horn Medicine Wheel. Star positions are for the present epoch, with additional positions at century intervals (AD 1000 to the present date) for Aldebaran, Rigel, Betelgeuse (B), and Capella (C), which is circumpolar. The position of the solstice sun has been taken as a fixed reference point in plotting the changing positions of these five brightest stars. Other stars participate in the same relative change, so that the constellation patterns remain essentially the same with time. Declinations (from Table 1) corresponding to cairn alignments EO, FA, FB, and FO are shown as lines of total width 1.5° adjacent to the probable alignment objects.

summer solstice in earlier epochs it rose earlier, in a darker sky, and at one time it rose just as the pre-dawn sky brightened to extinguish it. This momentary appearance at rising defines the heliacal rising of a star, which, because of the significant daily motion of the sun in right ascension, is a fairly well-defined point in time. The heliacal rising of the brightest stars was used for calendar reference by other early peoples. Perhaps the most commonly acknowledged use was in early Egypt, where at one time the heliacal rising of Sirius occurred at the summer solstice, which was also the season of the annual rise of the Nile.[25] Hipparchus reportedly made use of the changes in the heliacal rising times of certain stars to determine the difference between the lengths of the solar year and the sidereal year.[26]

At the Medicine Wheel the heliacal rising of Aldebaran—the only bright star to rise heliacally near the time of summer solstice—would have provided the only other celestial signal for the summer solstice;[27] moreover, with experience the observer, using the heliacal method, might increase the precision of the sunrise-azimuth method, since the daily solar motion in right ascension is greater than in declination. The heliacal method requires horizon pointers such as the cairns but not precise placements, since all that is needed is a rough indication of the place of rise to identify the star.

The FA cairn alignment could thus serve to point out the approximate place of rise of the one star whose momentary appearance near dawn would signal the solstice. Later in the summer, from the same backsight cairn, alignments FB and FO would point in similar fashion to the rising points of brighter Rigel and Sirius. Rigel, in familiar Orion, is the brightest object to rise heliacally after Aldebaran. It did so, at the latitude of the Big Horn Mountains, almost exactly one lunar month after Aldebaran's heliacal rise at solstice. Bright Sirius then rose heliacally about another lunar month later. It may be significant that there are 28 spokes in the Medicine Wheel, which could be used as day counters for these lunar-month intervals. It is at first surprising that stars were marked whose heliacal rising occurred after the solstice. Indeed, logical use might suggest instead the marking of presolstice events, to serve as warnings from which solstice countdowns could proceed. No such objects are marked, but there are two reasons why we should not expect them. One is the particular absence of bright stars in the region of the sky west of Aldebaran where presolstice heliacal rising would occur (Fig. 1.8). The other is the normally severe weather on the mountain in late spring, which would surely discourage use of the Medicine Wheel much before mid-June.

Why should Rigel and Sirius be marked? Perhaps simply because they were the brightest stars that rose in the summer dawn. Or, more logically, because their dates of heliacal rising would mark off the warmest moons of the year, the 2-month period after the solstice when the weather on the mountain was least severe. The heliacal rising of Sirius would be a good warning to leave the Medicine Wheel. We should also consider whether the Sirius, Rigel, and Aldebaran markers could have been laid down at different times by a sequence of different users, as the aspect of the sky changed with the precession of the equinoxes. Each of these stars at one time in the past rose heliacally at the solstice there—first Sirius, then Rigel, and, last of all, Aldebaran—but the periods between are measured in millennia and this is not consistent with the generally homogeneous construction of the Medicine Wheel or with reasonable estimates of its age.

An alternate interpretation of the peripheral cairn alignments does not require alignments on other than the sun and Aldebaran. It may be that cairn F was used only for the FA Aldebaran line, and that cairn B was sighted from cairn E, the solstice sunrise station, as another line to the same heliacal solstice star (Fig. 1.7). The line along EB is not as good an Aldebaran marker as the line along FA, but it is within 4°, close enough to permit identification (line 9 of Table 1.1). In this case, on days before the solstice, an observer at cairn E would note the daily shift of the sunrise point toward the EO line. From the same station, in the period before dawn, he could watch for the momentary appearance of a bright star (Aldebaran) along line EB. On the day of the solstice he would see both phenomena, about 1½ hours apart, in what would have been an impressive reaffirmation of the temporal coincidence of these two singularities of daybreak. At each subsequent dawn the ever-increasing persistence of Aldebaran in the predawn sky would confirm that the solstice was past, as would the slow shift of the sunrise point away from the line EO.

Cairn D remains the unused piece of the puzzle, for its relation to other cairns seems to fit no plausible sky alignment. It could be a crude attempt (11.5° error) to mark the celestial pole—a feat more difficult than the marking of objects that rise and set. In size and orientation this cairn appears anomalous, and it may have been a later addition, perhaps unrelated in purpose or of nonauthentic origin. The general appearance of the rest of the Medicine Wheel suggested to Grey[28] that at least some of the cairns may have been built before the rest of the structure. The uses proposed here make that seem likely, with phases of construction reflecting stages of refinement in its astronomical use. Cairns O, E, and C may once have stood alone, laid out by a first user to mark the rising and setting of the solstice sun, possibly on a lone pilgrimage or vision quest to the mountaintop. Cairns B, A, and F could have been added later by the same or different users for the added precision of Aldebaran's heliacal rise, and possibly for the mystical marking of Rigel and Sirius. The reported affinity of Amerindians for round objects[29] may explain the rim and spokes, which were likely added last. The 28 spokes were probably chosen to equal the days of the moon, for day counting or simply numerology. It should not surprise us that the wheel is not round (if it was built after the cairns), or that it meanders and flattens as though to lend significance to the more important cairns.

PROBABLE DATE OF CONSTRUCTION

The declination of the sun at solstice is a measure of the obliquity of the ecliptic, which changes with time. Since this is a well-known function, one might presume to establish the date of construction of a solstitial alignment such as the Medicine Wheel by making a sufficiently accurate measurement of alignment directions. An indicated azimuth of rise or set corresponds to a unique celestial declination, or obliquity, which in turn corresponds to a unique date in the terrestrial precession cycle. A number of workers[30] have cautioned against using such a procedure for sunrise or sunset markers. The change in the obliquity is so slow—about 1° in 8000 years—that the uncertainty in refraction and in the user's definition of sunrise introduces a chronological uncertainty of literally thousands of years, even for the most precise alignment determination. The situation is further degraded in the case of the Big Horn Medicine Wheel alignments by the crude nature of the cairns and our ignorance of their exact points of reference.

However, the implied use of Aldebaran as a heliacally rising star at summer solstice gives us a dating device which is not subject to these faults. This is a potentially powerful dating tool, since the change in Aldebaran's position with time is appreciable; moreover, the method is insensitive to the positional accuracy or stability of the cairns, once we are convinced that Aldebaran was so used. The absence of other bright candidate stars for the FA cairn alignment and the unique position of Aldebaran relative to that of the solstice sun leave little doubt that this was indeed the case.

In this dating method one uses the long-term change in the position of Aldebaran with respect to the solstice point, which comes about chiefly through precession of the equinoxes. Precession is such that at present the elongation angle between Aldebaran and the sun at solstice is decreasing at the rate of about 1.4° per century. Currently the separation is about 22°, such that, at the latitude of the Big Horn Mountains, Aldebaran rises about 9° before the sun on 21 June. As mentioned earlier, this is too close to permit one to observe the rising of a first-magnitude object, and Aldebaran at solstice now rises unseen in the predawn glow. Some time ago, when it rose approximately 11° before the sun, it was just far enough from the sun to rise heliacally on the solstice. Before that time, it rose even earlier, persisting longer before dawn and becoming less and less useful as a heliacally rising solstice marker. There is thus a period of time, centered on the time of nominal separation for heliacal rising, during which the star would have been useful. Neither the width of this time nor the separation angle for precise heliacal rising is an exact quantity, since they depend somewhat on the acuity of the observer's vision and on the variable brightness of the predawn sky. In taking 11° as a nominal separation angle for the heliacal rising of a first-magnitude star, I have followed the work of Lockyer.[31] A reasonable choice for an allowable tolerance about the value is ±1°.

Included in Figure 1.8 are the positions of Aldebaran (and of Rigel, Sirius, Betelgeuse, and Capella) from AD 1000 to the present in 100-year steps, from the calculations of Hawkins.[32] Figure 1.8 also shows the 11° heliacal horizon (the almucantar which lies 11° above the solar almucantar at summer solstice) and a portion of the 10° heliacal horizon for reference. Both of these horizon lines are drawn to include the effects of refraction and the measured dip angle in the FA cairn direction. Aldebaran was on the 11° line (for nominal first-magnitude heliacal rise) in about AD 1700. The uncertainty of ±1° in the value of the separation angle corresponds to an uncertainty of about ±200 years in date. Thus the use of Aldebaran as a heliacally rising solstice star implies that the FA cairn alignment was made in the period from AD 1500 to AD 1900 with a most probable value AD 1700. This agrees reasonably well with Grey's dendrochronological earliest date of AD 1760.[33]

AUTHENTICITY

The Big Horn Medicine Wheel is a crude structure of loose, unmortared stones. As with any archeological site, its authenticity is open to question and so is the possibility of serious modification since its discovery. It could easily have been built by a single person in a day, if it were no more than a randomly oriented pattern. In a recent popular article Ransom[34] has suggested that the entire structure was redesigned and relaid by U.S. Forest Rangers, using different stone, sometime between 1931 and 1955. Although this can be shown to be an invalid and unsubstantiated suggestion,[35] the

Table 1.2 Comparison of surveys of cairn positions.

Cairn	Azimuth of cairn center measured from the central cairn O (deg)			Predicted value (deg)
	Stockwell (1917)	Grey (1958)	Eddy (1972)	
A	014	015	017	
B	080	084	084	
C	121	126	129	125.4 (solstice sunset)
D	171	168	169	
E	235	232	235	235.5 (solstice sunrise)
F	292	293	295	295.8 (Sirius rise)

possibility remains that, since the original discovery of the Medicine Wheel, visitors could have significantly altered its form. It was not protected, other than by isolation, until about 50 years ago when the U.S. Forest Service built a low stone wall around the wheel. Although high steel fences have since been substituted, it is still relatively unprotected and unstabilized.

Before any conclusions can be drawn from the apparent astronomical alignment of the structure, it is clearly necessary to answer the question of its veracity.[36] The classification of the Medicine Wheel as a genuine artifact of early plains inhabitants must rest on evidence of its age, as established by the methods of archaeology and the inferences of astronomy invoked here. The apparent alignments of the cairns on a set of logically related celestial events of the summer solstice add a strong statistical argument that whoever laid down the cairns knew the annual path of the sun and the important stars of the summer dawn.

Definite evidence against any major modification of the wheel in this century is found if one compares historical photographs of the site with the present structure. Grey[37] made such a comparison as part of his study of the Medicine Wheel and concluded that the site in 1958 was in no basic way changed from its appearance in 1903. I carried out a similar study, with the help of the Western History Research Library, University of Wyoming, and the U.S. Forest Service, Medicine Wheel District. From a file of photographs dating from 1905, it can be established that the general appearance of the wheel has not changed in this period; more to the point, the cairn positions appear in early photographs just as they do in Figures 1.1 and 1.2.

In addition I have made a comparison of the three extant surveys of the Medicine Wheel: that of Stockwell in 1917,[38] Grey in 1958,[39] and the work presented here.[40] Grey compared his survey with the Stockwell map and, after correction for an acknowledged plotting error in the 1917 map, concluded that there was no evidence for change in the structure. My findings corroborate Grey's conclusion and further establish that the azimuths of the centers of the six peripheral cairns are the same in the three surveys, as best these reference positions can be determined. The comparison is given in Table 1.2. It would seem to establish that the cairn positions have been stable since at least 1917 and that the conclusions presented here concerning their alignment were valid at least at that time.

The number and arrangement of stones in each cairn have undoubtedly been altered through the years, by accident, removal, or intentional probing, so that their original dimensions and precise positions will never be established. We know, for example, that during the Grey survey the interior of each cairn was excavated. But the evidence is that their general positions and hence their relative alignments have been preserved.

CONJECTURE

It may be that the solstitial alignment of the Big Horn Wheel and its resemblance to the plan of a Medicine Lodge are related. The Medicine Lodge, or Sun Dance Lodge, was commonly built for the conduct of that important ceremony. The Sun Dance was performed traditionally in June—at the time of the summer solstice,[41] "when the sun is highest and the growing power of the world is strongest."[42] The cairns of the Medicine Wheel could have served to fix the time for that ceremony; moreover, certain aspects of the Sun Dance ritual, including the layout of the Medicine Lodge, could logically have been patterned after the Medicine Wheel. The Wheel and its mountain might then be considered sacred places, endowed with the ability to receive the message of midsummer from the sun itself. It would not seem unlikely that in the development and spreading of the Sun Dance ritual the pattern of the Medicine Wheel would have been copied, at least schematically. This suggested sequence—that the Medicine Lodge was patterned after the Medicine Wheel—is the reverse of that usually invoked.

We can imagine that the Big Horn Wheel may have been originally used by a knowledgeable few who climbed to the site in June to mark the day of the summer solstice for an ensuing Sun Dance ceremony. The ceremony itself, which often involved entire tribes, might then have followed, not atop the inhospitable mountain, but some lower site when the observers came down the mountain with the message from the sun to start. In time, particularly if the secrets of the cairns were known to only a few, this way of initiating the ceremony could have faded from use, and the Medicine Wheel could have been left behind, in the veil of mystery, as an original but unappreciated pattern. This conjecture seems reasonably consistent with the earlier limits of estimated dates for the structure and with reports that local Indians apparently knew nothing of the origin or purpose of the Medicine Wheel when it was first found. It would also add meaning to the Crow legend that the sun built it to show how to make tipis.

Of no less interest is the identity of the architects who first laid out the cairns. Did the nomadic tribes of the northern plains know enough of astronomy to perform this feat? The answer is surely yes. There should be little doubt that any people who lived by the sun would intimately know the dawn, and that any who lived the mercy of the seasons would know as well the solstices. The next step, of marking these phenomena with stones, would seem to follow naturally or, conceivably, could have been learned in trade and travel from the more advanced Pueblo people to the south, who, in turn, in earlier times, learned these things from the inhabitants of Mexico, if we are to accept the hypothesis of the northward infusion of astronomical culture.[43]

Why would a nomadic people wish to mark the solstice? Historical accounts of the Plains Indians do not emphasize this practice, or, to my knowledge, even acknowledge it, and indeed the custom is more commonly associated with agricultural societies. Still, there are other reasons for wanting an annual calendar reference, which include ritual, as cited above, and a basic need to plan for colder weather. These needs would survive into the historical era, although the methods of meeting them might change. With the encroachment of a white civilization on the northern plains in the 19th century the Indian's requirement for a natural calendar could have vanished, taking with it certain astronomical traditions.

ACKNOWLEDGMENTS

I am indebted to R. Williams, U.S. Forest Service, Medicine Wheel District, and to C. G. Roundy, formerly research historian, University of Wyoming, for providing historical records and illustrations, and to D. A. Breternitz, D. Mihalas, and R. Levine for helpful comments on the manuscript. I thank G. C. Frison, O. Stewart, R. Lister, and. D. Grey for providing useful information in initiating this study. The National Center for Atmospheric Research is sponsored by the National Science Foundation.

NOTES

1. D. Grey, *Plains Anthropology* 8 (1963):27.
2. C. Simms, *American Anthropologist* 5 (1903):107; B. Grinnell, ibid. (1922):24–299.
3. Grey, *Plains Anthropology*, carried out a dendrochronological study of a piece of tree limb found lodged between the two lowest courses of stone in cairn F (Fig. 1.2). Based on core samples taken from living trees and trees that had recently died in the Medicine Wheel area, he established a tree death date of AD 1760 for the limb. If the limb had been placed in the cairn as dead wood at the time of construction, this implies a construction date of about AD 1800 for cairn F. If the wood had been added later, the date becomes a latest date for this particular cairn.
4. W. R. Wedel, *Prehistoric Man on the Great Plains* (Norman: University of Oklahoma Press, 1961). Tipi rings (typically 2 to 13 meters in diameter) are often found in clusters and are generally believed to have served to hold down tipi covers.
5. T. F. Kehoe and A. B. Kehoe, *Journal of American Folklore* 72 (1959):115.
6. See, for example, W. M. Husted, *Plains Anthropologist* 8 (1963):221.
7. T. F. Kehoe, *Journal of the Washington Academy of Sciences* 44 (1954):133; H. A. Dempsey, ibid., 46 (1956):177; L. A. Brown, *Plains Anthropology* 8 (1963):225.
8. G. B. Grinnell, *American Anthropologist* 16 (1963):245.
9. Grinnell, *American Anthropologist*, 24–299.
10. R. Yellowtail, unpublished deposition given to P. L. Heaton, forest supervisor, Big Horn National Forest, 1952.
11. Grey, *Plains Anthropology*, 27.
12. Here, and as defined by A. Thom [*Vistas in Astronomy* 7 (1966):1], the backsight is nearer the observer, as with the sights on a rifle.
13. I. B. Taylor, Casper (WY) *Herald Tribune*, March 16, 1941.

14. H. H. Thompson, Sheridan (WY) *Post*, March 11, 1923.

15. E. C. Baity, *Current Anthropology* 14 (1973):389; A. Thom, *Megalithic Lunar Observatories* (Oxford: Clarendon, 1971); G. S. Hawkins, *Nature* (London) 200 (1963):306; A. Thom, *Megalithic Sites in Britain* (Oxford: Clarendon, 1967).

16. J. E. Reyman, thesis, Southern Illinois University (1971).

17. W. L. Wittry, *Explorer* 12:4 (1970):14.

18. W. R. Wedel, *American Antiquity* 32 (1967):54.

19. E. C. Parsons, *Pueblo Indian Religion* (Chicago: University of Chicago Press, 1939).

20. Simms, *American Anthropologist*, 107.

21. Grinnell, *American Anthropologist*, 24–299.

22. Copies of the map by A. G. Stockwell are kept in the historic file of the U.S. Forest Service, Medicine Wheel District, and in the Western History Research Library, University of Wyoming. As Grey (*Plains Anthropology* 8 [1963]:27) has pointed out, it displays distances erroneously as a result of a plotting error.

23. Grey, *Plains Anthropology*, 27

24. Thom, *Megalithic Sites in Britain*.

25. J. N. Lockyer, *The Dawn of Astronomy* (London: Cassell, 1894; reprinted by MIT Press, Cambridge, MA, 1964).

26. H. N. Russell, R. S. Dugan, J. Q. Stewart, *Astronomy* (New York: Ginn, 1945).

27. Reyman (thesis, 1871) has compiled lists of celestial features known to have been recognized and used in ceremony in Mesoamerica and in southwestern pueblos. The lists consist chiefly of planets, constellations, and asterisms. Only six individual stars appear, and one of these, which was known both in Mesoamerica and in Hopi ceremony, was Aldebaran. It may be that the reason for the importance of this star to early people was its heliacal rise at solstice.

28. Grey, *Plains Anthropology*, 27.

29. See, for example, J. G. Neihardt, *Black Elk Speaks* (Lincoln: University of Nebraska Press, 961; originally published by Morrow, New York, 1932).

30. See, for example, G. S. Hawkins, *Vistas in Astronomy* 10 (1968):45.

31. Lockyer, *Dawn of Astronomy*.

32. Hawkins, *Vistas in Astronomy*, 45.

33. As is evident in Figure 1.8, the precessional change in the declination coordinate is slight for objects in this part of the sky. If we use the deduced AD 1700 epoch to recompute in Table 1.1 the indicated stellar alignments for the lines FA, FB, and FO, the azimuth differences change (column 10) by no more than about 0.5°, which is a negligible correction considering the nature of the cairns. The amount of the change and its direction (favorable for Sirius only) may be seen in Figure 1.8 where the apparent declinations of the present cairn alignments are shown as lines alongside the presumed stellar object.

34. J. E. Ransom, *American West* 8:2 (1971):16.

35. Ransom (ibid.) based his conclusion on his personal recollections of how the wheel appeared to him as a youth of 7 and 17 years. He recalled that in 1921 and 1931 the wheel was made of nonlocal quartz and that it had only 18 spokes—a number which fits his theory of an Aztec origin. Abundant evidence in the form of historical photographs and surveys and an historical site record maintained by the U.S. Forest Service clearly establishes the general stability of the structure and negates Ransom's claim.

36. We are fortunate that at the present time the U.S. Forest Service, Medicine Wheel District, and the Department of Anthropology, University of Wyoming, are studying the archeological basis of the structure. Their work represents the first professional study of the

Medicine Wheel since the brief report by Grinnell in 1922 (*American Anthropologist*, 1903) and the volunteer study led by Grey in 1958 (*Plains Anthropology*, 1963).

37. Grey, *Plains Anthropology*, 27.

38. Copies of the map by A. G. Stockwell are kept in the historic file of the U.S. Forest Service, Medicine Wheel District, and in the Western History Research Library, University of Wyoming.

39. Grey, *Plains Anthropology*, 27.

40. Stockwell (see note 22) was concerned chiefly with the azimuths of the spokes and the shape of the rim of the wheel; in his map the cairns are shown only schematically. Since the spokes are not linear and according to photographic evidence apparently never were, the usefulness of this survey is limited. I found, however, that, within the limits of straight-line approximation, the number and individual positions of spokes in the Stockwell map agree with those in Grey's survey.

41. C. Wissler, *North American Indians of the Plains* (New York: American Museum of Natural History, 1920); J. E. Brown, ed., *The Sacred Pipe: Black Elk's Account of the Seven Rites of the Oglala Sioux* (Norman: University of Oklahoma Press, 1953).

42. Neihardt, *Black Elk Speaks*.

43. Reyman, thesis, Southern Illinois University (1971).

CHAPTER TWO

Geometry and Astronomy in Prehistoric Ohio

Ray Hively and Robert Horn

INTRODUCTION

Prehistoric geometric earthworks more extensive than Avebury, and requiring an effort in design and construction comparable to Stonehenge, once covered an area of 10.4 square kilometres near the present city of Newark, Ohio. (Fig. 2.1 is a 1934 aerial photograph showing a portion of the earthworks which still survives. Fig. 2.2 shows the full extent of the works as they appeared in 1847.) The Newark earthworks have been the subject of amateur excavation and widespread curiosity for two centuries. Recent archaeological research has begun to build an understanding of the social and environmental context of Newark and other comparable Ohio sites. There has been much speculation, especially in the literature of archaeoastronomy, concerning the possible geometric and astronomical significance of these and related works.[1] Much of

---

From *Archaeoastronomy* 4 (*JHA* 12) (1982):S1–S20. Reproduced by permission of the authors.

2.1 This photograph taken in 1934 by Reeves (1936) shows the most impressive part of the Newark earthworks which still survives, a circle (321 m in diameter and known as the Observatory Circle) connected by an avenue to an Octagon (with sides 190 m in length). This figure encloses 61 acres.

this speculation has been based upon nineteenth-century surveys which are known to be erroneous in quantitative details. Thus the aim of this paper is to provide a detailed analysis of the possible geometric and astronomical significance of the Newark earthworks, utilizing accurate survey data. Our analysis shows that the earthworks conform to a simple but precise geometric construction based on a single unit of length. Our work has also revealed substantial evidence that the earthworks may have been laid out at least in part as a lunar observatory.

BACKGROUND

Since 1800, when settlers in the Licking Valley first noticed large geometric figures drawn in heavy earthen embankments at the confluence of Raccoon Creek and the South Fork of Licking River, the Newark earthworks have occupied an anomalous position in American archaeology. The still surviving walls of the large geometric figures extend for close to 4 kilometres. The original extent of the walls may have been five times as great. Yet there is a "paucity of information" about the works and their archaeological context.[2] If the Newark works had been the only Native American artifacts of their kind in the drainage system of the Ohio River, they would almost certainly have been more closely studied and perhaps better known today. They were, however, merely the largest and northernmost of many similar structures on rivers tributary to the Ohio. Other extensive geometrical works were to be found on the banks of the Scioto River north of Chillicothe, Ohio, on Paint Creek near its confluences with the Scioto, and at the confluence of the Scioto with the Ohio River near Portsmouth. Works on a similar scale were located at the confluence of the Little Miami River with the Ohio, near Cincinnati, and at the confluence of the Muskingum River (fed by the Licking River) with the Ohio, near Marietta. Altogether, more than one hundred Ohio sites involving large earthworks, many of them geometrically regular, were known by 1900.[3] Unfortunately, only the Newark site has been preserved in a manner which allows its main features to be seen. The other geometric works have been damaged by erosion, the plough, and the bulldozer, until only a few features are still visible on the ground or on aerial photographs.

Geometric earthworks on the scale of Newark were the most spectacular evidence of complex social and economic organization found in Eastern North America in the period covering about 500 BC–AD 500. The archaeological evidence for the millennium shows hitherto isolated villages accustomed to a life of fishing, hunting and gathering, taking on traits of social stratification and cooperation perhaps related to the spread of maize as an important addition to village food resources. In many places, on the broad flood plains of glacial outwash streams, hierarchical social organization appeared in the village. Exchange of raw materials and manufactured goods gave otherwise independent villages the veneer of a common "culture." These common features are characteristic of an assembly of social, economic, and religious practices now called "Hopewell". They include, in addition to local hierarchy and inter-village exchange, the acquisition and redistribution of rare raw materials such as copper, marine shells, mica, galena, pipestone, obsidian, meteoritic iron, and jewel-quality flint used in the manufacture of ceremonial implements. In addition some "luxury" items such as mica mirrors, conch shell vessels, cold-hammered metal awls, and non-utilitarian celts were manufactured and traded. These exotic items both displayed the prestige of their owners and were essential features in a mortuary cult which consumed much of the labour and wealth of its adherents. What has been called the "Hopewell Interaction Sphere" covered much of the East, from the Great Lakes to the Gulf Coast. Raw materials were secured from areas as remote as the Grand Tetons. Some elaborate earthworks appear to have been ceremonial centres for the practice of the mortuary cult and, in some cases centres for the redistribution of exotic goods. The exchange-system may have made the redistribution of food in times of scarcity possible.[4] There is no evidence that the earthworks were themselves village locations or were even inhabited for long periods of time. The few traces of settlement in or near them appear to have been transient dwellings of construction crews or caretakers.[5] The Newark earthworks represent the acme of "Hopewell" in Ohio. They belong to a period, probably no earlier than AD

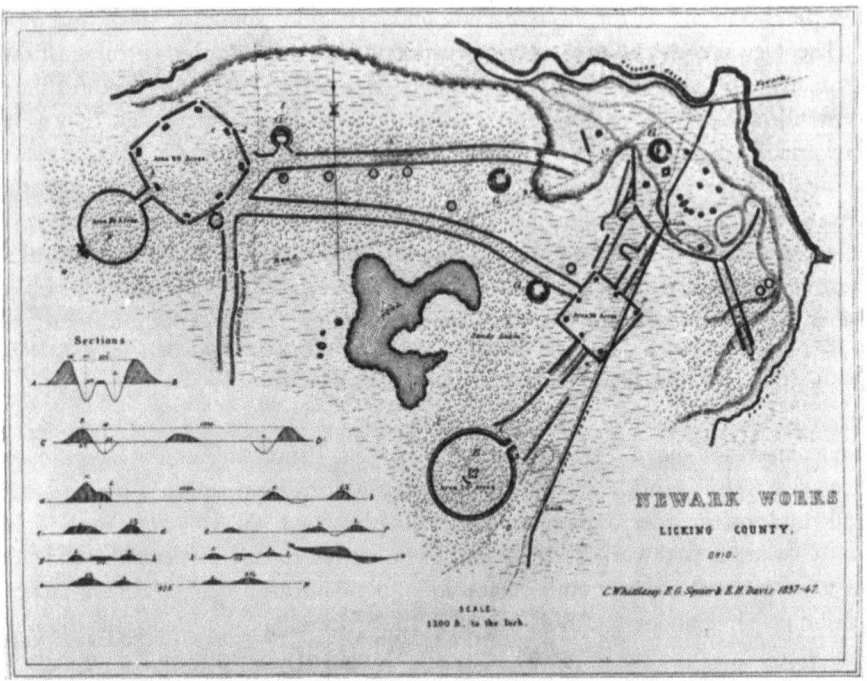

2.2 A qualitative plan showing the original extent of the Newark earthworks taken from Squier and Davis (1847). Only the Observatory Circle, the Octagon, the small circle south of the Octagon, the Fairground Circle, and a small part of the Square survive today.

250, when other similarly intricate and extensive works were built on the Ohio River, near the present cities of Portsmouth, Marietta, and Cincinnati. Since no radiocarbon dates exist for Newark, the date suggested has been reached by comparing ceramics and, for example, the style of platform mounds in the Newark octagon with similar sites where radiocarbon dates have been secured.[6]

DESCRIPTION

The Newark works are so complicated that a detailed description cannot be given here. Perhaps the best impression of the works as a whole can be gained from a study of the map in Figure 2.2. We will give a cursory description of some of the major surviving figures in order to establish the terminology to be used in the detailed analysis which follows.

The figures shown on the map are formed by large embankments of earth typically measuring 12 m at the base and some 1.7 m high. Perhaps the most striking feature is the quite symmetrical octagon-circle combination (marked F in Fig. 2.2) which encloses some 248,000 m^2 (61 acres). At the end of the circle directly opposite the avenue, the builders interrupted the wall, and extended it to 30.5 m outside its perimeter to form a large flat-topped mound which commands a view of the entire work. This structure, which has been called "Observatory Mound" for close to two centuries, stood 3.4 m high in 1847 and had a length of about 52 m; the associated circle will hereafter be referred to as the Observatory Circle. A large square enclosure encompassing over 81,000 m^2 (20 acres) is found southeast of the Octagon. The surviving part of this figure is now known as the Wright Square. We will refer to it as the Square. Another imposing figure (marked E in Fig. 2.2), a somewhat imperfect circle enclosing about 103,000 m^2 (25 acres), is located about 2 km southeast of the Observatory Circle. As this enclosure once housed the Licking County Fairgrounds it has taken its name from them. We will refer to it as the Fairground Circle.

Originally, the Octagon, Square and Fairground Circle each contained one or more earthen mounds. In the Octagon there are eight oval flat-topped mounds situated opposite the vertices. Each is placed about 30 m inside its vertex. Near the centre of the Fairground Circle lies a group of low mounds joined to form a figure which has often been described as a bird-effigy. The wingspread of the bird is 61 m; behind the effigy, opposite the entrance to the circle, is another low mound, crescent-shaped, as long altogether as the wingspread of the effigy.[7] The Octagon, the Observatory Circle, the Fairground Circle, a small portion of the Square, and a small circle southeast of the Octagon have been preserved. The remainder of the original earthworks shown on the 1847 map has been destroyed, although faint traces of additional features have been discerned on aerial photographs (see Fig. 2.1).[8] A probable method for the construction of these, and other Hopewellian geometric earthworks, using stakes and fibre or hide ropes, has been described by Morgan.[9]

SURVEY OF THE EARTHWORKS

The essential prerequisite for a reliable analysis of geometrical and astronomical content in an ancient work is an accurate plan of the original structure. An on-site survey of the Newark earthworks can no longer provide this information by itself, because the present structure is in part the product of at least two restorations. The Ohio Archaeological and Historical Society and the National Park Service undertook a major restoration project during 1933–36.[10] This project involved: (1) a major reconstruction of the walls of the Fairground Circle to restore it to a condition as described by the nineteenth-century survey of Squier and Davis;[11] (2) a reconstruction of part of the wall near the northern corner of the large Square; (3) cosmetic restoration of the Observatory Circle and Octagon involving such things as the elimination of footpaths and the removal of tree roots. A far more ambitious restoration of the Observatory Circle and Octagon was undertaken during 1893–96 by the Ohio National Guard. The State Adjutant General describes the restoration in his annual report of 1893:[12]

> Employing a surveyor, the original lines were ascertained.... The work at the start was very cautiously proceeded with, and all information which would assist in restoring the grounds to their early condition was eagerly sought for.

We have been unable to find, however, documentation, blueprints, or plans which give details of this restoration. Thus, no reliable conclusions can be formed from current surveys alone.

Three maps of the Newark earthworks, based on extensive surveys, were made prior to any recorded restoration: (1) a map published in 1820 by Atwater[13] giving the first detailed description of the site; (2) a map published in the voluminous and much celebrated work of Squier and Davis (Fig. 2.2); (3) a map based on a careful and accurate survey by Middleton in 1887–88 and published by Cyrus Thomas of the Smithsonian Institution's Bureau of Ethnology.[14] The Smithsonian survey was undertaken to ascertain the degree of geometrical regularity in Hopewellian earthworks and to test the accuracy of the earlier surveys. This survey demonstrated that the earlier surveys contained numerous and large quantitative errors in the size, shape, and orientation of the earthworks. Thus, the maps of Atwater and of Squier and Davis are useless for precise quantitative work. However, the description of the Newark site by Squier and Davis does assure us that as late as 1847 much of the site (including the Fairground Circle, the Observatory Circle, and the Octagon) was still covered by forest and had apparently retained its original form. By the time of the Middleton survey (1887) the effects of cultivation and the expansion of the city of Newark had destroyed large portions of the earthworks. The only parts of the site which Middleton could survey accurately were the Fairground Circle, the Observatory Circle, the Octagon, and parts of the Square. However, with the exception of certain parts of the Square, the Middleton field notes show that all parts of the earthworks were distinctly traceable and that the walls were never less than 0.75 m in height.

The accuracy of the Middleton survey, which will form the basis of our analysis, may be judged in several ways. Fowke, an archaeologist who participated in the survey, described the care with which the measurements were carried out:[15]

> Greater care was taken in getting bearings and distances than is usually employed in railway or canal surveys. Middleton and I, who did the work, stand by our figures, and with all the more reason, too, that in some cases they completely upset our antecedent ideas and opinions.

An independent survey of the Observatory Circle was carried out by Holmes[16] in 1892. He obtained a mean diameter of 321.3 m for the Observatory Circle, precisely the same value obtained by Middleton. Finally, during the summer of 1980 we undertook a survey of the site utilizing a transit and steel tape to judge the accuracy of the Middleton survey and to infer the nature of the State restoration. The results of our survey are compared with the measurements of Middleton in Table 2.1 using the notation defined in Figure 2.3. Distances and azimuths were all measured along the midline of the embankments. The comparison in Table 2.1 shows that the Middleton results agree with our results to within 30' of arc for azimuth measurements and to within 1 m for distance measurements. The accuracy of the Middleton azimuths can best be judged from a comparison of the azimuths for the Octagon walls *AB* and *BC*, where the two surveys agree within 15'. According to the Middleton field notes these

Table 2.1 Comparison of surveys of the Observatory Circle and Octagon.

Alignment	Azimuth		Distance (m)	
	Middleton (1887–88)	Hively-Horn (1980)	Middleton (1887–88)	Hively-Horn (1980)
KN	52°0'	51°50'	322.5 m	322.5 m
JK	52°4'	52°9'	89.9	89.0
LM	51°53'	51°38'	89.3	90.2
AB	130°19'	130°13'	189.6	489.7
BC	64°18'	64°31'	190.3	190.5
CD	39°50'	39°19'	190.5	191.0
DE	334°32'	334°20'	189.6	189.2
EF	308°28'	308°0'	189.3	188.4
GF	64°40'	65°38'	186.8	186.7
HG	39°15'	39°30'	189.4	189.3
AH	334°20'	334°29'	190.0	191.0

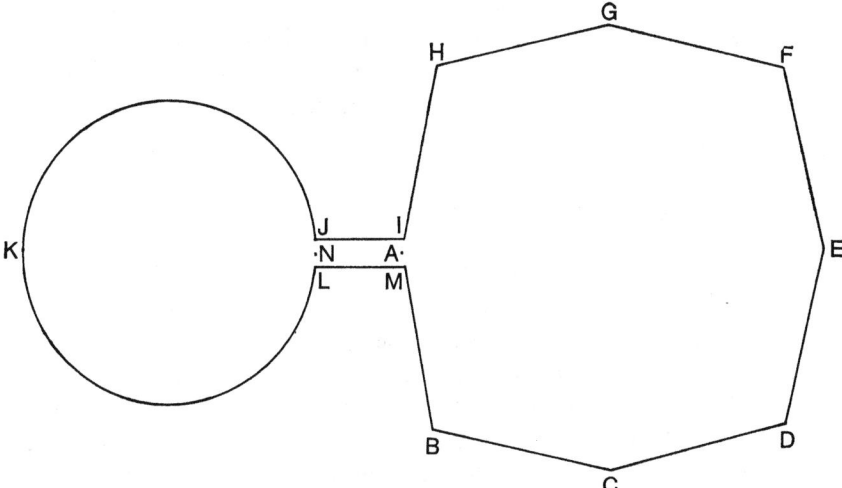

2.3 A schematic plan of the geometry defined by the midline of the embankments of the Observatory Circle–Octagon combination shows the lettered points used to define the azimuths given in Table 1. Point *A* is a vertex of the Octagon; Points *K* and *N* define the diameter of the circle passing through the centre of the avenue entrance.

walls were uninjured in 1888 and were still covered by the original forest. Thus, it is unlikely that the State restoration in 1893 could have altered their configuration significantly. On reading the manuscript of this paper in December 1980, Professor John Eddy indicated that he had completed a transit survey at the Newark site in 1978, and that his measurements (unpublished) agree with ours (personal communication).

There are, however, two significant discrepancies between the current configuration of the Octagon and the 1888 survey. Both the Thomas reports and the unpublished Middleton field notes[17] show the wall HA as being 168.2 m in length, some 9.2 m shorter than the other walls which ranged from 177.4 to 182 m in length. The height of wall HA in 1888 was 1.1 m, and Middleton records no evidence that any part of it was obliterated or difficult to trace. Presently the length of wall HA is 177.4 m, the same as the length of wall AB. If we accept the Middleton data on this point, we must conclude that the State reconstruction "restored" wall HA to a length and symmetry it did not originally possess. An examination of the soil in wall HA might resolve this question. The present locations of the centres of the mounds interior to the Octagon (opposite the vertices) differ from the positions recorded by Middleton. The centres of the mounds as restored are separated by distances ranging from 162 m to 164 m whereas Middleton records the separations as ranging from 162 m to 168 m. Each of the Octagon walls originally contained about 2400 m³ of earth, the great majority of which was in place for each wall in 1888. Thus, it is unlikely that the State would have significantly altered the alignment of the Octagon walls, and this is borne out by the close agreement of the surveys in Table 2.1. This agreement suggests that the primary effect of the State reconstruction of the Octagon was to increase the height of the walls to about 1.7 m. It must be admitted that the accuracy of the present positions of the ends of the walls and the interior mounds is less secure. Their positions could easily have been altered by moving relatively modest amounts of earth. Fortunately, our subsequent analysis will not depend in any important way on the position of the mounds.

The Square northeast of the Fairground Circle can no longer be surveyed on the ground, as it has been obliterated by construction. Fortunately, we do have the Middleton survey data for the size and shape of the Square. The location of the Square relative to the remainder of the earthworks can be determined from the position of the eastern corner of the Square which is visible on aerial photographs of the site taken by Dache Reeves in 1934.[18] The location so determined is consistent with the location of the part of the Square near the north corner (which has been reconstructed) and with the size as given by Middleton. The estimated uncertainty in our knowledge of the position of the Square is about 5 m. All the available evidence suggests that the Middleton survey has sufficient accuracy (azimuths with a maximum error of ±15') to provide a basis for a reliable analysis of the earthworks for geometrical regularity and astronomical orientation. Unless otherwise noted, the analysis of the following sections will be based on the Middleton data.

THE GEOMETRY OF THE EARTHWORKS

The first point to be noted regarding the geometry of the earthworks is their remarkable regularity. Consider the Observatory Circle. The midline of the embankment walls deviates by no more than 1.2 m at any place from a circle of diameter 321.3 m. A perfect circle of this diameter would have a circumference of 1009.4 m whereas the actual circle had a circumference of 1008.6 m. Thus, it is evident that the Observatory Circle very closely approximates a true circle. The configuration of the Octagon

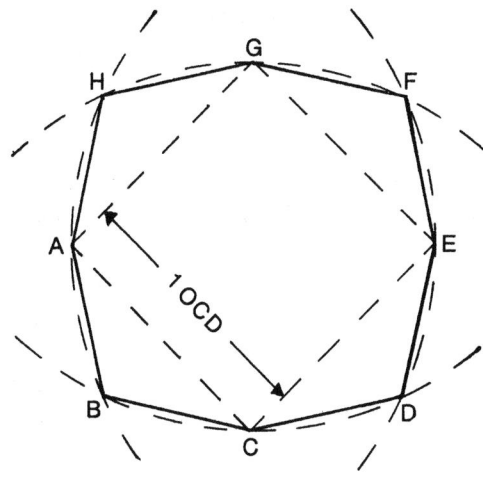

2.4 The dotted lines show a method which may have been used to construct the Newark Octagon. First a square (*ACEG*) of side 1 OCD (321.3 m) is laid out; then circular arcs of radius √2 OCD are drawn with centres at the vertices *A*, *C*, *E* and *G*. The intersections of the circular arcs locate the remaining vertices *B*, *D*, *F* and *H*.

demonstrates a similar degree of regularity and precision. For example, the angle of intersection of the diagonals *BF* and *DH* (see Fig. 2.3) differs by only 10' from a right angle, and the angle of the intersection of the lines *AE* and *CG* differs by only 2' from a right angle. The greatest departure from symmetry comes from the position of vertex *F*. It is located 6 m closer to the centre of the Octagon than its ideal position, and this results in side *FG* being about 2 percent shorter than the other sides.

We will now show that the Octagon can be laid out quite precisely using a method of geometrical construction based on only one length, the diameter of the Observatory Circle (321.3 m and hereafter denoted by the abbreviation OCD). First, it should be noted that the vertices *A*, *C*, *E*, and *G* form a square with sides *AC* = 318.6 m, *CE* = 320.0 m, *EG* = 321.0 m and *GA* = 320.3 m. The average deviation of these lengths from 1 OCD is 0.4 percent. Thus, the Octagon can be constructed in the following manner (see Fig. 2.4): first, lay out a square (*ACEG*) of side 1 OCD; secondly, draw circular arcs (of radius equal to the diagonal of the square) through each of the vertices *A*, *C*, *E*, and *G* using the opposite vertex as the centre of the circular arc. The intersection of these circular arcs locates the positions of the vertices *B*, *D*, *F* and *H*. With the exception of the angles and lengths connected with the position of vertex *F*, this construction reproduces the azimuths of the Octagon walls within 0°.6 and the lengths of the walls within 0.6 percent. The displacement of vertex *F* from its ideal position causes the azimuths of sides *FG* and *EF* to deviate by about 1°.5 from the predicted azimuths. An explanation of this asymmetry will be offered in the next section.

The conclusion suggested by the geometry of the Observatory Circle–Octagon combination is that both figures have been carefully and skillfully constructed using the same fundamental length, the OCD. Further evidence for this comes from the positions of the interior mounds as given by Middleton. The square formed by the centres of the mounds *B*, *D*, *F* and *H* has sides of length 1.002 OCD, 1.004 OCD, 1.007 OCD and 1.002 OCD respectively. Further support for the special significance of the OCD is found in the relation between the Fairground and Observatory Circles and between the Octagon and the Square. The distance between the centres of the Fairground Circle and the Observatory Circle is 5.99 OCD, and the distance between the centres of the Octagon and the Square is 6.02 OCD. Thus, both distances are 6 OCDs within better than 0.4 percent accuracy. These distances were determined from aerial photographs (taken in 1972) supplied by the Engineer's Office of Licking County, Ohio, and the distances are believed to be accurate to about 5 m. It should also be noted that the distance between the centre of the Octagon and the centre of the Observatory Mound differs by only 0–5 percent, from 2 OCD. In addition, the diameter of the small circle south of wall *BC* is 1/7 OCD within 0–3 percent. If the length of 1 OCD had a special significance for the builders of the earthworks, one would expect to find supporting evidence at other sites, and indeed this is the case. A Middleton survey as reported by Thomas[19] shows that the only other circle-octagon combination known to have been constructed by the Hopewell (the so-called High Bank Works near Chillicothe, Ohio) conforms to a geometric pattern based on a fundamental length of 0.998 OCD.[20]

Detailed survey data with respect to the Fairground Circle and Square are to be found in the Thomas reports, and we will only require a brief description here. It should be noted that the Fairground Circle is not a perfect circle, having maximum and minimum diameters differing by about 10 m. Thus, the position of the centre of the circle of best fit (with a diameter of 361.2 m) has an uncertainty of about 2 m. The Square has a geometrical precision comparable to that of the Octagon. The dimensions of the Fairground Circle and Square are not simply related to the OCD.[21]

The azimuthal orientations of the Fairground Circle and Square with respect to the Observatory Circle and Octagon display a curious relation which will prove to be important when considering the possible astronomical alignments of the site. Consider diameters of the Fairground and Observatory Circles oriented along the centres of the avenues leaving these circles as shown in Figure 2.5. A line between the northeast ends of these diameters has an azimuth of 129°.8. A line connecting the southwest ends has an azimuth of 129°.5. Thus, the Fairground Circle avenue axis is parallel displaced from the corresponding axis in the Observatory Circle along a mean azimuth of 129°.7. There are three reasons not to dismiss this as an unremarkable coincidence: (1) the distance between the centres of the two circles (6 OCD) suggests they may have been carefully placed with respect to one another; (2) the Square bears a similar kind of relation with respect to the Octagon; and (3) the azimuth of displacement, as shown in the next section, has a special astronomical significance. If lines are drawn from the Octagon vertices *A, C, E,* and *G* to vertices of the Square as shown in Figure 2.5, one finds that the Square is displaced from the Octagon along the azimuths $AO = 116°.2$, $CR = 116°.2$, $EQ = 117°.5$ and $GP = 117°.0$. Thus, the vertices of the Square are parallel displaced from the vertices of the Square *ACEG* along the mean azimuthal direction of 116°.7. As shown in the next section, this azimuth too has astronomical significance. These azimuths were determined from aerial photographs and have an uncertainty of ±0°.25.

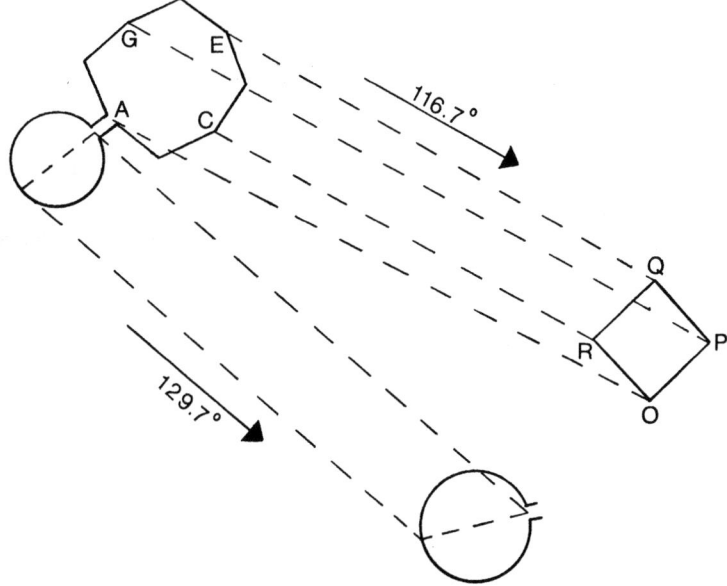

2.5 A sketch shows the parallel-displaced relationships between the Fairground and Observatory Circles and between the Square and the Octagon. The azimuths of the displacements mark southern extreme lunar rising points.

ASTRONOMICAL ALIGNMENTS

When examining the Newark site for conscious astronomical alignment it is essential to be conservative in method and interpretation. The site is very complex, and the number of alignments formed by connecting arbitrary points in the structure is so great that the likelihood of generating chance alignments with any astronomical phenomenon is disturbingly large. Consequently, we have only considered alignments between points for which independent geometric evidence exists to suggest that the points may have been consciously associated. For instance, we have considered alignments along linear embankments, axes of symmetry, and between special points (such as centres of geometric figures) separated by an integral number of OCDs. We have not considered alignments between mounds, between mounds and vertices, or between mounds and centres of geometrical figures. With only two exceptions we have rejected all potential alignments which did not have an azimuthal accuracy of better than 1°. The two exceptions occur because they are found in a context of similar alignments of much greater accuracy. It should also be remembered that a large portion of the original works shown in Figure 2.2 was destroyed before any accurate survey data could be obtained; thus, any astronomical information associated with that part of the site has been lost.

We have not considered stellar alignments, because uncertainties in the date of the site (several hundred years) are so great as to make the effects of stellar precession

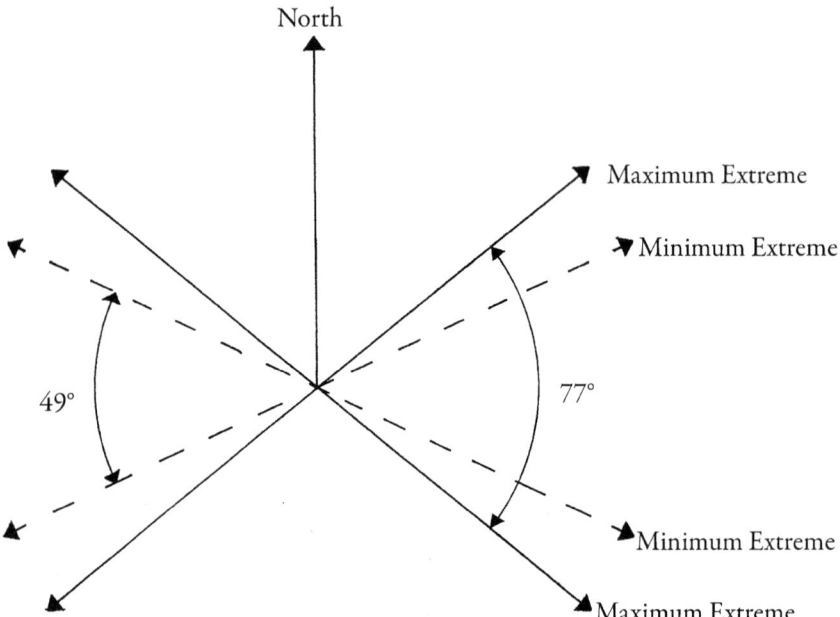

2.6 The extreme azimuths of the lunar rise-set points at Newark are shown by the arrows. The azimuths shown are approximate in that the effects of horizon altitude and refraction have been ignored. When the Moon is at a position for maximum extremes, the extreme north and south moonrises in a given month are separated by 77°; at the position for minimum extremes, the extreme moonrises are separated by 49°.

unacceptably large. In contrast the rise and set points of the Sun, Moon and planets show a negligible variation over a time scale of a few hundred years. Therefore, we have tested for astronomical alignments involving only these objects. We have carried out this test by comparing measured azimuths associated with features in the earthworks with the azimuths for the rise and set points of astronomical objects as given by the tables computed by Aveni.[22] For definiteness we have used the astronomical azimuths computed for AD 250. We have found no convincing evidence for any alignments with the Sun or planets. We have, however, found a substantial amount of evidence for lunar alignment.

The rising point of the Moon as marked along the horizon oscillates between a northerly and southerly extreme during each sidereal month (27⅓ days). Due to a slow precession of the Moon's orbit, these extreme northerly and southerly rising points oscillate between two fixed azimuths with a period of 18.61 years. A similar variation occurs in the setting point of the Moon as shown in Figure 2.6. A careful observer of the 18.6-year lunar cycle would therefore notice eight significant directions along the horizon (four moonrise points and four moonset points) where the Moon periodically reaches a maximum or minimum extreme. We have defined the rising point as the azimuth of the first gleam where the Moon's leading edge just comes over the horizon; similarly the setting point is defined as the azimuth of last gleam. The precise azimuth of a moonrise or moonset depends on the altitude of the horizon. The distant horizons along the astronomically significant directions at Newark are obscured by buildings and vegetation. Thus, we computed the horizon altitudes from a topographic quadrangle map prepared by the US Geological Survey in 1961. This procedure was checked in the field for directions in which the natural horizons were visible and was found to yield an accuracy of better than 0°.1.

The physical structure of the walls of the Octagon is well suited to marking astronomical phenomena on the horizon. The great length (179 m) of the walls enables the midline of the walls to define a precise azimuth to within 0°.25. The uncertainty in azimuth arises from the slight asymmetry of the walls which makes it impossible to locate the midline with perfect accuracy. The original height of the walls was about 1.7 m;[23] an adult standing at the end of a wall could sight with ease along the entire midline. First it should be noted that the avenue axis of the Octagon (*KNAE* in Fig. 2.3) points to the maximum northern extreme rising point of the Moon with an error of 0°.2. Our measurements on this point confirm an earlier conjecture by Eddy based on the Squier and Davis map.[24] The midline of the Octagon walls *AB, CB, EF* and *FG* are found to align closely with extreme rise and set points of the Moon. As shown in Table 2.2, the avenue axis and four sides of the Octagon mark five of the eight extreme lunar rise-set points with a mean accuracy of 0°.5. Moreover, the observation points for these alignments occur at the vertices *A, C, E,* and *G*, which (as shown in the last section) probably had a special role in the construction of the Octagon. It should also be noted that the four sides of the Octagon which are not astronomically aligned form closely parallel pairs as required by symmetry. Walls *CD* and *HG* differ by 0°.6 in azimuth, and walls *AH* and *DE* differ by only 0°.2. This stands in sharp contrast to the pairs of walls aligned to the Moon. Walls *AB* and *FE* differ in azimuth by 1°.9, and walls *BC* and *GF* differ by 1°.4. This suggests that the symmetry of the astronomically significant walls may have been deliberately distorted to achieve the lunar alignment.

In assessing the probability of these alignments being intentional there are a number of facts to be considered. First it should be noted that the requirements of (1) octagonal symmetry and (2) alignment with lunar extrema uniquely define the Newark octagon. Of the infinity of possible octagons which could have been constructed at this site, the one we find is precisely the one which matches the lunar extrema most closely. In fact we have been unable to design an equilateral polygon with eight or fewer sides which incorporates the extreme lunar points more efficiently and accurately than does the Newark octagon. When this fact is combined with the apparently intentional distortion of the Octagon in the direction of more accurate lunar alignments, the hypothesis of deliberate alignment must be taken very seriously. If these alignments are not due to chance, one would expect to see three kinds of supporting evidence: (1) alignments of comparable accuracy for the three remaining lunar events; (2) a repetition of the lunar alignments in other parts of the earthworks; and (3) a similar set of geometrical and astronomical patterns at other Hopewellian sites. Significantly, there is indeed supporting evidence of all three kinds.

Table 2.2 List of Lunar Alignments at Newark (AD 250).

Alignment	Azimuth (measured)	Horizon Elevation (computed)	Lunar Event	Lunar Azimuth (computed)	Error
Avenue axis of Octagon	52°.0	0°.51	max. north rise	51°.8	0°.2
Side *AB* of Octagon	130°.3	0°.61	max. south rise	130°.0	0°.3
Side *CB* of Octagon	244°.3	0°.59	min. south set	244°.6	0°.3
Side *EF* of Octagon	308°.5	1°.48	max. north set	307°.2	1°.3
Side *GF* of Octagon	65°.7	0°.46	min. north rise	66°.1	0°.4
Diagonal of avenue leading to Octagon (Figure 2.7)	65°.8	0°.35	min. north rise	66°.0	0°.2
Parallel displacement of Fairground Circle with respect to Observatory Circle (Figure 2.5)	129°.7	0°.60	max. south rise	130°.0	0°.3
Parallel displacement of Square with respect to Octagon (Figure 2.5)	116°.7	0°.75	min. south rise	115°.5	1°.2
Vertex F–north end of wall HA–tangent to Observatory Circle (Figure 2.8)	230°.5	0°.53	max. south set	230°.1	0°.4
North end of wall *HG* to south end of wall *DE* (Figure 2.8)	116°.3	0°.86	min. south rise	115°.6	0°.7
Centre of small circle to south end of wall *AH* (Figure 2.7)	293°.4	1°.43	min. north set	293°.0	0°.4
Centre of small circle to north end of wall *AH* (Figure 2.7)	308°.0	1°.45	max. north set	307°.2	0°.8
Observatory Mound centre–tangent to Fairground Circle	130°.1	0°.60	max. south rise	130°.0	0°.1
Northwest entrance to Square to north end of southeast wall of Square	116°.3	1°.18	min. south rise	116°.0	0°.3
Avenue axis of Fairground Circle	66°.6	0°.11	min. north rise	65°.7	0°.9
Centre of avenue entering northwest entrance of Square	306°.8	0°.97	max. north set	307°.7	0°.9
Diagonal of Square (*RP*, Figure 2.5)	92°.8	0°.98	rise-set points along straight line	92°.0	0°.8

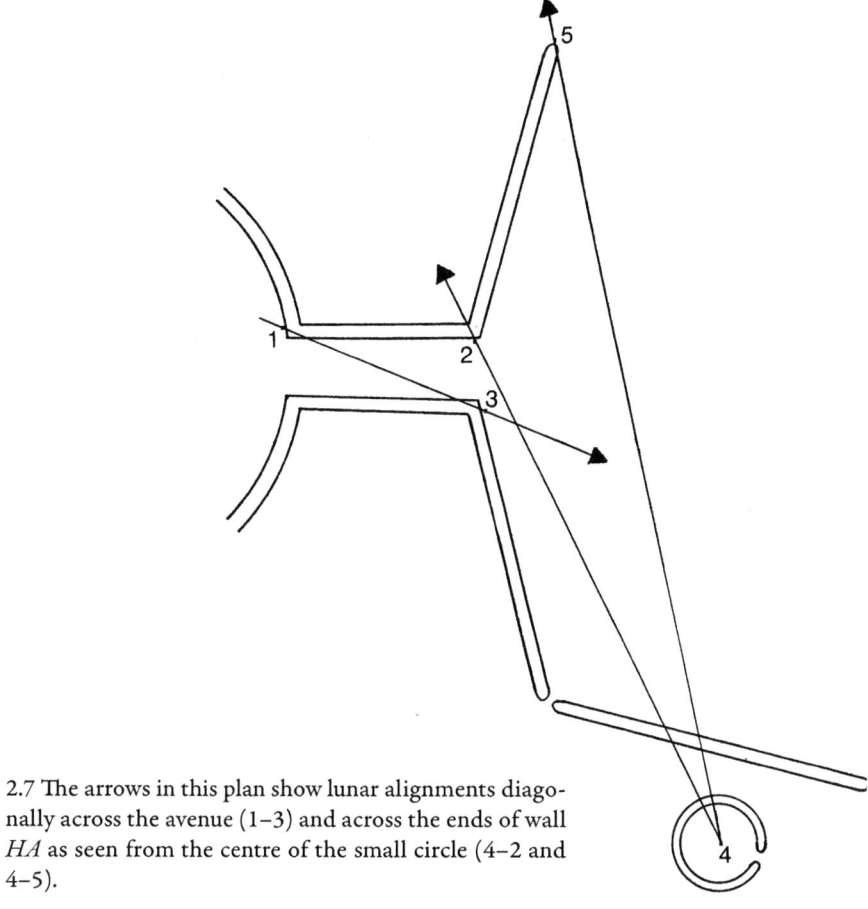

2.7 The arrows in this plan show lunar alignments diagonally across the avenue (1–3) and across the ends of wall *HA* as seen from the centre of the small circle (4–2 and 4–5).

Consider the avenue connecting the Octagon to the Observatory Circle. An observer sighting along the southern parallel of the avenue would be looking at the maximum northern extreme lunar rise point within 0°.1; an observer sighting along the northern parallel would be looking at the same lunar event within 0°.3. An observer sighting diagonally across the avenue (from Point 1 to Point 3 in Fig. 2.7) would be looking at the minimum northern extreme rise point within about 0°.2. Thus, an observer standing at the entrance to the avenue where it leaves the Observatory Circle would see the Moon rise in the avenue only when it was between the maximum and minimum northern extremes. Moreover, the 18.6-year cycle could be easily and precisely monitored by noting the position of the monthly northern extreme rise point as viewed from the south end of the avenue. As seen by an observer at Point 1 in Figure 2.7, the rise point would appear to swing wide of the avenue when the Moon reached the maximum and minimum extremes. This feature, if intentional, would account for the length to width ratio of the avenue.

Next we will show that the three "missing" alignments can be incorporated into features associated with the Octagon, features which appear inexplicable purely in geometrical terms. If one examines the gaps which appear at the vertices of the Octagon, there are two which are significantly different from the others. The bases of the Octagon walls come together at vertex F to form a very narrow gap or "observation point". Wall HA is also 9 m shorter than the other walls (according to Middleton), producing an unusually large gap at vertex H. A line from vertex F through the north end of wall HA passes through the north edge of mound H and is tangent to the midline of the Observatory Circle (see Fig. 2.8). This line also marks the maximum southern extreme set point within $0°.4$. Thus, this alignment contains four significant features and if intentional would account for the narrow gap at F, the short wall HA, the position of mound H, and the length of the avenue connecting the circle and Octagon. The minimum southern extreme rise point is marked within $0°.7$ by a line running from the north end of wall HG along the north edge of mound G and across the southern end of wall DE (see Fig. 2.8). The best argument for this being an intentional alignment comes from the position of mound G; its position and size are such that the north edge of the mound helps define the alignment without obscuring it. Moreover, this alignment, if intended, would offer an explanation for the sizes of the gaps connected with vertices G and E.

Recall that from Position 1 in the avenue (see Fig. 2.7) it was possible to monitor continuously the 18.6-year lunar cycle by observing the position of the extreme northern rise point. If this was an important function of the earthworks, one might expect to find a point from which the corresponding set points could be monitored in a similar fashion. Such a point is indeed marked by the centre of the small circle (of diameter 46 m) located south of wall BC. It should be noted that the position of this circle has no obvious geometrical relation to the rest of the structure. A line from the centre of the small circle thruugh the south end of wall HA marks the minimum northern extreme set point within $0°.4$. A line from the same centre to the northern end of wall HA marks the maximum northern extreme set point within $0°.8$. These two alignments, if deliberate, would uniquely determine the position of the small circle. As viewed from the circle centre the northern extreme set point always appears behind wall HA and reaches the ends of the wall only when the Moon is at the maximum or minimum extreme. Thus, an observer located at the circle centre could monitor the lunar cycle by observing the monthly northern extreme set point in relation to wall HA. Evidence that the circle and Point 1 in the avenue could have been associated in the minds of the builders comes from the fact that the distance between Point 1 and the midline of the circle wall is 1 OCD within 1 percent.

It must readily be acknowledged that the last group of four alignments we have proposed have an *ad hoc* character if considered alone. The argument for deliberate alignment is not as strong as for the alignments previously proposed. However, we believe there are two reasons not to dismiss these alignments as speculative or accidental without further consideration. First these alignments are precisely the ones predicted by the lunar context established by the more convincing alignments associ-

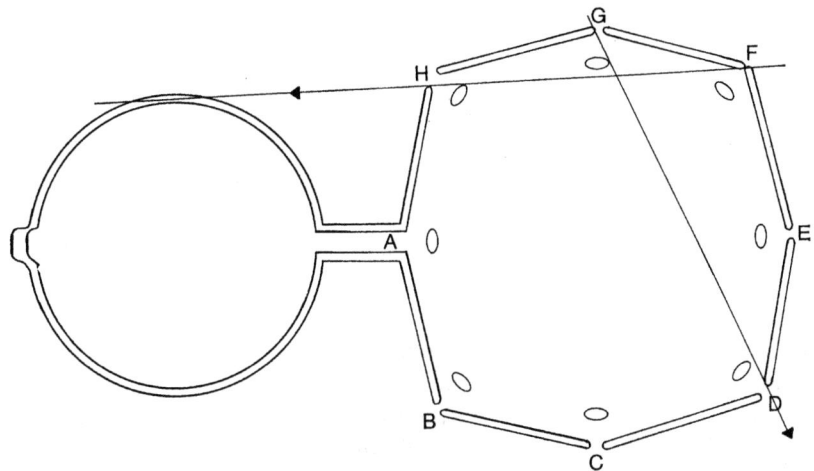

2.8 A sketch shows two of the "missing" lunar alignments which are unmarked by the sides of the Octagon. If these alignments were deliberate, they would account for the narrowness of the gap at *F*, the relatively short length of wall *HA*, and the positions and sizes of mounds *H* and *G*.

ated with the Octagon and its connecting avenue. Secondly, these alignments offer possible explanations for otherwise inexplicable features: the size of the gaps at the vertices; the position and size of the mounds; the location of the small circle.

Stronger evidence suggesting deliberate lunar alignments is found in the azimuthal directions of the parallel displacements of the Fairground Circle and the Square (Fig. 2.5). The displacement of the Fairground Circle with respect to the Observatory Circle aligns with the maximum southern extreme rise point to within 0°.3. It may also be significant that a line from the centre of the Observatory Mound tangent to the Fairground Circle has an azimuth which points to the maximum southern rise to within 0°.1. Therefore, when the Moon was at the maximum extreme in the 18.6-year cycle, an observer on the Observatory Mound would see the northern extreme rise occurring along the Octagon axis, and would see the southern extreme rise occurring the same month tangent to the Fairground Circle. The axis of the avenue leading from the Fairground Circle (equidistant from the midline of the avenue walls) points within 0°.9 to the minimum northern extreme rise point. Curiously, however, the avenue axis does not pass through the centre of the Fairground Circle. A line from the centre of the circle to the centre of the avenue entrance misses the lunar rise point by more than 3°.

The displacement of the Square with respect to the Octagon aligns with the minimum southern extreme rise point to within 1°.2. The same alignment is repeated more accurately inside the Square. A line from the centre of the northwest entrance to the Square (in the avenue leading from the Octagon in Fig. 2.2) to the north end of the southeastern wall of the Square points to the minimum south rise within 0°.3. This northwest entrance is not located at the centre of the side of the Square as is customary for squares constructed by the Hopewell.[25] Rather, the entrance is displaced about 30 m north of the centre of the northwestern side, probably for a conscious purpose and possibly to achieve the indicated lunar alignment. The other entrances to the Square were obliterated prior to the Middleton survey, and no reliable survey data are available for them. The avenue leading from the Octagon to the Square becomes straight for 160 m prior to joining the Square. The centre of this avenue points within $0°.9$ to the maximum northern set point. No reliable survey data exist for the other avenues which entered the Square.

The diagonal of the Square (RP in Fig. 2.5) also has a possible lunar significance. As the rising point of the Moon moves from south to north during the monthly cycle, there is a unique azimuth for which the Moon can be observed to rise and set on the same day in opposite directions along the same line. Due to the Moon's rapid apparent motion this azimuth is not 90° as it is for the Sun at the equinoxes; rather, for the Moon it is $92°.0$.[26] The azimuth of RP is $92°.8$. Thus, an observer could identify the midpoint of the monthly lunar cycle by noting the day when the rising and setting of the Moon most nearly occurred along the line RP. The fact that the rising and setting points of the Moon defined a straight line through the observer in this direction may well have had a special significance for the Hopewell, considering their obvious preoccupation with linear geometric figures.

The seventeen independent lunar alignments we have proposed for Newark are summarized in Table 2.2. Not all of these alignments have an equal probability of being intentional, and some may well be due to chance. We believe the first eight listed in the table are most likely to have been deliberate. These eight alignments have a mean error magnitude of $0°.53$. The mean error magnitude for all seventeen alignments is $0°.56$. Some perspective on the possible significance of this mean error may be gained by comparing the Newark alignments to the thirteen best lunar alignments proposed by Hawkins for Stonehenge which have a mean azimuthal error magnitude of $1°.3$.[27] It is well known that a precise marking of lunar extrema can be used in principle to predict lunar eclipses.[28] The average accuracy associated with the alignments we have proposed is sufficient to allow the prediction of the year when lunar eclipses occur near the equinoxes or solstices. This accuracy is not sufficient, however, to permit the prediction of all lunar eclipses. While eclipse prediction could have been a motivation for the Hopewell, our data do not justify any conclusions on this point.

EVIDENCE FROM OTHER SITES

Ultimately the geometrical and astronomical hypotheses outlined here must be tested at other Hopewellian sites. The degree of geometrical and astronomical sophistication implied by these hypotheses is unlikely to have been unique to Newark, but almost certainly would have been developed over a great length of time during which it would be manifested at other sites. Those sites which would provide suitable tests for these hypotheses would satisfy two criteria: (1) accurate, reliable survey data must be available; and (2) the sites should exhibit evidence (such as geometrical regularity) of having been laid out with great care. The criterion of reliable survey data restricts us to those sites surveyed by Middleton. Of the sites which Middleton surveyed, he noted that four of the Ohio sites possessed significant geometrical regularity: the Newark earthworks; the circle-octagon combination known as the High Bank Works; the square at Liberty Township; and the square at the Baum Works.[29] Space does not permit a detailed analysis of these other sites here. It is significant to note, however, that the size and orientation of all of these sites can be explained using geometrical and astronomical hypotheses similar to those outlined for Newark.

The diameter of the circle at High Bank is virtually identical with that of the Observatory Circle at Newark. The octagon can also be constructed using a method similar to that at Newark. The octagon at High Bank is oriented so that one of its principal axes (the one perpendicular to the short avenue connecting the circle and octagon) is aligned with the northern extreme rise point of the Moon within 1°. One side of the octagon is aligned to the summer solstice sunrise within about 1°. These alignments, if intentional, would completely determine the orientation and structure of the High Bank octagon. The lengths of the diagonals of the Baum and Liberty squares are both equal to three times the radius of the Observatory Circle within better than 1 percent. The Liberty square has exactly the same orientation as the Newark square to within 0°.3, which suggests some kind of astronomical orientation. Two sides of the Baum square are aligned with the summer solstice sunrise within 1°. This kind of analysis needs to be extended to other sites for which accurate survey data exist or for which such data can be obtained, perhaps by aerial photography.[30]

There is additional evidence suggesting celestial and perhaps lunar consciousness among the Hopewell. Specifically, six crescent-shaped mounds are present in the Newark earthworks. Five of these mounds are recorded on the Squier and Davis map, and one was discovered by Middleton.[31] The presence of numerous crescent-shaped mounds and the discovery of additional crescent-shaped artifacts associated with other Hopewell works have led to previous speculation that a lunar cult may have existed among the Hopewell,[32] although this speculation has found no champions among current archaeologists. A possible knowledge of the celestial origin of meteors has been attributed to the Hopewell by Prufer, based on evidence that they collected meteoritic iron.[33]

CONCLUSIONS

Our data and analysis show that the Newark earthworks were carefully conceived and constructed to exhibit a remarkable degree of symmetry, precision, and geometrical harmony, apparently based on a single length (the OCD). Further, we have shown that the structure of the Newark earthworks could be used for a relatively precise monitoring of the monthly and 18.6-year lunar cycles. The important and difficult question is whether the structure was consciously laid out by the Hopewell for this purpose. Surely it must be admitted that some of the lunar alignments may be due to chance. The difficulty in making objective, quantitative assessments of the probability of chance alignments is well known and has been widely discussed.[34] The large number of precise lunar alignments at Newark (especially those associated with the Octagon) weighs heavily against dismissing these alignments as mere coincidence and as unworthy of further investigation. Most of the main features related to the orientation, shape, relative size, and asymmetry of the surviving earthworks can be accounted for with the single hypothesis of deliberate lunar alignment. It should also be emphasized that all the alignments we have found correlate with a single object, the Moon. The intellectual power, tenacity of purpose, continuity, and desire for precision that would be implied by conscious lunar alignment is certainly no greater than that required in the construction of the earthworks themselves, a feat which is not in dispute. The geometric regularity of the works shows clearly that the Hopewell had a strong concern for geometrical harmony, and it is not surprising that they might record celestial harmonies (perhaps essential to their calendar) in the same structure.

ACKNOWLEDGMENTS

We wish to thank Anthony Aveni, John Eddy and Martha Potter Otto for critical comments on a draft of this paper; Olaf Prufer and Martha Potter Otto for assistance with archaeological materials and literature; Evan Farber and the Earlham College Library staff for help in securing manuscripts and aerial photographs; the Earlham College Professional Development Fund for financial support; and Rosemary Wesler for editing and typing the manuscript.

NOTES

1. See G. S. Hawkins, "The New World," in *Beyond Stonehenge* (New York: Harper & Row, 1973), 173–192; T. M. Cowan, "Effigy Mounds and Stellar Representation: A Comparison of Old World and New World Alignment Schemes," in *Archaeoastronomy in Pre-Columbian America,* ed. A. F. Aveni (Austin: University of Texas Press, 1965), 217–235; J. A. Eddy, "Archaeoastronomy of North America: Cliffs, Mounds and Medicine Wheels," in *In Search of Ancient Astronomies,* ed. E. C. Krupp (Garden City, NY: Doubleday, 1977), 133–163; W. Sullivan, "Ancient Mounds Taken as Clues to Advanced Cultures," *New York Times* (19 June 1979); S. Hagar, "The Portsmouth Works," *Popular Astronomy* 41 (1933):2–21; J. Marshall (as told to J. B. Carlson), "Geometry of the Hopewell Earthworks," *Early Man* (Spring 1979):1–5; S. D. Peet, "The Lunar Cult and the Calendar System," *American Antiquarian* (1896):116–125.

2. J. E. Bernhardt, "A Preliminary Survey of Middle Woodland Prehistory in Licking County, Ohio," *Pennsylvania Archaeologist* 46 (1976):39–54. Newark has received surprisingly scant attention in the literature of professional archaeology. The Licking Valley and its tributary streams have not been examined systematically. Current study of the earthworks will not be fully satisfactory until much more is known about the prehistory of the region. Bernhardt's survey presents evidence that Newark was the distribution centre for highly prized jewel-quality flint from nearby Flint Ridge.

3. W. K. Moorehead, "Report of Field Work in Various Portions of Ohio," *Ohio Archaeological and Historical Society Publications* 7 (1900):110–203 (Ohio Archaeological and Historical Society, Columbus, Ohio).

4. See M. F. Seeman in *Hopewell Archaeology: The Chillicothe Conference*, ed. D. S. Brose and N. Greber (Kent, OH: Kent State University Press, 1979), 39–46, and M. F. Seeman, *The Hopewell Interaction Sphere: The Evidence for Interregional Trade and Structural Complexity* (Indianapolis: Indiana Historical Society, 1979). See also Olaf H. Prufer in J. R. Caldwell and R. L. Hall (eds.), *Hopewellian Studies (Illinois State Museum Scientific Papers)* 12 (1964):37–83 (Illinois State Museum, Springfield, Illinois).

5. Prufer in Caldwell and Hall, *Hopewellian Studies*, 70.

6. Ibid., 50. Most archaeologists place Newark in the late Hopewell period, for which radiocarbon dates suggest a range from AD 0 to 500.

7. None of the mounds within the works described has been systematically explored. The bird-effigy was opened under the direction of H. C. Shetrone in 1928, after earlier amateur digging may have destroyed some evidence. Shetrone's workers found two copper artefacts, a crescent and a small, three-dimensional figure of a beaver or manatee. (Unpublished field notes in the possession of the Archaeological Division of the Ohio Historical Society.)

8. D. M. Reeves, "A Newly Discovered Extension of the Newark Works," *Ohio Archaeological and Historical Quarterly* 45 (1936):189–193.

9. R. Morgan, "Ohio's Prehistoric Engineers," *Ohio State Engineer* 20 (1937):2–5.

10. A blueprint in the possession of the Landscaping Division of the Ohio Historical Society dated 19 November 1936 shows reconstruction at the Fairground Circle and Square as well as "cosmetic" work at the Observatory Mound and in the Octagon.

11. E. G. Squier and E. H. Davis, *Ancient Monuments of the Mississippi Valley* (Washington, DC: Smithsonian Institution, 1847), 67–72.

12. Adjutant General, State of Ohio, *Annual Report*, 1893 (Columbus, 1893).

13. C. Atwater, "Description of the Antiquities Discovered in the State of Ohio and Other Western States," *American Antiquarian Society, Archaeologia Americana, Transactions and Collections* 1 (1820):109.

14. See C. Thomas, "The Circular, Square and Octagonal Earthworks of Ohio," *Smithsonian Institution Bureau of Ethnology*, Bulletin 10 (Washington, DC, 1889); ibid., "Report on Mound Explorations of the Bureau of Ethnology," *Smithsonian Institution Bureau of Ethnology, Twelfth Annual Report,* 1890–1891 (Washington, DC, 1894), 1–730.

15. G. Fowke, *Archaeological History of Ohio* (Columbus, OH: Ohio State Archaeological and Historical Society, 1902), 171.

16. W. H. Holmes, "Notes upon Some Geometric Earthworks, with Contour Maps," *American Anthropologist* 5 (1892):363–373.

17. Field Notes of J. Middleton, National Anthropological Archives, MS. no. 2400, Box 5, Smithsonian Institution, Washington, DC (unpublished).

18. Reeves, "Newly Discovered Extension," 189–193. See also the Reeves collection of aerial photographs (8" × 10" neg.) of Newark taken in January and October 1934: National Anthropological Archives, Smithsonian Institution, Washington, DC, negatives no. 124–141, 188–199.

19. Thomas, "Circular, Square and Octagonal Earthworks of Ohio," 22f.

20. Marshall, "Geometry of the Hopewell Earthworks," has sought to understand the geometry of the Hopewellian earthworks in terms of a design composed of right triangles with integral sides and using a unit of measurement of 57 m. The evidence from Newark alone does

not provide a conclusive case for this suggestion. The published data and analysis for other sites are as yet too meager to allow an evaluation of the suggestion.

21. But the following coincident *areas* should be noted. The area of the Square differs from that of the Observatory Circle by 0.36 percent. The area of the Fairground Circle differs from that of square *ACEG* within the Octagon by 0.22 percent. The total area of the Octagon doubles the area of the Observatory Circle and the Circle-Octagon avenue with an error of 0.19 percent.

22. A. F. Aveni, "Astronomical Tables Intended for Use in Astroarchaeological Studies," *American Antiquity* 37 (1972):531–540.

23. Cf. Squier and Davis, *Ancient Monuments of the Mississippi Valley*, 67ff.

24. Eddy, "Archaeoastronomy of North America," 150.

25. See the works by Thomas cited in note 14.

26. Due to variations in the Moon's apparent speed and in the orientation of the Moon's orbit, the precise value of this azimuth varies by $\pm 0°.3$ about an average of $92°.0$.

27. Hawkins, "The New World," 61.

28. A. Thom, *Megalithic Lunar Observatories* (London, Clarendon Press, 1973); E. C. Krupp, "The Stonehenge Chronicles," in *In Search of Ancient Astronomies,* ed. Krupp, 81–132.

29. Thomas, "Circular, Square and Octagonal Earthworks of Ohio," 32.

30. The authors are currently studying the geometry and astronomy of the sites noted.

31. Thomas, "Report on Mound Explorations of the Bureau of Ethnology," 460.

32. Cf. Hagar, "Portsmouth Works," 3f., and Peet, "Lunar Cult and the Calendar System," 116ff.

33. O. H. Prufer, "Prehistoric Hopewell Meteorite Collecting: Context and Implications," *Ohio Journal of Science* 41 (1961):341–352.

34. P. R. Freeman and W. Elmore, "A Test for the Significance of Astronomical Alignments," *Archaeoastronomy* (supplement to *Journal for the History of Astronomy* 1) (1979):S86–S96.

CHAPTER THREE

Archaeoastronomy at Machu Picchu

D. S. Dearborn and R. E. White

INTRODUCTION

In this paper, we will describe some of the results of an Earthwatch-sponsored expedition to study the old Inca citadel at Machu Picchu for sites of astronomical observation. The expedition took place in the summer of 1980; two groups of volunteers assisted the expedition leaders, Dearborn and White, in surveying and general data collection.

The Machu Picchu Citadel was chosen for this study because of its relatively unmolested condition. Unlike Cuzco, the Inca capital, the Spanish conquerors did not occupy Machu Picchu, and so did not have an opportunity to destroy this site of "pagan" worship. Perhaps even more significant is the fact that the citadel was voluntarily abandoned by its inhabitants. Since it was not continuously occupied, as was

From *Ethnoastronomy and Archaeoastronomy in the American Tropics*, ed. A. Aveni and G. Urton, Annals of the New York Academy of Sciences 385 (1982).

Cuzco, the original structures have not been pulled down to make way for modern buildings.

While our time at Machu Picchu was spent in a systematic routine of data gathering, there were two principal types of artifact for which we were searching. Our first interest was in finding structures or monuments designed for making precise astronomical observations associated with either an astronomically significant event (solstice, equinox, lunar excursion, etc.) or a cultural event. In addition, we hoped to find sites that may have been used for crude astronomical observations in conjunction with ceremonies.

While post-Conquest chroniclers leave us no direct information on Machu Picchu, their accounts do give us some idea of the observations that interested the Inca. Garcilazo de la Vega describes the preparations and ceremony attendant to watching sunrise on the day of the June solstice.[1] At Machu Picchu, we have found a structure that is well designed for use as a solstitial observatory. Because the terrain is quite different from Cuzco, it does not use the system of horizon-marking pillars described by Aveni[2] and Zuidema.[3] The flexibility of the Inca astronomers in adapting their observing techniques to the local environment suggests that they were interested in the astronomical event and not simply in making a ritual observation.

The results we wish to report on here involve the structures that Bingham called the "Torreon" and the "Intihuatana Stone."[4] We will present evidence that we believe demonstrates that the Torreon was useful for the observation of the June solstice, the solar zenith passage, and possibly constellations. The Intihuatana Stone itself does not seem to have been designed as an astronomical instrument (at least in any way we have fathomed), but its site may have been of great importance. Observations of sunsets made with reference to apparently artificial structures on the nearby ridge of San Miguel (located across the gorge of the Urumbamba) may have been used in calendric observations.

THE TORREON

A complete description of our findings on the Torreon has been submitted elsewhere (Dearborn and White, 1981), so the discussion here will be brief. The Torreon or "watch tower" is a temple structure that sits on a rock promontory with a clear view of the eastern horizon. Bingham noticed several similarities between the Temple of the Sun in Cuzco and the Torreon, including the fine masonry, the curved "semicircular" wall (Fig. 3.1), and the existence of water flowing nearby.[5]

Two windows penetrate the curved wall; one faces northeast, the other southeast. The interior is dominated by a large carved rock (or "altar"), which extends from the northeast window westwards through the center of the temple. This altar is part of the rock that supports the Torreon itself. It has been cut flat on top, with the exception of a small raised section or low spire in the southwest quadrant. One edge of the raised section runs along the center of the altar and points out of the northeast window.

3.1 The curved wall of the Torreon sits on a rock projection over the Royal Mausoleum with a clear view of the eastern horizon. The southeast window and the serpents door are seen from this view. Photograph by D. S. Dearborn.

Theodolite measurements, accurate to 3 seconds of arc, were made to determine the orientation of this edge. It was found to lie within 2 ± 5' of the direction of the rising point of the sun on the June solstice, which makes it a very precise sighting device for naked eye astronomy. The terrain at Machu Picchu is much more vertical than that in the Cuzco region. For the most part, the eastern horizon is more distant and less accessible. The technique of using pillars to mark the horizon would not be a suitable one over much of the horizon. A shadow-casting method using the solstice-pointing edge of the raised section of the altar would, however, be quite practicable.

Beginning in May, the sunrise illumination of the northeast window (or, more precisely, the shadow of a plumb-bob hung in the window) will lie across the altar and touch the solstice-pointing edge. Each day, as the sunrise point moves farther north, the angle between the window illumination (or plumb-bob shadow) and the solstice-pointing edge decreases until they become parallel on the day of the solstice.

Garcilaso de la Vega describes the preparations for the solstice festival, which involved a three-day fast.[6] This suggests that Inca astronomers did not simply observe the solstice, but made observations to predict it. Given a fiducial mark such as the solstice-pointing edge of the altar's truncated "spire," the date of the solstice can be predicted by measuring the angle between the plumb-bob shadow and the solstice-pointing edge. Such a measurement can be made accurately in a number of ways. For illustration, Dearborn and White (1981) describe a method they tried involving a cord stretched across the window. In any case, the orientation of the window and the carved "altar" stone form a usable system for making precise observations (and predictions) of the June solstice.

The other window faces the ridge of Machu Picchu to the southeast, near the place where the Inca Trail crosses that ridge. Its orientation is such that the sun, moon, and planets would never be observed to rise directly in the window: the window is oriented too far below the ecliptic. We, therefore considered the stars that would appear to rise in the strip of sky viewed from the window.

This strip of sky contains a large segment of the Milky Way, and includes several Inca constellations.[7] The constellation "Collca" (involving the tail of Scorpio) was found to lie just above the horizon at sunset on the day of the June solstice. Urton has described how modern Quechua-speaking groups observe the tail of Scorpio along with the Pleiades (also called "Collca") to signal the time of the solstice.[8] Since the Pleiades have approximately the same declination as the June solstice point, they are observed to rise through the window facing northeast. The event occurs just before sunrise as the solstice approaches. The two windows may then have been used together to observe a system of constellations.

Following the suggestion of J. Molloy, we examined the orientation of the two windows for another use. The widest possible observing angle through a window is obtained by viewing across the window jambs. The technique of using window jambs to define a position on the horizon and make astronomical observations has been used at Casa Grande[9] as well as Chichén Itzá.[10]

In the Torreon, we found that cross-jamb views made from the two windows overlap for only a 4° segment of the southeastern horizon, centered on the position of sunrise on the zenith passage date (February 14, October 29). During most of the year (February–October), the light at sunrise enters the Torreon through the northeast window. From October through February, however, it enters through the southeast window. Only for a five- or six-day period on either side of the zenith passage date will light enter both windows at sunrise.

As described here, the windows of the Torreon do not form an instrument for observing the precise date of zenith passage; they could be used together as a precise observing instrument, however, by observing the portions of the interior illuminated at sunrise. The Torreon, then, may have been used to define the period of time in which zenith passage occurs, and the precise date of zenith passage may have been observationally determined elsewhere.

In summary, the Torreon is designed for use as a precise instrument for observing the June solstice. In addition to this, it could be used to observe constellations and the approach of the zenith passage date.

THE INTIHUATANA STONE

The so-called "Intihuatana Stone" is one of the highest points of the citadel; it sits atop a terraced pyramid on the western edge of the plaza. Its altitude would allow obser-

3.2 Looking west across the Intihuatana Stone, a short pylon or spire projects above the base. Photograph by D. S. Dearborn.

vations of both the eastern and western horizons, but a wall constructed on the top of the pyramid restricts the view of the eastern horizon as seen from the Intihuatana Stone. With the exception of the ridge of San Miguel to the northwest, the western horizon is at a great distance and is unsuitable for constructing pillars to mark sunset positions.

The stone itself (Figs. 3.2 and 3.3) is intricately shaped. The base is not quite rectangular, with approximate dimensions of 2×1.6 meters. The top of the base descends in a slow helix around a central tapering spire or pylon. The pylon is truncated at a height greater than 50 cm above the base; the top surface of the pylon is beveled at such an angle that it appears to lie in a plane with the distant southwest horizon. While the faces of the pylon are not precisely flat, the azimuthal orientations are approximately 126° and 299° (narrow faces) and 40° and 225° (broad faces). The section of sky observed to set over the spire (when facing it square on) is centered on the declination strip −45°, which is 21½° south of the ecliptic plane; therefore, the sun, moon, and planets would never be observed to set within this portion of the sky. Perhaps coincidentally, however, it is the same strip of sky that would be viewed to rise through the southeast window of the Torreon.

While the stone itself does not seem to have any use as an observing instrument (at least, none that we have been able to determine), the site may have been important. Because of its position and altitude, an observer on the Intihuatana Stone pyramid has an unobstructed view of the ridge of San Miguel. This is one of the few segments of horizon near enough to the citadel to make the construction of pillars a suitable method for marking a horizon position.

3.3 The Intihuatana Stone as viewed from the northeast. Photograph by D. S. Dearborn.

While examining the ridge with a small telescope, Mr. Bernard Bell, one of the expedition members, observed what appeared to be two pillars that stuck up above the jungle and that appeared to be artificial. Slightly further north, he observed a natural rock outcrop that appeared to have a window through it (Fig. 3.4). Near sunset, when there was great contrast between the sky and the ridge, the window was clearly seen through the "backlighting" condition.

Dr. Manuel Silva, an anthropologist from the Instituto de Cultura who was accompanying our expedition, told us that he believed the pillars (and two mounds between them) to be parts of an old Huaca. On questioning some local farmers, he reported that, while a trail up the flank of San Miguel used to exist, it had not been maintained or used in over ten years. Reaching the San Miguel ridge would, then, require one to climb more than 1100 meters at an inclination of 65° to 70° through chest-high brush or jungle containing large numbers of poisonous snakes (fer-de-lance). Not being prepared at that time for such a trek, we were unable to reach the ridge to examine these structures closely and to determine whether or not they were artificial. We did, however, calculate the date on which the sunset occurs behind each of the structures as viewed from the Intihuatana Stone.

The window (which we dubbed "Pilar" for ease of referral) is located about 2.3 km from the Intihuatana Stone, and the opening is approximately 30 × 18 cm. Observations of the window from different locations suggest that at least one side is two meters deep. The strip of sky that passes beyond the window is at declination strip

3.4 A view of the ridge of San Miguel as seen from Huana Picchu. To the left (south) the pillars (covered in foliage) rise above the jungle with two mounds or pyramids between. On the right, a rock outcropping is penetrated by a hole or window about 2 meters below its crest. As viewed from the Intihuatana Stone, the sun will shine through this window on May 11 and August 2. Photograph by D. S. Dearborn.

17.7°. This gives two dates (May 11 and August 2) on which the sun sets behind the rock outcropping and shines through the window for a few extra seconds. Neither of these dates has any astronomical significance. The August date corresponds to the beginning of the period that is now called Herranza, which is associated with the time when the earth becomes fertile (the beginning of the planting season).[11]

There is also a possible association between Pilar and the observations in the Torreon. The dates on which the sun shines through Pilar closely bracket the dates on which the portion of the altar stone containing the solstice-pointing edge is illuminated at sunrise. Observations can be made to observe the solstice during this time.

Finally, Urton has pointed out to us that the time between solar passages behind Pilar closely corresponds to three sidereal lunar months, a period that Zuidema believes to have been an important unit of time in the Inca calendar.[12] While windows in rock are not common natural formations, they do exist; proof that Pilar is artificial must await closer examination of the crestline of the San Miguel mountain.

The "Huaca" pillars are located slightly over 2° and 3° north of Pilar. The northernmost pillar is thinner and, as viewed from the Intihuatana Stone, is covered with foliage. The dates when the sun would appear to set beyond the pillars are April 29 and May 3 in the late fall, and August 10 and 14 in the late winter. These dates correspond closely with the Festival of Santa Cruz (May 1–3),[13] and occur in a period associated with the harvest season. Again, proof that the pillars were used for the purpose of horizonal astronomy and to determine festival dates must await closer inspection.

Any expedition to San Miguel Mountain should examine the positions corresponding to sunset on other astronomically significant dates (solstice, nadir, passage). Since most of the ridge is covered in jungle impenetrable even to infrared film, a close inspection in person is mandatory.

CONCLUSION

The Torreon and the system including the Intihuatana Stone and the San Miguel Ridge are suitable for making astronomical calculations. The precision with which these Inca artifacts can be used strengthens our belief that they were so used.

In addition to the Torreon and Intihuatana Stone, a number of carved stones in the citadel area were examined for orientations suitable for astronomical observations. In one instance a crude alignment was found, but nothing with the precision of the Torreon.

Perhaps the most interesting structure requiring additional work is the site Bingham called "the Temple of the Moon."[14] While it is located on the north flank of Huana Picchu, where the South Celestial Pole is not visible, it is oriented along the cardinal directions, with walls facing due north, east, and west. Part of the structure is a narrow hall or corridor with high walls, which is oriented due north-south. Such a design would be particularly useful in observing the daily change in the altitude of the sun as it transverses the meridian. In this way, solar zenith passage could be observed very accurately.

We believe that our expedition was successful in finding evidence concerning the types of observations that interested the Inca. It has, however, opened our eyes just enough to see that much more study is needed. Additional work must be done on the ridge of San Miguel, at the Temple of the Moon, as well as other places in the citadel.

NOTES

1. G. de la Vega, in *The Incas,* ed. Alain Gheerbrant (New York: Avon Books, 1961).

2. A. F. Aveni, "Horizonal Astronomy in Incaic Cuzco," *Lateinamerika Studien* (München) 10 (1982): 175–193.

3. R. T. Zuidema, in *Native American Astronomy,* ed. A. F. Aveni (Austin: University of Texas Press, 1977).

4. H. Bingham, *Inca Land* (New York: Houghton Mifflin, 1922).

5. Ibid.

6. De la Vega, in *The Incas.*

7. G. Urton, *Ethnology* 17 (1978):157.

8. G. Urton, "The Astronomical System of a Community in the Peruvian Andes," Diss. University of Illinois, Urbana, chap. 7.

9. J. Molloy, R. E. White, T. P. Culbert, and Kayzer (Preprint, 1970).

10. A. F. Aveni, in *Archaeoastronomy in Pre-Columbian America,* ed. A. F. Aveni (Austin: University of Texas Press, 1975).
11. B. J. Isbell, *To Defend Ourselves* (Austin: University of Texas Press, 1978).
12. Zuidema, in *Native American Astronomy*.
13. Isbell, *To Defend Ourselves*.
14. Bingham, *Inca Land*.

PART II

Acquiring Cultural Context

In this section I have selected several articles that respond to the important critique that unless one seeks cultural context to validate claims, the results will never carry much weight. These pieces attempt to explain the relationship among astronomy, astronomical orientations, and the calendars developed by native people of the Americas. The contributions by Michael Coe and Miguel León-Portilla (on Mesoamerica) are largely overview pieces. Those written by Michael Zeilik (on Hopi-Zuni-Anasazi), R. T. Zuidema, and Anthony Aveni (on Andean Quechua societies) move closer to using cultural information to generate archaeoastronomical hypotheses. More recently published papers by Aveni (Chapter 10) and David Dearborn and Katharina Schreiber (Chapter 6) go even further.

Studies in Inca astronomy were initiated in the important paper titled "The Inca Calendar," an abridged version of which is reproduced here. It is the earliest exposition of anthropologist Tom Zuidema's many years of work on astronomy and calendar in the *ceque* system of the Inca capital of Cuzco, Peru, and its relation to social structure.

The ceques are an organization of invisible radial lines delineated by *huacas*, or sacred places, where people made offerings to the earth mother Pachamama at designated times. According to Zuidema, the ceque system is the organizing principle behind the social, political, geographic (especially hydrological), and cosmographic structure of the great city. Statements by the Spanish chroniclers about the ceques and their huacas have served as the basis for alignment studies carried out in the field by Zuidema and Aveni (see also Zuidema 1982). The first of these papers, "Horizon Astronomy in Incaic Cuzco," is reproduced here. Work on the ceques was updated by Bauer and Dearborn (1995), who have been critical of some of the alignments postulated by Aveni and Zuidema (especially zenith and anti-zenith sun alignments). Dearborn and Schreiber's interpretation of the ceque system and the responses by Aveni, Zuidema, and Hyslop are also reprinted here. Meanwhile Bauer's 1998 book on the ceque system, which gives better archaeological documentation of some huaca locations, has shown that the ceques are not as straight as Zuidema had portrayed them (cf., e.g., Zuidema's fig. 15.7 with Bauer's map 11.1). On the other hand, I have argued that astronomy works even in a crooked ceque system (Aveni 1996). That the role of astronomy in the ceque system of Cuzco remains controversial, I believe, is one of the soundest reasons for interested students to engage it.

Erected between 200 and 300 BC, the hilltop center of Chankillo, located on the north coast of Peru, offers one of the earliest examples of an Andean monumental calendar. As Ivan Ghezzi and Clive Ruggles demonstrate, thirteen towers that create a sawtooth profile, when sighted from a pair of artificial sighting points located due east and west of the ridge on which they are positioned, turn out to delimit the horizon solstitial positions. This collaborative study is significant because it extends the tradition of the use of towers as astronomical markers cited in the Inca chronicles backward nearly two millennia.

Astronomer Michael Zeilik's paper on archaeoastronomy in the U.S. Southwest is a model of how to apply an ethnographic research base to the comprehension of the place of astronomy in Pueblo life. It is also important because it underlines the social necessity of making *anticipatory* naked-eye observations and it demonstrates that setting up the ritual side of the agricultural calendar can be one of the principal motives for practicing skywatching.

Anthropologist/archaeologist Michael Coe's keynote address at the first international conference on archaeoastronomy, held in Mexico City in 1973, remains a classic introduction to the database for studies in Mesoamerica. Coe draws from a large body of resources, including ethnohistory, ethnography, the codices, and the monumental inscriptions, offering us a valuable perspective on Mesoamerican skywatching. Coe's primer on Aztec constellations has motivated a number of investigators to study star patterns in Native American cultures. Indeed, his closing paragraph has helped inspire a whole new generation of ethnoastronomers.

Next follow two synthesizing pieces that demonstrate the application of theories of culture to the study of ancient astronomy. A number of similar papers are listed in the bibliography. In "The Role of Astronomical Orientation in the Delineation of World View," I attempt to place our understanding of urban planning that incorporates astronomical alignments within the context of a social model that relates the center of a city to its periphery. I use Maya Copan, Aztec Tenochtítlan, and Inca Cuzco as examples.

Arguments about astronomically aligned sacred structures often can be greatly strengthened when ethnohistorical records actually cite them. Such is the case for the Templo Mayor of Tenochtítlan, the great Aztec temple of sacrifice. For example, one of the Spanish chroniclers tells us that King Moctezuma needed to tear this building down and reconstruct it because it was misaligned with the intended solar orientation (at the equinox). The piece by Ivan Šprajc, which builds on earlier work cited in the bibliography, shows, however, that the historical record can be as subject to interpretation as the alignments. Šprajc proposes that, placed in its mountain setting, the Templo Mayor was actually a solar calendar that utilized numbers known to have been significant in the written pre-Columbian manuscripts that survive.

Was the location of a building we find to be astronomically aligned deliberately chosen and are its architectural nuances purposely altered to accord with the local land- and skyscape? Šprajc proposes that this was certainly the case for the Templo Mayor, and in the article on Palenque's Temple of the Sun, which combines the efforts of three archaeologists, one of whom is a native Tzeltal Maya, Alonso Mendez and colleagues argue that this was surely the case. Their detailed study of structural alterations inside the temple (they also propose a blueprint of how they were executed in classical times) may have been related to accessing solar and lunar light at key times in their cycles.

The existence of a pre-contact written record in both monumental and manuscript form offers a great advantage to those who would attempt to understand the place of astronomy in ancient Maya culture. Current studies raise issues of Maya cultural identity and the role of astronomy in political ritual (cf., e.g., Rice 2004, 2007). Perhaps most important in this regard are the Maya E-Groups, specialized assemblages consisting of a radial pyramid (which can be accessed from each of the four directions) on the west side of an open plaza and three low-lying buildings situated on the east. In a number of instances, the latter mark the positions of sunrise at the solstices and equinoxes. But are we dealing with performative ritual space that incorporates astronomical phenomena or precise observations utilized by special skywatchers? Continuing studies (cf. Aveni and Hartung 1989; Aveni, Dowd, and Vining 2003) show that the situation is far more complex. The piece by James Aimers and Prudence Rice reproduced here represents the most thorough recent attempt to link the enigmatic E-Group set to Maya astronomy and geopolitical structure.

Mexican humanist and ethnohistorian Miguel León-Portilla's inspiring essay, delivered at the second Oxford conference, held in Merida, Yucatan, in 1986, is about the "ethos" of Mesoamerican astronomy. It constitutes yet another plea for establishing cultural context. I put it last in the sequence because of its breadth. León-Portilla not only places the understanding of the heavens in a religious context but also encourages us to think about the meaning of time in other cultures. In particular, his explanation of the cyclic calendar and its destinies offers a stark contrast to the Western view of time as a mechanism that exists apart from human concern.

THINGS TO THINK ABOUT

1. How do you know when to plant flowers out of doors? Do you think people really *need* celestial cues? If not, then why bother making astronomical alignments?

2. According to Zuidema, how did the ceque system of Cuzco contribute to the social and political cohesion of the Inca? Can you draw parallels to other cultures discussed in these articles?

3. Use the bibliography along with other articles in this section to follow up on how critics responded to Zuidema's suggestion that the Inca used horizon pillars in Cuzco to demarcate important times in the seasonal calendar.

4. Suppose you want to set up an observatory somewhere in Mesoamerica to chart the daily motion of the sun at the horizon. Select a pair of sticks, a foresight and a backsight, and position them 100 yards apart. Sink the foresight into the ground in a vertical position and line up the backsight so that the setting sun is positioned over both markers. (You will need help from someone to hold the stick because you need to step well back along the sight line to get an accurate alignment.) Next drive a stake in the ground at the point where the backsight touches the ground. This marks the sunset alignment on day one. A day or two later come back to the same place and set up the backsight again. You should note that this time the sun at horizon has shifted, and as a result, the stake marking the spot will fall a short distance to the north or south of the first position (depending on what time of year it is). Table II.1 gives the amount of lateral shift you can expect, in inches and feet for angles of one degree, 1/2°, and one minute of arc, or 1/60°, for a 100-yard baseline. First suppose you watch the sun's movement when it lies close to the equinoxes. Table II.2 shows the daily shift in azimuth around the time of the equinoxes, when it moves the fastest. By how much (in inches) would the backsight be shifted from day to day? Next, suppose similar observations are made around the times of the solstices, when the sun moves very slowly. Again consult Table II.2 for the shift as you attempt to determine the daily amount by which you would expect to move the backsight to fit the new alignment. Can you devise a plan to circumvent the obvious difficulty in charting the day-to-day slow motion of the sun around the solstices? This is a fascinating experiment you can easily try out anywhere, including, for short baselines, your backyard (if you have an unobstructed view of the horizon) or the roof of a building. You will come away amazed at how much precise astronomy you can do with a minimum of technology!

5. Do a brief survey of all the alignments mentioned in this section, paying close attention to the actual date of each proposed astronomical passage and the day or days this phenomenon supposedly works. From this survey, do you think astronomy was closely studied in order to *anticipate* certain events? Or were the phenomena (e.g., the passage of Venus in the window of Temple 22 at Copán)

Part II: Acquiring Cultural Context

Table II.1 Conversion of linear to angular azimuth error* for a 100-yard baseline

Angular Displacement	Approximate Equivalent in Linear Measure
1/60°	3/4 inch
1/2°	2 feet
1°	4 feet

* Note that one minute of arc (1/60°) is less than the daily change of position of sun at horizon within approximately two days of the solstices. Thirty minutes of arc corresponds approximately to the angular diameter of the sun or moon as well as to the approximate daily change of position of sun at horizon around the time of the equinoxes (see Table II.2).

Table II.2 The shift in the azimuth of sunrise for an observer in latitude 20°N*

Time	Daily Change in Azimuth in Minutes of Arc
Equinoxes	25.3
Equinoxes ± 1 day	25.2
Equinoxes ± 2 days	25.1
Solstices	0.0
Solstices ± 1 day	0.3
Solstices ± 2 days	1/2

* For observers in higher latitudes, the differences will be slightly larger because the sun crosses the horizon obliquely (see Aveni 2001:344n4).

part of the event itself? Or maybe both? Use concrete evidence from these articles to corroborate your arguments.

6. Compare León-Portilla's article on the Mesoamerican ethos with Coe's "Native Astronomy in Mesoamerica." What points do they agree make up the basic compendium of Mesoamerican archaeoastronomy? Where are there differences? Knowing that León-Portilla is a humanist and historian and Coe an archaeologist, how might these differences be explained?

7. In his widely referenced keynote paper, Coe states that the moon "travels at different times of the year. Lacking gravitational theory, the ancients could never satisfactorily account for the variation in the length of the synodic month" (p. 232). How is this specific quotation evidence that Mayanists still have far to go in understanding astronomy in other cultures?

8. Choose one of the articles in the following bibliography and research it. Discuss how it advances cultural astronomy over the articles reproduced in Part I.

SUGGESTIONS FOR FURTHER READING

Ashmore, W. 1989. "Construction and Cosmology: Politics and Ideology in Lowland Maya Settlement Patterns." In *Word and Image in Maya Culture: Exploration in Language, Writing, and Representation,* ed. W. Hanks and D. Rice, 272–286. Salt Lake City, University of Utah Press.

Aveni, A. 1996. "Astronomy and the Ceque System." *J. Steward Anthropological Society* 24(1, 2):157–172.

Aveni, A. 2000. "Out of Teotihuacan: Origins of Celestial Canon in Mesoamerica." In *Mesoamerica's Classic Heritage: From Teotihuacan to the Aztecs,* ed. D. Carrasco, L. Jones, and S. Sessions, 253–268. Niwot, University Press of Colorado.

Aveni, A. 2001. *Skywatchers: A Revised and Updated Version of Skywatchers of Ancient Mexico.* Austin, University of Texas Press.

Aveni, A., A. Dowd, and B. Vining. 2003. "Maya Calendar Reform? Evidence from Orientations of Specialized Architectural Assemblages." *Latin American Antiquity* 14(2):159–178.

Aveni, A., and H. Hartung. 1989. "Uaxactun, Guatemala, Group E and Similar Assemblages." In *World Archaeoastronomy*, ed. A. Aveni, 441–460. Cambridge, Cambridge University Press.

Aveni, A., H. Hartung, and J. C. Kelley. 1982. "Alta Vista (Chalchihuites), Astronomical Implications of a Mesoamerican Ceremonial Outpost at the Tropic of Cancer." *American Antiquity* 47(2):316–335.

Bauer, B. 1998. *The Sacred Landscape of the Inca: The Cuzco Ceque System*. Austin, University of Texas Press.

Bauer, B., and D. Dearborn. 1995. *Astronomy and Empire in the Ancient Andes*. Austin, University of Texas Press.

Carlson, J. 1993. "Venus Regulated Warfare and Ritual Sacrifice in Mesoamerica." In *Astronomies and Cultures,* ed. C. Ruggles and N. Saunders, 202–252. Niwot, University Press of Colorado.

McCluskey, S. 1987. "Science, Society, Objectivity, in the Astronomies of the Southwest." In *Astronomy and Ceremony in the Prehistoric Southwest,* ed. J. B. Carlson and W. J. Judge, 205–217. Papers of the Maxwell Museum of Anthropology, no. 2, Albuquerque.

Rice, P. 2004. *Maya Political Science: Time, Astronomy and the Cosmos*. Austin, University of Texas Press.

Rice, P. 2007. *Maya Calendar Origins: Monument, Mythistory, and the Materialization of Time*. Austin, University of Texas Press.

Williamson, R. 1984. *Living the Sky: The Cosmos of the American Indian*. Boston, Houghton Mifflin.

Zuidema, R. T. 1964. *The Ceque System of Cuzco*. Leiden, Brill.

Zuidema, R. T. 1982. "The Role of the Pleiades and of the Southern Cross and Alpha and Beta Centauri in the Calendar of the Incas." In *Ethnoastronomy and Archaeoastronomy in the American Tropics,* ed. A. Aveni and G. Urton, 203–229. Annals of the New York Academy of Sciences 385.

CHAPTER FOUR

The Inca Calendar

R. T. Zuidema

Most Spanish chroniclers in the sixteenth and seventeenth centuries report that the calendar of the Incas consisted of twelve months and was, in their own words, very similar to ours. Some authors in the nineteenth and early twentieth centuries indulged in highly fantastic and unfounded calendric schemes that were reputed to be Incaic, but these consisted more of the dross of simplistic Western ideas about primitive and archaic calendars.

Some students made a serious attempt; Nordenskiold (1925), for example, analyzed the "quipus" (knotted strings) and their calendrical content. In a sense even this was all guesswork as he had no ethnohistorical data to support his conclusions. Most students from the seventeenth century on stuck to the twelve month names and left it at that. Some modern students (e.g., Muller 1929, 1972) have tried to check out the

---

First published as "The Inca Calendar," by R. T. Zuidema, in *Native American Astronomy*, ed. A. Aveni. © 1977 by the University of Texas Press. All rights reserved.

Inca data on solar risings and settings, but as they did not know the place from where these were observed and as they hypothesized the wrong center of observation, they could not arrive at reliable reconstructions of the observational system.

Notwithstanding the apparent simplistic calendrical system of the Incas, there are many loose data, inconsistencies, and contradictions that do not fit into it. The well-informed indigenous chronicler Felipe Guaman Poma de Ayala (1936, ff. 893, 894 [pp. 883, 884]) is adamant about the fact that the Incas had an elaborate and precise calendar; he even gives us the picture, along with the name and address, of a calendar specialist—Juan Yumpa, from Uchucmarca in the province of Lucanas, whom he calls an "astrologuer-poet, who knows the round of the Sun and the Moon, eclipse and of stars and comets, hour, sunday, and month and year and of the four winds of the world for sowing the food since of old." Jose de Acosta (1954) in his famous chronicle, "Natural and moral history of the Indies," compares the Peruvian calendar favorably to the Mexican one (book 6, chap. 3). The work of recent scholars on the modern Andean agricultural and ceremonial calendar (Brownrigg 1973; Escobar et al. 1967) also demonstrates a high complexity, even if it has to be computed now by way of the Christian calendar.

Acosta had a direct knowledge of the complexity of the Inca calendar, as he refers (book 6, chap. 19) to the *ceque* system of Cuzco. This consisted of 328 *huacas* ("sacred shrines") in and around Cuzco which, in groups of an average of eight *huacas*, were aligned into forty-one directions (the *ceques*, or lines) toward the horizon. The *ceque* system was used for different social and ritual purposes. An explicitly expressed purpose (according to the priest Cristobal de Molina and the lawyer Juan Polo de Ondegardo, who did a painstaking research of recording it in the decade of 1560–1570 [Lohman Villena 1967]) was, however, also as a counting device for the calendar, each *huaca* standing for one day in the year, and as a system of sight lines for observing astronomical events on the horizon. The *ceque* system as a calendar was also mentioned by other students who knew about it—like Juan de Matienzo, the son-in-law of Polo de Ondegardo. At the moment we do not need to analyze why these students, or why the Jesuit Bernabe Cobo who preserved it in his chronicle, did not further analyze the *ceque* system as a calendar, as was done in the case of the Mexican and Maya calendars. One reason may be that in the Inca case the calendar was recorded on quipus in terms of an abstract theory of political organization, which has to be analyzed first, and not in terms of an elaborate system of gods, animals, and color directions, as it was preserved in Mexican pre- and postconquest codices. If we want to study the pre-Spanish Andean calendar, we will have to make the jump into insecurity through a study of the Inca theory of political organization and of an abstract *ceque* system.

In my rereading of the Spanish chroniclers for a study of Inca symbolism in religion and rituals and for a study of the iconography of pre-Spanish art, I also concentrated my attention on an analysis of the calendrical aspects. However, I want to treat most of these data as consequences of the calendar, to be explained by it, and not as information for its reconstruction. In the case of the preconquest data, I am aware so far of only two textiles that represent intricate calendrical accounts of all the days in a full solar year. I will analyze these first, as they set the stage for the kind of numerical computations that we also find in the *ceque* system as an Inca calendar. Here I have concentrated on three aspects that are perhaps necessary in any calendar as a description of astronomical events.

The first aspect involves the existence of schematic time units as defined by Neugebauer (1942) in the case of the Egyptian calendar. A schematic time unit is any fixed amount of days that repeats itself through time *independently* and that can be used to describe and measure astronomical time units. Such schematic time units are, of course, of importance to the administration of a centralized political organization. I will suggest that a first application of the *ceque* system was concerned with the recording of these schematic time units.

The second aspect involves the measurement of astronomical time units. In this case, the *ceque* system was used differently as a quipu for recording the measured astronomical time units.

The third aspect involves the use of the *ceque* system for observing astronomical events in terms of risings and settings of the sun, the stars, and possibly the moon. This, then, was still another use of the *ceque* system, independent from the two others.

Earlier fieldwork done in 1953 together with Dr. Manuel Chavez Ballon (see Zuidema 1964), as well as ongoing fieldwork started in the summers of 1973 and 1975, was primarily concentrated on exactly locating as many *huacas* as possible, especially those that were used as markers *(sucanca)* for astronomical observations. This work has advanced considerably and much use has been made of modern toponyms, data on old and new maps, data from chronicles and documents in general pertaining to the Cuzco area, and mythological and ritual data on Inca culture in general. Certain results of this work, for example, pertaining to the directional sequence of the *ceques*, will be used in the present paper. Nonetheless, much work still has to be done and a full discussion pertaining to the location of the *huacas* with their religious, ritual, and mythological significance must await another occasion. For a more elaborate discussion of the *ceques* and *huacas* used for the observation of sunrise and sunset during the solstices, see Zuidema (1976). Additional reasons for concentrating in this paper on numerical values in the *ceque* system are that this analysis can be done mostly independently from the location of the *huacas* and that arguments developed here may be helpful in locating the *huacas*. The discussion of Inca month names and their monthly rituals will also be kept to a formal statement of the problem as, again, a fuller understanding is dependent on an analysis of the numerical problems involved in the *ceque* system.

The most important reason, however, for choosing this approach in opening a discussion of Andean calendrical and astronomical problems lies in the nature of the *ceque* system itself. Knowing its functions as a day-to-day account of the whole year and as a system of astronomical observation, we would expect to have 365 or 366 *huacas* and 36 *ceques,* not 328 *huacas* and *41 ceques.* These and other numbers are squarely based on sidereal lunar computations and not on solar ones, whereas "solar" numbers are only secondarily embedded into this lunar system. The problem of lunar astronomical observations is far more complex than solar because of the more erratic movements of the moon. A precise location of the *huacas* and an understanding of the Inca techniques of astronomical observation are therefore also dependent on a better understanding of the Inca "theory of the moon" and this is what this paper will mostly deal with.

THE *CEQUE* SYSTEM

The *ceque* system has been compared to a giant quipu, laid out over the Cuzco valley and the surrounding hills that served in the local representation of the Inca cosmological system, in its spatial, hierarchical, and temporal aspects. The *ceque* system was used at different times of the year for different purposes and by different classes of people for recording superimposed cycles of ritual events. To understand the different calendrical cycles that were represented and integrated into one system, it will be necessary to distinguish and separate the layers of meaning attached to each event and its cycle. Not only can the *ceque* system be metaphorically compared to a quipu but every local group did in fact record its *ceque* system, that is, its political, religious, and calendrical organization on a quipu (Matienzo, p. 119), and the local keepers of these records, the *huacacamayoc* or *villcacamayoc* (specialists of the *huacas* and *villcas*, i.e., of the sacred shrines and objects) (Cook, book 13, chap. 22; Molina 1943, pp. 69–77), had to give account of them to the national administration.

The meaning of the *ceque* system can be explained best by way of the feast of the Capac hucha, when children were sacrificed (Molina, pp. 69–77) in Cuzco during the two solstitial feasts (Guaman Poma, ff. 247, 259, 262, 263). At that time, the ritual and calendrical (Matienzo, p. 120) value and rank of all the *huacas* in the empire, of every village, town, province, were reassessed. The representatives of these sent presents and children to the Inca in Cuzco. Some were sacrificed there while others were sent back and sacrificed at home. These *capac hucha* (also called *cachahui* or *cachahuaco,* from *cacha*, "messenger") followed a straight line (*ceque*) on their way home and did not go by way of the roads (Duviols 1967, pp. 26, 27, 38; Molina 1943; Rostworowski 1970, pp. 23–24).

The *huacas* of the *ceque* system in Cuzco were also recorded on a quipu. On the basis of this information, Molina wrote "Relacion de las huacas" (an account of the *huacas*) of Cuzco for the bishop of Cuzco. Polo de Ondegardo gave a similar account, from a quipu (Matienzo, pp. 119, 120), to the viceroy. Both accounts are from around 1570 and it is probable that their information, as preserved in the chronicle of Cobo, derived from the same quipu. Many of the place names and *huacas* mentioned by Molina in his description of the calendrical feasts of Cuzco coincide with those of the *ceque* system.

Guaman Poma mentions as a name for secretaries *quilla uata quipoc* (f. 359) "the quipu specialist for months and years" and as a name for accountants and treasurers *hucha quipoc* (f. 351), using the same word, *hucha,* as that for the human sacrifices.

4.1. The quipu specialist for the months and years. (From Guaman Poma, f. 360.)

Gonzalez Holguin also has a similar name, *hucha yachak,* for secretary. For us, the most important fact is the calendrical reference in the names given by Guaman Poma. In a drawing of the *hucha quipoc* (f. 360) he shows a man with a quipu and a tablet of 20 squares with black and white points in each of the squares (Fig. 4.1). The tablet was used for counting numbers and for "setting down feasts and sundays and months and years." We will argue later that the tablet may have had a calendrical function.

The reason why the *capac hucha* walked in a straight direction was probably because they had to follow a specific *ceque*. Since the *ceques* were considered to go out from the Temple of the Sun and since specific *huacas* also had a function in terms of astronomical observation, we may assume that the *ceques* were used as sight lines for the observation of solar, lunar, and stellar risings and settings. For the purpose of this paper it is of great importance that the *ceques* going straight N, W, and E could still be determined, because the *huacas* on these *ceques* are still precisely known.

The Order of the *Ceques*

The *ceque* system consisted of 41 *ceques* (lines) radiating from the Temple of the Sun in Cuzco and organizing 328 *huacas* in and around the city. Both numbers are of calendrical importance as $8 \times 41 = 12 \times 27\frac{1}{3} = 328$, 8 being the number of days in the Andean week, 12 the number of solar months in the year, and $27\frac{1}{3}$ the number of days in the sidereal lunar month. One *ceque* consisted of two lines running parallel to each other and not, as Cobo claims, one being the continuation of the other. However, for different calendrical reasons they were counted either as one or two. Finally, in his conclusion, Cobo does not say that there were 41 or 42 *ceques,* but 40.

The number 40 was important in the Inca theory of political organization. Local groups were matched, in an ascending order of size, to a hierarchical order of 100, 500, 1,000, 5,000, 10,000 and 40,000 (not 50,000!) families. The province of Cuzco, compared to a unit of 40,000, consisted of the city of Cuzco, the valley of Cuzco, and an outer zone of (probably 40) different non-Inca kingdoms that had received the honorary rank of Incas-by-privilege. In a special rite during the equinoctial feast of Cituay in September, 400 warriors had to "drive out the illnesses from Cuzco," each group of 10 following one *ceque* (Cobo, book 13, chap. 29). In this context we may understand why Polo de Ondegardo once mentions "more than 400 *huacas*" (p. 55) and at another place (p. 43) specifies the number as 340: 12 *sucancas* (observation posts on the horizon) and 328 *huacas*. Therefore, our first conclusion is that to the 40 or 41 *ceques* could be attached the schematic number values of either 8 ($8 \times 41 = 328$) or 10 ($10 \times 40 = 400$).

The distribution of the 40 (41, 42) *ceques* was not regular in the case of one of the four divisions of Cuzco. We will follow step by step how the Incas arrived at the actual distribution of the *ceques* since it is important for an understanding of the calendar.

The area surrounding Cuzco was first divided into a northern upper moiety called Hanan-Cuzco, and a southern lower moiety, Hurin-Cuzco. These moieties were themselves halved to produce four parts, *suyu*. A straight W-E line was formed

R. T. Zuidema

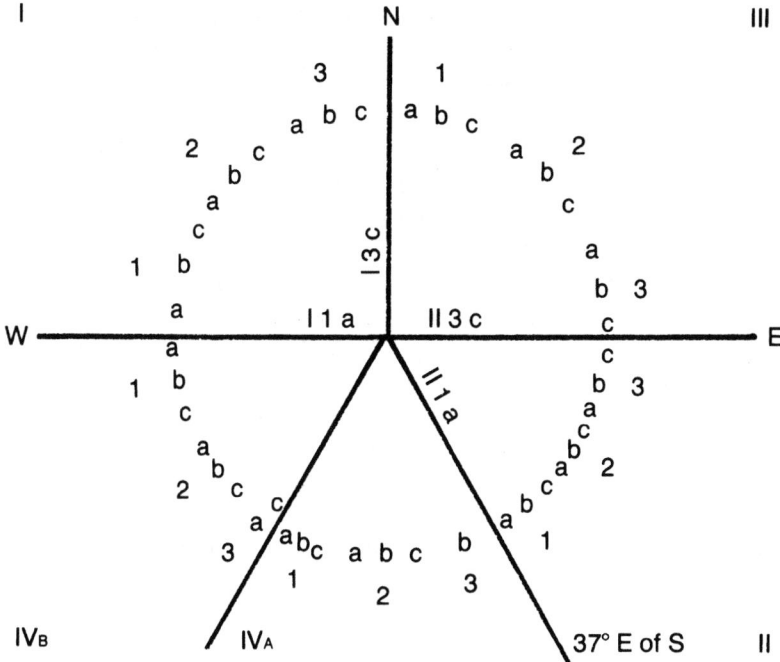

4.2. Scheme of the *ceque* system.

by the first *ceque* of Chinchaysuyu (NW) and the last *ceque* of Antisuyu (NE). The last *ceque* of Chinchaysuyu went straight north, separating this quarter from Antisuyu. The last *ceque* of Collasuyu, going 37° E of S, divided Hurin into two unequal parts: a small Collasuyu section (SE) and a large Cuntisuyu section (SW). Each *suyu* had groups of 3 *ceques* with the individual called either:

Collana, a, "principal, first"
Payan, b, "second, middle"
Cayao, c, from *calla* "origin"

A similar hierarchical order was applied to the groups of *ceques* (1, 2, 3) and to the *suyus* I, II, and III. In the latter context of primarily local and not hierarchical organization, Cuntisuyu (IV) represented an extra group in opposition to the first three. Cuntisuyu reversed the internal hierarchical order of the other *suyus*. The scheme is delineated in Figure 4.2.

The *huacas* and their *ceques* are enumerated in Cobo's "Relacion de los ceques" in the following order:

I 3 c b a, 2 c b a, 1 c b a
III 1 a b c, 2 a b c, 3 a b c
II 3 c b a, 2 c b a, 1 c b a
IVB 3 b, 2 c b a, 1 c b a, IVA 3 c-a, 2 c b a, 1 c b a

The Inca Calendar

In four cases, the generic name of the *ceque* was replaced by the name of the group associated with it (see IV g, V a). The fourth *ceque* of Cinchaysuyu (I 2 c) was called Payao and the fifth one (I 2 b) Cayao. Probably Cobo here made an error, the fourth *ceque* being Cayao and the fifth Payan. I corrected this error.

If we accept that Cobo's order of enumeration is based on Inca information, then we might conclude that the *huacas* and *ceques* of I and II are enumerated in ascending order. In IV the order may have been descending because of a reversed hierarchical value of the terms Collana, Payan, and Cayao (see V b). The Incas arrived at the number of 14 or 15 *ceques* in Cuntisuyu by first dividing this *suyu* into two parts, A and B. A comparison to the political organization of nearby San Jeronimo clearly demonstrates this intention (Zuidema 1964, pp. 222–223, 241–242). The Incaic organization there was ranked as one of 500 families. The 5 *pachacas* (*pachaca* = a group of a hundred) of San Jeronimo corresponded to the 4 *suyus* of Cuzco with Cuntisuyu, IV, matching the two last *pachacas*. From this comparison we might expect that the two halves of Cuntisuyu would also contain 9 *ceques* each, making a total for Cuzco of 5 × 9 = 45 *ceques*. A first reduction to 42 was obtained by *not* redoubling *ceques* 3 a b c. The final reduction to 41 resulted from joining *ceque* b to part B of IV and *ceques* c and a, taken as one, to part A. The suggested reduplication and reduction enable us to reconstruct the original situation:

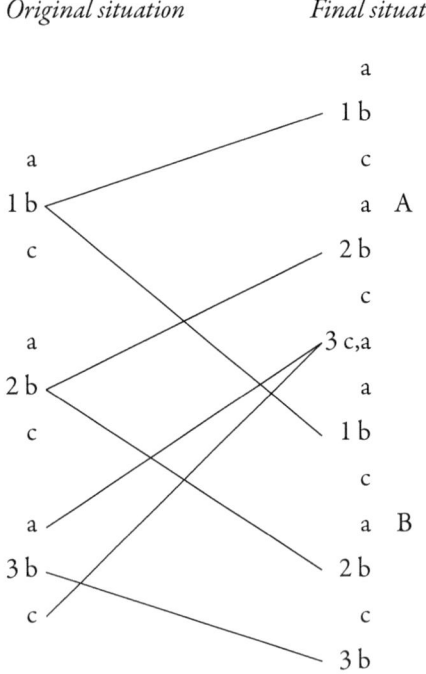

Later when I use the subscripts A and B (e.g., IVA 1 b) I will be referring to the final situation unless specific reference to the original situation is made.

The number 42 (or 41 or 40) was mentioned elsewhere in Peru in similar contexts. In two cases we can also study a similar reduction from 45.

Copacabana

The Incaic Temple of the Sun in Copacabana, on Lake Titicaca, was served by 42 "nations" from all over the empire (Ramos Gavilan 1621). The list consisted of:

1. One descent group of the Inca high nobility, called Sucsupanaca, to which always belonged the High Priest of the Sun.

2. The two groups of Hanan- and Hurin-Cuzco.

3. The four groups of Inca-by-privilege, Huaro, Quihuar, Papre, and Chilque.

4. Thirty-eight other nations.

Altogether, there were 45 groups although Gavilan specifically mentions the number 42. One reduction he applied was by joining Huaro and Quihuar and also Papre and Chilque. A similar reduction was used in similar contexts and on the same groups by Guaman Poma (ff. 84, 85, 750, 740) and in certain colonial documents. A further reduction could be achieved by taking Sucsupanaca as one group and Hanan- and Hurin-Cuzco as another. The total then would be one group of Inca high nobility, one group of Inca low nobility, and 40 other non-Inca groups. Of these 42 groups, 41 served the *panaca* (sisterhood or the grouping or household of the sisters and female cousins of a certain man) to which the High Priest of the Sun belonged, probably in calendrical ritual. Cobo may have referred to a similar situation in a village near Lake Titicaca when he says (book 14, chap. 17): "Once when I was present at the procession of Corpus Christi [i.e., the Catholic feast that replaced the old June solstitial celebration] in a village of Collao, I counted 40 dances, each different from the other, in which they imitated by way of their dress, their singing, and their dancing the different nations to which they belonged."

The Ritual of First Plowing in Hurin-Haucaypata

There were two "plazas of the Sun" in Cuzco: one of Hanan-Cuzco, Hanan-Haucaypata ("the plaza of rejoicing of Hanan"), in the center of town, and one of Hurin-Cuzco, Hurin-Haucaypata. The latter was close to the Temple of the Sun but outside the ceremonial part of the town within the confluence of two rivers. Calendrical festivities, with the Inca present, were carried out in both plazas. The second one was close to the field where the rituals of first harvesting, in May, and of first plowing, in August, were carried out. We will describe the latter feast (Rostworowski 1970b, pp. 165–166).

The Inca sat on his throne of gold (a pyramid) with his 4 councilors as chiefs of the 4 *suyus* of the empire behind him. They were described as chiefs of 100,000 families. Behind the 4 councilors were seated the chiefs of 10,000 families; there were 10 of these in each *suyu*. In the last position were seated the ambassadors of the provinces, who were apparently also divided into 40 groups. First, the Inca plowed a certain distance alone; next the 4 councilors plowed twice that distance; finally, the 40 chiefs with the rest of the people plowed till the end of the morning. In the afternoon, the 45 dignitaries (Inca, 4 councilors, 40 chiefs) who each had a golden digging stick played games while the other people continued plowing. The hierarchy involved was the same as in Copacabana and a similar type of reduction of the number of dignitaries was probably applied, the 4 councilors counting as one.

The Organization of Priests in Allauca

In an article on the ancestral organization of the village of Allauca (Zuidema 1973, from Hernandez Principe 1622) in the department of Ancash, central Peru, I analyzed the following organization of male and female priests distributed over its 4 *ayllus* (a vague or extended family by common agreement):

	First *ayllu*	Second *ayllu*	Third *ayllu*	Fourth *ayllu*
Male priests	8	8	4	4
Female priests	12	16	6	8

The numbers in the third and fourth *ayllus* are, for both men and women, half of those in the first and second *ayllus*, respectively. The first and third *ayllus* together and the second and fourth *ayllus* together each have 12 male priests. In a similar example from Cuzco, the Inca had 12 councilors: 4 from I, 2 from III, 4 from II, and 2 from IV (Guaman Poma, ff. 183, 365; Santacruz Pachacuti, p. 256). In the case of female priests of Allauca, the first and second *ayllus* together have 28 (28 is the number of days in a sidereal month). Together with the number of priests in the third and fourth *ayllus*, there are 42 female priests, as many as the number of *ceques* in Cuzco. The description of Allauca is all the more important for our purposes as it is given in the context of human sacrifices *(capac hucha, cachahui, ceque)*. We can conclude, therefore, that we are dealing here with the only specific description of a *ceque* system in Peru with a calendrical reference (numbers 12, 28, 42) similar to the *ceque* system of Cuzco.

The Kingdom of Nampallec in Lambayeque

The last example is from the pre-Inca and pre-Chimu kingdom of Nampallec (Cabello Valboa, book 3, chap. 17) in the north coast valley of Lambayeque. This organization consisted of 12 towns, including the capital, which were originally governed by the 12 sons of the second king in a legendary dynasty of 12 kings. These kings were served by 40 court officials who were each sent as a specialist by a different village.

R. T. Zuidema

THE POLITICAL SUBDIVISIONS OF CUZCO, THE GROUPS OF *CEQUES,* AND THE MEASURING OF ASTRONOMICAL TIME UNITS BY WAY OF *CEQUES* AND *HUACAS*

The Political Subdivisions of Cuzco

So far we have discussed the formal organization of the *ceque* system and we have suggested certain correlations between this formal system and another one consisting of schematic years, months, and weeks. Only after establishing such a formal and schematic system can one describe the astronomical time units which were measured by it. This fact is just as true for the modern student who wants to reconstruct the Andean calendar as it was for the Inca administration who wanted to use the same calendar in their whole empire. A mythical explanation says that it was for this reason that one Inca king divided the year into 12 months, because "in this way the generally accepted order would be such, from Quito to Chile, that they never would lose track of time" ("Discurso" in Maurtua, p. 150).

Astronomical values like the half-year periods between when the sun is in perihelion and aphelion, synodical months, solar months, sidereal lunar months, and possibly the Venus cycle were expressed by way of the numbers of *huacas* in the groups of *ceques* and combinations thereof. These combinations were explained in terms of the political and religious organization of Cuzco. Certain facets of this organization will have to be discussed before we can tackle the astronomical problem.

The anonymous chronicler best introduces the organizational problem when he mentions how one king "divided the population of Cuzco into 12 parts and ordered that each part would take up the names of its month and of the occupation carried out then, and that at the beginning of its month the group would come out on the central plaza, announcing its month and playing trumpets in order that everybody would know" ("Discurso" in Maurtua, p. 150).

Cobo also refers to this fact. On the one hand, he mentions the names of these 12 groups, each being associated with a certain group of 3 *ceques* of which it took care; on the other hand, he associates each group of 3 *ceques* with a specific *sucanca* on the horizon of Cuzco which indicated sunrise or sunset at the beginning of a specific month. Nonetheless, it is difficult to match the *ceque* system to these statements of Cobo, as the latter refer to a solar calendar and the totals of *ceques* and *huacas* fit only into a lunar calendar. So far, we can accept that certain associations were made between political groups and groups of *ceques* and months; however, they were not one-to-one associations. We become aware of this problem when discussing the association of the two non-Inca towns in the valley of Cuzco with the *ceques* and months.

Besides the Inca city of Cuzco at the west end of the valley, the town of Sanu (today San Sebastian), situated in the center of the Cuzco valley, represented the pre-Inca kingdom of the Ayarmaca, which was north of Cuzco. The town of Oma (today San Jeronimo), at the east end, represented in the valley an original kingdom of which the towns of Urcos and Huaro (some 40 km SE of Cuzco) formed the center. People from these towns had the special occupation, carried out outside of Cuzco, of being

87

priests and confessors to the Inca high nobility. Sanu and Oma are not mentioned on any *ceque,* but they are situated between III 3 c and II 3 c. The Ayarmaca of Sanu gave their name to *ceque* III 3 b which was close to the town, and, since the people of Oma worshipped as their ancestress a Mama Anahuarque (Rostworowski 1962, p. 137), we may assume that they took care of *ceque* IVb 3 b which was also called Anahuarque. This *ceque,* however, was not in the direction of the town of Oma. Whereas the high nobility of Cuzco celebrated its initiation feasts in December (Capac Inti raymi), the Ayarmaca did so in November (Ayarmaca raymi), and the people from Urcos (including those from Oma) had their initiation feast in October (Oma raymi) (C. Molina 1943, pp. 46–48).

These data, although helpful, still leave us completely confused about the association of political divisions with groups of *ceques* and months. The high nobility of Cuzco, who belonged to Capac ayllu ("the royal descent group") and who celebrated Capac Inti raymi, took care of the group of *ceques* I 1. We can relate Ayarmaca to III 3 and Oma to IV 3. But only the name Anahuarque suggests the association of the month Oma raymi, called after the town of Oma, to IV 3. In the two other cases we will have to suggest later, on astronomical grounds, that the month Ayarmaca was related to IV 2 and the month Capac Inti raymi to IV 1. Drawing a positive conclusion from a negative result, we must reaffirm that there was not a one-to-one relation of a political division with a group of *ceques* and with a specific month.

The problem of the association of the other 10 political divisions with groups of *ceques* is extremely complex and we will mention here only the results of our research.

The rank and function of each of these divisions were indicated in three different ways:

1. By the rank of its suyu and its group of *ceques*
2. By the mythical association to one of the former Inca kings
3. By the specific religious cult related to each group and its ancestor

The situation in Hanan-Cuzco was as follows:

I 1 (Collana)	10 Sun	8 Viracocha	(Collana) III 1
I 2 (Payan)	9 Thunder	7	(Payan) III 2
I 3 (Cayao)	6	Ayarmaca	(Cayao) III 3

The descending hierarchy was from I 1 to III 3 and is shown in correlation with the numbers of the last (10) to the first (6) king of Hanan-Cuzco. We must consider this distribution as regular in comparison to other similar examples. The first three groups were related respectively to the cult of the three major gods of the Inca pantheon. The cult of the visible Sun was of primary concern to Inca government. Thunder and Venus, considered as sons of Sun, were worshipped as gods of war and agriculture. Viracocha, the creator god, as the real, invisible sun, was considered ancestor to the priestly hierarchy. The three gods, then, were part of a solar cult within a solar calendar.

The following two political divisions of lesser importance had chthonic associations. The last group was non-Incaic. Because of the numbers of *huacas* involved, I will henceforth call I 1, I 2, and III 1 the major groups of *ceques* and III 2, I 3, and III 3 the minor ones.

The association in Hurin-Cuzco of groups of *ceques* with political divisions was not regular, although the latter derived their origin from a dynasty of five kings with a similar internal structure as those of Hanan-Cuzco. Some of the changes were due to the reduplication and reduction of *ceques* in IV and their reversed hierarchy. Of the major groups of *ceques:* II 2 now belonged to the descendents of the fifth king, corresponding to the tenth king (II) in Hanan-Cuzco. III now belonged to the descendents of the fourth king, corresponding to the ninth (I 2) in Hanan-Cuzco. IV 3, containing *ceque* Anahuarque (IVB 3 b), did not belong to an Inca group, but probably to the town of Oma. Of the minor groups of *ceques:* II 3 belonged to the descendents of the third king, corresponding to the eighth in Hanan-Cuzco. In Cuntisuyu we only know that IVB 1 belonged to the descendents of the first king. We have no data on the second king; therefore, we cannot draw any conclusions about the original situation in IV, except for the fact that these two other political divisions belonged to minor groups of *ceques* (IV 2 and IV 1).

We already mentioned the council of 12 nobles to the Inca king. Comparing the situation in Cuzco to that of Allauca, it seems most reasonable to suggest that only the 6 major groups of *ceques* were related to these councilors. That is:

4 from I 1 and I 2
2 from III 1
4 from II 2 and II 1
2 from IV 3

We will discover later that it was this group of *ceques* that defined the solar year by their number.

The Numbering of Huacas in the Ceque System

Cobo and Polo de Ondegardo mention explicitly 328 *huacas* of the *ceque* system. There were, however, some extra *huacas*. Here I will discuss to what extent the extra *huacas* might have affected the total number of *huacas* for the calculation of astronomically important numbers.

First, there were the 12 *sucanca* (Herrera, book 4, chap. 5; Polo), or 14 (Cobo), pillars used for astronomical observation. Both chroniclers mention these in addition to the total of *huacas,* although they do not give their names and locations in a special list. Of some *sucanca* we know, however, that they also functioned as regular *huacas* in the list of 328. Molina (pp. 26, 27) mentions two other *sucanca,* but these are located beyond the *ceque* system. Therefore, there is no reason to suppose that the sucanca were also used for counting extra days.

Second, the Temple of the Sun itself was counted as an extra *huaca*, although it was "not placed in the ceque system" (Cobo, p. 186). I want to propose that the temple could be counted, when necessary, as an extra *huaca* of IVB. Different data support this suggestion. Santillan (p. 47) includes the whole of Cuzco, as the center of Inca political organization, in Cuntisuyu. Guaman Poma (f. 183) relates the high priest of Cuzco, serving in the Temple of the Sun, to Cuntisuyu. Of its two subdivisions, IVA (37 *huacas*) and IVB (43 *huacas*), the latter one has more *huacas*, indicating its relatively higher importance. Whereas in the other three *suyus* some *ceques* started at the foot of the Temple of the Sun and others were farther removed from it, it was only in IVB that three *ceques* (1 a, 1 b, and 3 b) started from *huacas* within the Temple of the Sun. In IVA all *ceques* started on the edge of Cuzco. Finally, *ceque* IVB 3 b, called Anahuarque, is the longest single *ceque* of the whole system. For this reason I propose, too, that the descending hierarchy of the *ceques* and groups of *ceques* in IV, or at least in IVB, was inverted in opposition to their order in the other *suyus*.

Third, at the end of the description of the *ceques*, Cobo mentions 4 extra *huacas* that belonged to "different *ceques* that were not placed in the same order as the others when the investigation was made." Apparently (Cobo, book 13, chap. 28; Molina, pp. 64, 65), these *huacas* had a special function in relation to the beginning and the end of the irrigation system. One *huaca*, Tocoripuquio, should be found on *ceque* I 2 a but was forgotten: Cobo says that *ceque* I 2 a has 11 *huacas* although he mentions only 10. The name of another "extra" *huaca*, Quiquijana, was in fact also mentioned on *ceque* II 2 a.

I propose, therefore, that the last two extra *huacas*, not mentioned here, also corresponded to 2 of the 328 *huacas* and that we can take this number as having been intended by the informants in their description of the *ceque* system to Polo de Ondegardo. In the next paragraph I will show that there is also an internal consistency of the total of 328 *huacas* by way of the distribution of the numbers of *huacas* belonging to each group of three *ceques*.

The last problem we have to deal with here is the distribution of the *huacas* in Cuntisuyu. Although the number of *ceques* was augmented here, as was the space given to this *suyu*, the number of *huacas* was not. The different order of the *ceques* did, however, redistribute the *huacas* into other groups of other numbers. In the following diagram of the number of *huacas* in the *ceque* system, I will include for Cuntisuyu:

1. The original number of *huacas* in a situation of 3 × 3 *ceques*
2. The number of *huacas* in the later situation

The number of *huacas* in the original situation conforms completely to the pattern in the other *suyus* and I will analyze this distribution in the next paragraph. In the later situation the *huacas* are redistributed into the two major groups, IVB and IVA, with the number of *huacas* 43 and 37 respectively. Here I want to interpret the possible calendrical meaning of this secondary redistribution.

We noticed before that Cuntisuyu played a double role within the organization of 4 *suyus*. First, it is a *suyu* like the 3 others with a distribution of 3 × 3 = 9 *ceques* and with number values for the *huacas* of the 3 groups of *ceques* similar to those of the other *suyus*. Second, comparing the organization of 4 *suyus* in Cuzco to that of the 5 *pachacas* in San Jeronimo, we discovered that Cuntisuyu alone corresponded to the fourth and fifth *pachacas* of this non-Inca town, worshipping as their ancestress Mama Anahuarque. Similar 5-fold models were integrated in Cuzco into the 6-fold divisions of the *ceque* system in, respectively, Hanan-Cuzco and Hurin-Cuzco. The *panacas* of the 5 kings in each moiety were related to the first 5 groups of *ceques* and the non- and pre-Inca population to the sixth group. Apparently, a similar 6-fold geographic division of the whole Cuzco area was also known. Around Cuzco, 4 other temples of the sun were recognized, corresponding to the 4 *suyus*; but there were 6 sacred mountains: one in each of the 4 *suyus*, a fifth in Antisuyu, and a sixth in Cuntisuyu. The last two were close to the E-W line dividing Hanan- and Hurin-Cuzco (Sarmiento, chap. 31). The two southern mountains, Huanacauri in Collasuyu and Anahuarque in Cuntisuyu, were related to the pre-Inca population of Cuzco and to their Inca conquest. Elsewhere I have argued that the 6-fold division of Hanan-Cuzco can be interpreted as the imposition of one 4-fold division upon another in the following way:

```
          I 1      │  III 1
          (10) │ (8)
   I 3 ───────────┼─────────── III 3
   (6)              │        (Ayarmaca)
          I 2      │  III 2
          (9)  │ (7)
```

and

I 1, I 2	III 1, III 2
I 3	III 3

My argument is now that the data on Cuntisuyu also reveal a superimposition of one 4-fold model upon another one and that describing them within a 6-fold model the situation is this:

I	II
III	IV
IVa	IVb

According to this model we can consider the following totals of *huacas* in the *ceque* system:

1. I + II + III represent a unit similar to that of the three major groups of *ceques* in the organization of Hanan-Cuzco (see also next section). The total of the *huacas* is 85 + 85 + 78 = 248. This number was used in Babylonian astronomy,

and in other systems derived from it, as a close approximation of the period of 9 anomalistic months ($9 \times 27.55 = 247.95$), useful for the prediction of lunar eclipses (van der Waerden 1974, pp. 244, 245).

2. I + II + III + IV represent a system of 12 sidereal months ($328 = 12 \times 27\frac{1}{3}$) corresponding to another system of 12 solar months (see following sections).

3. I + II + III + IV + IVA, counting the *huacas* of IVA a second time, backward, gives a total of $328 + 37 = 365$, the number of days in the solar year. Perhaps a division of this total into 248 (I + II + III) and 117 (IV + IVA) may help us to understand the division of the year into 3 seasons of 4 months each as given by Guaman Poma. He claims that the 4-month season of August to November is governed by the moon and that then the nights are longer than the days. One conclusion of the analysis in the next section will be that the Incas had 4 months of 29 days and 8 months with a medium length of 31 days. The division mentioned here could have been another useful way to relate IV with its 117 huacas (IV + IVA) to 4 synodical months and I + II + III to the 8 solar months (8×31), being equal to 9 anomalistic months.

4. I + II + III + IVA + IVB, containing $365 + 43 = 408$ *huacas*, is close to that counted on the second textile of the Inca period: 407 squares ($32 + [7 \times 28] + 37 + [4 \times 28] + 30$). The squares in the corners of the textile are, however, difficult to count and the total may therefore be either 406 or 408. This total may explain why Polo de Ondegardo and Acosta mention 328 as well as "more than 400" *huacas* for the *ceque* system. We remember, moreover, that in the Situay raymi (September equinox) 400 warriors would follow the 40 or 41 *ceques* (10 warriors for each *ceque*). Four hundred eight days is close to the period of 15 nodical months ($15 \times 27.2 = 408$).

The Astronomical Significance of the Number of *Huacas*

Introduction. Professor T. Barthel, of Tubingen, suggested to me (personal communication) that in different cases the sum of a group of *ceques* with a high number of *huacas* (a major group) and a group with a low number (a minor one) indicates the value of a double sidereal lunar month. This suggestion led me to the discovery of an overall system involving sums of *huacas* of two or more groups of *ceques*.

The more important *suyus* (I and II) each have 2 major groups of *ceques* with values of 29 or more *huacas*. III and IV each have only one major group. In I, II, and III, these are higher-ranked groups. In IV, group 3 has the highest number. The numbers are those of a synodical month (29 or 30) or some days more (31 or 33), as suggested by Guaman Poma (f. 260) for adapting a synodical month to a calendrical one. The numbers of the minor groups of *ceques* are significantly lower but also show a regular pattern: III 2 and IV 2 each have 24 *huacas*, I 3 and III 3 of Hanan-Cuzco have 23 *huacas,* and II 3 and IV 3 of Hurin-Cuzco have 26 *huacas* (Table 4.1).

The perihelion and aphelion. The total of the higher numbers in Hanan-Cuzco ($33 + 29 + 31$) is 93; that in Hurin-Cuzco ($30 + 29 + 30$) is 89. These numbers could be

Table 4.1 Arrangement of *Huacas* in the *Ceque* System.

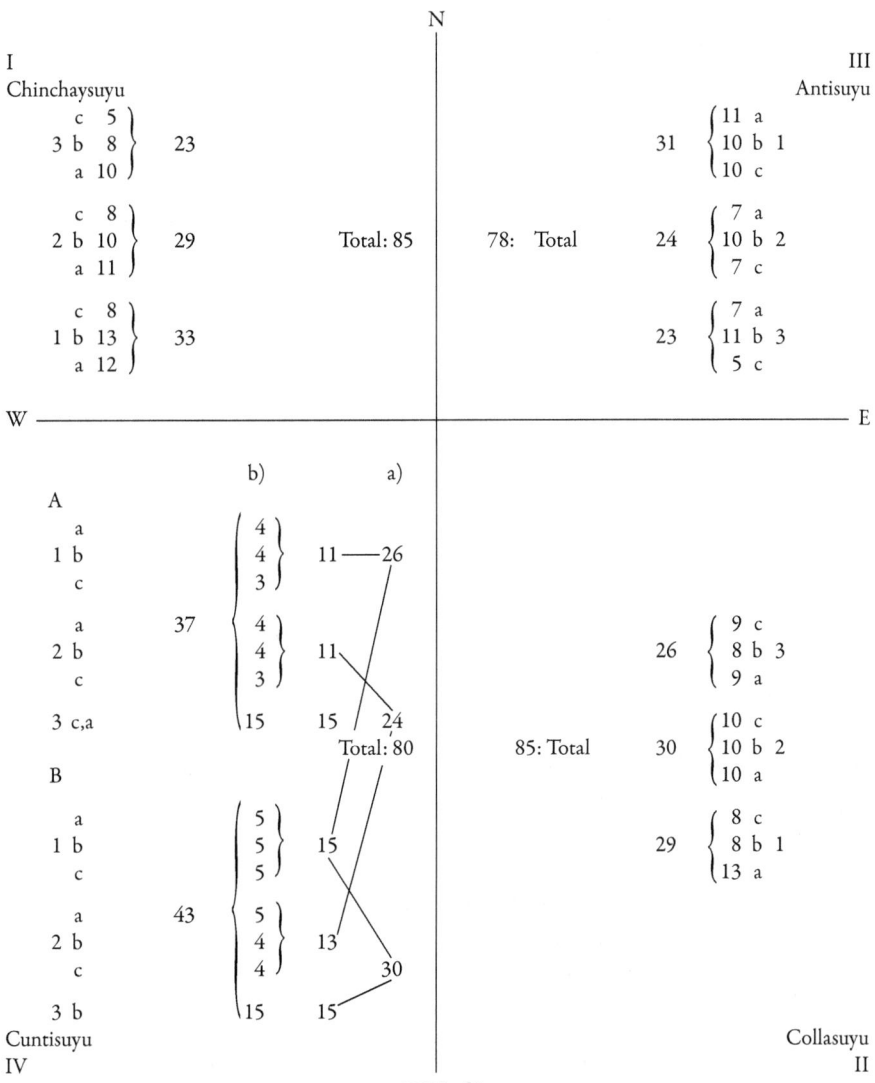

used for calculating the half year from March equinox over June solstice to September equinox when the sun is in aphelion (186 days) and the other half year when the sun is in perihelion (179 days). If we accept that the solstices fall halfway between the equinoxes (the solstices are difficult to fix on a specific day since the sun rises and sets at the same place on the horizon during different days, whereas during the equinoxes it moves fastest from one day to another), then we could say that the numbers 93 and 89 were used for calculating from equinox to solstice and from solstice to equinox back again. This would give us the values of: $(2 \times 93 = 186) + (2 \times 89 = 178) = 364$.

The Inca Calendar

In this case, the Temple of the Sun could have been added as an extra *huaca* to make a full solar year (364 + 1 = 365). The number 186 belonged to Hanan (I and II), the northern moiety of Cuzco. During that time of the year the sun also rises and sets north of the E-W line. The number 179 was in the same way related to Hurin (II and IV) and the south. We remember that it was probably also the 6 major groups of *ceques* that gave to the Inca his council of 12 persons. (Each representing one solar month?) From these data we conclude that each of the 6 major groups of *ceques* may have served 2 months in the calendar of 12 solar months. Analyzing first the case of Hanan-Cuzco I suggest that:

1. I 1 (33 *huacas*) indicated the two longest months around the June solstice (June 21), that is (in terms of the Christian calendar), from May 20 to July 24. These months were dedicated to the state cult of the Sun.

2. I 2 (29 *huacas*) indicated the months from April 21 to May 19 and from July 25 to August 22 and was dedicated to the cult of the Thunder and ritually to the first harvesting and first plowing, respectively.

3. III 1 (31 *huacas*) indicated the months from March 21 (equinox) to April 20 and from August 23 to September 22 (equinox), dedicated to the cult of Viracocha and ritually to water and irrigation.

We remember that according to Molina the year started "around the middle of May" and according to Guaman Poma the sun "sat still" in his solstitial seat till Santiago (July 25), which date he took as the first one of the month of August. The two days in May and July are close approximations in defining the period of the two solstitial months dedicated to the cult of the Sun. Apparently, Molina took the first day for starting his description of the ritual and solar year while Guaman Poma took the last day for describing the agricultural cycle in the year.

The perihelion and the synodic months. In Hurin-Cuzco we found as a value for the 6 solar months the number 178. That is, one day longer than 6 synodic months ($6 \times 29.5 = 177$) and one day short of the half year when the sun is in perihelion. The values of the *huacas* in the major groups of *ceques* of Hurin-Cuzco (30, 29, 30) could be used to describe the solar months in terms of synodic months. Because of the irregularities mentioned in Hurin-Cuzco, it is more difficult to decide upon the exact relationship of each group of *ceques* to two months. I would suggest the following relationships:

1. II 2 (30 *huacas*) indicated the two months around the December solstice from November 21 to January 20. The name of the fifth king, whose descendents took care of the *ceques* in II 2, was Capac Yupanqui, a synonym of Tupac Yupanqui (tenth king) and a name employing Capac ("royal") as do the two solstitial months of December (Capac Inti raymi and Capac raymi Camay quilla). Capac Yupanqui also had a specific relationship to the cult of the Sun (Guaman Poma, ff. 100, 101; Murua, book 1, chap. 11).

2. II 1 (29 *huacas*) indicated the two months from January 21 to February 18 and from October 23 to November 21. The *ceques* here were taken care of by the descendents of the fourth king who occupied in Hurin-Cuzco the same position as those of the ninth king in Hanan-Cuzco. The number of *huacas* in both cases was also the same.

3. IV 3 (30 *huacas* and 31 *huacas*) indicated the two months from February 19 to March 20 and from September 23 to October 23.

In conclusion, we might say that the particular phase of the moon that governed the September equinox in a specific year would recur also in the next December solstice and the next March equinox. After that day, this relationship of phase of the moon to solar month would change rapidly and so, after the following solstitial month of July, it would be very different. This fact might explain why we do find more month names during this season of 4 solar months (April to July) than there are months, since this would be the time for making necessary adjustments. We remember also the statement of the anonymous chronicler that each of the 12 political divisions in Cuzco had its name in common with a month. We found possible evidence for that statement only in one case: that of the town of Oma with the month Oma raymi and of the group of *ceques* to which *ceque* Anahuarque (IV 3) belonged. Our present explanation of the relationship between solar months and synodic months supports that statement.

The sidereal period of 73 nights (?). Having discovered the regularity and significance of the number of *huacas* in the major groups of *ceques*, we might ask if a case could be made for any regularity in the numbers constituting the minor groups.

In IV and I together and in III and II together there are, in each case, 73 $(26 + 24 + 23)$ *huacas* belonging to the minor groups of *ceques*. In this case the axis of division is not the W-E line but a N-S line $(5 \times 73 = 365)$. The period of 73 days could have been used to correlate the solar year with the sidereal lunar year (a double sidereal lunar year, including the Temple of the Sun once: $328 + 329 = 657 = 9 \times 73$) and the Venus cycle $(8 \times 73 = 584)$. Rauh has made a case for the Incaic use of a 73-day period, observing that Inca wooden bridges were rebuilt every 8 years, making the 8-year period a significant interval (Rauh, quotes in Thompson and Murra 1966). In section III c, I mentioned the possible importance of the period of 16 years for integrating solar and Venus rituals into one system.

I want to suggest in the next section that a 73-night period (possibly divided into periods of 36 and 37 nights) was used in sidereal time reckoning, especially in relation to the sidereal lunar calendar. In opposition to the major groups of *ceques* with solar relationships, the minor groups had chthonic relationships, that is, to the heaven at night.

The sidereal lunar months and their possible relationship to the solar year. In the data from the *ceque* system $(8 \times 41 = 12 \times 27⅓ = 328$ *huacas*), the sacerdotal organization of Allauca $(28 + 14 = 42$ female priests), the Inca period textile, and other pre-Spanish examples in Peruvian art and architecture, we find convincing evidence

THE INCA CALENDAR

that the sidereal lunar month was important in pre-conquest Andean culture. But we do not have any ethnohistorical or modern anthropological information on how such a calendar might have been used. One reason is that probably it was an exclusively female calendar. The following hypothetical suggestions are therefore based only on the significance of the numbers in the *ceque* system with the hope that this line of reasoning might be conducive to the discovery of new data confirming or rejecting it.

After 27⅓ days the moon will take the same position against the stars as before. The time at night, however, will be different. An 82-day period (3 × 27½ = 82) would have been convenient to see the moon at the same time of night against the same stars. The number of *huacas* in the four *suyus* are, however, only rough approximations:

    I: 85 *huacas*
    II: 85 *huacas*
    III: 78 *huacas*
    IV: 80 (or 81) *huacas*

Following the suggestion of Dr. Barthel, each major group of *ceques* can be combined with a minor one and will give a very close approximation of 2 sidereal months in their combined number of *huacas*. The most elegant combinations are:

$$\left.\begin{array}{l}31+23=54\\33+23=56\end{array}\right\}\ 110$$

$$\left.\begin{array}{l}30+24=54\\29+26=55\end{array}\right\}\ 109 \qquad \text{Total: 328}$$

$$\left.\begin{array}{l}30+24=54\\29+26=55\end{array}\right\}\ 109$$

These periods describe the sidereal lunar year as a completely independent cycle in terms of double sidereal lunar months (Fig. 4.3).

The numbers of these double months are built up from one solar month unit and another unit considerably lower. It is therefore reasonable to suppose that the sidereal lunar calendar was also integrated into the solar year. Taking into account (1) the double use of each of the 6 high numbers 33, 31, 30, 30, 29, 29 around a W-E axis and (2) a similar double use of the 6 low numbers 26, 24, 23, 26, 24, 23, but around an N-S axis, we develop the model in Figure 4.3. In each of the following periods the moon would be in the same sidereal position at the beginning and at the end (the dash indicating that three letters under it represent one calendrical position):

Q$\overline{\text{Pp}}$ab:	30 + 24 = 54
RQ$\overline{\text{Pp}}$abc:	29 + 54 + 26 = 109
SRQ$\overline{\text{Pp}}$abcd:	30 + 109 + 23 = 162
qp$\overline{\text{PA}}$B:	24 + 31 = 55
rqp$\overline{\text{PA}}$BC:	26 + 55 + 29 = 110
srqp$\overline{\text{PA}}$BCD:	23 + 110 + 33 = 166

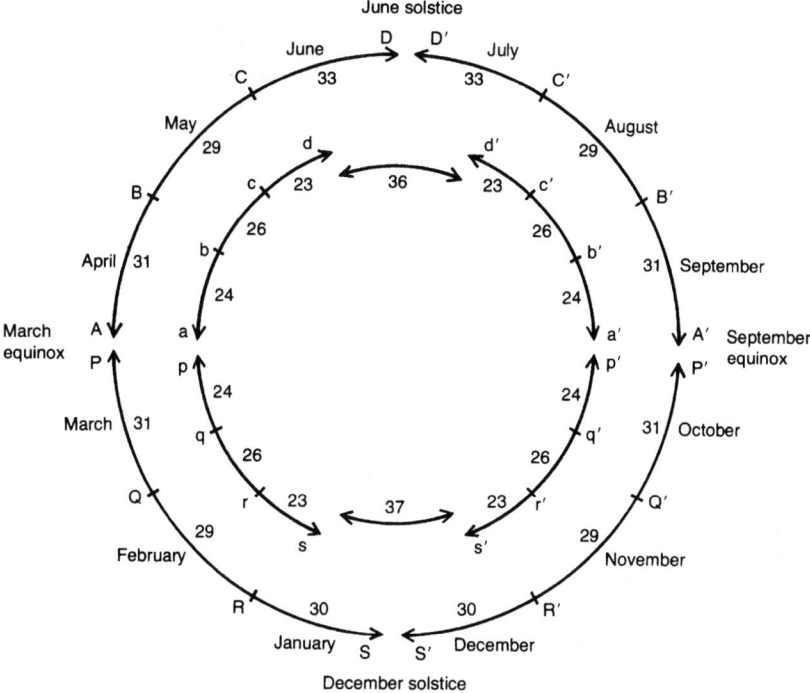

4.3. Hypothetical yearly cycle showing lunar months integrated into solar year.

These periods are calculated for the half year from the December to the June solstice. The same numbers would account for the half year from the June to the December solstice. Using the double sidereal lunar months in this way, within the context of the solar year and its solar months, would leave a gap in the sidereal time reckoning of 36 days around the June solstice and 37 days around the December solstice. Together, they are equal to any of the other four periods (73 days) of the sidereal year. A double lunar cycle, based on $328 + 329 = 9 \times 73$ nights, could therefore be left intact and would also be combined with a solar year in a cycle of 12 solar months.

We remember how a Venus cycle of 10 Venus years can be described in terms of 16 solar years. Similarly, 10 double lunar cycles of 657 (328 + 329) days equal 18 solar years. The periods of 3 years (Las Casas 1958, chap 140; Murua 1968, book 2, chap. 18) are mentioned for the renewal of the *acllas* (the virgins of the Sun) in their *acllahuasi* (house of the *acllas*); it may be that these years were calculated on the basis of lunar cycles. Six lunar cycles are equal to 5 solar years, the normal period used in the Inca age-class system.

Representation of a double sidereal month in Guaman Poma and in the Huari textile. In the introduction to the *ceque* system, I mentioned that Guaman Poma (f. 361) discusses a *quipucamayoc* ("*quipu specialist*") who was concerned with the assessment of local *ceque* systems (as indicated by his name *hucha quipoc*) and who was called (f. 359) a *quilla uata quipoc,* the quipu specialist for months and years. He was responsible for counting numbers "for setting down feasts and sundays and months and years." This man is shown in a drawing (f. 360) with a quipu and a kind of abacus (Fig. 4.1). Wassen (1940) has made a strong case saying that only the abacus was used for calculations like adding and subtracting and that the quipu served for recording the final numbers arrived at. However, since it is explicitly said that this man was concerned with the calendar, I want to suggest here another possible use of the abacus.

In a total of 20 squares, that is, 4 columns and 5 rows, there are $5 \times (5 + 3 + 2 + 1) = 55$ circles: 32 of these are white and 23 are black. The numbers are of the same magnitude as those found in the *ceque* system for respectively:

1. A double sidereal month.
2. The number of *huacas* in a major group of *ceques* representing a solar month (of 30 days) to which, in the words of Guaman Poma (f. 260), could be added one or two days according to the moon.
3. A minor group of *ceques,* having chthonic associations.

As the number 55 is the best whole number approximation to a double sidereal lunar cycle ($2 \times 27\frac{1}{3} = 54\frac{2}{3}$), we could imagine that white and black pebbles or kernels of corn were used for making the adjustments that Guaman Poma mentions in the correlation of a solar month to a sidereal cycle by changing the ratio of the two types of pebbles. A suggestion for such possible adjustments was made in the preceding section.

Earlier I indicated the importance of the period of 73 days, with its subdivisions of 26, 24, and 23 days, for relating a calendar consisting of solar months to one of sidereal lunar months. A similar relationship may be detected also in the Huari textile. On the basis of the organization of the circles in groups of 5 diagonals each, we discovered the importance of the number 72 (being the total of circles in all the first diagonals of each group of 5, of all the second diagonals, etc.). Because of certain irregularities in the color scheme of the diagonals, the total of 72 in four of the five cases was divided into two groups of $57 + 15$, $59 + 13$, $51 + 21$, and $52 + 20$ circles respectively; that is, the average division was of $54\frac{3}{4}$ and $17\frac{1}{4}$ circles, the first value being a very close approximation to the number of nights in 2 sidereal lunar months.

It is premature to guess how in the case of the textile the numbers $54\frac{3}{4}$, 72, and 73 (the total of the frontal figures and faces) exactly were used. The important thing is that these totals were represented in a calendar that otherwise stresses the existence of a year of 12 months of 30 (3×10) days and that a similar use of those numbers could have been made as in the *ceque* system.

THE *CEQUE* SYSTEM AND ASTRONOMICAL OBSERVATION: A PRELIMINARY DISCUSSION

Cobo and Polo mention that each group of 3 *ceques* was related to a *sucanca,* a pillar on the horizon as observed from Cuzco, which indicated the beginning of a month

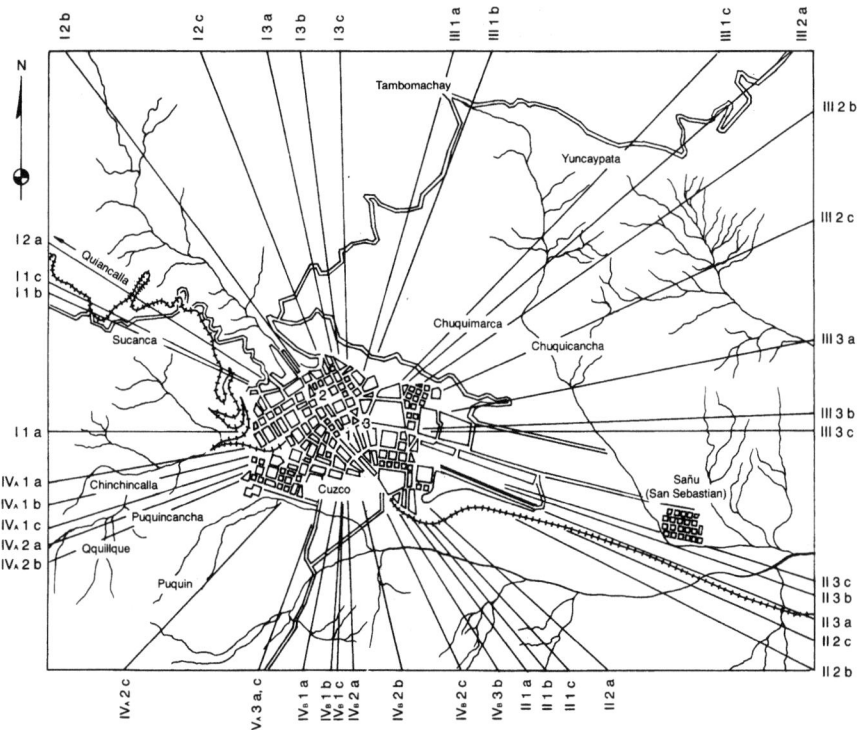

4.4. Map of the city of Cuzco with *ceques* superposed. (1) Coricancha (Temple of the Sun); (2) Hanan-Haucaypata; (3) Hurin-Haucaypata (Limapampa).

by the rising or setting of the sun at that place. In the case of three *huacas* Cobo also specifies their calendrical function as a *sucanca*. Our primary concern in the fieldwork carried out during the summer of 1975 was to find these *huacas* and the *ceques* on which they were located. For different reasons, however, we were obliged to try to locate all the *ceques*. The map of the *ceque* system (Fig. 4.4) is a first draft. For a general idea, I am confident that it is correct. I could locate *huacas* as place names from almost all the *ceques* with the help of old and modern maps, with data from other chroniclers, and by going into the field and questioning modern inhabitants.

Tentatively, I have drawn all the *ceques* as straight lines from the Temple of the Sun for the following reasons:

1. Except for *huacas* in or very near Cuzco, most of the *huacas* found could be considered as lying on straight lines.

2. Many of the place names referred to a general area (hacienda, mountain, etc.) and not to a specific rock or other exactly definable place. One of the exceptions was the next-to-the-last *huaca* of *ceque* II 1 a, called Guarmichaca. (This was the ninth *ceque* of Collasuyu in the sequence of *ceques* as used by Cobo. I will

indicate this sequence also when referring to specific *huacas*.) There are, at this place, ruins of a small Inca site, still known under that name, from which one can see Cuzco. The Temple of the Sun (now the church of Santo Domingo) was 37° E of magnetic north and its observation enabled me to define the first *ceque* of Collasuyu and its direction.

3. In the section dealing with the *capac hucha,* we observed that a child, sent home to be sacrificed, followed a *ceque* as a straight line. In modern indigenous agricultural practices this concept of *ceque* is still used to indicate the area from one line to the next. In some cases, when a *huaca* did not lie on its *ceque* in the *ceque* system, Cobo says so. While keeping the concept of *ceque* as a straight line, *huacas* of one *ceque* could be found to the right or left with some margin of freedom.

4. For the purpose of this research I was more interested in the *huacas* near the horizon (as seen from Cuzco) than in the nearby *huacas*. Therefore, the first ones have influenced my choice in the direction of a *ceque* as a straight line. In future research I hope to refine the whole map of the *ceque* system and give more exact locations of *huacas*.

In the following sections I will discuss the general data on Inca observatories and then the problem of location in Cuzco. In our fieldwork so far we were more successful in locating the *ceques* on which the solstitial *sucancas* were located than in locating the observation post(s). The latter cannot be identified with the Temple of the Sun as the center of the *ceque* system.

It will be understood from the data in this section that the *ceque* system was used in Inca astronomy quite independently from its use as a quipu in calendrical computations.

The Observatories

There are two traditions about the observatories in Cuzco: the first is that of Polo and Cobo and to this tradition we have already alluded. The other tradition is that of Betanzos (chap. 15) and Sarmiento (chap. 30), who both have very similar data, and of the anonymous chronicler ("Discurso" in Maurtua, p. 151) and Garcilaso (book 2, chap. 22). I will first analyze the second tradition.

All four chroniclers mention a group of four pillars for making one observation of a specific solar rising or setting. They all give some measures, in *pie, paso, vera, estado,* or *legua,* which I will convert into meters. The central pillars were smaller and were used for an exact observation; the outer pillars were bigger and helped to predict the approaching event.

According to Sarmiento (1947), the inner pillars were 16.72 meters apart and the outer ones 50.16 meters ($3 \times 16.72 = 50.16$) from each other. Betanzos gives the measurements as 8.36 meters for the inner ones and 41.80 meters for the outer ones. They do not give the distance from the center of observation to the posts; but, perhaps, this can be estimated. The height of the outer pillars was 4 meters.

According to the anonymous chronicler, the inner pillars were 69.65 meters apart, the outer ones 278.6 meters, and the distance of the center of observation 11,144 or 16,716 meters, that is, either 40 × 278.6 or 60 × 278.6. The angle of view of the outer posts would be about 1°. These distances between the pillars are satisfactory for good observations of one solar rising or setting. But the anonymous chronicler claims that when the sun arrived at one of the outer pillars (and he mentions these *sucanca* in relation to sowing in August), at that time the people living higher in the mountains would start sowing, whereas the people down in the valley started their sowing when the sun moved between the inner pillars.

Finally, Garcilaso mentions that the inner pillars were 50 meters apart and the outer pillars 150 meters. The pillars were observed on the skyline, but he does not know from where they were observed; he suspects the Temple of the Sun. His measurements seem to be of the same magnitude as those of the anonymous chronicler; both mention much higher values than do Betanzos and Sarmiento.

In fact, the latter authors seem to be describing a different system. Observations were not from Cuzco to the horizon, but from a flat place outside the city and higher up in the mountains where the Incas could set up a system of poles and observe them from a nearby place (perhaps from some 2,000 m, comparing their measurements to those of the anonymous chronicler). The wooden poles were used for actual observations. On the basis of these observations, a more elaborate and durable system with towers of stone was set up.

The chroniclers also do not agree on the day in the year for which the pillars were set up. The anonymous chronicler mentions only one set of pillars in the west for observing the opening of the agricultural year in August. Sarmiento says that the beginning of sowing was observed with pillars in the west and the beginning of harvest with pillars in the east. The dates would be around May and August. Probably the two sets of pillars could be used in relation to both events. Such a hypothesis could mean that both sets were used for one and the same event in May and in August, that is, for two days equally distant from the June solstice; or the data could mean that one set was used for indicating the beginning of May (April 21) and the end of August (August 23), while the other set was used for the end of May (May 20) and the beginning of August (July 25). In the case of the latter hypothesis, I am using the *ceque* system as a quipu, counting days by *huacas*. (See previous section.)

Garcilaso mentions four sets of four pillars for observing the rising and setting of the sun at the time of the two solstices.

None of the chroniclers mention *sucanca* for observing the equinoxes, but Garcilaso refers to a system of two gnomons for this purpose. In this case, the shadow of the gnomon itself was observed, so there was no need of any kind of horizon observation.

Despite the difficulties and insecurities of interpretation, I think that these four sources give us a better interpretation of the system of observation used by the Incas than do Polo and Cobo. Polo indirectly mentions ("Errores," p. 43) that there were 12 *sucanca* (i.e., the total of *huacas* which he mentions, 340, minus the total mentioned by Cobo, 328, equals the 12 *sucanca*). Since he claims (p. 16) that there was one pillar for each month and that the one for the beginning of "winter," the December solstice, was called Puncuy (= Puccuy) *sucanca* and the one for the beginning of "summer" Chirao *sucanca*, I concluded that there were 12. Cobo, probably basing his statement on the data of Polo (or Molina?), mentions 14 *sucanca*. To arrive at this number, he may have argued in three different ways:

1. With the expansion of groups of *ceques* in Cuntisuyu, he took 14 groups as corresponding to 14 *sucancas* and 14 months.
2. He excluded Puccuy *sucanca* and Chirao *sucanca* (mentioned also by him) from the 12 indicating the beginnings of the months.
3. Using a system of pillars as mentioned by Polo and Cobo, and observing both sunrise and sunset, one needs a system of 14 and *not* 12 pillars.

June solstice

Chirao	Chirao
July	June
August	May
September	April
equinox	equinox
October	March
November	February
December	January

December solstice

| Puccuy | Puccuy |

(In my earlier paper on the calendar, I used the 12 pillars as mentioned by Polo, in a similar diagram. Professor F. Lounsbury suggested to me that in fact I would need the 14 pillars.)

However, at the moment I have little confidence that these data or interpretations of Polo and Cobo are correct. My doubts are based on the facts that:

1. Both authors mention only one pillar for one month, although we know that for each observation 4, or at least 2, pillars are needed.
2. Cobo mentions in the *ceque* system only three *huacas* that served as *sucanca* and in each case these consist of 2 pillars.
3. These three *sucanca* are all located on the western horizon; his data agree here with those of the anonymous chronicler and Cieza *(Cronica,* chap. 92), who mentions that only in the west were artificial pillars as *sucanca* set up.

We conclude, therefore, that the informants of Polo (and, indirectly, of Cobo) may have thought of a system of 12 (3 × 4) pillars on the western horizon, consisting of three *sucanca*, each of two smaller pillars, to which were added in each case a higher northern and southern pillar. These higher pillars were probably situated on the two *ceques* next to the one of the *sucanca* and for this reason we may not have detected them yet. Furthermore, it is logical to suppose that *sucancas* only were set up on the western horizon. Only here, as the anonymous chronicler observes, were the mountains and the horizon near to Cuzco. We will notice from the data in the *ceque* system that the observation points in the east, beyond the valley, consisted of prominent natural features of the mountains.

THE *SUCANCAS* AND OTHER *HUACAS* OF ASTRONOMICAL SIGNIFICANCE AS MENTIONED IN THE *CEQUE* SYSTEM

Garcilaso remembers how he had seen the *sucancas* in his youth but that later they were destroyed by the Spaniards. He did not know from which point in Cuzco they were observed; but he said that once the *sucanca* are located it is easy to reconstruct the center of observation. For us, it was easier to locate the *ceques* on which the *sucanca* were situated than the center of observation. Therefore, I will discuss first the *sucanca* on the western horizon and then the other *huacas* of astronomical significance.

Cobo calls only one of the three *huacas* that served as *sucanca* by this name; but he says of this and two other *huacas* on the western horizon that they consisted of two artificial pillars and he indicates their calendrical function. We could not exactly locate the *huacas* themselves, as they were destroyed by the Spaniards, but, especially in the first two cases, we were quite successful in locating all the other *huacas* of their *ceques*. Therefore, on the basis of a description of the *sucanca*, of the direction of its *ceque* and of the adjacent *ceques*, and of the fact that, seen from Cuzco, it had to be on the horizon, we could indicate the area where each *sucanca* had to be located (Fig. 4.4).

1. Quiancalla (12 b, sixth *ceque*, ninth *huaca*), which means "from where one sees *(callan)* dawn *(quia)* or sunrise," indicated the June solstice ("the beginning of the summer" in the words of Cobo).
2. Sucanca (I 1 b, eighth *ceque*, seventh *huaca*) indicated that "then the people had to start sowing maize."
3. Chinchincalla (IVB 1 b, thirteenth *ceque*, third *huaca*) "was a big mountain with two pillars; when the Sun arrived, it was time to sow."

In the case of the latter two *sucancas*, we do not know if they indicated the beginning of a month or any other specific day within the agricultural cycle. Perhaps the name *chinchi* can help us. Gonzalez Holguin (1952) and Lira (1944) refer to a shrub, Chinchircuma, as "a yellow flower that is like feathers," and Lira also refers to a fruit of a thorny shrub, *Chinchi chinchi,* which was used against fevers. We know that the Incas started sowing when, in August, "a thorny plant was in bloom with yellow flowers" ("Discurso" in Maurtua, p. 149). Therefore, we might bring the *huaca* Chinchincalla in relation to this "thorny plant with yellow flowers," and the possibility exists that this *sucanca* indicated the beginning of the month of August (although such an interpretation would bring us additional difficulties).

Another *ceque* that we can consider in relation to the three discussed is the first *ceque* of Chinchaysuyu (I 1 a). All its *huacas* outside Cuzco are near the W-E line, as are the *huacas* of the last *ceque* of Antisuyu (III 3 c).

If Quiancalla served for observing the June solstice, we should expect the *huaca* used for observing the December solstice to be in the opposite direction. In fact, we do find mentioned in Collasuyu a *huaca* (II 2 a, fourth *ceque*, eighth *huaca*), Guancarcaya ("from where one observes the drum"?), "which is a canyon like a gate ... It was dedicated to the Sun and they offered him children in certain feasts that they had there." Because of its closeness to other *huacas* like Intipampa (II 2 b, fifth *ceque*, eighth *huaca*), "the plaza of the Sun," and Omotourco (II 2 b, fifth *ceque*, tenth *huaca*), which is also mentioned by Molina (p. 26) in relation to a *sucanca* and a solstitial ceremony (pp. 26, 48), we may conclude that Guancarcaya itself serves as a *sucanca* for observing sunrise during the December solstice.

Both solstitial *huacas*, Quiancalla and Guancarcaya, are situated, in their respective *suyu*, on the fourth *ceque* from the W-E line. This coincidence seems to be intended. Therefore, we may pay more attention to the fact that in Antisuyu a *huaca* was mentioned on the sixth *ceque* (i.e., on the fourth *ceque* north of the W-E line) Chuquicancha (III 2 c, sixth *ceque*, third *huaca*), as "a temple of the Sun." Nearby were two places, Susurpuquio (Molina, p. 20; or Susumarca, III 2 b, fifth *ceque*, eighth *huaca*) and Callachaca (Santacruz Pachacuti, pp. 2377 247; III 2 a, fourth *ceque*, third *huaca*; III 2 b, fifth *ceque*, eighth *huaca*; III 2 c, sixth *ceque*, second *huaca*), both mentioned in relation to the most explicit Inca myth of the appearance of the sun as a god. Not far from Chuquicancha was another Temple of the Sun bearing a similar name, Chuquimarca (III 1 c, third *ceque*, fourth *huaca*), where the Inca went during the June solstice. I conclude, therefore, that Chuquicancha had a function in relation to sunrise during the same solstitial feast in June. Moreover, I would propose that this temple had a role during the December solstice.

In the case of Quiancalla and Guancarcaya, I suspect that they were used not only for observing, respectively, the sunset in June and the sunrise in December from a central point in Cuzco but also for observing these events from each other as centers of observation toward the other *huaca* as *sucanca*. We may suspect a similar role for Chuquicancha in relation to another *huaca*, Puquincancha (IVa 2 b, tenth *ceque*, second *huaca*), as Temple of the Sun with a special function during the December solstice (Molina, pp. 60, 67). Using Puquincancha for observing a sunrise behind Chuquicancha during the June solstice and Chuquicancha for observing sunset behind Puquincancha during the December solstice might also explain another mythical reference to Chuquicancha in Callachaca. Once, during a severe drought (a reference to the dry season in June), rains (a reference to the wet season in December) only fell over Callachaca and only there produced a good crop (Santacruz Pachacuti, p. 247). (For a more detailed discussion of the *huacas* mentioned in this paragraph, see Zuidema 1976.)

THE CENTER(S) OF ASTRONOMICAL OBSERVATION

The anonymous chronicler says explicitly that the *sucanca* were observed from the middle of the plaza in Cuzco where a stone pillar was erected on top of an elevation, called Osno (= Ushnu). It is clear, then, that the center of astronomical observation was not in the Temple of the Sun as the center of the *ceque* system. However, after having located the *huacas* of *ceque* I 2 c to which the *huaca* Quiancalla belonged as the *sucanca* of sunset during the June solstice, we realized that, while some suburbs of Cuzco can be seen from there, the Incaic part of the city cannot be seen. On the other hand, there is an open view in the solstitial direction toward the Inca fortress of Sacsahuaman, which is situated in a plain to the northeast above Cuzco. Checking this information out, we observed sunset during the June solstice from Sacsahuaman, that is, from a round Inca structure there called Muyucmarca ("the round building") which was a Temple of the Sun. The point observed was within the area as described for the *huaca* and *sucanca* Quiancalla.

During our fieldwork, we made astronomical observations to various points on the horizon from the Temple of the Sun in Cuzco (Coricancha), from different points on the plaza, from atop the cathedral, and from Muyucmarca on Sacsahuaman. We worked under the assumption that all three could have been used by the Incas for astronomical observation. Only later did I realize that neither Coricancha nor the plaza near the cathedral was intended by the anonymous chronicler to be taken as the center of observations.

Cuzco had two plazas called Haucaypata ("Plaza of rejoicing"); both had an Ushnu and both were used for state rituals. Hanan-Haucaypata was in the upper half of Cuzco where the plaza of the king was also situated. Hurin-Haucaypata, belonging to the lower half (Hurin), was not far from Coricancha but was outside the ceremonial part of Cuzco. Today, as in colonial times, this plaza (or part of it) is known as Limapampa. Nearby were the sacred fields where the king did his ritual first plowing and first harvesting. The field was called Sausero (II 3 b, second *ceque,* third *huaca*); Limapampa was the first *huaca* of this *ceque* (II 3 b, second *ceque,* first *huaca*). Another field nearby, Guanaypata (III 3 a, third *ceque,* fourth *huaca*)—where according to the Inca origin myth a golden bar had sunk into the ground indicating that the city should be built there (Sarmiento, chap. 13)—had a "wall that had been built there by the Sun." Probably this wall, but certainly the Ushnu on Hurin-Haucaypata, was used for astronomical observations (III 2 b, fifth *ceque,* first *huaca*) (Cobo, book 13, chap. 32; Molina, pp. 29, 30, 36).

The conclusion reached here can be checked out in future fieldwork in two ways:

1. The position of Hurin-Haucaypata can be observed in relation to the crossing of the solstitial lines from Quiancalla to Guancarcaya and from Chuquicancha to Puquincancha.
2. Sarmiento and Betanzos mention two other observatories in flat areas above and outside Cuzco. Possibly, Chuquicancha and Muyucmarca on Sacsahuaman,

which were both temples of the sun, served this function. The position of Hurin-Haucaypata in relation to both these places can also be studied. Guaman Poma (ff. 261, 262, 263, 894) mentioned that solar observations were made from temples of the sun with windows. His description would probably fit structures like Muyucmarca.

Once the astronomical value of the places mentioned here is well assessed, we will then be able to study the calendrical importance of Sucanca (I 1 b) and Chinchincalla (IVB 1 b) and probably of many other *huacas* on the horizon as observed from Hurin-Haucaypata.

GENERAL CONCLUSIONS

A general problem of ethnohistorical studies on the Incas is that most Spanish chronicles were written forty years or more after the conquest. They reflect less what the Spaniards really saw of an empire in its full glory than what they interpreted from the elaborate and idealized memories of their older Inca informants who lived in a colonial society.

The *ceque* system, because of its intricate and systematic character, could be proven to be the most important exception to this general situation. Comparing it to the two pre-Spanish textiles, we find the same numbers discussed: 8; 27 or 28; 54, 55, or 56; 40, 41, or 42; 72 or 73. The Huari textile gives greater stress to the solar calendar with lunar numbers embedded into it; the Inca textile and the *ceque* system stress a sidereal lunar system containing solar and synodical lunar numbers. On the basis of a common astronomical theory, there seems to be a rather undogmatic concern, comparing the three calendars, in translating the astronomical data into a regular calendar. For instance, both the Huari textile and the *ceque* system started from a base of 45 diagonals, or *ceques*. In the first case 27 diagonals keep their full numbers of circles and 18 (2 × 9) diagonals have their numbers reduced, making 8 the average total of circles on a diagonal and 360 the general total of circles in the textile. In the case of the *ceque* system, 27 *ceques* also keep their original significance. But now, first, the number of the 18 other *ceques* themselves is reduced to 15 or 14 and, second, the number of *huacas* on each of these *ceques* is reduced, bringing down the total from a hypothetical 360 (45 × 8) to 328 (41 × 8). Concluding the comparison: the Huari textile starts from a situation of 45 × 8, reducing it to 41 × 8.

Ethnohistorical data have been supportive for the assumption of an 8-day week and a possible 10-day week also for the significance of a schematic month of 30 days, a synodic month of 29 or 30 days, and a solar month of 30 to 32 days. There is, however, no written evidence of the use of a sidereal month. The analysis of this paper may help to discover how Andean peoples actually observed this event. Here the data analyzed in another paper (Zuidema 1976) might be worth considering. The *ceque* indicating the setting of the sun during the June solstice contained also a *huaca* (Capipa[c]cha "the waterfall in the river Capi"; i.e., the "root" [capi] of Cuzco) where the Inca would bathe at new moon before June solstice and where another ritual dedicated to the moon was carried out at full moon after the December solstice. This and other *ceques* may have been used, then, in a special way to measure the movement of the moon along the horizon. Another possibility is that the Incas had knowledge of some form of a zodiac with "lunar mansions." Modern data on indigenous constellations collected recently by Urton (to be published) bring more order to the data on Inca constellations.

In this paper it has been my intention to establish a basis for the numerical analysis of the textiles and the *ceque* system. Later it may also be possible to interpret the number of *huacas* belonging to the individual *ceques*. Besides locating the *huacas* more precisely in the field, further research should also analyze the exact association of *panacas* (the "royal *ayllus*") with groups of *ceques* or to individual *ceques,* the similar association of the non-royal *ayllus* in the organization of Cuzco, and the relationship of both kinds of groups to the Inca month. Then a better understanding of various mythological and ritual data can also be reached and their importance to the calendar better understood.

ACKNOWLEDGMENTS

Support for this study, which is gratefully acknowledged here, was received from the National Science Foundation during the summer of 1973, when I carried out fieldwork in Cuzco, and during my year of sabbatical leave of 1973–74. During the summer of 1975, 1 received further support for carrying out fieldwork in Cuzco from the American Philosophical Society.

REFERENCES

Acosta, J. de. 1954. *Historia natural y moral de las Indias* (1590) *B.A.E.* 73. Madrid.
Brownrigg, L. A. 1973. "A Model of the Andean System of Time Dispersed and Congregated Ritual Calendars," Symposium 607, Andean Time: Ritual Calendars and Agricultural Cycles. Paper presented at the 72nd meeting of the American Anthropological Association, New Orleans.
Cobo, B. 1956. *Historia del Nuevo Mundo* (1633). *B.A.E.* 91–92. Madrid.
Discurso de la sucesion y gobierno de los Yngas. In Maurtua, *Juicio de limites entre el Peru y Bolivia,* vol. 8. Lima, Chunchos.
Escobar, G., R. P. Schaedel, and O. Núñez del Prado. 1967. *Organizacion social y cultural del sur del Peru*. Mexico City, Inst. Interamericano.
Guaman Poma de Ayala, F. 1936. *El primer nueva coronica y buen gobierno* (between 1584 and 1614). Paris.
Herrera, P. 1916. *A punta cronologico de las abroas y trabajos del cabildo a municipalidad de Quito desde 1534 hasta 1714 (primera epoca),* vol. 1. Quito.
Herrera y Tordesillas, A. de. 1944–1947. *Historia general de los Lechos de los Castellanos en las isles i tierra firme del mar oceano* (1601–1615). Reproduction of edition of 1726–1727. Asunción, Paraguay.

Holguin. See Gonzalez Holguin, D. 1952. *Vocamulario de la lengua . . . Quichua* (1608). Lima.
Las Casas, B. de. 1958. *Apologetica historia* (1564). *B.A.E.* 105–106. Madrid.
Lohman Villena, G. 1967. See Matienzo, J. de.
Matienzo, J. de. 1967. *Gobierno de Peru* (1567). Edition et étude préliminaire par Guillermo Lohman Villena. Paris-Lima.
Molina, C. de. 1943. *Relacion de las fabulas y ritos de los Incas* (1573). Lima.
Müller, R. 1929. Intiwatana (Sonnenwarten) im alten Peru. *Baessler Arkiv* 13, parts 3–4, pp. 178–187, Berlin.
Müller, R. 1972. *Sonne, mond und sterne uber dem reich der Inka.* Berlin, Heidelberg, and New York, Springer Press.
Murua, M. de. 1962. *Historia general del Peru* (1613). Madrid.
Neugebauer, O. 1942. "The Origin of the Egyptian Calendar." *Journal of Near Eastern Studies* 1 (January–October).
Nordenskiold, E. 1925. "The Secret of the Peruvian Quipus: Calculations with Years and Months in the Peruvian Quipus." *Comparative Ethnographical Studies* 6:1–2. Göteborg.
Polo de Ondegardo, J. 1916. Los errors y supersticiones de los indios (1571). Ed. Urteaga y Romero. *Col. Libr. Doc. Ref. Hist. Peru* 3:1–43, Lima.
Ramos Gavilan. 1621. *Historia del celebre santuario de nuestra senora de Copacabana, y sus Milagros invencion de la cruz de Carabuco.* Lima.
Rostworowski de Diez Canseco, M. 1970a. Mercaderes del valle de Chincha en la epoca perhispanica: Un documento y unos comentarios. *Revista Espana de Antropologia Americana* 5:135–178, Madrid.
Rostworowski de Diez Canseco, M. 1970b. Etnohistoria de un valle costeno durante el Tahuantinsuyu. *Revista del Museo Nacional* 35. Lima.
Santacruz Pachacuti Yamqui, J. de. 1950. *Relacion de antiguedades desde reyno del Piru* (1613). Ed. Jiménez de la Espada. Asunción, Paraguay.
Sarmiento de Gamboa, P. 1947. *Historia de los Incas* (1572). Buenos Aires, Biblioteca Emece.
Zuidema, R. T. 1964. *The Ceque System of Cuzco: The Social Organization of the Capital of the Inca.* Leiden, Ed. Brill.
Zuidema, R. T. 1966. El calendario Inca. In *Actas del XXXVI Congreso Internacional de Americanistas* 2:25–30. Seville.
Zuidema, R. T. 1973a. Kinship and Ancestor Cult in Three Peruvian Communities: Hernandez Principe's Account in 1622. *Bulletin Institut Français des Etudes Andines* 2 (1):16–33. Lima.
Zuidema, R. T. 1973b. La parente et le culte des ancêtres dans trois communautés péru-viennes: Un compte-rendu de 1622 par Hernandez Principe. *Signes et Langages des Amériques: Recherches Amerindiennes au Quebec* 3 (1,2):129–145.
Zuidema, R. T. 1973c. La quadrature du cercle dans l'ancien Péru. *Signes et Langages des Amériques: Recherches Amerindiennes au Quebec* 3 (1,2):147–165.
Zuidema, R. T. 1976. La imagen del sol y la huaca de Susurpuquio en el sistema astronomico de los Incas en el Cuzco. *Journal de Société des Americanistes* 43 (1974–1976): 199–230.

CHAPTER FIVE

Horizon Astronomy in Incaic Cuzco

Anthony Aveni

INTRODUCTION

It is often casually stated that the Inca of Peru could scarcely be expected to have developed a sophisticated system of astronomy like that of the Maya because of their inability to produce a written record. One tends to think of them as New World Romans—too occupied with war and conquest to turn to matters as esoteric as star-gazing. But their establishment of an empire stretching over nearly 30° of geographic latitude gave them reason enough to engage in calendrical matters, which can be anything but superfluous when one is faced with the task of regulating civil, agricultural, and religious dates on a national basis. When we assemble all the data from the post-Conquest chronicles alluding to the astronomical methods of the Inca, we are staggered by both the quantity

---

Reproduced from *Archaeoastronomy in the Americas*, edited by R. Williamson (Los Altos and College Park, Ballena Press and Center for Archaeoastronomy, 1981), by permission of the Malki Museum, Inc., 11-795 Fields Road, Banning, CA 92220, (951) 849-7289.

and quality of sky-watching which took place in and around Cuzco, the Inca capital. We are even more astonished by Incaic astronomical accomplishments when we realize both that the Inca empire began less than three centuries before Pizarro's entry into Cuzco, and that the Spaniards who wrote the history of the proceedings (like those who produced the "true" story of the conquest of Mexico) were not particularly disposed to believe that their subjects could have been sophisticated enough to pay much attention to the heavens. But we should realize that a long tradition of Andean cosmological thought had its roots in the earlier Andean cultures of Huari, Nasca, and Chavin. In this paper we employ the rich historical record left for us by the Spanish chroniclers, together with three seasons of archaeoastronomical field work in Cuzco, to derive specific astronomical sight-lines relating to solstitial and zenith-antizenith solar observations. The underlying framework for our study is the arrangement of *ceque* lines and *huacas* which united many aspects of life in the ancient capital.

THE *CEQUE* SYSTEM

As interpreted by Zuidema (1977a), the *ceque* system was an organization of 41 radial lines (or *rayas* as the Spanish called them), which emanated from the Temple of the Sun (Coricancha) in the center of Cuzco and crossed the landscape. Used for different purposes related to social organization (Zuidema 1964), irrigation (Zuidema 1978a), wind-directions (Zuidema 1977–8), and astronomy (Zuidema and Urton 1976), these imaginary lines were traced by a number of *huacas* or sacred places (of which there were 329 in all, counting the Temple of the Sun). Many of these *huacas* were rocks, springs, valleys, or mountains. Bernabé Cobo (1956:167–186) gives a full description of the location of the *huacas* and the situation of the *ceques* within the four principal quarters of the city. A large part of our work derives from our success in locating these *huacas*, tracing the *ceque* lines, and reconstructing the astronomical arguments given by the chroniclers based upon this system.

The chronicler F. Guaman Poma (1936:353–354) suggested that the Inca "had observatories with windows" to catch the first and last rays of the sun at the horizon. These observations were employed to know when to shear the llamas, sheep, and alpacas, and when to sow and harvest crops—events occurring at different seasons of the year. In other passages (Poma 1936:883–884) where this chronicler describes the duties in office of the state astrologer, he refers quite explicitly to the changing aspect of the sun in the local environment. The sun "sits in his chair one day and rules from that principal degree (of the December solstice). Then he sits in another chair where he rests and rules from that degree (of the other solstice)." From one seat to the other "he moves each day without resting"—a reference to the successive rising and setting points on the horizon. During the solstices "he rests for more than a day in his chair"—on those days the motion of the sun from day to day on the horizon becomes imperceptible.

Guaman Poma's statements, though brief, serve to establish both a purpose and a methodology in Inca observational astronomy; they also imply that the Inca had developed a sophisticated system of positional astronomy based upon accurate horizon observations. We shall see that evidence from other chroniclers expands and elaborates upon that of Guaman Poma. We will use this evidence together with other statements from the historical record and basic data acquired in the region about Cuzco, to deduce the extent and location of Inca astronomical observatories in the environment of the holy city. The system, as it is revealed to us, turns out to have been mostly practical and functional, and may not have constituted an astronomy as ritualistically based as that practiced in Pre-Columbian Mesoamerica.

While Guaman Poma mentions the use of windows, there is even better evidence relating to the use of pillars to mark the solar course. These pillars, called *mojones* (singular *mojon*) by the Spaniards, were used as border markers (called *sayhua* in Quechua) and high points of Inca mountain roads; a *ceque* line often terminated with a *mojon*. In Guaman Poma's pictorial treatise, pillars appear explicitly in drawings relating the history of the Inca (1936: 352, 354). In the first picture (Fig. 5.1) we see a *mededor* carefully measuring off blocks, which he supplies to the *mojonador* or pillar builder; both craftsmen work with great care as they fashion a *mojon* together. A completed model, taller than a man and with a rounded top, appears in the background. In the second picture (Fig. 5.1), three *mojones* adorn the scenery; one of them, apparently an important landmark, is next to a labeled road. A painting in the Church of Santo Domingo on the present site of the Temple of the Sun depicts a number of worshippers in front of a cathedral (Fig. 5.1). The painting is in the Flemish style of the late sixteenth century, but the environment is clearly that of Cuzco, although the horizon is somewhat distorted in comparison with the view which actually exists. If one looks to the left of the church, the careful viewer can see two square pillars appearing like towers perched on the Cuzco skyline; from all indications such structures must have been both prominent and abundant in the landscape.

The chronicler Bernabé Cobo (1956) is our principal source for much of our information on these towers. Cobo employs several pages to give readers a meticulous word-map of the *ceques* and *huacas* comprising the four quarters of the city.[1] We have extracted two passages buried in his ordered description which allude to the use of these *mojones:*

> Chinchaysuyu—*Ceque* 6 (Zuidema Number I 2a), *Huaca* 9: "A hill called Quiangalla that is on the road to Yucay where there were two monuments or pillars that they had for signs and when the sun arrived there it was the beginning of summer" [p. 172].
>
> Cuntisuyu—*Ceque* 13 (Zuidema Number IV A 1b), *Huaca* 3: "Chinchincalla is a large hill where there "were two monuments at which, when the sun arrived, it was time to sow" [p. 185].

Let us take Cobo's second statement first because it is simpler to analyze. As Figure 5.2 illustrates, the Chinchincalla twin pillars probably were fashioned to frame the setting sun at the December solstice (which is the time to *finish* planting). The most likely observation point would have been the Coricancha; the supporting evidence

5.1 *Mojones* or pillars in the environment of Cuzco: (top left) Inca tower builders. Example of their completed work in background (F. Guaman Poma, f. 352); (top right) towers mark important points in environment of ancient Cuzco (F. Guaman Poma, f. 354); (bottom) towers adorn background scenery of Cuzco in 16th century painting in Church of Santo Domingo in Cuzco.

Anthony Aveni

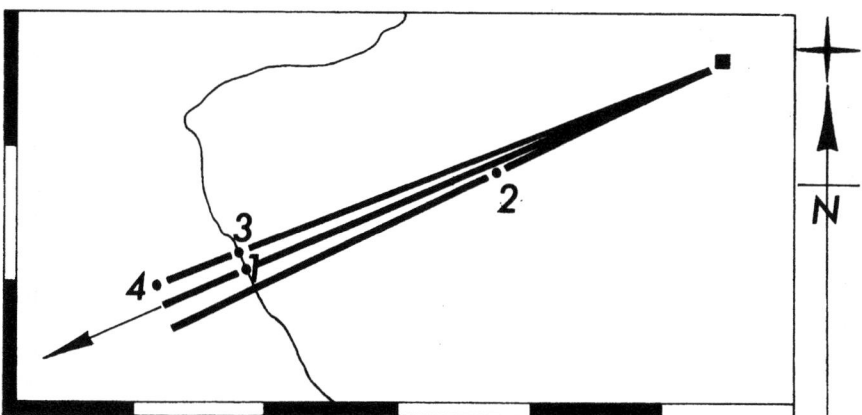

5.2 December solstice sunset alignment (*ceque* lines in dark ink). Curved line is the visible horizon as seen from the Coricancha (filled square), the observation point. Bars on the rim of the figure represent 1 km. (1) Location of pillars at Chinchincalla on Cobo's *ceque* 13 of Cuntisuyu; (2) Puquincancha (a *huaca* on *ceque* 10); (3) Ravaypampa (a *huaca* on *ceque* 14); (4) Pantanayoc (a *huaca* on *ceque* 14).

for this argument is quite strong. The line from the Coricancha to Chinchincalla can be drawn accurately from Cobo's description of the *huacas* on three closely spaced *ceques* in this region of Cuntisuyu quadrant. The line thus drawn is found to coincide precisely with the direction of the setting sun at the December solstice (the first day of summer in the southern hemisphere), as Figure 5.2 shows. Chinchincalla is delimited on the north by our precise determination from local informants of the location of Pantanayoc, a *huaca* on the adjacent fourteenth *ceque* to the north which Cobo describes thus: "It is a large hill parted in the middle that divides the roads of Chincha and Condesuyu" (1956:185). Ravaypampa, the next *huaca* of *Ceque* 14 of Cuntisuyu in the direction of Cuzco, is described as a "terrace on the slope of the hill of Chinchincalla," while the *huaca* Puquincancha limits our angular view on the next *ceque* (No. 12) to the south. Now a straight line from Pantanayoc to the Temple of the Sun passes immediately over a steep slope which must be the northern flank of Chinchincalla hill. The pillars must have been located higher up on that same slope. Of greatest astronomical significance is our discovery, from measurements with the surveyor's transit, that the December solstice alignment as measured from the roof of the Coricancha lies precisely in this direction on *Ceque* 13, between *ceques* 10 and 14. As is the case for June solstice observations, only a pair of pillars is needed to frame the sun at the horizon, when it comes to rest before turning around. Nevertheless, it would be useful to look for signs of rocks, pillars, temples, or other markers on adjacent *ceques* to the north of Chinchincalla. Such archaeological remains might originally have served as devices to mark the solar slowdown.

With regard to Cobo's other reference to horizon pillars, we find that the hill called Quiangalla can be located fairly accurately on Cobo's *Ceque* 6 of Chinchaysuyu,

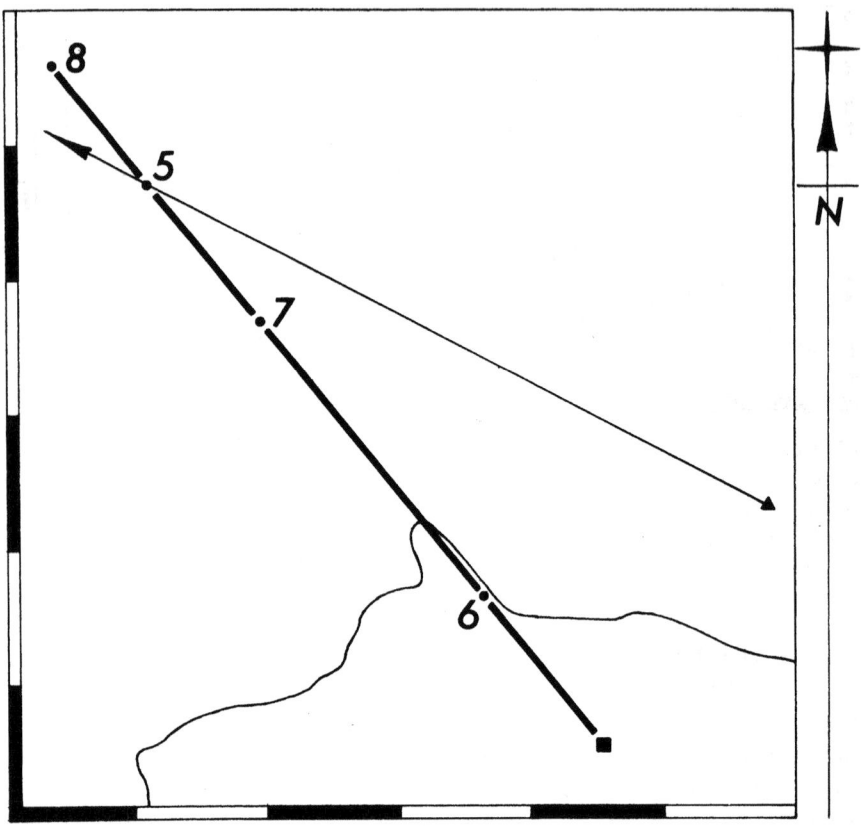

5.3 June solstice sunset alignment viewed from Lacco (filled triangle) to a pair of pillars on Quiangalla (5), a *huaca* on *ceque* 6 of Chinchaysuyu. Coricancha, visible horizon, and 1 km lengths as in Fig. 5.2. Other *huacas:* (6) Ticcicocha; (7) Quisco; (8) Guargua Illapuquiu.

on which it is the ninth *huaca* as one counts outward from the Temple of the Sun (see Fig. 5.3). But neither the Coricancha (Sun Temple) nor any other monument near it could have served as the observing station because not all parts of the hill Quiangalla are visible from the center of the city. By meticulously tracing the June solstice-sunset direction from Quiangalla back to Cuzco using the surveyor's transit, we found that it passed about 2 km north of the center of the city. If long-distance June solstice-sunset observations were marked by a pair of pillars on Quiangalla, the most likely observation point from which to observe the event would be 5 km to the southwest in the area of the *huaca* Chuquimarca (III 1c, No. 4). This was called a temple of the Sun; it was the one where the Inca spent his time celebrating the June solstice rituals. Chuquimarca was on or near a mountain called Manturcalla (III 1c, No. 6). Although neither name is used any more, by locating the surrounding *huacas* and using colonial documents that mention Manturcalla in the context of other place-names there, Zuidema has suggested that the most logical candidate for Chuquimarca is a complex today called Lacco. It consists of a sculptured rock with caves and ruins of buildings, terraces, stairways, and a canalized stream. If we suppose Lacco to have been the place, and then trace the June solstice-sunset line more exactly from it, we find that the astronomical sight-line crosses *Ceque* I 2a at a point that falls on a sttaight line passing through at least three other *huacas* of that ceque; i.e., we find that *Ceque* I 2a (*Ceque* No. 6 of Chinchaysuyu) is a straight line, and the Quiangalla pillars lay on it. Though the towers are long since destroyed, an archaeological probing for some remains might prove fruitful. The search could be limited to the 400 m^2 area within which our astronomical-ethnohistorical argument suggests the pillar remains must lie.

Having established a technique that illustrates how we can use astronomical measurements and calculations in conjunction with information on the placement of towers to mark the winter and summer solstices, we turn next to a more complex argument in which we show that the Inca devised a far more basic astronomical line that defined the zenith sunrise–antizenith sunset axis, an alignment that actually cut across the *ceque* system and incorporated several observing stations and a gnomon. The basic reasoning behind the establishment of this line, its connection with the aforementioned solstice lines, and the place of such linear schemes in the general context of Andean cosmological thought is developed further in a separate paper by Zuidema in *Archaeoastronomy in the Americas* (Williamson 1981). Here we elaborate only the technical argument.

We begin with another of Cobo's descriptions, in which he states that the seventh *huaca* of the eighth *ceque* of Chinchaysuyu (Zuidema Number I 1b), the northwest quadrant of Cuzco,

> ... was called Sucanca. It was a hill where the irrigation canal of Chinchero comes. On it were two pillars or monuments to signal that when the sun arrived there it was time to begin to plant maize. The sacrifice made there was addressed to the sun and they asked that it would arrive on time so that they would have good reason to plant ... [1956:173–174].

The earlier Anonymous Chronicler (probably from Maurtua, ca. 1570:150–152) mentioned four little pillars serving the same function, located in the same general direction. These were visible at about 2 to 3 leagues from the city and were used to regulate the sun. The chronicler says they were "high on a hill overlooking Cuzco from the west":

> When the sun passed the first pillar they prepared themselves for planting in the higher altitudes, as ripening takes longer.
> When the sun entered the space between the two pillars in the middle it became the general time to plant in Cuzco; this was always the month of August.
> And when the sun stood fitting in the middle between the two pillars they had another pillar in the middle of the plaza, a pillar of well worked stone about one estado high, called the *Ushnu*, from which they viewed it. This was the general time to plant in the valley of Cuzco and surroundings [author's italics].

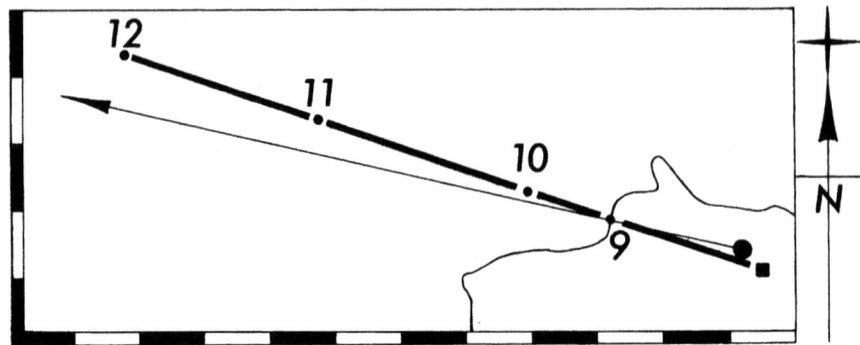

5.4 18 August antizenith sunset line viewed from the Ushnu (filled circle) to the 4 sucanca or pillars on Cerro Picchu (9) which represent one *huaca* on the 8th *ceque* of Chinchaysuyu. Other *huacas*: (10) Urcoscalla; (11) Poroypuquio; (12) Collanasayba.

The northernmost pillar (see Fig. 5.8a) evidently served as a device to warn that the planting season in the Cuzco valley was approaching. People who cultivated crops at higher altitudes, where growth occurred at a slower pace, would be allowed sufficient additional time to sow their seeds before planting commenced in the valley. Though the chroniclers provide no evidence, the southernmost pillar was perhaps used as a similar warning device during the harvest season when the sun, moving from left to right, passed it in April.

There is no doubt that these particular pillars resided on the slope of Cerro Picchu above the Carmenga district (located in the western section of the modern city and still retaining its old name). Another chronicler, P. Cieza de León, establishes this for us when he states that

> on the hill of *Carmenga* [i.e., Cerro Picchu] they have at *definite intervals*, small towers, which serve to keep track of the movement of the sun, which they regard as important and they had a plaza where they say a long time ago there was a swamp or lake at which the founders smoothed over the mortar and stone. . . . From this plaza the four royal roads go out [1973:214].

Cobo tells us that *Ceque* 8, radiating from the Temple of the Sun, passed over the hill of Carmenga and terminated at Sicllabamba. Different *huacas* on that *ceque* can be pinpointed exactly (Fig. 5.4) because today half of them can still be located and many still retain their old names. Urcoscalla was a place "where those who travel to Chinchaysuyu lose sight of Cuzco" (Cobo 1956:173–174). On the basis of sightings of Cuzco made from the hills west of Cerro Picchu, we equate the area where the city disappears from view with a place bearing the modern name of Arco Punco. Radially outward from the Coricancha along the same *ceque (Huaca* 12) we find Poroypuquio ("the well of Poroy"); Cobo says that the Spanish built a watermill there, and the ruins of the mill can still be seen near the village of Poroy. Collanasayba ("the prin-

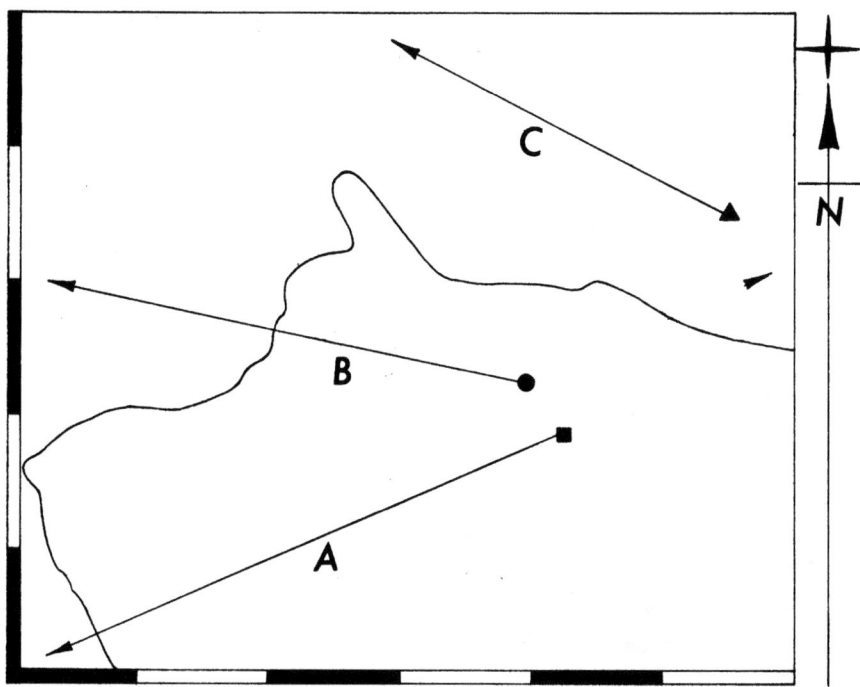

5.5 Summary schematic showing all 3 alignments derived for Cuzco: (A) December solstice sunset; (B) June solstice sunset; (C) Antizenith sunset.

cipal *mojon*"), located 13 km radially outward from the Temple of the Sun, is a rock outcrop about an acre in extent located at a prominent bend in the river coming from Poroy. Original Inca walls were found when the modern road to Chinchero was constructed, and the hacienda there still bears the original name Sicllabamba; the hill which belonged to Sicllabamba is also called Collanasayba. The house of that hacienda still stands adjacent to this, the last *huaca* of its *ceque*. The place must have been ideal for the ritual purpose of throwing offerings (brought from Cuzco) into the river at the end of the *ceque*, a practice referred to often by the chroniclers.

Using the surveyor's transit, we traced the *ceque* as a straight line over the mountain between Collanasayba and the Coricancha in Cuzco. The goal of this exercise was to approximate the position on Cerro Picchu where the straight line would cross. Since Sucanca lay on the *ceque* line but was astronomically observed from the *Ushnu*, we saw a means of fixing the important sunset line and the corresponding date in the agricultural calendar, thereby locating the site of the *Ushnu* (known to have been somewhere in Hanan Haucaypata—the present Plaza de Armas) as well. Other chroniclers such as Betanzos (1968) and Sarmiento de Gamboa (1942) discuss these same pillars, and Garcilaso de la Vega (1969:116–118) also mentions having seen them standing in

5.6 Monroy (1653) painting of the Plaza Hanan-Haucaypata of Cuzco. *Rollo* (at lower left), which Pizarro placed on the site of the Ushnu, once marked the center of a much larger plaza that can be visualized by removing the row of buildings passing horizontally across the picture and including the *rollo*. The plaza boundary also can be traced roughly in Fig. 5.7 if the Ushnu (B) is imagined as the center. For orientation, the cathedral just to the left of bottom center is at position C in Fig 5.7.

1560, though he could not be sure of the place from which they were supposed to be observed when the sun went down.

The position of the Picchu Sucanca thus having been located (we erected our own Sucanca on the spot by piling up stones), we set up the transit there and shot a series of lines into the Plaza de Armas. We sought to determine for what dates sunset could be viewed over the Sucanca from various points in the plaza. One line to which we paid particular attention was that corresponding to the 18 August sunset date (the broad arrow in the enlargement of the Plaza in Fig. 5.7), for it coincides with the sunset point on the day when the sun passes through the nadir (the point opposite the zenith). The importance of the passage of the sun through the zenith is mentioned frequently in the chronicles (e.g., see Garcilaso de la Vega 1969). The

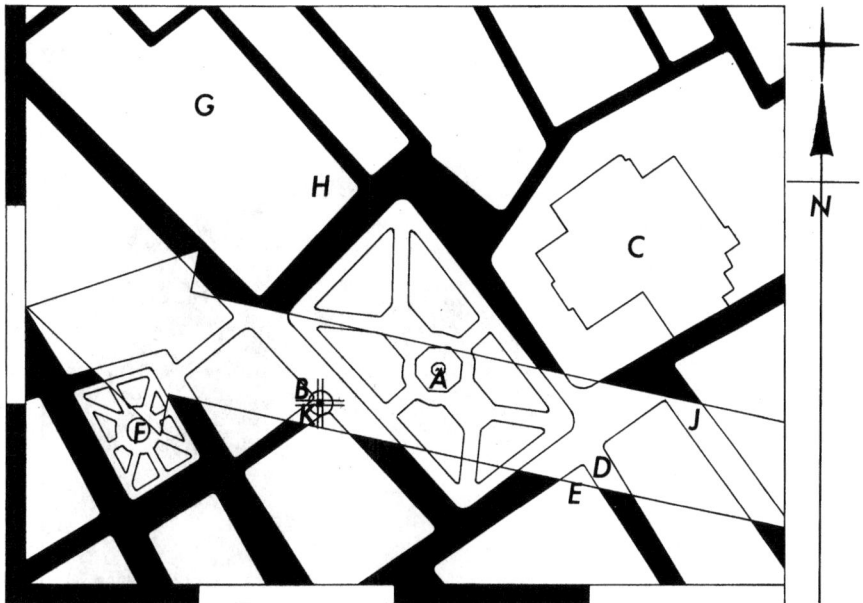

5.7 18 August antizenith line (arrow) traced backwards from the known position of the *sucanca* across the Plaza de Armas, the center of which today is marked by a fountain (A). The *rollo* or *picota* discovered in the Monroy painting (Fig. 5.6) was located at B. It was placed on the site of the Ushnu by Pizarro and is seen to fall within the predicted observational zone. Bars bordering the figure represent 100-meter lengths. Other features identifiable in Cuzco today: (C) Cathedral; (D) Calle Loreto; (E) Compania; (F) Plaza Regocijo; (G) Garden of Ticcicocha; (H) Casana Palace; (J) Calle Sta. Catalina Angosta; (K) Calle Medio.

zenith-antizenith direction accords very well with the vertical symbolism so prominent in Andean thought (Zuidema 1977–8; Isbell 1978), and with statements made by Guaman Poma (1936:883–884) that the earth opens up in February and August, the zenith-antizenith sun dates. On the antizenith sunset day the sun disappears over the western horizon at a point about 180° opposite the sunrise point on the day of the zenith passage. (Actual differential horizon elevations cause the line to deviate by as much as 0.5° from linearity, but the azimuth is, nevertheless, about 283°.)

By a fortunate circumstance, our assumptions and calculations were further aided in the summer of 1978 as a result of our examination of a 1653 painting by the artist Estrada Monroy (Fig. 5.6), located in the cathedral of Cuzco. It shows the city in the aftermath of the earthquake of 31 March 1651, barely a century after the conquest. The view shown is from the top of the cathedral looking toward the southwest (the same as the orientation of the map). Near the center of the picture (enlarged in the inset) we see a vertical column supporting a cross. This is the *rollo* (or "*picota*") which Pizarro was said to have placed on the site of the *Ushnu* "in the middle of the plaza"

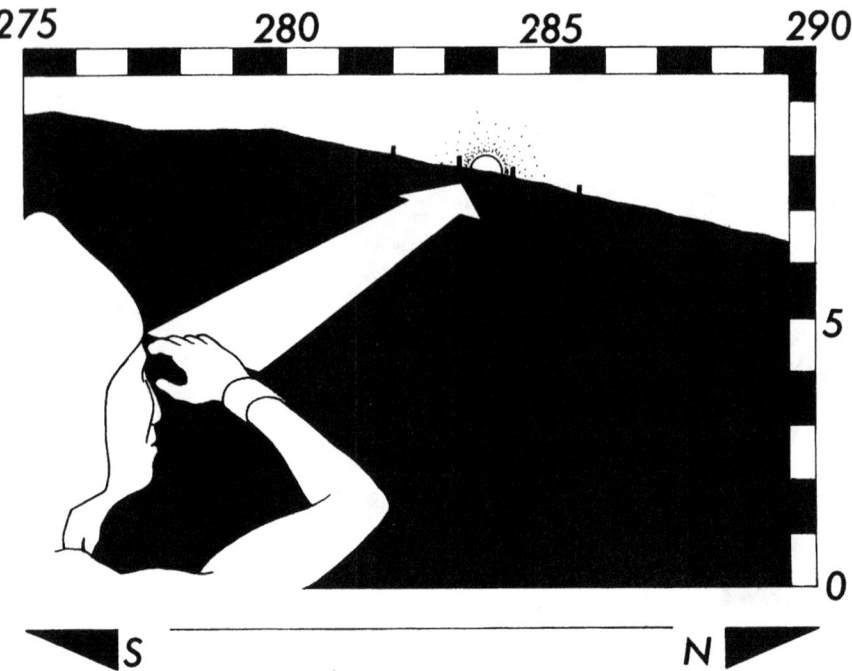

5.8a Observing the 18 August antizenith sunset on the western horizon of Cuzco. Observer is at the Ushnu 2 km from the pillars. The horizontal scale tabulates azimuth, the vertical scale, altitude, both in degrees. The size of the sun is exaggerated.

shortly after he conquered the city (Betanzos 1968:70). (The *Ushnu* itself is a very elusive concept; its relation to verticality in the Inca cosmos has been fully developed by Zuidema [1977–8], who has demonstrated that it was manifested in a multitude of material objects including an altar of sacrifice, a platform of stones, a *mojon,* and even a hole in the earth.) Note that if we join the Plaza de Armas (the old Hurin-Haucaypata) with the Plaza de Regocijo (the old Cusipata) to the southwest to form a single plaza, the stone would have lain near the center; the old layout can be imagined in Figure 5.7. The chronicler Santacruz Pachacuti Yamqui talks about the center of the double plaza Haucaypata-Cusipata, so we now understand that this was what other chroniclers meant when they referred to the "middle of the plaza." We note further that the *Ushnu* is situated comfortably within the 18 August viewing band as calculated from our measurements of the *ceque* line containing the Sucanca on Cerro Picchu.

While the Anonymous Chronicler mentions a chain of four pillars, presumably running perpendicular to the *ceque* lines, why does the description of Cobo include only two of them? Evidently Cobo mentions only those towers located on the *ceque* whose *huacas* he was delineating, as he did in the case of the solstice pillars. In Figure 5.8a we see the horizon geometry of the Picchu scheme and attach a photograph (Fig. 5.8b) that shows the situation at present as viewed from the plaza.

5.8b View as it appears today from close to the site of the Ushnu. (Rooftop blocks the view from the exact site today.) Antizenith sunset occurs behind the light pole, on left (see arrow sight-line of Fig. 5.7).

The line from the *Ushnu* to the pillars on Cerro Picchu turns out to be the tip of a rather large iceberg, of which I shall reveal only the basic outline. As Zuidema shows in an accompanying paper, this line turns out to be derived from observations taken in the other direction, that of the rising sun at the day of its passage across the zenith. This line passes from the *Ushnu* of Hanan-Haucaypata of the northern moiety through a second *Ushnu,* that of Hurin-Haucaypata of the southern moiety. The latter *Ushnu* is located on the site of the plaza Limapampa. This observational line is extended to point to the eastern horizon, where sun temples were situated. The concept of reversing the sight-line to view the antizenith event (which coincides with both the time of planting and the time the earth opens) became a guiding principle for the Inca, a brilliant stroke which also embodied the dualism of the vertical-antivertical axis so central to Inca and even contemporary Andean cosmic thought (e.g., see Isbell 1978:197–220). Lying along the same line (which crosses from one side of the Inca capital to the other) is the Sunturhuasi building and the two *Ushnu,* functioning as gnomon used to make measurements of the sun when it migrated to the overhead position. The sighting of the moon at the zenith, when the sun lay at the antizenith, may also have been involved. These ideas are merely presented here without proof, in order to indicate the direction of our research. The entire argument, together with the solstice-watching scheme, is summarized in Figure 5.5, which provides strong evidence concerning not only the position and function of the Cerro Picchu pillars, but also the reality, in some instances, of Cobo's straight-line description of the *ceques*. Archaeologists are encouraged to begin explorations of the predicted *Ushnu* site as well as of the sites of the pillars.

CONCLUSIONS

We have attempted to explicate here both the exactitude and the complexity of the most basic types of Incaic astronomical orientations. In this task we have employed ethnohistorical evidence, astronomical arguments, and architectural and topographic data gathered through appropriate instrumentation in the Cuzco environment. We have established beyond doubt the existence of three astronomical sighting directions which emanated from three different points in the Cuzco area: (1) a pair of towers which mark the June solstice-sunset point as viewed from Lacco, a complex of rock carvings on a hill north of Cuzco; (2) a pair of towers which mark the December solstices as seen from Coricancha, the center of the *ceque* system; and (3) four towers situated on Cerro Picchu which mark time for the planting season, centered about the place where the sun sets on the day of passage through the antizenith.

We have discussed only formative concepts relating to the last sight-line, but have done so in enough detail to make it clear that it was of great importance, serving as an embodiment of the up-down dualism which enters so frequently into Andean thought. The scheme of employing horizon observations to determine planting dates is similar to that reported by Urton (1978b) in his study of present-day Misminay, a village northwest of Cuzco.

We find a tricentric concept with unequal numbers of towers at only one horizon difficult to deal with because it is not what we expected to find. Western astronomers would never have operated this way, and Garcilaso's confused statement about the pillars (a description which we can no longer take too seriously) reveals how shallow the penetration of the Renaissance mind into Incaic astronomy was. To add to the complications, we are confronted with a basic astronomical concept—the zenith-antizenith passage of the sun—which has no analog outside the Tropics. Near the equator, vertical motion on the celestial sphere is dominant. By contrast, circular orbits pivoted about the celestial pole characterize the principal aspect of the heavens as seen in the temperate latitudes that served as the environmental backdrop for the more familiar astronomy of the Classical world. It will be interesting to see how this principle holds in other tropical zone astronomies, such as those of Polynesia, Java, Central Africa, and Mexico.

In spite of perceptual difficulties, this early foray into the fundamentals of Inca astronomy raises questions not only for astronomers, but also for agriculturists, ethnologists, and archaeologists. There are many hypotheses to test, and what once appeared (astronomically speaking) to be an unknown world sealed by the absence of the written word has opened into a rich field for future research.

NOTES

We are indebted to the organization Earthwatch (Belmont, Massachusetts) for supporting this research and to Stephen Fabian, Gary Urton, and William and Billie Jean Isbell for useful discussions relating to this problem. We thank Peter Dunham for doing the illustrations.

1. The city was divided into quadrants, two of which belonged to the upper moiety of Hanan Cuzco (Chinchaysuyu, the northwest, and Antisuyu, the northeast) and two to the lower moiety of Hurin Cuzco (Cuntisuyu, the southwest, and Collasuyu, the southeast). Curiously, the division between the two southern quadrants occurred at azimuth 146° and not 180°. The Temple of the Sun served as the center of coordinates, having been situated at the confluence of the two rivers which flowed through the valley of Cuzco.

REFERENCES

Betanzos, J. de. 1551. *Suma y narración de los Incas*. Madrid, Biblioteca de Autores Españoles, Cronicas de interes indigena.

Cobo, B. 1653. *Historia del Nuevo Mundo*. Madrid, Biblioteca de Autores Españoles (1964).

Garcilaso de la Vega, G. S. 1969. *Royal Commentaries of the Incas*, 2 vols., trans. Harold V. Livermore. Austin, University of Texas Press.

Guamán Poma de Ayala, F. 1584–1614. *El Primer Nueva Coronica y Buen Gobierno*. Paris, Musée de l'Homme (1936).

Isbell, B. J. 1978. *To Defend Ourselves*. Austin, Foundation for Latin American Studies.

Zuidema, R. T. 1964. *The Ceque System of Cuzco: The Social Organization of the Capital of the Inca*. Leiden, E. J. Brill.

Zuidema, R. T. 1977a. "The Inca Calendar." In *Native American Astronomy*, ed. Anthony Aveni, 219–259. Austin, University of Texas Press.

Zuidema, R. T. 1977b. "Mito y Historia en el Antiguo Peru." *Allpanchis Phuturinqa* 10: 15–52.

Zuidema, R. T. 1978a. Lieux sacrés et irrigation: Tradition historique, mythes et rituals au Cuzco. *Les Annales*. Paris.

Zuidema, R. T. 1978b. Shaft tombs and the Inca Empire. *Journal of the Steward Anthropological Society* 9:1–2 (Fall 1977 and Spring 1978).

Zuidema, R. T. 1980. El Ushnu. *Revista de la Universidad Complutense*. Editorial de la Universidad Complutense de Madrid, 28 (117):317–362.

Zuidema, R. T. n.d.a. El Sunturhuasi. (Ms.)

Zuidema, R. T. n.d.b. Llamas and the Inca Calendar. (Ms.)

Zuidema, R. T., and G. Urton. 1976. La Constellation de la Llama en los Andes Peruana. *Allpanchis Phuturinga* 9:59–119.

CHAPTER SIX

Here Comes the Sun
The Cuzco–Machu Picchu Connection

David S.P. Dearborn and Katharina J. Schreiber

PROLOGUE

In January 1986 at the Second Oxford Conference on Archaeoastronomy held at Mérida, Mexico, Anthony Aveni suggested to Dearborn that recent research at Machu Picchu should be compared to what was known about Inca observations in Cuzco. This suggestion provided the inspiration for the present article which attempts to do just that: compare what we have found at the Inca site of Machu Picchu with the work of Aveni and Tom Zuidema in and around Cuzco.

We divide this comparison into three parts: evidence for June solstice observations, evidence for December solstice observations, and evidence for observations of the zenith and antizenith dates. In order to make this comparison we have had

---

First published as "Here Comes the Sun: The Cuzco–Machu Picchu Connection," by David S.P. Dearborn and Katharina J. Schreiber, from *Archaeoastronomy* 9:15–37. © 1987 by the University of Texas Press. All rights reserved.

to summarize the data from numerous articles by Zuidema and Aveni, and to evaluate this data. We are not always in agreement with Zuidema and Aveni about significance and reliability of their results, but we trust we have not misrepresented their interpretations.

Our work at Machu Picchu stems from three seasons of fieldwork. In 1980 work by Dearborn and White focused on the structure known as the Torreón, and its association with the June solstice and perhaps the zenith (Dearborn and White 1982, 1983). In 1982, work by Dearborn, along with White and Mannheim, attempted to locate points of observation of the antizenith. And in 1984, Dearborn, Schreiber and White studied a structure known as Intimachay and found that it was associated with the December solstice (Dearborn, Schreiber and White 1987). These results are herein summarized, and compared with what is "known" about Inca astronomy in the Cuzco area.

INCA RELIGION AND CULTURE

Beginning sometime in the early 15th century, the Inca began the conquest that established an empire extending south from Cuzco into Bolivia, Chile and Argentina, and north through Peru and Ecuador to Colombia. According to the oral tradition recorded by the Spanish, this conquest was begun by Sapa Inca Pachacuti Yupanqui. He is also given credit for having introduced the fine cut stone masonry for which imperial Cuzco is known, and he formalized much of the religion that the Inca spread throughout their empire.

Among the principal Inca deities were the Sun, Moon, and stars. While astrology seems to be a common Andean interest, they included astronomical objects in their origin myth. Their descent from Inti, the Sun god, was used as a reason for their success in conquest, and was developed into the state religion. Conquered peoples were not required to give up their own gods, but had to acknowledge the superiority of Inti, and his Inca descendants. Cobo (1983) records that Pachacuti ordered temples be built to the sun in all conquered lands. At sites like Pachacamac, these were major construction projects intended to impress all who came of the importance and power of the Inca. Because of the Inca interest in the Sun, as well as the Moon and stars, archaeoastronomical studies serve to help define the place that astronomy had in the intellectual and cultural organization of the Inca empire which they called Tawantinsuyu (Land of the Four Quarters).

Sources like Garcilaso de la Vega (1960), Felipe Guaman Poma de Ayala (1936), Bernabé Cobo (1983, and Rowe 1979), Cristóbal de Molina (1943), and Juan de Polo de Ondegardo (1916) give us some information on when the Inca celebrated fiestas and what events were of an astronomical nature like the solstices or zenith passage dates, and what events included observations to indicate times of local importance, such as when to plant. The observations themselves can be considered both true astronomy, such as monitoring the motion of the Sun, as well as of a symbolic sort like simply observing the sunrise on a particular date.

Interestingly, these chroniclers also give us some indication of the technology used in these observations. They indicate that the different techniques used by the Incas might include horizon observations, windows, and shadow casting. For example, in the case of horizon observations, Cobo's list of *huacas* around Cuzco states clearly that some of them were pillars or markers along the horizon that were used for observing the movement of the sun:

> [Ch-6:9] la nouena Guaca era un cerro llamado, Quiangalla, que esta en el camino de yucay, donde estauan dos *mójones,* o pilares que tenian por señal que llegando alli el sol era el principio del verano (Cobo, in Rowe 1979: 24).

> [Ch-6:9] The ninth *guaca* was a hill named Quiangalla which is on the Yucay road. On it were two markers or pillars which they regarded as indication that, when the Sun reached there, it was the beginning of the summer (translation by Rowe 1979: 25).

The use of windows for observing the rays of the sun is described by Guaman Poma:

> ... y lo ciguin el senbrar la comida en que mes y en q. dia y en que ora y en punto por donde anda el sol lo miran los altos serros y por la manana de la claridad y rrayo q. apunta el sol a la uentana por este rreloxo, cienbra y coxe la comida del año in este rreyno (Guaman Poma 1936: 235).

> In the sowing of the crops, they follow the month, the day, the hour, and the point where the Sun moves; they watch the high hills in the morning, the brightness, and the rays that the Sun aims at the window; by this clock they sow and harvest each year in this domain.

And Garcilaso describes the way in which the Incas observed shadows in order to define the day of the equinox and zenith:

> Para verificar el equinoccio telan columnas de piedra riquisimamente labradas, puestas en los patios o plazas que habia ante los templos del sol; los sacerdotes cuando sentian que el equinocco estaba cerca, tenian cuidado de mirar cada dia la sombra que la columna hacia. Tenian las columnas puestas en el centro de un cerco redondo muy grande que tomaba todo el ancho de la plaza o del patio; por medio del cerco echaban por hilo de oriente a poniente una raya, que por larga experiencia sabian donde habian de poner el un punto y el otro. Por la sombra que la columna hacia sobre la raya, velan que el equinoccio se iba acercando; y cuando la sombra tomaba la raya de medio a medio, desde que salla el sol hasta que se ponia, y que a mediodia bañaba la luz del sol toda la columna en derredor sin hacer sombra a parte alguna, decian que aquel dia era el equinoccial (Garcilaso 1960: 73).

> To verify the equinox they had columns of richly worked stone, placed in the patios or plazas in front of the temples of the sun; when they sensed that the equinox was near, the priests took care to look every day at the shadow that the column made. They had the columns put in the center of a circle which was very large and took the whole width of the plaza or patio; in the middle of the circle they made a line from east to west with string, because of long experience they knew where to put each point. By the shadow that the column made on the line they saw that the

equinox was approaching; and when the shadow bisected the line, from where the Sun rose to where it set, and at noon the light of the Sun bathed the entire column all around without making a shadow on any part, they said that that day was the equinox.

The mestizo chroniclers, Guaman Poma and Garcilaso, talk of windows and shadow casting from pillars. These are techniques that would have been useful in or near the confines of a temple where access was restricted and only a select group might be present, such as the priests or nobility. The horizon pillars, on the other hand, which were described by several Spanish chroniclers, were public features visible to all. The Spanish could hardly miss the pillars as they represented a public manifestation of a pagan religion. On the other hand, they could easily have been unaware of the more private features and may not have observed these other techniques.

The accounts must, however, be used with care. They were often politically or religiously motivated, and seldom give a systematic description of events as one expects in a modern history. Ethnographic studies by Urton (1981) and others have examined some associations between these ethnohistorical data and modern practices giving us a better understanding of the meaning of those records.

Zuidema has used this ethnohistorical and mythological data to develop hypotheses defining the place that astronomy had in Andean culture and thought. Zuidema's hypotheses often address questions that are interesting in a human, cultural context, but are not amenable to finding that physical data to support or deny them. His work with Aveni (1981) has, however, led to some evidence for alignments and observations not evident in the historical records. We will present some of the ethnohistoric documentation that forms the basis of these hypotheses and we examine the physical evidence obtained in Cuzco and Machu Picchu (Dearborn and White 1982, 1983; Dearborn, Schreiber and White 1987) where it supports or is at least consistent with these hypotheses.

MACHU PICCHU

The bulk of our own research has been at Machu Picchu. Machu Picchu is an ideal site at which to work given the good state of preservation of the architecture there. Features such as windows are frequently not preserved at Inca sites; Machu Picchu offers the rare opportunity to study a site that is relatively intact. Rowe (1987) has analyzed an early colonial document that lists Inca property holdings; he finds that Pachacuti Inca held an estate named Picchu (named after the hill near Cuzco) approximately where Machu Picchu is located. This discovery gives an understanding for the extensive ceremonial areas that exist at Machu Picchu. It was a planned community where Pachacuti could relax and carry out the duties of the imperial religion that he had substantially developed.

Our research at Machu Picchu has shown that the Incas were capable of precise observations and could have defined the exact dates not only of astronomical events, but also days leading up to them, if they desired to. Some of the structures that we examined at Machu Picchu may be analogous to those reported in Cuzco. Machu Picchu was a planned community (as opposed to one that evolved over time) so it contains all the temples and other features that Pachacuti wished to have at his estate. The intentional destruction of Inca sites and the continuous habitation of Cuzco have obscured much relevant evidence in the Cuzco area and thus the discernment of actual observation features there is somewhat more difficult than at Machu Picchu.

INCA CUZCO AND THE CORICANCHA

Cuzco was the capital of Tawantinsuyu and as such contained the holiest places in the empire. Cuzco was rebuilt by the Inca Pachacuti not just to be a home to the Children of the Sun, but to impress visitors (allies or otherwise). Among the most important and impressive shrines in Cuzco was the Coricancha, the enclosure of gold, in which the Temple of the Sun was located. This was among the most sacred places to the Inca.

Rowe (1944), in his study of the Coricancha, described its significance to the Inca as being akin to the church of the Holy Sepulchre to Christians. According to Garcilaso, it contained temples to the Sun, Moon, and Stars (in particular the Pleiades and Venus), Rainbow and Thunder. The Coricancha was also the ceremonial center of Cuzco and the point of origin of the *ceque* system, a series of lines that radiated out from Cuzco (discussed in more detail below).

While much of the Coricancha still stands, the Hall of the Sun, itself, was destroyed some time between 1560 and 1638 (Rowe 1944: 62). Whatever remains must exist under what now is the church of Santo Domingo. We may never know what the true observational capabilities of this structure were, although we do know from the account of Garcilaso that the equinox/zenith was observed there, as described above.

Zuidema (1982) found the west wall of the Coricancha to face near the point on the horizon where the Pleiades rise. Using his published data we find this wall oriented to the rise of objects having a declination of 21.1 degrees. While the Coricancha is one of the oldest Inca shrines, Pachacuti is supposed to have rebuilt it when he introduced the fine masonry to Cuzco. The walls that remain today were probably constructed around 1450. At that time the declination of the Pleiades was about 22.5 degrees. While 1.4 degrees is too large a difference for this to have been a true observing instrument (the eye could easily notice the difference if sighting along a wall), it is as close as one might expect for constructing a structure merely oriented to a star group. This is especially true when considering that the azimuth of a stellar rise appears to vary due to atmospheric clarity and lunar position.

Given the importance of the Pleiades as a constellation, the suggested orientation is not unreasonable. Urton has found that even today some people observe the first morning that the Pleiades rise before the Sun as a herald of the solstice, and for omens of the coming year. Zuidema, however, has suggested another possible intent for the Coricancha orientation.

The Coricancha orientation could mark the rise point of the sun around May 25. The full Moon that follows this date will be the month that includes the Inti Raymi festival. This idea is supported by Molina's description of the month of May:

> They commenced to count the year in the middle of May, a few days more or less, on the first day of the moon; which month, being the first of their year, was called *Huaca* and Llusque, and they performed the following ceremonies, called Yutip-Raymi, or festivals of the Sun (Molina 1873: 16).

While Molina is the only chronicler to claim that the year "began" with Inti Raymi, he may have been talking about the commencement of a separate (from the solar year) lunar calendar, or have simply chosen to begin his account of the months with the most important public festival.

Shifting from the Julian to Gregorian calendar shifts Molina's date to late May; the first full moon following May 24 will include the feast of Inti Raymi. If the orientation of the Coriancha was intended to mark the date to begin watching for the full moon, then its alignment is quite precise.

Before turning from the Coricancha, we must discuss briefly the *ceque* system (Fig. 6.1). The *ceque* system consisted of a large number of *huacas* (sacred places) arranged in 41 "lines," that radiated outward from the Coricancha. Cobo described 328 of these *huacas* in a list that was translated by Rowe (1979). He lists the *huacas* by *suyu* (quarter) and *ceque* (e.g., CH-9:4 is the 4th *huaca* of the 9th *ceque* line of Chinchaysuyu).

Suyu	Direction	Ceques	Huacas
Chinchaysuyu	Northwest	9	85
Antisuyu	Northeast	9	78
Collasuyu	Southeast	9	85
Cuntisuyu	Southwest	14	80

Some of these *huacas* were used in making observations of the Sun, and we will refer to them frequently. These observations were generally made from one *huaca* to another. Suggestions have been made also of some alignments of *ceque* lines.

An important property that is often assumed for these *ceques* (lines of *huacas*) is that they are straight. Aveni (1981) supports this assumption with a statement that three *huacas* of CH-6 lay on a straight line. He recently expanded that statement to say "within a few degrees" (Aveni 1987). Cobo gives us evidence that the *huacas* did not necessarily lie in an absolute straight line when he says:

> [Ch-9:4] la quarta Guaca era una fuente llamada Pomacucho, que estaua algo apartada deste *ceque* ... (Cobo, in Rowe 1979: 28).

> [Ch-9:4] The fourth *guaca* was a fountain named Pomacucho which was somewhat separated from this *ceque* (translation by Rowe 1979: 29).

This variation from a straight line must be considered as a source of uncertainty in locating *huacas* on a straight line between two other *huacas*.

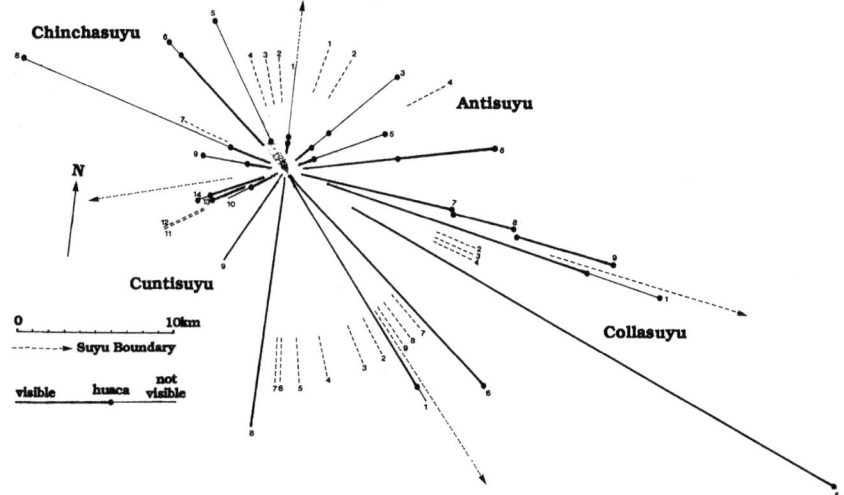

6.1 The *huacas* of the *ceque* system were divided into four *suyus* and extended out to four leagues from Cuzco. The locations of the *ceques* drawn as solid lines follow Zuidema (1982), and the approximate locations of the remaining *ceques* are indicated by dotted lines and were taken from Zuidema (1978).

INTI RAYMI: THE JUNE SOLSTICE

Ethnohistoric Documentation

According to Garcilaso (1960, book 6, chapter 20) the Inti Raymi festival was the most important celebration of the year. It marks the date when the Sun has moved as far north as it travels and turns around to begin the journey back to the south. This was not just an Inca celebration. People from all parts of the empire paid homage to "the father of all the Incas."

As described by Garcilaso, the festival was preceded by three days of fasting and on the eve of the feast day sacrifices were made. Gifts were given to the Sun by ambassadors from all of the participating nations. On sunrise of the feast day, the Inca and his relatives watched the Sun rise from the great plaza of Haucaypata (today the Plaza de Armas). Peoples of non-Inca blood did the same from the plaza Cusipata separated from Haucaypata by the river Huatanay. After drinking to the Sun, a bowl of *chicha* was poured into a channel that flowed from the plaza to the house of the Sun. The Inca then led a procession to the Temple of the Sun in the Coricancha where additional sacrifices were made. This was followed by feasting, drinking and honoring people who had distinguished themselves during the year. The festival lasted for nine consecutive days.

While Garcilaso de la Vega was writing forty years after he himself had left Peru in 1560, the importance of Inti Raymi as a festival worshipping the Sun is echoed by many chroniclers. Inti Raymi was one of the "three principal feasts of the year" as

6.2 A view across the Plaza de Armas in Cuzco from the steps of the cathedral (which extends out into the original plaza of Haucaypata). The buildings on the opposite side of the square lie near the path of the Huatanay river which separated the plaza Haucaypata from Cusipata. The *Usno* was located somewhere in this plaza, and the pillars that marked the beginning of planting were on Picchu, the hill to the right.

described by Molina (1873: 36). In describing the ceremonies associated with Inti Raymi, he says that, in addition to various ceremonies at the Coricancha and elsewhere, "on every other day of this month they went to burn sheep and the other offerings at the following places." He then names 21 locations where these sacrifices were made.

The sites named by Molina form a circuit some distance to the southeast of Cuzco, and Zuidema suggests that they were visited sequentially by a procession (Zuidema 1981). The first location was named Succanca and the second was Omoto-yanacauri. The last two were Quispicancha and Sulcanca. The similarity of these names to the *huaca* named Sucanca (which was used as a solar marker) leads Zuidema to suspect that the first and last locations were part of the system of such markers. (The Sucanca will be discussed further below in the sections dealing with the December solstice and zenith-antizenith alignments.)

It is clear from all descriptions of Inti Raymi that it was an extremely important and sacred event associated with the worship of the god, Inti, and of making observations of his travels through the sky.

The June Solstice at Cuzco

In addition to the primary Temple of the Sun at the Coricancha, there were a number of other temples of the Sun in and around Cuzco. For example, Cobo's list of *huacas* includes several shrines that were temples of the sun, or were at least dedicated to the sun. These include:

> AN-3:4 *Chuquimarca*. It was a temple of the Sun on the hill of Mantocalla, in which they said that the Sun descended to sleep. For this reason, in addition to other things, they offered it children.[1]
>
> AN-6:3 *Chuquicancha*. It is a well-known hill which they held to be a house of the Sun. On it they made very solemn sacrifice to gladden the Sun.
>
> CO-4:8 *Guancarcalla* is a ravine like a gateway which is next to the hill mentioned above [Raurao Quiran]. It was dedicated to the Sun, and they offered it children in certain festivals which they held there.
>
> CU-10:2 *Puquincancha*. It was a house of the Sun which was above Cayocache. They sacrificed children to it.

We also know of many markers, some of which were used to observe the Sun's motion along the horizon:

> CH-6:9 *Quiangalla* is a hill which is on the Yucay road. On it were two markers [*mojónes*] or pillars which they regarded as indication that, when the Sun reached there, it was the beginning of the summer.
>
> CH-8:7 *Sucanca*. It was a hill by way of which the water channel from Chinchero comes. On it there were two markers [*mojónes*] as an indication that when the Sun arrived there, they had to begin to plant the maize. The sacrifice which was made there was directed to the Sun, asking him to arrive there at the time which would be appropriate for planting; and they sacrificed to him sheep, clothing, and small miniature lambs of gold and silver.
>
> CH-8:13 *Collanasayba*. It was a marker [*mojón*] which is on a hill at the beginning of Sicllabamba, as the end and limit of the *guacas* of this *ceque*.
>
> CO-8:3 *Mudca*. It was a stone pillar which was on a small hill near Membilla. They offered it only ground up shells.
>
> CU-13:3 *Chinchincalla* is a large hill where there were two markers [*mojónes*]; when the Sun reached them, it was time to plant.

Aveni (1981) had discussed the possible association of two of these *huacas*, Chuquimarca and Quiangalla, with a June solstice observation (Fig. 6.3).

Chuquimarca was a temple of the Sun where the Sun descended to sleep. This suggests a setting Sun. However, as this structure is located to the east of the Coricancha, one cannot observe a sunset by sighting along this *ceque* line from its origin at the Coricancha. Zuidema suggests that, instead, a sunset observation was made *from* Chuquimarca on the June solstice. The best candidate for the point that was observed from there is the *huaca* Quiangalla, on which two markers were erected in order to observe the beginning of summer. (It should be noted that although the

Here Comes the Sun

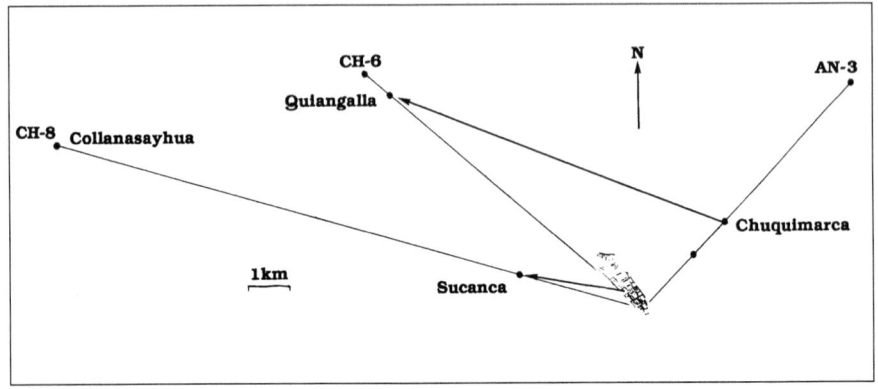

6.3 Sunset on the June solstice was probably observed from the temple of the Sun, Chuquimarca on AN-3 to a pair of pillars called Quiangalla located on CH-6. An observation to begin planting around Cuzco was also made from the Usno, located somewhere in the main plaza, to the central pair of four pillars located on the hill Picchu. These pillars were probably the *huaca* Sucanca.

June solstice is the beginning of the austral winter, this dry season was often referred to as summer.)

In order to identify the location of Quiangalla, Aveni (1981) and Zuidema (1981) defined the path of *ceque* CH-6 by locating three of its *huacas* (unspecified in their reports), and defined the location of Quiangalla as the point where this *ceque* passed over a high ridge. By sighting back toward the *ceque* line, AN-3, on which Chuquimarca was located, they were able to specify its probable location, if a solar observation were made on the June solstice. They believe that Chuquimarca was on a small hill on which is located a ruin now known as Lacco. If these remains are truly Chuquimarca, then the June solstice sunset crosses the horizon across the valley to the northeast in a well-defined, restricted area (400 square meters, according to Aveni). The pillars of Quiangalla should have been located near there.

There is no longer any surface evidence of the existence of the pillars, but it is not unlikely that they were intentionally destroyed by the Spanish in their quest to obliterate objects of pagan worship. Perhaps future archaeological investigation will be able to determine their actual location and demonstrate their utility for marking the beginning of "summer" as viewed from Lacco (or some nearby site). In sum, while few actual physical remains have been found, Aveni and Zuidema have made a convincing case, and it is very encouraging that the sunset direction from Lacco crosses the *ceque*, CH-6, along the horizon.

The June Solstice at Machu Picchu

When we began our study of Machu Picchu, among the first structures examined was the Torreón. It was constructed on a large rock prominence with a clear view

6.4 The hill on which Lacco is located. Zuidema and Aveni argue that the Temple of the Sun, Chuquicancha, must have been near here. The Antisuyu road passes below the hill (through the center of the photograph).

to the east. The walls are of the finest stone work to be found anywhere. One wall is curved to conform to the contour of the rock foundation and has two windows facing northeast and southeast. Bingham (1948: 148) compared a number of superficial characteristics of the Torreón to the Coricancha and suggested that it was a temple of the Sun. In addition to the curved wall of fine masonry, similarities included a door with holes that Bingham compared to the niches in the Coricancha used for the placement of jewels and golden idols.

The Torreón also contains features that are unique among the buildings at Machu Picchu. The two windows in the curved wall have pegs projecting from the wall exterior very close to the upper and lower corners of the window. These pegs are flat on the top and rounded on the bottom. The interior of the structure is dominated by an irregular platform carved from the bedrock. The Torreón was clearly a special structure at Machu Picchu.

It was found that the northeastern window is aligned to a declination of +21.6 (±0.6) degrees (providing an interesting similarity to the Coricancha). A significant feature of the Torreón is, however, the straight edge that was cut into the bedrock platform that dominates its interior. This edge does not point through the center of

HERE COMES THE SUN

6.5 The Torreón at Machu Picchu (*right of center*). Its foundation is a stone prominence with a clear view to the east. The two windows with the exterior pegs are clearly visible in the curved wall.

the window, but at a slight angle. It is oriented very precisely to the rising point of the Sun on the June solstice. The solstitial alignment of the central stone is accurate to approximately 2 arc minutes, a precision comparable to the best naked eye astronomical measurements (including the effects of refraction and the change in the obliquity of the ecliptic). This orientation is of sufficient precision to strongly argue that it was intentional.

While the building was capable of being used to determine the solstice with no additional features, it would have been slightly awkward. An observer located at the rear of the stone could sight along the edge only to see the wall below the window. Lifting one's eye to the horizon to sight the Sun loses the accuracy with which the edge was constructed. However, if one were to cast a shadow through the window and onto the stone, a more accurate observation could be made. We therefore turned to the pegs located near the upper corners of the window.

Projections or pegs are common enough on Inca masonry, but these are quite distinct. Almost all projections are found near the base of large stones and are flat on the bottom. It is believed that they were used for manipulating the stone while it was being worked. Their locations on particular stones seem almost random. The Torreón

6.6 Sunlight entering the window of the Torreón a few minutes after sunrise on a day near the June solstice. Just at sunrise, the light actually reaches a little further back on the rock.

is constructed with coursed masonry of finely cut small stones, not the large stones that usually have projections. Furthermore, the projections are specifically related to the windows, not randomly located along the wall. It seems apparent that the pegs are shaped to support something in the window, not for lifting the stones. We reasoned that anything supported on these pegs casts a shadow and decided to investigate the utility of such shadow casting. As described above, shadow casting was a principle known to, and used by, the Inca.

To investigate this possibility, a plumb bob was hung from a stick supported by the upper pegs. While Bingham (1979: 193) found plumb bobs at Machu Picchu, we do not suggest that this was necessarily the instrument used with the window. Any object hung or supported on the window at sunrise will cast a shadow on the central stone. It was demonstrated that the shadow of a vertical object acted almost like a hand on a clock to indicate not only the day of the June solstice, but also individual dates for a range of time around the solstice. Beginning in May, the light from the rising Sun will enter the window and illuminate the edge in the stone that points to the solstice rising position. The shadow of a plumb bob (or any other object) will cross the

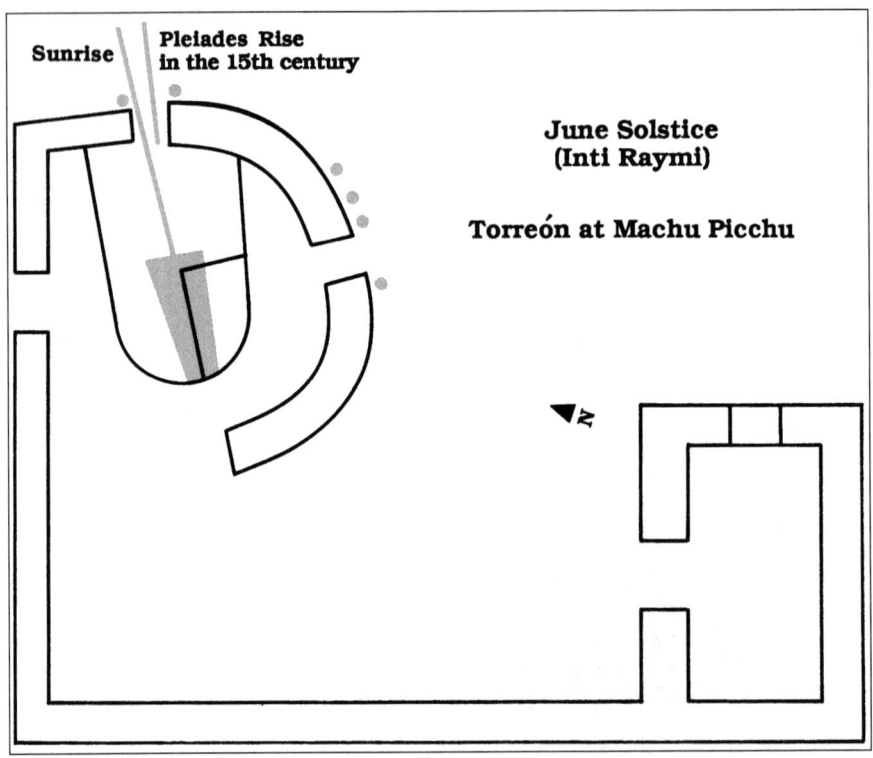

6.7 A schematic plan of the Torreón.

stone forming an angle with the cut edge that points toward the solstice. Each day as the sun approaches the solstice rising position, this angle will decrease until the day of the solstice, when the shadow will become exactly parallel with the edge. In this way, one could utilize the accuracy of the edge carved into the rock and determine the date well in advance of the solstice.

So, it is clear that the window of the Torreón was used in conjunction with the edge cut in the large interior stone platform to indicate the sunrise on the day of the June solstice. We suggest further that a shadow casting device was used in the window, probably making use of the pegs near the window corners. Whether or not this was a plumb bob or some other device must remain speculative at present.

Finally, the orientation of the window in the Torreón might serve another purpose: the observation of the first morning rise of the Pleiades. Because the rise position of the Pleiades is very close to that of the Sun on the day of the solstice, a window oriented to one naturally included the other. In any case, the similarity between the orientation of this window of the Torreón and the orientation of the Coricancha is interesting.

Subsequent investigations in the "Intihuatana Barrio" at Pisac revealed a building with remarkable similarities to the Torreón. The Pisac structure is in much poorer

6.8 The edge cut into the rock foundation of the Temple of the Sun in the Intihuatana *barrio* at Pisac is damaged, but like that in the Torreón, is aligned to the June solstice.

condition, but, like the Torreón, it was founded on a rock prominence that had a clear view to the east. The masonry is of very fine coursed stones that form a curved wall conforming to the shape of the rock as in the case of the Torreón. The walls are no longer high enough to tell if windows with pegs ever existed, but an edge is carved into the foundation stone, which, like that in the Torreón, is oriented to the June solstice.

In conclusion, the Torreón is a special building that was constructed with a very precise June solstice alignment and was constructed in a manner that allowed observations to follow the motion of the Sun through a broad range of dates. Similar structures, like the one at Pisac, may have been located at other Inca sites as well.

Evidence for June Solstice Observations

From the construction of the Torreón, we know that the Incas were able to build structures with observing capabilities accurate to minutes of arc and could have easily determined individual days from observation there. The structure at Pisac reassures us that the alignment is not fortuitous. The similarities in orientation between the Torreón and the Coricancha also support the possibility that the orientation to a declination of 21 degrees was intentional. Given the utility of shadow casting in the Torreón, we prefer the interpretation that the orientation referred to a particular sunrise (near May 25) rather than to the Pleiades, but these are not mutually exclusive options.

The Chuquimarca-Quiangalla alignment, suggested by Aveni and Zuidema for marking the June solstice sunset, is a well formulated proposition. They have probably identified the correct area for each site, but the pillars at Quiangalla no longer exist (on the surface) and the exact location of the observing site might be open to debate. Further investigation, possibly involving archaeological excavation, to either locate some remains of the pillars or to identify the nature of Zuidema's site near Lacco, could establish a more solid link between these sites and a solstice observation.

Finally, the number of temples of the Sun that existed in Cuzco show that (at least here) the Inca established different sites for the worship of Inti. This suggests that the observations needed to establish the time to begin the Inti Raymi festival may have been made from a variety of different locations.

CAPAC RAYMI: THE DECEMBER SOLSTICE

Ethnohistoric Documentation

While Inti Raymi was a festival for everyone to show reverence to the Sun (and the Inca), Capac Raymi, associated with the December solstice, was quite a different cultural event. As described by Molina, it was the ceremony at which the Inca youth (Capac Churi) became adults (were knighted) and wore the large earspools for which the Spaniards called the Incas "*orejones*." Molina (1873: 36–47) described 23 days of preparatory rituals during which there were processions carrying images of the Sun, sacrifices and the presentation of weapons to the youth about to be knighted. (In particular, these included slings, and the youth were reminded that the first Incas carried slings when they emerged from the cave at Pacariqtambo.) These young men fasted, competed in races, were flogged and had their ears pierced.

Several *huacas* listed by Cobo are associated with ceremonies carried out during Capac Raymi. These include:

> CH-3:8 *Calispuquiu*, a fountain . . . which was below the said house of Tupa Inca. All those who were made *orejones* in the festival of Raymi went to wash in it.

> CH-4:5 *Guamancancha,* near the fortress on a small hill of this name. It was an enclosure inside of which there were two small *buhios* designated for fasting when *orejones* were made.

> CH-5:7 *Chacaguanacauri*. It is a small hill on the way to Yucay where the young men who were preparing themselves to be *orejones* went for a certain grass which they carried on the lances.

> CH-9:3 *Quinoacalla*, a hill . . . in Carmenga, where it was ordained that the *orejones* should rest in the festival of Raymi.

> CH-9:6 *Apuyauira*, a stone . . . on the hill of Picho [Piccho]. They believed that it was one of those who emerged from the earth with Huanacauri, and that after

having lived for a long time he climbed up there and turned to stone. All the *ayllos* went to worship at it in the festival of Raymi.

AN-5:1 *Usno,* a stone . . . which was in the plaza of Hurin Aucaypata; this was the first *guaca* to which those who were being made *orejones* made offerings.

CO-6:7 *Huanacauri,* the most important shrine in the kingdom. It was here that the Inca received the sign from Inti that they should settle in the Cuzco area. As part of their preparation for Knighthood, sacrifices were offered here.

CU-2:4 *Rauaraya.* It is a small hill where the Indians finished running on the feast of the Raymi, and here a certain punishment was given to those who had not run well.

In Molina's description of Capac Raymi he says that on the 23rd day of the festival an image of the Sun was carried to a house of the Sun called Puquinque (Puquincancha). They remained there for a number of days and a variety of sacrifices was made. The image of the Sun was then returned to its normal resting place to conclude the ceremony and the month (Molina 1943: 60).

December Solstice at Cuzco

Sunrise Observations. The timing of the Capac Raymi celebration associates it with the December solstice. There was elaborate preparation that culminated with a procession and vigil at Puquincancha. Puquincancha is the temple of the Sun that Cobo lists as:

CU-10:2 *Puquincancha.* It was a house of the Sun which was above Cayocache. They sacrificed children to it.

If Capac Raymi was a normal length month of 30 days, the vigil there may have lasted a week ending in late December. Some sort of observation or counting of days must have been done to determine when to commence this ceremony. Zuidema (1981) has suggested that Puquincancha is the likely site of such observations, and we can now examine the physical evidence as to what observations were made.

In 1967, excavations during a construction project partially exposed a structure with fine stone masonry (under 2 m overburden) near the streets of Belén and Santiago in Cuzco (Rowe, personal communication). The location is approximately correct for this ruin to be the remains of Puquincancha. From this location, Zuidema (1981), working with Aveni, has determined the direction of the December solstice sunrise to be towards a hill now known as Mutu (approximately 25 km from Puquincancha). Zuidema identifies this hill as the last *huaca* on CO-5:

CO-5:10 *Omotourco,* a small hill . . . which is opposite Quispicancha in the puna or páramo. On top of it were three stones to which they offered sacrifices.

He further identifies Omotourco as being Omoto-ynacauri, the second place on Molina's list of sites where sacrifices were made during Inti Raymi celebrations. The first site, which was probably nearby, was named Succanca, a name similar to that used for a pair of the pillars located around Cuzco that were used for solar observations. Zuidema believes that, in addition to marking the first place for sacrifices during Inti Raymi, Mutu may have marked the sunrise on Capac Raymi as viewed from Puquincancha.

Mutu (Omotourco) is a long distance from Puquincancha and a marker there could not be seen against a rising Sun. For instance, at this distance (25 km), a pillar 15 meters tall has an angular size of 2 arc minutes and would be lost in the glare of the sun. The observation must therefore have been simply the sunrise against the hill. This type of observation is very similar to the types of solar observations described by Urton (1981) in Misminay. Unless the horizon has many features, it is less accurate than observation against a pillar. But because of the great distance involved, the markers discussed by Zuidema could serve no part in the actual observation. At most, they served to mark an important place.

Puquincancha was clearly a temple of the Sun associated with Capac Raymi. Since the marker on Mutu was not observable, it is not surprising that there is no ethnohistorical reference to such use. The ethnohistorical association between Capac Raymi and Puquincancha is clear, however. It is difficult to evaluate the archaeological investigation of Puquincancha. Was the Sun simply observed to rise over a hill, or was there some other feature or mechanism involved, as in the Torreón?

Sunset Observations. A second alignment has been proposed by Aveni (1981) involving a sunset alignment of the *ceque* Cuntisuyu 13 (Fig. 6.9). While there are no historical references to observations along *ceque* lines, astronomical orientation of such lines is clearly a proposition worth testing. CU-13 is interesting in this regard judging from Cobo's description of the *huacas*:

> CU-13:3 *Chinchincalla* is a large hill where there were two markers (*mojónes*); when the sun reached them, it was time to plant.

The direction of the adjacent *ceque*, CU-14, was determined by identifying the following two *huacas*:

> CU-14:3 *Ruauypampa*. It was a terrace where the Inca lodged which was on the slope of the hill of Chinchincalla.
>
> CU-14:4 *Pantanaya*, a large hill cleft in the middle which divides the Roads of Chinchaysuyu and Condesuyu, or Cuntisuyu.

This information permitted Aveni to identify the hill of Chinchincalla. The location and direction of CU-13 could then be further restricted from the identification described above of Puquincancha on CH-10. However, because Puquincancha is relatively close to the Coricancha and its full extent has not been determined from excavation, the direction of CU-10 is not very restrictive. The diagrams presented by Aveni show a 4 degree angle between CU-10 and CU-14. He concludes that this angle includes the December solstice sunset point, and that CU-13 was constructed as a solstice marker.

David S.P. Dearborn and Katharina J. Schreiber

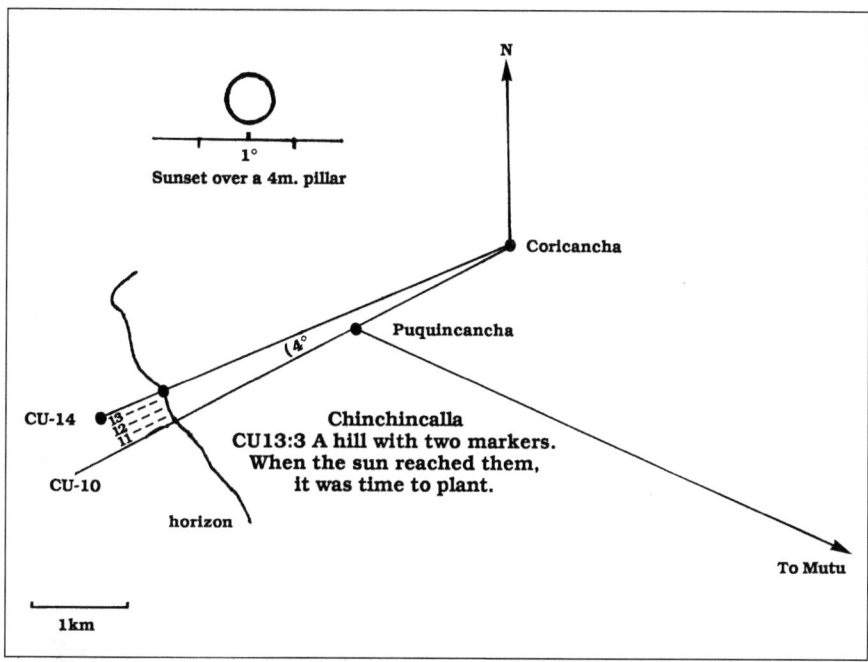

6.9 Puquincancha was a house of the Sun that was used during the Capac Raymi ceremony, associated with the December solstice. From there the solstice sunrise could be viewed over the distant hill Mutu, which Zuidema believes was the last *huaca* of CO-5. Aveni (1981) has claimed that the *ceque* CU-13 was aligned to the sunset on the December solstice. The coarseness in the determination of this alignment gives two other *ceques* at least as good a claim to this distinction. The insert shows the appearance of the Sun setting over a large pillar as viewed from the Coricancha.

No information on the azimuths determined for CU-14, or CU-10 were presented by Aveni in his published account, but Zuidema (1982) says CU-14 has an azimuth of 247 degrees. This places CU-10 at about 243 degrees. The horizon altitude in this area ranges from 6 to 8 degrees leading to a December solstice sunset azimuth of 247.5, about half a degree (a full solar diameter) north of CU-14. The sunset azimuth for a flat horizon is 245.5 degrees.

If the quoted azimuth is correct, then CU-13 (and even CU-14) is too far south for the pillars to have been used for an actual observation from the Coricancha on the December solstice. Cobo's account does not refer to a Capac Raymi observation, but rather states that these pillars were used for a solar observation to determine the time for planting. Based on planting and harvest dates from the modern village of Misminay, located near Cuzco in a similar environmental zone (Urton 1981), local planting times range from August through December. A late December date for the observation of this *huaca* is inconsistent with the beginning of the planting season.

6.10 The exterior of Intimachay. The door on the right is recessed nearly two meters behind the window to the left.

Zuidema (1981) suggests that perhaps Cobo was referring to the end of the planting season rather than the beginning.

As Zuidema has shown, the locations from which the horizon markers around Cuzco were observed were not necessarily in the Coricancha. Like the observations described for the June solstice, the pillars could well have been observed from elsewhere. The existence of these pillars on CU-13 therefore does little to indicate that this direction was observed from the Coricancha or to give CU-13 a better claim to a solstice alignment than CU-11 or CU-12. In fact, the first *huaca* of CU-12 may have a better claim, since it has an association with the festival of Raymi:

CU-12:1 *Cunturpata*. It was a seat on which the Inca rested on the way to Raymi.

None of the *huacas* on CU-13 list any Raymi connection. One might as well suggest that these *ceque* lines are coarsely clustered towards the solstice but are not intended for observations. However, Zuidema's (1982) map of the *ceque* lines shows lines clustering in non-astronomical directions as well.

In sum, the evidence for a December solstice observation from the Coricancha along *ceque* CU-13 does not seem entirely plausible to us, at least if we believe that the Incas were concerned with making precise observations. The towers along this *ceque* are certainly in the general direction of the December solstice sunset as viewed from the Coricancha, but the uncertainty in the direction is large and the ethnohistoric data (taken literally) are inconsistent. As we have seen that the Incas were capable of very precise observing techniques, this apparent imprecision in the case of a very important ritual event seems inconsistent.

The December Solstice at Machu Picchu

In 1984 we examined a feature we call Intimachay at Machu Picchu (Dearborn, Schreiber and White 1987). It is a small cave, faced with good coursed masonry, and a window carved partially from a boulder that forms part of the front wall. The interior of the cave has several broad steps or platforms carved into the bedrock and a wall of coursed masonry with two niches. Two sides of the window had been carved through 2.2 meters of solid rock.

While a window is a very reasonable addition to a cave, the full view of this window contains only two degrees of horizon. The view from the interior of the cave is further limited by a stone that baffles the window from inside the cave, resulting in a view of much less than a degree. This window is ineffectual for illuminating the interior of the cave except in a very special way. It was precisely aligned with the rising of the Sun on the date of the December solstice. Because of the finite diameter of the Sun, and the view of the window, light from the rising Sun is capable of entering the cave for about 10 days before and after the solstice.

While it is not possible to reduce the number of days on which the Sun enters the cave below 8, observing the precise location of the spot where the light hits the back wall of the cave might allow one to determine the precise date. Furthermore, given that the December solstice occurs during the rainy season, the ability to function over a range of dates is a benefit. The view of the window was remarkably well aligned and collimated, and was constructed with some considerable effort to admit light to the cave only for this brief period.

One might wonder why Pachacuti Inca chose to use a cave to observe the December solstice rather than some other structure. Polo (1873: 153–4) cites the Inca creation myth in which "seven men and women had come out of a cave that they had called Paccari-tampu, five leagues from Cuzco, where a window was carved in masonry in most ancient times." This was the first Inca, Manco Capac, with his brothers and sisters. They wandered the Earth until they arrived at Huanacauri, a hill south of Cuzco, where the Sun, their father, gave them a sign that they should settle there.

Huanacauri was one of the most sacred Inca shrines and played a prominent place in the Capac Raymi festival. It is clear from the description of the Capac Raymi ceremony that part of initiating young men into manhood involved reminding them of their origins. Additional evidence of such indoctrination comes from a statement of Molina (1873: 4) when he says "in a house of the Sun called Poquen Cancha (Puquincancha), which is near Cuzco, they had the life of each one of the Yncas, with the lands they conquered, painted with figures on certain boards, and also their origin."

On this basis we might associate the Capac Raymi celebration with the Inca origin myth. As some versions of this myth say that the original Incas emerged from a cave, we might then suggest an association between Capac Raymi and caves. The use of a window at Intimachay might then be seen as similar to the window described at the cave at Pacaritambo. These associations suggest a possible explanation for why Intimachay is a cave rather than some other form of structure.

Evidence for December Solstice Observations

At Machu Picchu we have identified a feature, Intimachay, that was constructed to admit light at sunrise for only a brief period around the December solstice. In Cuzco, it is clear that Puquincancha was used in the Capac Raymi celebration. Historical accounts suggest that they spent several days, perhaps on the order of a week, watching there. A case has been made by Zuidema and Aveni that the December solstice sunrise was observed from Puquincancha to rise over a distant hill (with an unobservable marker). Without further investigation of Puquincancha, it is difficult to know if the observation was simply of sunrise over the hill Mutu or if Puquincancha contained some means of observing the Sun. (In the Torreón, while the *Sun* rose over a mountain, it was the *stone* that was observed.)

The record is less clear about observing December solstice sunsets (and since this falls in the rainy season, the afternoon sky is also less clear). The work of Aveni shows that there are five *ceques* within a few degrees of the solstice sunset point. While the claim has been made the CU-13 is oriented to the solstice, no evidence has been presented to give it a superior claim over the others, except for the existence of the pillars of Chinchincalla. These pillars were used to define a planting time and were probably observed from somewhere other than the Coricancha.

THE ZENITH AND ANTIZENITH DATES

Ethnohistoric Documentation

As discussed above, Garcilaso (1960) described the observation of the March and September equinoxes by the Incas. They used a fine stone column and the shadow of this column was observed along an east-west line to determine the equinox. While the description is quite specific as to the equinox, it says that the column is "fully lighted without any shadow," indicative of the Sun passing through the zenith. It then proceeds to say that as the Incas moved towards Quito they noticed that all shadow disappeared near the days of the equinox. They considered this to be quite special.

Garcilaso's description of the equinox observations in Cuzco suggests that the zenith passage dates (February 13–14 and October 30) were recognized there. Furthermore, they clearly noticed that the zenith passage date of the Sun became coincident with the equinoxes near Quito. If the Inca recognized and observed the zenith, it stands to reason that they conceived of an opposition to the zenith passage, or an antizenith passage of the Sun, as the Inca perceived their world in terms of dual oppositions (Zuidema 1981, 1982). The antizenith passage of the Sun should probably correspond to the Sun's passage through the nadir (April 27 and August 18 in Cuzco), but its precise dates depend on exactly what observation is used to define the phenomenon, since the nadir cannot be observed directly.

Duality, a pervasive concept in the Andes, is a view in which the world is divided (or sorted) into dual oppositions (for example, Hanan-Hurin, the upper-lower division of geographic space). The Sun and Moon were perceived to be in opposition to each other and were given masculine and feminine attributes respectively. While there is no ethnohistoric reference to an antizenith opposition to the zenith passage, the concept may have been too alien for the Spanish to record in a direct manner. Zuidema (1981, 1982) has presented a range of circumstantial evidence to suggest that antizenith (and zenith) passages were of comparable importance to the solstices and equinoxes.

The basis of this suggested importance seems to have been that the zenith-antizenith passages provided an alternative to using solstices and equinoxes in the quartering of the year, resulting in a more natural set of seasons. At the correct latitude, the zeniths and antizeniths quarter the year nearly evenly, but in Cuzco they result in alternating "seasons" of 72 and 110 days. Zuidema has examined the recorded activities and festivals of the year for evidence of this quartering. The most suggestive correlation comes from the proximity of the nadir passage dates to the beginning of the planting season (August) and harvest season (May). Planting and harvest times are based on the environment and were determined in the Cuzco area many generations before Inca times. If they recognized an antizenith passage date and noticed that it fell near these agricultural times, they might have associated it with agricultural pursuits.

The zenith passage dates in Cuzco fall in the modern months of February and October, and while this is a phenomenon that the Inca are recorded to have observed, the activities of these months seem less fundamental than planting and harvesting. In his description of the activities of February, Guaman Poma (1936: 239) mentions no festivals, but says that during this month, the Inca sacrificed gold, silver and animals to the Sun, Moon, and Stars, and many *huacas*. It was a time of much rain and hunger when many of the young and old died. Molina (1873: 51) calls the month Atun-pucuy, and states that they had no special festival in it.

The other zenith passage date falls at the end of October in a month that they called Aya Marca Raymi. Guaman Poma (1936: 257) says that they held a feast in this month (Aya Marcai Quilla, in his terms) to honor the dead. Molina (1873: 35) describes this as a month when the youth of the village of Ayamarca were knighted and the youth of Cuzco began preparations for Capac Raymi. He also mentions that food was offered to the *huacas*. The descriptions of the activities of these months are much shorter and have less detail than the months containing the solstices.

The Zenith-Antizenith at Cuzco

The zenith passage of the Sun is an observable phenomenon, and Garcilaso's description leaves little doubt that it was noticed in shadow-casting observations in the Coricancha in Cuzco. If the antizenith passages were recognized, their dates depend on exactly what observation was used to define this phenomenon. The observation would have involved some reversal or opposition of a zenith passage observation, as a true nadir passage is not directly observable.

One possibility, for reversing an observation of the Sun's passage through the zenith, might be the observation of the full Moon passing through the zenith, which occurs roughly at the time of the solar antizenith. This phenomenon does not occur every year (though it will pass within 10 degrees of the zenith each year) and can occur over a broad range of dates (a full month). Such an observation lacks the precision that would be desirable for marking the beginning and end of the agricultural cycle, and thus in providing the important quartering of the year.

Another possibility for reversing the zenith is to observe the point on the horizon where the Sun rises on the zenith, then reverse this direction to define the point of sunset on the day of the antizenith. Zuidema (1981, 1982) has discussed several possible zenith-antizenith axes, but the best case has been described by Aveni (1981). It involves the proposed observation of the antizenith from the *Usno* in the main Plaza against the four pillars on what must be the hill Picchu (Fig. 6.11). The Anonymous Chronicler described their use for making a solar observation that determined the time to plant in and around Cuzco. This was part of a regular set of observations for measuring the progress of the Sun. He described some pillars and the *Usno* saying:

> ... e hizieron que en la serrania más alta, á vista de los ciudad del Cuzco, á la parte del Poniente, hicieron quarto pilares á manera de torrecillas, que se pudian sojuzgar de á dos y tres leguas en paraje de ducientos pasos desde el primero al postreros, y los dos en medio auia cincuenta pasos del uno al otro, y los dos de cabos rrepartidos por su quenta á propósito de sus fines de manera que entrando el Sol por el priver pilar, se apercebian para los sementeras generales, y comencaun á sembrer legumbres por los altos, por ser más tárdio; y entrando el Sol por los dos Pilares de en medio, era el punto y tiempo general de sembrar en el Cuzco, y era siempre por el mes de Agosto. Es ansi, que, para tomar el punto del Sol entre los dos pilares de en medio tenian otro pilar en medio de placa, pilar de piedra muy labrada, de un estado en alto, en un paraje señaldo al proposito, que le nombrauan Osno, y desde alli tomauan el punto del Sol en medio de los dos Pilares, y estando ajustado, hera el tiempo general de sembrar en los valles del Cuzco y su comarca (Anonymous Chronicler 1906: 151).

> ... in the highest mountains to the west as seen from Cuzco, they made four pillars in the form of small towers that can be distinguished from 2 or 3 leagues, with a distance from first to last of 200 *pasos*, and between the middle ones of 50 *pasos*, and the separation of these two cylinders was proper for their intended use; with the result that when they became ready for sowing in general and they started to sow in the higher places as the plants are later here, and when the Sun entered the central pillars it was time for sowing in Cuzco, and this was always the month of

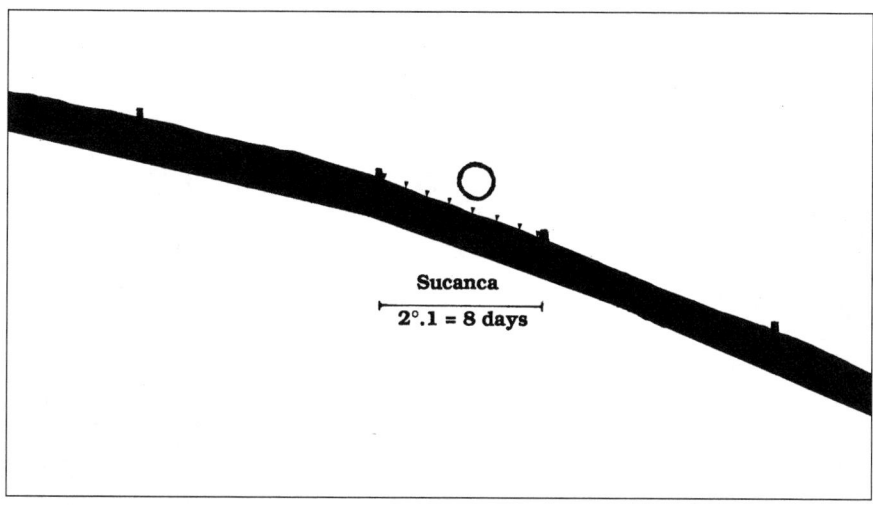

6.11 The sizes of the pillars on the hill Picchu are not known, but this shows to scale the Sun setting beyond pillars four meters tall (as viewed from the plaza). It took the Sun approximately 8 days to move between the inner pillars (the *huaca* Sucanca). The last gleam of the Sun would occur between the pillars for seven days, and the full disk pass between them for 5 or 6 days. These sunsets defined a range of time when it was proper to begin planting around Cuzco.

> August. The place for taking the Sun to enter the central two pillars was another pillar in the middle of the plaza, a pillar of well dressed stone, one *estado* high, in the middle of the plaza, in a place for that purpose was singled out, that they call *Osno*, and they measured the Sun between the two pillars, and when it was exactly there, it was the time for sowing in the valleys of Cuzco and around (translation by Zuidema 1981, with revision by Patricia J. Lyon).

These pillars are probably correctly identified as the *huaca*:

> CH-8:7 *Sucanca*. It was a hill by way of which the water channel from Chinchero comes. On it there were two markers as an indication that when the Sun arrived there, they had to begin to plant the maize. The sacrifice that was made there was directed to the Sun, asking him to arrive there at a time which would be appropriate for planting, and they sacrificed to him sheep, clothing, and small miniature lambs of gold and silver.

These quotes unambiguously state that an alignment was constructed using the sunset to specify an agricultural date. Aveni has examined this alignment to support the zenith-antizenith hypothesis. The general orientation of Chinchasuyu-8 (see Fig. 6.3) was determined by locating the last *huaca* of this *ceque*:

> CH-8:13 *Collanasayba*. It was a marker *[mojón]* on a hill at the beginning of Sicllabamba, as the end and limit of the *guacas* of this *ceque*.

HERE COMES THE SUN

6.12 The Intihuatana stone sits on a small terraced hill with construction on three sides and an open view to the west and north.

Aveni and Zuidema found an hacienda approximately 10 km from Cuzco in the appropriate *suyu* bearing the name of this *huaca*. Assuming that the *ceque* is straight, the place where the *ceque* line crosses the ridge of Picchu can be located and tentatively identified as the *huaca* Sucanca.

The two inner pillars that were probably the *huaca* were 50 paces apart. Zuidema believes these were roman (two step) paces (about 1.4 m each). The pillars were therefore separated by 70 meters, or as viewed from the *usno* in the main plaza 1.9 km away, about 2.1 degrees. The diameter of the Sun is 0.5 degrees, and through mid-August the Sun moves about 0.3 degrees per day along the horizon. The entire disk of the Sun sets between these pillars for 6 days, and the last gleam of the Sun for 7 days. It requires 26 or 27 days to pass between the two outer pillars. Either these inner pillars were meant to define a range of days (not a specific date), or perhaps only the day that the Sun passed one of the pillars is significant.

We cannot know if these inner pillars were centered on the point where Chinchasuyu-8 crosses the horizon or if just one of the pillars resided there. This results in a 7-day uncertainty in the absolute date being specified by this alignment. If the *ceque* lines are only straight to "a few degrees," another week's uncertainty in the dates being specified could be included. Additional uncertainty is added from questions of where the *Usno* was located in the plaza.

Standing on Picchu, one can observe the sunrise on the zenith passage date in the direction of Tipón, an Inca site to the southeast of Cuzco. Projecting this line down through Cuzco produces a path through Haucaypata along which a reverse, antizenith, observation could be made. Aveni does this, and produces a path which lies within 30 meters of the location that he proposes for the *Usno* (from other evidence). If we take this as a measure of the uncertainty in the *Usno*'s location (instead of 100 meters, from the width of the plaza) we obtain another 3 days' uncertainty in the actual date that the Sun was seen to cross between the inner pillars.

The resulting uncertainty in the date of this agricultural alignment is at least 10 days. It occurs in August, as the Anonymous Chronicler said, but the uncertainty is too large to unambiguously associate it with an antizenith passage observation. While the physical evidence is inadequate, Zuidema (1981) has offered support in the ability of an antizenith concept to link several activities and myths of the Incas.

As described by Molina (1873: 36), in the Inti Raymi celebration, sacrifice was made at a marker called Sulcanca near Quispicancha (today's Tipón). Molina does not describe Sulcanca as being used astronomically (they were the markers of the first and last stop of the Inti Raymi procession), but its similarity to the name (Sucanca) recorded by Cobo for the *huaca* on Picchu does suggest a possible relation between the places. If Sulcanca were on the hill above Tipón, it would mark the zenith passage dates' sunrise as observed from Sucanca on Picchu. As with Succanca near Mutu, any pillar located near Tipón is too far away to be observed from Sucanca at Cuzco. The zenith sunrise observed from Sucanca (not described historically) is simply over a distant mountain. The reversal of this observation, also suggested by Zuidema, could form an antizenith passage observation, but is difficult because Picchu does not form the horizon from Tipón.

As a final line of evidence, Zuidema (1981) offers mythical connections. He states that Quispi (as in Quispicancha) can mean crystal or translucent. In one Inca myth, the (soon to be) Inca Pachacuti threw a crystal into a spring just before a battle with the Chanka. The Sun god came forth to talk with him. This provides a link between crystals, the rising Sun, and the name of the Inca site near Sulcanca. This type of evidence is presented to indicate an interest in observing sunrise in the vicinity of Tipón. Such evidence is difficult to evaluate quantitatively but must be considered.

In sum, Zuidema and Aveni have presented an interesting case, but no clear-cut substantive evidence to support it. The zenith sunrise could have been observed over Tipón, but the suggestion that it was viewed from the Sucanca at Cuzco has not been demonstrated.

The Zenith-Antizenith at Machu Picchu

The Torreón. Returning to Torreón, we did find some evidence of a zenith passage date observation. The Torreón has two windows, one of which was used to observe the June solstice sunrise, and we should consider here the reason for the other window. The view directly out of the second window includes a region of sky that includes several Inca constellations, such as the tail of Scorpio, called Collca (storehouse) by the Incas. However, the Sun could never be observed straight out of this window. Urton (1981) shows that the tail of Scorpio was perceived to be in opposition to the Pleiades, but a window by itself does not form a precise observing instrument for observing stars. This window has pegs at the four corners much like the other window through which the June solstice sunlight entered, suggesting that perhaps something sat in this window. Stars could be viewed through something set there, but this forms a good observatory only if a place from which to sight is also well defined.

Examining the full range of view of the two windows, we found that the cross-jamb views intersected for about four degrees of horizon. The position is centered on the rising point of the Sun on the dates of zenith passage. Only for a period of about five days on either side of the zenith passage date does the Sun enter both windows at sunrise. Light or shadows cast by something set or hung in the window could again be used to make a precise observation and perhaps to determine precise dates for a range of time around the zenith passage date. In any event, the period around the zenith passage date is easily and dramatically indicated in the Torreón by the first morning light entering both windows.

While this suggests that a horizon observation was made to monitor the approach of the zenith passage date at Machu Picchu, this observation could not be reversed from the Torreón to observe the antizenith. The light that enters the windows on these dates hits the walls. It would be very difficult to take the direction defined by this entering light and project it up to the horizon. Sunset from the Torreón is over a group of buildings on the hill above and behind it.

The Intihuatana as Usno. We considered the possibility that there was an *Usno* at Machu Picchu, and, as Aveni has suggested for Cuzco, that the antizenith might be observed from it. The most likely candidate for an *Usno* at Machu Picchu is the so-called Intihuatana stone. It is located on top of a small hill on the western edge of the city. The hill has been terraced all around and the top leveled. Walls of fine masonry are found to the south and east of the stone. They contain several windows including two that appear to have once been doors. What appears to be a seat is carved into the Intihuatana stone, similar to the description of the Capac *Usno* in Cuzco. A rectangular pillar approximately 60 cm high sits on an uneven bench on top of the Intihuatana stone. In addition, there are many facets cut into the stone and a peg that projects to the northeast. It is a complex shape, difficult to describe adequately, but one thing must be realized. It is intentional and it is the shape that its markers wanted it to be. In the Torreón, the precision to which the Inca stonemasons could work has been demonstrated. One can hear and read many stories about the Intihuatana, some of them demonstrably wrong, some merely untestable.

The orientation of the step in the bench was examined as were the edges of the pillar, but no solar alignment was found. Furthermore, the edges, while smooth, are not truly straight. The stone does not seem suitable as an observing instrument by itself. As has been suggested, it could have been covered with other materials, and

6.13 Looking west across the Intihuatana stone, the nearby ridge of San Miguel is seen to rise to the right. The stone has what appears to be a seat as well as many edges and projections.

could have been used with partially destroyed walls nearby, but to date nobody has been able to offer evidence of such use.

If the Intihuatana were analogous to the *Usno* in Cuzco, it might have been used with a marker on the horizon. Albornoz described the *Usno* as "a pillar of gold in the plaza," and Inca *Usnos* more generally, saying:

> There is another general *guaca* on the royal roads and in the plazas of the towns, which they call *uznos*. They were shaped like a ninepin, and made of many different types of stone or of gold and silver. All of them had structures in the places I have said, made like towers of very handsome stonework, as in Bilcas and in Pucara and in Guanaco el Viejo and in Tiaguanaco. The lords sat on the said *uzno* to drink to the Sun and they made very many sacrifices to the Sun (translation by Rowe 1979: 74).

Cieza de Leon describes it as "shaped like a sugar loaf, well encased and full of gold" (Rowe 1979: 74). Machu Picchu, in all likelihood, had an *Usno* that was built by the same Inca who placed the one in the plaza in Cuzco. In addition, the description of Inca *Usnos* is consistent with the Intihuatana at Machu Picchu.

In 1980, Bernard Bell, one of the Earthwatch volunteers participating in the project, sighted what appeared to be a pillar projecting out of the jungle on the San Miguel ridge near Machu Picchu. This ridge is a little over 2 km northwest of the Intihuatana stone, a distance very similar to that separating the *Usno* and the ridge of Picchu in Cuzco. A few degrees away is a natural rock prominence that contains a hole through which sky could be seen. The date that the Sun sets behind the prominence and gleams for a moment through the hole is approximately August 2; this is very near the beginning of the traditional agricultural year, and when Zuidema (1981) believed the Sun passed the outer pillar on the ridge of Picchu at Cuzco. The pillar on the San Miguel ridge was located where the Sun sets on May 1 and August 10. Thus it misses the true antizenith passage date by a week and misses the reverse of the zenith passage sunrise by more. However, it is within the limits of the *Usno*-Sucanca alignment.

In the 1982 field season, Bell and Dearborn climbed the ridge near Machu Picchu to see if the possible pillar and hole in the rock were artificial. While logistical difficulties prevented examining the whole ridge (including the equinox and solstice sunset points as viewed from the Intihuatana), they did examine (walk is not an accurate description) the portion of the ridge beyond which the Sun sets through August, and early September, but did not locate any pillars. If any pillars were constructed to mark a sunset date in August they would have been located in the area examined. Unlike Cuzco, there is no indication of the intentional destruction of *huacas* around Machu Picchu, so the absence of such features cannot be attributed to the zealousness of the Spaniards. The features observed on the ridge were entirely natural. (Many of the *huacas* that Cobo describes in the *ceque* system were natural features and moreover, Cobo does not say that the markers on Picchu were artificial. The Anonymous Chronicler does, however, say that they were man-made.)

In sum, there is no evidence that the Inca paid any attention to this natural rock pillar on the ridge near Machu Picchu; and even if they did, it was not associated with an antizenith sunset. Hence, the use of the Intihuatana remains a mystery to us.

Evidence for Zenith-Antizenith Observations

Garcilaso's description of shadow casting to determine the equinox leaves little doubt that the Inca were aware of the days when the Sun passes through the zenith, and in some instances, considered them special. It is odd, however, that Molina does not list any special feast in the month of February when this occurs.

While there is no historical reference to an antizenith passage date, the Spanish may not have recognized the concept and hence did not record it in those terms. The concept of duality or opposition is strong enough in the Andes to propose that, in addition to the zenith passage date, they recognized some opposition to it.

The ethnohistoric evidence that a solar observation was made to commence planting is unambiguous. Ethnographic studies of Urton at Misminay find that solar observations are sometimes used today to determine the beginning of the planting season (Urton 1981). He found informants who indicated solar horizon positions typically 11 to 12 degrees north of the east-west line, corresponding to a date about a week after the August antizenith passage date. Furthermore, Urton does not indicate that these modern informants associate this observation with an opposition to the zenith passage date. Planting times, in such instances, must be based on a pragmatic consideration of when the last frost will occur. Sites at higher altitudes will be forced to a later starting date.

We found evidence at Machu Picchu for a horizon observation of the rising Sun around the zenith passage dates. The reversal of such an observation could lead to an "antizenith" sunset but not as viewed from Torreón. The physical evidence in Cuzco, linking the passage of the Sun between two pillars on the Picchu ridge to the antizenith dates, is uncertain to at least 10 days. Claims that this must be an antizenith passage observation are based on the idea that such an alignment must exist and was observed along the horizon. As a measure of the uncertainty in this association, Zuidema has modified his original hypothesis such that he now suggests that antizenith passage dates occurred when the Sun passed the northernmost (outer) pillar. The inner pillars then correspond to dates in late August and early September (Zuidema, personal communication). These new dates are more in line with contemporary planting times in Cuzco.

Finally, we have found no analog to the Cuzco agricultural (August) observation at Machu Picchu. Because Machu Picchu is considerably lower than Cuzco and in an entirely different ecozone, the planting cycle was entirely different. Furthermore, Machu Picchu was not a large public place, so a public ceremony to begin the planting season may have been less important there.

As discussed in the beginning of this section, an antizenith period could be defined by observing the full Moon's passage through the zenith. Because of the inclination of the Moon's orbit to the Sun, this can occur (in Cuzco) when the Sun has a declination from 8.5 to 18.5 degrees, so this defines a broad period, and not a particular date like the Sun passing a pillar does. Such a full Moon can occur from August 1 to September 2, and April 12 to May 14. The Incas may well have noticed that the full Moon passed through the zenith during periods associated with planting and harvest. Whether such an observation was made and perceived by the Incas as a reversal of the sun at the zenith is a proposition that may never be demonstrated.

Before we can accept the notion that the August agricultural observation described in Cuzco was an opposition to the zenith passage date, more supporting data must be obtained. Other Inca sites must be examined for evidence of an antizenith alignment used by the Inca. Antizenith and agricultural alignments might be disentangled by considering sites with different altitudes and latitudes. This will result in a separation of the antizenith passage date from local time to begin planting.[2]

CONCLUSIONS

We have attempted to compare the results of work at Machu Picchu to what is "known" about Inca astronomic observations in and around Cuzco. In doing so we have had to re-evaluate the evidence presented by Zuidema and Aveni, if for no other reason than to be sure we were looking for the same sorts of observations and evidence. In some cases we regard the evidence from Cuzco to be quite convincing, in other cases somewhat less so.

Regarding observations of the June solstice, we find a convincing case made by Zuidema and Aveni that the Chuquimarca-Quiangalla alignment was used in Cuzco for observing sunset on the June solstice. At Machu Picchu the evidence is also convincing that the Torreón was used for observing sunrise on this solstice.

Regarding the December solstice, a reasonable case has been made for the observation of sunrise from Puquincancha in Cuzco. The case for a sunset observation along *ceque* CU-13 we do not find compelling, if we believe that the Incas were capable of and concerned with making precise observations. At Machu Picchu we have identified a structure, Intimachay, that seems to have been created expressly for monitoring sunrise on and around the December solstice. Interestingly, we feel the case for such an observation is made stronger based on the data from Machu Picchu than it is on the data from Cuzco.

Finally, it is clear that the Incas observed the zenith passage dates in Cuzco, based on the account of Garcilaso. Likewise at Machu Picchu, the Torreón may have been used to observe sunrise on the zenith passage dates as well. A strong logical case for the probable existence of antizenith observations has been made by Zuidema, but the physical evidence connecting such an observation to the beginning of planting has not been clearly demonstrated in Cuzco. Likewise at Machu Picchu, we have as yet found no clear alignments used for observing the antizenith. In the absence of such physical evidence, the nature of an observation to reverse the zenith passage remains uncertain. By analogy to the solstices, it is possible that one zenith passage date opposed the other. This provides an opposition that is distinct from the nadir passage of the Sun and the planting season.

ACKNOWLEDGMENTS

We wish to thank Tony Aveni for the inspiration to write this article. Both he and Zuidema offered advice and made various suggestions to us in informal conversations as we were preparing this paper. John Rowe provided invaluable aid in helping us track down various ethnohistoric references, as did Pat Lyon who also aided in the translation of particularly (to us) obscure passages. Finally we wish to thank all the Earthwatch volunteers (especially Bernard Bell) who participated in field research at Machu Picchu, and our colleague Ray White who organized those projects.

NOTES

1. For this and each of the following references to *huacas* described by Cobo, we are using the translation by John Rowe, published in his 1979 transcription and translation of Cobo's list of *huacas* in the area around Cuzco. We will not repeat this citation for each one.

2. Work like that of Ziolkowski and Sadowski (1984) at Ingapirca in Ecuador is inadequate for making statements one way or another. They measured the azimuth of the axis of an important structure from a map. Without information on the correction to true north, the altitude of the horizon, the azimuth indicates the set position of objects with a declination of about 12 degrees. They conclude, based on work of Aveni, that this was meant to represent the antizenith passage date in Cuzco. The only way to transfer such an alignment to Ecuador would be by counting the days to know when this phenomenon occurs in Cuzco, then observing the sunset locally. The declination given by Ziolkowski corresponds to a date nearly a week after the antizenith passage date in Cuzco, and without the information on true north and horizon it is impossible to evaluate this claim further.

REFERENCES

Anonymous Chronicler. 1906. Discurso de la suscesion y gobierno de los Yngas (ca. 1580). *Juicio de Limites entre el Peru y Bolivia,* vol. 8. Lima, Maurtua.

Aveni, Anthony. 1981. Horizon Astronomy in Incaic Cuzco. In *Archaeoastronomy in the Americas,* ed. Williamson, 305–318. Los Altos, CA, and College Park, MD, Ballena Press/Center for Archaeoastronomy cooperative publication.

Aveni, Anthony. 1987. Inka Order in an Animate World: Astronomy and the Built Environment in the Andes (part 1). A lecture presented in the pre-Columbian session of the 40th annual meeting of the Society of Architectural History in San Francisco.

Bingham, Hiram. 1948. *Lost City of the Inca.* New York, Duell, Sloan and Pearce.

Bingham, Hiram. 1979. *Machu Picchu: A Citadel of the Incas.* New York, Hacker Art Books.

Cobo, Bernabe. 1983. *History of the Inca Empire,* trans. Roland Hamilton. Austin, University of Texas Press.

Dearborn, David S.P., and Raymond E. White. 1982. Archaeoastronomy at Machu Picchu. In *Ethnoastronomy and Archaeoastronomy in the American Tropics,* ed. Anthony F. Aveni and Gary Urton, 249–259. Annals of the New York Academy of Sciences, vol. 385. New York Academy of Sciences, New York.

Dearborn, David S.P., and Raymond E. White. 1983. The "Torreón" at Machu Picchu as an Observatory. *Archaeoastronomy: Journal for the History of Astronomy* 5:S37–S49.

Dearborn, David S.P., Katharina J. Schreiber, and Raymond E. White. 1987. Intimachay: A December Solstice Observatory. *American Antiquity* 52:346–352.

Garcilaso de la Vega. 1960. Primera Parte de los Comentarios Reales de los Incas. *Biblioteca de Autores Españoles,* vol. 133. Madrid.

Garcilaso de la Vega. 1961. *The Royal Commentaries of the Inca,* trans. Alain Gheerbrant. New York, Avon Books.

Guman Poma de Ayala, Felipe. 1936. *Nueva Corónica y Buen Gobierno (Codem Péruvien illustré). Travaux et Mémoires de l'Institut d'Ethnologie,* vol. 23. Paris, Institut d'Ethnologie.

Molina Cristobal de. 1873. An Account of the Fables and Rites of the Yncas. In *Narratives of the Rites and Laws of the Yncas,* trans. Clements R. Markham. London, Hakluyt Society.

Molina Cristobal de. 1943. Fabulas y Ritos de los Incas [1574]. In *Las Crónicas de los Molina, Los Pequeños Grandes Libros de Historia Americana,* serie I, tomo 4, segunda paginación. Lima.

Polo de Ondegardo, Juan de. 1916. Los errores y supersticiones de los indios sacados del tratado y averiguacion que hizo el Licenciado Polo, 195 L. *Colección de Libros y Documentos Referentes a la Historia del Perú,* vol. 3. Lima.

Polo de Ondegardo, Juan de. 1973. Report by Polo de Ondegardo. In *Narratives of the Rites and Laws of the Yncas,* trans. Clements R. Markham. London, Hakluyt Society.
Rowe, John. 1944. An Introduction to the Archaeology of Cuzco. Papers of the Peabody Museum of American Archaeology and Ethnology, vol. 27, no. 2. Harvard University, Cambridge, MA.
Rowe, John. 1979. An Account of the Shrines of Ancient Cuzco. *Nawpa Pacha* 17:1–80.
Rowe, John. 1987. Pachacuti's Royal Estate at (Machu) Picchu. Lecture presented at the annual meeting of the Institute of Andean Studies, Berkeley.
Urton, Gary. 1981. *At the Crossroads of the Earth and the Sky.* Austin, University of Texas Press.
Ziolkowski, Marius S., and Robert M. Sadowski. 1984. Informe acerca de las investigaciones arqueoastronómicas en el Area central de Ingapirca (Ecuador). *Revista Española de Antropologia Americana* 19:103–125. Madrid.
Zuidema, R. Tom. 1977. The Inca Calendar. In *Native American Astronomy,* ed. A. Aveni, 219–259. Austin, University of Texas Press.
Zuidema, R. Tom. 1981. The Inca Observations in Cuzco of the Solar and Lunar Passages through the Zenith and Anti-zenith at Cuzco. In *Archaeoastronomy in the Americas,* ed. Ray A. Williamson, 319–342. Los Altos, CA, and College Park, MD, Ballena Press/ Center for Archaeoastronomy cooperative publication.
Zuidema, R. Tom. 1982. Catachilla: The Roles of the Pleiades, and the Southern Cross, and Alpha and Beta Centauri in the Calendar of the Incas. In *Ethnoastronomy* and *Archaeoastronomy in the American Tropics,* ed. Anthony F. Aveni and Gary Urton, 203–229. Annals of the New York Academy of Science, vol. 358. New York Academy of Science, New York.

On Seeing the Light

A Reply to *Here Comes the Sun* by Dearborn and Schreiber

by Anthony Aveni

The philosophy that certain truth can be arrived at by accepting only that knowledge acquired from actual experiment and the observation of natural phenomena is what we call positivism. When 19th century philosopher Auguste Comte inculcated that doctrine as a way of understanding the natural world, he could not have predicted that the tentacles of his brand of logic would become a way of life for his 20th century progenitors and that it would pervade the human as well as the physical (sciences).

In the interdiscipline of archaeoastronomy, we dwell in a shadowy twilight zone illuminated both by the hard facts of the phenomenal world, the semi-hard facts of archaeology (some would disagree about where these fit on the hardness scale), and

---

First published as "Comments: On Seeing the Light: A Reply to *Here Comes the Sun* by Dearborn and Schreiber," by Anthony Aveni, from *Archaeoastronomy* 10:22–24. © 1992 by the University of Texas Press. All rights reserved.

the somewhat more interpretive body of knowledge acquired from the anthropological disciplines (which "hard" scientists like to call "squishy"). In such a situation there is a danger in overtipping the delicate balance of input data from the various disciplines that contribute to rational thinking in our field. Such a danger may allow certain truth to be decided by one of the disciplines. For a time this is what happened in Megalithic astronomy, and with Dearborn and Schreiber's (1986) (hereinafter "D and S") paper, we run the risk of seeing it happen once again.

Precision, exactitude, and quantitative methodology are biblical terms for the physical scientist, and when logical positivists rivet attention full-throttle on such concepts, one is not surprised to see them arrive via a stepwise process at conclusions that they characterize as either having been or not having been demonstrated.

But reality is surely more complicated than this, for as we have already suggested, there are different kinds of evidence that cannot be treated so rigorously.

Take ethnohistoric evidence, for example. True, it is physical in the sense that a manuscript is a thing; however, a manuscript can be filled with corruptible hearsay ideas which, since Comte's time, simply do not seem to carry for us the force of a single measurement of a cut rock surface or the direct observation of a sunrise. Still, for what it is worth, reference to horizon astronomical observation in Cuzco is given by at least eight chroniclers, the least reliable of whom is Garcilaso, the very chronicler to which D and S devote most of their attention. Is this choice mitigated by the fact that Garcilaso is the most readily available chronicler in English translation? A comparison of original references among the chroniclers to horizon pillars reveals that the fanciful Garcilaso, a man steeped in European Renaissance tradition who was relatively blind to the Andean view of the natural world, deviates radically from all the other chroniclers in his description of the number, dimensions, spacing, and general arrangement of the Cuzco sun pillars. ("Garcilaso's version of Inca history is largely fictitious; most of it seems to be a pious fraud perpetrated by Garcilaso himself . . ." J. H. Rowe in R. Hamilton [ed.] *History of the Inca Empire,* Austin, University of Texas Press 1979, p. x.) His arguments about clocks, sundials, shadow casting, and symmetrically placed groups of pillars come straight out of the Renaissance mentality. Why careful scientists who would take the extreme trouble to measure sunrises at Machu Picchu to arc-minute accuracy should commit such errors in the judgment of another kind of evidence concerning precision—historical precision—is not only inconsistent but also inexcusable.

What does ethnohistory say about Machu Picchu? Concerning that place there is not a shred of evidence in the ethnohistoric record of any astronomical/calendrical observation having been done there—not even so much as a mention of its existence. Therefore, there is no basis to employ this site as a standard of comparison for archaeoastronomical hypotheses. If this is true of Machu Picchu, then it must also be true for the Torreón and the Capac Raymi cave. Likewise, there is no basis for the hypothesis that the Intihuatana was an Usno, astronomical or otherwise. That it is a "thing" accords it no special status. Unlike Cuzco, the capital and center of the empire Tahuantinsuyu, where we have ample historical justification to argue astronomical alignments in detail, at Machu Picchu one is in much the same situation as the astronomer who confronts Stonehenge. There is no matrix to either generate or couch hypotheses. There are only buildings, alignments, a horizon, and the mind of the positivist. Yet this has been powerful enough stuff out of which to make a case, for as Hawkins (1965, p. vii) once opined, "If I can see any alignment, general relationship or use for the parts of Stonehenge, then these facts were also to the builders."

In this age of homage to "thingness," technology has as much a *sine qua non* as precision in the mental kit of the modern logical positivist who practices archaeoastronomy. The maxim he/she applies seems to be: since we use it, can we find it in them? This is where the solstice pegs and platforms in the Torreón come from, replete with quotations of accuracy of two arc minutes "(including the effects of refraction and change in the obliquity of the ecliptic)" (p. 23). Are the authors suggesting (without stating it) that the Inca cared about and deliberately strove for such accuracy?

What does the ethnohistoric record say about Andean astronomical technology? Practically nothing. There is no evidence that platforms were carved, that grooves rather than horizons were looked at, that frames, or otherwise, ever were hung on windows. Moreover, we already have an adequate hypothesis to account for knobs on Andean blocks. The authors' experiments, which show that if one suspends a suitable frame from the Torreón window one can get the shadow of a string to fall on the edge of a cut surface at a particular time, serve only to show how the modern technologically oriented mind seeks to explain things. Whether the contraption *did* work that way is pure conjecture. I can take the knife I used to butter my toast this morning and employ it to screw a hinge onto the door of my study. Does this demonstrate the purpose for which the knife was intended?

My intention in this reply is not to deal specifically with point-by-point criticisms of Zuidema and Aveni's published research on Cuzco astronomy (which will be done elsewhere), but rather to reveal basic methodological and interpretive differences between us and our critics. However, let me employ two specific cases of astronomically related *ceques* to amplify the difference between these outlooks. The *ceques* CU 10–14 are very closely bunched together, but there is no evidence they overlap. We know this because we have determined as best we can the location of *huacas* that delineate these *ceques*. These five *ceques* are gathered within about a four-degree zone on the visible horizon as seen from Coricancha. Chinchincalla, CU13-3, must have been the one incorporating the sun pillars; at least Cobo tells us so. Having positively located Pantanaya in the landscape on CU14 and having measured its 247° azimuth from Coricancha, we can say that Chinchincalla lies on a line directed slightly south of Az=247°. There is a flat-topped portion of a hill at the horizon that fits Cobo's description of the area, especially since the horizon slopes off steeply toward the north, fitting the description of Rauaypampa, CU 14-3, given by Cobo. The area of the flat top of the hill is more than 100 square meters, which subtends more than a degree as seen from Coricancha. Sunset over a 7½° horizon at 13° 30' S latitude (AD 1500) corresponds to approximately 247½° azimuth (incidentally, the sunset azimuth quoted [p. 28] for a "flat" horizon—I assume the authors' mean "of zero altitude"—is irrelevant). This is ½° north of our placement of CU-14 and between ½° and 1½° north of the ethnohistorically documented astronomical *ceque* CU13. All of this reasoning is dismissed by Dearborn and Schreiber as "too far south for the pillars to have been used for an actual observation from the Coricancha on the December solstice" (p. 28). Our conclusion remains that, regardless of how precisely the pillars function for the 20th century positivist, the weight of all the evidence suggests the Inca used them to mark a date in the calendar, most probably the December solstice.

The hypothesis that zenith-antizenith solar observations were conducted at the horizon of Cuzco owes its origin once again to the chronicles, Garcilaso not among them, contrary to what our critics seem to believe (pp. 30–1). That D & S devote only one paragraph consisting of 12 lines to reversible dualism (B. J. Isbell 1982) suggests that they have not really penetrated the concept of the use of these dates in the calendar and mythology, not to mention the landscape. All *ceque* lines are not straight "to a few degrees." Some are demonstrably straighter than others. Therefore, one cannot arbitrarily tack ±3° on to every *ceque* line related to an astronomical argument, as D & S (p. 33) do in evaluating how precisely the anti-zenith date can be pinpointed via location of the *ceque* line containing the sucanca on Cerro Picchu. The details of this argument become moot, however, for the authors in this particular instance are criticizing an idea that is already out of date. Based on arguments relating to the use of the *ceque* system as a calendar, Zuidema has proposed a revision in the positioning of the pillars on Cerro Picchu. Consequently, we no longer regard antizenith sunset being coincident with the middle pair on C. Picchu as an integral part of the astronomical argument (see e.g. most recently Zuidema 1988, p. 350, n 4). What all of this does for the location of *Usno* has yet to be determined by us. Such is the state of incompleteness of our work that it would be advisable for all potential reviewers to hold their tongues and their pencils on this one issue until the final, completed work is published.

There are many lessons to be learned from a careful reading of Dearborn and Schreiber's critique. A positive one among them is the need for one group of investigators to examine carefully the data and arguments of another group. For this and all the lavish attention paid our work by our colleagues we are sincerely grateful. However, one would like to suppose a careful re-examination would include a personal visitation to the sites under discussion—that is, to the remote *huaca* sites as well as the more easily accessible Coricancha and the Plaza de Armas. Who besides Zuidema and I have walked over and examined this area in detail with map and chronicler's description in hand? I would not attempt to criticize the astronomy of Machu Picchu had I not visited and measured there as well.

But the primary lesson, one that even a student of archaeoastronomy with no real interest in the Inca *per se* should be able to learn, lies in the realization that one cannot strictly apply the same scientific methodology that is used to study the stars to the various kinds of data that relate to human behavior. Indeed, the miscalculation of the power, the force, and the universality of application of quantitative and technological rigor to solve all our problems—those of the past included—may go down in history books of the 21st century as one of the great shortcomings of the 20th.

REFERENCES

Dearborn, D., and K. Schreiber. 1986. Here Comes the Sun. *Archaeoastronomy* 9:15–37.
Hawkins, G. 1965. *Stonehenge Decoded*. New York, Delta Dell.
Isbell, B. J. 1982. Culture Confronts Nature in the Dialectical World of the Tropics. In *Archaeoastronomy and Ethnoastronomy in the American Tropics,* ed. A. Aveni and G. Urton, 353–363. Annals New York Academy of Science, vol. 385, New York.
Zuidema, R. T. 1988. A Quipu Calendar from Inca. In *World Archaeoastronomy,* ed. A. Aveni, 341–351. Cambridge, Cambridge University Press.

Blinded by the Light
A Response to On Seeing the Light by A. Aveni
by David S.P. Dearborn

Aveni's reply presents an opportunity to examine the major differences in our approach to archaeoastronomy. He does not refute our representation of his work, stating that a "point-by-point" response will be done elsewhere, but instead discusses positivism. As part of this he then criticizes the "scientific methodology" that we bring to our study as out of place.

A "positivist" approach is one that emphasizes what can be derived from physical evidence, and rejects ideas that do not manifest themselves in a physical manner. Aveni instead has "seen the light" of an approach which asserts that the important questions involve the thoughts and beliefs of a people, and may not manifest themselves physically. In particular, he follows the work of Zuidema which falls into this category. Zuidema works hard and is creative in his efforts to seek the order of the Andean mind, but the basis in physical evidence of his hypotheses is sometimes very weak.

We begin our response by stating that we have never doubted the value of Zuidema's efforts, and have demonstrated our conviction by examining them critically instead of simply ignoring them as some have, or blindly accepting them. The question here is the level at which one requires such hypotheses to be verified by physical evidence before believing them, and this is a matter of judgment. Clifford Geertz said:

> The force of our interpretations cannot rest, as they are now so often made to do, on the tightness with which they hold together, or the assurance with which they are argued. Nothing has done more, I think, to discredit cultural analysis than the construction of impeccable depictions of formal order in whose actual existence nobody can quite believe.

First published as "Comments: Blinded by the Light (A response to *On Seeing the Light* by A. Aveni)," by David S.P. Dearborn, from *Archaeoastronomy* 10:24–27. © 1992 by the University of Texas Press. All rights reserved.

He was not arguing for a positivist approach, but to temper those elegant interpretations by requiring close examination of all evidence pertaining to the order imposed. This is what we have done. After a critical evaluation, we have found that we do not always agree with our colleagues, and at the heart of the difference is our inability to reject the utility of scientific methodology. We now turn from the main issue to deal with some of Aveni's specific criticisms.

Regarding Garcilaso and the Zenith Passage of the Sun

In his reply Aveni claims that we depend excessively on Garcilaso, and suggests that it was due to its availability in English. This is an example of the unfortunate lack of accuracy evident in the reply. In our article, we quote extensively from Cobo, Guaman Poma, Molina, Polo, Pachacuti Yamqui, an anonymous chronicler, and others, often giving both the Spanish and an English translation. We understand the problem with accepting any of the chroniclers as accurate, and included a caution to that effect in our article.

Aveni claims that our inclusion of Garcilaso was an inexcusable error in judgment. We included Garcilaso because he had been referred to by Zuidema, and contrary to Aveni's assertion in his reply, because it is the only direct source of ethnohistoric evidence for any interest in the Zenith passage by the Inca. If we had left it out, we would have been accused of ignoring the ethnohistorical documentation supporting the hypothesis of Zuidema and Aveni. In this and many other places in his reply, Aveni demonstrates just how casually he read both our article and the works of Zuidema.

In conversations with Zuidema during preparation of our article, we asked him if there was any other ethnohistoric documentation for the zenith passage, and he cited none. Since then, he has stated that the zenith passage interest of the Inca was well (but indirectly) documented in a description of a harvest festival that occurred in April. He claimed (at a meeting of The International Congress of Americanists, 1988) that the Inca traveled in a procession exactly along the direction of the zenith passage date sunrise as viewed from Sucanca (towards Tipon). If this is the festival that Molina (el Almagrista 1968:81–82) described, it was a celebration for the harvest of the maize in April of 1535. In it, the Incas went out of Cuzco, and conducted ceremonies for 8 or 10 days, which included much drinking, singing to the sun, and burning many "sheep." The only indication of where this occurred is Molina's statement that

> Sacaban en un llano, que es a la salida del Cuzco, hacia donde sale el sol en amaneciendo...
>
> They took (the effigies) to a plain that is just out of Cuzco, where the sun would rise at dawn...

In April, the sun rises to the northeast of Cuzco, nearly orthogonal to the direction of travel along a Sucanca-Tipon line. One might suspect that the festival was associated with the *huaca* described by Cobo (Rowe 1979):

AN-3:6 Mantocalla was a hill which was held in great veneration, on which at the time of shelling maize, they made certain sacrifices. For these (sacrifices) they placed on the hill many bundles of carved firewood dressed as men and women and a great quantity of maize ears made of wood. After great drunken feasts, they burned many sheep with the said firewood and killed some children.

This *huaca* is to the northeast of Cuzco on the same *ceque* as Chuquimarca, which Zuidema locates somewhere near Lacco. Again, this is far from the Sucanca to Tipon line. The "indirect" support then fails, leaving Garcilaso as the sole ethnohistoric evidence for the hypothesis of the zenith passage importance. While there is little ethnohistoric documentation on the zenith passage, there is none for its opposition, the "antizenith" passage. As we discussed in our paper, if the zenith passage were important, some opposition to it might be inferred from duality, a concept that we did not question in our paper. We also agreed that the Spanish Chroniclers might not recognize and record such an alien concept, and record an antizenith. Unfortunately, neither is it present in the writings of any of the native or mestizo chroniclers.

Finally, when Aveni published his work on the Usno-Sucanca alignment, a topic on which he has given many talks, it was presented as evidence for an August 17 (antizenith) alignment supporting the hypothesis of Zuidema. Now, he advises "potential reviewers to hold their tongues and their pencils on this one issue until the final complete work is published." How many secondary publications like that of Hyslop's (1985, see review in Archaeoastronomy Vol IX) must there be before we are permitted to examine the evidence?

Additionally, we are chided for addressing this "idea that is out of date," referencing an article of Zuidema that is due out soon. Zuidema has now modified his belief to locate the Usno-Sucanca alignment near September 2, nearly 15 days after the originally claimed date. They have not abandoned the antizenith concept, simply moved it to a different pillar. Our work then stands as a warning of how weak the archaeological support was that Sucanca marked the antizenith, or how little it now refutes Zuidema's new claim that one of the other pillars marks this date.

If one eliminates Garcilaso, as Aveni seems to prefer, and Aveni's archaeological evidence (the previously claimed Usno-Sucanca alignment) as Zuidema has done, little remains to support their hypothesis of a zenith-antizenith partitioning of the year. It is unclear what Aveni would have us base our belief on, and his response gives no additional material. The importance of the zenith passages as well as of any observation opposing them is then a matter of debate.

Ceques, Their Direction and Straightness

In locating *huacas* like Sucanca and Quiangalla, Aveni has used the assumption that the *ceques* are straight lines. This was supported by a statement (Aveni 1981) that several (unspecified) *huacas* on one *ceque* were located and found to lie on a straight line. In his response, we see for the first time from Aveni, a statement that "all *ceque* lines are not straight." Additionally, we did not "arbitrarily" assign an uncertainty of 3 degrees to their alignment. As discussed in our article, this uncertainty was inherent in the nature of the *huaca*, and the size of the plaza, and Zuidema's latest modifications in his hypothesis for the use of Sucanca, change its location by at least this amount.

While the importance of a researcher describing the location and means of identification of *huacas* for which he claims an alignment seems obvious, it apparently must be discussed. In a detailed study of Callachaca published by Niles (1987), she offers tentative identifications of *huacas* along the *ceques* AN-4, AN-5, and AN-6. The identifications are based on the surviving place names, *huaca* descriptions by Cobo, and relative positions along Inca roads (or paths). Precise locations, descriptions, and photographs are given to each candidate. She finds them to lie along paths which, as viewed from the Coricancha, would subtend nearly 25 degrees, and appear (again from the Coricancha) to overlap each other.

Niles' identifications of *huacas* on AN-5 are in direct conflict with Zuidema's (1982) claim that it aligns with the Pleiades rise. The places identified by Niles are generally farther south, and subtend an angle from the Coricancha through which nearly a quarter of the sky will rise. At the very least Niles' study shows the difficulty in identifying *huacas* and the necessity for a complete description of how they are identified before their identification can be accepted. In the absence of any information by Zuidema of how he identified the *huacas* on this *ceque,* it is difficult to judge whose identifications are more likely. This lack of detail in the papers of Aveni and Zuidema results in a great deal of trouble for anyone attempting to evaluate their work, instead of merely accepting it. The presentation of good data is not just the compulsive behavior of a positivist, but the foundation on which one's colleagues can base substantive judgments of one's interpretations.

It is not strange that Aveni believes we have not visited many of the sites around Cuzco discussed by him and Zuidema. In the absence of any precise description of where the sites were located it is difficult to do so. In spite of this, we have visited the sites of Tipon, Lacco, Callachaca, and many sites much more difficult to reach. True, we did not walk the area of Sucanca and Chinchincalla where, as Aveni and Zuidema admit in their papers, nothing remains. It is unclear what he expects us to learn when he states in his response that we should have visited those sites.

Machu Picchu, Precision, and Pillars

Aveni again raises the question of our use of structures at Machu Picchu as evidence that the Inca made precise astronomical observations. As we stated in our article there is little ethnohistoric information on the site, but Aveni's comparison of working there to working at Stonehenge is so fallacious that it is difficult to comprehend. We do know from documents (Rowe 1987) that Machu Picchu was an estate of the Inca Pachacuti, who was responsible for the initial expansion of the empire and for rebuilding much of Cuzco. Therefore, while the ethnohistoric data for Cuzco does not refer specifically to the site of Machu Picchu, it does refer to structures built by the same people, in the same style, and at the same time. Are we to believe that when

Pachacuti visited Machu Picchu, that he gave up all of his culture and religion? Also, the Torreón is not a unique structure as is Stonehenge. The same alignment was found in a very similar structure at Pisac.

We find the criticism of our interest in precision strange in any researcher. The precision that we found in the edges cut into rock at Machu Picchu and Pisac was such that any suggestion that both alignments were accidental requires a belief in an occurrence with a probability that is literally a million to one. Added to this, we know that the Inca had a strong interest in the June solstice (where we have no such knowledge of the builders of Stonehenge). His diversion to a discussion of the pegs at the windows has been answered elsewhere, and has no bearing on the point he was attempting to make in his reply. To argue that the precision we found at these sites was unimportant is to argue that the Inca did not really monitor the motion of the sun in direct contradiction to statements by chroniclers like Sarmiento (1965).

> Y para que el tiempo del sembrar y del cogar se supiese precisamente y nunca se perdiese, hizo poner en un monte alto al levante del Cuzco cuatro palos, apartados el uno del otro como dos varas de medir, y las cabezas de ellos unos agujeros, por de donde entras el sol a manera de reloj o astrolabio. Y considerando a dónde heriá el sol por aquellos agujeros al tiempo del barbechar y sembrar, hizo señal en el suelo, ... Y diputó personas que tuviesen cuenta con estos, relojes y notificasen al pueblo los tiempos y sus diferencias que aquellos relojes señalasen.
>
> And so that the time of sowing and reaping be known precisely and never lost, he caused the placement on a high hill to the east of Cusco four poles separated from one another by two measuring sticks and in the tops of them some holes through which the sun might enter like a clock or astrolabe and considering where the sun went through those holes at the time of cultivating and planting he made there marks on the ground ... he appointed persons to attend to these clocks and notify the people the times and intervals that those clocks might indicate.

While Sarmiento's description seems to confuse the towers on the hills with the gnomons in the temples, he is insistent on the precision. The accuracy that we found in the Torreón is a natural consequence of a person actually observing. To even a casual observer, a shift of a quarter of a degree in the position of a sunrise from some alignment is a clear miss. (E.g. the angular diameters of the sun and moon are approximately 0.5 degrees. If one of those objects gleams over the horizon a quarter of a degree from some marker, anyone with normal vision will clearly see that it misses the marker.)

In his response, Aveni has given the azimuth of CU-14 for the first time. Initially he claims it to lie on an azimuth of 247, coincident with the value that we accepted from Zuidema. In a later sentence he claims a *huaca* on this *ceque*, CU-14:3, was measured north of this at 247.5 and an altitude of 7.5 degrees. This direction is indeed that of the Solstice as viewed from the Coricancha. The pillars, Chinchincalla are then somewhere in a 4 degree range south of that ceque (Aveni suggests .5 to 1.5 degrees). In either case, this is outside of the area where the sun will be seen to set by either a "20th century positivist" with glasses or an Inca with normal vision. He then labels irrelevant our effort to make the 247 degree azimuth given by Zuidema (and here by Aveni) consistent with their claim that Chinchincalla did mark a solstice. In the absence of any altitude information, we considered a flat or zero altitude horizon the most favorable assumption to their case (through we noted the true horizon was 6 to 8 degrees) and even here, the alignment fails. We believe this discussion underscores the necessity of stating the measurements as well as the interpretation.

Finally, Aveni and Zuidema have made convincing arguments that the other pillars, like Sucanca and Quiangalla, were not observed from the Coricancha, but from other *huacas*. We then see no reason to accept the assertion that because the pillars were probably near the solstice set point as seen from the Coricancha that they were observed from there. Especially when a December solstice alignment is not consistent with Cobo's statement that the sunset between these Pillars marked a time to plant.

Conclusions

Returning again to the real issue in Aveni's response, what is the appropriateness of scientific methodology, and what level of evidence must we require before accepting interpretations? These are points of serious debate when addressing questions of cultural motivation, and we cannot offer to draw a simple line between what should be believed and what should not. Nevertheless, returning to Geertz, can we accept interpretations that rely solely on "the tightness with which they hold together, or the assurance with which they are argued"?

By considering his work, we have accepted the approach of Zuidema as a means of generating hypotheses, and originally began this study with a preference that they be substantially correct. Our article gave an orderly presentation of the archaeological, ethnohistorical, and ethnographic support for these hypotheses. We did not conclude that Zuidema's hypotheses were wrong, simply demonstrated that in some instances, the supporting data were very weak. Aveni has not disputed any of our presentation, but instead criticized our reliance on physical evidence and "scientific methodology." We respond that without some physical evidence how can we know that any of the concepts that have been proposed were ever in the Andean mind? We are not positivists who reject out of hand any hypothesis not supported (or supportable) by physical evidence as non-science. We see the purpose of such hypotheses as part of an order or framework for investigating interesting cultural questions. In the end, however, the acceptance of those hypotheses must be judged at some level by the supporting data.

REFERENCES

Aveni, Anthony. 1981. Horizon Astronomy in Incaic Cuzco. In *Archaeoastronomy in the Americas,* ed. R. Williamson, 305–318. Santa Barbara, Ballena Press.
Cobo, Bernabe. 1983. *History of the Inca Empire,* trans. Roland Hamilton. Austin, University of Texas Press.

Hyslop, John. 1985. *Inkawasi: The New Cuzco,* Cañete, Lunahuana, Peru, BAR International series 234, xii–147, Oxford.

Molina, Cristobal de (el Almagrista). 1965. *Relación de Muchas Cosas Acaescidas en el Peru.* Biblioteca de Autores Españoles, Tomo 209, 56–96, Madrid.

Niles, Susan A. 1987. *Callachaca: Style and Status in an Inca Community.* Iowa City, University of Iowa Press, 171–206.

Rowe, John. 1979. An Account of the Shrines of Ancient Cuzco. *Nawpa Pacha* 17:1–80.

Rowe, John. 1987. Pachacuti's Royal Estate at (Machu) Picchu. Lecture presented at the annual meeting of the Institute of Andean Studies, Berkeley.

Sarmiento de Gamboa. 1965. *Historia Indica.* Biblioteca de Autores Españoles, vol. 135, p. 236.

Zuidema, R. Tom. 1981. The Inca Observations in Cuzco of the Solar and Lunar Passages through the Zenith and Anti-zenith. *Archaeoastronomy in the Americas,* ed. R. Williamson. Santa Barbara, Ballena Press.

Zuidema, R. Tom. 1982. Catachilla: The Role of the Pleiades, and the Southern Cross, and Alpha and Beta Centauri in the Calendar of the Incas. *Ethnoastronomy and Archaeoastronomy in the American Tropics,* ed. A. Aveni and G. Urton. Annals of the New York Academy of Sciences 358:203–229.

Here Comes the Sun: The Cuzco–Machu Picchu Connection

A Reply to Dearborn and Schreiber

by R. T. Zuidema

Dearborn and Schreiber, in their recent article comparing Incaic astronomy in Machu Picchu and Cuzco, concentrate on the issue of precision. Given the importance of the problem, I want to respond to their article. It seems to me that they define precision wrongly; that they misrepresent the results that Aveni and I obtained in Cuzco; and that they do not come to interesting results of comparison.

The authors see precision in the first place as a problem of observation, not of calendar and its place in culture. While many of their references to Inca culture are anecdotal and thus irrelevant, they do not hesitate to warn the reader that "the accounts (of the chroniclers) must... be used with care." This attitude may have made them unaware—giving one example—that a crucial text of Molina (1573), quoted in an atrocious English version from 1873, does not read, as they claim, that the Incaic month of the June solstice began with a full moon after May 25 (Gregorian calendar), but with a new moon. Precision does not belong only to astronomy!

First published as "Comments: A Reply to Dearborn and Schreiber *Here Comes the Sun: The Cuzco–Machu Picchu Connection,*" by R. T. Zuidema, from *Archaeoastronomy* 10:28–29. © 1992 by the University of Texas Press. All rights reserved.

David S.P. Dearborn and Katharina J. Schreiber

I want to confine myself to one point of substance where their misconception of "precision" impedes comparison of our measurements of Coricancha, the temple of the Sun in Cuzco, with their measurements of the "Torreón" in Machu Picchu. In order to arrive at such a goal one should investigate first if alignments of both buildings could have served similar purposes. Let me repeat, therefore, my part of the argument of how precision in Coricancha was obtained with two kinds of observations, each made with different techniques.

First, calendrical exactitude was obtained by observing a specific sunrise for a date *not* during a solstice. Coricancha turns out to be aligned precisely to a sunrise on May 25 (and not to various sunrises during a vague period around the June solstice) (Zuidema 1982a pp. 214–5). Although the choice of this date seems arbitrary, as it is not for any easily detectable solstice, equinox or zenith passage, it serves to *calculate* the exact date of the June solstice and to define precisely the Full Moon related to that event. In addition, it is confirmed by the statement of Molina alluded to above (which I translated from the original manuscript with good transcriptions of its Quechua words).

The other observations from Coricancha concerned the Pleiades, in whose case, the same kind of precise alignment was not needed. I asked the question *if* the toponyms and the myths and rituals referring to the Pleiades could make such observations acceptable. I also asked *if* the Pleiades could have had a role in the calendar. Any calendar taking stars into account does so, *not* for their exact rising or setting points—stars are recognized within constellations for which no such exact points need to be given—*but* for the dates of their first and last appearance. In the case of the Pleiades, Coricancha would have served a purpose different from that for the Sun.

Although I gave a full discussion of these issues and I explained them in the sequence presented here, Dearborn and Schreiber chose to question our measurements, asking, first of all, *if* Coricancha was aligned *exactly* to the Pleiades. They misunderstood the problem.

One might have expected Dearborn and Schreiber to investigate a use of the Torreón in Machu Picchu similar to that of Coricancha, given that its northeastern window faces almost the same direction as Coricancha (a fact they observed) and that it also is *not* adapted to observing sunrise during a period around the June solstice. We are not told for which day of sunrise the alignment of the window might have been useful. It could have been the very same day for which the Incas in Cuzco had built Coricancha! Instead, we get the hypothesis of a minor alignment of a rock's edge towards June solstice sunrise, which serves no precise calendrical purpose, and we are asked to accept the highly improbable use of a plummet outside the window. One cannot build a theory of Inca astronomy and its calendar on such a hypothesis. No real comparison with Coricancha was made.

Our research on astronomy and the calendar in Cuzco is based on a full use of the chroniclers and is developed in more publications than the three on which Dearborn and Schreiber base their sweeping judgments (Zuidema 1966, 1973, 1976a, 1976b, 1977, 1978, 1980, 1981a, 1981b, 1982a, 1982b, 1982c, 1983, 1985 up to 1986). For an excellent use and overview of these studies, see Anders 1986). In our research, a late chronicler like Garcilaso, in whom Dearborn and Schreiber place such an undue trust, does not play a role. We might put to a better use one of their contentions and claim that Garcilaso's information is "not amenable to finding physical data to support him."

We know, from early chronicles and from our work, that the Incas did not use Coricancha for observing a June solstice sunrise. Their interest was in *predicting* the event. This they accomplished by combining three types of observation: 1) an exact observation of Sun (25 May); 2) an exact one of the Moon (New Moon after 25 May allowing to calculate for every year the Full Moon closest to the June solstice); and 3) one of the Pleiades (first heliacal rise in the morning some days after 3 June) (Zuidema 1982b pp. 214–8). We are not dependent on a precise alignment of the latter observation, which was obtained by way of the Sun *and* the Moon. Nevertheless, the Pleiades served calendrical precision by way of sidereal-lunar observations throughout the year. Overall precision was obtained by combining various types of observation and calculation and not by an isolated and unsupported precise result outside the general context of the culture and its calendar. We are reminded of what Aaboe and de Solla Price (1964) once said about "the derivation of accurate parameters from crude but crucial observations" for obtaining precise calendrical results. They showed, among other things, that the Greek determined the exact dates of the solstices, not by using these events themselves, but by observing sunrises and -sets before and after.

In the meantime, we are still waiting for the Sun to come, and to rise, from Machu Picchu.

REFERENCES

Aaboe, Asger, and Derek J. de Solla Price. 1964. Qualitative Measurement in Antiquity: The Derivation of Accurate Parameters from Crude but Crucial Observations. In *L'Aventure de la Science: Mélanges Alexandre Koyré* vol. 1:1–20.

Anders, M. 1986. "Investigation of Storage Facilities in Pampa Grande, Peru. Ph.D. thesis, Cornell University.

Zuidema, R. Tom. 1966. El calendario Inca. *Actas del XXXVI Congreso Internacional de Americanistas* 2:25–30. Seville.

Zuidema, R. Tom. 1973. Kinship and Ancestor Cult in Three Peruvian Communities: Hernandez Príncipe's Account of 1622. *Boletín del Instituto, Franegs de Estudios Andinos* 2:1 pp. 16–33.

Zuidema, R. Tom. 1976a. La imagen del Sol y la huaca de Susurpuquio en el sistema astronómico de los Incas en el Cuzco. *Journal de la Société des Américanistes* 62 (1974–1976):200–230.

Zuidema, R. Tom, and Gary Urton. 1976b. La Constelacién de la Llama en los Andes Peruanos. *Allpanchis Phuturinga* 9:59–119.

Zuidema, R. Tom. 1977. The Inca Calendar. In *Native American Astronomy,* ed. A. F. Aveni, 219–259. Austin, University of Texas Press.

Zuidema, R. Tom. 1978. Mito, rito, calendario y geografía en el antiguo Peru. *Acts du XLIIe Congrés International des Américanistes* vol. 4, pp. 347–359. Paris.

Zuidema, R. Tom. 1980. El Ushnu. In *Economia y Sociedad an los Andes y Mesoamerica,* ed. J. A. Alcina Franch. Revista de la Universidad Complutense de Madrid, vol. 28:117, pp. 317–362.

Zuidema, R. Tom. 1981a. Inca Observations of the Solar and Lunar Passages through Zenith and Anti-zenith at Cuzco. In *Archaeoastronomy in the Americas,* ed. R. Williamson, 319–342. Los Altos, Ballena Press.

Zuidema, R. Tom. 1981b. Comment on a review by J. H. Rowe. *Latin American Research Review* 16(3): 167–170.

Zuidema, R. Tom. 1982a. Bureaucracy and Systematic Knowledge in Andean Civilization. In *The Inca and Aztec States, 1400–1800,* ed. Collier, Rosaldo, and Wirth, 419–458. New York, Academic Press.

Zuidema, R. Tom. 1982b. Catachillay: The Role of the Pleiades and of the Southern Cross and Alpha and Beta Centauri in the Calendar of the Incas. In *Ethnoastronomy and Archaeoastronomy in the American Tropics,* ed. A. F. Aveni and G. Urton, 203–229. Annals of the New York Academy of Sciences, vol. 385.

Zuidema, R. Tom. 1982c. The Sidereal Lunar Calendar of the Incas. In *Archaeoastronomy in the New World,* ed. A. F. Aveni, 59–107. Cambridge, Cambridge University Press.

Zuidema, R. Tom. 1983. Towards a General Andean Star Calendar in Ancient Peru. In *Calendars in Mesoamerica and Peru: Native American Computations of Time,* ed. A. F. Aveni and G. Brotherston, 235–262. Oxford, BAR International Series vol. 174.

Zuidema, R. Tom. 1985. L'Organisation andine du savoir rituel et technique en termes d'espace et de temps. *Techniques et Culture* 6:43–66.

Stand and Deliver
A Comment on Zuidema's Reply to Dearborn and Schreiber
by David S.P. Dearborn

The crux of Zuidema's reply seems to be that our article did not correctly represent the thrust of his work. He then cites a long list of publications on the topic of Inca astronomy, presumably implying how much important material we left out. Many of those articles merely repeat the material contained in the ones that we cited in support of Zuidema's hypothesis that the *ceque* system was used as a calendar. We did not include a discussion of the calendar hypothesis for two reasons. First, our article was a comparison of the work in Cuzco and Machu Picchu, and the identification of a "*ceque* system" at Machu Picchu would be speculative at best. Second, Zuidema's ideas of how the *ceque* system may have functioned continue to evolve. We had intended to wait for his book on the topic before analyzing this hypothesis.

First published as "Comments: Stand and Deliver: A Comment on Zuidema's Reply to Dearborn and Schreiber," by David S.P. Dearborn, from *Archaeoastronomy* 10:29–30. © 1992 by the University of Texas Press. All rights reserved.

As the alignments and orientations that we examined in our paper are used as physical elements supporting Zuidema's hypotheses of how astronomy and calendrics related to Inca social order, we considered a critical examination of them essential as a means of evaluating the foundations of the hypotheses. We included all of the evidence that we thought pertinent to the alignments and orientations compared, and neither Zuidema nor Aveni has presented anything in their replies to add to the evidence that we presented. In fact Aveni wants to remove the only direct quote (by Garcilaso) supporting the existence of an observation that *they* claim.

Zuidema goes further stating that many of our references are anecdotal and thus irrelevant, yet nearly all of them are used by him (including Garcilaso, Molina, Cobo, Guaman Poma and others). It is possible to misunderstand the quotations of chroniclers, and here at least, Zuidema tries to give an example instead of merely accusing with generalities and value laden words. He states that "a crucial text of Molina (1573), quoted in an atrocious English version from 1873 does not read as they claim, that the Incaic month of the June solstice began with a Full Moon after the May 25 (Gregorian Calendar) but with a New Moon." The passage referred to says nothing of a Full Moon in either Markham's or Zuidema's translation, and the only place that we have ever seen such a claim was Zuidema's article (1982b—see his reference list) where he said "A month including 21 June has *to begin after May 25 with a Full Moon* following June 7." The critical part of the Molina quote is a claim that late May was a significant time, and this was the same in both translations. While the accuracy of Markham's translation is irrelevant, in this case it was a concern, and we checked all of the quotations that we used against the original Spanish. An example of where this was relevant is the reference in which Molina says that they stayed at Puquincancha for a "number of days" following the 23rd day of the festival. Markham left this out of his translation.

In the section that Zuidema is referring to, we were presenting the idea that he gives two paragraphs later, that the orientation of the Coricancha was set "to define the Full Moon related to that event (Inti Raymi)" or in our words "the Full Moon that follows this date will be the month that includes Inti Raymi." Perhaps we should have said "in the month," but we certainly never said "begin the month" as Zuidema does in his 1982 article or attributes to us in his reply.

Zuidema originated a number of creative hypotheses, and we believe he had many insights on Andean culture. Nevertheless, as we stated in the prologue of our paper, we found areas of agreement, as well as areas where we could not agree on the significance of their results. In those instances of disagreement, we had hoped that Zuidema and Aveni would respond with additional evidence supporting their claims, or at least restating the evidence that we presented, and this was not done. The defensive attitude shown by Zuidema to our examination is misplaced, and his one attempt to respond directly to our article is based on attributing something to us that we did not say. There is an almost grasping attempt to claim that our criticisms stemmed from lack of understanding of their arguments instead of presenting those arguments. In what seems like desperation they denigrate physical evidence like the principal (only) alignment carved in a stone dominating the center of the Torreón, and replicated in a similar structure in Pisac, because it shows an accuracy that they consider unimportant in their work. Zuidema concludes that he is still waiting for the light to shine from Machu Picchu. If he feels our article fell short, we would encourage him to write his own for *Archaeoastronomy*.

On Shadows of Doubt
A Reply to Dearborn's Review of *Inkawasi—the New Cuzco*
by John Hyslop

D.S.P. Dearborn's review (*Archaeoastronomy*, Vol. IX, Nos. 1–4, pp. 115–122, 1986) of Chapter 4 of my monograph *Inkawasi—the New Cuzco* (BAR International Series 234, Oxford, U.K., 1985) stimulates me to correct some information and raise questions about Dearborn's approach.

Dearborn criticizes several of the proposed alignments at Inkawasi for a lack of precision. His standard of evidence is a proposed observatory (the Torreón) at Machu Picchu. His often-published contention that this "observatory" is accurate to a few minutes of arc does not resolve the greater problem of demonstrating that a cut edge on a rock in the Torreón was in fact aligned with the June solstice sunset. His argument hangs on a hypothetical string or other device hung on one side of a window of the Torreón. As he and Schreiber have noted (*Archaeoastronomy*, IX, 1986: 23), without some alignment device, a viewer "could sight along the edge only to see the wall below the window." The alleged accuracy of Dearborn's Torreón observatory lacks the crucial physical evidence to create the alignment, as well as historical backing that strings or other devices in windows were used. The Torreón should not be used as a standard for judging Inka astronomical activities at Inkawasi or elsewhere.

The *Inkawasi* monograph proposes that several Inka astronomical interests were important in the layout of the site. A number of factors, discussed in my monograph (p. 70) but ignored by Dearborn, may cause variations in the Cuzco alignments when they were used in the layout of other Inka settlements. No one yet knows just how accurate Inka astronomy was when used in settlements away from Cuzco. For example, alignments may have been introduced into site planning to express ideas associated with them, but aligned compounds or site axes may not have served as observation devices. This and other factors suggest that precision is but one test (the main one used by Dearborn) of Inka astronomical influences on site design.

First published as "Comments: On Shadows of Doubt: A Reply to Dearborn's Review of *Inkawasi—the New Cuzco*," by John Hyslop, from *Archaeoastronomy* 10:30–32. © 1992 by the University of Texas Press. All rights reserved.

Dearborn examines several of the architectural alignments at Inkawasi which I propose were based on astronomical concerns. He criticizes using alignments proposed by Zuidema as "known" (particularly to star groups) when in fact they are poorly based. I only examine two alignments with star groups (both used by Dearborn and collaborators in previous works), one to the Pleiades rise azimuth and one to the area of the quasi-circumpolar constellations. In the latter case I am careful to point out (p. 62) that a possible *zeque* sight line (Anahuarque) to the region "is not a proven astronomical sight line." My justification for studying a possible alignment with that region of the horizon does not come primarily from Zuidema, but from ethnographic work by Gary Urton, who found modern Quechua-speakers with an Andean cosmology (possibly similar to the Inka) having a great interest in Andean constellations and the Milky Way which appear over a number of degrees on the southeastern horizon.

I agree with Dearborn that the alignment I propose to that region is not specific, but think premature his opinion that it "is inadequate," since I begin my discussion by noting that I am not dealing with a sight line to a point, but to an arc on the horizon. Dearborn increases the ineligibility of the fit by including measurements of architecture which, in the text (p. 64), I discount as possibly aligned.

Dearborn evaluates by two possible zenith sunset alignments (parallel sides of the same architectural compound) and points out correctly, I believe, that the one on the northern side of the compound is more adequate (half a degree off) than the other. He criticizes that this is not "the axis that Zuidema discusses for Cusco," a point clearly made in my discussion (p. 60). I also note that this alignment could achieve the same calendrical result required by Zuidema, a point ignored by Dearborn.

The most complex set of alignments at Inkawasi deals with those established by a trapezoidal plaza which opens to the northeast. Twice in his review Dearborn notes that the plaza is well off from the proposed alignments when viewed from the same plaza. It is, but I specifically argue that the plaza was not laid out from observations made at it. I reject Dearborn's creation and repetition of alignments with considerable error which I never proposed.

Dearborn argues that the lines of the side of the plaza do not converge on point D, as I maintain, but rather at a point some considerable distance before it. To reach his conclusion he had to extract measurements from the general site map. In the monograph (Appendix 3), I warn that the map is not sufficiently accurate for archaeoastronomical research (few archaeological plans of large sites are). It was made before the astronomical measurements. Because of the precision required for archaeoastronomical research, I returned to Inkawasi after the mapmaking and made the alignment study. I argue that the lines of the trapezoidal plaza do converge very near to point D, and this is illustrated in the photographs I publish on pages 67 and 68 (figures 33 and 34). The aerial photograph published here for the first time (Fig. 6.14) is supplementary evidence.

Dearborn suggests adjusting the alignments established by Inkawasi's trapezoidal plaza by nearly four degrees because of the horizon elevation when viewed from point D. This would still allow it to point to the June solstice sunrise and the Pleiades rise azimuth, but would exclude the northern lunar rise, an alignment we all consider poorly substantiated. His suggestion appears an option worth considering. I am surprised, however, that he did not report my logic for using a zero-degree horizon. In Appendix 3 (p. 133) I wrote that the trapezoidal plaza's alignments might have been established from the slope to the west of the main plaza where the observation of the horizon beyond the plaza is level with an observation point. In fact, I state that my proposed alignments would work "only if the observations were made from nearby heights."

The other bone I have to pick with Dearborn deals with his objections to my halving and quartering of Inkawasi. We apparently agree that Inka settlements may have been divided into two and four parts. Few researchers have attempted to define Inka settlements in such an Andean way, and I am pleased to have tackled this thorny subject.

Dearborn misses the point that the halves and quarters of the *physical plan* of Cuzco did not occupy exactly the same space that was partitioned by the *ceque* system. The Inkas divided each somewhat differently into two and four parts. The central sector of the Inka capital was divided into two parts, *hanan* (upper) and *hurin* (lower), by the road which ran along the southeastern side of the dual Inka plaza. The main Sun Temple at Cuzco, the center of the *ceque* system, was located within the *hurin* part of the city. Likewise, the proposed Sun Temple at Inkawasi is also within the *hurin* sector. Even given this fact, Dearborn argues that at Inkawasi the partitioning of the physical plan of the city should begin at the construction I propose as a Sun Temple. Such was not the case in the physical plan of Cuzco, and is not the case for my proposed divisions of Inkawasi.

My proposed Sun Temple compound at Inkawasi is based on its orientation, its position adjacent to a trapezoidal plaza, and its location within a specific quadrant, all similar to the main Sun Temple at Cuzco. I reject his contention that I "apparently did not measure it," since the compound's orientation was based on the measurement of one side of the trapezoidal plaza which also forms one side of the compound. There can be little doubt about the orientation of the compound I propose as a Sun Temple. Dearborn's argument that a Sun Temple should have a "beautifully curved wall" like Cuzco's and the Machu Picchu Torreón (who said it was a Sun Temple?) is contradicted by the shape of several other clearly identified Inka sun temples at Pachacamac, Vilcas Waman, the Island of the Sun, none of which has curved walls.

Dearborn then argues that my explanation of the placement of the Qollasuyu quadrant within Inkawasi on the basis of the proposed Sun Temple is circular. My halving and quartering of Inkawasi was based on more complex factors than Dearborn represents. Prime among them is the road pattern, the primary access to water of the *hanan* half, and the general placement of the *suyu* around Cuzco. A possible Sun Temple in Inkawasi's Qollasuyu quadrant is a useful confirmation of a partitioning scheme based on a number of arguments.

I propose that sector F1 at Inkawasi, which has fourteen radial architectural units, reflects spatial organization in the *zeque* system's Kuntisuyu quadrant, which has 14 lines. Dearborn doubts this, noting that there is "more to consider than a numerical coincidence of 14." I do consider much more. Thus I define a number of similarities between the *zeque* system's Kuntisuyu and sector F1. They include: the internal grouping of the units into groups of three, a generally similar orientation (noted by Dearborn but rejected for lack of precision), division of the rectangular radial units into two groups by one central non-rectangular unit, the arrangement of the units around a construction of ceremonial significance (a subsidiary *ushnu*), and the sector's location on the southwest side of the site. Dearborn questions, without further comment, whether Zuidema's groupings of *zeques* in Kuntisuyu is comparable to that of the units of sector F1. I include a discussion (p. 55) discussing just that.

I did not claim sector F1 was part of a *zeque* system, or that it should mirror the Cuzco *zeque* system's Kuntisuyu in every way. Differences always occur when ideas are transferred from one medium to another. I proposed (pp. 52–56) only similarities between F1's *spatial organization* and that of a quadrant of the *zeque* system. The system's spatial organization is useful for understanding some Inka architectural planning at Inkawasi and at other Inka settlements. In a recently completed study on *Inka Settlement Planning* (University of Texas Press, Austin, 1990), I discuss a number of other cases of radial layouts in Inka site designs. Those not intent on rejecting my interpretation of sector F1 will find the additional examples supportive.

I should like to add that a number of Dearborn's claims, such as "the basic structure of the city [Cuzco] evolved from its pre-imperial form," are without certain historical or archaeological confirmation. This remains a question for future archaeological and historical research.

Dearborn correctly notes that references (pp. 65–66) to a June solstice *sunset* were in error. It was a sunrise, mentioned casually as part of a line from the main *ushnu* to a peak on the northeastern horizon. It was not included in the tables since there was no evidence that it had any influence on Inkawasi's layout.

I am not certain that all I propose about Inkawasi's structural organization is as I propose. My presentation, as I predicted in the Preface, could be expected to raise questions. The hypothetical nature of some of the interpretations depends on many lines of reasoning, most non-quantitative, and perhaps perplexing to some physical scientists.

Finally, only one chapter of the *Inkawasi* monograph deals with the concerns discussed above. For most archaeologists and Andean scholars, the other chapters dealing with historical evidence, military aspects, and activity patterns may also be of interest.

Lengthening Shadows
A Response to Hyslop's Comments
by David S.P. Dearborn

Hyslop begins his response by questioning the results obtained at Machu Picchu as indicative of the precision of Inca astronomy. He echoes Aveni's concern on the proposed use of the pegs that exist on the windows of the Torreon, but at least recognizes that the critical element is the accuracy cut into the stone there and in Pisac. The Inca may have been very interested in constructing symbolic alignments, which were never meant to function for an observation, but it seems odd that they would do so if they never actually observed objects. The precision that Hyslop and Aveni wish to disavow is simply that which an observer would have if he tried to make an observation, and that is what we found at Machu Picchu.

In his book, Hyslop attempts to determine the motivation behind the layout of a planned Inca site, Inkawasi. In particular, he proposed an astronomical basis, founded substantially on the astronomical alignments that have been claimed by Aveni and Zuidema for Cuzco. Inkawasi is linked to Cuzco by a story recorded generations after its abandonment, claiming that it was named New Cuzco and that the districts and hills were given names the same as those in Cuzco. Very little is known about Inkawasi beyond this, and that it was built for Topa Inca to use during his Campaign in the Huarco region.

If the story is correct, the simplest explanation could be that as many towns in Massachusetts are named after villages in East Anglia, the areas around Inkawasi were named after places around Cuzco. Inkawasi was a planned site (that is, it did not evolve like a city which can grow without a common plan), and should go beyond the transference of familiar names, to have embedded in its structure any principles used by the Inca architects in city design. In his book, Hyslop claims to have found evidence of such organization in certain alignments (which were too inaccurate to be intended for observation).

The difficulty with "symbolic" alignments which function for an actual observation is not finding candidates; it is knowing which, if any, of these alignments were recognized by the builders. Confidence in the proposed alignments would be improved if such alignments were known to exist in Cuzco. For this, Hyslop accepts the alignments claimed for the Pleiades rise, Alpha and Beta Centauri rise, and an "antizenith" sunrise. Given the importance of these claims for his thesis, it would have been desirable if he had examined what these claims were based upon.

First published as "Comments: Lengthening Shadows: A Response to Hyslop's Comments," by David S.P. Dearborn, from *Archaeoastronomy* 10:32–34. © 1992 by the University of Texas Press. All rights reserved.

First, there is no ethnohistoric documentation claiming that any of the ceques in Cuzco were aligned to an astronomical object, either symbolically, or functionally, nor is there any mention of an "antizenith" passage of the sun. As we demonstrated in our article (Dearborn and Schreiber, *Archaeoastronomy*, Vol IX), the actual direction of the Usno-Sucanca alignment to the antizenith sunset is so uncertain that it provides no support to the antizenith hypothesis. Since then, Zuidema has arbitrarily changed the antizenith sunset to a different pillar, and Aveni has asked all reviewers to "hold their tongues and their pencils on this one issue until the final work is published." We assume that Aveni's admonition must then apply to colleagues basing their research on his previously published findings.

The two stellar groups for which Hyslop suggests alignments are clearly ones recognized by the Inca. While it is certainly possible, and even reasonable to suppose that the Inca constructed "symbolic" alignments, what is the evidence that they did? Unfortunately, Zuidema does not give us any usable information on how he determined the *ceques* AN-5 to be aligned to the Pleiades, and CU-1 to Beta Centauri's rise. However, as I point out in my reply to Aveni, the published identifications of Niles (1987) of the huacas of AN-5 are all south of the Pleiades, and define a region of the horizon through which nearly a quarter of the sky passes. This provides little confidence that a Pleiades alignment is what was intended. While it is entirely possible that the Inca did construct such symbolic alignments, it is difficult to accept them as known on the basis of the evidence that has been presented. The value of studies such as Hyslop's would be to find a related alignment at a site like Inkawasi. In this case it would have to originate from the building that he believes corresponded to the Coricancha. Not from the junction of one set of roads to a portion of mountain ridge viewed over a plaza.

However, in examining the alignments proposed by Hyslop, we found the alignment from his point D to the plaza most interesting, though data included in the book were not consistent with the convergence point at point D. The alignment is not precise, but neither is it suitable for a true observation. The Pleiades rise over a high ridge well above the plaza, so the lack of precision is understandable. The question that Hyslop must address is: was it intended? His case might have been improved if the viewing location had been an Usno, but it is still a possibility.

I included calculations of the orientation of the plaza itself (as viewed from the plaza) because Hyslop made his measurements from there and I checked all of the measurements that he included in the text. If allowed an accuracy of few degrees for the symbolic nature of the alignment, I can find December and June solstice sunsets, a Betelgeuse rise (an alignment also proposed by Zuidema), and a llama (dark cloud constellation) set (not proposed by Zuidema but equally well founded in the ethnography).

A more fundamental reason for determining the position on the sky marked by the southern edge of the Plaza (as viewed from there) is that he used this to orient the structure that he associates with the Coricancha. Zuidema has measured the orientation of one of the walls of the Coricancha and believes its proximity to the Pleiades rise to be significant. The orientation here does not match.

David S.P. Dearborn and Katharina J. Schreiber

Finally, the bulk of Hyslop's work involves a proposal for the partitioning of Inkawasi into quarters. The partitioning that he suggests is reasonable, but many alternative possibilities could be suggested with equal probability. For every way in which he claims that sector F1 mirrors Cuntisuyu, it is easy to cite three ways in which it does not. Perhaps that is telling us about what the Inca considered fundamental about Cuntisuyu (which would not include its orientation, or the number of *huacas*), or perhaps it is a large structure that was built to conform to the terrain.

Hyslop ends by suggesting that "the hypothetical nature of some of the interpretations depends on many lines of reasoning, most non-quantitative, and perhaps perplexing to some physical scientists." Perhaps if he knew more about the development of quantum mechanics, he would know that such fields often develop from instinct, and the mathematical rigor then follows. I know physical scientists and engineers who not only read but write novels, poetry, and music. They participate in politics, and other complex activities not easily quantified. In suggesting that they are incapable of dealing with non-quantitative lines of reasoning, Hyslop resorts to a simplistic stereotype (a short fellow with glasses, a pocket protector, and a calculator attached to his

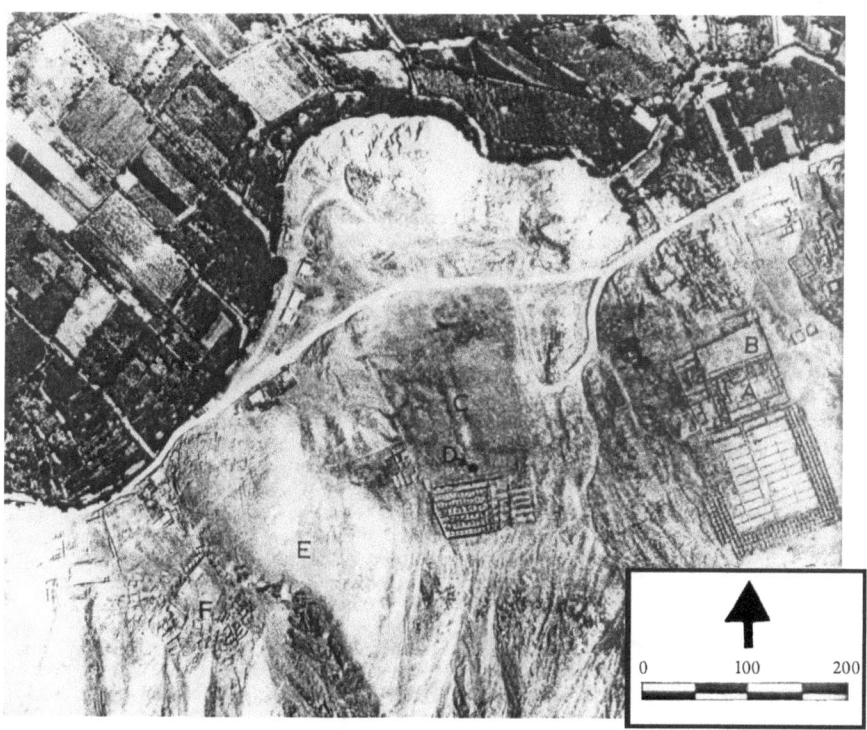

6.14 Aerial photograph of the Inka garrison, Inkawasi, in the Cañete Valley of Peru. Key: A, Compound proposed as a Sun Temple; B, Trapezoidal plaza; C, Main plaza; D, Point D, where several lines cross; E, Slope to west of main plaza; F, Sector F1 with radially arranged units. Scale approximate. From Servicio Aereofotográfico Nacional, Lima.

belt) instead of responding to valid questions of why we should accept or even consider his particular interpretations. He has begun an interesting study, but if a reader of *Archaeoastronomy* were to approach his book without a background in Andean archaeology, he might come away believing that more had been found than just one possibility. I hope Hyslop extends his study and finds similar order at other planned sites.

CHAPTER SEVEN

Chankillo

A 2,300-Year-Old Solar Observatory in Coastal Peru

Ivan Ghezzi and Clive Ruggles

The identification of places from which astronomical observations were made in prehistory, together with evidence on the nature and context of those observations, can reveal much about the ways in which people before the advent of written records perceived, understood, and attempted to order and control the world they inhabited.[1] Evidence of systematic observations of the changing position of the rising and setting sun along the horizon,[2] in particular, can provide information on the development, nature and social operation of ancient calendars.[3] Solar horizon calendars were certainly important among indigenous Americans, one of the best-known modern examples being at the Hopi village of Walpi.[4] In pre-contact Mesoamerica, systematic studies of the orientations of sacred buildings and city plans strongly suggest the existence of horizon calendars in which special significance was attributed to certain key dates. It has been argued that these include not only the solstices, but also the

---

From *Science* 315 (2007):120–124. Reprinted with permission from AAAS.

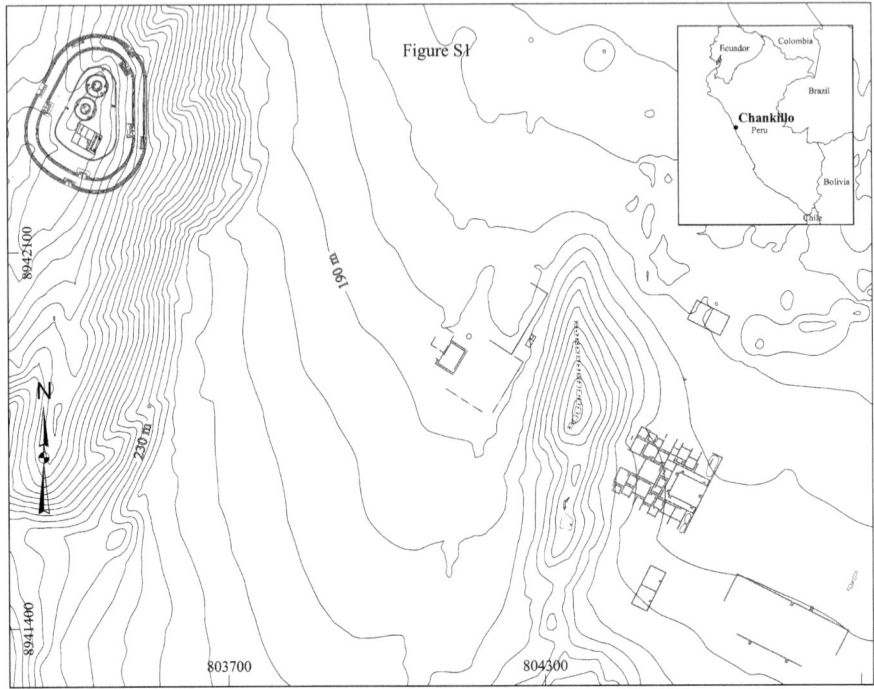

7.1 Map of the Chankillo site.

dates of solar zenith passage[5] and dates counted off from these at intervals significant in the intermeshing cycles of the Mesoamerican calendar round.[6] In South America, accounts going back to the 16th century record various details of pre-conquest practices relating to Inca state-regulated sun worship and related cosmological beliefs.[7] Various schemes of landscape timekeeping have been suggested, supported by a combination of historical evidence and analyses of the spatial disposition of sacred architecture—and in particular the system of shrines placed along lines (*ceques*) conceived as radiating out from the central sun temple, the Coricancha, in Cusco.[8] "Sun pillars" are described by various chroniclers as having stood around the horizon from Cusco and been used to mark planting times and regulate seasonal observances,[9] but all the Cusco pillars have vanished without trace and their precise location remains unknown. Here, we describe a much earlier structure in coastal Peru that seems to have been built to facilitate sunrise and sunset observations throughout the seasonal year.

The group of structures known as the Thirteen Towers is found within Chankillo, a ceremonial center in the Casma-Sechín River Basin of the coastal Peruvian desert (Fig. 7.1). Seventeen ^{14}C dates fall between 2350–2000 calibrated years before the present (B.P.) (Fig. 7.2), and point to the beginning of occupation at the site in the fourth century BC, during the late Early Horizon period.[10] The site contains multiple

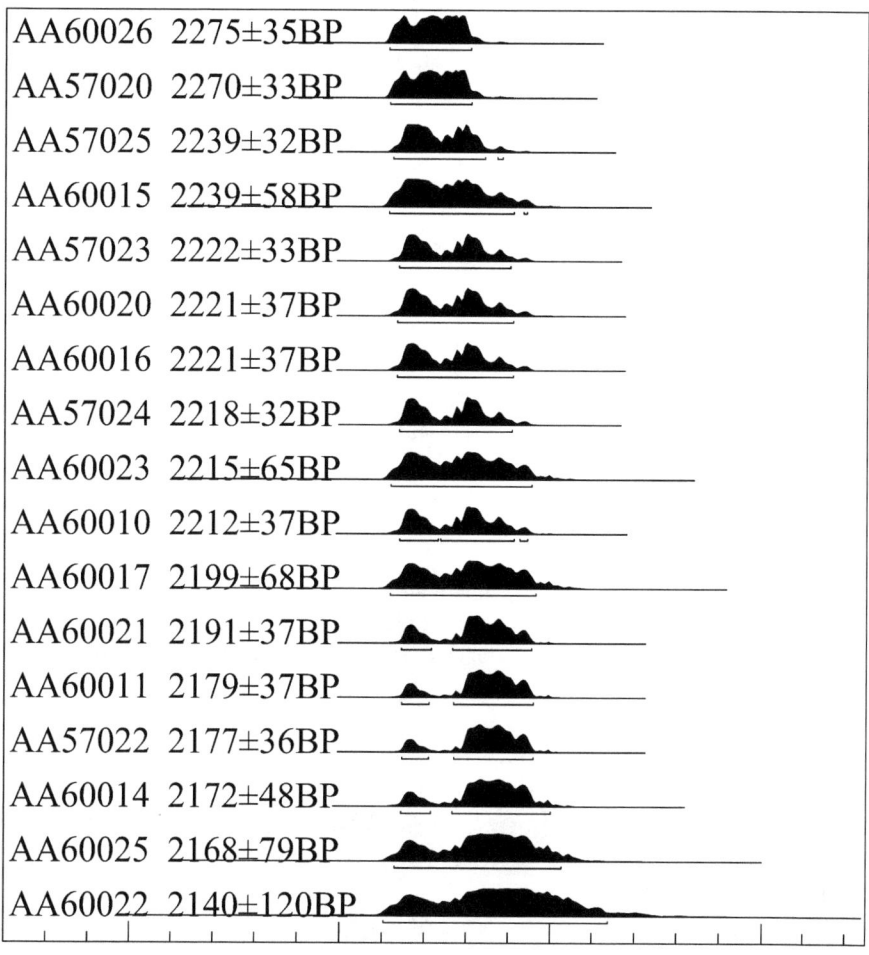

7.2 Calibrated years B.P. date ranges (±SE) for samples from Chankillo, prepared by means of the program OxCal version 3.10[11] with the use of Southern Hemisphere atmospheric data.[12] For each sample, the first column represents the laboratory (NSF-Arizona Accelerator Mass Spectrometry Laboratory) identification number. The shaded area refers to the probability distribution of possible intersection points with the calibration curve, and the horizontal line below represents the 2-sigma calibrated age range. Five dates (AA57020 to AA57025) were sampled following dendrochronological principles from the outer sapwood rings preserved under bark in algarrobo (*Prosopis* sp.) lintels found still plugged into the architecture; these give a firm date for the construction of the site. The rest were obtained from the remains (including seed and fiber) of plants with short life spans. Thus, the "old wood" problem, especially troubling on the coastal desert of Peru, was minimized. CalBC, calibrated years BCE; CalAD, calibrated years CE.

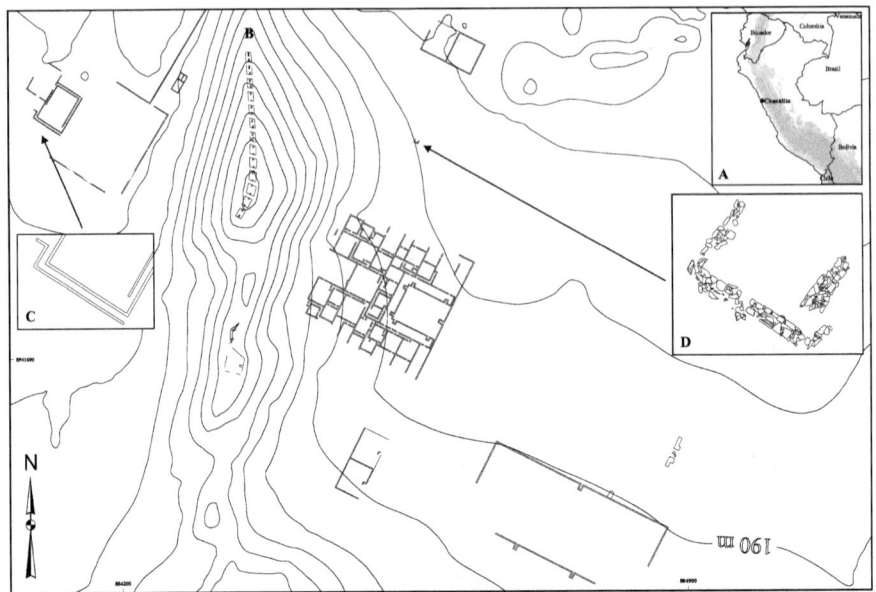

7.3 Plan of the Thirteen Towers and adjacent buildings in Chankillo. (A) Location within Peru. (B) The Thirteen Towers. (C) The external corridor and western observing point. (D) The eastern observing point.

7.4 The Thirteen Towers of Chankillo, as viewed from the fortified temple. Tower 1 is the leftmost tower in the image.

7.5 The fortified temple at Chankillo (Photo courtesy of Servicio Aerofotografico Nacional, Peru).

standing structures and plazas over approximately 4 km² of rock outcrops and sand ramps. It is oriented south of east (azimuth 118°). Its best-known feature is a 300-m-long hilltop structure built in a remote location and heavily fortified with massive walls, restricted gates, and parapets (Fig. 7.5). This famous structure has been discussed often as a fort, a redoubt, or a ceremonial center.[13] However, recent research supports an alternative interpretation as a fortified temple.[14] A lesser-known part of the site is a ceremonial-civic area to the east, which contains buildings, plazas, and storage facilities. The Thirteen Towers form the most outstanding feature within this area: a row of 13 cuboidal constructions placed along the ridge of a low hill (Fig. 7.3b). The towers run north-south, although Towers 11–13 are twisted around slightly towards the southwest. Seen from the buildings and plazas below this hill, on either side, the towers form an artificial toothed horizon with narrow gaps at regular intervals (Fig. 7.4).

The towers are relatively well-preserved; their corners have mostly collapsed, but enough of the original architecture survives to allow reconstruction (Fig. 7.3). They were flat-topped and rectangular to rhomboidal. Their size (75–125 m²) and height (2–6 m) vary widely. Nonetheless, they are regularly spaced: the gaps between the towers vary from 4.7 to 5.1 m. Each tower has a pair of inset staircases leading up to the summit on the north and south sides (Fig. 7.6). Most of the northern staircases

7.6. Oblique view of Tower 1 with excavated northern staircase.

are centered along this side, although not all are aligned with the general orientation of the tower. Most of the southern staircases are often offset toward the east. The staircases are narrow (1.3–1.5 m wide), but because the heights of the towers vary, they are of different lengths (1.3–5.2 m). Most of the tower summits are well preserved; no artifacts remain on these surfaces, though it is clear they were foci of activity.

A group of enclosures is found 200 m to the west of the towers (Fig. 7.3). The southernmost enclosure contains a building composed of two courtyards. The southeast courtyard is 53.6 m long and 36.5 m wide, and is well-preserved. Running along its southern side is a unique construction: a 40 m-long exterior corridor (Fig. 7.3c). The corridor, like the rest of the building, was carefully constructed, plastered, and painted white; however it never led into it. Instead, it connected a doorway on the northwest side, to which access was restricted by a blocking wall, with an opening on the southeast side that directly faced the towers 235 m away. The southeast opening, unlike every other doorway at Chankillo, did not have the typical barholds, or small niches where a pin was firmly tied into the stone masonry and presumably used to attach a wooden door.[15] We infer that the purpose of the corridor was to orchestrate movement from its restricted entryway to a doorless opening directly facing the towers. Considering the original height of the corridor walls, estimated at roughly 2.2 m, only when the opening was reached would there have been an unobstructed view of the full row of towers. Surrounding the opening, at floor level, archaeological excavations have revealed offerings of pottery, shell, and lithics. No other offerings were found associated with openings in excavations elsewhere at the site.[16] This suggests that significant elements of ritual were involved in the process of passing through the corridor and standing at the end of it to observe the towers. Consequently, we term this opening the "western observing point."

To the east of the towers (Fig. 7.3) is a large area (1.4 km^2) with several buildings, including an impressive complex of interconnected patios and rooms, *chicha* (corn beer) storage facilities, and a large plaza (0.16 km^2). In several places within the plaza there were surface offerings of ceramic panpipes and thorny oyster (*Spondylus princeps* sp.), and middens near the plaza contained remains of serving vessels, more ceramic panpipes, and abundant maize remains. This whole area was probably a setting for large ceremonial feasts.

From several locations around this ceremonial area, the Thirteen Towers command the landscape and could be used as solar horizon markers, but one isolated building is of particular interest (Fig. 7.3d). It is a small, isolated building in the middle of a large, open space. Its position in relation to the Thirteen Towers is almost an exact mirror of the western observing point: the two lie almost exactly on an east-west line, are at the same elevation, and are at roughly the same distance from the towers. When viewed from inside this building, the spread of the towers forms an artificial horizon as well.

Only an incomplete outline of a rectangular room, 6 m wide, is preserved. Like the corridor leading to the western observing point on the opposite side of the towers, this room had a doorway—in this case on the southeast side—that was restricted by a small blocking wall. We hypothesize that this structure was the eastern observing point, but the exact position cannot be known with the same certainty as the western observing point.

We determined the locations of the two observing points together with the corners of each tower using hand-held differential GPS equipment. This enabled each point on the false horizon formed by the towers, as viewed from each observing point in turn, to be defined in terms of its azimuth, altitude, and (astronomical) declination (Tables 7.1 and 7.2, for more detail, see Tables 7.3 and 7.4). Independent compass-clinometer determinations of azimuths and altitudes, calibrated using a direct observation of sunrise against the towers, provided consistency checks. By "altitude," we mean the vertical angle between a viewed point and the horizontal plane through the observer, "elevation" being the height of a location above sea level.[17]

Declinations of +23.75° and −23.75° correspond to the center of the sun at the extreme positions of sunrise and sunset in 300 BC, at the June and December solstices respectively, with the sun's disc extending between +23.5° and +24.0° (June) and between −24.0° and −23.5° (December).[18] Intermediate declinations correspond to sunrise and sunset on other dates.

As viewed from the two observing points, the spread of the towers along the horizon corresponds remarkably closely to the range of movement of the rising and setting positions of the sun over the year. This in itself argues strongly that the towers were used for solar observation. From the western observing point, the southern slopes of Cerro Mucho Malo, at a distance of 3 km, meet the nearer horizon (formed by the nearby hill on which the towers are constructed) just to the left of the northernmost tower, Tower 1, providing a thirteenth "gap" of similar width to those between each pair of adjacent towers down the line.

From the eastern observing point, the southernmost tower, Tower 13, would not have been visible at all, and the top of Tower 12 would only just have been visible (it is only partially visible now in its ruinous condition). From here, the December solstice sun would have been seen to set behind the left side of the southernmost visible tower, Tower 12, while the June solstice sun set directly to the right of the northernmost tower, Tower 1 (Fig. 7.7). In either case, once the sun had begun to move significantly away from either of its extreme rising positions a few days after each solstice, the various towers and gaps would have provided a means to track the progress of the sun up and down the horizon to within an accuracy of two or three days.

If we accept that the towers were used as foresights for solar observations, then does their disposition suggest anything about the way the year might have been broken down? The flat tops of the towers originally formed their own smooth, "false" horizon, their varying heights compensating to some extent for the slope of the hill on which they were built. This false horizon was broken at intervals by deep, narrow cuts formed by the gaps between the towers. When viewed from the western observing point, the sun rose for just one or two days in each gap. One possibility, then, is that critical sunrises were observed in the gaps. However, the regularity of the gaps argues against this, suggesting instead that the year was divided into regular intervals. The

Table 7.1. The annual movement of sunrise against the line of towers as viewed from the western observing point.

Observed point	Declination	Notes	In 300 BC, the center of the sun rose here: on [Greg date]	and [Greg date]	Dates of sunrise in the gaps	Interval (days)		Interval (days)
Cerro Mucho Malo (point on slope)	23.7	June solstice sunrise (dec +23.75°)	26-Jun*	16-Jun*				
Cerro Mucho Malo meets local horizon	22.1		14-Jul	31-May				
					15-Jul		30-May	
Tower 1 center top	20.2		24-Jul	21-May		16		16
					31-Jul		14-May	
Tower 2 center top	16.8		07-Aug	07-May		12		12
					12-Aug		02-May	
Tower 3 center top	13.4		18-Aug	26-Apr		11		12
					23-Aug		20-Apr	
Tower 4 center top	9.8	Antizenith sunrise (dec +9.6°)	28-Aug	15-Apr		10		10
					02-Sep		10-Apr	
Tower 5 center top	6.2		07-Sep	05-Apr		9		10
					11-Sep		31-Mar	
Tower 6 center top	2.0	"Temporal equinox" sunrise (dec between +0.6 and +1.0°)	17-Sep	25-Mar		11		11
					22-Sep		20-Mar	
Tower 7 center top	-1.8		27-Sep	15-Mar		9		9
					01-Oct		11-Mar	
Tower 8 center top	-5.4	Zenith sunrise (dec −6.9°)	06-Oct	06-Mar		10		10
					11-Oct		01-Mar	
Tower 9 center top	-9.1		15-Oct	24-Feb		9		10
					20-Oct		19-Feb	
Tower 10 center top	-12.8		*25-Oct*	14-Feb		10		10
					30-Oct		09-Feb	
Tower 11 center top	-16.6		06-Nov	02-Feb		10		10
					12-Nov		27-Jan	
Tower 12 center top	-20.3		20-Nov	19-Jan		16		17
					28-Nov		10-Jan	
Tower 13 center top	-23.7	December solstice sunrise (dec −23.75°)	15-Dec†	25-Dec†				

* The sun will also rise at this position on any date during the period 16 Jun to 26 Jun
† The sun will also rise at this position on any date during the period 15 Dec to 25 Dec

This table lists the declination of the center top of each tower, and the gap between each pair of towers. The following columns show the dates in the year (using the Gregorian calendar, taking the June solstice as June 21) when the sun would have risen at the point in question, and the intervals (numbers of days) between sunrises in successive gaps. The quoted dates and the intervals are only accurate to within ± 1 day at best, with larger errors possible near the solstices, where the daily change in the sunrise position is extremely small.

Table 7.2. The annual movement of sunset against the line of towers as viewed from the eastern observing point.

Observed point	Declination	Notes	In 300 BC, the center of the sun set here: on [Greg date]	and [Greg date]	Dates of sunset in the gaps		Interval (days)	Interval (days)
Tower 12 center top	-23.0	December solstice sunset (dec −23.75°)	06-Dec	02-Jan				
Tower 11 center top	-20.0		18-Nov	20-Jan				
Tower 10 center top	-15.9		03-Nov	04-Feb				
GAP	-13.7				28-Oct	11-Feb		
Tower 9 center top	-11.5		21-Oct	18-Feb			12	12
GAP	-9.5	Zenith sunset (dec −9.6°)			16-Oct	23-Feb		
Tower 8 center top	-7.3		10-Oct	01-Mar			12	12
GAP	-4.9				04-Oct	07-Mar		
Tower 7 center top	-2.9		29-Sep	13-Mar			11	12
GAP	-0.5				23-Sep	19-Mar		
Tower 6 center top	1.5	"Temporal equinox" sunset (dec between +0.6 and +1.0°)	18-Sep	24-Mar			11	11
GAP	4.1				12-Sep	30-Mar		
Tower 5 center top	6.4		06-Sep	06-Apr			11	12
GAP	8.3				01-Sep	11-Apr		
Tower 4 center top	10.3	Antizenith sunset (dec +9.6°)	27-Aug	16-Apr			11	12
GAP	12.4				21-Aug	23-Apr		
Tower 3 center top	14.4		15-Aug	29-Apr			12	12
GAP	16.3				09-Aug	05-May		
Tower 2 center top	18.2		02-Aug	12-May			16	16
GAP	20.3				24-Jul	21-May		
Tower 1 center top	22.1		14-Jul	31-May				
Tower 1 top right	23.7	June solstice sunset (dec +23.75°)	26-Jun*	16-Jun*				

* The sun will also set at this position on any date during the period 16 Jun to 26 Jun

The details are similar to Table 7.1. Note that Tower 13 was not visible, and that no gaps were visible between Towers 12 and 11 or between Towers 11 and 10.

sunrises in the gaps between the central towers, Towers 3 to 11, were all separated by time intervals of (or close to) 10 days, implying that a 10-day interval may have been a feature of the solar calendar. However, the time intervals are longer between the outer towers in the line, where the sunrise moves along more slowly. Furthermore, the situation is different from the eastern observing point (Fig. 7.8), since no gaps would have been visible between the southernmost towers in the line as far as Tower 10 (and

7.7 The Thirteen Towers as viewed from the western observing point, annotated with the positions of sunrise at the solstices, equinoxes, and the dates of zenith and antizenith passage in c. 300 BCE. Tower 1 is the leftmost tower in the image.

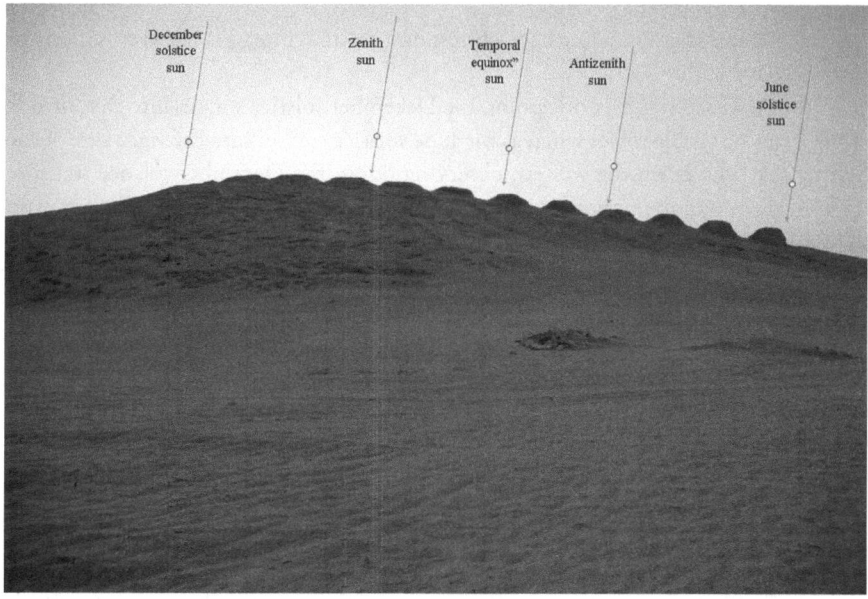

7.8 The Thirteen Towers as viewed from the eastern observing point, annotated with the positions of sunset at the solstices, equinoxes, and the dates of zenith and antizenith passage in c. 300 BCE. Tower 1 is the rightmost tower in the image.

7.9 June solstice sunrise between Cerro Mucho Malo and Tower 1, as viewed from the western observing point today. The position of the June solstitial sunrise has shifted rightwards by about 0.3° since 300 BC.

possibly 9), and the remaining gaps correspond to time intervals between sunsets of 11 or 12 days (Table 7.2).

From the eastern observing point, the December solstice sun set into the left side of the leftmost visible tower whereas the June solstice sun set into the right side of the rightmost tower. From the western observing point, the December solstice sun rose up from the top of the rightmost tower while the June solstice sun rose a little way up the slopes of Cerro Mucho Malo (Fig. 7.9). There is an evident symmetry here also, suggesting that this natural hill was perceived as the leftmost "tower" in this profile. Midwinter would have been the one time of year when the sun was seen to emerge from a natural hill rather than a human construction.

Equinoctial sunrise (declination 0.0°) occurred in the central gap directly between Towers 6 and 7. If Cerro Mucho Malo is included, so that there are thirteen gaps, then this is the central one. In the other direction, equinoctial sunset occurred just to the right of this same gap, which seen from the east is the central gap within the twelve visible towers. However, the applicability of the concept of the equinox outside a Western conceptual framework is highly questionable.[19] If, as here, there is clear evidence that a mechanism existed to help count off the days, then the mid-days between the solstices (the "temporal equinoxes" or "Thom equinoxes") are more likely to have been significant. In 300 BC, the sun's declination on these days was between +0.6° and +1.0°, and there is no evidence that these days were specially marked.

Table 7.3. The annual movement of sunrise against the line of towers as viewed from the western observing point.

Observed point		Az (true)	Alt	Dec	Notes	In 300 BC, the center of the sun rose here: on [Greg date] June solstice:		and [Greg date] June solstice:		Dates of sunrise in the gaps	Interval (days)		Interval (days)
Cerro Mucho Malo	Point on horizon profile	64.0	6.6	24.2									
Cerro Mucho Malo	Point on horizon profile	64.3	6.3	24.0									
Cerro Mucho Malo	Point on horizon profile	64.4	6.2	23.9									
Cerro Mucho Malo	Point on horizon profile	64.7	6.1	23.7	June solstice sunrise (dec +23.75°)	26-Jun*	5	16-Jun*	360				
Cerro Mucho Malo	Point on horizon profile	65.0	6.0	23.4		02-Jul	11	12-Jun	356				
Cerro Mucho Malo	Meets local horizon	66.5	5.25	22.1		14-Jul	23	31-May	344				
										15-Jul		30-May	
Tower 1 left	Base	66.8	5.5	21.8		16-Jul	25	29-May	342				
Tower 1 left	Top	66.8	6.5	21.5		18-Jul	27	28-May	341				
Tower 1 center	Top	68.2	6.75	20.2		24-Jul	33	21-May	334		16		16
Tower 1 right	Top	69.5	6.75	18.9		30-Jul	39	15-May	328				
Tower 1 right	Base	69.5	6.0	19.0		30-Jul	39	15-May	328				
										31-Jul		14-May	
Tower 2 left	Base	70.2	6.25	18.3		01-Aug	41	13-May	326				
Tower 2 left	Top	70.2	7.0	18.1		02-Aug	42	12-May	325				
Tower 2 center	Top	71.5	7.25	16.8		07-Aug	47	07-May	320		12		12
Tower 2 right	Top	72.8	7.25	15.6		11-Aug	51	03-May	316				
Tower 2 right	Base	72.8	6.75	15.7		11-Aug	51	03-May	316				
										12-Aug		02-May	
Tower 3 left	Base	73.5	7.0	14.9		13-Aug	53	30-Apr	313				
Tower 3 left	Top	73.5	7.5	14.8		13-Aug	53	30-Apr	313				
Tower 3 center	Top	75.0	7.50	13.4		18-Aug	58	26-Apr	309		11		12
Tower 3 right	Top	76.4	7.75	12.0		22-Aug	62	21-Apr	304				
Tower 3 right	Base	76.4	7.25	12.1		22-Aug	62	22-Apr	305				
										23-Aug		20-Apr	
Tower 4 left	Base	77.2	7.5	11.2		24-Aug	64	19-Apr	302				
Tower 4 left	Top	77.2	8.0	11.1		24-Aug	64	19-Apr	302				
Tower 4 center	Top	78.6	8.0	9.8	Antizenith sunrise (dec +9.6°)	28-Aug	68	15-Apr	298		10		10
Tower 4 right	Top	80.1	8.25	8.3		01-Sep	72	11-Apr	294				
Tower 4 right	Base	80.1	7.75	8.4		01-Sep	72	11-Apr	294				
										02-Sep		10-Apr	
Tower 5 left	Base	80.9	7.75	7.6		03-Sep	74	09-Apr	292				
Tower 5 left	Top	80.9	8.50	7.5		03-Sep	74	09-Apr	292				
Tower 5 center	Top	82.1	8.75	6.2		07-Sep	78	05-Apr	288		9		10
Tower 5 right	Top	83.3	8.75	5.1		09-Sep	80	02-Apr	285				
Tower 5 right	Base	83.3	8.25	5.2		09-Sep	80	03-Apr	286				
										11-Sep		31-Mar	
Tower 6 left	Base	84.9	8.5	3.6		13-Sep	84	29-Mar	281				
Tower 6 left	Top	84.9	9.0	3.5		13-Sep	84	29-Mar	281				
Tower 6 center	Top	86.4	9.0	2.0	"Temporal equinox" sunrise (dec between +0.6 and +1.0°)	17-Sep	88	25-Mar	277		11		11
Tower 6 right	Top	87.9	9.25	0.5		21-Sep	92	21-Mar	273				
Tower 6 right	Base	87.9	8.75	0.6		21-Sep	92	22-Mar	274				
										22-Sep		20-Mar	
Tower 7 left	Base	88.8	9.0	-0.3		23-Sep	94	19-Mar	271				
Tower 7 left	Top	88.8	9.5	-0.4		23-Sep	94	19-Mar	271				
Tower 7 center	Top	90.2	9.5	-1.8		27-Sep	98	15-Mar	267		9		9
Tower 7 right	Top	91.7	9.75	-3.3		30-Sep	101	12-Mar	264				
Tower 7 right	Base	91.7	9.25	-3.2		30-Sep	101	12-Mar	264				
										01-Oct		11-Mar	
Tower 8 left	Base	92.2	9.25	-3.7		01-Oct	102	11-Mar	263				
Tower 8 left	Top	92.2	9.75	-3.8		02-Oct	103	10-Mar	262				
Tower 8 center	Top	93.9	10.0	-5.4	Zenith sunrise (dec −6.9°)	06-Oct	107	06-Mar	258		10		10
Tower 8 right	Top	95.6	10.0	-7.1		10-Oct	111	02-Mar	254				
Tower 8 right	Base	95.6	9.5	-7.0		10-Oct	111	02-Mar	254				
										11-Oct		01-Mar	
Tower 9 left	Base	96.1	9.75	-7.5		11-Oct	112	01-Mar	253				
Tower 9 left	Top	96.1	10.25	-7.6		11-Oct	112	28-Feb	252				
Tower 9 center	Top	97.6	10.25	-9.1		15-Oct	116	24-Feb	248		9		10
Tower 9 right	Top	99.1	10.5	-10.6		19-Oct	120	20-Feb	244				
Tower 9 right	Base	99.1	10.0	-10.5		19-Oct	120	21-Feb	245				
										20-Oct		19-Feb	
Tower 10 left	Base	100.1	10.0	-11.5		21-Oct	122	18-Feb	242				
Tower 10 left	Top	100.1	10.5	-11.6		22-Oct	123	17-Feb	241				

The data given for the sides and center-top of each tower consist of the (true) azimuth, altitude, and declination, quoted to the nearest 0.1°. The following columns show the dates in the year (using the Gregorian calendar, taking the June solstice as June 21) when the sun would have risen behind the point in question, and the dates when it would have risen in the gaps between each pair of towers. The quoted dates, and the quoted intervals (numbers of days) between them, are only accurate to within ± 1 day at best, with larger errors possible near the solstices, where the daily change in the sunrise position is extremely small.

Table 7.4. The annual movement of sunset against the line of towers as viewed from the eastern observing point.

Observed point		Az (true)	Alt	Dec	Notes	In 300 BC, the center of the sun set here: on [Greg date] June solstice:	and [Greg date] # days after June solstice:		Dates of sunset in the gaps			
										Interval (days)		Interval (days)
Tower 12 left	Base	246.6	11.75	-24.7	INVISIBLE							
Tower 12 left	Top	246.6	12.25	-24.7	JUST VISIBLE WHEN COMPLETE? December solstice sunset (dec –23.75°)							
Tower 12 center	Top	248.4	12.5	-23.0		06-Dec	168	02-Jan	195			
Tower 12 right	Top	250.2	12.5	-21.2	HIDDEN BEHIND TOWER 11							
Tower 12 right	Base	250.2	12.0	-21.2	HIDDEN BEHIND TOWER 11							
Tower 11 left	Base	249.2	12.0	-22.1		29-Nov	161	09-Jan	202			
Tower 11 left	Top	249.2	12.75	-22.2		30-Nov	162	09-Jan	202			
Tower 11 center	Top	251.5	12.75	-20.0		18-Nov	150	20-Jan	213			
Tower 11 right	Top	253.9	12.75	-17.7	HIDDEN BEHIND TOWER 10							
Tower 11 right	Base	253.9	12.25	-17.6	HIDDEN BEHIND TOWER 10							
Tower 10 left	Base	253.7	12.25	-17.8		10-Nov	142	29-Jan	222			
Tower 10 left	Top	253.7	12.75	-17.9		10-Nov	142	28-Jan	221			
Tower 10 center	Top	255.7	12.75	-15.9		03-Nov	135	04-Feb	228			
Tower 10 right	Top	257.7	12.75	-14.0		29-Oct	130	10-Feb	234			
Tower 10 right	Base	257.7	12.0	-13.9		28-Oct	129	11-Feb	235			
										28-Oct		11-Feb
Tower 9 left	Base	258.1	12.0	-13.5		27-Oct	128	12-Feb	236			
Tower 9 left	Top	258.1	12.5	-13.6		27-Oct	128	12-Feb	236			
Tower 9 center	Top	260.2	12.5	-11.5	Zenith sunset (dec –9.6°)	21-Oct	122	18-Feb	242	12		12
Tower 9 right	Top	262.3	12.5	-9.5		16-Oct	117	23-Feb	247			
Tower 9 right	Base	262.3	11.75	-9.4		16-Oct	117	24-Feb	248			
										16-Oct		23-Feb
Tower 8 left	Base	262.3	11.75	-9.4		16-Oct	117	24-Feb	248			
Tower 8 left	Top	262.3	12.25	-9.5		16-Oct	117	23-Feb	247			
Tower 8 center	Top	264.5	12.25	-7.3		10-Oct	111	01-Mar	253	12		12
Tower 8 right	Top	266.7	12.25	-5.2		05-Oct	106	07-Mar	259			
Tower 8 right	Base	266.7	11.5	-5.1		05-Oct	106	07-Mar	259			
										04-Oct		07-Mar
Tower 7 left	Base	267.2	11.25	-4.6		04-Oct	105	08-Mar	260			
Tower 7 left	Top	267.2	11.75	-4.6		04-Oct	105	08-Mar	260			
Tower 7 center	Top	269.0	11.75	-2.9		29-Sep	100	13-Mar	265	11		12
Tower 7 right	Top	270.7	11.5	-1.2		25-Sep	96	17-Mar	269			
Tower 7 right	Base	270.7	11.0	-1.1		25-Sep	96	17-Mar	269			
										23-Sep		19-Mar
Tower 6 left	Base	271.9	10.75	0.1		22-Sep	93	20-Mar	272			
Tower 6 left	Top	271.9	11.25	0.0	"Temporal equinox" sunset (dec between +0.6 and +1.0°)	22-Sep	93	20-Mar	272			
Tower 6 center	Top	273.5	11.25	1.5		18-Sep	89	24-Mar	276	11		11
Tower 6 right	Top	275.1	11.25	3.1		14-Sep	85	28-Mar	280			
Tower 6 right	Base	275.1	10.5	3.2		14-Sep	85	28-Mar	280			
										12-Sep		30-Mar
Tower 5 left	Base	277.0	10.25	5.1		09-Sep	80	02-Apr	285			
Tower 5 left	Top	277.0	10.75	5.0		10-Sep	81	02-Apr	285			
Tower 5 center	Top	278.5	10.75	6.4		06-Sep	77	06-Apr	289	11		12
Tower 5 right	Top	279.9	10.75	7.8		02-Sep	73	10-Apr	293			
Tower 5 right	Base	279.9	10.0	7.9		02-Sep	73	10-Apr	293			
										01-Sep		11-Apr
Tower 4 left	Base	280.7	9.5	8.8		31-Aug	71	12-Apr	295			
Tower 4 left	Top	280.7	10.25	8.7	Antizenith sunset (dec +9.6°)	31-Aug	71	12-Apr	295			
Tower 4 center	Top	282.4	10.25	10.3		27-Aug	67	16-Apr	299	11		12
Tower 4 right	Top	284.1	10.0	12.0		22-Aug	62	21-Apr	304			
Tower 4 right	Base	284.1	9.25	12.1		22-Aug	62	22-Apr	305			
										21-Aug		23-Apr
Tower 3 left	Base	284.9	9.25	12.9		19-Aug	59	24-Apr	307			

The details are similar to Table 7.3. Note that Tower 13 was not visible, and that no gaps were visible between Towers 12 and 11 or between Towers 11 and 10.

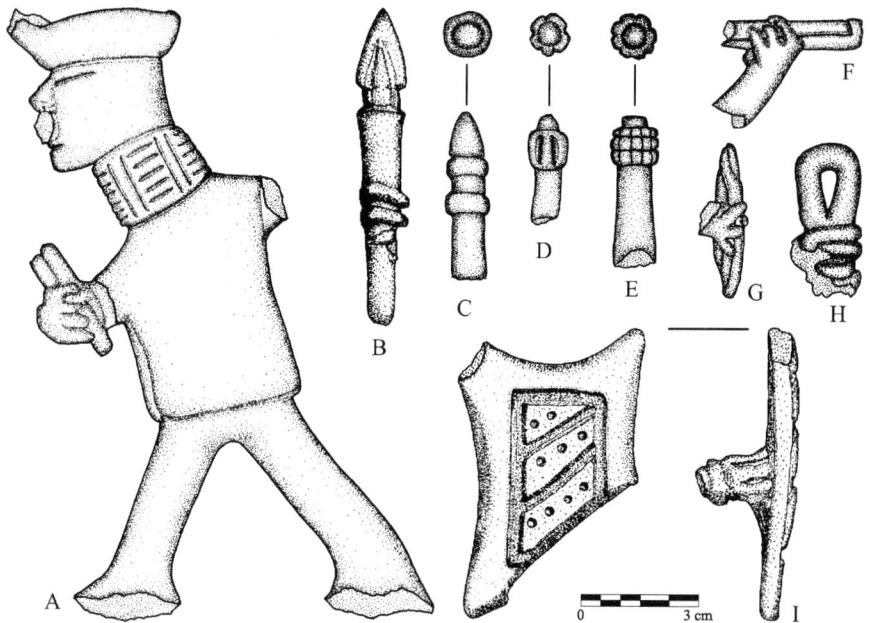

7.10 (A) Warrior ceramic figurine. Weapon types found at Chankillo: (B) spear; (C), (D) and (E) clubs; (F) spear-thrower; (G) darts; (H) sling; (I) shield.

A variety of evidence suggests that the date of solar zenith passage was significant to early cultures in the American tropics in general and in the Andes in particular.[20] It has also been suggested that the dates of solar antizenith passage might have been of significance in Inca Cusco,[21] although this idea has been debated.[22] However, there is nothing in the pattern of disposition of the towers to suggest that it was deliberately preconceived in relation to sunrise or sunset on these dates. Only zenith passage sunset falls close to (and even then, not exactly within) a gap between two towers.

Astronomical "explanations" can be fitted notoriously easily to preexisting alignments. Repeated instances of solar and lunar alignments can provide strong evidence of intentionality, as among many local groups of later prehistoric tombs and temples in Britain, Ireland and Europe.[23] However, at a unique site there is always a danger of circular argument if the judgment of what might have been significant to people in the past is made solely on the basis of the alignment evidence itself. Fortuitous stellar alignments are particularly likely, given the number of stars in the sky and the fact that their positions change steadily over the centuries owing to precession. The Chankillo towers, on the other hand, just span (to within a couple of degrees) the solar rising and setting arcs as seen from two observing points, each clearly defined by a unique structure with no other apparent purpose. Thus we are not selecting putative astronomical targets from innumerable possibilities but seeing direct indications of all four solstitial rising and setting points—astronomical "targets" whose broad significance across cultures is self-evident and widely attested.

It is uncontroversial to postulate direct observations of the annual movement of the rising or setting sun along the horizon for the purposes of regulating seasonal events such as religious festivals, or for maintaining a seasonal calendar. Nonetheless, evaluating the nature of the observations made and the social and ritual context within which they operated and derived their relevance is not simple. This point is well illustrated by recent debates concerning the function of the so-called E-group structures in the Mayan heartlands of the Peten, Guatemala.[24] In the case of the Thirteen Towers and nearby plazas, we can infer, they provided a setting for people participating in public rituals and feasts directly linked to the observation and interpretation of the seasonal passage of the sun. By contrast, the observing points themselves appear to have been highly restricted. Individuals with the status to access them and conduct ceremonies would have had the power to regulate time, ideology, and the rituals that bound this society together. Additionally, the excavations at Chankillo have uncovered ceramic warrior figurines holding a great variety of offensive (and defensive) weapons (Fig. 7.10).[25] The figurines wear signs of distinction, such as headdresses, shirts, and especially neck, chest, and nose ornaments. The artistic representation of these warriors, holding specialized weapons and wearing the symbols of their high status, indicates the possible rise of a class of war leaders and the centralization of power and authority in the hands of a few. Thus, sun worship and related cosmological beliefs at Chankillo could have helped to legitimize the authority of an elite, just as it did within the Inca empire two millennia later. And this, in its turn, implies that the towers were not a simple instrument for solar observation but the monumental expression of existing—and therefore by implication even older—knowledge.

There is increasing evidence that the sun cult, which as the official cult of the Inca empire, regulated calendrical ceremonies and supported the established social hierarchy, had precursors. For example, historically attested sunrise ceremonies at a sanctuary on the Island of the Sun in Lake Titicaca,[26] surrounding a crag regarded as the origin place of the sun, almost certainly had pre-Incaic roots.[27] Given the similarity between the solar observation device at Chankillo and the Cusco pillars documented some two millennia later,[28] it seems likely that similar practices were common within many of the great states that developed in the Andes prior to, as well as including, the Inca empire.

NOTES

We thank the numerous archaeologists and volunteers who participated in the Chankillo project, and especially J. L. Pino. We thank Yale University, Pontificia Universidad Católica del Peru, National Science Foundation, Wenner-Gren Foundation, The Field Museum, Schwerin Foundation, and Earthwatch Institute for support. R. Towner and K. Anchukaitis were instrumental in securing five samples for dendrochronological dating. The NSF funded all AMS radiocarbon dates. We thank the Asociación Cultural Peruano Británica in Lima, Peru, for logistical and financial support.

1. C.L.N. Ruggles in *Archaeology: The Key Concepts*, ed. A. C. Renfrew and P. G. Bahn (London and New York: Routledge, 2005), 11–16; C.L.N. Ruggles, *Ancient Astronomy: An Encyclopedia of Cosmologies and Myth* (Santa Barbara: ABC-CLIO, 2005).

2. A. F. Aveni, *Skywatchers* (Austin: University of Texas Press, 2001), 55–67.

3. Not all accurate sky-based seasonal calendars rely upon horizon observations of the sun: one exception is the traditional calendar of the Borana of Ethiopia and Kenya (M. Bassi, *Current Anthropology* 29 [1988]:619), which is luni-stellar.

4. S. C. McCluskey, *Journal for the History of Astronomy*, 8 (1977):174.

5. A. F. Aveni and H. Hartung, *Transactions of the American Philosophical Society* 76:7 (1986):1.

6. I. Šprajc, *Orientaciones Astronómicas en la Arquitectura Prehispánica de México* (México DF: Instituto Nacional de Antropología e Historia, 2001).

7. M. S. Ziótkowski and R. M. Sadowski, eds., *Time and Calendars in the Inca Empire* (Oxford: BAR International Series 479, 1989); B. S Bauer and D.S.P. Dearborn, *Astronomy and Empire in the Ancient Andes* (Austin: University of Texas Press, 1995).

8. R. T. Zuidema, *The Ceque System of Cuzco: The Social Organization of the Capital of the Inca* (Leiden: Brill, 1964); A. F. Aveni, *Stairways to the Stars* (New York: Wiley, 1997), 147–176; B. S. Bauer, *The Sacred Landscape of the Inca: The Cusco Ceque System* (Austin: University of Texas Press, 1998).

9. Bauer and Dearborn, *Astronomy and Empire in the Ancient Andes*, 67–100.

10. I. Ghezzi in *Andean Archaeology III: North and South*, ed. W. Isbell and H. Silverman (New York: Springer, 2006), 67–84.

11. C. Bronk Ramsey, *Radiocarbon* 37, 425 (1995); C. Bronk Ramsey, *Radiocarbon* 43, 355 (2001).

12. F. G. McCormac et al., *Radiocarbon* 46, 1087 (2004).

13. J. R. Topic and T. L. Topic in *Arqueologia, Antropologia e Historia en los Andes: Homenaje a Maria Rostworowski*, ed. R. Varon and J. Flores (Lima: IEP, 1997), 567–590.

14. Ghezzi in *Andean Archaeology III*, 67–84.

15. I. Ghezzi, *Proyecto Arqueológico Chankillo: Informe de la Temporada 2003* (Lima: Instituto Nacional de Cultura, 2004).

16. Ibid.

17. C.L.N. Ruggles, *Astronomy in Prehistoric Britain and Ireland* (New Haven: Yale University Press, 1999), ix.

18. Ibid., 18, 24, 57.

19. C.L.N. Ruggles, *Archaeoastronomy* 22 (supplement to *Journal for the History of Astronomy* 28) (1997):S45.

20. A. F. Aveni, *Skywatchers*, 40–46, 265–269.

21. R. T. Zuidema in *Archaeoastronomy in the Americas*, ed. R. A. Williamson (Los Altos: Ballena Press, 1981), 319–342.

22. B. S. Bauer and D.S.P. Dearborn, *Astronomy and Empire in the Ancient Andes*, 94–98.

23. C.L.N. Ruggles, *Astronomy in Prehistoric Britain and Ireland*, 91–111; M. A. Hoskin, *Tombs, Temples and Their Orientations* (Bognor Regis: Ocarina Books, 2001).

24. A. F. Aveni and H. Hartung in *World Archaeastronomy*, ed. A. F. Aveni (Cambridge: Cambridge University Press, 1989), 441–461; A. F. Aveni, A. S. Dowd, and B. Vining, *Latin American Antiquity* 14 (2003):159; G. R. Aylesworth, *Archaeoastronomy: The Journal of Astronomy in Culture* 18 (2004):34.

25. Ghezzi in *Andean Archaeology III*, 67–84.

26. B. S. Bauer and C. Stanish, *Ritual and Pilgrimage in the Ancient Andes: The Islands of the Sun and the Moon* (Austin: University of Texas Press, 2001).

27. D.S.P. Dearborn, M. T. Seddon, and B. S. Bauer, *Latin American Antiquity* 9 (1998): 240.

28. Bauer, *Sacred Landscape*.

CHAPTER EIGHT

Keeping the Sacred and Planting Calendar
Archaeoastronomy in the Pueblo Southwest

Michael Zeilik

INTRODUCTION

The US Southwest defines a cultural area that is large in space and deep in time. During the 1000 years before the *entrada* of the Spanish, the Southwest harbored numerous, interacting cultures with possible connections to prehistoric Mesoamerica. Those cultures included the Patayan, Hohokam, Mogollon, and Anasazi. Archaeoastronomy in the Southwest aims to infer the content and the impact of astronomy in these prehistoric societies. This region has a great historical advantage—the people now living there have a cultural descent from prehistory. That cultural connection runs especially deep with the historic Pueblo people directly from their immediate protohistoric ancestors and on to their Anasazi (and Mogollon) ancestors who lived throughout the northern Southwest in the broad Anasazi region (now known as the Four Corners area).

From *World Archaeoastronomy*, ed. A. Aveni (New York: Cambridge University Press, 1989).

I will focus in this paper on the Anasazi, in part because of the rich cultural heritage of the present-day Pueblos (revealed in the ethnographic record) and in part because of the rich archaeological tradition of the Anasazi region. Here the past comes alive when a Turtle Dance takes place at San Juan Pueblo—perhaps a displaced winter solstice dance (A. Ortiz, 1985, private communication). I can easily imagine a similar dance—different in detail but with the same ritual intentions—in the sunny plaza of a prehistoric pueblo. That image contains the essence of the oft-maligned argument by analogy that I will use in a conservative way: as a culturally appropriate source of hypotheses for testing in a prehistoric context (Binford, 1972; Reyman, 1975). I will presume that those Puebloan concepts about astronomy that are most frequent within the diversity of the historic Pueblos mark cultural traditions that are the deepest in time. Specific cultural discontinuities certainly *did* occur between the past and the present—but aspects essential to survival and adaptation to the high desert stayed mostly intact. In particular, the sacred and planting calendars, kept by observations of the sun and the moon, control the life of the historic Pueblos—an essential integration of astronomy, agriculture, and ritual. The same was likely an astronomical practice of the Anasazi. As Cordell (1984, pp. 271–2) has put it:

> Agricultural people throughout the world are concerned with calendrical observations. These are crucial to the time of planting and harvesting outside of equatorial regions and to water management where supplemental watering is necessary to insure the success of crops. In Chaco Canyon, and in the San Juan as a whole, short growing seasons and inadequate rainfall are problems for agriculturists, so it is likely that *accurate* astronomical observations were important to the Chaco Anasazi [author's italics].

Overall, I find ethnographic data to be a mind-opening, wondrous window on Pueblo life. Ethnographic analogy, properly used, can flesh out the old bones of prehistoric ruins. Ethnographic analogy does not straitjacket our development of testable hypotheses. To the contrary, it illuminates rather than retards our insights to Anasazi life.

We shall consider three points about methods for practising archaeoastronomers. First, we must respect the archaeological record and the interpretations derived from it. We should work with an archaeological awareness for every site and culture and even generate hypotheses that the archaeologists can test for us! While we do so, we must remember that a major problem for the archaeologist is the *control of time* at a site; dating information might not be there or it has a large uncertainty. Second, I feel arguments for astronomical aspects of Anasazi life for which no ethnographic analogy exists have an additional burden of proof placed upon them. To demonstrate that a site "works" astronomically is not enough; it should "work" culturally, also. This means marshalling evidence from different aspects of the archaeological record so that any astronomical proposition makes cultural sense. Third, we need to involve expertise from diverse disciplines: art, art history, archaeology, anthropology, astronomy, ethnology, folklore, and history. This strategy requires that we form multiple hypotheses, involve multiple investigators (to avoid observer bias), and following multiple lines of evidence—for we know from the historic Pueblo example that cultural artifacts have multiple uses and meanings. This multivariate approach (Young, 1987b), however, must be conducted in the appropriate cultural context.

Finally, we must ask how *valid* our conclusions are, given our methods and ethnographic base from the historic Pueblos. Hedges has made some wise observations in this regard with reference to California archaeoastronomy. His points (Hedges, 1985, pp. 35–7) also hold true, in my opinion, for the Pueblo Southwest, especially his plea that "we must develop a coherent, logically consistent line of reasoning in the interpretation of archaeoastronomical sites" (p. 35).

ASTRONOMY IN THE HISTORIC PUEBLOS

We owe a dubious debt to ethnographers who visited the Pueblos before the economic encroachment of Anglo society in the 1930s. These people—Jesse W. Fewkes, Frank H. Cushing, John G. Bourke, Alexander M. Stephen, Matilda C. Stevenson, Elsie C. Parsons, and Edward S. Curtis—left us detailed records of many aspects of Pueblo life. So many and so detailed are they that the astronomical unity of the calendar gets lost—in part, because it was so transparent to the Pueblo people. (See *Sun Chief* [Simmons, 1942] and *Pueblo Indian Journal* [Parsons, 1939], for wonderful examples of the "right time" for religious and planting activities.)

Few attempts have been made to gather this material in one place and make sense of the often contradicting pieces. Reyman (1971, pp. 114–22) made a substantial effort for his pioneering study of kiva orientations at Mesa Verde and Chaco in the context of the influence of Mexican ceremonialism on the prehistoric Southwest. (His work influenced that of Williamson *et al.*, 1975, for astronomical alignments at Chaco.) Ellis (1975) revealed her intimate association with Rio Grande ceremonialism in an attempt by ethnographic analogy to place the hypothetical "AD 1054 supernova" pictograph at Chaco (Brandt *et al.*, 1975; Brandt and Williamson, 1977) in the context of the Pueblo calendar watch. For Hopi First Mesa, McCluskey (1977, 1981, 1982) broke new ground in a historical understanding of the synchronization of solar- and lunar-timed ceremonies. M. Jane Young (1987a) and Barbara Tedlock (1983) have given insights about Zuni from their own experiences there. (See also Young and Williamson, 1981, for information on Zuni constellations.) Feeling the need for coherent, comprehensive information, I have gathered material from 19 historic Pueblos on sun- and moon-watching for calendrical purposes (Zeilik, 1983b, 1985a, 1986a) and am continuing work on miscellaneous sky watching (planets, stars, comets, and meteors).

The prehistoric observation of and possible rock art record (Miller, 1955) about the 1054 supernova (which formed the supernova remnant now called the Crab Nebula in the constellation of Taurus the Bull) played a key role in sparking interest in Southwestern archaeoastronomy—at least among astronomers. The pictograph in Chaco on the canyon wall below Penasco Blanco (Fig. 8.1) is one of the better known

8.1 A portion of the rock art panel, below Penasco Blanco in Chaco Canyon, which may or may not represent the AD 1054 supernova in conjunction with a waning crescent moon (see Chapters 27 and 28) (M. Zeilik).

of the rock art sites (Brandt and Williamson, 1979; Koenig, 1979; Mayer, 1979). Most of the *astronomical* objections to this rock art site as representing the 1054 supernova can be eliminated simply by flipping the view (Gaustad, 1980); but the nagging doubt remains that this configuration represents Venus next to the crescent moon rather than the supernova (Ellis, 1975, pp. 86–7). Other sites seem to have to be carried along a very thin line of reasoning to be interpreted as the 1054 supernova (Hedges, 1985, p. 26). Workers now have moved away from the search for the supernova; yet we need to recall that it motivated much early work, especially at Chaco.

In general, we should distinguish between astronomical *purposes* and the astronomical *practices* (that derive from the purposes). In the Pueblo Southwest context, astronomy serves the purposes of establishing and validating (1) sacred directions and cosmic patterns, (2) cosmic mythology, (3) certain ritual sites and shrines, (4) the ritual and planting calendar, and (5) times for hunting and gathering. These desired ends prompted the development of horizon calendars, light and shadow markers, and lunar phase counts for tracking the calendar. The main task of the calendar watch centers on methods to *anticipate* festival dates (McCluskey, 1982; Zeilik, 1986a). Pueblo ceremonies must be announced ahead of time so that ritual preparations can be properly carried out, and the meshing of solar/lunar ceremonies worked out by the proper intercalation (McCluskey, 1977). The ceremonial cycle typically spans a year, and solar and lunar observations conducted by the appropriate religious officials set the timing for the rituals that are laid out in a sequence, where the end of one ceremony marks the start of the next.

Anticipation provides a culturally correct way to understand the term "precision" or "accurate," when combined with historical information about festival dates as has been done by McCluskey (1977) for Hopi and Lyon (1985) for Shalako at Zuni. Sun priests historically at Hopi and Zuni were able to fix solstice ceremonies to within the astronomically correct day (or two) by making anticipatory observations about two weeks ahead of time (Zeilik, 1985a). In contrast, lunar ceremonies appear to be moveable festivals, with a scheduling decision under the power of the person responsible for setting the date (Frigout, 1979; Sekaquaptewa, 1983; Zeilik, 1986a). And the date of a particular lunar phase (such as full) may have been established to within a day (or two) of the astronomically correct phase (Bunzel, 1932b, pp. 702–3).

An emphasis on calendric anticipation switches our attention from directions to rate of change of direction (angular speed) of the sun. This mental realignment should not, however, overlook the importance of direction to the Pueblo sense of sacred space. The special directions are usually the four solstitial ones (winter and summer rising and setting), as well as the zenith and nadir, such as embodied in the placement of corn ears in the Chief kiva during the Hopi Powamu ceremony (Stephen, 1936, p. 228, fig. 140). Special colors are also linked with each of the sacred directions, often with several animal associations, too (Parsons, 1939). Such sacred directions may result in buildings with special orientations in historic and prehistoric times. The point here is that cycles such as the sun's seasonal swing along the horizon involve *motions* to achieve the extreme positions. The Pueblos embodied both in their astronomical practices: the directions tend to be cosmic; the motions, calendrical.

Finally, we need to consider how the calendrical observations were made. For the sun, a horizon calendar was most commonly used, and we have good maps for some of them (Stephen, 1936, maps 4 and 12). Astronomically, such horizons need to be fairly distant (a few kilometers at least) with some, but not necessarily dramatic, relief. The sun priest also goes to a special sun-watching station, which is typically not marked by rock art (Zeilik, 1985d; Young, 1987a). The ethnographic analogy then raises a key archaeological problem: What would be the material evidence for a sun-watching station? At Zuni, the historic sunwatching station at the old village of Matsakya was well marked (Fig. 8.2), but it no longer exists today.

The second calendrical technique involves the use of light and shadow cast through windows/portals against a wall with markers (Lange, 1959, p. 56); this method has more hope for archaeological remains. At Zuni, the famous passage of Cushing (1979, p. 117) indicates markers of some kind on the wall. The diaries of John G. Bourke tell us what was going on (Bourke, 1881, entry for 19 November):

> After breakfast, Cushing, Pedro Pino [the Governor of Zuni] and myself went to the upper story of one of the highest houses on the Eastern side of the Pueblo; here in the West wall was an old blue china plate fixed there, so the head of the house said, in the time of the Spaniards, to conceal a painting of the Sun, which faced a small rectangular aperture in the eastern wall. When the sun shone through the

8.2 Sun-watching station at Matsakya, one of the old Zuni villages. Pekwin used this location for horizon observations prior to the summer solstice (Stevenson, 1904, pp. 148–9). Note the upright stone with the sun face. Photo by M. C. Stevenson (Smithsonian Institution photo no. 84-7550).

aperture farthest to the North, Spring had come and the season of planting had arrived; the more Southerly aperture allowed the rays of the Sun to fall upon the center of the plate (in ancient times upon the face of the sun picture) about the period of the Autumnal equinox—and when the light struck a certain point in the wall, it was the time of the Winter Solstice.

This statement reveals the practices of solar calendars at Zuni at a time before Anglo culture had much impact. Sunlight hits the sun face during harvesting season; planting and the winter solstice were also noted. And we also know that at the same time, Pekwin (Fig. 8.3), the sun priest (or "cacique of the Sun," as Bourke called him) did a horizon watch (Bourke, 1881, 19 November) as well as kept a wall calendar in his own house (according to his brother). The half of that calendar that marked the time from spring planting to the summer solstice consisted of a horizontal line of scratches onto which sunlight fell through an east-facing window. The half of the year to the winter solstice was marked by strings of abalone shells, hanging on the wall. From this information and the argument by analogy, we can tell archaeologists to look for rooms that have windows situated so as to cast sunlight (probably in the morning) against a wall. Any markers on these walls, however, would probably not have lasted 1000 years.

Michael Zeilik

8.3 Pekwin at Zuni in 1896. Photo by M. C. Stevenson (Smithsonian Institution photo no. 2250).

I find these Zuni examples especially curious because we are told that Pekwin keeps both a horizon watch (Stevenson, 1904, pp. 109, 117–18) and also one involving a wall calendar. What was the cultural reason for this redundancy? Did Pekwin use both to anticipate and confirm important times? For instance, one major sequence of the wall calendar includes the time of planting, winter solstice and summer

Keeping the sacred and planting calendar

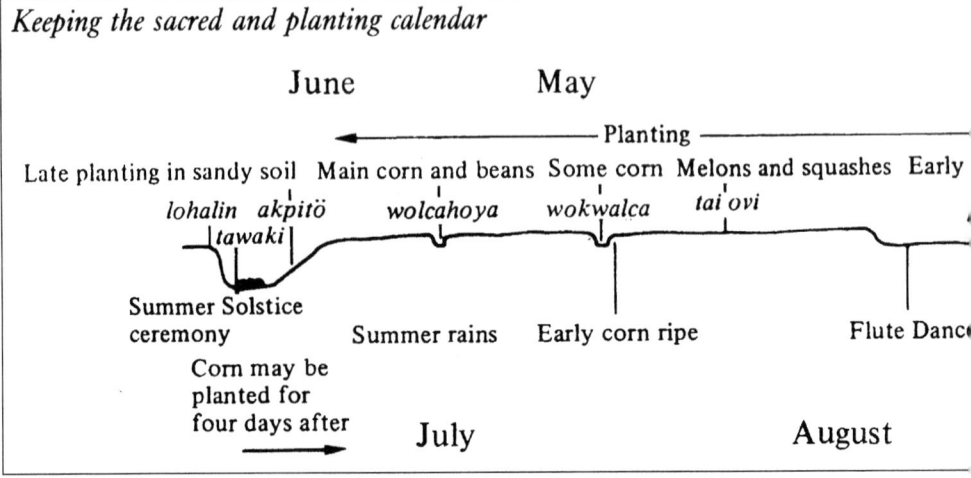

8.4 Planting and ritual horizon calendar, from winter to summer solstice, for the Hopi village of Shimopovi (after Forde, 1931, fig. 6B). The interval from December to June is read from right to left along the top; July to December from left to right along the bottom.

solstice (Bourke, 1881, 3 September). Pekwin's horizon calendar involves similar time sequences. Did these two parallel calendars both support the degree of precision that Pekwin was supposed to achieve? Bourke (1881, 25 November) states that

> Cushing took me to call upon the Priest of the Sun. This official, in the different Pueblos, is supreme during time of peace, that is to say his edicts can only be revoked by a council of all the Caciques. The present incumbent of the office in Zuni is a very young man and has but recently entered upon his duties. I have noted elsewhere [20 November 1881] that *by a mistake in his orientation* he ante-dated the festival of the Winter Solstice by some (15) or eighteen (18) days, a mistake very summarily corrected by the Council of the Caciques [author's italics].

Note that this passage tells us that other religious officials kept watch on Pekwin's work, and could track the calendar as well as he to insure that "correct" times were announced. Also, note that Pekwin's mistake in anticipatory observations was far too large to be, acceptable to the council.

I have discussed Zuni in some detail as a specific example of pan-Pueblo practices and as a case where we have substantial information—most of it consistent!—predating the 20th century. The practices then may reflect the survival of an astronomical tradition that traces back to prehistoric times.

In summary, the ethnographic evidence about the historic Pueblo calendar indicates:

(1) attention to a one-year *seasonal cycle* with the winter solstice usually as the "middle";

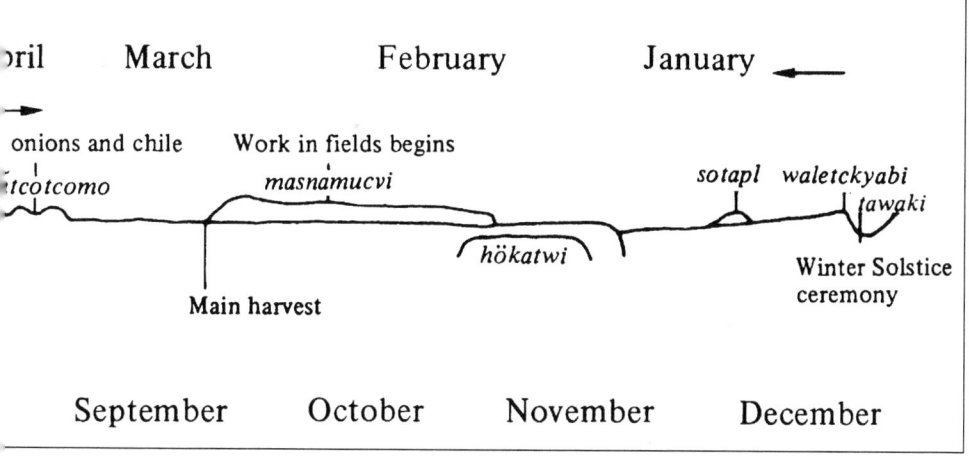

(2) a *fixed sequence* of ceremonies, in which the completion of one sets the stage for the next;

(3) an *anticipation* of ritual times so that proper preparations can make the ceremonies effective;

(4) *astronomical observations* that set the calendrical schedule.

In what follows, I will use the winter solstice, planting season (mid-April up to the summer solstice as for the Hopi—see Forde, 1931; Beaglehole, 1937; Titiev, 1938), and the summer solstice as the order of importance for seasonal dates (Fig. 8.4). (The pueblos rarely mark the equinoxes as such; those that do correlate roughly with planting and harvesting. See Ortiz, 1969, p. 104, for explicit comments about the importance of the equinoxes to the Tewa, which marked the seasonal transfer of power between the summer and winter people.) I will also define *calendrical site* to mean one that could be used to achieve at least the same degree of accuracy of scheduling as attained by Puebloan calendar keepers in historic times.

Finally, my focus on the *Pueblo* Southwest should not result in the inference that only the Pueblo people paid attention to the sky. The Navajo and Apache did, also, in their own cultural context (see Farrer and Second, 1981, for examples from the Mescalero Apache). And the Indian peoples of North America in the greater Southwest (Spier, 1955) and all other regions had some type of calendrical and cosmological systems (see Cope, 1919, and Stirling, 1945, for general references; and see the journal *Earth and Sky* for California).

Table 8.1 Possible Anasazi horizon calendars.

Place	WS	VE	SS	AE	Anticipation
Chaco					
Wijiji	+	–	–	–	+ WS
Pueblo Bonito	–	+	+	+	+ except WS
Penasco Blanco	poor				
Fajada Butte	poor				
Rio Grande/Chama					
Tsiping Pueblo	–	+	–	+	+ equinoxes
Tsankawi	+	+	+	+	+ all
Kuaua	+	+	–	+	+ WS

WS = winter solstice; SS = summer solstice; VE = vernal equinox (planting); AE = autumnal equinox (harvesting). *Key:* "+" means the site works for the time given; "–" that it does not.

HORIZON CALENDARS

Because they are so prosaic, horizon calendars have received little attention—even though we have the most and best ethnographic information about them! They are also hard to substantiate archaeologically, for it is difficult to ascertain where a sun priest might have stood to make his observations. On the other hand, if the horizon is distant—ten or so kilometers—the choice of the standing spot is not crucial if it is within or close to a pueblo. Among the historic Pueblos, the sun priest rarely travels very far to make his observations—from the roof at Walpi (Stephen, 1936, p. 62), near the edge of the mesa for Hano (Yava, 1978, p. 72), or a few kilometers east from Zuni (Halona) to the deserted village of Matsakya (Stevenson, 1904, p. 118). Hence, by ethnographic analogy, one can search for a spot in or close to the archaeological site of a village that provides a view of the horizon to see if a calendar can be kept in this fashion. If not, a search should then be made for light/shadow markers. That conclusion can be reached immediately if no good horizon line presents itself for the appropriate spans of time.

Archaeological sites that have been tested for horizon watching are (Table 8.1): Penasco Blanco (Williamson, Fisher and O'Flynn, 1977; M. Zeilik, unpublished field notes), Wijiji (Williamson, Fisher and O'Flynn, 1977; Williamson, 1983; Zeilik and Elston, 1983), Fajada Butte (Williamson, Fisher and O'Flynn, 1977), and Pueblo Bonito (Zeilik, 1985e)—all in Chaco Canyon; and Tsiping Pueblo in the Chama valley of New Mexico (Zeilik, 1984a), Tsankawi (M. Zeilik, unpublished field notes) on the Pajarito Plateau, and Kuaua along the Rio Grande (J. Lowther and M. Zeilik, unpublished field notes). The Chacoan Great House sites range in dates from AD 900 to 1100 (those at Fajada Butte date to the 1200s); Tsiping is a later occupation, from AD 1250 to 1450, and a probable prehistoric Tewa site, as is Tsankawi (which is now part of the Bandelier National Monument), which may have been occupied up to the late 15th century. Kuaua was occupied into historic times and may have been visited by the Spanish (Dutton, 1963, pp. 3–18).

8.5 Sunset on the vernal equinox behind the sacred mountain of Pedernal as viewed from the amphitheater area at Tsiping Pueblo (M. Zeilik).

The view from Penasco Blanco, at the edge of the mesa where the prehistoric road leads to a point above the hypothetical "1054 supernova" site, spans the year but has little to offer for anticipatory markers or planting (Zeilik, unpublished field notes). The horizon from the Wijiji rock art site, which has a nearby rock pillar, can anticipate the winter solstice (Williamson, Fisher and O'Flynn, 1977; Williamson, 1983; Zeilik and Elston, 1983). The confirmatory alignment above the rock pillar on the winter solstice only is visible from a narrow range along the viewing ledge, because of the closeness of the pillar. The cultural context is confused between Anasazi and Navajo (Williamson, 1983); overall, a clear conclusion is hard to reach. The site does *not* work for a planting sequence or the summer solstice.

The horizon view from the east side of Pueblo Bonito provides a reasonable profile for summer solstice through planting, but it has only one possibility for an anticipatory marker for the winter solstice. The view from atop the canyon's rim, near the staircase, might be better and needs to be checked out. From a ceremonial area just outside Tsiping, the historically sacred mountain of Pedernal provides a profile (Fig. 8.5) for anticipation of and tracking of a planting/harvesting calendar at sunset. For the solstices, the horizon from Tsiping offers no clear anticipatory or confirmatory possibilities either at sunrise or sunset (M. Zeilik, unpublished field notes).

Tsankawi sports a stunning view of the Sangre de Cristo mountains, which lie about 50 km to the east, and a panorama of the Rio Grande Valley and Jemez range (Hewett, 1906, pp. 20–1). From the tip of Tsankawi Mesa (see Hewett, 1906, pl. VIIa), the view encompasses a horizon profile that easily can be used for winter solstice, summer solstice, and a planting calendar (Zeilik, unpublished field notes). Other prehistoric pueblos in the same region have roughly the same view, so they can have similar horizon calendars. We also have ethnographic evidence from the Tewa pueblo of San Ildefonso (which is the historic Pueblo closest to Tsankawi) that the same mountain range was used to anticipate (by 12 days) and confirm the summer solstice (Stevenson, n.d.).

Kuaua contains, in its square kiva, the famous ceremonial murals (Dutton, 1963). Some of these sacred figures depict the Sun Father (based on ethnographic interpretations) and relate to ceremonies at the winter and summer solstice, as well as general symbolism related to fertility. At the site today (part of Coronado State Monument), the tree line of the Rio Grande makes observations difficult from the ground. We therefore observed from the roof of the Museum, which would mimic observations from the second story of the pueblo. The Sandia and Ortiz Mountains to the east provide a clear horizon profile at sunrise from the equinoxes to the winter solstice (Lowther and Zeilik, unpublished field notes). We know that at Santa Ana Pueblo, the Sandia Mountains are used to track the approach of the winter solstice (White, 1942, pp. 92, 103), for which Masewi (the war priest and one of the Twin War Gods) begins the watch in September, just the time when the sun begins to climb up the slope of Sandia. Dutton (1963, p. 205) states that some of the Kuaua people may have joined those at Santa Ana.

I conclude that Anasazi sites need *far* more work in the search for potential horizon calendars. Our firmest ethnographic information about calendrical sun-watchers involves horizon profile sequences. By the ethnographic model—especially for the Hopi—we expect that each pueblo kept its own calendar by its own sun-watcher or watchers. Hence, each site should be checked carefully for a useful horizon track—or lack of one! If the horizon is not workable for certain intervals, we may then infer that light and shadow techniques might have provided an alternative procedure.

LIGHT AND SHADOW

Light and shadow phenomena draw our attention by their visual appeal, especially through the medium of photography. The three-slab site and other sites at Fajada Butte (Sofaer and Sinclair, 1987) and the petroglyph sites in the Petrified Forest (Preston and Preston, 1985, 1987) have sparked public attention to this aspect of Southwestern archaeoastronomy. The petroglyph site near Holly House in Hovenweep is also visually stunning (Williamson and Young, 1979; Williamson, 1979, 1981, 1984), but has a lower visibility in the public eye. Other light and shadow sites associated with window/portals in buildings include the corner openings at Pueblo Bonito; the northeast opening in Casa Rinconada in Chaco; and openings in add-on rooms at Hovenweep Castle, Unit Type House, and Cajon Group Ruins at Hovenweep.

A general claim is sometimes made that such sites work as intentional and accurate calendrical markers. To make such a case requires, in my opinion, that the site serves to anticipate the solstices within an accuracy of a day or two and to keep track of a planting calendar. This requirement brings up the issue of *practical resolving power*— the ability of an observer to interact with a site to mark reliably one-day differences in shadow positions. Note that an emphasis on anticipation shifts our attention from the *positions* of the light-shadows to the rate of change of position on a surface, i.e. *linear speed*.

Two physical elements enter into this resolving power issue. The first is the linear speed along the surface per day, as the shadow mimics the seasonal motion of the sun. Imagine the shadow as a lever with a fulcrum at the point of the shadow-casting edge. Then, the longer this throwing arm, the larger the linear motion for each degree, say, of solar motion. So better resolving power requires longer shadow throws. The second is the sharpness of the shadow edges. When the light source has a finite size, such as the sun (which subtends ½° in the sky), a shadow has a fuzzy edge, even if the shadow-casting edge is smooth. An empirical examination of the degree of fuzziness gives, with a lever arm of 1 m, a width of about 1 to 2 mm. The fuzziness increases in direct proportion to the length of the shadow throw. So, shorter is sharper.

The two physical elements work against each other. A larger linear motion requires a longer throw; for sharper edges, a shorter one. Given the eye's perceptive ability for edge contrast, I suspect that an optimal length exists that balances these two considerations. From my experience, I surmise that this length is about 1 m (though this optimum length needs to be established by experiment). My subjective judgement is that a linear motion of at least 0.5 to 1 cm/day is needed for reliable observations with a one-day resolving power.

The final element for shadow-casting reliability is the actual visibility of the shadow edge and ability to note its position from a selected observing position. That is, what you can do "being there" at the site, *not* what can be done after the fact with photographic records. (Photography increases the contrast along a shadow's edge.) To analyze the site correctly requires that we interact with it in the same way as a prehistoric observer might. So, the roughness of the shadow-casting edges (which will increase the fuzziness of the shadow), the texture and color of the surface on which the shadows fall, the sharpness of petroglyphs/pictographs on the surface, and the viewing position and angle all enter into the character of the observations. These are all specific to the site, so each site should be judged based on these details of the site's geometry. Claims for precision cannot exceed the practical resolving power that the site offers.

The above analysis is only *one* part of a site's interpretation. An essential piece is the archaeological context of the site—especially difficult for rock art images because they cannot be dated absolutely as can buildings by tree-ring or radiocarbon techniques. Again, the problem is the archaeological control of time. The archaeological

Table 8.2 Anasazi interior sites

Place	Average speed (cm/day)	WS	VE	SS	AE	Anticipation
Hovenweep						
Hovenweep Castle:						
WS port	1	+	–	–	–	+ WS
SS port	4	–	+	+	–	+ SS
Unit Type House:						
WS port	2	+	–	–	–	+ WS
E Port	3	–	+	–	+	+ E
SS port	2	–	+	+	–	+ SS
Cajon Ruin "Tower":						
WS port	?	?	?	?	?	?
E port	3	–	+	–	+	+ E
SS port	2	–	+	+	–	+ SS
Chaco						
Pueblo Bonito						
Room 228	3	+	–	–	–	+ WS
Casa Rinconada	6	–	–	+	–	+ SS

WS = winter solstice; SS = summer solstice; E = equinox.
Key: "+" means the site works for the time given; "–" that it does not; "?" that insufficient information is available. Average speeds are used, rounded off to one significant figure, to allow direct comparison among sites.

8.6 The southeast corner of Pueblo Bonito today. The second storey corner opening just to the right of center is that to room 225B; the corner opening to the far left is that to room 228B. The stepped stonework in front of the wall at bottom right may have served as a shrine for offerings of effigies (M. Zeilik).

Michael Zeilik

context then illuminates the cultural one, an important consideration at sites that may have been used by different cultures at different times.

I will now examine light/shadow sites, dividing them into two groups: interior (buildings) and exterior (rock formations with art).

Interior Sites

At Hovenweep, Williamson, Fisher and O'Flynn (1977) investigated a number of tower structures. Williamson (1981) elaborated on Hovenweep Castle and the Cajon Group ruins. These places and the Unit Type House have add-on rooms that contain portals apparently positioned to let in light for calendrical purposes. (Some portals are vents; others provide sights to other buildings in the canyons—not *every* one has an astronomical purpose.) Cajon lies about 10 km south of Hovenweep Castle and Unit Type House, which are within a kilometer of each other. Zeilik (1983a, 1987, and unpublished field notes) re-examined these sites and found that all could be used in an anticipatory fashion and that the average speeds of the light beams on the walls ranged from about one to a few centimeters per day (Table 8.2). Photos taken by Fewkes (1919) just prior to 1900 indicate that, for Unit Type House and Hovenweep Castle, the portals were intact before any stabilization. Hence, they were originally in the structures, although we are not sure to what extent the insides of the openings were plastered and so would restrict the view compared with today.

Cajon and Hovenweep Castle work calendrically at sunset; Unit Type House in early morning (about half an hour after local sunrise because of a nearby ridge that blocks the sunlight). In all three places, the horizon view does not work well for a calendar. The play of light within the rooms against the walls of each can anticipate the solstices, confirm them, and keep track of a planting calendar. So I judge them to be the best examples we have found to date of intentional calendrical structures in a large part because of the integrity of the structures. The next archaeological step would be to excavate these rooms to see if they contain any artifacts that might show a different ceremonial usage than other rooms also trenched for a controlled comparison.

In Chaco, Reyman (1976) noted that corner openings to rooms 225B and 228B at Pueblo Bonito are oriented to the winter solstice sunrise (Fig. 8.6). The horizon profile from Bonito contains little relief from the end of October up to the winter solstice. The openings to rooms 228 and 225 will work to anticipate and confirm the winter solstice (Zeilik, 1985e). In room 228, the sunlight first enters the room at the end of October and moves horizontally at an average speed of about 3 cm/day. Hence, the beam's motion could be used to anticipate the winter solstice to within an accuracy of one day.

The main problem with these corner openings is that we do not know whether they were exterior windows or interior doors! I have examined pre-reconstruction photos of Pueblo Bonito taken in 1920; they clearly show the opening to room 228 and a barely visible outline of the one to room 225 but these photos provide no clear way to tell if the openings were corner doorways—of which Pueblo Bonito has a number of examples—or actually were windows. Archaeologists could help here by providing information as to the nature of the openings. Lekson (1985, private communication) suggests that they are most likely doorways, because of their size, location above the floors, and their identical construction to other interior openings at the first-story level of Bonito. In contrast, Reyman (1976, and 1986, private communication) has noted that photographic plates from the 1890s (by G. Pepper and R. Wetherill) show the windows; and that stabilization reports and present conditions at the site indicate that the exterior walls perpendicular to the south walls were buttressing walls added for structural support.

Upon first examination, the northeast opening of the great kiva of Casa Rinconada at Chaco appears to be aligned to the summer solstice sunrise (Williamson, Fisher and O'Flynn, 1977; Williamson, 1981, 1982). The beam of light strikes one of the large niches about half an hour after local sunrise and makes a visually appealing display. Sunlight first enters the opening at the beginning of May (Zeilik, 1984b), so the horizontal motion along the wall can easily be used to anticipate the summer solstice to the day—the average speed is some 6 cm/day (Table 8.2). But the archaeological evidence argues against this aspect of Casa Rinconada as calendric. Firstly, photos prior to the reconstruction clearly show that the opening was in shambles; in particular, the right-hand edge (as viewed from the inside) was non-existent. So the position of that edge and hence the width of the window are a result of the reconstruction rather than original. Second, a low wall probably cut the opening off from a view of the sky. Third, even if all else worked correctly, the light beam would have struck the northwest roof support post rather than the niche, which was probably bricked in. All in all, I judge the evidence weighs against an intentional calendrical phenomenon at Casa Rinconada on the part of the Chacoan Anasazi.

Table 8.2 summarizes the known properties of most of the Anasazi interior sites, which may have had an analogous function to those at Zuni. Note that for most sites, the average linear speed of the sunlight beam is 2+ cm/day—enough speed to achieve one-day resolving powers. Clearly, this area will be a fruitful one for future work, if we can get around the reconstruction problem.

Exterior Sites

The three-slab site atop Fajada Butte in Chaco has the widest reputation among the Anasazi exterior sites (Sofaer, Zinser and Sinclair, 1979a, b). Sofaer and Sinclair (1987) have improved the knowledge of the ethnographic context and also have expanded the view to other petroglyph sites on the butte, which they claim mark local solar noon and the seasons. Sofaer and colleagues find particularly attractive the purported integration of the sites for sun/moon and season/noon observations. Newman, Mark and Vivian (1982) infer that the three-slab site is likely a natural rockfall, rather than a "construct."

I have re-evaluated the three-slab site in some detail (Zeilik, 1985b, c) and will not repeat those comments here. (Carlson, 1987, has also analyzed the functions of the site.) But let me make four general remarks. One, any claim for "precise" or "accurate" must be made explicit and should be put in a cultural context. For example, the shafts of sunlight at the three-slab site do not, in my judgement, permit the anticipation of the solstices to within a day or two—they do not have sufficient practical resolving power. Two, the Fajada Butte sites must be placed in a firmer archaeological context. I note that the small site 29 SJ 1360, which was occupied during the 10th and early 11th centuries, is located at the base of the butte. Could it have housed the sun priest who might have monitored the astronomical observations on the butte (McKenna, 1984, p. 389)? Three, the ethnographic information shows that the historic Pueblo people clearly distinguished between sun shrines (see Fewkes, 1906, for examples) and sun-watching stations for calendrical purposes—a distinction often confused in the archaeoastronomical literature. Four, calendrical sites tend to be accessible, since at times the sun-watcher needs to make daily observations. The three-slab site is a hard climb (especially in winter) from the occupation sites within the canyon. The butte does contain two small habitation sites, which, from an analysis of ceramics found associated with them, appear to have been occupied late in Chaco's history (Toll, Windes and McKenna, 1980), around AD 1175–1225. We cannot be certain whether these habitation sites have any temporal connection to the rock art; if so, we may be seeing a development on the butte *after* the Bonito phase, which ended around AD 1130.

In my judgement, the three-slab site fits much better with the attributes of historic sun shrines rather than calendrical observational places. If so, the butte may have been a sacred—even cosmological—place, but I think that its astronomical use for calendrical keeping was much less likely its purpose. That is, the sunlight motions at the three-slab site act as a *seasonal hierophany* that contains calendrical information but has limited practical resolving power (see Hedges, 1985, p. 27 for a similar conclusion). Put another way: I think that the site has more of a *commemorative* than a calendric function, if it were intentionally used by the Chacoan Anasazi.

Preston and Preston (1985, 1987) have worked on a number of so-called "solar observatory" sites in Arizona—most of which are in the Petrified Forest National Park. They propose that petroglyphs were used prehistorically in two ways for calendrical observations: (1) by light/shadow interacting with the rock art in "significant" ways, and (2) by using a glyph to position an observer's head for sighting a horizon feature at sunrise or set. (No ethnographic evidence has been found to date for the use of rock art as a head-positioning device for calendrical observations.) The Prestons define "significant interactions" as tangent/center shadows striking spirals/circles (and in other interactions for anthropomorphic figures). The light/shadow phenomena occurring here strongly resemble those at Fajada Butte, but rather than being concentrated in one locale they are spread out over many locations.

I have visited the site close to the Puerco ruin at the time of the summer solstice. The horizon view from the ruins would not make a good calendar. To the east and south of the ruin lie many boulders with rock art. The one noted by the Prestons is a small (10 cm wide, 12 cm high) spiral-like figure, whose pecked line is about 1 cm wide and very irregular. At mid-morning, a shaft of light (formed by a natural slit in the boulders to the northeast some 1.6 m away and so has noticeably fuzzy edges) moves down along the rock and crosses the spiral in about half an hour. Park Rangers at the site told me that the light first comes through the slit (which is fairly narrow—about ½° on the sky as seen from the petroglyph) roughly a week before the solstice.

These observations raise a number of questions. One, given the density of petroglyphs and rock edges at the site, what was the probability that the interaction was accidental? Two, to what extent does the fuzziness of the shadow edge and the irregularity of the pecking limit the practical resolving power of the site? The Prestons' sites need to have such questions asked about them. Preston (1984) has made one stab at the probability problem; he finds that the Cave of Life site is the only one where the statistical chances that the interactions are accidental are very small.

Finally, the Holly site at Hovenweep presents a dramatic display at (and before) the summer solstice. Then, about half an hour after local sunrise, two shafts of light move horizontally across a petroglyph panel that includes spiral and circle shapes. Only about seven minutes pass from the first appearance of the beams and their merging near a spiral. Williamson and Young (1979) have noted symbols on the panel that may relate to the Twin War Gods (Venus as Morning and Evening star) and a snake/serpent that may be a water symbol. The petroglyphs here have many motifs similar to those around the "1054 supernova" site below Penasco Blanco at Chaco (M. Zeilik, unpublished field notes). Williamson (1984, pp. 93–103) has concluded that the site may have been used to keep a planting calendar similar to that at Hopi—especially pertinent to the farming area in which it is located. He also suggests that the change in the positions of the light shafts would allow a practiced observer to anticipate the summer solstice to "within three or four days." (In contrast, the summer solstice portal in Hovenweep Castle would allow a trained observer to predict the summer solstice to within a day or two.)

In summary about the exterior sites: I concur with Young (1986) and Schaafsma (1985, pp. 264–6) that some Anasazi rock art served to interact with light and shadow. However, I think that a better case needs to be made that these interactions work *calendrically* in the appropriate cultural context. Because it works today does not guarantee that it was used that way prehistorically.

No ethnographic analogy has been found yet for exterior sites whose light/shadow interacts with rock art for calendrical purposes. (Of course, the physical principle is the same as that for light entering the windows of buildings.) The closest process comes from Zuni, where Pekwin visited a "small post of petrified or silicified wood" (Fewkes, 1891, p. 4). Fewkes states that this post ". . . is in certain particulars a gnomon"—an implication that the position of its shadows was used as a marker at sunrise. Bourke (1881) clarifies the situation a bit, when he writes that the stone pillar was a place for Pekwin to make sacrifices and watch the sun. Bourke and Cushing go with the Zuni governor, Pedro Pino,

> to the Eastern boundary of the Pueblo to an open field in the center of which was a vertical, four sided piece of petrified wood, put there, Pedro said, a long, long while

ago. In line with it away off to the East, six or seven miles on the crest of a mesa could be discerned another monument, pointed out by Pedro.

The implication to me is that Pekwin used the pillar in the field to position himself in the right alignment with the pillar on the mesa. This case does not much resemble prehistoric exterior light/shadow sites.

LUNAR OBSERVATIONS

Ethnographic information shows that some Pueblos paid close attention to the phases of the moon and used them as a timing device (among other functions, such as weather prediction) for some festival dates—most notably Shalako at Zuni (Zeilik, 1986a) and Powamu at Hopi (McCluskey, 1977). The regular phase cycle serves as a nightly marker that can be used to anticipate events that coincide with a specific phase. For example, at Zuni, the head of the Long Horns had to anticipate full moons (for preparation and planting of prayer feathers by the impersonators of the Shalako) by two days (Bunzel, 1932c, p. 963). He could have done so by spotting the first visible crescent and then counting for ten days (Zeilik, 1986a).

Because the Pueblos had a reliable seasonal calendar (based on solar observations), the use of the moon to schedule ceremonies brings up the ancient problem of synchronizing the solar and lunar calendars. This intercalation issue clearly arises at Zuni in the coordination of Shalako, the winter solstice ceremony, and the New Fire ceremony that marks the start of a new ritual year. Pekwin sets the date for the winter solstice ceremony by observations of the sun; Sayatasha (the head of the Long Horn Kachina group) establishes the Shalako schedule by observing the phases of the moon and has the final decision for the date (Bunzel, 1932c). Sayatasha consults with Pekwin to ensure that Shalako does not conflict with the Winter Solstice festival.

Meanwhile, we are also informed that Pekwin is supposed to arrange for the winter solstice ceremony to occur at the full moon in December (Bunzel, 1932a, p. 354; see also Tedlock, 1983)! Reyman (1980) has emphasized how troublesome this scheduling requirement will be for Pekwin, for even given a three-day range, at best the full moon/winter solstice coincidence can happen only every three years. However, Pekwin does *not* control the date of Shalako, which *is* based on lunar phases. The 49-day countdown to Shalako begins on the day of a full moon (Bunzel, 1932a, p. 523), and that is directed by Sayatasha. Another articulation of the sun and moon occurs at the winter solstice ceremony, for the first full moon that follows begins the count of ten full moons that is the first stage of the timing for Shalako. Overall, I do not see how Sayatasha and Pekwin kept the timings in line.

One hint comes from the sequence of Zuni months, in which the first six are named and the next six are nameless or repeated in names (Stevenson, 1904, p. 108). Parsons (1917, pp. 300–1) suggested that the non-naming of the November moon, in particular, provides the elasticity to reconcile the Shalako and winter solstice ceremonies. Shalako dates (Lyon, 1985) need to be compared carefully to winter solstice celebratory dates to illuminate this issue (as McCluskey, 1977, has done for the Hopi).

The Pueblos with sun- and moon-watching invited the intercalation problem. The months began with the observation of the new (crescent) moon, the seasonal cycle by sun-watching. The months tended to be named after key ceremonies within them (especially at Hopi; see Ellis, 1975, pp. 65–8 for a summary) or seasonal events (see Harrington, 1916, pp. 63–6 for the Tewa month names and a comparison to those at Jemez and Zuni). The Pueblos that tally and name months usually have 12 or 13 in a seasonal year (see Cope, 1919, pp. 151–3 for a summary). I suspect those with a fixed 12 months have adopted, at least in part, a European calendar. Those that have 13 months (or more) probably are the closest to the aboriginal calendar, for those allow intercalation by a "short month" technique. One Hopi, Abbott Sekaquaptewa (1983), has written in a Third Mesa newspaper that, for the case of the winter solstice ceremony and the next new moon (Pamuya), the adjustment involves "the elimination of one or a couple of the four day cycles after the final rites of the winter solstice ceremony." This procedure speeds up the arrival of the next lunar cycle and "allows a continuity of the normal ceremonial and spring planting times."

McCluskey (1977) has analyzed the actual dates for the winter bean ceremony (Powamu), where the full moon after the close of the winter solstice ceremony signals the planting of beans in kivas (Stephen, 1936, p. 162; Parsons, 1939, p. 505). Now, each year a given phase of the moon arrives 10.9 days earlier, so that, after three years, the moon phases are 32.7 days "ahead" of the sun. A simple intercalation scheme would occur every 2.71 years. McCluskey (1977) finds this periodicity in the Powamu dates. The inference is that one "month" is not counted every 2.71 years; it may be a very short interval. Unfortunately, similar evidence needs to be gathered for other Pueblos, and no general, consistent scheme of intercalation appears to have been used in historic times.

No firm information is yet available about an interest in or knowledge about the lunar 18.6-year standstill cycle. One hint comes from Curtis (1922, p. 156, n. 1) in his discussion of the Flute ceremony at Hopi: "It is said that at villages other than Walpi, the rising of the moon at a certain landmark on the horizon determines the date; and the date is not the same at Walpi as at the other pueblos." If this information is correct, then some Hopi religious officers observed the horizon position of the rising moon, which goes through a complete north-south azimuthal swing in a month. That swing is especially noticeable at major standstill, but it does not immediately follow that the standstill cycle was noted at Hopi. In fact, the context of the reference is the Flute/Snake ceremonies, which occur at the same time of year but in *alternate* years (Curtis, 1922, p. 156), rather than an 18.6-year (or 19-year) cycle.

Sofaer, Sinclair and Doggett (1982) have argued that the Anasazi marked the standstill cycle at the three-slab site in Chaco at moonrise (see Chapters 29 and 30). Such a cycle does not fit into the typical historic Pueblo ceremonial framework. That ritual round usually encompasses but one seasonal year in a fixed sequence of events.

8.7 Zuni calendar stick, drawn by J. G. Bourke (1881, 21 November).

Longer cycles are typically not counted. For example, Dutton and Marmon (1936, p. 5) state that

> ... the Lagunans have had no very accurate system of "year counts." Those who have not adopted use of modern calendars, keep track of time as did their ancestors, by particular events or phenomena, or by saying that such-and-such occurred when so-and-so was a boy. They have no eras, cycles, or periods of several or many years.

One exception to this framework of short cycles is given in an obscure reference in Stephen (1936, p. 1039, n. 1) in which, from early notebooks, he writes: "The kachinas reckon their feasts by the year of thirteen moons, consequently their Keli only coincides with the secular Keli once in thirteen years, which period is called a glad year." The context here is the Hopi way of reckoning the months. The "Keli" probably refers to the initiation into the Hopi kachina societies. The general tribal initiation occurs in November, and every four to six years (Curtis, 1922, pp. 107–8) takes place in a long form (16-day long). No other source contains information on a Hopi "glad year" cycle, nor does Stephen mention it elsewhere in *Hopi Journal*. It may have been bad information, or it may have come from a Hopi clan that died out during Stephen's stay there.

If the Fajada three-slab site did intentionally mark the lunar extremes, questions naturally follow based on ethnographic analogy. What ceremony did the major or minor standstill trigger? How were these ceremonies anticipated? Let me speculate that they were anticipated within an accuracy of a year. Can an observer reliably see this motion in moonlight, limited to northern declinations, with the given fuzziness of the shadow's edge? A practical test of lunar shadow watching needs to be made to ascertain if that site is usable for tracking the standstill cycle. (I note here that Zeilik 1985b and 1985c contain an error in the rate of motion of the moon's shadow over the standstill cycle; the numbers have been corrected—see Sofaer and Sinclair, 1986a,b. The corrections do not negate my point about the limited resolving power of the motion of the edge of the moon's shadow to anticipate the time of the standstills.) In other words, the moonlight shadows at the site may have provided a *cyclical hierophany* with a modest resolving power for calendrical purposes (if the lunar standstill cycle did in fact matter to the Chacoan Anasazi).

Lunar calendars kept by phases (with the first visible crescent starting the month) present a serious archaeological problem: What would be the material remains? We have two hints. At San Ildefonso, Stevenson (n.d.) reported that the moon-watcher kept a tally for the year by notching the edge of a flat stone on which was drawn a moon face. These stones were then deposited in a ceremonial chamber. At Zuni and Hopi, we have evidence of the use of calendar sticks to track the months. Fewkes

8.8 Zuni religious officials in the 1880s, photographed in Boston in 1882. From left to right: Nanahe, a Hopi who married in Zuni, Nayuchi, the Head Bow Priest; Layutsailunkia, foster father to Frank Cushing; Pahlowahtiwa, governor of Zuni; and Kiasi, Junior Bow Priest. (Courtesy Photo Archives, Museum of New Mexico, negative no. 89333, Benjamin W. Kilburn, photographer.)

(1892, p. 151) described the one at Hopi, and Alexander Marshack, at my suggestion, found the same stick in the Smithsonian Institution. The Hopi stick has a total of 28 grooves, which could be used to count or record a month of phases if the times of

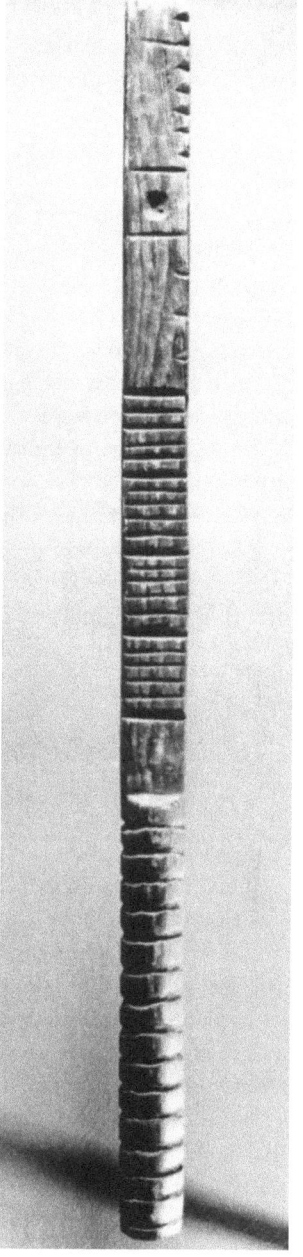

8.9 Probable Hopi calendar stick, collected by Fewkes. (Photo by Alexander Marshack.)

invisibility are not counted. The Zuni stick (Fig. 8.7) is described and drawn by Bourke (1881, 21 November), based on information acquired in a meeting with the Zunis Nanahe and Nayuchi (Fig. 8.8), who was Head Bow Priest at the time), who implies that it was used to tally a year, starting with ". . . the moon of the Fire Festival, or winter solstice." By Fire Festival, Bourke means the New Fire ceremony, which takes place right after the Zuni winter solstice ceremony. (I note that the Zuni stick more closely resembles the Winnebago one described by Marshack, 1985, than does the Hopi one.) Fewkes (1892, p. 151) states that he has seen calendar sticks at Walpi and Hano. He describes them as follows:

> These sticks are about a foot or a foot and a half long, and are divided into two parts, one section being round, the other flattened on one side. The round section is girt by fifteen shallow parallel grooves, and occupies about a third of the whole length of the stick. The remaining two thirds of the stick has a number of parallel grooves or notches cut upon the flattened surface. Five of the latter grooves, which are situated at equal distances, are deeper than the remaining, and between each pair there are four smaller parallel grooves arranged at equal distances. The space in which these grooves are cut occupies about one half of the flat portion of the stick. The remaining half, or that more distant from the round section, is divided into two parts, which are separated by a rectangular space, in, the centre of which there is a depression called the "*nā-tā'l-tci.*" On one side of the depression there are three notches, on the other, seven.

The stick found by Marshack (Fig. 8.9) matches this description quite well, and was probably collected by Fewkes at Hopi.

Sticks or rocks (even rock art?) with similar markings could be inferred to be Anasazi devices for tracking months, if found in the right archaeological context. More important, the mode of keeping such tallies would reveal the Anasazi solution to the problem of lunar calendar keeping.

THE MESOAMERICAN CONNECTION?

Archaeologists have debated for many years the possible influences of Mesoamerican cultures on the prehistoric societies of the US Southwest. The interpretations tend to fall into two camps: those that argue for a strong (controlling?) Mexican influence, specifically from the Toltec-Chalchihuites cultures (Kelley and Kelley, 1975), and those that claim that the Southwest—specifically Chaco—developed in essentially a cultural isolation (McGuire, 1980). I do not want to be caught in the cross-fire between these two groups but do want to propose one possible solution to their feud. It lies in a careful comparison of prehistoric Southwestern and Mesoamerican astronomical practices (Reyman, 1971)—a comparison that I now feel we can begin to make with some reliability.

A general comparison of historic Pueblo practices shows that, when compared with calendrical activities from Mayan and Central Mexican cultures, the Southwest *lacked*: (1) written calendars, (2) a numerical system with long counts, (3) 260-day ritual "years," (4) close attention to conjunctions of Venus, (5) a system of year bearers, and (6) zenith passages of the sun.

Some Southwestern investigators make the claim that the kachina cult is an outgrowth of Mexico that worked its way up the Rio Grande valley in the 14th century (Ellis and Hammack, 1968), especially on the basis of a change in the rock art (Schaafsma and Schaafsma, 1974). Sometime after AD 1300, a new rock art style pervades the Rio Grande valley (and extends as far west as Hopi)—a style that includes masked figures similar to the kachina dancers. The motifs derive from the Jornada Mogollon of southern New Mexico and include representations of the Mexican deity Quetzalcoatl. One of the most striking elements is that of Venus as Morning or Evening star (Schaafsma, 1980). Another Mexican influence relates to pecked-cross figures, which are abundant in the Mogollon region (Zeilik, 1980), but which are less common in Anasazi territory.

Reyman (1971), Ellis (1977) and Young (1989) have taken steps in a cosmological context by comparing Hopi and Zuni kachinas with Mesoamerican deities—specifically for Ellis and Young through parallels in iconography that may derive from parallel functions of supernatural beings in different (but perhaps intertwined) cultural contexts. Fundamental to the Puebloan attitude towards the gods is the notion of *reciprocity* between human beings and supernaturals so that the cosmic order can be sustained. That reciprocity shows up strongest in the sun/water cult relating to fertility—and that lays the base for sacred directions in both Puebloan and Mesoamerican world views. But, we still need to reveal the connections to astronomy, if a cross fertilization has indeed occurred.

One striking area of similarity relates to sacred directions (Reyman, 1971; Broda, 1982; Vogt, 1985; and Young, 1989). The important parallels here include the connection of specific animals and colors to each direction, as well as mountain shrines as the physical embodiments of the sacred space. But the most promising aspect to me is that of the *solstitial* points as cardinal points, rather than the European notions of north, south, east and west at right angles to each other. When seen in the Pueblo context, they may represent cultural borrowings or a misunderstanding by an investigator from a European background who "expected" the "proper" cardinal directions. Zenith passages do integrate particular directions and times in Mesoamerica (Broda, 1982, pp. 90–3); they do not occur in the Southwest. I do find it intriguing that the Isleta ceremonies of drawing down the sun (Parsons, 1932, pp. 290–3) and the moon (Parsons, 1932, p. 342) involve the use of sunlight and moonlight at meridian passage—perhaps an analogy to Mesoamerican zenith tubes (Broda, 1982, pp. 90–3) even though the sun does not transit the zenith at Southwestern latitudes.

I suggest that one way to clear up the cloud around the Southwestern-Mesoamerican connection requires a close look at Venus in the Southwestern context—a job yet to be done. We have many tales of the Pueblo Twin War Gods, who are related to Venus as Morning and Evening star (Young and Williamson, 1981). In the Rio Grande Pueblos, the two War Chiefs (or Outside Chiefs) are the two aspects of Venus. Curiously, only they can criticize the actions of the cacique—the religious head of the Pueblo—who does the sun-watching. As Bandelier noted in 1880:

> If the cacique shows unfavorable character, is not peaceable, quiet, or is ambitious, then the capitan de guerra [War Chief] calls a meeting at the estufa [kiva], which meeting is attended by all those who are regarded as capable of understanding the matter and worthy of participation in it... (Lange and Riley, 1966, p. 179).

The cacique represents Earth Mother, who must know that the people of the Pueblo are doing the proper rites at the proper time. The War Chief, in contrast, represents Father Sun—a position that fell to him when the older position of War *Priest* disappeared after the Spanish contact with the Pueblos (F. H. Ellis, private communication). The War Priest and the cacique together headed the Pueblo, with equal but complementary duties as earth's representatives of Father Sun and Earth Mother. The War Priest had the responsibility of keeping the Pueblo safe from invaders and witches; his two assistants in this task were the War Chiefs, sons of the Sun. When the office of the War Priest died out, the "elder" War Chief took over some of the War Priest's duties and carries them out today. Hence, one aspect of Venus has acquired some of the Sun's power in the Pueblo symbolic context. Needed here is a detailed investigation of Pueblo recordings of observations of Venus, a comparison to Venus configurations on those dates, and the search for a cycle (five or eight years, for example) in the historical record.

CONCLUSIONS AND SUGGESTIONS

Workers in the Southwest have come a long way since the concerted efforts to find rock art that may represent the 1054 supernova (Miller, 1955; Young, 1986). That field work drew attention to the issue of astronomy in the prehistoric Southwest. The next step involved possible kiva alignments (Reyman, 1971) and the search for sun-watching stations (Williamson *et al.,* 1975; Williamson, Fisher and O'Flynn,

1977) and the first instances of light and shadow interactions. That stage culminated in the recognition of the sunlight interactions at the three-slab site on Fajada Butte by Jay Crotty and Anna Sofaer and the subsequent long-term efforts there by Sofaer and co-workers, especially Rolf Sinclair at that site and others on the butte. We have recently begun to appreciate the possibilities of concrete expressions of cosmic structure by the Anasazi—such as with Casa Rinconada (Williamson, 1982, 1987)—based on available ethnographic information of Puebloan cosmology (such as described for Hopi Third Mesa by Titiev, 1944, and by Parsons, 1939, more generally). However, such possibilities will be very hard to demonstrate conclusively (Reyman, 1971, p. 321, rejected this notion based on his field work).

We are now at a new stage in the search for Anasazi astronomy—that of trying to place sites and phenomena into a specific cultural context. To do so successfully requires that we draw the archaeologists into the arena as active participants. They will tell us about the problems of the control of time with which we must now grapple in order to connect at the right level in the cultural matrix.

How to entice the archaeologists, most of whom have so far stood at the sidelines? I propose that we focus on the *material correlates of astronomical activities*. We need to demonstrate that the focus on the sky results in material remains that differ from those developed for other ritual activities. I have mentioned some so far: windows/portals that admit light onto walls with calendrical markings; calendar sticks or stones; light/shadow interactions on rock art at key seasonal times (solstices, planting); offerings from rituals that occur seasonally.

What can we do for ourselves? First, we need more *cooperative* work to make effective use of our energies that must span a wide geographic area. Second, in contrast, we also need more *independent* study of the same sites to avoid observer bias. Third, we must *share* our results, even in preliminary form, faster so that we can develop effective ventures in new directions. Fourth, we should develop a *broader* context with the Southwest and Mesoamerica. Fifth, we need *intensive* work in specific areas, such as a field school at Chaco Canyon. Sixth, light/shadow sites need an *objective* evaluation of their practical resolving power. Finally, we must hammer out appropriate methodologies so that we can trust our interpretations and conclusions, which we should view as tentative insights into Anasazi astronomy and cosmology.

ACKNOWLEDGMENTS

I thank Rolf Sinclair, Anna Sofaer, M. Jane Young and Steve McCluskey for sending me copies of pertinent papers. Ray Williamson, Tony Aveni, Jonathan Reyman, Curt Schaafsma, Polly Schaafsma, John Carlson, Steve McCluskey and M. Jane Young read over and commented upon early drafts of this paper. Tom Windes and Joan Mathien made substantial comments on a later draft, as did Florence Ellis, whose knowledge of the Pueblos is deep and intimate.

REFERENCES

Beaglehole, E. 1937. Notes on Hopi Economic Life. *Yale University Publications in Anthropology* 15.

Binford, L. R. 1972. Methodological Considerations of the Archaeological Use of Ethnographic Data. In *An Archaeological Perspective*, 59–67. New York, Seminar Press.

Bourke, J. G. 1881. Diary. West Point, NY: United States Military Academy Library, Special Collections and Archives. Photofacsimile in Special Collections, Coronado Room, Zimmerman Library, University of New Mexico. Entries for November 1881.

Brandt, J. C., S. P. Maran, R. A. Williamson, R. S. Harrington, C. Cochran, M. Kennedy, W. J. Kennedy, and V. D. Chamberlain. 1975. Possible Rock Art Records of the Crab Nebula Supernova in the Western United States. In *Archaeoastronomy in Precolumbian America*, ed. A. F. Aveni, 45–58. Austin, University of Texas Press.

Brandt, J. C., and R. A. Williamson. 1977. Rock Art Representations of the AD 1054 Supernova: A Progress Report. In *Native American Astronomy*, ed. A. F. Aveni, 171–177. Austin, University of Texas Press.

Brandt, J. C., and R. A. Williamson. 1979. 1054 Supernova and Rock Art. *Archaeoastronomy Supplement to the Journal for the History of Astronomy* 1:S1–S38.

Broda, J. 1982. Astronomy, Cosmovision, and Ideology in Pre-Hispanic Mesoamerica. In *Ethnoastronomy and Archaeoastronomy in the American Tropics*, ed. A. F. Aveni and G. Urton. Annals of the New York Academy of Sciences, vol. 385.

Bunzel, R. L. 1932a. Introduction to Zuni Ceremonialism. *Bureau of American Ethnology 47th Annual Report*, pp. 467–544.

Bunzel, R. L. 1932b. Zuni Ritual Poetry. *Bureau of American Ethnology 47th Annual Report*, pp. 611–835.

Bunzel, R. L. 1932c. Zuni Katcinas. *Bureau of American Ethnology 47th Annual Report*, pp. 837–1086.

Carlson, J. P. 1987. Romancing the Stone, or Moonshine on the Sun Dagger. In *Astronomy and Ceremony in the Prehistoric Southwest*, ed. J. B. Carlson and W. J. Judge. Albuquerque, Maxwell Museum Press. Papers of the Maxwell Museum Press, no. 2, pp. 71–88.

Cope, L. 1919. Calendars of the Indians North of Mexico. *University of California Publications in American Archaeology and Ethnology* 16 (4):119–176.

Cordell, L. S. 1984. *Prehistory of the Southwest*. New York, Academic Press.

Curtis, E. S. 1922. The Hopi. *The North American Indian* 22.

Cushing, F. H. 1979. My Adventures in Zuni. Reprinted in *Zuni*, ed. J. Green, 46–134. Lincoln, University of Nebraska Press.

Dutton, B., and M. A. Marmon. 1936. The Laguna Calendar. *The University of New Mexico Bulletin, Anthropological Series*, 283 (2):1.

Dutton, D. P. 1963. *Sun Father's Way: The Kiva Murals of Kuaua*. Albuquerque, University of New Mexico Press.

Ellis, F. H. 1975. A Thousand Years of the Pueblo Sun-Moon-Star Calendar. In *Archaeoastronomy in Precolumbian America*, ed. A. F. Aveni, 59–87. Austin, University of Texas Press.

Ellis, F. H. 1977. Distinctive Parallels between Mesoamerican and Pueblo Iconography and Deities. Paper delivered at Guanajuato, Mexico.

Ellis, F. H., and L. Hammack. 1968. The Inner Sanctum of Feather Cave: A Mogollon Sun and Earth Shrine Linking Mexico and the Southwest. *American Antiquity* 30:25–44.

Farrer, C. R., and B. Second. 1981. Living the Sky: Aspects of Mescalero Apache Ethnoastronomy. In *Archaeoastronomy in the Americas,* ed. R. A. Williamson, 137–151. Los Altos, CA: Ballena Press/Center for Archaeoastronomy.

Fewkes, J. W. 1891. A Few Summer Ceremonials at Zuni Pueblo. *Journal of American Ethnology and Archaeology* 1:1–61.

Fewkes, J. W. 1892. A Few Summer Ceremonials at the Tusayan Pueblos. *Journal of American Ethnology and Archaeology* 2:1–161.

Fewkes, J. W. 1906. Hopi Shrines Near East Mesa, Arizona. *American Anthropologist* (NS) 8:346–375.

Fewkes, J. W. 1919. Prehistoric Villages, Castles, and Towers of Southwestern Colorado. *Bureau of American Ethnology Bulletin* 79:plates 14a, b, c; 19c; 32b.

Forde, C. D. 1931. Hopi Agriculture and Land Ownership. *Journal of the Royal Anthropological Institute of Great Britain and Ireland* 61:357–405.

Frigout, E. 1979. Hopi Ceremonialism. *Handbook of North American Indians,* ed. A. Ortiz, 9:564–576. Washington, DC, Smithsonian Institution.

Gaustad, J. E. 1980. The Chaco Canyon Supernova Pictograph—A Reorientation. *Archaeoastronomy* 111 (4):33–34.

Harrington, J. P. 1916. *Bureau of American Ethnology 29th Annual Report,* pp. 29–618.

Hedges, K. 1985. Methodology and Validity in California Archaeoastronomy. In *Earth and Sky,* ed. A. Benson and M. Hoskinson, 25–39. Thousand Oaks, Slow Press.

Hewett, E. L. 1906. Antiquities of the Jemez Plateau. *Bureau of American Ethnology Bulletin* no. 32.

Kelley, J. C., and E. A. Kelley. 1975. An Alternative Hypothesis for the Explanation of Anasazi Cultural History. In *Collected Papers in Honor of Florence Hawley Ellis,* ed. T. R. Frisbie. Papers of the Archaeological Society of New Mexico, no. 2, pp. 178–223.

Koenig, S. H. 1979. Stars, Crescents, and Supernovae in Southwestern Indian Art. *Archaeoastronomy Supplement to the Journal for the History of Astronomy* 1:539–550.

Lange, C. H. 1959. *Cochiti: A New Mexico Pueblo Past and Present.* Austin, University of Texas Press.

Lange, C. H., and C. L. Riley. 1966. *The Southwestern Journals of Adolph F. Bandelier: 1880–1882.* Albuquerque, University of New Mexico Press.

Lyon, L. 1985. Chronology of the Zuni Sha'lak'o Ceremony. In *Southwestern Culture History: Collected Papers in Honor of Albert H. Schroeder,* ed. C. H. Lange, 233–249. Papers of the Archaeological Society of New Mexico, no. 10.

McCluskey, S. C. 1977. The Astronomy of the Hopi Indians. *Journal for the History of Astronomy* 8:174–195.

McCluskey, S. C. 1981. Transformations of the Hopi Calendar. In *Archaeoastronomy in the Americas,* ed. R. A. Williamson, 173–182. Los Altos, CA, Ballena Press/Center for Archaeoastronomy.

McCluskey, S. C. 1982. Historical Astronomy: The Hopi Example. In *Archaeoastronomy in the New World,* ed. A. Aveni, 31–57. Cambridge, Cambridge University Press.

McGuire, R. H. 1980. The Mesoamerican Connection in the Southwest. *The Kiva* 46:3–38.

McKenna, P. J. 1984. The Architecture and Material Culture of Chaco Canyon, New Mexico. In *Reports of the Chaco Center,* ed. W. J. Judge. Albuquerque, Division of Cultural Resources, National Park Service, no. 7.

Marshack, A. 1985. A Lunar-solar Year Calendar Stick from North America. *American Antiquity* 50:27–51.

Mayer, D. 1979. Miller's Hypothesis. *Archaeoastronomy Supplement to the Journal for the History of Astronomy* 1:S51–S74.

Miller, W. C. 1955. Two Prehistoric Drawings of Possible Astronomical Significance. *Astronomical Society of the Pacific Leaflet* 3:14.

Newman, E. B., R. K. Mark, and R. G. Vivian. 1982. Anasazi Solar Marker: The Use of a Natural Rockfall. *Science* 217:1036–1038.

Ortiz, A. 1969. *The Tewa World*. Chicago, University of Chicago Press.

Parsons, E. C. 1917. Notes on Zuni. *Memoirs of the American Anthropology Association* 4 (3).

Parsons, E. C. 1925. A Pueblo Indian Journal 1920–1921. *Memoirs of the American Anthropological Society* 32.

Parsons, E. C. 1932. Isleta, New Mexico. *Bureau of American Ethnology 47th Annual Report*, pp. 193–466.

Parsons, E. C. 1939. *Pueblo Indian Religion*. Chicago, University of Chicago Press.

Preston, A. L., and R. A. Preston. 1985. The Discovery of 19 Prehistoric Calendric Petroglyph Sites in Arizona. In *Earth and Sky,* ed. A. Benson and M. Hoskinson, 123–133. Thousand Oaks, Slow Press.

Preston, R. A. 1984. Calendrical Petroglyph Sites in Arizona: New Evidence and Statistical Studies. Paper presented at the 1984 International Conference on Prehistoric Rock Art and Archaeoastronomy, Little Rock, Arkansas.

Preston, R. A., and A. L. Preston. 1987. Evidence for the Calendric Function at 19 Prehistoric Petroglyph Sites in Arizona. In *Astronomy and Ceremony in the Prehistoric Southwest,* ed. J. B. Carlson and W. J. Judge. Albuquerque, Maxwell Museum Press. Papers of the Maxwell Museum of Anthropology, no. 2, pp. 191–204.

Reyman, J. E. 1971. Mexican Influence on Southwestern Ceremonialism. Ph.D. dissertation, Southern Illinois University, pp. 114–122.

Reyman, J. E. 1975. The Nature and Nurture of Archaeoastronomical Studies. In *Archaeoastronomy in Precolumbian America,* ed. A. F. Aveni, 205–215. Austin, University of Texas Press.

Reyman, J. E. 1976. Astronomy, Architecture, and Adaptation at Pueblo Bonito. *Science* 193:957–962.

Reyman, J. E. 1980. The Predictive Dimension of Priestly Power. In *New Frontiers in Archaeology and Ethnohistory of the Greater Southwest,* ed. C. L. Riley and B. C. Hendrick, 40–59. Transactions of the Illinois Academy of Science 72:4.

Schaafsma, P. 1980. *Indian Rock Art of the Southwest*. Albuquerque, University of New Mexico Press.

Schaafsma, P. 1985. Form, Content, and Function: Theory and Method in North American Rock Art Studies. *Advances in Archaeological Method and Theory* 8:237–277.

Schaafsma, P., and C. F. Schaafsma. 1974. Evidence for the Origins of the Katchina Cult as Suggested by Southwestern Rock Art. *American Antiquity* 33:535–545.

Sekaquaptewa, A. 1983. Out of Phase with the Moon Phase. In *Quatoqi*. Arizona, Oraibi.

Simmons, L. W. 1942. *Sun Chief: The Autobiography of a Hopi Indian*. New Haven, CT: Yale University Press.

Sofaer, A., and R. Sinclair. 1987. Astronomical Markings at Three Sites on Fajada Butte. In *Astronomy and Ceremony in the Prehistoric Southwest,* ed. J. B. Carlson and W. J. Judge. Albuquerque, Maxwell Museum Press. Papers of the Maxwell Museum of Anthropology, no. 2, pp. 43–67.

Sofaer, A., and R. M. Sinclair. 1986a. Letter. *Science* 231:1057–1058.

Sofaer, A., and R. M. Sinclair. 1986b. Appraisal. *Archaeoastronomy Supplement to the Journal for the History of Astronomy* 10:S59–S66.

Sofaer, A., R. M. Sinclair, and L. E. Doggett. 1982. Lunar Markings on Fajada Butte, Chaco Canyon, New Mexico. In *Archaeoastronomy in the New World*, ed. A. F. Aveni, 169–181. Cambridge, Cambridge University Press.

Sofaer, A., V. Zinser, and R. M. Sinclair. 1979a. A Unique Solar Marking Construct. *Science* 206:283–291.

Sofaer, A., V. Zinser, and R. M. Sinclair. 1979b. A Unique Solar Marking Construct of the Ancient Pueblo Indians. In *American Indian Rock Art*, ed. F. G. Bock, K. Hedges, G. Lee, and H. Michaelis, vol. 5:117–25. El Toro, CA, American Rock Art Association.

Spier, L. 1955. Mohave Culture Items. *Museum of Northern Arizona Bulletin*, no. 28.

Stephen, A. M. 1936. *Hopi Journal*, ed. E. C. Parsons. New York, Columbia University Press.

Stevenson, M. C. 1904. The Zuni Indians: Their Mythology, Esoteric Societies, and Ceremonies. *Bureau of American Ethnology 23rd Annual Report*.

Stevenson, M. C. N.d. Material on the Tewa, Harrington Papers. Washington, DC: Smithsonian Anthropological Archives.

Stirling, M. W. 1945. Concepts of the Sun among American Indians. *Annual Report of the Smithsonian Institution*, pp. 387–400.

Tedlock, B. 1983. Zuni Sacred Theater. *Native American Quarterly* 7 (4):93–110.

Titiev, M. 1938. Dates of Planting at the Hopi Pueblo of Oraibi. *Museum Notes of the Museum of Northern Arizona* 11 (5):39–42.

Titiev, M. 1944. Old Oraibi: A Study of the Hopi Indians of Third Mesa. *Papers of the Peabody Museum, Harvard University* 22:1.

Toll, H. W., T. C. Windes, and P. J. McKenna. 1980. Late Ceramic Patterns in Chaco Canyon: The Pragmatics of Modeling Ceramic Exchange. In *Models and Methods in Regional Exchange*, ed. R. E. Fry. Washington, DC: Society for American Archaeology.

Vogt, E. Z. 1985. Cardinal Directions and Ceremonial Circuits in Mayan and Southwestern Cosmology. *National Geographic Society Research Reports* 21:487–496.

White, L. A. 1942. The Pueblo of Santa Ana, New Mexico. *Memoirs of the American Anthropological Society*, no. 60.

Williamson, R. A. 1979. Field Report Hovenweep National Monument. *Archaeoastronomy* 2 (3):11–12.

Williamson, R. A. 1981. North America: A Multiplicity of Astronomies. In *Archaeology in the Americas*, ed. R. A. Williamson, 61–80. Los Altos, CA, Ballena Press/Center for Archaeoastronomy.

Williamson, R. A. 1982. Casa Rinconada: Twelfth-Century Anasazi Kiva. In *Archaeoastronomy in the New World*, ed. A. Aveni, 205–218. Cambridge, Cambridge University Press.

Williamson, R. A. 1983. Sky Symbolism in a Navajo Rock Art Site, Chaco Canyon. *Archaeoastronomy* 6 (14):59–65.

Williamson, R. A. 1984. *Living the Sky: The Cosmos of the American Indian*. Boston: Houghton Mifflin.

Williamson, R. A. 1987. Light and Shadow, Ritual, and Astronomy in Anasazi Structures. In *Astronomy and Ceremony in the Prehistoric Southwest*, ed. J. B. Carlson and W. J. Judge, 99–119. Albuquerque, Maxwell Museum Press. Papers of the Maxwell Museum of Anthropology, no. 2.

Williamson, R. A., H. J. Fisher, and D. O'Flynn. 1977. Anasazi Solar Observatories. In *Native American Astronomy*, ed. A. Aveni, 203–217. Austin, University of Texas Press.

Williamson, R. A., H. J. Fisher, A. F. Williamson, and C. Cochran. 1975. The Astronomical Record in Chaco Canyon, New Mexico. In *Archaeoastronomy in Precolumbian America*, ed. A. F. Aveni, 33–43. Austin, University of Texas Press.

Williamson, R. A., and M. J. Young. 1979. An Equinox Sun Petroglyph Panel at Hovenweep National Monument. In *American Indian Rock Art,* ed. F. G. Bock, K. Hedges, G. Lee, and H. Michaelis, vol. 5:70–80. El Toro, CA, American Rock Art Research Association.

Yava, A. 1978. *Big Falling Snow.* Albuquerque, University of New Mexico Press.

Young, M. J. 1986. The Interrelationship of Rock Art and Astronomical Practice in the American Southwest. *Archaeoastronomy Supplement to the Journal for the History of Astronomy* 10:543–558.

Young, M. J. 1987a. The Nature of the Evidence: Archaeoastronomy in the Prehistoric Southwest. In *Astronomy and Ceremony in the Prehistoric Southwest,* ed. J. B. Carlson and W. J. Judge. Albuquerque, Maxwell Museum Press. Papers of the Maxwell Museum of Anthropology, no. 2, pp. 169–190.

Young, M. J. 1987b. Issues in the Archaeoastronomical Endeavor in the American Southwest. In *Astronomy and Ceremony in the Prehistoric Southwest,* ed. J. B. Carlson and W. J. Judge. Albuquerque, Maxwell Museum Press. Papers of the Maxwell Museum of Anthropology, no. 2, pp. 219–232.

Young, M. J. 1989. The Southwest Connection: Similarities between Western Puebloan and Mesoamerican Cosmology. In *World Archaeoastronomy,* ed. A. Aveni. Cambridge, Cambridge University Press.

Young, M. J., and R. A. Williamson. 1981. Ethnoastronomy: The Zuni Case. In *Archaeoastronomy in the Americas,* ed. R. A. Williamson, 183–191. Los Altos, CA, Ballena Press/Center for Archaeoastronomy.

Zeilik, M. 1980. Pecked-cross-like Petroglyphs in New Mexico. *Archaeoastronomy* 3 (1):21.

Zeilik, M. 1983a. Anticipation in Anasazi Astronomy. Paper presented at the 56th Annual Pecos Conference, August 1983.

Zeilik, M. 1983b. Historic Puebloan Sun Watching. Paper presented at the First International Ethnoastronomy Conference, Washington, DC, September 1983.

Zeilik, M. 1984a. A Possible Equinoctial Sun-sighting Station at Tsiping, New Mexico. *Archaeoastronomy* 7 (1–4):70–75.

Zeilik, M. 1984b. Summer Solstice at Casa Rinconada: Calendar, Hierophany, or Nothing? *Archaeoastronomy* 7 (1–4):76–81.

Zeilik, M. 1985a. The Ethnoastronomy of the Historic Pueblos. 1. Calendrical Sun Watching. *Archaeoastronomy Supplement to the Journal for the History of Astronomy* 8:S1–S25.

Zeilik, M. 1985b. The Fajada Butte Solar Marker: A Reevaluation. *Science* 228:1311–1313.

Zeilik, M. 1985c. A Reassessment of the Fajada Butte Solar Marker. *Archaeoastronomy Supplement to the Journal for the History of Astronomy* 9:S69–S85.

Zeilik, M. 1985d. Sun Shrines and Sun Symbols in the US Southwest. *Archaeoastronomy Supplement to the Journal for the History of Astronomy* 9:S86–S96.

Zeilik, M. 1985e. Keeping a Seasonal Calendar at Pueblo Bonito. *Archaeoastronomy* 8.

Zeilik, M. 1986a. The Ethnoastronomy of the Historic Pueblos. 2. Moon Watching. *Archaeoastronomy Supplement to the Journal for the History of Astronomy* 10:S1–S22.

Zeilik, M. 1986b. Reply. *Science* 231:1058.

Zeilik, M. 1986c. Response. *Archaeoastronomy Supplement to the Journal for the History of Astronomy* 10:S66–S69.

Zeilik, M. 1987. Anticipation in Ceremony: The Readiness Is All. In *Astronomy and Ceremony in the Prehistoric Southwest,* ed. J. B. Carlson and W. J. Judge, 25–41. Albuquerque, Maxwell Museum Press. Papers of the Maxwell Museum of Anthropology, no. 2.

Zeilik, M., and R. Elston. 1983. Wijiji at Chaco Canyon: A Winter Solstice Sunrise and Sunset Station. *Archaeoastronomy* 6 (1–4):66–73.

CHAPTER NINE

Native Astronomy in Mesoamerica

Michael D. Coe

INTRODUCTION

If any one trait can be said to be distinctive of the native cultures of prehispanic Mesoamerica, it is a deep concern with the heavenly bodies and the passage of time as marked by the apparent movements of these objects. We know of the Mesoamerican obsession with the sun, the moon, and the night sky from the testimony of carved stone monuments, of the surviving native books, and from statements of the native intelligentsia set down after the Spanish conquest. For instance, we are told by the historian Torquemada of the meritorious qualities of Nezahualpilli, king of Texcoco, in the following terms:

First published as "Native Astronomy in Mesoamerica," by Michael Coe, in *Archaeoastronomy in Pre-Columbian America,* ed. A. Aveni. © 1975 by the University of Texas Press. All rights reserved.

> It is said that he was a great astrologer; that he was much concerned with understanding the movement of the celestial bodies. Inclined to the study of these things, he would seek in his kingdoms for those who knew of these things, and he would bring them to his court. He would communicate to them all that he knew. And at night he would study the stars, and he would go on the roof of his palace, and from there he would watch the stars, and he would discuss problems with them. (León-Portilla 1963, p. 142)

There is abundant evidence for much of Mesoamerica that the study of the heavens was the province of specialists, generally the priests as among the Aztecs, and we even have a Nahuatl word for "astrologer," *ilhuica tlamatilizmatini*, "the wise man who studies heaven."

It comes, then, as a surprise to learn how poor, scanty, and misleading our information is on native astronomy in Mesoamerica. In Spanish writings and compilations of data on central Mexico and the Maya area on the eve of the Conquest, the subject is hardly mentioned at all, and then in the most equivocal terms. Even the great Sahagun devotes only a few pages to it. And yet these men were contemporaries of Copernicus, and lived during a time when all of Europe was astounded and perplexed by the revolution in our knowledge of the universe that was then taking place. I am inclined to think that this revolution largely bypassed Spain and the soldiers and missionaries that she sent to the New World. These were men who, if they thought about astronomy at all, thought about it in terms of the judicial astrology then in vogue, and were inclined to dismiss with contempt native concepts of the heavens. I will later point out how utterly confusing the Spanish accounts really are. The failure of the Spaniards to properly record or to understand resulted in one of the greatest intellectual losses in all history.

Our present knowledge of Mesoamerican astronomy thus comes largely from the study of codices and monuments, particularly those from the Maya area. Since the nineteenth century, scholars such as Forstemann, Seler, Nuttall, Spinden, Thompson, and Caso have devoted themselves to this pursuit, and have revealed a world of knowledge unsurmised or ignored by the Spaniards.

But there are obviously entire areas of native astronomical concepts and practices not clearly visible in the reliefs and surviving books. As an example, large numbers of deities in the central Mexican and Maya manuscripts have stellar attributes. What does this mean? In the present state of our knowledge, we are not at all certain. What instruments did the specialists use to observe and measure the heavens (Nuttall 1906)? We can only guess at this stage. What constellations and asterisms were important to them, and why? Did they have a zodiac, and if so, was it solar or lunar?

There is a final source of information that has usually been overlooked: the American Indians of Mesoamerica and adjacent regions whose cultures have largely survived the continuing onslaught of European civilization. This area has hardly been touched, and the reasons are not hard to find. In the first place, there is scarcely an ethnologist or social anthropologist who can identify anything other than the moon and the Big Dipper in the night sky; the so-called natives are a great deal wiser. As a

Huichol told Carl Lumholtz (1900, p. 59), who was recording information about a native planisphere, "People think we Indians don't know anything, but we know more than the whites." Secondly, the local subculture of the social anthropologists who have worked in Mesoamerica has generally ignored the problem of survivals of nature culture in favor of acculturation and community studies, which are for the most part of little or no interest to archaeologists.

I am convinced that there is still much to be learned from modern ethnoastronomical research on these peoples, particularly on those who are still relatively isolated from the processes of ladinoization. The fragmentary data suggest that many groups retain native constellations and names for the bright stars, and this is an area which sorely needs further research. It is true that the great specialists in astronomy, those who had deep scientific and esoteric knowledge of the heavens, were effectively eliminated by the Spanish overlords after the Conquest. But specialists (on what might be called the "folk" level of organization) probably still survive in remote areas. One must not expect "the man in the street" to have much of this knowledge, any more than one could find out from the modern New Yorker how a television set works. In what is probably the most complete study ever made of the starlore of any American Indian group, Father Berard Haile (1947) found that the average Navajo is unacquainted with this body of knowledge; not even most singers know anything about it. Singers who wish to reach a high degree of proficiency in the star-gazing art must lie out under the stars night after night, and through the seasons, with older practitioners to memorize the heavens, and then this person must be able to reproduce what he has seen in colored sands within the hogan. Navajo specialists of high degree are able to prescribe the proper ceremonial for sick clients by viewing the refracted colors of first-magnitude stars through crystals or glass.

In this brief survey, I am going to leave the extremely important subject of archaeoastronomy aside, and concentrate only upon native astronomy as revealed to us in the documents, whether pre- or posthispanic. It is my hope to be able to single out those celestial phenomena that seem to have been of most significance to the native Mesoamerican mind, so that those investigating the possible orientation of ancient buildings and cities, and perhaps their complete layout, might know which correlations are more likely than others, and which ones do or do not conform to the Mesoamerican mental set. As Burland (1952, p. 26) has so aptly put it, "Each correction to past work is a step nearer the truth, and we must, if we wish our researches to progress, give up our preconceptions and try to understand native American cultures by 'thinking Indian' and seeing nature as they saw it—simply and with a respect for its mystery."

THE MESOAMERICAN UNIVERSE

To understand Mesoamerican astronomy, one must study their conception of the cosmos in which the heavenly bodies acted out their role. For this, we have abundant evidence from both the central Mexicans and the Maya, well analyzed by Thompson

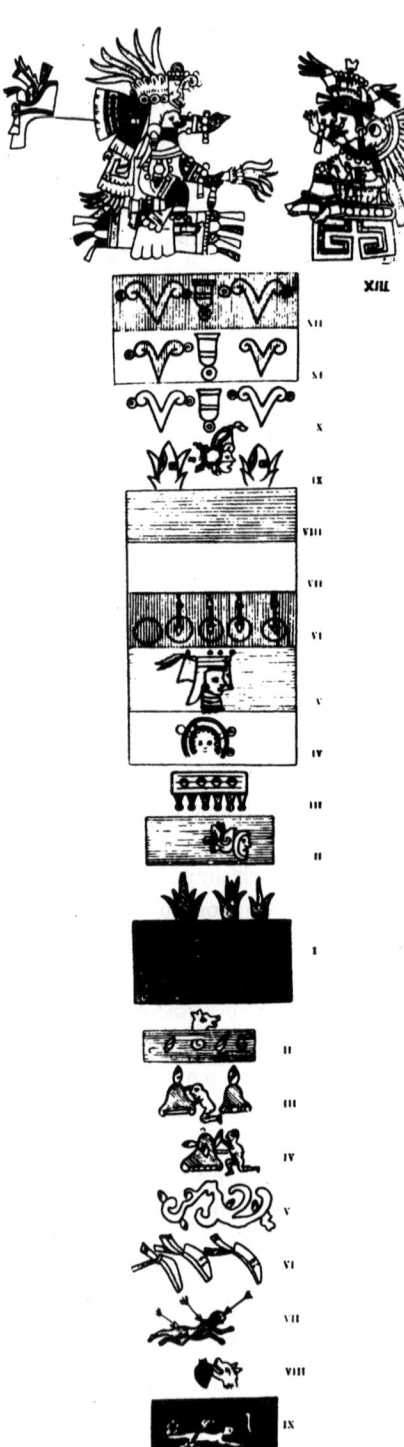

9.1 The heavens, the earth and the levels of the underworld as depicted on page 2 of the Codex Vaticanus A (Rios) (after Nicholson 1971, Fig. 7).

(1950, 1970), Soustelle (1940), Caso (1954), León-Portilla (1963), and Nicholson (1971). The picture that has been built up is almost Ptolemaic in its scope. In place of the concentric spheres of Ptolemy, in which astral objects fulfilled their geometric roles, the Mesoamericans conceived a layered universe, well illustrated by two pages of the Codex Vaticanus A (Fig. 9.1). The earth itself was conceived of at times as a large wheel or disk, at other times as a figure resembling our four-leaved clover; apparently this formed the back of an enormous saurian lying in water, surrounded by water-lilies and other aquatic vegetation.

The water surrounding the earth was called by Nahuatl speakers *teoatl* or "divine water," and *ilhuica atl*, "celestial water," since the seas which bordered the land extended up to the heavens on the horizon. These heavens were conceived of as thirteen in number, although it is obvious from our codex that the earth was counted as the first layer. Each heaven contained celestial objects as well as certain gods. Through the second and lowest, *Ilhuicatl metzli*, traveled the moon on its course, and from this layer were suspended the clouds. Above it was *Citlalco*, the place of the fixed stars and the abode of *Citlallicue*, She of the Starry Skirt, an astral goddess who seems to have been a female aspect of the dual Creator Deity. The sun traveled its diurnal road in the fourth layer, *Ilhuicatl Tonatiuh*;

while Venus, usually given a name meaning "great star," inhabited the next layer up. The sixth layer was *Ilhuicatl Mamalhuazocan*, the Heaven of the Fire Drill, a constellation the identity of which is debatable; in it were seen the comets, called "stars that smoke." Also in this heaven were the Fire Serpents, reptiles whose duty was to conduct the sun each day from the eastern horizon to the zenith. The seventh layer was the blackish or green heaven in which were winds and storms, while dust filled the eighth or blue heaven. Nicholson (1971, table 2) thinks it possible that thunder came from the ninth heaven, called *Itztapatl Nanatzcayan*, "where the stone slabs crash together." Layers ten through twelve were associated with the colors white, yellow, and red. Finally, in the thirteenth layer, *Omeyocan*, resided the dual, male-female, creator god whose all-embracing role as the progenitor of time, space, the gods, and all things has been so well worked out by León-Portilla.

Of course, this scheme has been largely derived from central Mexican sources, but there is strong reason to believe that it also applies to the Maya. Furthermore, both Maya and Mexicans seem to have had a nine-layered underworld, beginning with the earth again as the first layer.

Another extremely important concept in the Mesoamerican world-view is that of color directions. Each cardinal point was associated with a color (to the Aztecs these were east–red, north–black, west–white, south–blue), a tree, a bird, and (according to Vaticanus A), a part of the human body. Most importantly, to each direction was assigned a day in the 260-day count and a division of the same. Thus, space and time became inextricably intertwined in a kind of all-embracing mechanism.

Perhaps most important of all in their cosmological thinking was the calendar itself. At its heart was the sacred 260-day count, the origin of which remains obscure. This was based upon the permutation of thirteen numerical coefficients with a sequence of twenty days, the names of which are largely of animal origin, although plants and natural forces also enter into the scheme. At the time of the Conquest, almost all prognostications were based upon this calendar, a fact which caused Sahagun to utter the testy comment:

> For the art of judiciary astrology, common among us, is founded upon natural astrology, which is in the signs and planets of the heavens and in their courses and aspects. But this art of soothsaying followeth, or is founded upon, some characters and numbers in which no natural foundation existeth, but are only an artifice made by the devil himself. Nor is it possible that any man could have made or invented this art. For it hath no foundation in any science nor in any natural order. Rather it appeareth to be a thing of fraud and deceit than rational or ingenious. (Sahagun 1957:145)

A contrary view has been advanced by several students of the subject (summarized in Thompson 1950, pp. 98–9), namely that there *is* a rational basis for the count, if at one time, as among some modern Maya, the count was fixed rather than cyclical. It so happens that at about 15° north, which is the latitude of Copan and many Maya sites, the two passages of the sun through the zenith take place at an interval of 105 days; this is also the interval between the two planting dates in this area. If the 260-day count began as an immovable segment of the agricultural year, but at some later point in the distant past was set rolling as a perpetual cyclical count, then this *might* explain its adoption in the first place. I consider these arguments extremely tenuous. At any rate, since it was associated with the color-direction concept, with the gods, and with the affairs of men, this ritual count was the most significant mental construct in Mesoamerica.

Detailed information from central Mexico and the Maya area shows that most or all Mesoamericans conceived of a dynamic universe, one in a constant state of change. The cosmos had been initiated with an old male-female couple, situated in the navel of the world and in the center of the thirteenth heaven; this dual divinity was at times manifest as the Old Fire God, who acted as the lord of the household hearth, of time, and of the solar year. To start the universe on its space-time course, the male-female divinity produced four offspring, each assigned to the four directions and appropriate colors; among the Mexicans, these were the four Tezcatlipocas, among the Maya the quadripartite deities known as the Bacabs. The struggle between these offspring, particularly between the Black Tezcatlipoca, the ruler of the north, and the White Tezcatlipoca of the west, called Quetzalcoatl, resulted in a series of cyclical creations and destructions known as "suns." Each of these creations was, however, imperfect, and it was only in the fifth sun, our own, that the Sun, Moon, and men were created as we know them. Interestingly enough, the stars had been formed in earlier creations and thus pre-existed the Sun.

In summary, the Mesoamerican cosmos was one in constant flux, in which space and time were co-terminous, in which the heavenly bodies moved in fixed layers, and which was in constant peril of cataclysm. It was this world-view which guided native astronomy.

THE SUN AND THE YEAR

Among the central Mexicans, the sun was deified as a young, red-visaged personage with the name Tonatiuh (Fig. 9.2), often symbolized in eagle form. To the Maya, he was an old god with large eye and Roman nose (Fig. 9.3), and his avian counterpart was the scarlet macaw. According to Aztec mythology, once the perfect, fifth sun had been created in Teotihuacan, the ancient capital city of highland Mexico, it had to be fed with hearts of brave captives, and mankind was specifically created for this sanguinary end.

The so-called solar calendar among both Mexicans and Maya numbered exactly 365 days, divided into eighteen 20-day periods (or *veintenas,* in Spanish) totaling 360 days, with five "days without name" at the end. This has been aptly termed the Vague Year, since, in spite of assertions in some of the ethnohistoric records, there is no evidence that the Mesoamericans ever intercalated days or leap days. This Vague Year permutated with the 260-day count to produce a Calendar Round of 51 Vague Years of 18,960 days, a time span which to them contained some remarkable numerical properties. Among the Maya, each Vague Year was named from the 260-day count position

Michael D. Coe

9.2 Tonatiuh, the Aztec Sun God, on page 12 of the Codex Cospi.

9.3 The Maya Sun God, from page 11a of the Dresden Codex.

on which its first day fell, and among the Aztec (according to Caso 1971, p. 34b), from the 260-day count position of the final day of the eighteenth month. These four days were the so-called Year Bearers.

The fact that the length of the solar year is actually 365.2422 days means that the Vague Year calendar was constantly gaining on the seasons by a factor of thirteen days every 52 Vague Years. It was once advanced by the astronomer John Teeple (1930, pp. 70–85) that the Maya had a system of "determinants" by which the Maya expressed the accumulated error since the inception of their Long Count Calendar, which was based upon a 360-day count. It has been conclusively shown by Proskouriakoff that these alleged "determinant" dates are actually historical events occurring at irregular time intervals. Thus results a fundamental problem in Mesoamerican astronomy: the Vague Year was constantly running ahead of the sun and the stars. On the other hand, their calendrical system was an eminently rational one which, by avoiding lunations for the most part, was similar to that of the Egyptians and was a prototype of the system of Julian days by which modern astronomers record time intervals. And furthermore, we know from their lunar calculations that the Maya, at any rate, had a remarkably accurate knowledge of the true length of the tropical year.

The beginning of the Vague Year therefore changed by about a quarter from year to year. There has been some dispute, based on conflicting evidence, as to the *veintena* which actually began the Vague Year among Nahua speakers. According to Caso (1971, p. 341), this was Izcalli, the first day of which fell on 24 January in the year 1521; Nicholson (1971, table 4) would make it Atlcahualo, which commenced on 13 February in that year. Landa, writing of the Yucatec Maya, correctly begins their Vague Year with the first of Pop, and has it fall on 16 July 1533. There is, therefore, not the slightest justification for the following statement in the Histoire du Mechique:

> They counted the year from the spring equinox, when the sun makes a straight shadow, and as soon as it was felt that the sun was rising, they counted the first day, and [thence] the days by twenties. (Garibay 1965, p. 69)

In fact, I have not been able to discover any other references to the solstices and equinoxes in the early data, whether archaeological or ethnohistorical, although these would have been easy to calculate using instruments as simple as a gnomon, sighting sticks, and horizon landmarks. I have no doubt, however, that they were important. Rafael Girard (1966) has discovered some remarkable information from the contemporary Chorti Maya of the Guatemala-Honduras border. The Chorti priests calculate the beginning of the Vague Year on 8 February, after the sun has completed an extensive "rest" in the winter solstice. The night of 30 April–1 May they take to signal the first passage of the sun across the zenith, based upon the position in the night sky of Orion's Belt, the Southern Cross, and the Pleiades; according to his informants, it is the position of at least some of these following sunset that is important. The summer solstice is marked on 21 June, 52 days following zenith passage. After another 52 days occurs the second passage of the sun through the zenith, heading south toward the equinox. All of these points are marked by important festivals, including a winter solstice celebration which takes place from 19 to 27 December, at which time the native priests say that the Pleiades and Orion's Belt rise at sundown and vanish at dawn.

Girard's discovery of the importance of these solar observations (which unfortunately he does not describe in enough detail), and their correlation with the celestial sphere, is so far unique for Mesoamerica. But he goes one step further, by showing that for the Chorti Maya, the four directions are not the cardinal points but the solstices, a finding that has been confirmed among the highland Maya of Chiapas. How widespread this concept was in ancient Mesoamerica remains unknown.

A further problem relating to the solar calendar is the point at which the Mesoamericans began their day, and the division of it into so-called "hours." Caso (1971, p. 345) suggested that the central Mexicans began the day at noontime, this point in time being easy to observe (presumably with a gnomon). But for the Classic Maya, Thompson (1950, pp. 174-4) feels that it started at sunrise, although the modern Jacaltec and Ixil Maya begin it with sunset. At any rate the data are far from conclusive.

Seler claimed that the well-known sequence of Nine Lords of the Night and Thirteen Birds of the Day, which accompany in unvarying succession the 260 days of the ritual calendar, proved that the day had thirteen hours and the night nine. No other scholar has followed him in this. There is evidence that the Aztec priests divided our 24-hour day into nine ritualistic periods, four in the day and five by night

(Thompson 1950, p. 177); one of the principal duties of the priests was said to be the observation of the stars so as to tell the nocturnal divisions, and to sound trumpets and drums for the observance of the appropriate rites. The Zapotec supposedly had nine divisions for the day, and nine for the night.

Now, one can devise simple means to tell the time of day. The question is, how did they divide the night accurately into equal parts without timepieces? There is absolutely no evidence that the Mesoamericans had water clocks or any other kind of non-solar mechanism. One thus is led to doubt statements that they could determine midnight or any other night hour except in the vaguest way; even then the divisions would have been of different lengths as the seasons advanced.

THE MOON AND THE LUNISOLAR CALENDAR

The moon was almost generally a female deity, although in the central Mexican accounts of the creation of the Fifth Sun at Teotihuacan, the god seems to have been a male. As a female, the lunar orb was for the Mesoamericans the very embodiment of the fair sex. The young, waxing moon was seen as a beautiful woman, forming part of a complex of youthful goddesses associated with sexual love; her image can be seen in the central Mexican codices and, above all, on many pages of the Maya Dresden Codex (Fig. 9.4). As the moon waned and gradually slipped back towards the eastern horizon, she became an old and somewhat malevolent deity, with snakes in her hair or on her skirt, or with spindles placed in her headdress as an indication of her role as patroness of weaving. Another domain over which she ruled was that of childbirth. Again, she apparently formed part of a larger complex of aged goddesses and merged in many ways with some of these, particularly with the female half of the dual Creator God.

The association between the Moon Goddess and water was a close one. The heaven in which she traveled was that of the Rain God, and her symbol among the central Mexicans was a kind of cross-sectioned vessel—or womb—in which water can be seen. This concept persists today among the Chorti Maya, who explain the waxing and waning moon as a pot being filled and then gradually emptied of water. Another strong association is the rabbit (Fig. 9.5), a creature which all Mesoamerican people see on the face of the moon; the rabbit is the symbol of drink and drunkenness, and it is likely that the moon's domain extended to the complex of pulque deities.

It is only for the Maya that we have information on the role that the moon played in the calendar, and on lunar observations. Landa (Tozzer 1941, pp. 133–4) states of the Yucatec Maya, "They divide [the year] into two kinds of months, the one kind of thirty days and called *U*, which means 'moon,' and they counted it from the time at which the new moon appeared until it no longer appears"; his other kind of "month" was the twenty-day *veintena*, called *uinal* in Maya. Notwithstanding Landa's claim, neither the Maya nor any of the Mesoamericans attempted to construct a grand lunisolar calendar for civil and religious purposes. In failing to do this, they are probably unique among the early civilizations, but they were lucky. The moon has the unfortu-

9.4 The Young and the old Moon Goddess, on pages 16c and 43b of the Dresden Codex.

9.5 The goddess Tlazolteotl as Moon Goddess, standing before the rabbit in the moon. From the Codex Borgia, page 55.

nate property of following a path similar to the sun's and of being the same apparent size as the sun; these facts, and religious considerations, led ancient astronomers in the Old World to try to correlate the two main heavenly bodies into one grand scheme, an attempt which led to centuries or even millennia of confusion.

The problem is that the moon's orbit is elliptical, as is the earth's; this means that it travels at different speeds at different times of the year. Lacking gravitational theory, the ancients could never satisfactorily account for the variation in the length of the synodic month, which can be as short as 29.26 days, or as long as 29.80 days.

Nonetheless, the Maya kept a very close account of synodic lunations over a very long period of time, beginning at least as early as AD 300. In spite of claims to the contrary, the only astronomical calculations which surely are present on Maya monuments of the Classic Period are lunar; these are given following the initial Long Count date at the beginning of an inscription. Since the majority of dates in Classic times are now recognized as historical, or as Period Ending dates which occurred in the life of a ruler and were celebrated by him, it is clear that the moon was felt to exert a powerful influence on terrestrial events. The lunar data are presented in a passage of up to six glyphs, and include the following information: (1) the age of the current moon, (2) the number of moons already completed in a lunar half-year, and (3) the length of the present lunation, either 29 or 30 days. Parenthetically, neither on the monuments nor in the codices is there any indication that the Maya reckoned the sidereal period of the moon, which averages 27.32 days.

Of greatest interest to modern scholars are the attempts by the Maya to correlate synodic months with the solar calendar. At first, each center had its own formula for correlating the two by groupings of moons. As Teeple was able to show, by AD 682, Copan began using the formula 149 moons equals 4,400 days, which means that the average length of a lunation was given the remarkably accurate value of 29.53020 days. This system was rapidly adopted all over the Maya area, but lack of uniformity again appeared after AD 756.

Among the lunisolar formulae adopted by the Maya was one which appears to have been in use at Palenque, in which 405 lunations equal 11,960 days (Thompson 1950, p. 246). This apparently foreshadows the famous eclipse tables on pages 51–58 of the Dresden Codex, believed by Thompson to have been compiled in the 12th century. Among the properties of the number 11,960 which must have seemed remarkable to the Maya was the fact that, in addition to correlating the solar round with the moon, it also contains exactly forty-six 260-day counts. This coincidence is taken advantage of in the eclipse tables. Thompson (1950, pp. 245–6) has summarized the workings of these tables as follows:

> The arrangement of the groups of moons within the table in Dresden is such that there is no doubt whatever that the cycle of 11,960 days had been divided in such a way as to give a series of days, at intervals of 177 (occasionally 178) and 148 days, on which eclipses might, but not necessarily would, occur. After each occurrence of a five-lunation group of 148 days there is a picture. Most of these carry symbols indicative of an eclipse or at least of conjunction. . . . It has been suggested that these pictures may indicate lunar eclipses between two partial eclipses of the sun one lunation apart.

The nodes, when the path of the moon and that of the sun cross, occur every 173.31 days, the eclipse half-year, as the Maya astronomers well knew, for eclipses could only take place within about 18 days from the node. It so happens that three eclipse half-years are exceedingly close to twice the 260-day count, another coincidence that the Maya exploited in this table.

Some remarkably acute astronomical records must have been kept over a period of time to work out such a table, although Alexander Pogo (1937) had the feeling that a knowledgeable native astronomer could have worked it out successfully from observations of lunar eclipses stretched out only a third of a century. The question still remains whether they were aware of the regression of the nodes, but Thompson feels that even if they had no knowledge of this event, the Maya astronomers at least had observed their effect and devised means of periodically correcting the eclipse table.

We are not exactly sure how the Maya reckoned the age of the moon. Spinden once advanced the idea that they counted the days of a lunation from full moon, but Thompson (1950, pp. 236–7) has effectively disposed of this idea. Although Landa's testimony states that the count was from first appearance after conjunction, Thompson feels that linguistic evidence indicates that disappearance or conjunction (astronomical "new moon") was the more likely starting points. Sahagun's description, on the other hand, favors the Landa hypothesis:

> When the moon is born anew, it seems like a delicate little arch of wire, no radiance does it emanate; little by little, it begins to grow. (León-Portilla 1963, p. 49)

Finally, interesting though the seasonal movements north and south of the risings and settings of the moon might be to investigators of ancient orientations, all of our sources remain silent on the subject.

THE CELESTIAL WANDERERS

If the vagaries of the moon must have been puzzling to the ancient Mesoamericans, those of the planets must have been equally so. The apparent loops or retrograde motions in the orbits of the brightest planets must have been apparent to careful observers like the Maya astronomers, and they must have begun compiling records of their motions at an early date. Curiously, vocabularies of native languages give no indication that the planets were viewed as different in any way from the fixed stars. This can be seen in the Maya Books of Chilam Balam, compilations of pre- and post-Spanish materials, in which the Spanish term *planetob* (pluralized in Maya fashion) is employed in the astrological sections.

While a profound concern with the moon seems to have been effectively confined to the Maya, the Venus cult was pan-Mesoamerican. All of the peoples of our area realized that with the Morning and Evening Star, they were dealing with the same heavenly body. All of them seem to have calculated its synodic period as 584 days, the nearest whole number to its actual average value, 583.92 days. It so happens that $5 \times 584 = 8 \times 365$, so that five synodic periods of Venus exactly correspond to eight Vague Years. It is an even more remarkable coincidence that in 104 Vague Years, or two times the Calendar Round which coordinates the 260-day count and the Vague Year, there are exactly 146 260-day counts and 65 Venus periods. Among the heavenly bodies, only the moon could not be coordinated into this grand system. Small wonder that the Mesoamericans considered their calendar to be divine.

In almost all lexicons, Venus as Morning Star is glossed by a compound word which can be translated as "Great Star," although the Maya Motul Dictionary (1929) also gives the term *xux ek,* meaning "wasp star." In the central Mexican and Maya codices, the Venus period began with its heliacal rising in the east as Morning Star. There

9.6 The Venus God spearing the Water Goddess, from the Codex Borgia, page 53.

is abundant evidence that this event boded ill for the inhabitants of the earth. At each appearance with the dawn sun at 584-day intervals, the Venus regent threw his spear at a victim symbolizing an aspect of Mesoamerican daily life: at a water goddess, signifying impending drought (Fig. 9.6); at a jaguar throne, symbol of the rulers; at various deities; at the jaguar warriors, i.e. the soldiery; and at the Maize God, indicating starvation.

In the central Mexican codices, the table is laid out in five sections to present 65 Venus periods or 104 Vague Years, and there are no subdivisions of the full synodic period. In the Maya Dresden Codex, and in the newly discovered Grolier Codex which can be ascribed to the Maya-Toltec (Coe 1973), the 584 days are separated into four subperiods: 1) Morning Star (236 days), 2) Superior Conjunction (90 days), 3) Evening Star (250 days), and 4) Inferior Conjunction (8 days). The days in the 260-day count which initiated these subperiods are outlined in tables which also cover 104 Vague Years.

The problem that arises with these Venus tables is the small error that accumulates over the centuries, gradually displacing the true heliacal rising of Venus at the start of the 104 Vague Year count from its official position, the day 1 Flower in the 260-day count. Teeple, followed by Thompson (1950, pp. 226-7), has shown how this correction was made, by subtractions of small numbers of days at the end of 57 and 61 Venus periods.

Venus was enormously important in Mesoamerican religion and mythology. A large body of myth relates to the apotheosis of Quetzalcoatl-Kukulcan, the Feathered Serpent, as the Morning Star, and he and the Evening Star were conceived of as a pair of Hero Twins. At other times and for other purposes, at least in central Mexico, the Morning Star was coterminous with Mixcoatl-Camaxtli (Nicholson 1971, pp. 426-7), a god complex associated with the northern hunters known as the "Chichimecs," and, especially with the sacrifice of captives, emphasizing the basically malevolent character of this great heavenly body.

There is still debate about other members of our solar system that might have been consistently observed by Mesoamerican astronomers. It has long been recognized that certain bands appearing in Maya reliefs and codices represent the sky, or at least a segment of it. Recognizable signs or glyphs in these celestial bands include the sun, the moon, and Venus, but there are several other signs yet untranslated. From this alone one would draw the conclusion that the band represents the ecliptic, and that the remaining signs are planets other than Venus.

On pages 22-23, 43-45, 58 and 59 of the Dresden Codex are tables giving multiples of 78 and 780 days; the accompanying pictures show a strange monster with upturned snout and cloven hoofs descending from a celestial band. The synodic period of Mars being 779.936 days, Ernst Forstemann (1906, pp. 215-6) and the astronomer R. W. Willson (1924, p. 30) drew the conclusion that these tables deal with Mars, and that the creatures depicted were the "4 Mars beasts." Teeple and Thompson (1950, pp. 257-8) have doubted this assertion, but the coincidence seems to me to be convincing, especially since the synodic period of Mars would have been easy to observe for Maya specialists experienced in the recording of celestial events. More problematical is a table on pages 30-33 which has a base of 117 days, very close to the average synodic period of Mercury of about 116 days, for that planet is not easy to observe. I continue to be puzzled at the absence of any plausible reference to the synodic period of Jupiter, 398.88 days. At any rate, I think that the data do suggest the Maya were deeply interested in the planets, above all, in Venus.

THE FIXED STARS

There are said to be some 3,000 stars which the average observer can pick out with the naked eye at any one moment on a clear night (Moore 1965, p. 1). It could certainly be expected that at least all of the first and second magnitude stars visible in the latitudes of Mesoamerica as well as the Milky Way would be of considerable interest to their astronomers. Furthermore, there are certain asterisms and perhaps even constellations which would form significant groupings no matter who the observer, such as the Pleiades, Orion's Belt, the Hyades, Castor and Pollux, the Southern Cross, possibly the Northern Cross in Cygnus, the Big Dipper in Ursa Major, and Cassiopeia. At any rate, native ways of classifying and grouping the fixed stars comprise one of the most interesting areas of the young study called ethnoscience.

I have mentioned the poverty-stricken nature of the Spanish or Spanish-influenced sources on native astronomy, particularly as these apply to Mesoamerican asterisms and constellations. Just about all the information which Sahagun wishes to impart on this subject is presented on two pages of the unedited Madrid manuscript, accompanied by drawings of the sun, the moon, Venus, comets, and apparent Aztec constellations. I will return to these drawings later. But to point up the pitfalls involved in using these sources, let me quote from the account given us by Tezozomoc (1944, p. 396) of the admonitions given to Moctezuma Xocoyotzin on his election as emperor. He was especially to make it his duty to rise at midnight (and to look at the stars):

> ... at *yohualitqui mamalhuaztli*, as they call "the keys of Saint Peter" among the stars in the firmament; at the *citlaltlachtli*, the north and its wheel; at the *tianquiztli*, the Pleiades (Spanish, "las cabrillas"); and the *colotlixayac*, the constellation of the Scorpion, these mark the four parts of the world, governed by heaven. Toward morning he must also carefully observe the constellation *xonecuilli*, which is the Cross of St. James (Spanish, "la encomienda de Santiago"), which appears in the southern sky in the direction of India and China; and he must carefully observe the morning star, which appears at dawn and is called *tlahuizcalpan tecutli*. (translation adapted from Seler 1904, p. 355, with emendations)

Now, just what does this source mean by such terms as "the keys of Saint Peter" and "the Cross of Saint James?" Ever since the Venerable Bede, Christian scholars had been attempting to substitute personages drawn from the Bible and Christian hagiography for the pagan deities of Classical mythology as designations for the European constellations; this effort reached its culmination in 1627, with the star atlas of Julius Schiller (Allen 1963, p. 28). Apparently, such piecemeal changes had crept into popular parlance in Spain during the 15th and 16th centuries, and it is unfortunate that our scanty sources usually use this terminology rather than the Classical names of serious astronomers, astrologers, and navigators.

Sahagun's figure of the Pleiades (Fig. 9.7a) is surely that, since it closely conforms to the appearance of that asterism in the sky. The Nahuatl name given by Tezozomoc, *tianquiztli* or "marketplace," is matched by words in Maya languages meaning "a multitude" or "something heaped up." Perhaps more significant is the Yucatec and Lacandon Maya term, *tzab* or "rattlesnake rattle," for there is a clearcut association between the asterism, the native term, and the supreme Maya creator god, Itzamna. There are strong reasons for believing that Itzamna was the counterpart of the dual creator god of the central Mexicans, and of his avatar, the Old Fire God, the lord of time. Throughout much of the native New World, the Pleiades seem to have played a role closely connected with the creation of the night sky, with the hearth and its fire, and with the agricultural cycle (for its role in the determination of the seasons, see Levi-Strauss 1964, pp. 222–45). I suspect that it, and not Polaris, was thought of as the center of the firmament. Among the well-documented Navajo, the Fire God or

Native Astronomy in Mesoamerica

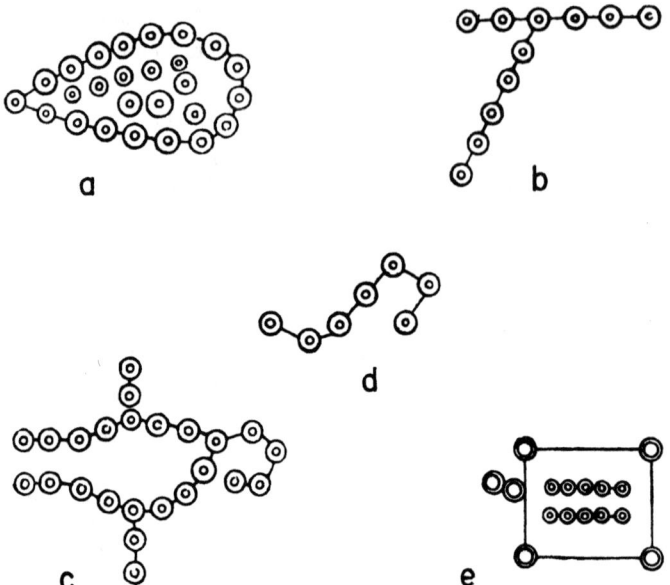

9.7 Aztec constellations as given by Sahagun. a, Pleiades. b, Mamalhuaztli. c, Citlalcolotl. d, Xonecuilli. e, unnamed constellation.

Black God created all of the stars in an orderly fashion, only to have some of them strewn about by a trickster Coyote; his own symbol was the Pleiades, which hopped from his foot up to his face, where it remained permanently lodged.

Among the Aztecs, the most celebrated feast was the Toxiuhmolpilia, the Binding of the Years, a ceremony which took place every 52 Vague Years. According to Sahagun (1957, pp. 143–4), all fires were put out at the end of this period, which equalled a Calendar Round. The priests and temple servants of the great temple proceeded to the Hill of the Star which they reached before "midnight."

> Having reached there, they looked at the Pleiades to see if they were at the Zenith, and if they were not, they waited until they were. And when they saw that they had now passed the Zenith, they knew that the movements of the heavens had not ceased and that the end of the world was not then, but that they would have another fifty-two years, assured that the world would not come to an end.

This ceremony took place at 52-year intervals in the year 2 Reed and in the month Panquetzaliztli, which in the time of the conquest lasted from 21 November to 10 December (Nicholson 1971, table 4).

As Burland (1952, pp. 23–6) has noted, there are several problems with this account. The first is, how did they know when midnight occurred? Secondly, if no intercalations were made, by the end of 52 years the midnight culmination of the Pleiades would have been 13 days off. And finally, because of the effect of precession, the Pleiades would cross the zenith at midnight about one day earlier every 69 years.

Informed Spanish observers could have questioned the native priests on these matters, but they failed to do so.

It is obvious that the Vague Year never could have effectively functioned as an agricultural calendar, and it could be expected that Mesoamerican farmers would have used striking asterisms like the Pleiades to mark points in time for their agricultural round; this seems to be the case for the Chorti, and also for the Lacandon, for whom Baer and Baer (n.d., p. 136) state, when the Pleiades have reached the tops of trees by the dawn, the milpa must be burned and the corn planted.

Another Sahaguntine constellation is called *Mamalhuaztli* (Fig 9.7b), or the Fire Drill, and interpretation of this group has given rise to great confusion. Sahagun says:

> Hazia esta gente particular reverencia y particular sacrificios a los mastelexos del cielo que andan cerca de las cabrillas que es en el signo del toro. (Sahagun 1953, p. 60)

It will be remembered that Tezozomoc calls the Fire Drill "the keys of Saint Peter." In their translation of Sahagun, Dibble and Anderson have uncritically identified it with Castor and Pollux in Gemini, which is unlikely on the face of it since these stars are nowhere near the "sign of Taurus." Their reasoning is probably based on an alternate term, *astillejos*, which Sahagun gives for the group; this is a word which, like *mastelexos*, might be translated as "little sticks," not an unreasonable description of the group as it is shown, or the fire-making apparatus itself. However, the Diccionario de la Real Academia, 1970, derives *astillejos* from the Latin *aster* or star, and following earlier editions, glosses the word as Castor and Pollux. Suspecting this identification, I looked in the Nebrija Dictionary, published in 1493 or 1495, and probably available to Sahagun; there we find *astilejos* defined as the constellation Orion.

I am convinced that the Fire Drill is in fact the Belt and Sword of Orion, which form a figure closely conforming to the picture, and which certainly do "march near the Pleiades"; the latter, of course, are part of Taurus. Just to show, however, that there is always room for doubt, Seler (1904, p. 357), following Jose Ma. Vigil and the Julius Schiller atlas, saw "the Keys of St. Peter," or the Fire Drill, as the Hyades in Taurus. We may never track this down, which is unfortunate since to the central Mexicans the Fire Drill was of great importance, being addressed as "Lord of the Night" in nocturnal offerings and sacrifices. Sahagun (1953, p. 60) informs us that the figure of the constellation was even burned onto the wrists, lest, after one went to the land of the dead, fire might be kindled on those parts.

Less problematical is *Citlalcolotl*, the Scorpion (Fig. 9.7c). Both ethnohistoric and ethnological evidence (i.e. from the Huichol and Maya) leads me to believe that this is our constellation, Scorpius. It will be seen that as Sahagun's figure is inverted, it is remarkably close to Scorpius as we know it, with the curved tail prominently displayed. Either this group naturally looks so much like a scorpion that it has received this name independently in both hemispheres, or it was in some way diffused to the New World.

The identity of Sahagun's *Xonecuilli* (Fig. 9.7d) or "twisted" is insoluble at present. He calls these the "S-chaped stars in the mouth of the trumpet (Spanish, *bocina*)," which would presumably indicate that the group formed our Little Dipper in Ursa Minor, an identification accepted by Zelia Nuttall (1901, p. 33). However, Tezozomoc glosses *Xonecuilli* as "the Cross of St. James, which appears in the southern sky in the direction of India and China," from which Seler (1904, p. 358) concluded that this is the Southern Cross, a constellation recognized by many peoples in Mesoamerica. Later on, however, in his commentary on the Borgia Codex, Seler (1963, 1, pp. 193–4) suggests that *Xonecuilli* might be certain stars in Hercules and Draco, an identification for which he has little supporting evidence.

One further constellation is shown by Sahagun (Fig. 9.7e) but not described; Seler thinks this is Tezozomoc's *Citlaltlachtli*, the Star Ballcourt, and ascribes it to stars which circle the Pole, but this is extremely tenuous.

I cannot go into individual stars which were, and in some places still are, given special names in Mesoamerica, such as Polaris. Information from the Tzeltal (gathered by my former student Allen Turner) and from the Lacandon is very suggestive. The data given by Bruce et al (1971, p. 15) for the Lacandon are extremely rich: certain asterisms like "turtle" (also given in the Motul) are not identified, but these people have names for Jupiter, Venus, Orion's Belt ("Peccary"), Rigel ("Woodpecker"), Betelgeuse ("Red Dragonfly"), Ursa Minor ("Alligator"), and Sirius (large species of woodpecker). Turner's brief research with a Lacandon informant suggests that further work will throw light upon such questions as what the Maya had in mind when they refer to a "turtle" constellation, which the Motul Dictionary places in Gemini.

The Milky Way was and is of universal significance in Mesoamerica. The Nahuatl word for the Galaxy was *Citlallicue* (Starry Skirt), which is in fact, the name of the goddess who created the stars. The Yucatec Maya term was *Tamacaz*, which is also the name of the fer-de-lance, and associations with snakes seem to be universal in our area. In a modern dictionary of the Quiche Maya language (Leon 1954), there are two terms for the Milky Way, one (*sac bey*) for summer, and one (*xibal bey*) for the winter, when it is bifurcated; the bifurcation is identified with the Underworld, and it is quite probable that the Maya, like many other American Indians, thought of the Milky Way as the road of the souls journeying to that region.

Fragmentary allusions suggest that many of the Mesoamerican gods had their abode in the heavens. The Historia de los Mexicanos por sus Pinturas says that Tezcatlipoca and Quetzalcoatl, after creating the Milky Way, now live in the Galaxy, and Tezcatlipoca is specifically associated with Ursa Major by a myth having to do with his expulsion from heaven by his rival Quetzalcoatl (Garibay 1965, p. 30). Other deities are said to have once lived in the sky, but later descended to earth, such as *Yacatecuhtli*, the Merchant God, and *Mixcoatl*, the god of hunting. Probably all of the gods had stellar associations.

THE QUESTION OF A ZODIAC

Stansbury Hagar (1912), a confirmed diffusionist, was the first to advance the claim for a solar zodiac in the Maya area, based on evidence which has later been critically

Michael D. Coe

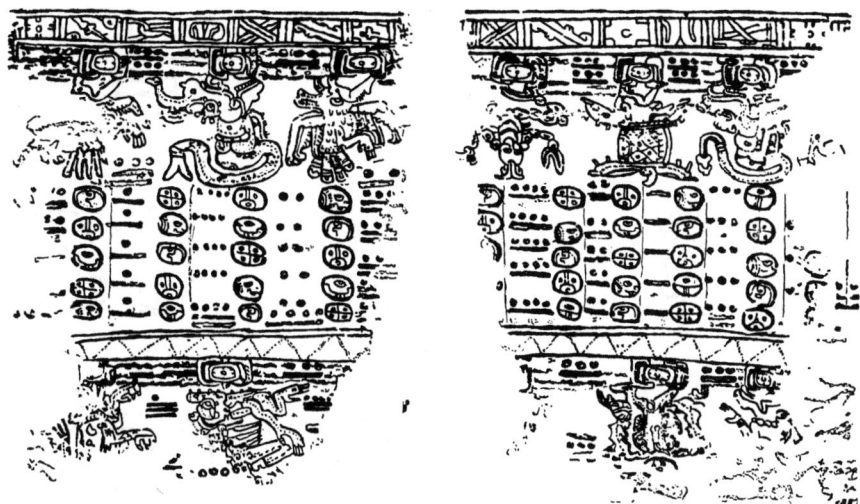

9.8 Probable asterisms in the Paris Codex, pp. 23–4.

examined by Spinden (1916). In the Maya Paris Codex, on pages 23 and 24, are two horizontal celestial bands, beneath which are pendant thirteen animals each connected with a sun symbol (Fig. 9.8). Hagar says that this is a zodiac. Because at least some of the figures might be star groups (i.e., rattlesnake—Pleiades, turtle—Gemini?, jaguar—Tezcatlipoca), Hagar wanted to see this as a zodiac. Ever since Forstemann (1903) tried to work out the meaning of the numbers and day names appearing in this table, there has been much discussion of what this is all about. Recently, my colleague Floyd Lounsbury has reached the conclusion that these pages comprise an eclipse table; the animals are in the process of eating the solar eclipse symbols themselves, recalling the Yucatec Maya word for eclipse, *chi'ibal kin,* "to eat the sun."

Just as dubious as a zodiac is a stellar band on the east wing of the Monjas at Chichen Itza (Fig. 9.9), in which animal and other figures connected with star symbols are placed at intervals between sky signs. By an ingenious rearrangement of the stones forming the relief, Seler (1910) was able to match up the sequence of figures with those in the Paris. Needless to say, however, nobody has been able to prove either of these to be zodiacs.

There is some curious information in the text and illustrations of Duran (1971) which *might* be interpreted as an indication of a kind of zodiac. According to him, each "month" or *veintena* in the Vague Year had its own "planet" or constellation, some of which he shows in the sky surrounded by clouds. These are also known as *veintena* symbols in other sources, such as the pierced bird for the "month" Tozoztontli.

This is clearly a subject which needs more research before it is dismissed. The native zodiac (if it existed at all) may not have been like ours, i.e. constellations in

9.9 Sequence of probable asterisms in the stellar band on the East Wing of the Monjas, Chichen Itza, Yucatan.

a band extending on either side of the ecliptic, but a system of lunar mansions like the *hsiu* of China, which extend from pole to pole and which are designated by star groups which can occur anywhere within them.

CONCLUSIONS

Certain generalizations can be made about native astronomy in Mesoamerica. Firstly, they seem to have dealt exclusively with synodic periods rather than sidereal. Secondly, all of the calendars which we have, including the Venus table of the Maya, are *official* calendars, codified long ago in the past and only seldom offering the native priests or astronomers the opportunity to build in necessary corrections. As Spinden (1916, p. 66) has said, "It has long been recognized that interpolated corrections would vitiate the elaborate calculations of the Maya where solar, lunar, and Venus periods are correlated over vast stretches of time." Numerology ruled supreme in Mesoamerica, allying their astronomy much more closely with that of Mesopotamia than with the Greeks, whose obsession was geometry.

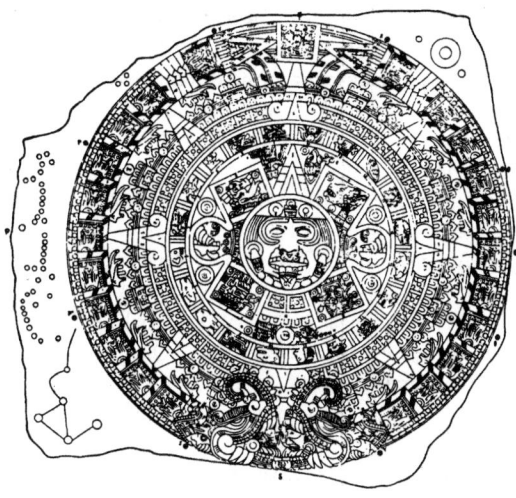

9.10 Probable constellations on the border of the Aztec Calendar Stone (after Nuttall 1901, Fig. 56).

It is certain that Mesoamerican astronomy was far more complex and advanced than our fragmentary data indicate. Some aspects of it may have been the result of diffusion. Kelley (Moran and Kelley 1969) has made the ingenious suggestion that the sequence of twenty named days may be a reduction of an oriental system of 28 lunar mansions, and Needham (1959, p. 407) has commented upon the unusual coincidence that the Maya astronomers and those of the Han Chinese worked with an eclipse calendar of 11,960 days. Finally I might point out that the practice of indicating constellations by circles connected with straight lines, to be found not only in the Primeros Memoriales (the first version of the Florentine Codex; see Sahagun 1905) but also on the borders of the Aztec Calendar Stone (Fig. 9.10), has no antiquity whatever in Europe, but goes back as far as the Han Dynasty in China. In fact, it does not appear in the western part of the Old World until 1785, as a conscious effort of French astronomers to reconcile their star maps with those of the Chinese (Deborah Jean Warner, personal communication).

I will end with a plea to everyone concerned to collect ethnoastronomical data from the surviving native peoples of Mesoamerica; there is not much time left to salvage this body of information, which must surely help answer many of the problems raised here.

REFERENCES

Baer, P., and M. Baer. N.d. Materials on Lacandon Culture of the Petha Region. *Microfilm Collection of Manuscripts on Middle American Cultural Anthropology* 28, Chicago.
Bruce, R., C. Robles U., and E. Ramos C. *Los Lacandones 2, Cosmovision Maya*. Instituto Nacional de Antropologia e Historia, Departmento Investigaciones Antropologias, Publicaciones 26. Mexico.
Burland, C. 1952. The Toltec-Style Calendar of Mexico. *Thirteenth International Congress of Americanists* 23–26. Cambridge, England.
Caso, A. 1954. *El Pueblo del Sol*. Mexico, Fondo de Cultura Economica.
Caso, A. 1971. Calendrical Systems of Central Mexico. In *Handbook of Middle American Indians* 10:333–348. Austin, University of Texas Press.
Coe, M. 1973. *The Maya Scribe and His World*. New York, Grolier Club.

Duran, F. Diego. 1971. *Book of the Gods and Rites and the Ancient Calendar.* Norman, University of Oklahoma Press.
Förstemann, E. 1906. Commentary on the Maya Manuscript in the Royal Public Library of Dresden. *Papers of the Peabody Museum, Harvard University* 6:2. Cambridge.
Garibay, A. M. 1965. *Teogonia e Historia de Los Mexicanos.* Mexico, Editorial Porrua.
Girard, R. 1966. *Los Mayas.* Mexico, Libro Mex.
Hagar, S. 1912. Zodiacal Symbolism of the Mexican and Maya Months and Day Signs. *17th International Congress of Americanists,* Mexico, pp. 140–159.
Haile, B.O.F.M. 1947. *Starlore among the Navajo.* Santa Fe, Museum of Navajo Ceremonial Art.
Leon, J. de. 1954. Diccionario Quiche-Español. Guatemala, Guatemala Editorial Landivar.
León-Portilla, M. 1963. *Aztec Thought and Culture.* Norman, University of Oklahoma Press.
Levi-Strauss, C. 1964. *Le Cru et le Cuit.* Paris, Librairie Plon.
Lumholtz, C. 1900. Symbolism of the Huichol Indians. *American Museum of Natural History Memoirs* 3, New York.
Moran, H. A., and D. H. Kelley. 1969. The Alphabet and the Ancient Calendar Signs. *Daily Press,* Palo Alto.
Needham, J. 1959. Mathematics and the Science of the Heavens and the Earth. *Science and Civilization in China* 3. Cambridge, Cambridge University Press.
Nicholson, H. B. 1971. Religion in Pre-Hispanic Central Mexico. In *Handbook of Middle American Indians,* 10:395–446. Austin, University of Texas Press.
Nuttall, Z. 1901. The Fundamental Principles of Old and New World Civilizations. *Archaeological and Ethnological Papers of the Peabody Museum, Harvard University,* 2, Cambridge, pp. 1–602.
Nuttall, Z. 1906. The Astronomical Methods of the Ancient Mexicans. *Boas Anniversary Volume,* ed. B. Laufer, 290–298. New York, Stechert.
Pogo, A. 1937. Maya Astronomy. *Carnegie Institution of Washington, Year Book* 36, Washington, DC, p. 2435.
Ruppert, K. 1953. Florentine Codex, Book 7, trans. C. E. Dibble and A.J.O. Anderson. Santa Fe, School of American Research and the University of Utah.
Ruppert, K. 1957. Florentine Codex, Books 4 and 5, trans. C. E. Dibble and A.J.O. Anderson. Santa Fe, School of American Research and the University of Utah.
Seler, E. 1904. Venus Period in the Picture Writing of the Borgian Codex Group. *Bureau of American Ethnology Bulletin* 28, Washington, DC, pp. 355–391.
Soustelle, J. 1940. *La Pensée Cosmologique des Anciens Mexicains.* Paris, Hermann et Cie.
Spinden, H. J. 1916. The Question of the Zodiac in America. *American Anthropologist,* n.s. 18:53–80.
Teeple, J. D. 1930. Maya Astronomy. *Carnegie Institution of Washington, Publication* 403, Contribution 2, Washington, DC.
Tezozomoc, H. A. 1944. *Cronica Mexicana.* Mexico, Editorial Leyenda.
Thompson, J.E.S. 1950. Maya Hieroglyphic Writing: An Introduction. *Carnegie Institution of Washington, Publication* 589.
Thompson, J.E.S. 1970. *Maya History and Religion.* Norman, University of Oklahoma Press.
Wilson, R. W. 1924. Astronomical Notes on the Maya Codices. *Papers of the Peabody Museum, Harvard University* 6:3, Cambridge.

CHAPTER TEN

The Role of Astronomical Orientation in the Delineation of World View
A Center and Periphery Model

Anthony Aveni

INTRODUCTION

While this symposium consists of individuals who might define themselves on job application forms as archaeologist, art historian, astronomer, or historian of religion, the common ground that binds most of us together is that we all study the behavioral characteristics of a culture. For we are really Americanists—better, Mesoamericanists. That label and the tendency to employ it, now with perhaps greater validity than in the past, may reflect the broadening that has begun to occur in our respective disciplinary-based studies and to which our host has referred in his organizational statement. Davíd Carrasco also has suggested that this interdisciplinary broadening is guided, at least in part, by New World imagery, such as page one of the Codex Fejérváry Mayer,

---

From *The Imagination of Matter: Religion and Ecology in Mesoamerican Traditions*, ed. D. Carrasco, British Archaeological Reports International Series 515 (1989). Reproduced by permission from the author.

in which we find a cognitive map that unites the most vital elements of a Central Mexican system of belief. Its quadripartite form encompasses a system of classification that tries to organize and explain in general how time and space are related, and specifically, where such seemingly diverse entities as plant and animal forms, deities, the sun, death, the parts of the human body, and the sensation of color fit into the relationship.

Colored ourselves by a reductionist way of viewing things, we usually validate knowledge by isolating and examining its component parts in detail. The notion of a constellation of ideas (like the Fejérváry World Diagram) that interrelates sky objects, plants, and human bodily parts seems to lack the complexity and sophistication that satisfies us, the precision that we believe can be articulated only when one selects and dissects a given segment of human concern quite apart from all other concerns. We identify as primitive any scheme of ideas that falls short of this intellectual program. But primitive implies only the first stage on an evolutionary path which, like the Way of the New Testament religion will lead to the ultimate resolution of all concerns, if followed without deviation or abnormality.

The persistence in the belief that our writing, our astronomy and science, our politics and religion, our very human values, perhaps even our biological species, are the closest approximation to the idealized and ultimately realizable end product of human and social evolution provides an obstacle along another Way—the way of understanding the ideas of other cultures.

Rather than regarding world views of other people as deficiency-oriented systems that lack one or another characteristic that keep them from being like our own, it might be useful to examine some losses on our side of the ledger. The separation of earth and sky in western cosmology did not occur until the Medieval Period when maps of the world took on decidedly religious connotations. Prior to that time, the bond between heaven and earth was enforced by the knowledge and practice that one needed to observe the stars in order to produce a faithful map of the environment. The principle connecting star and mountain was associative rather than causative. The place of the mountain on the world map was determined by the visual observation of how high above it the Pole Star appeared.

When we look at the role of astronomy in the developed cultural nuclei of America, we find that as we try to study its aspects isolated from all other spheres of human inquiry, the matter or substance of it dissolves away into nothingness. It is not until we take up the question of the relation between the stars and the land, water, agriculture, kingship, and kinship that we can even begin to make sense of anything we read in Maya inscriptions, the Mexican temple alignments, or time-marking pillars on the horizon of the capital city of the Inca. Furthermore, I find that when we begin to examine very closely the way in which precise knowledge about the celestial environment is expressed in the ceremonial and civic architecture, we discover that such knowledge is tied to each culture's comprehension of what we call the biological and social environments and that it incorporates what we call religious belief as well.

Following leads that began initially with the formulation of a particular astronomical problem, I recently began to discover some cross-cultural patterns that led me to formulate a working model that I would dare to propose as the basic thesis for my paper:

Among highly organized, bureaucratic societies, the politico-religious philosophy of the ruling body often is expressed or articulated physically through a large-scale apparatus or plan that incorporates and joins information about the terrestrial and celestial environment within major architectural elements both in the ceremonial center and at the visible edge of space surrounding the center. In each case, the accommodation of the built environment to the natural one consists of more than just a simple functional alignment (such as sunrise on the patron saint's day). Usually, environmental time and space are intimately linked to sacred time and space through a set of principles that seem to reinforce one another repeatedly. In the following sections, I would like to argue in favor of this physical center-periphery model by discussing and synthesizing three case studies from the pre-Conquest Mexica world of Central Mexico, the Classic Maya culture, and imperial Cuzco in the Andean highlands.

THE TEMPLO MAYOR OF TENOCHTITLAN AND ITS ENVIRONMENT

Summarizing his extensive excavations at the Great Temple, Matos (1984, p. 78) has remarked: "In short, the Great Temple of Tenochtitlan symbolized to the Mexicas water and war, life and death, Tlaloc and Huitzilopochtli" and (p. 70): "The prominence of these two deities reflects the fundamental needs of the Mexicas: their economy was based on agricultural production (hence the importance of water and rain) and on tribute collected by conquest (hence the importance of war). Thus, we expected that all the elements associated with the Great Temple, such as offerings and sculptures, would in some way be related to two fundamental themes." In a recent paper Calnek, Hartung, and I (Aveni, Calnek and Hartung 1988) have used myth as the starting point in an attempt to explicate one aspect of the nature of the complex symbolism that might underlie some of the practical motives associated with the construction of the temple that Matos has spoken about.

Thus, we follow the guiding assumptions that in the real world of the Mexica, religious and economic themes were intimately related and that these people harbored specific and quite complex reasons for their actions, including the emplacement of the architectural components that made up their ceremonial space. Specifically, we argued that the builders of the great Mexica ceremonial center of Tenochtitlan possessed certain cosmological predilections related to the myth of the founding of their city, that they sought to replicate in their sacred space in a variety of ways. As is often the case in Mesoamerica, the myth was embedded in a system of horizon astronomy linked to mountain worship.

The involvement of the mountain cult with the civic calendar is strengthened by the observed orientational properties of a shrine on Mt. Tlaloc which is located directly in front of the Templo Mayor at a distance of 44 km. Sun and mountain together offered a visual reminder and presented a live enactment of the tradition that told of the circumstances of Mexica origination. The view from the summit of the Templo Mayor and in general, anywhere in the precinct, also reinforces the belief that time itself (encapsulated in the sun's motion) was meted out in perfect harmony with events that led to the establishment of their regime. At a more specific level, the horizontality of the environmental time-keeping scheme stresses the concept of pivoting the count of the days about the equinox and dividing time intervals into periods of 20 days, marked by prominent places reached by the setting sun on the horizon, an idea that seems concordant with the written record concerning the calendar. While a synopsis of our procedure is in order, details should be gleaned from a review of our paper.

Our approach was initiated as a result of some attention we paid to Motolinia's (1971 [1541]) statement about the twisted nature of the Templo Mayor and its relation to observations said to have been made at the time of the equinox. It was further conditioned by the strong preponderance of a *locative* element in the foundation myth, and the centrality of the cardinal direction principle in the ethnohistoric writings, a theme which is well known over a long period of Mesoamerican history (e.g. Duran 1971, p. 78).

The founding myth offers a very detailed recording of esoteric ritual procedures employed to determine how, when, and where the highly dramatic public viewing of the eagle (or sun, said to indicate the hierophany of the founding) would take place.

Despite its richly symbolic language, Tezozomoc's (1975 [1723]) colorful account of the founding myth, which we have basically followed, includes some rather specific place names, orientations, and associated ritual processes:

(A) The *locations* singled out include: (1) Chapultepec; (2) Tepetzinco; (3) Tlalcocomolco; and (4) the Nopal foundation site where the wandering Mexicas were told to look for the eagle perched upon the cactus. Of these, the latter three were marked *point-locations*: Tepetzinco by the decapitated head of Copil, alter ego of Huitzilopochtli, and the Nopal site by his heart.

Tlalcocomolco also is marked to be the place mentioned by Torquemada (1975) where Chimalpopoca (third ruler, ca. 1415–1426) was said to have dedicated an important religious monument/sacrificial stone. Unfortunately, there is no present means by which to identify its actual location.

(B) The *orientations* singled out are: (1) east-looking and (2) north-looking, but toward a cave and mountain. These cross at some unspecified location.

Other interpretations gleaned from the text suggest that:

(1) The nopal was viewed by priests who made east and north observations.
(2) When the assembled people saw the eagle, it was "as if from afar," so at least at some physical distance.
(3) We may observe, somewhat conjecturally, that both "eagle" and "Huitzilopochtli," here linked as a single being, reasonably were identified with the solar disk, and since observation seems to have occurred in early morning, with the rising sun.

These ethnohistoric data led us to a general orientation hypothesis which we reexamined in the light of recent archaeological evidence: *that the founding of Tenochtitlan is related to the observations of sunrise at a position and an alignment determined by the crossing of east-looking and north-looking observations, which stipulate that the sun was observed atop the nopal cactus, and that the crossing point lay to the west of that location.*

While we have not yet assembled all of the pieces of the orientation puzzle together perfectly, we have proposed a tentative building sequence that best fits the archaeological and ecological data:

(1) Reflecting the orientation principle so central to their foundation myth, the Mexica probably laid out their earliest rudimentary structures in the east-west direction. Peaks on the mountain ranges to the east or west may have served as the original reference points.

Recent archaeological discoveries provide strong circumstantial evidence linking the degree of skewing found in Tenochtitlan's city plan to the Huitzilopochtli cult, e.g. note that a line from the raised altar at the rear wall of the Temple of Huitzilopochtli extended westward along the direction of alignment of the building, passes through no less than five highly distinctive archaeological features. From west to east, these features are: (1) a small plaster head in the center of the top step beneath the platform; (2) a sacrificial stone; (3 & 4) buried offerings, No. 33 & 34; and (5) a raised altar against the rear wall of the temple, where a statue of the deity is likely to have been placed. (Aveni, Calnek and Hartung 1988, fig. 3)

The half of the Templo Mayor occupied by Tlaloc is asymmetrically displaced to a position north of the east-west axis, suggesting an iconographic if not theologically expressed subordination to the chief tutelary god of the Mexica. This line also crosses the intersection of the north side of Calle Argentina and the west side of Calle Guatemala at a distance of 100 omitl (55 m) from the altar (very close to the intersection point of the four causeways—point 0 in fig. 3 of our 1988 paper—as best we can determine it).

(2) When they wished to incorporate the equinoctial alignment as an enduring principle into the three-dimensional enclosed environment of their ceremonial space, they laid out the Templo Mayor at 6¾° S of E (as we still observe and measure it today) so that the sun could be observed to rise in the notch between the temples or perhaps over the Temple of Huitzilopochtli at the equinox.

(3) The rebuilding of the temples at different heights must have presented something of a problem and this may be the underlying source of Motolinia's famous statement about the building being twisted and his need to rebuild it. But the general conformity of the alignments of the later phases may be taken to imply either that the difference of linear height between observer and sun disk always was kept constant in the engineering problem or that the desire to preserve the equinox orientation, once established, was simply abandoned because of other (non-astronomical) concerns.

The extension of the perpendicular to the front of the Templo Mayor along the East avenue from the intersection point of the four causeways intersects the eastern horizon at a point close to the place where the sun rises about 20 days before spring equinox. We can still view the eastern horizon today by looking along the modern Calle Guatemala toward Cerro Tlaloc. This street appears to retain almost exactly the same orientation as the Templo Mayor; this would have been the case when it ended in the canoe harbor, later Atarazanas, and eventually the Plaza de San Lazaro. Thus, there would have been an unobstructed view of these hills in pre-conquest times.

Now, the foundation myth suggested an eastward-looking observation around the time of the equinox and the architecture accords with that general orientation. Accordingly, we inquired into what environmental reference points might be found in the landscape surrounding the city. Thus, if we prolong the mean axis of the Templo Mayor eastward (azimuth $96^{3}/_{4}°$) to the most remote point of the valley from which the Templo Mayor still can be seen, we find that the line intersects the "puerto" between two rather prominent and culturally important hills situated along the eastern horizon of Mexico City at a range of 44 km. The hill to the north is Cerro Tlaloc (elev. 4130m), regarded by the chroniclers as perhaps the most important mountain in the landscape, for it was called "Tlalocan," the house of the rain god. In the Mexica religion, this mountain was said to be the origin of rain, mist, clouds, and snow. Pomar (1981) describes an idol of Tlaloc that was worshipped on the top of the mountain. Ixtlilxochitl, the anonymous chronicler of the Anales de Quauhtitlan, Muñoz Camargo, and Torquemada, also refer to it as a highly venerated place that was visited by many of the important Mexica rulers. (For details, see Broda 1982, p. 47.) Duran had described a rather extensive ruin on top of the mountain consisting of a large courtyard where there took place an annual feast on 29 April (9 May New Style) (during the month of Huey Tozoztli). According to Wicke and Horcasitas (1957, p. 85), who ascended the peak in the mid-1950s, this was the time maize seeds first sprouted. "It was attended by priests, rulers, and lords from both the Valley of Mexico and that of Puebla, for whom provisional shelters of branches and straw were built on the slopes below the courtyard." Duran describes the sacrifice of young children that was performed there. Once the ceremonies had concluded, everyone hurried to the Valley of Mexico where on the same day, a second ceremony in honor of Chalchiuhtlicue, goddess of the waters, also took place. This statement offers a direct physical connection between the main ceremonial precinct and this particular segment of its periphery.

Following an ascent to the summit, Wicke and Horcasitas, as had Rickards (1929) some years before them, provided sketches and descriptions of the place. The rectangular enclosure at the summit is about 40m (east-west) by 50m (north-south) and consists of massive fallen walls made of roughly hewn stone that must have been several feet high (to fit with Duran's description). Among several smaller structures within the rectangular enclosure is a rectangular pit (about a meter square) carved out of rock. When I participated in a climb up Mt. Tlaloc in October 1984, my attention was directed to the fact that the pit contained water. Whether this water-filled pit served a sacrificial purpose or was merely a part of the rather scant water supply on the mountain, I cannot say. The nearest water source, an aqueduct, commences below the hill.

Access to this rectangular enclosure is provided by an impressive 125m-long walled pathway which, according to Wicke and Horcasitas, entered from the north. However, our transit measurements yielded an alignment of 281°30' or 11°30' N of W. We note that the orientations of the causeway and enclosure are different. Furthermore, we found the causeway to be skewed by 8° relative to the enclosure itself, the latter actually facing 273°30' (3°30' N of W) or *directly on line to the Templo Mayor*.

Richard Townsend (P.C.) has remarked that the causeway also served a social purpose; namely, it controlled traffic. For the ritual of the making of rain, the procedure involved processional representatives from four different cities, each of whom was delegated the responsibility of dressing different portions of the Tlaloc idol. The causeway served the function of bringing these factions together and funnelling them into and out of the ceremonial enclosure.

The view confronting one who exits the causeway is quite dramatic (Aveni, Calnek and Hartung 1988, fig. 7): One sees nearby Cerro Huepango in the foreground, and, to the left, the most remote sighting that one can make of it from an easterly direction, the Templo Mayor (scarcely visible today because of the perpetual haze of modern Mexico City).

Sunset on 18 April in the current calendar would have occurred along the direction of the axis of the causeway, a date which falls almost exactly one Mexica month (actually 21 days) short of the 9 May (New Style) date on which the great feast honoring Tlaloc at his birthplace was conducted. Therefore, one wonders whether the causeway was misaligned deliberately with the axis of the enclosure in order to accommodate an orientation to the setting sun during the month that housed this particular ritual.

All of these observations suggest that Mt. Tlaloc, the fabled Tlalocan, on the periphery of the valley containing the great ceremonial center, may have played a role in the planning and orientation scheme of the Templo Mayor. Viewed from the Templo Mayor, Cerro Tlaloc stands 2°26' above the horizon at azimuth 93° while Cerro Telapon, the hill to the south of the "puerto" and at an altitude of 2°23', lies at azimuth 99°. In between, at azimuth 96°, the horizon dips to 1°54'. The sun would have risen there as seen from the center of the city about 14 days before the vernal and after the autumnal equinox 9 March, 6 October, having risen over Cerro Tlaloc at the north edge of the puerto on 28 September and 16 March (equinox ±5 days).

The actual projection of the mean axis of all measured phases of the Templo Mayor to the horizon strikes a point on the upslope of Cerro Yeloxochitl, also called Cerro Colotepec, that corresponds to sunrise 16 days before equinox. The axis of the large Phase V facade projects almost to the top of C. Colotepec (17–18 days before equinox). However, the sun would have arrived at C. Telapon, the highest point of this horizon feature on 1 March and 14 October, about 20 days from the equinox positions. The choice of marking a 20-day interval from the equinox along the horizon makes sense, given the concept of the Mesoamerican year, which, as we have already pointed out, consisted of individually named 20-day months, plus the five *nemontemi* or "meaningless days."

There is another directional aspect of the myth described in Tezozomoc that turns out to be in accord with this particular type of orientation calendar. Recall that Tepetzinco had been mentioned in connection with the sacrifice of Copil and that his decapitated head had been placed atop that outcrop in Lake Texcoco less than five km. from Tenochtitlan. Though it is not known to be marked architecturally (and we have not really explored the possibility), Tepetzinco continued to play an important ceremonial role up to the time of the Spanish conquest (Sahagun 1950–52, Book 2). Whether by chance or design, its orientation viewed from the temple precinct is toward azimuth 82½° or 7½° N of E. Taking account of horizon elevation, sunrises in this direction would have occurred on 8 April and 6 September, the first of which falls 19 days *after* the spring equinox. Therefore, in a spatial sense, Tepetzinco becomes the *calendrical reciprocal* of Telapon-Yeloxochitl. If this relation is not simply coincidence, we would be obliged to conclude that early Mexica priest-astronomers, or if one prefers, the god Huitzilopochtli himself, were primarily concerned with equinox-related sunrise observations spaced at equal intervals defined by the vigesimal counting of whole day periods required by their calendrical system.

In our future joint research we will investigate the degree to which the calendar was incorporated in city planning. We must investigate whether the calendar in Sahagun can be made to fit with the named point locations both within and without the ceremonial center where he tells us various celebrations were carried out, i.e., were other sunrise and sunset observations involved with these points and do the directions fit the dates?

As we demonstrated in our 1988 paper, all of these alignments would have constituted an ideal match only during the first years after the institution of a recalibrated calendar; however, citations from Ixtlilxochitl imply not only that equinox observations were employed to recalibrate calendars, but also that the 365-day calendar was not routinely corrected via leap-year intercalations. When, and of course if, astronomical phenomena were marked by monumental structures such as the Templo Mayor, the purpose was almost certainly to ritualistically commemorate crucial events. The celebration of the founding of Mexico-Tenochtitlan unquestionably belongs in that category.

There is some evidence that the orientation principles evident in the architecture of Tenochtitlan are the proudest of a tradition that began in the highlands at least as early as Teotihuacan. The care and precision in planning and laying out pre-Columbian cities appears to have been duplicated all over the Valley of Mexico, even down to the clockwise (east of north) skew and the equinox orientation.

Building on ideas developed by Dow (1967) and Millon (1973), on several occasions (Aveni, Hartung & Buckingham 1978 and references therein), we have argued that astronomical factors played a role in the orientation of Teotihuacan and that certain cross petroglyphs pecked into floors and rock outcrops at the site functioned as markers for long-distance sightlines that were used in setting up the orientation as well as re-orienting the city.

The argument that the equinox is prominent in the Teotihuacan plan is discussed at length by Aveni & Hartung (1982); for example, equinox sunset is marked by a line passing from the top of the Pyramid of the Sun through a pair of petroglyphs at ranges of 6½ and 7½ km, respectively, that lie on the horizon precisely in line with the pyramid and at approximately the same elevation as its summit. Here then, is at least one example of a possible historical precedent for the equinoctial orientation alluded to in the Templo Mayor.

If Teotihuacan tradition influenced Mexica myth, why should not one expect the orientational properties of the ceremonial center to be carried forward as well? Even the identity of sacred places in the landscape may have endured. In this connection, we note that the Tlaloc-like statue located in front of the National Museum of Anthropology in Mexico City was found near Coatlinchan in a cañada in the foothills of Cerro Tlaloc. Might this indicate that, already in the time of Teotihuacan, this particular mountain belonged to Tlaloc?

COPAN: THE CENTER (TEMPLE 22) AND THE PERIPHERY (THE STELA 10-12 BASELINE)

In the great southern Maya city of Copan, as in much of the Maya world and in Tenochtitlan as well, astronomical events also can be interpreted as a means of the validation of the power of rulership. In the past decade, the dramatic advances in the decipherment of Maya script and the revelation of dynastic sequences and historical events have been accompanied by the realization that celestial phenomena played a direct physical role, not only in the symbolic process but also as visible triggering mechanisms to promote social action, such as the conduct of accession rituals and even war.

Closs, Aveni, and Crowley (1984) have argued that the worship and observation of the planet Venus in connection with rain in the milpa agriculture may have been a special feature of Temple 22 at Copan, one of three structures at the site of which the associated iconography possesses Venus symbols. The key to our argument linking the center and periphery of Copan, however, rests upon an observational relationship between the position of the planet on the horizon and the direction of the single narrow window located on the west side of the structure. Looking through this window, one could have sighted Venus above the western horizon during the months of April and May while it was on its way to or returning from one of its eight-year horizon extremes.

Specifically, at the time of the building of Temple 22, in the year prior to a great extreme, the first day of visibility of Venus always fell within or very close to an eight-day period ranging between 25 April and 3 May and in a year following a great extreme, the last day of visibility of Venus possessed essentially the same property. These extremes occurred at the same time of the year as the traditional period associated with the onset of the rainy season. We have used the ethnohistoric and ethnographic record to argue for a ritual connection among rain, maize planting, and Venus. In this scenario the windows would have served as structures to aid in marking the time when Venus would reappear in the sky following its last appearance in the west during the rainy season.

The Role of Astronomical Orientation in the Delineation of World View

The importance to the Maya of doing things on the proper day is reflected in the very existence of the codices, which, we can be sure, were intended to be carried from place to place by the priests, whose function it was to give the correct date for a civic, social, or agricultural ceremony or an event based on what was happening in the natural world. And we know that the codices are based at least in part upon real astronomical observations.

What do we know of the relationship between Venus and the ecology of the Maya region? In the Maya area, the origins of Quetzalcoatl-Venus probably stem from the Late Classic period. The Venus-rain relationship is further strengthened in Mopan Kekchi and Chol myths relating to the discovery of maize in songs to Xipe Totec and even in the Dresden Codex. Lahun Chan, the manifestation of Venus as Morning Star on Dresden 47, is depicted with the nose of the rain god Chac. God L, the manifestation of the Morning Star on Dresden 46, appears as a protagonist in the celestial downpour scene on Dresden 74. In the same vein, a black god closely affiliated to and possibly identical with God L is the protagonist in scenes showing heavy rainfalls on Madred 32a-b (see Closs, Aveni, and Crowley for details).

The beginning of the rainy season is the single most important climatic factor in the agricultural year of the Maya. Its actual occurrence is variable from year to year and region to region, but its ritual celebration is not. Morley and Brainerd (1956: 134) write: "Planting is begun immediately after the first rains, which all Maya believe will fall on the Day of the Holy Cross (3 May)." While this is something of an overstatement, it still seems to be generally applicable in the Maya lowlands.

Redfield and Villa Rojas (1962: 84, 110), speaking of the Yucatecan village of Chan Kom, write: "The most important festal day of the year is the Day of the Cross, the third of May." They also write: "Novenas may be offered to the Holy Cross at any time; but during April and May, and especially on May 3 ... the Holy Cross is honored by novenas and jaranas in all the villages of the region." Thompson (1970: 268) describes celebrations and offerings made by the Tzotzil on May 3 for abundant rains and bountiful crops, etc. Other rituals in the Copan area focus rather narrowly on the period of 25 April–3 May, which corresponds precisely to the time when Venus appears in the window.

The sun-Venus connection, both in the sky and in the iconography of Temple 22, offers a logical explanation for the observed fact that the alignment of the window of Temple 22 also marks the place where the sun set on precisely the same dates that it set along the solar-related baseline between Stelae 12 and 10, which are located on opposite sides of the Copan valley and which constitute the only pair of a number of Copan outlying stelae that are intervisible (Aveni n.d.).

That Copan outliers may have constituted territorial boundary markers is not inconsistent with the hypothesis that these two outlying stelae also formed a calendric baseline used to anticipate the time of the initiation of the milpa agriculture as had been suggested long ago by Morley. Indeed, given generally bad weather at the time of milpa, the Maya would have had a great need to anticipate this time of year by establishing a solar baseline that formally marked a time well in advance of the start of the rainy season.

Now, we had also noted that the dates indicated by the 12-10 baseline fit ideally into an orientation calendar consisting of 20-day periods symmetrically pivoted about the dates of passage of the sun across the zenith (Aveni 1977). Though similar in that it spatializes time intervals in a vigesimal calendar, this orientation calendar stands in marked contrast to the Mexica orientation calendar, discussed earlier, in which the equinox seems to serve as the base date. We also pointed out that this 7 km baseline lies parallel to and cuts across the south end of the acropolis where Temple 22 and its associated structures are located. Actually, when we demonstrated this fact, based on measurements taken in the field, we had not realized the possibility that Venus might have played a role in the orientation scheme. Indeed, Stela 10, at the western horizon, would have been visible in the window of Temple 22 in the ceremonial center, from which one could have sighted Venus, bringer of rains, on the same evening that one also viewed the sunset from the periphery of the site. Thus, there is an implied physical connection (a kind of mutual reinforcement) between sun and Venus, and between Temple 22 and the baseline. They are connected by a concrete set of observations, and were likely accompanied by an attending ceremony.

What kind of ceremony? Miller (1988) interprets the Main Acropolis of Copan, and particularly Temple 22, to be the residence of the royal family. She argues that the ruler literally sat over the glyphic text of Temple 22, elevated and enframed by the bicephalic dragon as he appeared within the interior chamber of the temple. Here, youthful rulership is celebrated and symbolized as one of the stages of the growth of the maize plant.

The ruler enters through the mouth of the monster represented by the doorway of Temple 22, which Miller suggests was not unlike the typical Chenes facade. He becomes enthroned as if planted like a kernel of corn; then the Maya ruler sprouts and grows. This metaphor of maize as the flourishing career of the sun-god ruler is consistent with our hypothesis that attempts to connect celestial representations of this king with the act of sowing maize and specifically the apparition of Venus in Temple 22 with the time of maize planting. Recalling the way it was employed at Bonampak, Miller (1988, p. 27) argues that the Venus imagery associated with the Jaguar Staircase suggests that the building might have been employed as a place for ceremonies related to the sacrifice of captives, a setting in which the image of Venus itself could well have played a role.

Temple 22 seems to be related to the re-enactment and re-affirmation of kingship. The ceremony, like the calendar of which it is a part, has both its public and private aspects. That part of the scheme visible to the public consists of the king on his throne, Venus in the sky, and the sun coming to the horizon. The secret knowledge, known only to the priests, is encapsulated in the view that appeared only to them through the window of the temple and along the sight line between stelae 12 and 10. We have, on the one hand, in *real time,* the visual phenomenon of the sun attached to Venus, in effect following or preceding it, as the two are observed twisting about in the western sky (Aveni, n.d.) and, on the other hand, in *mythological time,* the Venus and sun symbols appearing in stucco carvings at opposite ends of the bicephalic monster draped over the door of Temple 22. The eye, viewing at the appropriate time, is confronted with two different kinds of visual imagery that really have the same meaning.

Astronomically, the bicephalic serpent that has Venus symbols at one end and the sun sign at the other, illustrates the visual reality that exists between sun and Venus—their inextricable bonding in the pre-dawn or early evening sky. We realize the power of this relationship only when we go out-of-doors and confront these events as the Maya saw them. Only then do we understand that Venus and the sun bear a unique as well as highly prominent visual relationship to each other in the sky. Wherever the sun goes, so, too, goes bright Venus, never very far from him, always suggesting his presence.

It is interesting that for the Maya of Copan, Venus was implicated with solar time. Perhaps this is not so surprising given their penchant for cyclicity. We do not know whether this was also the case for Mexica Tenochtitlan, perhaps because we have not really looked for it. We know of no Venus symbols on the Templo Mayor to motivate us. However, we do know of the importance assigned to Venus in the Central Mexican codices and even that its disappearance of eight days was noted in the Annales de Cuauhtitlan. Is this enough to lead us to take seriously the notion that this celestial luminary is present in orientation schemes relating center and periphery?

Schele and Miller in *The Blood of Kings* (1986, pp. 106–7) suggest that social cohesion was the purpose of symbolic arrays like Temple 22 and its iconography. "The Maya solution to social crisis was not to manipulate economics or intensify agricultural technology; instead, they adjusted ideology"—adapted it to fit their changing concept of social reality. These architectural hierophanies and their associated myth, diverse in symbolism, grand in scale and commanding public space, were intended to convince large numbers of people of one emerging fact of Maya life in the Classic that was to last for hundreds of years: that "differential social ranking and a ruling elite are the natural order of existence ordained by the gods."

PAIRED OPPOSITES IN THE CELESTIAL AND TERRESTRIAL ENVIRONMENT OF CUZCO: TAHUANTINSUYO

In the Inca mentality, social cohesion was achieved through a highly organized system of belief that centered on ancestor worship and that enjoined practically all spheres of social and natural activity into a coherent whole. The formal statement of the basic tenets of that belief system consisted of an organizational map of the capital city that lay over the city itself and that was comprised of both natural and man-made elements.

We can think of the ceque system of Cuzco, the Inca capital, as a highly ordered cognitive map that resembles a quipu. Many details of the organization and classification of information encompassed by the system have been passed down by the Spanish chroniclers, most notably P. Bernabe Cobo (1956), and it is Zuidema's 1977 interpretation of Cobo's description that I shall follow throughout. The ceque system consists of 41 invisible radial lines (ceques) crossing the landscape and emanating from the Coricancha, a temple for worship of ancestors lying near the center of Cuzco. These lineations were marked or traced by 328 huacas or sacred places that consisted of rocks, springs, or other water sources, valleys, mountains, and various man-made shrines. One huaca, for example, was a great stone, one of a sculpted pair, called Collaconcho, that stood in the fortress Sacsahuaman.

> They declare that, bringing it for the structure, it fell three times and killed some Indians. The sorcerers, in questions they put to it, said that it had replied that, if they persisted in wanting to put it in the structure, all would have a bad end, apart from the fact that they would not be able to do it. From that time on, it was considered a general guaca to which they made offerings for the strength of the Inca. (Rowe 1979, p. 21)

The ceques are divided among the four principal quarters (suyus) which, themselves, reflect the hierarchical sub-division of a moiety system based on water and irrigation. The northern pair of suyus constitutes the up-river half of the moiety and the southern pair the down-river sector, the Coricancha, having been built at the confluence of the two rivers that define the valley of Cuzco. Within each suyu the ceques are grouped in threes according to which particular kinship group was assigned to care for them. The system of ceques and huacas seems to have functioned largely as a mnemonic scheme that encoded and interrelated information about Inca dynastic history, social organization, irrigation, wind direction, calendar, and astronomy. (For details see Zuidema 1977.)

The ecology or more precisely, hydrological, basis of the ceque system cannot be underestimated. It is the subject of a recent study by J. Sherbondy (n.d.). She suggests that the ceque lines served as boundaries between *ayllus* or "groups of people who considered themselves related by descent from a common ancestor and who hold corporate rights to the lands and water of the ayllu." In effect, then, the ceque system defines irrigation districts; it is a map of the distribution of water rights. The *sacred* huacas that delineated these ceques also served the *secular* purpose of marking borders between ecological zones, regions of different land use, etc. The offerings made at the huacas really were directed to the ancestors within the earth who were the original owners of the precious water supply.

As in the case of Copan where certain outlying stelae functioned as boundary markers and astronomical indications at the same time, so, too, were astronomical considerations present in the layout of Cuzco. Astronomical restrictions had to compete with hydrological restrictions just as the conduct of ritual and astronomical patters must be made to go together in the *Dresden Codex*. In the case of Cuzco, we cannot assume all astronomical sight lines emanated from one point. In Cuzco the June solstice sunset observation was cared for by the ayllus of Hanan (upper) Cuzco, i.e. from a place in their district, while its opposite, the December solstice sunrise, was made from a "counterpoint" in Hurin (lower) Cuzco. We might wonder: Did the Inca decentralize their observations so as to give a sense of autonomy to the members of the two moieties in an otherwise rigid system?

Of particular interest to us in the context of a center-periphery model is an astronomical sighting line utilized by the Inca to formalize the planting season. It employs one of the huacas (the seventh) of one of the ceques (the eighth of the Chinchaysuyu [NW] quadrant). According to Cobo, this huaca

> was called Sucanca. It was a hill by way of which the water channel from Chinchero comes. On it there were two markers as an indication that when the sun arrived there, they had to begin to plant the maize. The sacrifice which was made there was directed to the sun, asking him to arrive there at the time which would be appropriate for planting, and they sacrificed to him sheep, clothing, and small miniature lambs of gold and silver. (Rowe 1979, p. 27)

During the course of our fieldwork on the location and mapping of the ceque system (Aveni 1981, Zuidema 1981) we determined that this pair of pillars was the central group of a larger system described by an Anonymous Chronicler (1906) and said to have been visible at about two to three leagues from the city where they were used to regulate the sun:

> When the sun passed the first pillar, they prepared themselves for planting in the higher altitudes, as ripening takes longer. When the sun entered the space between the two pillars in the middle it became the general time to plant in Cuzco; this was always the month of August. And when the sun stood fitting in the middle between the two pillars they had another pillar in the middle of the plaza, a pillar of well-worked stone about one estado high, called the Ushnu, from which they viewed it. This was the general time to plant in the valleys of Cuzco and the area surrounding it.

We found it interesting that the formal date of planting in the Cuzco valley corresponded rather closely to the day of the year that could be regarded as the reciprocal in time to the day of the passage of the sun across the overhead position (the zenith). In spatial terms, on the "anti-zenith" day (18 August, 23 April), the sun disappears over the western horizon at a point 180° opposite the sunrise point on the days of zenith passage (30 October, 13 February). Based upon his study of the ceque system as a calendrical counting device, Zuidema (1988) has recently proposed a slight shift in the date when the sun arrived at the central and outer pillar. This shift does not affect the argument made in this paper.

Verticality can be tied to the fertility theme, for the chronicler Guaman Poma makes several statements about the earth mother opening up to receive seed and then life-giving water in February and August, the zenith/anti-zenith sun dates (see references in Aveni 1981, p. 306).

We have walked and measured the line from the Ushnu to the pillars on Cerro Picchu and we extended it for several kilometers in either direction. We have argued that the formalized anti-zenith sightline in Inca ceremonial space actually was derived from observations taken in the other direction, that of the rising sun at the day of its passage across the zenith. This line passes over the Ushnu of Hanan-Haucaypata (the so-called northern moiety) and then extends 26 km to a point on the eastern horizon about the town of Quispicanchis where other structures, now badly ruined, once were situated. Lying along this same zenith/anti-zenith line, which crosses from one side of the Inca capital to the other, is the Sunturhuasi, a tall structure that may have functioned as a gnomon. It symbolically represented the upward projection that complemented the downward directed Ushnu, perhaps in the same way the image of the Pyramid of the Sun in positive space at Teotihuacan was reversed in the depressed space of the Ciudadela, both lying along the same Teotihuacan/N-S axis. Thus, it is possible that both Ushnu and Sunturhuasi functioned to mark the sun when it migrated to the overhead position. It may be added that the chronicler Guaman Poma has specifically associated the Sunturhuasi with astronomical sightings (see Aveni 1981).

The concept of the reversal of the horizontal sight line to view the anti-zenith event (which coincides with both the time of planting and the time the earth opens) became a guiding principle for the Inca, a brilliant stroke which also embodied the dualism of the vertical/anti-vertical axis so central to Inca and even contemporary Andean cosmological thought (see e.g. Isbell 1978: 197–220).

The Coricancha, center of the ceque system, also manifests at least one aspect of this orientation calendar. Its walls are essentially aligned with pairs of ceques that go out in the solstice directions, but we believe they are slightly and deliberately deviated from this direction, possibly to include stellar groupings in the time-marking scheme (e.g. the Pleiades, the heliacal risings of which were employed to begin the count of the seasonal year and were also intended to anticipate the solstice timings [Zuidema 1982]). So many principles and ideas deemed important by the Inca are skillfully unified in the zenith/anti zenith sightline that runs horizontally from one end of the valley to the other, cutting through the heart of the city. As in the case of the other schemes of which we have spoken that tie together the natural and ceremonial environments, it is difficult to delineate an evolutionary building sequence, but whatever it was, we believe it began with the recognition that the most important time of year, the planting season, coincided with the time when the sun lay at its most remote position in time from the day when it passed overhead. Given the evidence, we believe it is reasonable to hypothesize that the architects of the city first sighted the place of zenith sunrise from the vicinity of Cerro Picchu and six months later made the reverse sighting from the top of the mountain above Quispicanchis. As in the case of Tenochtitlan, the transformation of this direction in ecological space-time into the relatively more confined ceremonial environment was probably not a simple task because Cerro Picchu, which lies immediately west of the center of Cuzco, looms 9° high above the horizon and produces an azimuthal shift or skew in the sunset position relative to what it would be along the relatively level horizon that can be viewed along the 26 km long baseline that transcended the populated portion of Cuzco. Perhaps, as in the case of the Templo Mayor, "things became a little twisted," as Motolinia says. But these are technical details, the solution to which we need not argue here. The problem is essentially the same in each instance—to transform environmental events to ceremonial events.

CONCLUSIONS

It is now time to "ascend from the cases" and hopefully come to some general truths. Responding to the thought-provoking guidelines that Davíd Carrasco has provided for this symposium, I decided to review and rethink three case studies associated with the research in which I had participated on the delineation of calendrical systems via orientation principles. In the context of this conference, I have attempted to entertain the question of whether these diverse, disconnected cultures exhibited anything in common, both formally and processually in the way they have gone about classifying, expressing, or using their astronomy and relating it to other categories of human knowledge.

It does not require much deep thought to see that there is plenty of general common ground here. Each of the building schemes I have outlined seems to have involved deliberate planning by a centralized bureaucratic authority. But, as Charles Long has warned in his commentary: Whose knowledge are we dealing with? Archaeoastronomy may serve the purpose of discerning one concrete aspect of the ritual behavior of these Native American people, but we must be careful about how we interpret what these data imply about social behavior and systems of religious belief.

I believe we can state with some certainty that the elaborate detailed planning we have uncovered comes only with the formation of a state and can hardly be expected to be a part of a less organized or sophisticated polity. We may be witnessing the regularity of nature's events being assigned a place in the social order by (at least partially) a theocratic ruling body, one that associated itself with the awesome and impressive forces of nature as a means of establishing and sustaining its power. Thus, the Moctezumas of Tenochtitlan, Eighteen-Jog of Copan and Pachacuti Inca of Cuzco all have something in common, at least at a general level. But we must be concerned with the similarities and differences in the details of the formal plan and the questions raised by such a study that can then be posed for the students of anthropology and religion.

As is often the case when one attempts to synthesize information, I have found it necessary to produce my own taxonomic scheme, which I display in Table 10.1. The table is self-explanatory. First of all, note that for all three case studies there are two kinds of orientations, one in the ceremonial space of the center involving the most important building complex and the other relating built structures at the visible horizon. Moreover, the latter are physically connected by deliberate alignments to the central building complex. For the case of Cuzco, the connection is an indirect one embedded within the ceque system.

I found it meaningful that in every case the time period of the calendar that becomes the focus of all the skywatching, aligning, and concomitant ritual is *the period in the dry portion of the year immediately preceding the rainy season*. In the case of the Templo Mayor, the supposed place of origination of water, the very Tlalocan, is tied directly to the observational scheme in a redundant, reinforcing way. Clearly, one building could have done as well to fix the date; but there are two. In Copan, the

Table 10.1 Toward a center-periphery model of the ceremonial center that incorporates astronomy and calendar.

	Formal Arrangement				Comments		
	Center		Periphery				
Site	Architectural Alignment	Associated Astronomical Event	Architectural Alignment	Associated Astronomical Event	Social Principle	Ecological Principle	Outcome of Ritual	Associated Iconography
Copan	Temple 22, West-Facing Window.	Venus at Horizon, Late April–Early May. Year Preceding Great Standstill.	Stela 12–Stela 10 (7 km Baseline).	Sunset 12 Apr, or 20 Days Before Zenith Passage.	Differential Social Ranking is the Natural Order of Things. The Ruler is Directly Linked to Celestial Events.	Fertility; Coincidence of Rain-Maize Via Venus-Maize-Rain Complex.	Good Maize Crop Assured.	Bicephalic Serpent, Young Corn Iconography, Possibly Identified With Ruler.
Cuzco	Ushnu to Pillars on C. Picchu (2 km Baseline). Coricancha Walls.	Anti-Zenith Sunset. Solstice, Heliacal Rise of Pleiades.	C. Picchu to Hill Above Quispicanchis (26 km Baseline).	Zenith Sunrise–Anti-Zenith Sunset.	It is Important to Pay Tribute to Ancestors Who Live in the Earth And From Whom the People Inherit Their Land And Water by Divine Right.	Fertility of Pachamama. Earth Opens Up; Water & Irrigation.	Planting in Different Altitudes, Apparently Regulated.	Architectural Forms of Ushnu & Sunturhuasi.
Tenochtitlan	Templo Mayor.	Equinox Sunrise.	Templo Mayor to Horizon Including Tlaloc Shrine and Tepetzinco (44 km Baseline).	Equinox Sunrise ±20 Days.	The Organization of The Aztecs as a People. Celestial, Visual Re-Enactment of the Foundation Myth.	Origin of Rain; Fertility. Maize Sprouting. Maize-Rain Connected To Child Sacrifice.	Rain Assured.	Symbolism of Tlaloc.

creation of a calendrical orientation scheme based on solar horizon observations that harmonizes perfectly with key events in the agricultural cycle can only be characterized as ingenious. The timing of the sun observations 20 days (one Mesoamerican month) away from the equinox (in Tenochtitlan) or solar zenith passage (in Copan) would have impressed the commoner as much as the pronouncements of an Einstein or Newton, two gods of the modern dynasty of Science, impress us. And Cuzco's calendar could be employed nowhere else but in Cuzco. The timing would be thrown off either by latitude, climate, or altitude anywhere else. Of the many calendrical-environmental connected sets that could be discovered, the Inca seemed to achieve one that worked remarkably well.

In all three cases there had developed a tradition out of the universal cosmological conviction that binds human social and environmental order to the movement of things in the heavens. We must face it: the sky is the most dependable. In Central Mexico we can trace the cosmic orientation principle to Teotihuacan, and in Cuzco, the organizational principle of the ceque system may have had its roots in time at least as far back as the Late Intermediate Period on the coast, where we find hints of it in the lines of Nazca (Aveni 1986). There is evidence that the Inca tried to propagate these concepts of order in detail, though likely not with much success, to other parts of the foreign world they administered; e.g. to Huanuco Pampa where we have evidence for bi-, tri-, and quadripartite ceque symbolism and to Incawasi in the highlands, as well as to Tambo Colorado on the south coast of Peru (and possibly in Tiahuanaco).

Given the wide variety of iconographic symbols that we can identify as being associated with these center and periphery sighting schemes, we can only imagine the number of icons and artifacts whose association with the calendar still lies hidden. The same holds for the myriad astronomical phenomena for which no evidence in the material record has yet been uncovered. We were fortunate enough to discover the Venus reference at Copan only because the Venus Table in the Dresden Codex has been deciphered and we know what a Maya Venus symbol looks like. For me, the mental exercise of synthesizing already published data is helpful in organizing thoughts and redirecting questions for future study and I am grateful for being provoked to do so in the context of an interdisciplinary symposium such as this one.

The most obvious question I now ask: Just how universal is this center-periphery calendrical model? I would like to subject some of the other archaeoastronomical studies to my "formative" Table 10.1 to see whether various elements of those investigations fit or alter the model. To the anthropologist, I would ask: Can these considerations help in the problem of understanding the formation of states? Have we underplayed the role of the use of environment, architecture, and hierophany in propagating knowledge of the social order? The historian of architecture or Mesoamerican art might inquire about whether there is a place for orientation among the many parameters used to delineate the evolution of style. In a recent monograph on the Puuc sites of the Yucatec Maya (Aveni and Hartung 1986), we have tied orientational shifts and trends to studies of settlement patterns and the testing of various models that try to account for the interrelations among the different Maya cities. Perhaps future inquiries might focus on some of the broader aspects of city planning implied by the model offered here.

REFERENCES

Anonymous Chronicler. 1906 [16th cen.] Discurso de la Sucesion de Gobierno de los Yngas. In *Juicio de Limites entre el Peru y Bolivia, Prueba Peruana,* ed. V. Maurtua. 8:149–165.

Aveni, A. 1977. Concepts of Positional Astronomy Employed in Ancient Mesoamerican Architecture. In *Native American Astronomy,* ed. A. F. Aveni, 3–20. Austin, University of Texas Press.

Aveni, A. 1980. *Skywatchers of Ancient Mexico.* Austin, University of Texas Press.

Aveni, A. 1981. Horizon Astronomy in Incaic Cuzco. In *Archaeoastronomy in the Americas,* ed. R. Williamson, 305–318. Los Altos and College Park, Ballena and the Center for Archaeoastronomy.

Aveni, A. N.d. The Real Venus-Kukulcan in the Maya Inscriptions and Alignments. Paper read at VI Mesa Redonda de Palenque, Palenque, Mexico, June 1986.

Aveni, A. 1986. The Nazca Lines: Patterns in the Desert. *Archaeology* 39 (4):32–39.

Aveni, A., E. Calnek, and H. Hartung. 1988. Myth, Environment, and the Orientation of the Templo Mayor, *American Antiquity* 53 (1988): 287–309.

Aveni, A., and H. Hartung. 1982. New Observations of the Pecked Cross Petroglyph. *Lateinamerika Studien* 10:25–41.

Aveni, A., and H. Hartung. 1986. The Maya City and the Calendar. *Transactions of the American Philosophical Society* 76 (7):1–87.

Aveni, A., H. Hartung, and B. Buckingham. 1978. The Pecked Cross Symbol in Ancient Mesoamerica. *Science* 202:267–279.

Broda, J. 1982. Astronomy, Cosmovision and Ideology in Pre-Hispanic Mesoamerica. In *Ethnoastronomy and Archaeoastronomy in the American Tropics,* ed. A. Aveni and G. Urton, 81–110. New York, Annals of the New York Academy of Sciences, 385.

Closs, M., A. Aveni, and B. Crowley. 1984. The Planet Venus and Temple 22 at Copan. *Indiana* 9:221–247.

Cobo. 1956 [1653]. *Historia del Nuevo Mundo.* Madrid, Biblioteca de Autores Espanoles, V. 91–92.

Dow, J. 1967. Astronomical Orientations at Teotihuacan: A Case Study in Astroarchaeology. *American Antiquity* 32:326–334.

Duran, D. 1971 [1570]. *The Book of the Gods and Rites and the Ancient Calendar,* trans. and ed. F. Horcasitas and D. Heyden. Norman, University of Oklahoma Press.

Isbell, B. J. 1978. *To Defend Ourselves.* Austin, University of Texas Press Foundation for Latin American Studies.

Matos, E. 1984. The Great Temple of Tenochtitlan. *Scientific American* (August):70–79.

Miller, M. 1988. "The Meaning and Function of the Main Acropolis, Copan. In *The Southeast Classic Maya Zone,* ed. G. Willey and E. Boone, 149–194. Washington, DC, Dumbarton Oaks.

Millon, R. 1973. *Urbanization at Teotihuacan, Mexico,* Vol. 1, *The Teotihuacan Map,* Part 1, Text. Austin, University of Texas Press.

Morley, S., and G. Brainerd. 1956. *The Ancient Maya.* Stanford, Stanford University Press.

Motolinia, T. 1971 [1541]. *Memoriales o Libro de las Cosas de Nueva España y de las Naturales de Ella,* ed. E. O. Gorman. Mexico, UNAM.

Pomar, J. B. 1981. *Relacion de Texcoco,* Nueva Collecion de Documentos para la Historia, de Mexico, ed. J. Garcia Icazbalceta, 111:1–69 (2nd ed. Mexico 1941).

Redfield, R., and A. Villa Rojas. 1962. *Chan Kom: A Maya Village.* Chicago, University of Chicago Press.

Rickards, C. 1929. The Ruins of Tlaloc, State of Mexico. *Journal de la Société des Américanistes de Paris* 21:197–199.

Rowe, J. 1979. An Account of the Shrines of Ancient Cuzco. *Nawpa Pacha* 17:1–79.

Sahagún, B. de. 1950–1982. *Florentine Codex: General History of the Things of New Spain* (12 vols. and intro.), trans. C. Dibble and A.J.O. Anderson. Sante Fe, Monographs of School of American Research.

Schele, L., and M. Miller. 1986. *The Blood of Kings.* Fort Worth, TX, Kimbell Art Museum.

Sherbondy, J. 1980. Water in the Ritual Calendar of Cuzco. Paper presented at the 1980 American Anthropological Association meeting, Washington, DC, 1980.

Tezozomoc, A. 1975 [1609]. *Cronica Mexicana.* Mexico City, Editorial Porrua.

Thompson, J.E.S. 1970. *Maya History and Religion.* Norman, University of Oklahoma Press.

Torquemada, J. 1975 [1723]. *Monarquia Indiana.* Mexico, Editorial Porrua.

Wicke, C., and F. Horcasitas. 1957. Archaeological Investigations on Mt. Tlaloc, Mexico. *Mesoamerican Notes* 5:83–97.

Zuidema, R. T. 1977. The Inca Calendar. In *Native American Astronomy,* ed. A. Aveni, 219–259. Austin, University of Texas Press.

Zuidema, R. T. 1981. Inca Observations of the Solar and Lunar Passages through Zenith and Anti-zenith at Cuzco. In *Archaeoastronomy in the Americas,* ed. R. Williamson, 319–342. Los Altos and College Park, Ballena and Center for Archaeoastronomy.

Zuidema, R. T. 1982. Catachillay: The Role of the Pleiades and of the Southern Cross and α and β Centauri in the Calendar of the Incas. In *Ethnoastronomy and Archaeoastronomy in the American Tropics,* ed. A. Aveni & G. Urton, 203–230. New York, Annals of the New York Academy of Sciences, 385.

Zuidema, R. T. 1988. The Pillars of Cuzco: Which Two Dates of Sunset Did They Define? In *New Directions in American Archaeoastronomy,* ed. A. Aveni, 143–169. B.A.R. International Series, 454.

Astronomical Alignments at the Templo Mayor of Tenochtitlan, Mexico

Ivan Šprajc

INTRODUCTION

Systematic archaeoastronomical research carried out during the last few decades has revealed that architectural orientations in Mesoamerica exhibit a clearly non-random distribution and that civic and ceremonial buildings were mostly oriented on the basis of astronomical considerations, particularly to the sun's positions on the horizon on certain dates of the tropical year.[1] While the alignments to sunrises and sunsets on the solstices and equinoxes have been found on various archaeological sites, the most frequent orientational groups correspond to other dates whose significance is less obvious. According to various hypotheses put forward thus far, the solar dates recorded by the orientations can be interpreted in terms of their relevance in the agricultural cycle and in the computations related to the calendrical system. It has been suggested, for

From *Archaeoastronomy* 25 (*JHA* 31) (2000):S11–S34. Reproduced by permission of the author.

example, that the dates indicated by the alignments are separated by calendrically significant intervals. The most elaborate model of this type has been proposed by Tichy,[2] who contends that these dates mark intervals of 13 and 20 days and multiples thereof; on the other hand, he also suggests that the orientations are spaced in accordance with a geometrical system based on a 4.5° angular measurement unit. Some authors reconstructed possible horizon calendars for particular sites, on the assumption that prominent peaks of the local horizon served as natural markers of sunrises and sunsets on relevant dates.[3]

In order to test such hypotheses, I undertook precise measurements of alignments at 37 Preclassic, Classic and Postclassic archaeological sites in central Mexico. This involved measuring not only the orientations of civic-ceremonial structures but also the alignments to prominent mountains on the local horizon, placed within the angle of annual movement of the sun. The analyses of the data obtained show that the dates of sunrises and sunsets both along the architectural orientations and above the prominent hills on the local horizon exhibit consistent patterns, being separated by intervals that are predominantly multiples of 13 and 20 days and are, therefore, significant in terms of the Mesoamerican calendrical system. Furthermore, the most frequently recurrent dates, registered at a number of sites, apparently marked crucial moments of a ritual agricultural cycle. The regularities detected strongly suggest that the important ceremonial structures were constructed on carefully selected places, in order to employ certain surrounding peaks as natural markers of horizon calendars. Both the orientations embodied in the monumental architecture of a particular site—occasionally dominating the entire urban layout—and the prominent features of the local horizon allowed the use of an observational calendar that, in view of the lack of permanent concordance of the calendrical and tropical years, was necessary for predicting important seasonal changes and for an efficient scheduling of the corresponding agricultural activities. It is also obvious, however, that this practical function of observational calendars was deeply embedded in the ritual and intimately related with social organization, religion and political ideology.[4]

The results of my research in central Mexico agree with some general ideas formerly expressed by other authors, but differ in important details which concern the principles underlying the orientational patterns, and the use of observational calendars. While some of Tichy's models,[5] for example, do have a real basis—even if his specific hypotheses are not corroborated—his geometrical orientational scheme can hardly be sustained.[6]

The Templo Mayor of Tenochtitlan, one of the structures included in my study, exemplifies the observational and calendrical function of the alignments at central Mexican sites from the Preclassic onwards.

ARCHITECTURE AND CHRONOLOGY

The remains of the Templo Mayor of Tenochtitlan (Fig. 11.1) are located in Mexico City's historical centre, immediately northeast of the Metropolitan Cathedral (longi-

11.1 Remains of the Templo Mayor of Tenochtitlan, with its various structural phases (view to the north).

tude: 99°07'51" W; latitude: 19°26'03" N; altitude above sea level: 2240 m[7]). The earliest vestiges of a settlement in the area occupied in later times by the Templo Mayor ceremonial precinct date from the Early Postclassic.[8] However, the greater part of architectural remains discovered so far belong to the Late Postclassic, including the various structural stages of the Templo Mayor, the main building of the sacred precinct of the Mexica capital. Even if there is no agreement about the details concerning the chronological sequence of the Templo Mayor's construction, it seems that Phase II can be dated, according to several propositions, to the fourteenth century;[9] it is thus probable that the earliest temple (nowadays covered by the construction called Phase II) was built in the same century or even in the previous one.

The research accomplished so far[10] has made it possible to distinguish seven principal building stages of the Templo Mayor. Each of the known superimposed structures, all of them similar in shape, is characterized by a double stairway on the west side. Upon the platform of Phase II the remains of upper twin sanctuaries are also preserved, dedicated to the gods Tlaloc and Huitzilopochtli. Not only the Contact-period historical sources but also an enormous amount of offerings and other archaeological finds provide information as to the ritual activities and complex symbolism associated with the Templo Mayor.[11]

Table 11.1 Data on the orientations of the Templo Mayor of Tenochtitlan.

Structure	A	h	δ	Dates
Templo Mayor Phase II	97°42' ± 30'	2°02' ± 5'	−6°39' ± 30'	Mar 3, Oct 10 ± 1^d
	277°42' ± 30'	2°07' ± 3'	7°54' ± 30'	Apr 9, Sep 1 ± 1^d
	6°30' ± 1°			
Later phases	95°36' ± 30'	1°55' ± 5'	−4°43' ± 30'	Mar 9, Oct 5 ± 1^d
	275°36' ± 30'	2°22' ± 5'	6°00' ± 30'	Apr 4, Sep 7 ± 1^d
	6°40' ± 30'			

ARCHITECTURAL ORIENTATIONS AND ALIGNMENTS TO PROMINENT HORIZON FEATURES

The data on architectural orientations at the Templo Mayor of Tenochtitlan are listed in Table 11.1. The mean east-west and north-south azimuths (with estimated margins of error) appear in the second column (A), whereas the corresponding horizon altitudes are given in the third column (h). The astronomical declinations calculated for each azimuth and horizon altitude, taking into account the effects of atmospheric refraction, appear in the fourth column (δ),[12] while the dates on which the sun had these declinations are listed in the fifth column.[13]

The east-west orientation azimuth of Phase II is based on the azimuth of the narrow passageway that separates the upper twin sanctuaries (Fig. 11.2), because the latter probably reproduces the intended orientation of the temple with particular fidelity: the drawing of the Templo Mayor in the Tenochtitlan map attributed to Cortés shows a face representing the sun flanked by the two upper sanctuaries, thus suggesting that the observations were made precisely along the passage between them.[14] Even if this is not an indisputable proof that the orientation of the passage is the most relevant one, it does seem significant, on the one hand, that other east-west lines measured on the Phase II structure exhibit very divergent azimuths and, on the other, that their mean is very close to the present-day azimuth of the passage (cf. infra).

The remains of the various construction stages of the Templo Mayor are nowadays considerably displaced from their original position, due to differential settlements that the architectural complex has undergone through the centuries[15] and which must have also resulted in horizontal movements. At present, the azimuth of the axis of the passageway between the twin sanctuaries of Phase II is 97°32', but originally it must have been a little larger, because the structure is strongly inclined, its southeast extreme exhibiting the highest elevation. Measuring relative heights of various points on the upper platform,[16] I was able to determine the approximate inclination angles in the north-south and east-west directions and to calculate, on these grounds, the probable magnitude of horizontal movements. The calculations, presented in detail in the Appendix, indicate that a small rotation movement in the horizontal plane must have accompanied the process of settling of the structure and that the east-west architectural alignments originally had slightly greater azimuths than they have nowadays. Since the magnitude of this horizontal skew may have been between 0 and 20 min-

Ivan Šprajc

11.2 View to the east along the passageway between the upper sanctuaries of Tlaloc (left) and Huitzilpochtli (right) of Phase II of the Templo Mayor of Tenochtitlan.

utes of arc, depending on the sequence of the movements, I added to the measured azimuth of the passage (97°32′) the mean value of 10′. Although the estimated margin of error of the azimuth thus obtained is, according to these calculations, ±10′, it seems reasonable to consider a larger value: on the one hand, the calculations are valid for a rigid body, whereas the building most surely has not moved uniformly in all of its parts; on the other hand, we can suppose that telluric movements, which are so common in the region and whose effects may have been intensified by the characteristics of the swampy ground, triggered some additional and irregular horizontal dislocations that cannot be reconstructed. Furthermore, it should be recalled that the value 97°42′ corresponds to the azimuth measured along the passageway between the twin sanctuaries and corrected for the estimated horizontal rotation, while we have no compelling evidence that this was, indeed, the most relevant alignment for observations. The mean azimuth of all of the east-west lines measured on Phase II is 97°24′; this value is, significantly, very close to the present-day azimuth of the passage, but it also has a margin of error, since the individual azimuths diverge considerably. The margin of error of ±30′ assigned to the east-west orientation azimuth of Phase II of the Templo Mayor (Table 11.1) is based on these considerations.

Aveni et al. and Ponce de León give for the passage of Phase II the azimuths 97°46′ and 97°25′, respectively.[17] Exploring the effects of the structure's sloping, Ponce de León[18] measured the axis of the passage projected to the present ground level, and concluded that the azimuth of 98°48′ he established for this virtual axis must be considered as very close to the original azimuth of the passageway. Even if Ponce de León's

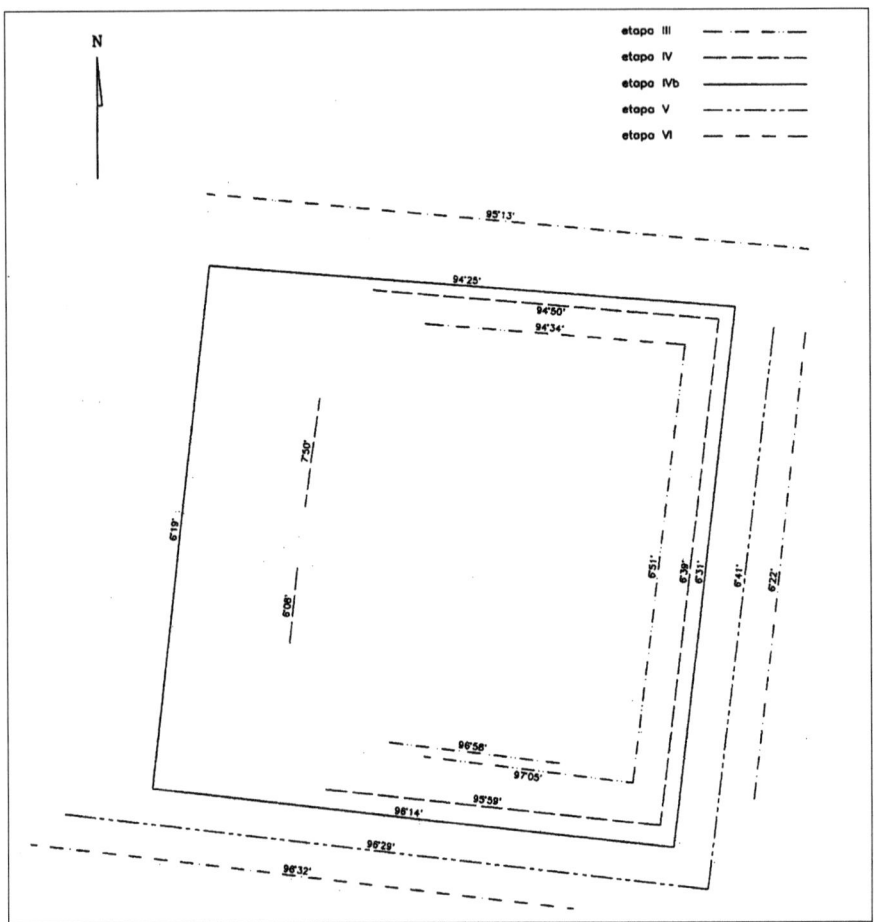

11.3 Azimuths of the lines measured on late phases of the Templo Mayor of Tenochtitlan.

analysis is detailed and careful, it should be pointed out that the azimuth obtained by his procedure is most probably too large: by projecting the axis of the passage to the actual ground level, along the plane perpendicular to the upper platform of the structure,[19] we get a line connecting two points which—located on the front and rear façades—originally were not on the same level, if we consider the inclination of the structure, whose southeastern extreme is nowadays its most elevated part. The azimuth of this alignment does not necessarily reproduce the original orientation of the passageway, since it depends on the position of the axes around which the structure rotated and on the sequence of these movements.[20]

The results of my measurements show that the orientation of Phase II, at least in the east-west direction, differs from the one incorporated into the later superimposed structures. Measuring the alignments between the corners of the preserved slanted faces *(taludes)* of the later phases—or between the points near the corners that are not exposed or preserved—I obtained the azimuths shown in Figure 11.3;[21] the mean values appear in Table 11.1. The azimuths of the alignments may nowadays, due to settlements, slightly differ from the original ones, but the formula discussed in the Appendix and derived with the purpose of estimating possible horizontal movements of Phase II cannot be applied to the case of later phases, since the latter have not moved as rigid bodies. The degree of subsidence observable at different points is directly proportional to their distances from the central part of the construction mass, which is the most elevated one, because the compressibility of underlying clays was reduced by the pressure of the first superimposed buildings.[22] As it is obvious, therefore, that the settlements did not produce uniform horizontal skews, it can be assumed that by averaging the extant azimuths, the eventual errors of individual values cancel out.

Ponce de León[23] also inferred that Phase II, on the one hand, and the later superimposed buildings, on the other, had different orientations. For the line connecting central points of the stairways of the late phases he obtained the azimuth of 96°02'. Assuming also for these structures a skew similar to the one detected on Phase II, he added to the measured azimuth the value 1°23'—i.e. the difference between the existing (97°25') and the original azimuth (98°48') he determined for the passageway of Phase II—and concluded that the value obtained, 97°25', must be considered as the original orientation azimuth of the structural phases later than the second one. In view of the argument presented above, however, the conclusion seems hard to accept, both because the correction value determined for Phase II (1°23') is excessive and because the Phase II structure tilted in a relatively uniform manner, while the differential settlements of subsequent phases caused different parts of the structures to incline in different directions. It can be observed that the azimuth measured by Ponce de León, without correction (96°02'), is quite close to the mean value based on the *taludes* (95°36': Table 11.1). However, the line measured along the central points of the stairways of the superimposed buildings does not necessarily reproduce with precision the orientation of each of them, because it could never be visually controlled by the builders. On the other hand, we can recall that the successive stages of the contemporary Tenayuca pyramid share the same orientation, but their central east-west axes move progressively towards the south.[24]

The data displayed in Table 11.1 show that the north-south azimuths of Phase II and of the late phases are practically equal. Furthermore, the listed values, the result of my own measurements, agree with the mean of 6°42'± 23' established by Aveni, Calnek and Hartung[25] and based on the north-south lines. Observing that the latter do not exhibit notable divergences, Aveni et al. concluded that all of the structural stages possessed very similar orientations.[26] However, the east-west azimuths of the late phases are consistently smaller than those measured on Phase II, their mean values being 95°48' (Phase III), 95°25' (Phase IV), 95°19' (Phase IVb) and 95°52' (Phase VI) (cf. Fig. 11.3). Since these values do not differ from each other in a significant and systematic way, it is likely that the mean value based on them and given in Table 11.1 represents the intended orientation of the late phases of the Templo Mayor with reasonable accuracy.[27] This conclusion is supported by the fact that various adjacent structures contemporary with the last phases of the Templo Mayor[28] exhibit comparable orientations. For example, the azimuths of the east-west axes of Structures C and F, located immediately to the north and south of Phase VI, are 95°47' and 95°04', respectively. The pronounced inclinations of both structures suggest that their original orientations were quite similar to those of the late phases of the Templo Mayor: since Structure C, to the north, presents the greatest elevation in its southwest corner, its original east-west azimuths must have been slightly smaller than nowadays, while those of Structure F, alternatively called Red Temple and situated to the south, were probably greater, because the most elevated part of this building is its northwest corner. The east-west azimuth of Structure B located immediately west of Structure C is 95°23', while the south face of Structure E, also known as House of the Eagles and occupying the extreme north of the excavated area, aligns with an azimuth of 95°06'. It seems, then, that the orientation of the Templo Mayor was reproduced in the contemporary neighbouring buildings.

It has been commonly held that the streets in the historical centre of Mexico City follow the orientation of the Templo Mayor and associated structures.[29] This opinion is reflected also in the reconstruction plans of the sacred precinct of Tenochtitlan.[30] It should be pointed out, however, that the orientations of the greater part of the buildings that have been excavated are slightly skewed counterclockwise relative to the present urban layout. As the plan of Vega Sosa shows, the structures excavated in the area of the nearby Metropolitan Cathedral exhibit such a deviation with regard to the ground plan of the church, whose axes agree with the orientation of the surrounding streets.[31] The fact that the colonial urban layout corresponds rather with the orientation of Phase II of the Templo Mayor[32] suggests that this alignment, even though in later times it no longer prevailed in the ceremonial precinct, had been dominant in the early period of Tenochtitlan and persisted in certain buildings and streets, or even in the greater part of the prehispanic urban layout, until the Conquest, when it was adopted by the colonial town.[33]

Considering that the orientation of the Templo Mayor changed, beginning with Phase III, it can be recalled that the latter, according to various authors, belongs to the reign of Itzcóatl.[34] The modification can thus be understood as a part of the ambitious programme of reforms for which this ruler is particularly well known. The orientations of the structures excavated in the area of the cathedral have not been measured with precision, but Structure A (Temple of Tonatiuh) appears to reflect the same change:[35] while the first construction phase follows the orientation of the cathedral and, therefore, of Phase II of the Templo Mayor, the superimposed buildings exhibit a skew in the same direction (counterclockwise) as the late phases of the Templo Mayor. However, whereas Structure A is late, some other buildings that share the same deviation, notably Structures C and D,[36] belong, according to Vega Sosa,[37] to the early periods of occupation of the site (c. AD 950–1350). Consequently, it is possible that the new orientation incorporated into the Templo Mayor after Phase III had forerunners, but became dominant in the sacred precinct only in the late periods of Tenochtitlan.

Table 11.2 Data on the eastern horizon of the Templo Mayor of Tenochtitlan.

Mountain	A	h	δ	Dates
Cerro Tláloc	93°11'	2°17'	−2°19'	Mar 14, Sep 28
Cerro Tlamacas	74°40'	0°58'	14°40'	Apr 29, Aug 13

The results of my analysis of the alignment data referring to prominent horizon features at a number of archaeological sites in central Mexico suggest that, in the case of the Templo Mayor of Tenochtitlan, the mountain tops Tláloc and Tlamacas, visible on the eastern horizon, must have been particularly important: they marked sunrises on the dates that, together with those recorded by architectural orientations, composed observational calendar schemes comparable to those reconstructed for other sites.[38] The azimuths (A), altitudes (h), declinations (δ) and sunrise dates corresponding to the two mountains are listed in Table 11.2.[39]

OBSERVATIONAL CALENDARS

Employing the data presented in Tables 11.1 and 11.2, it can be calculated that various dates recorded by architectural alignments and certain mountain peaks on the horizon are separated by intervals that are, or approach, multiples of 13 and of 20 days. Cerro Tlamacas demands particular attention, because the dates it registers divide the year into intervals of approximately 105 and 260 days. The "ideal" dates would be April 30 and August 13, which are commonly marked by architectural orientations and prominent horizon features at various sites.[40] In fact, Cerro Tlamacas could have recorded these dates if the last contact of the solar disk with the horizon was observed, i.e. if it was the tangent position of the sun upon the mountain that was relevant for determining the dates corresponding to the alignment. The declination of the sun required for seeing its lower limb aligned with the top of Cerro Tlamacas, when observing at the Templo Mayor, is 14°45'.[41] If for a 4-year period in the mid-fourteenth century—assuming the site for the construction of the Templo Mayor was chosen around that time—we examine solar declinations calculated for the moments of sunrise on relevant dates, we find that Gregorian dates on which Cerro Tlamacas was aligned with the centre of the sun and with its lower limb were those listed in Tables 11.3 and 11.4, respectively.[42]

It can be observed that the intervals separating the dates registered by the centre of solar disk behind the summit of Cerro Tlamacas are 105 or 106 and 259 or 260 days. However, if the dates on which the sun's disk was seen tangent to the mountain were relevant, the short interval was 105 or, once in the four years, 106 days, while the long interval was always 260 days.[43]

Assuming that the interval of 260 days was particularly important, because it separated the same dates of *tonalpohualli* (the sacred 260-day calendrical count), it can be concluded that *the dates of the observational calendar of the Templo Mayor were recorded by tangent positions of the sun on the horizon along the alignments*.[44]

Table 11.3 Dates recorded by the solar disk's center aligned with Cerro Tlamacas, and intermediate intervals, for a period of four years in the middle of the 14th century.

Year	Date	Interval (days)
1341	Apr 29	
		106
	Aug 13	
		260
1342	Apr 30	
		105
	Aug 13	
		260
1343	Apr 30	
		106
	Aug 14	
		259
1344	Apr 29	
		106
	Aug 13	
		259
1345	Apr 29	

Table 11.4 Dates recorded by the solar disk's lower limb aligned with Cerro Tlamacas, and intermediate intervals, for a period of four years in the middle of the 14th century.

Year	Date	Interval (days)
1341	Apr 30	
		105
	Aug 13	
		260
1342	Apr 30	
		105
	Aug 13	
		260
1343	Apr 30	
		105
	Aug 13	
		260
1344	Apr 29	
		106
	Aug 13	
		260
1345	Apr 30	

The interval of 46 days between the dates marked by the Tláloc and Tlamacas peaks also demands attention, because the sunset dates corresponding to the orientation of the Templo Mayor's Phase II subdivide it in intervals of 26 or 27 and 20 or 19 days (cf. Tables 11.1 and 11.2). In the late fourteenth century, when Phase II was probably erected,[45] the sun's lower limb aligned with the Tláloc and Tlamacas mountain tops on March 14 or 15 and April 29 or 30, respectively, but the intermediate interval was predominantly 46 days. Supposing the accuracy of the observational calendar was more important in spring, before the onset of the rainy season, it is likely that the orientation of Phase II recorded sunsets on April 9 or 10, ideally separated by the exact intervals of 26 (2 *trecenas*) and 20 days (1 *veintena*) from those marked by Cerro Tláloc and Cerro Tlamacas, respectively.[46] The structure could have registered these dates if the tangent position of the sun on the horizon was determinant and, moreover, if the original east-west orientation azimuth of Phase II was approximately 97°50' (declination required: 8°06'), i.e. about 8' greater than the one given in Table 11.1 (97°42'). The latter has been determined from assessment of the magnitude of horizontal skew originated by settlements, by adding the mean correction value of 10' to the present azimuth of the passageway between the twin sanctuaries (97°32') *(vide supra,* and the Appendix). However, according to the argument presented in the Appendix, the original azimuth could have been, indeed, up to 20' greater than it is nowadays.[47]

It also seems significant that the interval between the sunrise dates corresponding to the orientation of Phase II approaches 39 days (3 *trecenas*). However, the spring interval (from March 3 to April 9), though presumably the more important, is 37 days. If the original orientation azimuth was about 97°50', as suggested above, and if the tangent position of the sun on the horizon was observed, the date of sunset along the axis of Phase II was April 9 or 10, while the sunrises occurred on the same date of March 3, which means that the interval between the two dates did not reach 39 days. The "ideal" date would have been March 1/2, 13 days before the one recorded by Cerro Tláloc, and April 9/10, 20 days before the sunrise above Cerro Tlamacas, but these dates, given the horizon altitudes, could not be recorded by one and the same orientation.[48]

It can be hypothesized that sunrises on March 1 or 2 (13 days before sunrise above Cerro Tláloc and 39 days before sunset in the axis of the temple) were marked by other orientations. While the idea that two slightly different alignments were embodied in the same Phase II of the Templo Mayor is not supported by the measured alignments,[49] it is not impossible that some other neighbouring building(s) recorded the relevant sunrise dates, which composed an observational calendar in combination with the dates of sunset in the axis of the Templo Mayor.[50] The hypothesis obviously has no support until a required orientation is found, incorporated into a structure contemporaneous with Phase II.

The available evidence suggests that the primary concern of the builders of Phase II was to orient the structure toward the point on the western horizon where the sun set 26 days after it had risen above Cerro Tláloc and 20 days before the same phenomenon occurred above Cerro Tlamacas. Table 11.5 presents the dates and intervals of the observational calendar that could have been in use in the late fourteenth century, if the east-west orientation azimuth of the structure was approximately 97°50' and if tangent positions of the sun upon the horizon were relevant for determining the dates. As one can see, in the spring half of the year the interval between the sunset in the axis of the structure and the sunrise above Cerro Tlamacas is invariably 20 days, while the distance between the sunrise above Cerro Tláloc and the sunset marked by the building is 26 days, except in 1382, when it is 27 days. It may be noted, again, that the long interval separating the sunrises above Cerro Tlamacas is always 260 days. Also significant might be the fact that the long interval between the dates of sunset in the axis of the structure (e.g., from 1380 August 31 to 1381 April 9) is constantly 221 days, i.e. 17 *trecenas*.

As for the late orientation of the Templo Mayor, the underlying astronomical and calendrical motives seem to be clear: the intervals composing the observational calendar that can be reconstructed are, or approximate to, multiples of *trecenas*. The shortest intervals between the sunrises and sunsets in the axis of the structure are 26 or 28 days, while the consecutive sunrise/sunset dates are separated by intervals of 155/156 days; furthermore, the sunset dates recorded by the temple's orientation fell 25 days before and after the sunrises above Cerro Tlamacas (cf. Tables 11.1 and 11.2). An "ideal" scheme of intervals would have been the one shown in Table 11.6, where the short intervals between the consecutive dates of both sunrises and sunsets in the axis of the structure are always 156 days (12 *trecenas*), while the spring intervals from the sunrise to the sunset marked by the structure, as well as from the latter to the

Table 11.5 Possible observational calendar related to Phase II of the Templo Mayor, for a 4-year period in the late 14th century.

Alignment	Date	Interval (days)	Date	Interval (days)
	1380		*1382*	167
Cerro Tláloc, sunrise	Mar 14		Mar 14	
		26		27
Templo Mayor, Phase II, sunset	Apr 9		Apr 10	
		20		20
Cerro Tlamacas, sunrise	Apr 29		Apr 30	
		105		105
Cerro Tlamacas, sunrise	Aug 12		Aug 13	
		19		19
Templo Mayor, Phase II, sunset	Aug 31		Sep 1	
		28		27
Cerro Tláloc, sunrise	Sep 28		Sep 28	
	1381	167	*1383*	168
Cerro Tláloc, sunrise	Mar 14		Mar 15	
		26		26
Templo Mayor, Phase II, sunset	Apr 9		Apr 10	
		20		20
Cerro Tlamacas, sunrise	Apr 29		Apr 30	
		106		105
Cerro Tlamacas, sunrise	Aug 13		Aug 13	
		19		19
Templo Mayor, Phase II, sunset	Sep 1		Sep 1	
		27		27
Cerro Tláloc, sunrise	Sep 28		Sep 28	
	1382	167	*1384*	168
Cerro Tláloc, sunrise	Mar 14		Mar 14	

sunrise above Cerro Tlamacas, are 26 days (2 *trecenas*). Calculations show that this scheme could have been achieved if the declinations corresponding to the east- and west-working orientation of the building were about −4°27' and 5°55', respectively. In the fifteenth century, the sun had these declinations when its lower limb "touched" the east and west horizon of the Templo Mayor at azimuths 95°25' and 275°25'. Consequently, the ideal dates of the observational calendar could, indeed, be recorded by one and the same architectural orientation, but *only if tangent positions of the sun on the horizon were relevant* and, at the same time, if the orientation azimuth was about 95°25', i.e. 11' smaller than the one given in Table 11.1 (95°36'). Since the latter derives from the azimuths measured on the preserved segments of the lowest wall faces of the late structural stages, it is obvious that the margin of error that has to be allowed for exceeds the correction of 11' necessary for obtaining the ideal value. It is thus very likely that, starting with Phase III, the azimuth of the intended east-west orientation of the Templo Mayor was about 95°25'.[51]

In the light of comparative evidence from other sites it is unlikely that the alignments to the mountains Tláloc and Tlamacas were fortuitous. While information

Table 11.6 Possible observational calendar related with the late phases of the Templo Mayor of Tenochtitlan.

Alignment	Date	Interval (days)		Date
		156		
Templo Mayor, late phases, sunrise	Mar 9			Oct 4
		26	27	
Templo Mayor, late phases, sunset	Apr 4			Sep 7
		26	25	
Cerro Tlamacas, sunrise	Apr 30			Aug 13
		105		

on the eventual importance of Cerro Tlamacas in prehispanic times seems to be lacking, the symbolic and ritual significance of Cerro Tláloc is amply documented in early colonial written sources and corroborated by archaeological remains on the mountain's summit.[52] Several historical sources mention that the selection of the site for the construction of the Templo Mayor was conditioned by the presence of caves, rocks and water springs.[53] On the other hand, Mazari et al. and Mazari, analysing the settlements of the Templo Mayor in terms of the soil mechanics, argue that no natural island had ever existed on the spot and that the temple was built upon a huge artificial platform some 11m in height, submerged approximately 6m below the lake surface.[54] This interpretation, if correct, may give further support to the idea that the site, apparently hardly appropriate for building a temple, was chosen on astronomical grounds, because it allowed the use of an observational calendar in which some significant dates were marked by the sun's positions over certain prominent horizon features.

DISCUSSION OF SOME PREVIOUS HYPOTHESES

Aveni, Calnek and Hartung[55] also ascribe astronomical motives to the location of the Templo Mayor of Tenochtitlan. They observe that the sun rises over the peaks of Telapón and Tepetzinco (Peñón de los Baños) about 20 days before and after, respectively, the spring equinox.[56] Their inference about the importance of equinoxes is based on indirect data,[57] but it might be significant that the dates of sunrise over Cerro Tepetzinco are close to the sunset dates marked by the orientation of Phase II of the Templo Mayor.[58] Even if Cerro Tepetzinco, with its summit lying below the actual skyline, does not seem appropriate for exact astronomical observations, it may have had a symbolic influence on the location of the Templo Mayor of Tenochtitlan. We may recall the mythical significance of Tepetzinco, the place where Copil's head was deposited, as well as the argument of González Aparicio that this rocky outcrop had an important role in the urban planning of Tenochtitlan.[59]

Aveni et al.[60] find allusions to the observation of the sun relative to the mountains in the myth about the founding of Tenochtitlan, as narrated by Alvarado Tezozómoc in his *Crónica Mexicáyotl*. They comment that the scene with the eagle perched on top of a cactus was seen, according to the story, from far away and that the eagle, identical to Huitzilopochtli, must refer to the sun, probably the rising sun. Since the myth also mentions that the Mexica recognized the site prophesied by Huitzilopochtli when they saw rocks and caves to the east and north, Aveni et al. conclude that the founding of Tenochtitlan must have been related to the observation of sunrise at a position where relevant alignments to the east and to the north intersected. If the story reflects the importance of the mountains to the east as calendrical markers, the reference to the elevation to the north might be associated with the Guadalupe mountain range and its highest peak, Cerro Cuauhtepec, currently also known as Pico Tres Padres;[61] on the other hand, the text might refer to Cerro Chiquihuite, which for an observer at the Templo Mayor marks the direction to the astronomical north.[62]

Ponce de León[63] mentions another alignment that may have been involved in considerations about the placement of the Templo Mayor of Tenochtitlan: the western extension of the solstitial axis of the pyramid and urban layout of Cholula crosses Cerro Tehuicocone, in the mountain ridge north of Iztaccíhuatl, and reaches the sacred precinct of Tenochtitlan. While Cholula is not visible from the Templo Mayor, the alignment to Cerro Tehuicocone may not be fortuitous: though little prominent, the peak marked winter solstice sunrises.[64]

In his attempt to reconstruct the observational calendar of the Templo Mayor, Drucker suggests that at both Teotihuacan and Tenochtitlan observational schemes composed of 20-day periods were in use, with a "core interval" of 180 days, from September 22 to March 20. Drucker calculates that the Templo Mayor azimuth of 97°06' (measured by Aveni) corresponded in the mid-fourteenth century to sunrises on March 1 and October 12, and to sunsets on April 8 and September 2, and concludes that these dates, except April 8, represent initial days of three of the 20-day periods composing his observational calendar scheme.[65] Drucker's hypotheses must be rejected because, in the first place, his calculation procedures are erroneous.[66] Therefore, the dates he determines do not correspond to the azimuth of 97°06' and, even less so, to the azimuths of 97°42' and 95°36', which actually represent the orientations of the Templo Mayor (Table 11.1). Furthermore, to my knowledge there is no unequivocal evidence ascribing a special importance to the dates September 22 and March 20, central dates of Drucker's scheme.

Galindo[67] remarks that, according to Sahagún, the feast of Yoaltecuhtli was celebrated in the sign called Nahui Ollin, which was the day 203 of the count of *tonalamatl*. Considering that Sahagún places the beginning of the prehispanic year on February 2, or February 12 in the present calendar, Galindo observes that the day 203 of the calendar falls exactly on September 2, the day when the sun set in the axis of the Templo Mayor. It must be pointed out, in the first place, that the number "203" represents an inadequate translation of the Nahuatl term used in the *Florentine codex*. Anderson and Dibble corrected the error in their second edition of the work: the text relates simply that the feast was celebrated every 260 days,[68] without mentioning any relationship with the beginning of the calendrical 365-day year. Furthermore, the date September 2 referred to by Galindo is based on the azimuth of 97°25' determined by Ponce de León[69] for the late phases of the Templo Mayor; as I have argued above, this azimuth approximately corresponds to Phase II, while the superimposed buildings—

including Phase VII, i.e. the temple seen by Spanish conquerors—had a different orientation.[70] Likewise, the day March 4 associated by Galindo[71] with sunrises in the axis of the Templo Mayor and with the first day of the month of Tlacaxipehualiztli, according to Sahagún's correlation, corresponds to the azimuth 97°25' and, therefore, could not be recorded by the orientation of the Templo Mayor at the time of the Conquest.

Galindo[72] also refers to the dates March 27 and December 12 mentioned by Durán and associates as the first one with the sunset behind Cerro La Malinche; in Durán's scheme, the two dates correspond to the days 4 Ollin of *tonalpohualli*. However, beside the fact that the coincidence of a certain date of *tonalpohualli* with one and the same date of the tropical year, recurring only at 42-year intervals,[73] can hardly be considered as relevant for explaining the significance of the alignments, it should be recalled that the calendar of Durán is fictitious—or a "model calendar" because its indigenous year starts arbitrarily with 1 Cipactli and 1 Cuahuitlehua (Atlcahualo), corresponding to March 1 of the Julian calendar.[74]

THE ORIENTATION OF THE TEMPLO MAYOR AND THE COMMENT OF MOTOLINIA

Finally, let us examine the hypotheses that have been put forward with respect to the famous statement of Fray Toribio de Motolinía, that the feast of Tlacaxipehualiztli "fell when the sun was in the middle of *Uchilobos,* which was the equinox."[75] The text, evidently referring to the Templo Mayor of Tenochtitlan, owes its importance to the fact that it seems to be the only documentary reference relating a Mesoamerican temple with astronomical observations. No wonder, then, that there have been various attempts at reconciling Motolinía's comment with the archaeologically attested layout of the Templo Mayor.

Aveni and Gibbs,[76] finding that the temple's orientation does not correspond to the equinox sunrises on the natural horizon, suggested that the observations of the equinoctial sun could have been made at the Temple of Quetzalcoatl, situated, according to some sources, west of the Templo Mayor: due to the height of the latter, the sun would have appeared in the notch between the twin sanctuaries only after having moved considerably southwards on its oblique daily path, and reaching the azimuth corresponding to the orientation of the Templo Mayor.[77]

Aveni, Calnek and Hartung[78] further elaborated the hypothesis, taking into account the most recent archaeological data. They proposed that the Mexica laid out their earliest temple structures in the east-west direction, i.e. to the equinox sunrise, but as the altitude of the successive superimposed buildings was growing, they skewed the orientation to the south, so that the equinoctial sun could be observed along the passageway between the upper sanctuaries from some point located in front of the building and along its extended axis. According to Aveni et al., "the general conformity of the alignments of the later phases, however, may be taken to imply either that the differences of linear height between observer and sun disk were always kept constant in the engineering problem, or that the desire to preserve the equinox orientation, once established, simply was abandoned."[79]

Tichy[80] argues that the hypothesis forwarded by Aveni et al. is unlikely and that the orientation of the structure must be explored relative to the sun's positions on the horizon. Even if the possibility that some prehispanic structures contained oblique alignments, referring to astronomically significant positions at considerable altitudes, cannot be discarded, the azimuthal distribution patterns exhibited by Mesoamerican architectural orientations indicate that the latter, indeed, recorded astronomical phenomena on the horizon.[81]

Quoting Motolinía's comment about the coincidence of the feast of Tlacaxipehualiztli with the equinox, Aveni et al.[82] mention that the month of Tlacaxipehualiztli began, in Sahagún's correlation, on March 4 of the Gregorian calendar, so that the feast, usually celebrated at the end of the month, would have occurred about March 23, very close to the equinox. Sahagún's correlation, which makes the first day of Tlacaxipehualiztli occur on March 4, Gregorian, is based on information compiled in his time[83] and thus cannot be relevant for interpreting Motolinía's statement, which refers to an astronomical phenomenon related to the Templo Mayor: even if the structure was not destroyed immediately, its ritual and astronomical function did not survive beyond the Conquest. Furthermore, Motolinía says that, when the Spaniards conquered the land, the natives of New Spain started their year at the beginning of March, the first month being Tlacaxipehualiztli, while Sahagún affirms that the indigenous year began in early February with the month Atlcahualo,[84] so that the first day of the following month Tlacaxipehualiztli, although it coincided with March 4, Gregorian, fell in *February* of the Julian calendar, as Aveni et al. also observe.[85] This means that, if we rely on Sahagún's correlation and, at the same time, accept as correct Motolinía's statement about the feast of Tlacaxipehualiztli (last day of the month) falling on or near the equinox, we are forced to reject as false the information given by the same Motolinía about the beginning of the month Tlacaxipehualiztli in March, which seems arbitrary. As can be seen immediately, Motolinía's data quoted above are internally coherent[86] and, moreover, perfectly congruent with the orientation that has been determined for the fate phases of the Templo Mayor (Table 11.1).

Both Motolinía's comment and the drawing of the Templo Mayor in the map of Tenochtitlan attributed to Cortés have been interpreted as references to the observation of *sunrises* between the twin sanctuaries.[87] In fact, Motolinía's text,[88] having it that the feast of Tlacaxipehualiztli "fell when the sun was in the middle of *Uchilobos*," is not explicit and may well refer to the *sunset* in the axis of the building.[89] Indeed, in 1519 the last day of Tlacaxipehualiztli fell, according to the correlation established by Caso,[90] on March 25 of the Julian calendar, equivalent to April 4 of the Gregorian calendar, which was precisely *the date of sunset along the axis of the late stages of the Templo Mayor*. Consequently, Motolinía's statement can be understood as a reference to the sunset in the structure's axis on the specified date. This interpretation agrees not only with Caso's correlation and his argument,[91] based on various sources and supported by Prem,[92] that the main feast of every month was celebrated on its last day, but also with the comment of Motolinía[93] himself that the last day of the month was "solemn and very festive among them."

Even the fact that Motolinía correlates the feast of Tlacaxipehualiztli with the equinox is only apparently contradictory. As mentioned above, in the Julian calendar, which was in use in Motolinía's times (until the Gregorian reform, adopted in Mexico in 1583[94]), the feast (and the sunset in the axis of the Templo Mayor) fell in 1519 on March 25; the friar's information becomes entirely understandable and accurate, if we recall that this day, the Feast of the Annunciation on which Jesus Christ's conception was celebrated, was in the Middle Ages commonly identified with the vernal equinox.[95] It seems, then, that Motolinía did not refer to the astronomical equinox but rather only made note of the correlation between the day of the Mexica festival, which in the last years before the Conquest coincided with the solar event in the Templo Mayor, and the date in the Christian calendar that corresponded to the traditional day of spring equinox.

Considering that the offerings found at the Templo Mayor and other types of data reflect the enormous importance of the ceremonies carried out in Tlacaxipehualiztli,[96] it is not impossible that the temple's orientation had some relationship with this month, though the correspondence was more symbolic than calendrically precise and stable. It can be pointed out that the date of the spring sunset recorded by the late orientation of the Templo Mayor (April 4, Gregorian) fell on a day within the month Tlacaxipehualiztli during a period of some 80 years; even if it may be fortuitous, it is nonetheless a fact that the date of sunset in the axis of the Templo Mayor coincided with the first day of Tlacaxipehualiztli in the late forties of the fifteenth century, i.e. precisely in the period of Itzcóatl, the ruler responsible of the construction of Phase III,[97] which is the first one that has the new orientation. In this context it seems significant that, according to the written sources, the ceremonies of consecration of the *Huey Teocalli,* intertwined with the Tlacaxipehualiztli rites, acquired importance during the reign of Motecuhzoma Ilhuicamina (1440–69), Itzcóatl's immediate successor on the Mexica throne.[98]

POSSIBLE OBSERVATIONAL TECHNIQUES

It seems fairly certain that the Templo Mayor, like other architectural orientations in central Mexico, recorded astronomical phenomena on the horizon, but we can only speculate about the possible observational methods. The sunrises may have been observed along the passageway between the twin sanctuaries (Fig. 11.2), as the drawing in the early colonial map of Tenochtitlan suggests.[99] In this case the dates corresponding to the orientation could have been determined with ease and better precision if the observations were made from a distant point. Moreover, if the observation point was at the natural ground level, it necessarily had to be located relatively far from the temple: as the height of the latter was growing (by each superimposed building), the distance had to increase.[100] If the observations were carried out from the upper part of a building situated along the axis of the Templo Mayor, the distance could have been

11.4 Sanctuary of Huitzilpochtli on the upper platform of Phase II of the Templo Mayor of Tenochtitlan (view to the east).

smaller. For the moment, however, we have no evidence suggesting the location of the eventual observing point.

On the other hand, it is worth noting some architectural elements of Phase II that may have allowed observations of the sun or light-and-shadow effects in the upper sanctuaries. Recalling Hartung's suggestion, based on illustrations in some codices, that astronomical observations could have been carried out from the interior of the temples,[101] I measured the imaginary line connecting the centre of the sacrificial stone, found in situ in front of the sanctuary of Huitzilopochtli, and the centre of the small rectangular pedestal built upon the bench abutted to the interior east wall (Fig. 11.4). The alignment does not seem to be astronomically significant, because the corresponding azimuth, 99°37', coincides with none of the others that have been measured in the building.[102]

Between the jambs of the entrance to the sanctuary of Tlaloc and two abutted pillars there are vertical slits that could have facilitated the observation of solar rays projected upon the interior east wall of the chapel on certain dates, a few moments before the sunset. To the idea expressed by Hartung,[103] that the temples' jambs possibly incorporated astronomical alignments, it can be added that the sufficiently narrow slits, allowing the passage of solar rays on certain dates only, certainly could have served as very appropriate devices for precise astronomical and calendrical observations. However, the slits of the Tlaloc sanctuary would not have allowed high accuracy, because each of the two, defined by rather irregular wall faces, is approximately

1.20m long and between 2cm and 5cm wide. The observational hypothesis is further weakened by the fact that the two slits, one to the north and the other to the south of the entrance (Fig. 11.2), have very divergent azimuths (94°35' ± 30' and 98° ± 30', respectively).

The adjacent Huitzilopochtli's sanctuary has no comparable masonry pillars abutted to the jambs but rather two low walls, which flank the access to the inner sanctum (Fig. 11.4). Vestiges of stucco, framing rectangular spaces upon the two walls, as well as remains of wood found on both of them during excavations, indicate that wooden pillars were placed on top of the low walls and abutted to the jambs of this sanctuary.[104] It seems significant that the jamb faces are much smoother and more parallel to each other than those of the Tlaloc sanctuary: the azimuths of north and south jambs are 98°48' ± 30' and 97°40' ± 30', respectively. It should be pointed out, however, that the measured lines are, again, short[105] and that the original azimuths cannot be accurately determined, because their exact values depend on the thickness of the stucco that covered the jambs and which is preserved in fragments. Moreover, the surfaces without stucco on the jambs are of roughly the same width as those on the abutted walls, suggesting that the two wooden pillars were not separated from the jambs.[106] In other words, the idea that slits, comparable to those of the adjacent Tlaloc shrine, existed between the jambs and wooden pillars of Huitzilopochtli's sanctuary must remain, in the light of the currently available evidence, merely a speculation.

Since the alignments discussed differ notably, the corresponding sunset dates would have fallen several days before and after those listed in Table 11.1 and recorded by the azimuth of the passageway between the twin sanctuaries. Even if the possibility that certain alignments were astronomically significant and intentional cannot be discarded, it would be too venturous to speculate along these lines, because some of the measured azimuths may differ from the original ones, both because it is impossible to reconstruct the original thickness of the stucco layers and because of possible measurement errors arising from the shortness of the lines. Furthermore, no alignments of this type that could serve as comparative data are preserved in other sites.

It is not impossible, of course, that the sanctuaries originally had some architectural elements, now lost, that permitted the observation of the projection of the sun's rays on relevant dates (e.g. openings, such as those of the Temple of the Seven Dolls at Dzibilchaltún, Yucatán[107]). If light-and-shadow effects were observed in the west-facing sanctuaries of the Templo Mayor at sunset, we can suppose that some adjacent buildings, sharing the same orientation but facing east (like Structures C and F, contemporary with the late phases of the Templo Mayor; see above) may have served for observing this type of phenomena in the morning, when the sun rose in the axis of the Templo Mayor.

To conclude this discussion on possible observational practices, let us return, once more, to the quoted statement of Motolinía. Commenting upon the feast of Tlacaxipehualiztli and the associated solar phenomenon at the Temple of Huitzilopochtli, the author adds that the building was a little twisted, and that "*Mutizuma* wanted to tear it down and set it straight."[108] The remark, brief and apparently insignificant, reveals nothing about the observational methods employed, but it does suggest that the orientation of the temple was not merely symbolic but also functional. Considering that the mean east-west azimuths of the late phases do not exhibit significant differences *(vide supra)*, the referred imprecision could not be large; if in spite of that it was detected and, moreover, became a matter of concern of the supreme Mexica lord, it seems obvious that the observations were made continuously and that the function of certain structural elements was to mark astronomically relevant alignments with precision. Why was the building twisted? Aveni et al.[109] consider that the skew may have been a consequence of the difficulties the architects had to face, as they wished to preserve the equinoctial alignment in different building stages, each one with a greater height (cf. supra). Another possible explanation is related with the phenomenon whose effects have been analysed above: the archaeological evidence indicates that settlements represented a serious problem already for the Mexica builders, forcing them continuously to strengthen, correct and re-level their temples.[110] As I have argued, the settlements were accompanied by slight movements of the alignments in the horizontal plane; could not it be that this was the cause of the imperfection that Fr. Motolinía alludes to?

FINAL REMARKS

In the light of comparative evidence from other central Mexican archaeological sites,[111] it can be concluded that the Templo Mayor of Tenochtitlan was constructed on a spot that was deliberately chosen, with the purpose of employing some prominent peaks on the local horizon as natural markers of the sun's position on certain culturally relevant dates of the tropical year, whereas the architectural orientations were laid out to pinpoint dates that were in a meaningful relation to those marked by the horizon features. The observational schemes were composed of calendrically significant and, therefore, easily manageable intervals. It is more than likely that observational calendars had practical uses, allowing an efficient scheduling of agricultural and associated ritual activities in the annual cycle. While some dates recorded by the alignments probably marked crucial moments of a canonic or ritualized agricultural cycle, others must have had "auxiliary" functions. Since the intervals that separated them were multiples of basic periods of the calendrical system, it was relatively easy to predict the most important dates, knowing the sequence of the intervals involved and the mechanics of the calendar: it should be recalled that the days separated by multiples of 13 days had the same *trecena* numeral, whereas the phenomena separated by multiples of 20 days occurred on the dates that had the same *veintena* sign of the 260-day count. This anticipatory aspect of observational calendars must have been of major significance. Important dates, supposing they were related to subsistence activities, had to be announced ahead of time, because the ceremonies officially inaugurating certain stages of agricultural cycle had to be prepared with due anticipation; on the other hand, direct observations on relevant dates may have been obstructed by cloudy weather.[112] Notwithstanding, it should be recalled that astronomical alignments at

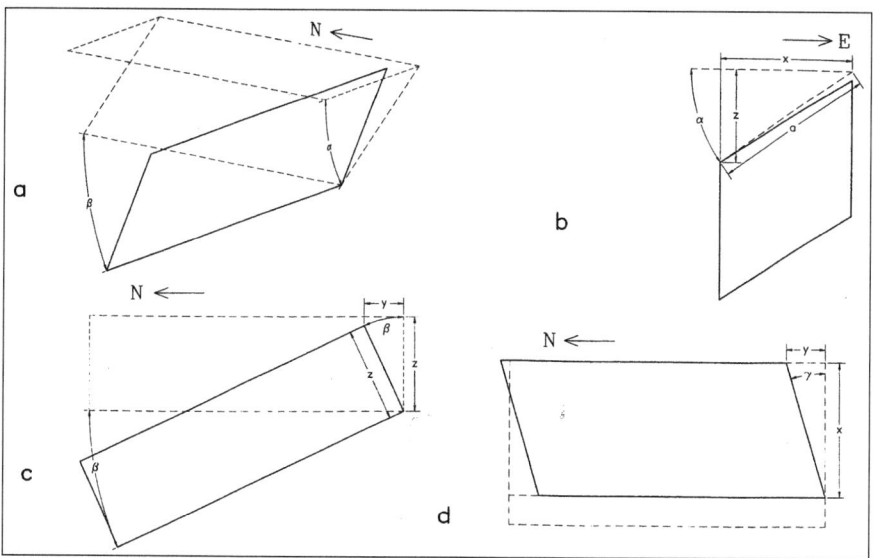

11.5 Schematic representation of one possible sequence of movements that resulted in the existing inclination of Phase II of the Templo Mayor of Tenochtitlan: (a) perspective view to the northeast; (b) side view to the north; (c) side view to the east; (d) plan.

the Templo Mayor of Tenochtitlan, as well as at other Mesoamerican sites, are associated with the most important civic and ceremonial buildings, obviously revealing that astronomical practices had a paramount role in social, religious and even political life of prehispanic societies.

APPENDIX: POSSIBLE HORIZONTAL SKEWS RESULTING FROM SETTLEMENTS OF PHASE II OF THE TEMPLO MAYOR OF TENOCHTITLAN

The southeast corner of the strongly tilted second structural stage of the Templo Mayor of Tenochtitlan is, at present, its most elevated part. By measuring relative heights of various points on the upper platform, I was able to determine the approximate inclination angles along the north-south and east-west axes of the structure, and to calculate, on these grounds, the magnitude of probable horizontal movements caused by settlements. Though the ground surface supporting the architectural masses of the Templo Mayor is estimated to have undergone settlements of up to 11m,[113] it can be assumed, for the purpose of this calculation, that only west and north parts of the structure subsided. The situation is shown schematically in Figures 11.5 and 11.6.

The rectangle outlined in each of these figures with a bold line represents the inclination of the base of Phase II, as observed nowadays, though intentionally exaggerated, in order to facilitate visualization of the movements and to illustrate the derivation of the expression for calculating the range of horizontal skews. We can

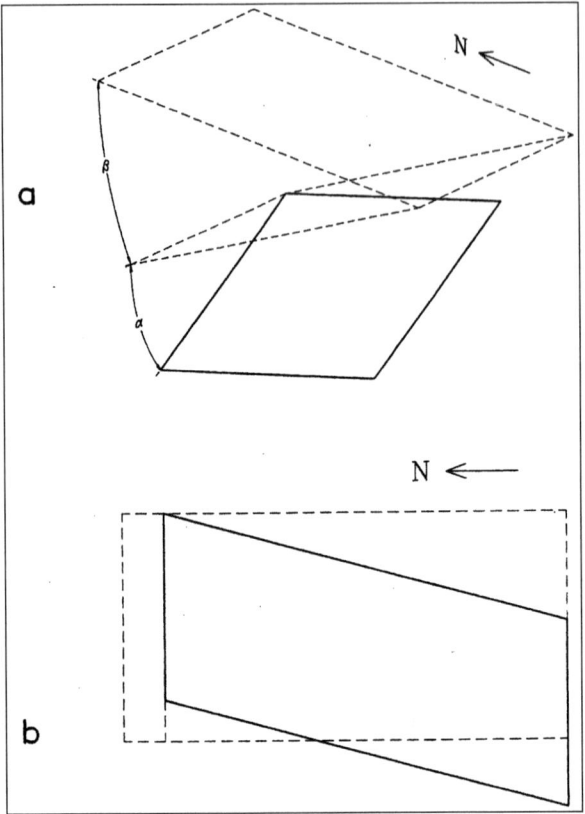

11.6 Schematic representation of another possible sequence of movements that resulted in the existing inclination of Phase II of the Templo Mayor of Tenochtitlan (cf. Fig. 5): (a) perspective view to the northeast; (b) plan.

imagine that the rectangle represents the base of the building, though it may also correspond to the upper platform or to whatever parallel section of the structure, considering that uniform movements that characterize the behaviour of rigid bodies will be assumed. Ideally, the movements that have resulted in the extant inclination of the structure can be separated in two components: those having a horizontal rotation axis in the north-south direction provoked a greater settlement of the west part, whereas the north part of the structure subsided as a result of the movements around an east-west horizontal rotation axis. The exact location of the axes around which the structure really rotated is irrelevant for the calculation, because the developed expression involves only the inclination angles, which are in any case equal. Supposing these horizontal axes were always placed along the east and south sides of the building's base, two ideal sequences of settlements can be reconstructed.

The first case is illustrated in Figure 11.5. If the structure first suffered a settlement of its west part (i.e. rotation around the eastern axis) and later of its north part (rotation around the southern axis), we can observe that the north-south azimuths remain equal, while the east-west azimuths diminish to an extent depending on inclination angles (Figures 11.5a and d): if the building inclined first by a vertical angle (α in the east-west direction and, afterwards, by a vertical angle β in the north-south direction (Figure 11.5a), the azimuths of the east-west lines decreased by a horizontal angle γ (Figure 11.5d). Figure 11.5b shows that

$$x/a = \cos \alpha, \text{ and so } x = a \cos \alpha, \qquad (1)$$
and
$$z/a = \sin \alpha, \text{ and so } z = a \sin \alpha, \qquad (2)$$

while Figure 11.5c implies that

$$y/z = \sin \beta, \text{ and so } y = z \sin \beta. \qquad (3)$$

Eliminating z between (2) and (3), we have

$$y = a \sin \alpha \sin \beta. \qquad (4)$$

Since from Figure 11.5d it follows that

$$\tan \gamma = y/x,$$
we have
$$\tan \gamma = (a \sin \alpha \sin \beta) / (a \cos \alpha) = \tan \alpha \sin \beta. \qquad (5)$$

The angle γ represents the decrease in the azimuths of east-west lines, if the movements that provoked the inclination of the body occurred as shown in Figure 11.5a.

The effects of the inverted sequence of movements are illustrated in Figure 11.6: if we consider that the first movement, provoking the settlement of the north part, occurred around the south axis and was followed by one around the east axis, resulting in subsidence of the west part of the structure (Figure 11.6a), we can observe that the north-south azimuths increased, while the east-west azimuths remained equal (Figure 11.6b). The increase of the north-south azimuths can be calculated by the same Equation (5), interchanging the values of α and β.

It should be emphasised that these are, of course, two *ideal* sequences of movements. There is no doubt that Phase II of the Templo Mayor subsided gradually; however, particular moving sequences must have been comparable to those described, having combined effects that resulted in the skew of *all* horizontal alignments within the ranges that can be calculated. Equation (5) allows the estimation of the *maximum* values of deviation in the horizontal plane of the lines incorporated into the structure. Since the maximum values of α and β, which define the inclination of the upper platform of Phase II, are approximately 8°30' and 2°15', respectively,[114] it follows that the east-west/north-south alignments may have suffered an azimuthal decrease/increase of up to approximately 20'. It should be reiterated that these are the maximum values calculated for one or the other group of alignments, and that gradual settlements with different sequences of the structure's movement may have resulted in slightly smaller azimuthal variations, though both in east-west and north-south alignments. Consequently, the mean correction value of 10' considered for diminishing/increasing the existing north-south/east-west azimuths measured on Phase II of the Templo Mayor seems to be sufficiently realistic.

NOTES

1. Anthony F. Aveni, "Possible Astronomical Orientations in Ancient Mesoamerica," in *Archaeoastronomy in Pre-Columbian America*, ed. A. F. Aveni (Austin, University of Texas Press, 1975), 163–190; idem, "Concepts of Positional Astronomy Employed in Ancient Mesoamerican Architecture," in *Native American Astronomy*, ed. A. F. Aveni (Austin, University of Texas Press, 1977), 3–19; idem, *Skywatchers of Ancient Mexico* (Austin, University of Texas Press, 1980); Anthony F. Aveni and Sharon L. Gibbs, "On the Orientation of Precolumbian Buildings in Central Mexico," *American Antiquity* 41 (1976):510–517; Anthony F. Aveni and Horst Hartung, *Maya City Planning and the Calendar (Transactions of the American Philosophical Society* 76:7; Philadelphia, American Philosophical Society, 1986); Franz Tichy, *Die geordnete Welt indianischer Völker: Ein Bespiel von Raumordnung und Zeitordnung im vorkolumbischen Mexiko* (Das Mexiko-Projekt der Deutschen Forschungsgemeinschaft, 21; Stuttgart, 1991).

2. Tichy, *Die geordnete Welt indianischer Völker*.

3. For example, Arturo Ponce de León, *Fechamiento arqueoastronómico en el Altiplano de México* (Mexico City, 1982); A. F. Aveni, E. E. Calnek, and H. Hartung, "Myth, Environment, and the Orientation of the Templo Mayor of Tenochtitlan," *American Antiquity* 4 (1988):287–309; Tichy, *Die geordnete Welt indianischer Völker*, 159ff; Johanna Broda, "Astronomical Knowledge, Calendrics, and Sacred Geography in Ancient Mesoamerica," in *Astronomies and Cultures*, ed. C.L.N. Ruggles and N. J. Saunders, 253–295 (Niwot, University Press of Colorado, 1993), 258ff; Stanislaw Iwaniszewski, "Archaeology and Archaeoastronomy of Mount Tlaloc, Mexico: A Reconsideration," *Latin American Antiquity* 5 (1994):158–176; Jesús Galindo Trejo, *Arqueoastronomía en la América antigua* (Mexico City, 1994), 129ff; Rubén B. Morante L., "Evidencias del conocimiento astronómico en Xochicalco, Morelos" (unpublished M.A. thesis, Mexico City, 1993); idem, "Evidencias del conocimiento astronómico en Teotihuacan" (unpublished Ph.D. dissertation, Mexico City, 1996).

4. The detailed argument and the supporting evidence are exhaustively presented in: Ivan Šprajc, "Orientaciones en la arquitectura prehispánica del México central: Aspectos de la geografía sagrada en Mesoamérica" (unpublished Ph.D. dissertation, Mexico City, 1997).

5. Tichy, *Die geordnete Welt indianischer Völker*.

6. Šprajc, "Orientaciones," 39ff, 70ff.

7. This altitude above the sea level was reconstructed by Luis González Aparicio, *Plano reconstructivo de la región de Tenochtitlan* (Mexico City, 1973), 17ff, for the level of the lake of Texcoco in prehispanic times; it is thus probable that it approximately corresponds also to the level of the ground on which each of the successive structural phases of the Templo Mayor was built, even if nowadays they are situated at lower and differing attitudes, due to settlements in the marshy subsoil.

8. Constanza Vega Sosa, "La cronología relativa de México-Tenochtitlan," *Mexicon* 12 (1990):9–14; Leonardo López Luján, pers. comm. (May 1997).

9. Leonardo López Luján, *Las ofrendas del Templo Mayor de Tenochtitlan* (Mexico City, 1993), 73–77, fig. 14.

10. For research history and bibliography, see ibid., 19ff.

11. Ignacio Marquina, *El Templo Mayor de México* (Mexico City, 1960); idem, *Arquitectura prehispánica*, 2nd ed. (Mexico City, 1964; 1st ed. 1950), 180–204; Eduardo Matos Moctezuma, *Una visita al Templo Mayor de Tenochtitlan* (Mexico City, 1981); idem (ed.), *El Templo Mayor: Excavaciones y estudios* (Mexico City, 1982); idem, "Los edificios aledaños al Templo Mayor," *Estudios de cultura náhuatl* 17 (1984), 15–21; idem, *The Great Temple of the Aztecs: Treasures of Tenochtitlan* (London, 1988); J. Broda, D. Carrasco, and E. Matos

Moctezuma, *The Great Temple of Tenochtitlan: Center and Periphery in the Aztec World* (Berkeley, University of California Press, 1987); Elizabeth Hill Boone (ed.), *The Aztec Templo Mayor* (Washington, DC, Dumbarton Oaks Research Library and Collection, 1987); López Luján, *Las ofrendas del Templo Mayor*.

12. The refraction factors used in these calculations (taken from Gerald Hawkins, "Astro-archaeology," *Vistas in Astronomy* 10:45–88 [1968], 52, table 1; A. Thom, *Megalithic Lunar Observatories* [Oxford, Clarendon Press, 1971], 28ff, table 3.1; Aveni, *Skywatchers*, 128) were corrected for the altitude above the sea level, employing Formula 7 of Hawkins ("Astro-archaeology," 53).

13. The dates are given in the (proleptic) Gregorian calendar, which provides the closest approximation to the tropical year. Due to precessional variations in the obliquity of the ecliptic, on the one hand, and in the heliocentric longitude of the perihelion of the Earth's orbit, on the other (the latter element determining the length of astronomical seasons), one and the same solar declination does not necessarily correspond in any time span to exactly the same date of the tropical (Gregorian) year. The dates in Table 11.1 have been determined on the basis of the sun's positions given in the tables of Bryant Tuckerman, *Planetary, Lunar, and Solar Positions: AD 2 to AD 1649* (Philadelphia, American Philosophical Society, 1964) (the procedure is described in detail in Šprajc, "Orientaciones," 30f); the dates corresponding to Phase II and to later phases are valid for the fourteenth and fifteenth centuries, respectively.

14. Cf. Marquina, *El Templo Mayor*, 30, 113, fig. 1; idem, *Arquitectura prehispánica*, 183, fig. 6 bis; Aveni and Gibbs, "On the Orientation of Precolumbian Buildings," 514, fig. 3; Matos, *The Great Temple*, 146, fig. 115.

15. Marcos Mazari, Raúl J. Marsal, and Jesús Alberro, "Los asentamientos del Templo Mayor analizados por la mecánica de suelos," *Estudios de cultura náhuatl* 19 (1989):145–182; Marcos Mazari M., *La Isla de los Perros (con un apéndice)* (Mexico City, 1996).

16. I am grateful to Leonardo López Luján for his help in these measurements, as well as in other works I carried out at the Templo Mayor. I also wish to thank Eduardo Matos Moctezuma, director of the Museo del Templo Mayor, who kindly authorized all the measurements I made on various occasions at this archaeological site.

17. Aveni, Calnek, and Hartung, "Myth, Environment, and the Orientation," 296; Ponce de León, *Fechamiento arqueoastronómico*, 54.

18. Ponce de León, *Fechamiento arqueoastronómico*, 54ff, plates 12 and 13.

19. Cf. ibid., plates 12 and 13.

20. The alignment described and measured by Ponce de León would correspond to the one originally incorporated into the passageway only if the existing inclination of the structure were the result of two successive rotations only: the first around a north-south axis and the second around an east-west axis. There is no doubt, however, that the movements were gradual and in different directions; after the first subsidence of the northern part of the building, any subsequent settling of its western part—the structure rotating around a horizontal north-south axis—increased the azimuth of all of the east-west lines projected to the horizontal plane along the planes perpendicular to the base (already inclined) of the structure. Considering that the tilt of the building is particularly pronounced in the east-west direction, it is highly probable that the azimuth of the virtual axis measured by Ponce de León exceeds the original azimuth of the passageway.

21. I am indebted to José Guadalupe Orta B. and Pascual Medina M., topographers of the Dirección de Registro Público de Monumentos y Zonas Arqueológicos, INAH, Mexico, who kindly helped me in these measurements, carried out with a total station and GPS receivers.

22. Mazari, Marsal and Alberro, "Los asentamientos del Templo Mayor," 169f; López Luján, *Las ofrendas del Templo Mayor*, 70; E. Ovando-Shelley and L. Manzanilla, "An Archaeological Interpretation of Geotechnical Soundings under the Metropolitan Cathedral, Mexico City," *Archaeometry* 39 (1997):224f.

23. Ponce de León, *Fechamiento arqueoastronómico*, 31, 56f, plate 13.

24. Marquina, *Arquitectura prehispánica*, 168, plates 49 and 50; Šprajc, "Orientaciones," 230ff, fig. 5.16.

25. Aveni, Calnek, and Hartung, "Myth, Environment, and the Orientation," 294, table 2.

26. Ibid., 295.

27. The mean value 95°36' given in Table 11.1 has been calculated on the basis of the mean east-west azimuths of Phases III, IV, IVb, and VI; because on Phase V only the south face could be measured (Fig. 11.3), its azimuth has not been taken into account in this calculation.

The azimuth 97°06' obtained by Aveni (*Skywatchers*, 314; Aveni and Gibbs, "On the Orientation of Precolumbian Buildings," 512, table 1) was not measured on Phase VII (Aveni, Calnek, and Hartung, "Myth, Environment, and the Orientation," 294) but rather at the southwest extreme of Phase IV, which had been exposed before the extensive excavations directed by Eduardo Matos Moctezuma began in the area (Leonardo López Luján: pers. comm., June 1997). The azimuth exceeds considerably the mean given in Table 11.1, most probably because it was measured along a relatively short section of the south face's west part: due to differential settlements, the preserved faces or *taludes* are nowadays slightly convex; moreover, the azimuths of the south faces are consistently greater than those of the north faces (see Fig. 11.3).

28. Matos, *Una visita*, 37, 41; idem, "Los edificios aledaños al Templo Mayor"; López Luján, *Las ofrendas del Templo Mayor*, 78ff.

29. Cf. Ponce de León, *Fechamiento arqueoastronómico*, 30f, photo 7; Aveni, Calnek, and Hartung, "Myth, Environment, and the Orientation," 303.

30. For example, Marquina, *Arquitectura prehispánica*, 185, plate 54; Alejandro Villalobos Pérez, "Consideraciones sobre un plano reconstructivo del recinto sagrado de México-Tenochtitlan," *Cuadernos de arquitectura mesoamericana* 4 (1985):62, fig. 5.

31. Constanza Vega Sosa, "El Templo del Sol, su relación con el glifo chalchíhuitl; el Templo de Ehécatl-Quetzalcóatl," in *El recinto sagrado de México-Tenochtitlan: Excavaciones 1968–69 y 1975–76*, ed. C. Vega Sosa (Mexico City, 1979), 75–86, Plan 1. These skews have been corroborated by recent explorations (Alvaro Barrera: pers. comm., May 1997).

32. For example, the azimuth of Calle Guatemala is approximately 97°20', whereas the streets Tacuba and Donceles have azimuths around 98°10' (cf. similar values in Aveni, Calnek, and Hartung, "Myth, Environment, and the Orientation," 296, table 3).

33. George Kubler, *Mexican Architecture of the Sixteenth Century* (Westport, CT: Greenwood Press, 1972; 1st ed., 1948), 102, mentions that Mexico City still reveals the form of the Aztec capital and that many central streets follow the pattern of prehispanic canals. In fact, the archaeological information about the course of prehispanic avenues in the immediate vicinity of the Templo Mayor is lacking, so that we do not know for sure whether parts of urban layout of Tenochtitlan are, indeed, preserved in modern streets (and, if so, to what extent and how accurately). To give a concrete example, some archaeological data support the opinion first expressed by Marquina (*El Templo Mayor*, 32) that the modern street of Tacuba, assumed to be a survival of the easternmost part of the causeway to Tlacopan, actually runs a trifle south of the latter: Margarita Carballal: pers. comm., June 1997; Margarita Carballal Staedtler and María Flores Hernández, "Las *calzadas* prehispánicas de la *Isla de México*: Algunas considera-

ciones acerca de sus funciones," *Arqueologia: Revista de la Dirección de Arqueología del INAH*, 2a época, no. 1 (1989), 76.

34. Matos, *The Great Temple*, 73; López Luján, *Las ofrendas del Templo Mayor*, 73ff, fig. 14.

35. Vega Sosa, "El Templo del Sol," Plan 1.

36. It must be pointed out that here we are dealing with the structures excavated in the area of the cathedral, because the same letters were assigned to other buildings in the immediate neighbourhood of the Templo Mayor.

37. Vega Sosa, "La cronología relativa de México-Tenochtitlan," 13f.

38. Šprajc, "Orientaciones."

39. The data on the rest of the prominent features on the horizon of the Templo Mayor are given in Šprajc, "Orientaciones," 305f, tables 5.4.20.2 and 5.4.20.3. As for the methodological criteria employed for the selection of the horizon features considered in my comparative analyses, see ibid., 16f. The dates in the last column of Table 11.2 are valid for the fourteenth century (cf. supra: note 13), because it was probably at that time that the place for the construction of the Templo Mayor was selected.

40. Šprajc, "Orientaciones," 74ff.

41. Cf. ibid., 27, 94ff.

42. The dates listed in Table 11.3 are those on which the sun's declination at sunrise was equal or closest to 14°40', required for the centre of solar disk to be aligned with Cerro Tlamacas (cf. Table 11.2), while on the dates in Table 11.4 the declination of the sun was closer or equal to 14°45', necessary for the alignment of the mountain top with the sun's lower limb.

43. It may be worth noting that the interval from 1343 August 14/13 to 1344 April 29, in Tables 11.3 and 11.4, is 259/260 days, because 1344 was a leap year; in other years the interval between the same dates is one day shorter.

44. It should be pointed out that the four-year patterns of exact dates of solar phenomena (i.e. of certain declinations of the sun) exhibit gradual variations through time (of ±1 day) which derive from the system of intercalations used in the Gregorian calendar. However, the patterns of intervals remain constant during longer periods. In Tables 11.3 and 11.4 the dates for a 4-year span in the mid-fourteenth century are given, but the schemes of intervals would be practically identical if reconstructed for the thirteenth or fifteenth century. The dates registered by tangent positions of the sun would always tend to separate 260-day intervals. Even though the interval of 260 days, too, would sometimes inevitably diminish or increase for 1 day, the "irregularities" of this kind would be much less frequent than in the scheme of the dates recorded by the centre of solar disk.

45. López Luján, *Las ofrendas del Templo Mayor*, 73ff.

46. Spanish terms *trecena* and *veintena* are commonly employed for basic 13-day and 20-day periods, respectively, of Mesoamerican calendar.

47. As shown in Appendix, the horizontal skew of the east-west lines could have reached the maximum value if the structure suffered, first, a major subsidence of its west part and, afterwards, minor settlements of the north part. Indeed, such sequence of movements is likely, considering that the most intensive settlements of the Templo Mayor seem to have been provoked by the weight of the stairways: cf. Mazari, Marsal and Alberro, "Los asentamientos del Templo Mayor," 168f, 178f; López Luján, *Las ofrendas del Templo Mayor*, 70.

48. The dates March 2 and April 10, separated by 39 days, could have been registered with an azimuth of approximately 98°05', which is, however, hardly reconcilable with the argument concerning the probable effects of settlements, because it implies a horizontal skew of more

than ½° (recall that the existing azimuth of the passageway between the twin sanctuaries is 97°32'), whereas the results of calculations make values in excess of 20' unlikely (see Appendix). Furthermore, the alignment of 98°05' actually would not have provided a basis for an ideal observational calendar: since the date of sunset at the azimuth of 98°05' would have been, in the late fourteenth century, invariably April 10, while the sunrises above Cerro Tlamacas occurred on April 29 or 30, the interval between the two phenomena would have varied from 19 to 20 days; on the other hand, the sunrise above Cerro Tláloc, falling predominantly on March 14, would not have subdivided the 39-day span between the dates of sunrise and sunset in the axis of the structure in ideal intervals of 13 and 26 days.

49. The east-west lines of the northern half of the building have smaller azimuths than those of the southern half. However, since the alignments on each of the two halves do not tend to be parallel to each other (instead, the azimuths increase progressively from north to south, the extreme values being 94°08' and 100°00'), it seems that this peculiarity of construction was not a result of the purpose of incorporating two different orientations into the same building. It may also be mentioned that the twin sanctuaries of Structure I of Teopanzolco, which is the only comparable case known at the moment, evidently share the same orientation.

50. This is the way the orientations of the Templo Mayor and the Calendrical Temple of Tlatelolco, on the one hand, and of Structures I and II of Teopanzolco, on the other, seem to have functioned: Šprajc, "Orientaciones," 268ff, 291ff.

51. Significantly, the declinations (dates) corresponding to the orientation of the church of San Luis at Huexotla (Edo. de México), apparently built upon the ruins of the main temple of the prehispanic town, are almost identical: Šprajc, "Orientaciones," 249.

52. Aveni, Calnek, and Hartung, "Myth, Environment, and the Orientation," 298ff; Johanna Broda, "Las fiestas aztecas de los dioses de la lluvia: Una reconstrucción según las fuentes del siglo XVI," *Revista española de antropologia americana* 6 (1971): 277ff; eadem, "Cosmovisión y observación de la naturaleza: El ejemplo del culto de los cerros en Mesoamérica," in *Arqueoastronomía y etnoastronomía en Mesoamérica*, ed. J. Broda, S. Iwaniszewski, and L. Maupomé (Mexico City, 1991), 475f; idem, "The Sacred Landscape of Aztec Calendar Festivals: Myth, Nature, and Society," in *To Change Place: Aztec Ceremonial Landscapes*, ed. D. Carrasco, 74–120 (Niwot, University Press of Colorado, 1991), 95; Stanislaw Iwaniszewski, "La arqueología de alta montaña en México y su estado actual," *Estudios de cultura náhuatl* 18 (1986):256f, 260; idem, "Archaeology and Archaeoastronomy of Mount Tlaloc"; Šprajc, "Orientaciones," 255ff.

53. López Luján, *Las ofrendas del Templo Mayor*, 88ff; Ovando-Shelley and Manzanilla, "An Archaeological Interpretation," 222. While vestiges of water springs have, indeed, been found in the area of the Templo Mayor ceremonial precinct (López Luján, *Las ofrendas del Templo Mayor*, 88f; Ovando-Shelley and Manzanilla, "An Archaeological Interpretation," 222, 232), the allusions to caves and large rocks are not reconcilable with the geological and geomorphological lacustrine environment (Ovando-Shelley and Manzanilla, ibid., 232f).

54. Mazari, Marsal and Alberro, "Los asentamientos del Templo Mayor," 155, 168, 177; Mazari, *La Isla de los Perros*, 11ff.

55. Aveni, Calnek, and Hartung, "Myth, Environment, and the Orientation."

56. Ibid., 302; Anthony F. Aveni, "Mapping the Ritual Landscape: Debt Payment to Tlaloc during the Month of Atlcahualo," in *To Change Place: Aztec Ceremonial Landscapes*, ed. D. Carrasco (Niwot, University Press of Colorado, 1991), 67.

57. Aveni, Calnek, and Hartung, "Myth, Environment, and the Orientation," 289f, 304f, 307.

58. Ibid., 302; cf. Galindo, *Arqueoastronomía*, 166.

59. Aveni, Calnek, and Hartung, "Myth, Environment, and the Orientation," 292, 302; Broda, "The Sacred Landscape," 186ff; González Aparicio, *Plano reconstructivo*, 47f, 53.

60. Aveni, Calnek, and Hartung, "Myth, Environment, and the Orientation," 292f.

61. Ibid., 304. Aveni, "Mapping the Ritual Landscape," 63, mentions various archaeological sites that seem to exemplify the symbolic importance of the mountain located to the north of a ceremonial centre. It may be added that the north-south axes of the structures examined at central Mexican archaeological sites align in more cases with a mountain to the north than to the south: Šprajc, "Orientaciones," 38.

62. Ponce de León, *Fechamiento arqueoastronómico*, 58.

63. Ibid.

64. Šprajc, "Orientaciones," 313f.

65. R. David Drucker, "A Solar Orientation Framework for Teotihuacan," in *Los procesos de cambio (en Mesoamérica y áreas circunvecinas): XV Mesa Redonda*, ii (Guanajuato, 1977), 281ff, fig. 3.

66. To obtain the date corresponding to a certain declination of the sun in the past, Drucker (ibid., 278) multiplies the present declination value with a constant derived from de Sitter's formula. However, the formula developed by de Sitter for calculating the obliquity of the ecliptic in any epoch (Thom, *Megalithic Lunar Observatories*, 15), while it makes possible to determine the maximum/minimum declinations of the sun (attained at solstices), is not sufficient for establishing the *exact dates* on which the sun, in a given period, had certain declinations, since the corresponding moments of the year depend not only on the obliquity of the ecliptic but also on the length of the seasons, which varies as a function of the secular movement of the perihelion-aphelion line of the Earth's orbit: cf. Šprajc, "Orientaciones," 30f. Moreover, the declinations and dates determined by Drucker ("Solar Orientation Framework," 282) as corresponding to the azimuth 97°06′, allowing for the horizon altitude of 2°10′ (both for east and west), actually do not derive from the formula presented by himself (ibid., 278).

67. Galindo Trejo, *Arqueoastronomía*, 166f.

68. Arthur J.O. Anderson and Charles E. Dibble, *Florentine Codex: General History of the Things of New Spain: Fray Bernardino de Sahagún, Book 2—The Ceremonies*, 2nd ed., rev. (Monographs of the School of American Research, no. 14, Part III; Santa Fe, 1981), 216. I am indebted to Leonardo López Luján and Alfredo López Austin for calling my attention to this fact.

69. Ponce de León, *Fechamiento arqueoastronómico*, 31.

70. The orientation of Phase VII has not been determined directly by measurements, but the remains of this structure clearly show that it was erected on top of the former Phase VI, preserving its orientation: López Luján, *Las ofrendas del Templo Mayor*, 72, and pers. comm., June 1997.

71. Galindo Trejo, *Arqueoastronomía*, 167.

72. Ibid., 167.

73. Elzbieta Siarkiewicz, *El tiempo en el tonalámatl* (Warsaw, 1995), 94.

74. Cf. Hanns J. Prem, "Das Chronologieproblem in der autochthonen Tradition Zentralmexikos," *Zeitschrift für Ethnologie* 108:1 (1983): 143ff, table 3.

75. Fray Toribio de Benavente o Motolinía, *Memoriales o libro de las cosas de la Nueva España y de los naturales de ella*, ed. E. O'Gorman (Mexico City, 1971), 51.

76. Aveni and Gibbs, "On the Orientation of Precolumbian Buildings," 513ff.

77. Ibid., 515, fig. 4; Aveni, *Skywatchers*, 245ff, fig. 81.

78. Aveni, Calnek, and Hartung, "Myth, Environment, and the Orientation," 294ff.

79. Ibid., 297.

80. Tichy, *Die geordnete Welt indianischer Völker*, 94.

81. Šprajc, "Orientaciones," 9; idem, "La astronomía en Mesoamérica."

82. Aveni, Calnek, and Hartung, "Myth, Environment, and the Orientation," 291.

83. Šprajc, "Orientaciones," 106f. In fact, this correlation appears only in the *Florentine codex* and *Historia general;* in his other texts, Sahagún gives slightly different correlations: Renate Bartl, Barbara Göbel and Hanns J. Prem, "Los calendarios aztecas de Sahagún," *Estudios de cultura náhuatl* 19 (1989):13–82.

84. Motolinía, *Memoriales*, 44f; Fray Bernardino de Sahagún, *Historia general de las cosas de Nueva España*, 6th ed. (Mexico City, 1985), 77ff.

85. Aveni, Calnek, and Hartung, "Myth, Environment, and the Orientation," 291.

86. It should be clarified that the data relevant in the present context and quoted above are coherent, but apparently do not belong to the same author: Motolinía (*Memoriales*, 44f) himself mentions that the year began in March with the month of Tlacaxipehualiztli and that the feasts took place on the last day of each month, whereas the reference to the solar phenomenon in the Templo Mayor is part of an interpolation that does not pertain to the text of *Memoriales* (ibid., 50).

87. Cf. A. Maudslay, "A Note of the Position and Extent of the Great Temple," in *Trabajos Arqueológicos en el centro de la Ciudad de México*, 2nd ed., ed. E. Matos Moctezuma, 269–272 (Mexico City, 1990; orig. publ. in 1912), 272; Aveni and Gibbs, "On the Orientation of Precolumbian Buildings," 513; Aveni, *Skywatchers*, 248.

88. Motolinía, *Memoriales*, 51.

89. The fact that Marquina, *El Templo Mayor*, 113, paraphrasing Motolinía, mentions the sun "in front of Huichilobos" shows clearly that the text is ambiguous.

90. Alfonso Caso, *Los calendarios prehispánicos* (Mexico City, 1967), 58, table 4.

91. Ibid., 39, 51.

92. Hanns J. Prem, "Los calendarios prehispánicos y sus correlaciones: Problemas históricos y técnicos," in *Arqueoastronomía y etnoastronomía en Mesoamérica*, ed. J. Broda, S. Iwaniszewski, and L. Maupomé, 389–411 (Mexico City, 1991), 395.

93. Motolinía, *Memoriales*, 45.

94. Caso, *Los calendarios prehispánicos*, 98f.

95. Robert R. Newton, *Medieval Chronicles and the Rotation of the Earth* (Baltimore: Johns Hopkins University Press, 1972), 27; Stephen C. McCluskey, "The Mid-quarter Days and the Historical Survival of British Folk Astronomy," *Archaeoastronomy* 13 (1989):S2; idem, "Astronomies and Rituals at the Dawn of the Middle Ages," in *Astronomies and Cultures*, ed. Ruggles and Saunders, 110f, 114. Even if the canonical date of ecclesiastical equinox established in AD 325 by the Council of Nicaea was March 21, the Roman tradition correlating the equinox with March 25 (VIII Kal. Aprilis) survived, as well: Newton, *Medieval Chronicles*, 22–27. Newton mentions two medieval calendars—one of them recorded by Bede—which attest to the coexistence of both traditions, because in each of them the equinox is annotated for both 21 and 25 of March (ibid., 26f). Incidentally, Bede is one of the authors Motolinía (*Memoriales*, 46) quotes in his discussion on various calendars.

96. Johanna Broda, "Tlacaxipeualiztli: A Reconstruction of an Aztec Calendar Festival from 16th Century Sources," *Revista española de antropología americana* 5 (1970):197–274; López Luján, *Las ofrendas del Templo Mayor*, particularly 270–289.

97. Matos, *The Great Temple*, 73; López Luján, *Las ofrendas del Templo Mayor*, 73ff, fig. 14.

98. López Luján, *Las ofrendas del Templo Mayor*, 272.

99. Cf. Marquina, *El Templo Mayor*, 30, fig. 1; idem, *Arquitectura prehispánica*, 183, fig. 6 bis; Aveni and Gibbs, "On the Orientation of Precolumbian Buildings," 514, fig. 3; Matos, *The Great Temple*, 146, fig. 115; Aveni, *Skywatchers*, 247, fig. 81b.

100. For example, if the upper platform of the last structural phase was about 30m high (cf. Marquina, *El Templo Mayor*, 44), the observer had to stand at a distance of more than 800m if he wanted to see the sunrise on the natural horizon and, at the same time, between the two upper sanctuaries.

101. Horst Hartung, "A Scheme of Probable Astronomical Projections in Mesoamerican Architecture," in *Archaeoastronomy in Pre-Columbian America,* ed. A. F. Aveni, 191–204 (Austin: University of Texas Press, 1975), 193, figs. 3 and 4.

102. It could be speculated that, a few moments before sunset on certain dates, the sacrificial stone's shadow was observed, projected onto the pedestal, which probably supported a statue of Huitzilopochtli (López Luján, *Las ofrendas del Templo Mayor*, 71). However, as the pedestal is wider than the sacrificial stone, the phenomenon would have occurred on several consecutive days. Particular dates could have been determined if the pedestal or the bench had had some markings, no traces of which, however, have been detected.

103. Hartung, "Scheme of Probable Astronomical Projections," 196.

104. Francisco Hinojosa: pers. comm., May 1997.

105. The inner face of each jamb is trapezoidal, being its maximum width, which diminishes upwards, about 185cm (along the intersection with the upper horizontal face of the abutted wall).

106. I am indebted for this caution to Francisco Hinojosa, May 1997.

107. Cf. Ivan Šprajc, "El Satunsat de Oxkintok y la Estructura 1-sub de Dzibilchaltún: Unos apuntes arqueoastronómicos," in *Memorias del Segundo Congreso Internacional de Mayistas* (Mexico City, 1995), 585–600.

108. Motolinía, *Memoriales*, 51.

109. Aveni, Calnek, and Hartung, "Myth, Environment, and the Orientation," 297.

110. López Luján, *Las ofrendas del Templo Mayor*, 70; Francisco Hinojosa: pers. comm., May 1997.

111. Šprajc, "Orientaciones."

112. Ibid., 114ff; cf. Michael Zeilik, "The Ethnoastronomy of the Historic Pueblos, 1: Calendrical Sun Watching," *Archaeoastronomy* 8 (1985):S1–24.

113. Mazari, Marsal and Alberro, "Los asentamientos del Templo Mayor," 155.

114. Relative heights of different points on the platform do not render in all parts exactly the same inclination angles α and β, which indicates that the structure, undergoing differential settlements, has not moved strictly as a rigid body.

CHAPTER TWELVE

Astronomical Observations from the Temple of the Sun

Alonso Mendez, Edwin L. Barnhart, Christopher Powell, and Carol Karasik

INTRODUCTION

Maya architecture is a repository for ancient astronomical knowledge. Investigating the celestial alignments of Maya architecture, archaeoastronomers have identified dozens of structures that were oriented to the Sun, stars, and planets rising and setting on the horizon. Astronomical observations formed the basis of the Classic Maya calendar, which eventually integrated the cycles of the Sun with the movements of the Moon and five visible planets. The calendar supported a religious system that linked the heavens with seasonal cycles and the agricultural rituals associated with them (Milbrath 1999:1). Decipherments of carved inscriptions reveal that royal ceremonies and accessions were also timed to coincide with significant stations of the Sun

---

Reproduced by permission from *Archaeoastronomy: The Journal of Astronomy in Culture* 19 (2005):44–73.

or with rare planetary conjunctions (Aveni 2001:163–214). The role of astronomy in agriculture, politics, and religion exemplifies the Maya penchant for interweaving nature, human society, and the divine. The night sky, with its infinite population of souls, gods, and monsters, presented a mirror image of the hidden underworld below. Alignments to celestial bodies expressed the bonds between earth and the many levels of the cosmos.

The ancient Maya exhibited their scientific and spiritual understanding of the cosmic realms through astronomical hierophanies (Aveni 2001:220–221). As defined by the religious historian Mircea Eliade (1958:11), a hierophany is the manifestation of the sacred in an object or event in the material world. Archaeoastronomers have adopted the term to describe phenomena of sunlight and shadow that play across architectural features during important stations of the Sun. If accompanied by public ceremonies, these dazzling displays must have generated awe and religious fervor among the populace and confirmed the power of the divine ruler. Such spectacles rely on the alignment of monumental buildings with the Sun. In the Maya region the most renowned example takes place at Chichén Itzá during equinox when, in a dramatic play of light and shadow, the triangular pattern of a serpent appears on the balustrade of the pyramid El Castillo. Other hierophanies depend on the position of the Sun as seen from a meaningful vantage point. At Dzibilchaltún, the rising Sun at equinox, when viewed from the main causeway, fully illuminates the central door of the Temple of the Seven Dolls (Chan Chi and Ayala 2003). Both Yucatecan sites demonstrate the precise interplay between monumental architecture and the Sun during key stations of the year (Figs. 12.1 and 12.2).

Precedents for solar-oriented structures exist at Early Classic Maya sites. Known as "Group E" complexes, after the architectural complex identified as Group E at Waxaktun, numerous examples have been found throughout the Maya area (Aveni 2001:288–292). Characteristically, the Group E complex contains a single temple, used for sighting, that stands directly west of three buildings that mark winter solstice, summer solstice, and equinox, the midpoint between those two solar extremes. Because Group E complexes were primarily used to record the known positions of the Sun rather than to obtain new astronomical information, they cannot be considered genuine observatories. Instead, the complexes served as stages for "ritual observation" that may have been the focus of public ceremonies (Krupp 1983:249).

Calendrical observations were closely tied to religious observances among the ancient cultures of the American Southwest. Pueblo Sun Priests still conduct solar observations prior to the solstices, which are crucial phases in the Pueblo agricultural and religious cycles. In fact, the main responsibility of the Pueblo Sun watchers is to anticipate the exact days of the solar stations in order to set the dates for major rituals (Zeilik 1985:S3).

At Palenque, Aveni (1980:284, 2001:295) sees a subtle distinction between astronomically oriented structures designed for ritual purposes and structures that served a symbolic function—to manifest astronomical hierophanies. To date, archaeoastronomers (Aveni and Hartung 1978; Carlson 1976) have identified the celestial orienta-

12.1 Equinox at El Castillo, Chichén Itzá, Yucatán (photograph by Christopher Powell).

12.2 Equinox sunrise at the Temple of the Seven Dolls, Dzibilchaltún, Yucatán (photograph by Felipe Chan Chi).

tion of a number of buildings; namely, the alignment of the Temple of the Count to Sirius; the Temple of the Foliated Cross to Capella; and House A and the eastern side of the Palace to the Moon at maximum elongation. Considering the solar alignments of major buildings at the site, Aveni and Hartung (1978) noted that Temple XIV is aligned with the winter solstice and that the western side of the Palace is aligned with the setting Sun at zenith passage. John Carlson (1976) was the first to hypothesize that the Temple of the Sun is oriented to face the rising Sun at winter solstice, a theory supported by other scholars (Aveni 2001; Aveni and Hartung 1978; Milbrath 1999).

Following the discovery of these alignments, only a few researchers have witnessed hierophanies at the site.

Anderson et al. (1981) documented a series of solar events that occurred in the Tower of the Palace. Standing inside the Tower, the three investigators noticed that at sunset on April 30, the Sun's rays passed directly through the T-shaped window on the western facade and struck an interior wall of the viewing chamber. With the approach of summer, the image of the T-shaped window moved progressively to the east. On June 22, the investigators saw the complete image of the T glowing on the wall. After the summer solstice, as the northerly position of the Sun decreased, the projected image shifted until only a fraction appeared on the wall. By August 12, a few days after zenith passage at Palenque, the Sun's rays were aligned perpendicularly with the western wall of the Tower; consequently, the light entering the window did not project an image on the angled wall. For the three investigators, these observations demonstrated the existence of specially designed interior spaces, oriented with extreme precision, which made it possible for Maya astronomers and calendar keepers to monitor the Sun at solstice and zenith passages.

At the sites of Monte Albán and Xochicalco, zenith passages were observed through zenith sighting tubes. At Palenque, where no zenith tubes have been found, the Tower evidently served this purpose and thus functioned as a working observatory. Anderson et al. (1981) suggest that observations made from the Tower allowed astronomers to divide the solar year into two periods: the 105-day agricultural cycle and the 260-day ritual cycle. In reality, the 105-day period, from zenith passage on April 30 to zenith passage on August 12, occurs only at 15° latitude. At higher latitudes, this inter-zenith period was more symbolic than practical, an ideal cycle designed by the Maya hierarchy to fix the length of the relatively arbitrary growing season, from the first sprouting of corn to its maturity, as well as to commemorate the anniversary in the solar year of the date of Creation, 13.0.0.0.0 4 Ahaw 8 Kumk'u. As we have recently discovered, alignments to zenith passage may be seen in the relationship between the Temple of the Sun and the Temple of the Cross, where its importance is bound up with the recorded creation myth and dynastic history.

Three hierophanies, witnessed during the solstices, have been associated with rites of divine kingship. While standing in the Tower of the Palace, Linda Schele (in Carlson 1976:107) observed that the "dying" Sun at winter solstice, setting over the ridge directly behind the Temple of the Inscriptions, appeared to enter the earth through the royal tomb of Janahb Pakal. Schele interpreted this solar event as an annual reenactment of Janahb Pakal's descent into the underworld, as depicted in the iconography on the sarcophagus lid (Fig. 12.3).

The second hierophany described by Schele (in Carlson 1976:107) also occurred during winter solstice. Seen from the Tower, the Sun setting behind the Temple of the Inscriptions sent a shaft of light that slowly mounted the terraces of the Temple of the Cross, and as the base of the pyramid sank into shadow, a final beam of light entered the temple and illuminated God L, portrayed on the eastern doorjamb of the sanctuary. Schele speculated that this phenomenon symbolized the transfer of royal power

12.3 Winter solstice Sun setting behind the Temple of the Inscriptions, as seen from the Palace, Palenque, Chiapas (photograph by Alonso Mendez).

from Janahb Pakal to his son and heir Kan B'ahlam II, an event that occurred under the aegis of God L.

Another dynamic relationship between the Temple of the Inscriptions and the Temple of the Cross has been described by Anderson and Morales (1981). At sunset during summer solstice, they noted that the light entering the western window of the anterior corridor of the Temple of the Inscriptions aligned with the eastern window directly across the corridor and then highlighted the upper platform of the Temple of the Cross, where a major stela once stood. The researchers suggest that Stela I, believed to be a portrait of Janahb Pakal or Kan B'ahlam, marked a solar observation point. Anderson and Morales's report reveals not only the longitudinal orientation of the Temple of the Inscriptions to the summer solstice but also its remarkable alignment to the Temple of the Cross. Their observation also reinforces Schele's theory that the visual effects seen at summer solstice represented the transfer of royal power from Janahb Pakal to Kan B'ahlam.

Schele's theory begins to address the metaphysical connection between astronomical phenomena and historical events recorded in the hieroglyphic inscriptions. In fact, her on-site observations helped confirm her readings of two great historical moments: the heir-designation ceremony of the young prince, Kan B'ahlam, held during summer solstice of AD 641, and the death and burial of his father, Janahb Pakal, some 40 years later. As it turns out, Schele's winter solstice observation requires some correction; the Tower where she was standing was not erected until the eighth century (Hartung 1980:76) and was therefore not the correct stage for watching the setting Sun. House E of the Palace, where Pakal was crowned, is the proper vantage point for viewing the winter solstice Sun sink behind the Temple of the Inscriptions where Pakal is buried (see Fig. 12.3). The great pyramid was completed by Kan B'ahlam and dedicated on 9.12.16.12.19 10 Kawak 7 Pax, December 23, 688 (David Stuart, personal communication 2005), two days after winter solstice. We still see the drama of birth, death, and royal succession written in light, for the last rays of the winter solstice Sun illuminate the Temple of the Cross, built by Kan B'ahlam to commemorate his accession to the throne. Years after his father designated him as heir on the summer solstice, Kan B'ahlam continued to honor his father's interest in solstitial alignments.

Father and son also shared an intense preoccupation with the planets. Aldana's (2001:131–132) reading of the texts in the Temple of the Inscriptions suggests that Janahb Pakal's fascination was largely oracular. Many scholars (Aldana 2004; Aveni and Hotaling 1996; Lounsbury 1989:253–254) have noted that Pakal's katun-ending ceremonies were synchronized with multiple astronomical events, particularly the appearance of Venus at maximum elongation. As will be seen in the final section of this article, the Moon and major planets also played a role in the timing of rituals conducted by Kan B'ahlam. Moreover, sometime during his reign, astronomers perfected the 819-day calendar, which took into account the cycles of Saturn and Jupiter (Aldana 2004; Lounsbury 1978; Powell 1996). Given the numerous allusions to astronomical phenomena in the art and literature, considerable speculation has gone into equating the rulers and patron gods of Palenque with specific planets (Kelley 1980; Lounsbury 1985; Schlak 1996).

In sum, on-site observations make up a small part of the multidisciplinary inquiries into the astronomical knowledge buried in the inscriptions, art, and architecture of Palenque. Aside from studies made by Milbrath (1988), little recognition has been given to the role of the anti-zenith, or nadir, passages in the Maya calendar. Venus, Jupiter, and Saturn have received enormous attention, but the Moon has been mysteriously slighted. Progress has been made in identifying the orientations of major buildings and their possible ritual and calendrical significance, but there has been little headway in discovering the overall cosmological scheme of the ceremonial center. Numerous observations from a significant vantage point are needed for a fuller appreciation of the alignments at Palenque. Our ongoing investigations show that the Cross Group, the Temple of the Inscriptions, *and the Palace* exhibit astronomical alignments that are fundamental to understanding their design, function, and interrelationships.

OBSERVATIONS IN THE TEMPLE OF THE SUN

The Temple of the Sun is the westernmost building in the Cross Group, a complex of three temples erected on three hills rising above a small plaza (Baudez 1996:121–124) (Fig. 12.4). Completed by Kan B'ahlam in AD 692, the Temples of the Cross, Foliated

Cross, and Sun represent shrines to the three patron deities of Palenque, GI, GII, and GIII, respectively. Each temple contains tablets with texts that tie the history of Kan B'ahlam's lineage to those gods of Creation. Each temple also commemorates major events in Kan B'ahlam's reign. With the Temple of the Cross on the tallest hill to the north, the Foliated Cross nestled at the base of the eastern mountain, and the Sun on a low mound in the west, the temples represent a cosmogram of the Upper, Middle, and Lower Worlds. At the same time, the temples represent the three "hearthstones of creation" located at the center of the universe in the constellation of Orion (Freidel et al. 1993:65–69). Sharing architectural and artistic styles as well as textual cross-references, the temples of the Cross Group are also interrelated in their alignments to the Sun and Moon and to one another.

The Temple of the Sun is the most intact structure in the Cross Group. After reconstruction in the 1950s and the replacement of the lintels, the temple is now the most reliable focus for on-site observations. The hierophanies that occur inside the temple are characterized by thin rays of light that cross the temple floor at well-defined angles, which, we propose, were determined by architectural features and by the position of the Sun.

The following descriptions of these solar events are based on naked-eye observations, corroborated by topographical measurements from the latest map of Palenque (Barnhart 2000) (see Fig. 12.4). Data for building azimuths are taken from Carlson (1976) and Aveni and Hartung (1978); solar azimuths and lunar information for Palenque's latitude, from Starry Night Deluxe (Andersen et al. 1997).

Winter Solstice

It has long been held that the Temple of the Sun was oriented to face the rising Sun on the morning of the winter solstice (Aveni and Hartung 1978:175; Carlson 1976; Milbrath 1999:69). John Carlson (1976:110) recorded the winter solstice orientation of the Temple of the Sun as 119°46′. Estimating a horizon line of 9° altitude, Carlson predicted that on winter solstice the first rays of the Sun would enter the central doorway of the temple at a "directly perpendicular" angle.

Our eyewitness observations did not confirm Carlson's hypothesis. At 9:23 AM on December 21, the rising Sun broke the horizon at approximately 130′ azimuth directly above El Mirador, the mountain under which the Cross Group was built (Fig. 12.5). At a vertical angle of 30°, the first rays shone through the central doorway of the temple at an angle 10° south of the transverse axis of 119°46′ (Fig. 12.6). At that moment the light lined up along the back edges of the medial walls (Fig. 12.7). That is the farthest the sunlight now reaches into the temple's interior during winter solstice.

On a flat horizon at Palenque's latitude of 17°28′N, the winter solstice sunrise occurs at an azimuth of 114°23′E and sunset at 245°36′W. As noted, the observed sunrise emerging over the peak of El Mirador is closer to 130°. The mountain completely blocks the Sun's visibility at a 9° altitude and prevents the early-morning rays from entering the temple at a perpendicular angle. However, the fact that the Sun breaks

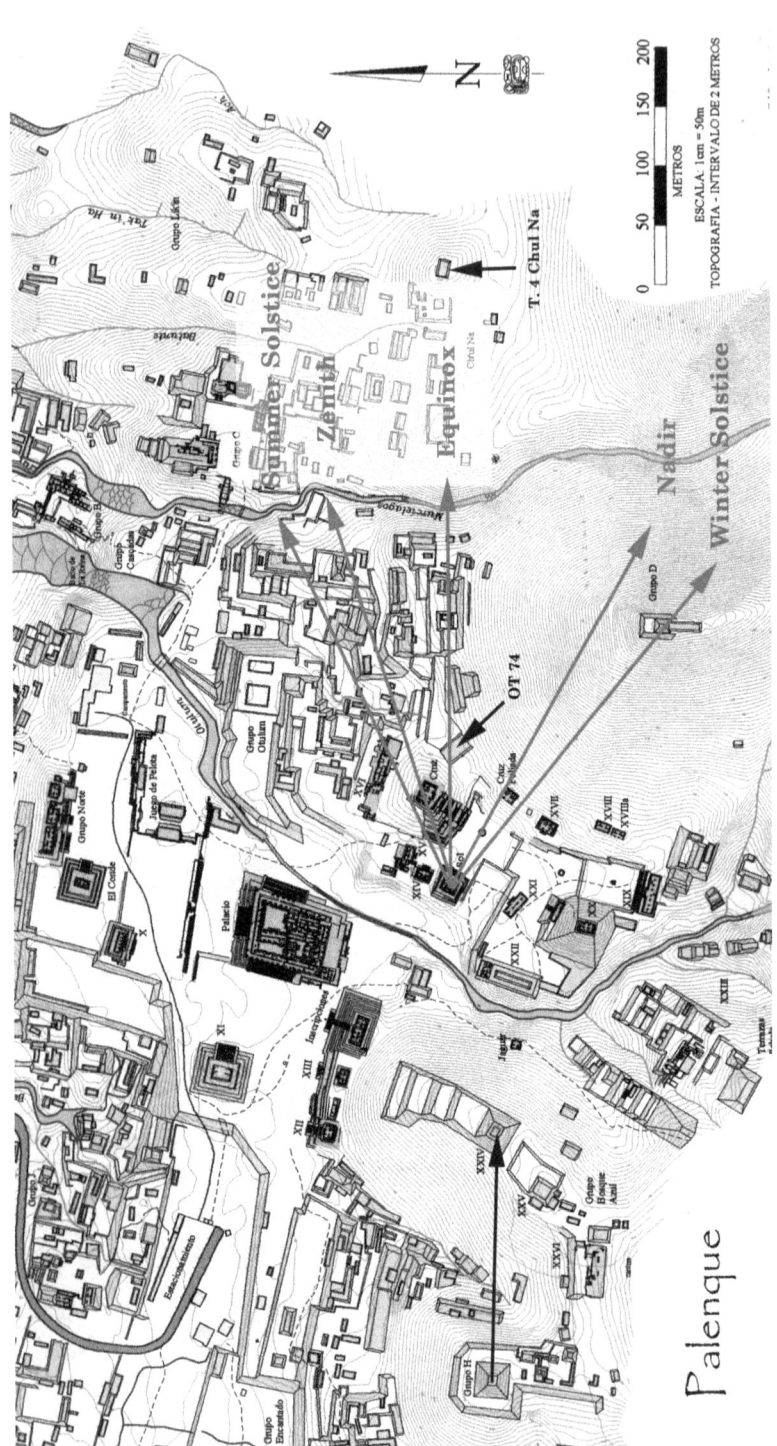

12.4 Map of Palenque showing solar alignments from the Temple of the Sun (Edwin L. Barnhart and Alonso Mendez).

Alonso Mendez, Edwin L. Barnhart, Christopher Powell, and Carol Karasik

12.5 Winter solstice sunrise over El Mirador, as seen from the Temple of the Sun (photograph by Alonso Mendez).

12.6 Plan view of the Temple of the Sun illustrating the angle of light during winter solstice sunrise (drawing by Alonso Mendez).

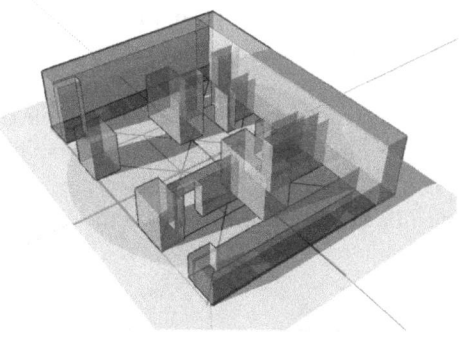

from the peak of El Mirador is consistent with a recognized pattern of solstice orientations seen at numerous Mesoamerican sites with prominent topographical irregularities such as mountains or clefts between mountains (Malmström 1997).

The problem remains: How do we account for the 119°46'E–299°46'W alignment of the Temple of the Sun? One possible explanation is that the transverse axis of the temple closely matches the maximum elongations of the Moon. Like the Sun, the Moon has its maximum northern and southern rising and setting points. Called lunar standstills, they occur on the solstices, at full or new Moon, and repeat every 18⅔ years (Aveni 2001:72–73). At Palenque's latitude, the Moon at its maximum southern extreme rises at 120° and at its maximum northern extreme sets at 300° on a

12.7 Interior of the Temple of the Sun showing sunlight on the medial wall during winter solstice sunrise (photograph by Alonso Mendez).

flat horizon. The Moon rising at its maximum southern extreme on the eastern horizon would not have been visible from the central doorway of the Temple of the Sun; however, from the Temple of the Foliated Cross, the Moon at its maximum northern extreme would have been seen setting over the roof comb of the Temple of the Sun (Fig. 12.8).

This correlation between the Temple of the Sun and the Moon lends weight to lunar interpretations of the iconography in the Tablet of the Sun (Bassie-Sweet 1991:192–198). But as we shall see, sunrise at nadir passage provides a decisive answer regarding the solar orientation of the temple.

Equinox

Equinox marks the midpoint between the solstices and corresponds to the time of year when the Sun rises and sets on the celestial equator. As a result, the length of the day is equal to the length of the night. During vernal equinox, on March 21, the Sun passes from the southern to the northern hemisphere; at autumnal equinox, on September 20, the Sun retreats to the southern hemisphere. Sunrise on both days occurs at an azimuth of 90° due east and sunset at 270° due west.

Viewed from the Temple of the Sun on March 21, the Sun rises at a low point on the horizon, between El Mirador and the Temple of the Cross, at an azimuth of

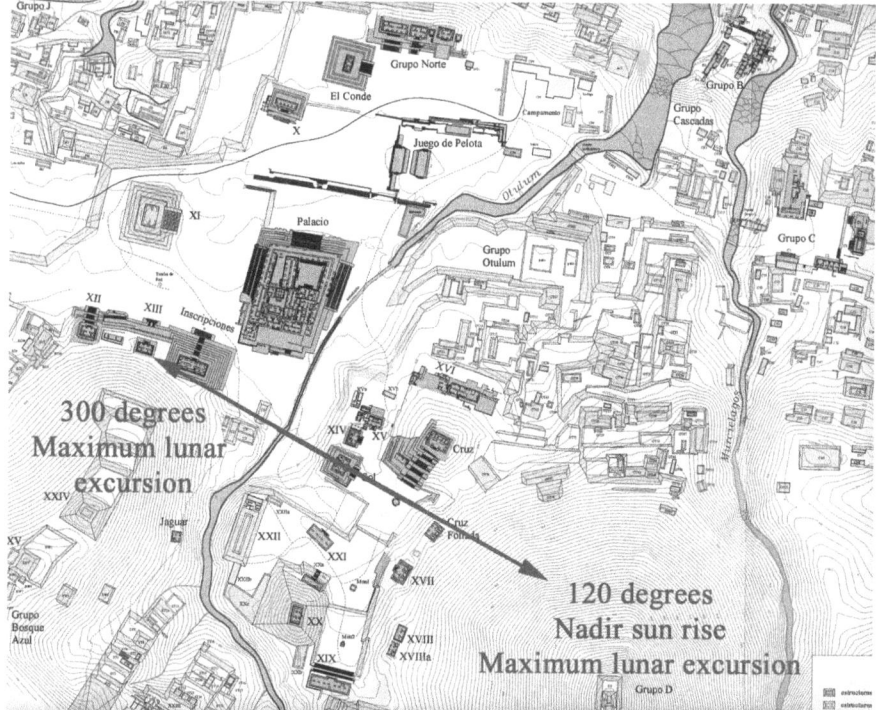

12.8 Map of Palenque showing the orientation of the transverse axis of the Temple of the Sun to nadir sunrise and to the maximum southern and northern lunar excursions (Edwin L. Barnhart and Alonso Mendez).

91° (Fig. 12.9). At 6:50 AM, sunlight enters the middle doorway of the temple at an oblique angle of 29° north of the transverse axis. The medial wall and sanctuary wall narrow the first ray until it becomes a thin knife of light reaching into the southwestern corner of the central posterior chamber (Figs. 12.10 and 12.11). The light then retreats from the corner, and by 7:30 AM, it disappears completely.

This corner, formed by secondary wall B and the back wall (see Fig. 12.11), was apparently added to define the angle of sunlight at equinox. By observing the alignment of diagonal shafts of light in the far corner of the Temple of the Sun, Maya astronomers would have been able to recognize the exact days of vernal and autumnal equinoxes. This knowledge may have served to fix the dates for agricultural activities in the solar calendar (Aveni 2001:293–294).

As mentioned earlier, equinox sightings have been documented at sites with flat horizons, such as Chichén Itzá, Dzibilchaltún, and the Group E architectural complexes found at Waxaktun and numerous other sites throughout the Maya area. The Cross Group at Palenque does not fit the Group E pattern. From the Temple of the Sun, the low notch between El Mirador and the Temple of the Cross defines the horizon at equinox. Our topographical surveys indicate that much of this low notch was

12.9 Equinox sunrise as seen from the Temple of the Sun (photograph by Edwin L. Barnhart).

human made. Pending further exploration, structure OT74 and the terraces of the Otolum Group could prove to be the true markers for equinox sightings from the Temple of the Sun (see Fig. 12.4). Because they are related to the Cross Group and show alignments to equinox, these structures may eventually illustrate a concerted effort by Palencano builders to establish corridors of sightings.

Given the hilly terrain, it is likely that astronomers made initial long-distance sightings of the Sun from high observation points. On the peak of El Mirador, the platform of Group D has a cardinal orientation, possibly toward equinox, but the buildings are too badly deteriorated for this to be said with any confidence. Smaller hills crowned with standing structures do show definite equinox orientations to one another (e.g., between the Temple of the Cross and Temple IV in the Ch'ul Na Group and between Temple XXIV and Group H; see Fig. 12.4). Such long-distance sightings would have decreased any margin of error in reading the slight differences in azimuth during the two successive days of equinox (Aveni 2001:65–66). Established readings taken from lofty elevations could then be transferred to plazas or groups lacking clear views of the horizon. This was probably the case for the Temple of the Sun; modifications were made to the low notch as well as the interior of the temple so that the building would better interact with the Sun at equinox. Future excavations as well as further investigations of ancient surveying techniques will no doubt shed light

Alonso Mendez, Edwin L. Barnhart, Christopher Powell, and Carol Karasik

12.10 Interior of the Temple of the Sun showing equinox sunlight reaching the southwestern corner of the central chamber (photograph by Edwin L. Barnhart).

12.11 Plan view of the Temple of the Sun illustrating the angle of light during equinox sunrise (drawing by Edwin L Barnhart).

on the complex interrelationship between elevated sightings and plaza orientations.

Summer Solstice

Summer solstice corresponds to the time of year when the Sun rises and sets at its maximum position in the north. At Palenque's latitude of 17°28'N, sunrise on a flat horizon occurs at an azimuth of 65°14'E and sets at 294°44'W.

At 7:00 AM on June 21, the Sun, when viewed from the interior of the Temple of the Sun, rises from its northernmost point on the horizon, grazing the northwestern

317

corner of the Temple of the Cross (Fig. 12.12). Light enters the Temple of the Sun at an oblique angle of 50° north of the transverse axis, or approximately 70° azimuth (Fig. 12.13).

The diagonal light entering the northeastern doorway continues to steal across the temple floor. As it pierces the dark interior, the broad ray, blocked by consecutive wall edges, grows increasingly narrow until it becomes a thin beam of light striking the corner of the southwestern chamber (Fig. 12.14). By 7:30 AM the rays recede from the temple.

The angle of light seen within the temple appears to be the direct result of significant architectural details that suggest that this solstice alignment was intentional. The northeastern corner is set back 10 cm from the rest of the facade, a notable difference that allows the Sun to penetrate the interior at the desired angle.

Because the temple faces El Mirador, early-morning rays can enter the front doorway only at azimuths between 90° and 65°. The ancient architects left evidence that implies close attention to those particular angles of sunrise. Their concern is apparent in the addition of two interior walls (A and B) with doorways that precisely frame the light. Given the fact that the other secondary wall (C) does not have an opening, we must assume that the architects chose to capture the light that radiates from the northern part of the horizon. Careful alignment of the doorways to the medial wall

12.12 Summer solstice Sun rising at the northwestern corner of the Temple of the Cross, as seen from the Temple of the Sun (photograph by Alonso Mendez).

12.13 Summer solstice sunlight entering the northeastern door of the Temple of the Sun (photograph by Alonso Mendez).

12.14 Summer solstice sunlight striking the southwestern interior corner of the Temple of the Sun (photograph by Alonso Mendez).

permitted the ray of light to pass through the temple. By capturing sunlight through doorways at a diagonal, observers were also able to confirm the position of the Sun on the horizon on significant dates with greater precision.

A final point needs to be made concerning the intentionality of design as well as the function of the temple. Even though the diagonal of the temple (66°14′) is only one degree off the true azimuth for summer solstice sunrise (65°14′), it is the *visible light* entering the temple at 70° and its relationship to the transverse axis that mark the angle of the summer solstice (Fig. 12.15). This indicates knowledge of the solstitial azimuth prior to the construction of the temple; later design modifications reaffirm this knowledge both visually and conceptually. Based on these factors, we conclude that the temple functioned as a space for ritualized astronomical observations.

Observations of the Sun deep inside the temple would have been conducted by priests, astronomers, or rulers, whereas the diagonal light entering through the northeastern doorway would have been visible to a larger group of observers. A person standing directly in the center of the building is fully illuminated by the dazzling morning rays (Fig. 12.16). This powerful lighting effect may have been employed during public rituals that took place during summer solstice.

On the evening of the summer solstice, at 6:18 PM, we made our principal observation from the small "altar stone" near the base of the stairway of the Temple of the Cross. At an azimuth of 290°, the Sun sinks behind the Inscriptions Prospect, and the last rays of light pierce the center of the roof comb of the Temple of the Sun (Fig. 12.17).

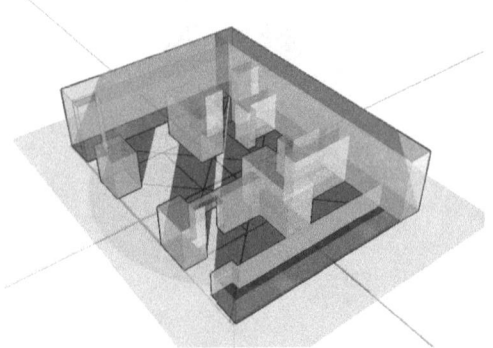

12.15 Plan view of the Temple of the Sun illustrating the diagonal orientation of the temple and the angle of sunlight entering the temple during summer solstice (drawing by Alonso Mendez).

12.16 Summer solstice sunlight illuminating a figure standing in the center of the Temple of the Sun (photograph by Christopher Powell).

Although our observation point was not on the transverse axis of the temple (119°46'E–299°46'W), there is nevertheless a noteworthy relationship between the setting Sun at summer solstice and the temple as viewed from the plaza. With an unobstructed horizon, the visual effects of this alignment may have been more striking.

In a reconstructed drawing of the roof comb (Robertson 1991:47–48), a seated stucco figure surrounded by a double-headed serpent and sky bands holds a double-headed serpent bar (Fig. 12.18). Four *Bacabs* support the sky serpents. The figure is seated on a *Kawak,* or Earth Monster (Robertson 1991:47), perhaps an implicit reference to the Sun setting over the mountain behind the Temple of the Sun (Milbrath, personal communication 2004).

On the roof frieze below, a smaller figure, whom Robertson identifies as Kan B'ahlam in the guise of God K, is seated on a throne in the company of two kneeling figures holding what appear to be the Jester God/*Tok Pakal* and God K manikin. The related image depicted on the Tablet of the Sun shows the "sun shield" looming above a more elaborate throne. In both the roof frieze and the tablet, the 7 and 9 gods flank the central image, another parallel between this seated figure and the shield on the tablet.

Alonso Mendez, Edwin L. Barnhart, Christopher Powell, and Carol Karasik

12.17 Summer solstice Sun setting behind the roof comb of the Temple of the Sun, as seen from the Cross Group Plaza (photograph by Susan M. Prins).

12.18 Reconstructive drawing of the roof comb of the Temple of the Sun (drawing by Merle Greene Robertson).

Zenith Passage

Zenith passage occurs only within the limits of the Tropics of Cancer and Capricorn and corresponds to the days when the Sun reaches a 90° vertical position from the horizon. At midday the Sun is directly overhead in the center of the sky. If

12.19 Zenith sunrise over the Temple of the Cross, as seen from the interior of the Temple of the Sun (photograph by Alfonso Morales).

a gnomon were planted in the earth, an observer would notice the total absence of a shadow at noon. Zenith passage had major significance in ancient Mesoamerican calendars, mainly marking the ideal beginning of the 105-day agricultural cycle in early May and harvest in early August. At 15° latitude, the second zenith passage corresponds with the mythical day of Creation, August 13, 3114 BC (Coggins 1996:21; Freidel et al. 1993:97; Malmström 1997:52).

Zenith varies according to latitude. At Palenque, the first zenith passage takes place on May 7 and the second on August 5, when the sun rises at an azimuth of 72°02' and sets at an azimuth of 287°07'.

12.20 Zenith sunrise over the Temple of the Cross, as seen from the central doorway of the Temple of the Sun (photograph by Alonso Mendez).

On the day of zenith passage, sunrise at Palenque occurs at about 6:30 AM, but direct light is not visible from the Temple of the Sun until 8:00 AM. Seen from the central doorway of the temple, the Sun rises directly over the roof comb of the Temple of the Cross in a spectacular display of architectural alignment between the two buildings (Figs. 12.19 and 12.20).

Inside the Temple of the Sun, a wide beam of light enters the northeastern doorway, as it does during summer solstice. Originally, doorway A had a lintel that limited the maximum extension of the morning light. With the aid of a plumb rod marking the edge of the doorway and approximating the height of the door, we were able to observe a thin ray of light, defined by the width of the door and the angle of entry

12.21 Zenith sunlight aligning to the southeastern corner of the sanctuary, Temple of the Sun (photograph by Alonso Mendez).

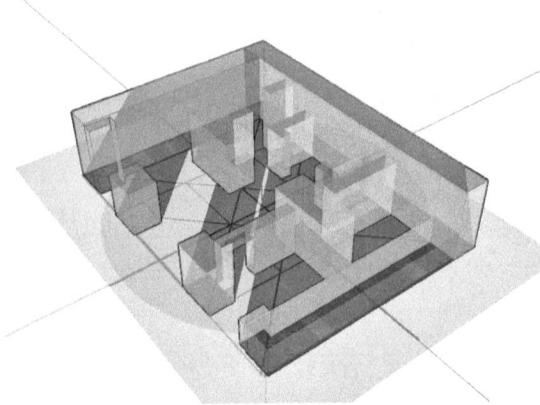

12.22 Plan view illustrating the angle of zenith sunlight in the Temple of the Sun (drawing by Alonso Mendez).

(approximately 45° north of the transverse axis), advancing toward the southeastern corner of the sanctuary (Figs. 12.21 and 12.22).

Nadir Passage

Nadir is the opposite of zenith. Like zenith passage, nadir passage varies according to latitude: the higher the latitude, the closer the distance between winter solstice and nadir passage, summer solstice and zenith passage. At the equator, nadir and zenith coincide with the equinoxes. At the 23° latitude of the tropics, nadir and zenith correspond with the solstices. At Palenque's latitude of 17°28′N, nadir passage occurs at midnight on January 29 and November 9, when the Sun passes at 90° below the horizon (Fig. 12.23).

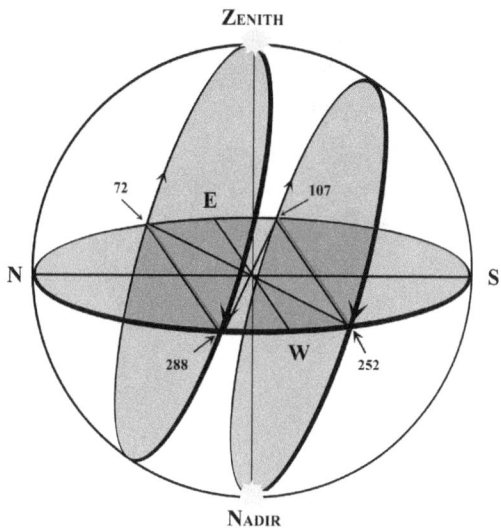

12.23 Diagram showing the azimuths of zenith and nadir passages for Palenque's latitude of 17°28'N (drawing by Alonso Mendez).

The nightly passage of the Sun under the earth is described in contemporary cosmology and folk tales (Gossen 1974:34; Karasik 1996:232, 273). Additionally, ethnographers have found that modern Maya languages equate our cardinal directions of north and south with zenith, "up" or "above," and nadir, "down" or "below" (B. Tedlock 1992:19–24). According to Coggins, the same was true for the ancient Maya (B. Tedlock 1992:19).

How did the ancient Maya determine the times of year when the Sun reached what they considered to be the center of the underworld? Nadir can be arrived at geometrically by measuring the angle between summer solstice and zenith sunrises and then transposing that angle to the known winter solstice azimuth and the presumed azimuth of nadir. Alternatively, a straight line can be extended from the point on the horizon line where zenith sunrise occurred through the observer's location to a point opposite on the western horizon. The arithmetical solution is as simple. Astronomers at Palenque may have counted the number of days from summer solstice to zenith and then counted the same number of days from winter solstice to nadir. The results would have been fairly accurate. Astronomers could then correlate their findings by observing bright stars on the horizon. Sirius, the brightest star in the sky, would have been the logical marker for nadir passage. During the Late Classic period, Sirius rose at 106°26' and set at 253°33', remarkably close to the azimuths for the rising and setting nadir Sun (107°45'E–252°5'W) in the seventh century. Aveni and Hartung (1978:176) have already noted that Palenque's Temple of the Count faces Sirius. It is possible that the temple also faced the rising Sun at nadir passage. The Pleiades passing through the zenith of the night sky also presaged the solar nadir passage, but only in November (Krupp 1983:205–209; Milbrath 1998:27).

Milbrath's (1988:26–28) identification of the east-west axis of El Castillo at Chichén Itzá is perhaps the first recognition of the relationship between zenith and nadir as seen in the alignment of a Maya temple. At Chichén Itzá's latitude of 20°, the temple's western facade is aligned to the zenith sunset on May 25 and July 20, whereas its eastern face is oriented to the nadir sunrise on November 22 and January 21. Milbrath mentions that the November nadir marked the beginning of the dry season and the January nadir marked the commencement of the agricultural season.

12.24 Nadir sunlight entering the interior of the Temple of the Sun (photograph by Alonso Mendez).

12.25 Nadir sunlight entering the interior of the Temple of the Sun (photograph by Alonso Mendez).

Milbrath proposes that in addition to its calendrical significance, the November nadir announced the period of warfare.

At Palenque, the nadir Sun now rises and sets at an azimuth of 107°40'E–252°32'W. Viewed from the Temple of the Sun at 9:15 AM on November 9, the sunrise breaks the horizon at a vertical angle of 23° and a horizontal azimuth of 120°. Light floods the temple at a direct perpendicular angle, illuminates the room to the south of the sanctuary (Figs. 12.24 and 12.25), and reaches the entrance to the inner sanctuary (Fig. 12.26). The photograph in Figure 12.27, taken from the threshold of the sanctuary, shows the centering of the Sun below the lintel of the middle doorway. This is the effect that Carlson had predicted for winter solstice. Our observations establish that the transverse axis of the temple is oriented to the rising Sun at nadir.

As mentioned earlier, the transverse axis of the temple also marks the southern maximum extreme of the rising Moon as well as the northern maximum extreme of the setting Moon (300°), events that occur during the solstices every $18^{2/3}$ years. Although the moonrise at 120° is not visible from the Temple of the Sun, the alignment of the temple with the lunar standstills merits attention (see Fig. 12.8).

AN ASTRONOMICAL BLUEPRINT FOR THE LAYOUT AND DESIGN OF THE TEMPLE OF THE SUN

This section presents a hypothetical discussion of surveying methods that may have been employed during the Classic period. Our experience during the Palenque Mapping

12.26 Nadir sunlight reaching the threshold of the inner sanctuary, Temple of the Sun (photograph by Alonso Mendez).

12.27 Nadir sunrise centered below the lintel of the middle doorway, Temple of the Sun (photograph by Alonso Mendez).

Project (1998–2000) gave us some practical insights into the challenges the ancient Maya faced when they were laying out and designing the Temple of the Sun, and the Cross Group as a whole.

A nucleus of bedrock lies at the core of the Cross Group and forms the mass of the temples' substructures. The ground-penetrating radar survey, conducted by the Proyecto de las Cruces, along with accompanying test pits that explored the terraces, revealed that the Temple of the Cross, northeast of the Temple of the Sun, rests on solid bedrock as far up as the sixth terrace (Hanna et al. 1996:5). This natural configuration of bedrock salients also lined the ravine that would become the Cross Group Plaza.

First, the deep ravine that ran between the major hills had to be leveled. Topsoil was removed in order to reach the bedrock carved by millennia of rains rushing down the mountainside. Then, thousands of cubic meters of clay were laid down to ensure that later flooding would never undermine the foundation of the Cross Group Plaza. Upon this foundation, construction of the Temples of the Sun, Cross, and Foliated Cross began. In this case, it was not necessary to move mountains but to "dress" them.

The builders of the Cross Group must have recognized the advantages of raising temples on natural elevations, one of which would face the rising Sun. While standing on the foundation for the future Temple of the Sun, the builders had observed the major stations of the Sun in relation to El Mirador. During winter solstice the Sun rose directly over the peak; at equinox the Sun rose from the cleft between El Mirador and the ridge that would become the Temple of the Cross; at zenith the Sun rose directly above that same ridge; and at summer solstice, just north of the ridge. This information was all the builders needed for their preliminary design (Fig. 12.28). The goal was to build two terraced platforms. The Temple of the Sun would eventually serve as the solar observation point; the Temple of the Cross would serve as the back-sight for those observations.

Initially the taller platform of the Temple of the Cross was used for taking sightings that would orient the entire group. From this high vantage, surveyors were afforded a clear view of the eastern and western horizons. The primary observation was made during summer solstice to establish the extreme northern position of the Sun. The surveyors drove a gnomon into the center of the stucco floor of the platform. Sunrise observations were recorded on the floor by tracing the shadow of the gnomon and connecting that painted line to a sighting stick aligned to the first rays of the Sun. The surveyors then waited until the Sun set in the west. A second line painted over the evening shadow of the gnomon ran toward a sighting stick aligned to the last rays of the Sun. In one day of observations, the surveyors had drawn two intersecting lines that crossed at approximately 50°, the angle of the solstices. By dividing these angles in half, they established the line of the equinox. This alignment was preserved in the diagonal orientation of the Temple of the Cross, which is less than one degree off due east/west (Fig. 12.29).

The longitudinal axis of the Temple of the Cross (119° E–301° W) coincides with the maximum excursions of the Moon. On a flat horizon at Palenque's latitude, the

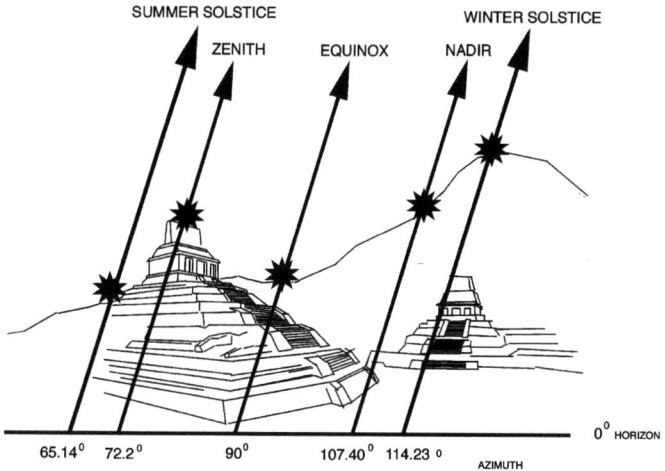

12.28 Solar alignments as seen from the Temple of the Sun (drawing by Alonso Mendez).

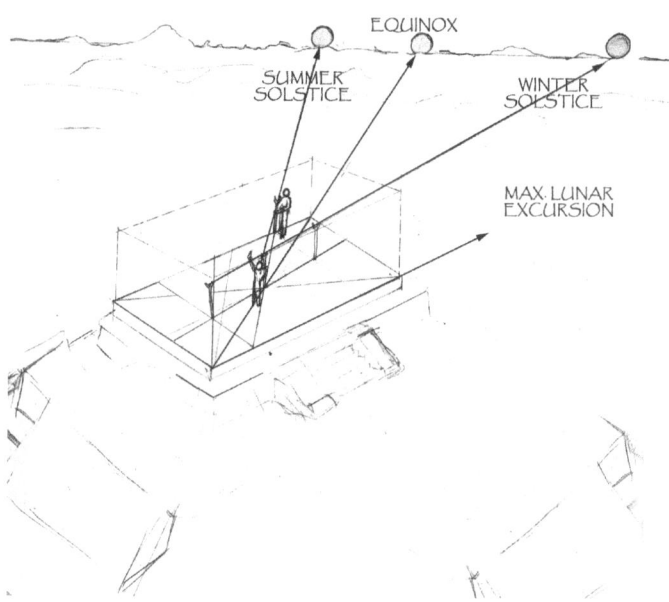

12.29 Hypothetical illustration of methods used in aligning the Temple of the Cross to the Sun and Moon (drawing by Alonso Mendez).

Moon, at its maximum southern position, rises and sets at an azimuth of 120° E–240° W and at 60° E–300° W at its maximum northern position. The lunar azimuths also match the transverse axis of the Temple of the Sun (119°46'E–299°46'W).

It was possible to project the measurements down to the platform of the Temple of the Sun with the aid of plumb bobs and sighting rods. Two crossed wooden sticks may have served as a rudimentary surveyor's transit. With this device, surveyors could project desired angles either vertically or horizontally (Aveni 2001:65). Adjustments could then be made to the height of the platform to bring the future temple into alignment with El Mirador, the low notch, and the Temple of the Cross.

After the solar angles were drawn on the platform of the Temple of the Sun, the builders began to lay out the geometric proportions of the temple. They probably relied on the same methods and the same tools traditionally employed in measuring houses made of wattle and daub (Anderson, personal communication 2004). Braided henequen rope and wooden stakes were used for squaring the building from corner to corner. In effect, the temple builders repeated the actions of the Maker, Modeler who laid out the cosmos (Tedlock 1985:72):

> The fourfold siding, fourfold cornering,
> measuring, fourfold staking,
> halving the cord, stretching the cord
> in the sky, on the earth,
> the four sides, the four corners, as it is said,
> by the Maker, Modeler...

The repetitive measurements of halving the cord, then stretching or "doubling" the cord, suggest that the Creators, like modern Maya house builders and farmers, began with a square (Christenson 2003:65; Vogt 1990:17–19). Next, the sides of the initial square were extended to produce a double square. This double square defined two larger squares, which, overlapping, produced a rectangle with a 3:4:5 ratio. The overlapping squares marked the two main piers of the temple facade, the width of the medial doorway, the facade of the sanctuary, and the secondary walls that framed the sanctuary. For the width of the walls, the builders stretched a cord from the center of the initial square to the outer rectangle and made a circle. The points where the radius intersected the major diagonals marked the interior corners of the temple. Three inner doorways eventually would allow sunlight to travel along the original sight lines that defined the interior space. Thus, a formula of progressive squares and rectangles produced a beautifully proportioned floor plan that also was in keeping with the principal solar orientations of the temple (Fig. 12.30).

The exterior dimensions of the temple were based on an integral right triangle with 3:4:5 proportions and interior angles of 90°, 53°, and 37° (Fig. 12.31). The 90° angle was inherent in the initial square, whose 45° diagonal was the angle between the observed sunrise at zenith and nadir passages. The 53° angle was the angle between summer solstice and nadir passage/maximum lunar excursion (Fig. 12.32). The observed sunlight entering the temple during summer solstice created a 50° angle with the transverse axis (see Fig. 12.16).

When the walls and vaults were raised, the roof comb was added. The side elevations conformed to the 3:4:5 principal proportion of the temple (Fig. 12.33).

During excavations conducted by M. A. Fernández (1991:239–241) between 1942 and 1945, three offerings were discovered in the floor of the temple. Offering

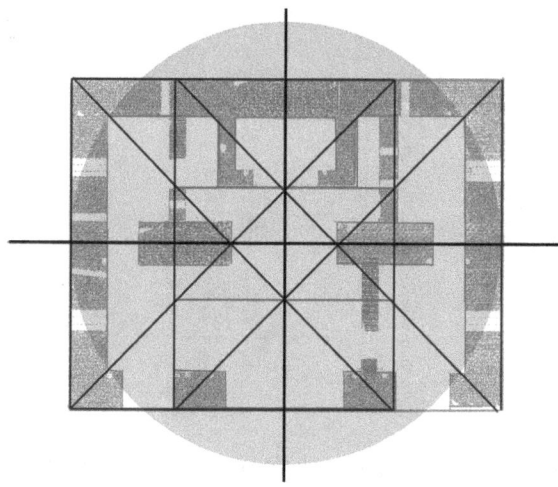

12.30 Plan view of the hypothetical geometric layout of the Temple of the Sun (drawing by Alonso Mendez after Merle Greene Robertson).

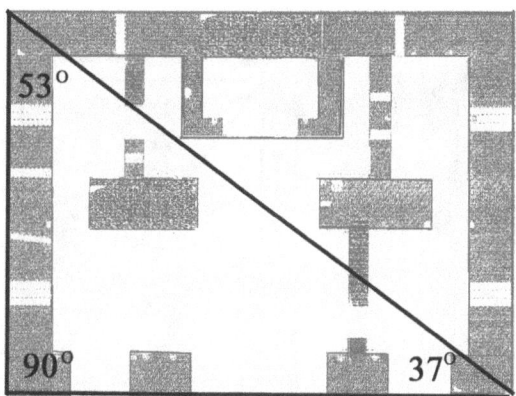

12.31 Plan view of the Temple of the Sun showing its geometric proportions and its respective angles (drawing by Alonso Mendez after Merle Greene Robertson).

1 (a cist containing two lidded vessels) and Offering 2 (a lidded vessel accompanied by jaguar phalanges) were buried on the transverse axis of the temple; Offering 3, a stucco mask representing a solar deity, aligned with Offering 2 along the 70° azimuth where the solstice light enters the building (Fig. 12.34).

The hieroglyphic texts tell us that after the temple was built, two dedicatory rites were held, one celebrating the completion of the temple on 9.12.18.5.17 3 Caban 15 Mol (AD July 24, 690), the second celebrating the completion of the inner sanctuary on 9.12.19.14.12 5 Eb 5 Kayab (AD January 10, 692). Sometime during those two years, walls B and C (see Fig. 12.11) were added to define the space of the sanctuary so that it would better interact with the diagonal light at equinox; doorway A defined summer solstice and zenith (see Figs. 12.15 and 12.22).

The Cross Group raised Kan B'ahlam's royal lineage above that of previous dynasties and rooted his legitimacy in the mythology of Creation (Aldana 2004; Baudez 1996:126; Freidel et al. 1993:283). Charged with religious symbolism, the Cross Group also embodied the movements of the Sun and Moon. The Temple of the Sun, with its back to the Plaza of the Inscriptions, became the new vantage for witnessing

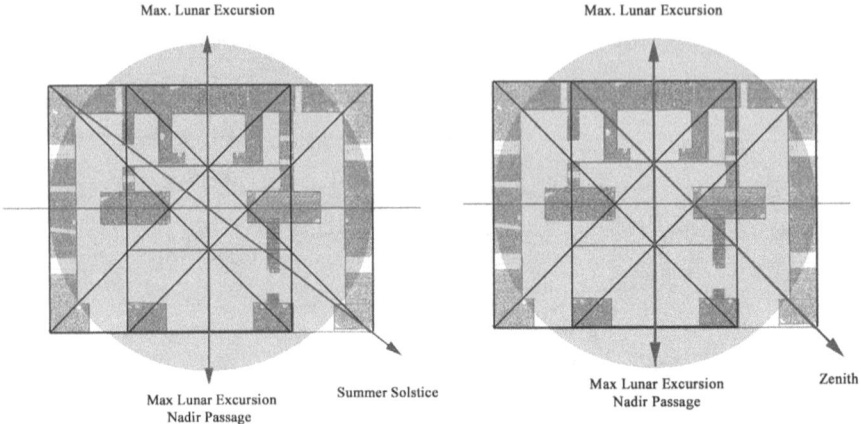

12.32 Plan view of the Temple of the Sun showing the relationship between the temple's astronomical alignments and geometric proportions (drawing by Alonso Mendez after Merle Greene Robertson).

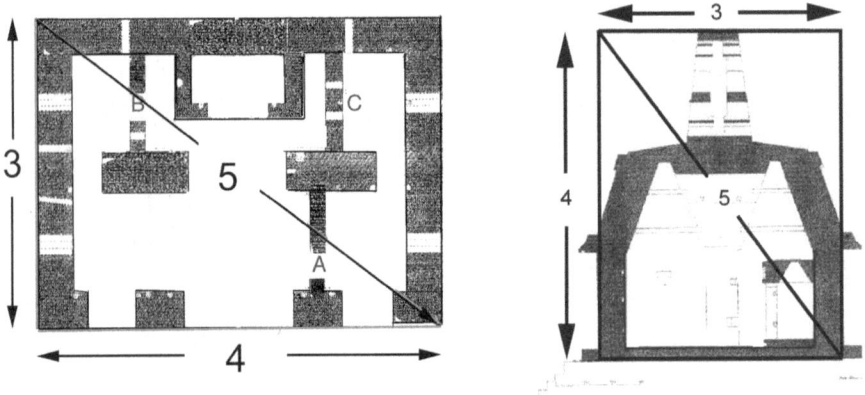

12.33 Diagrams illustrating the principal proportions of the plan and elevation, Temple of the Sun (drawing by Christopher Powell after Merle Greene Robertson).

the repeating cycles of hierophanies that would convey the spiritual and intellectual message of the new ruler.

ASTRONOMICAL ELEMENTS IN THE TEXT AND ICONOGRAPHY OF THE TEMPLE OF THE SUN

The Temple of the Sun was originally named for the prominent "sun shield" displayed on the carved tablet in its inner sanctum (Fig. 12.35). In addition to the imagery, the hieroglyphic inscriptions on the tablet and *alfardas* contain numerous direct, and

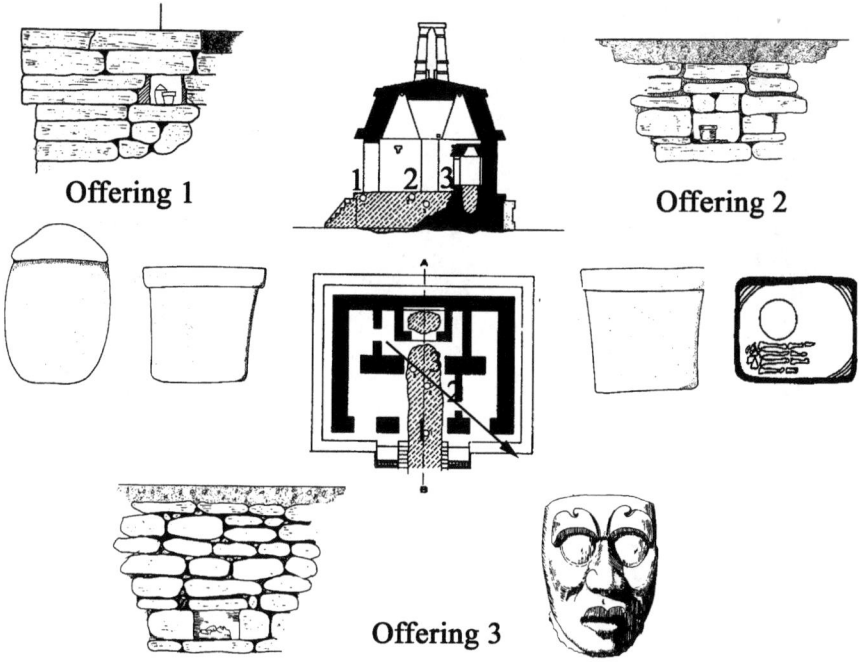

12.34 Drawing showing the distribution of three offerings buried in the Temple of the Sun (after Miguel A. Fernández).

oblique, astronomical references (see Table 12.1). Floyd Lounsbury (1978, 1989) has provided brilliant mathematical insights into the Jupiter-Saturn periods related to the 819-day count, the Venus and Mars cycles, and the 108-year tropical year drift cycle. Given the orientation of the temple, this section focuses on the transits of the Sun and Moon alluded to in the Tablet of the Sun. (For correlations between calendrical data and astronomical phenomena, we used the 584285 GMT+2 standard and the Maya Date–Maya Calendric Calculator [Bassett 1999].)

The text of the tablet begins with the birth of GIII on 1.18.5.3.6 13 Cimi 19 Ceh (A1-D6) (Gregorian: October 25, 2360 BC, Julian: November 12, 2360 BC). The second born of the Palenque Triad, GIII is named in the text as *Ahaw Kin,* "Lord Sun" (Lounsbury 1985:50–51), though he is currently called *K'inich Ahaw,* "Sun-Faced Lord" (Miller and Taube 1993:130; Montgomery 2002:28; Wald 1999:93). Long considered a solar deity, he is often associated with the Sun in the underworld (Kelley 1976:6). Dennis Tedlock (1985:368, 1992:264) identifies GIII with the younger Hero Twin, Xbalanque, or "little jaguar sun," and equates him with the full Moon. It is interesting to note that a full Moon rose on his birthday (Milbrath 1999:102).

The later passages of the tablet cite important dates in Kan B'ahlam's life; those events are keyed to major astronomical phenomena. The first historical date, 9.12.18.5.16 2 Cib 14 Mol (D16–O6) (AD July 23, 690), corresponds to a rare Jupiter/

12.35 Tablet of the Sun (drawing by Linda Schele).

Saturn/Mars conjunction. This major astronomical event, related to the dedication of the Cross Group, is also referenced in the texts of the Temples of the Cross and Foliated Cross. That night, Mars set at 252°45', less than a minute off the Sun's nadir position on the western horizon; Jupiter and Saturn set at 254°; and the Moon set at 250°22'. In other words, this conjunction was amplified by the fact that the major planets entered the underworld through the portal of the nadir Sun.

Kan B'ahlam dedicated the *K'inich B'ahlam Kuk Nah* building (Houston 1996), the Temple of the Sun, on 9.12.18.5.17 3 Caban 15 Mol (N7-N8) (AD July 24, 690). That morning Venus rose at 68°, within 2° of the Sun's summer solstice position. Like the Sun at that time of year, Venus would have been visible through the northeastern doorway of the Temple of the Sun.

Following the dedication of the temple, Kan B'ahlam performed a bloodletting rite to the gods, on 9.12.18.5.19 5 Cauac 17 Mol (N13-N16) (AD July 26, 690). Apparently Kan B'ahlam had waited for the Moon to reach its southern extreme and to begin its northern journey before conducting the bloodletting ritual. The anniversary of Janahb Pakal's accession to the throne, July 29, fell immediately on the full Moon and five days before the Sun's zenith passage. The night before, the full Moon passed through the Sun's nadir position, rising at 106°9' and setting at 251°43'. Aveni and Hotaling (1996:363) mention that Jupiter, Saturn, and Mars, which had come

Table 12.1 Dates from the Temple of the Sun and their proposed astronomical significance.

Glyph ID	L.C. Tzolkin and Haab	Context	GMT + 2 Correlation (Gregorian 854285)	Event	Proposed Astronomical Significance*
A1-B9	1.18.5.3.6 13 Cimi 19 Ceh	INSG Mythological Date Tablet of the Sun	October 25, 2360 BC (Jul.: Nov 12, 2360 BC)	Birth of GIII	Full Moon at zenith passage
D16-O6	9.12.18.5.16 2 Cib 14 Mol	Historical Date Tablet of the Sun	AD July 23, 690	Dedication Temple of the Sun	Jupiter/Saturn/Mars/Moon (Conjunction at nadir azimuth); Venus rises at 67° azimuth and A.M.G.E.†
N7-N8	9.12.18.5.17 3 Caban 15 Mol	Historical Date Tablet of the Sun	AD July 24, 690	Dedication Temple of the Sun	Venus Rises at 68° azimuth (diagonal of temple)
N13-N16	9.12.19.5.19 5 Kawak 17 Mol	Historical Date Tablet of the Sun	AD July 26, 690	Bloodletting Rite	Full Moon rises at nadir azimuth (July 29); Jupiter/Saturn/Mars conjunction at max. proximity
P6-Q6	9.10.8.9.3 9 Akbal 6 Xul	Historical Date Tablet of the Sun	AD June 17, 641	Heir Designation Kan B'ahlam	New Moon; Summer solstice Jupiter/Mars/Moon conjunction (June 25)
P14-Q15	9.10.10.0.0 13 Ahaw 19 Kankin	Historical Date Tablet of the Sun	AD December 6, 642	War Event 10 Tun Completion	Full Moon/partial lunar eclipse (December 14)
L1-M6	9.12.11.12.10 8 Oc 3 Kayab	Historical Date Tablet of the Sun	AD January 10, 684	Accession Kan B'ahlam	Jupiter in retrograde; Full moon sets at 299° (max. northern excursion); Moon/Mars conjunction
A1-G2	9.12.19.14.12 5 Eb 5 Kayab	Historical Date (alfardas)	AD January 10, 692	Dedication Sanctuary	Full Moon at 1° from Zenith azimuth (January 12)
H1-L2	9.13.0.0.0 8 Ahaw 8 Uo	Historical Date (alfardas)	AD March 18, 692	13th Katun Celebration	Vernal equinox; Saturn/Mars/Moon conjunction; Venus at nadir azimuth

*Astronomical data from Starry Night Deluxe (Julian).
† A.M.G.E = Greatest elongation as "Morning Star."

within 10° of longitude the night before the dedication of the Cross Group, were even closer (within 4°) on the anniversary of Janahb Pakal.

The inscriptions then move back in time to record Kan B'ahlam's heir-designation ceremony. The celebration, which began on 9.10.8.9.3 9 Ak'bal 6 Xul (AD June 17, 641), culminated five days later on the summer solstice. As the text states (Q5-Q10), *i-u-ti b'olon Ak'b'al Wak Tzikin k'alwani u-naah'tal Ook-te K'in K'inich Kan B'ahlam B'aakel Wayal yi-chi-nal GI*: "And then it happens, (9 Ak'b'al 6 Tzikin) after the fifth changover [day] he is bound as the first Pillar of the Sun, the radiant Kan B'ahlam, the Bone Spirit, in the presence of GI" (Stuart 2006:168). The heir-designation ceremony, culminating on the summer solstice, is perhaps the most critical of the solar allusions in the text, as it defines the transfer of divine status from Janahb Pakal to his son Kan B'ahlam.

On June 17, a new Moon, in conjunction with the Sun, set midway between the solstice and zenith at an azimuth of 292°. On June 21, the Moon, 3.8 days before first quarter, passed through the zenith. By June 25 the Moon had moved away from the Sun to join Mars and Jupiter in a conjunction that set directly on the Sun's equinox position.

The final date on the tablet, 9.10.10.0.0 13 Ahaw 18 Kankin (P14-Q16) (AD December 6, 642), relates to a war event that coincided with a ten *tun* anniversary. As was customary, the six-year-old heir designate, or his father, was required to capture and sacrifice nobles from a rival kingdom (Schele and Freidel 1990:236). On the same date, Smoking Squirrel of Naranjo and Ah Cacaw of Tikal engaged in battle with the cities of Ucanal and Calakmul, respectively (Schele and Freidel 1990:251). Why was this date so propitious for war? Astronomically, this date fell close to a full Moon, which entered into a partial eclipse on December 14, perhaps an ideal time for a beheading. That night the Moon rising at 66°40', slightly more than one degree from the azimuth of summer solstice, seemed to be in direct opposition to the Sun, which set in the southwest at 245°37', only 10 minutes away from the azimuth of winter solstice.

A brief text prominently displayed above the center of the "sun shield" on the tablet records the accession of K'inich Kan B'ahlam II on 9.12.11.12.10 8 Oc 3 Kayab (L1-M6) (AD January 10, 684). At the ripe age of 49, Kan B'ahlam waited 132 days after his father's death before he came to the throne (Schele and Freidel 1990:240). A possible explanation may lie in the astronomical significance of this date. According to Lounsbury (1989), the event was timed to Jupiter's retrograde motion. The Moon was equally prominent that evening. One day before full, and in conjunction with Mars, the Moon reached its maximum northern excursion, rising at 60° and setting at 299°39', less than one degree off the transverse axis of the Temple of the Sun. The placement of the accession text at the center of the tablet parallels the commemoration of this event in the central alignment of the temple. The visual and conceptual ramifications are profound. The Moon not only mirrored the position of the rising Sun during nadir passage but also echoed the setting Sun at summer solstice, the time of Kan B'ahlam's heir-designation ceremony (see Fig. 12.17).

12.36 Hieroglyphic texts from the *alfardas*, Temple of the Sun (after Linda Schele).

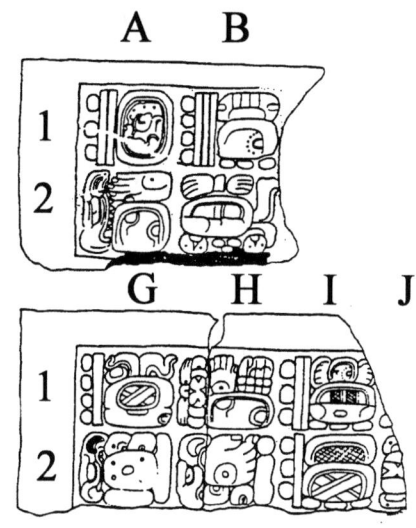

On the eighth anniversary of Kan B'ahlam's accession, he dedicated the sanctuary of the Temple of the Sun. According to a separate text recovered from the *alfardas* of the temple (Fig. 12.36), this ceremony occurred on 9.12.19.14.12 5 Eb 5 Kayab (A1-G2, *alfardas*) (AD January 10, 692). Two days later, a full Moon rose at an azimuth of 73°, only one degree from the Sun's position during zenith passage. Observed from the central doorway of the Temple of the Sun, the Moon would have been seen emerging directly from the center of the roof comb of the Temple of the Cross.

A few months later, Kan B'ahlam celebrated the completion of Katun 13 on the vernal equinox, 9.13.0.0.0 8 Ahau 8 Uo (H1-L2, *alfardas*) (AD March 18, 692). Saturn, Mars, and the Moon rose in conjunction. The planets would have been seen rising from the peak of El Mirador at the position of the rising Sun at winter solstice. Venus would have been visible from the Temple of the Sun directly centered in the doorway at 30° altitude and 120° azimuth, the Sun's nadir position on the horizon.

A series of dramatic lunar events surrounded Kan B'ahlam's death, recorded on the Zapata Panel as 9.13.10.1.5 6 Chikchan 3 Pop (AD February 20, 702). One month earlier, on AD January 20, 702, the Moon reached its maximum northern excursion, completing the cycle of 18⅔ years that began on the date of Kan B'ahlam's accession. Rising at an azimuth of 60° and setting at 300°, the full Moon would have been seen from the Temple of the Foliated Cross as it descended behind the Temple of the Sun, precisely along the line of the transverse axis. On the date of Kan B'ahlam's recorded death, the Moon, one day before full, rose in the sky near the zenith position of 72° azimuth and set near the position of the summer solstice Sun (285° azimuth). On AD April 20, 702, shortly before Kan B'ahlam's younger brother assumed office, a total lunar eclipse took place, and although it occurred just below the horizon at 103°, its red penumbra would still have been visible; by dawn the full Moon set at 254°, the nadir position of the Sun. Two and a half years later, Kan B'ahlam's spirit is said to have risen from the underworld, on 9.13.13.15.0 9 Ahaw 3 Kankin, or AD November 6, 705 (Apotheosis Panel, Temple XIV). That evening the Sun set within one degree of nadir. Three nights later the full Moon rose directly from the zenith position on the horizon, traveled across the height of the sky, and eclipsed the Pleiades before they descended into the horizon. Kan B'ahlam's life, death, and apotheosis were circumscribed by recurring patterns of the Sun and Moon (see Table 12.2).

Considerable debate surrounds the solar and lunar symbolism of the Tablet of the Sun. But first, a description of the iconography is in order. The dominant imagery on the tablet ostensibly refers to war, perhaps to the war event associated with Kan B'ahlam's heir designation or to the role of the king, upon his accession, as a warrior. On either side of the panel, two men face each other—the same two men who appear on each of the tablets in the Cross Group. The man on the right has been identified as K'inich Kan B'ahlam II. The smaller figure, draped in "winding sheets," probably represents Kan B'ahlam's dead father, K'inich Janahb Pakal (Schele and Freidel 1990:254), although some scholars identify the figure as the young Kan B'ahlam before his accession to the throne (Martin and Grube 2000:169; Milbrath 1999:233). He offers a personified eccentric flint with flayed face shield: the *Tok Pakal,* an icon that denotes royal lineage as well as warrior status (Freidel et al. 1993:305). The larger figure on the right offers a God K manikin, another symbol of royal lineage (Miller and Taube 1993:110–111) as well as blood sacrifice (Freidel et al. 1993:194). These same objects are offered by the two kneeling figures on the roof frieze of the temple.

Both figures stand on the backs of sun deities and pay homage to the tablet's central motif, the "war stack" (Schele and Freidel 1990:259): a shield in front of two crossed spears resting upon a platform adorned with the masks of a jaguar and two serpents. The face emblazoned on the shield has been identified as the Jaguar God of the Underworld, the Night Sun (Schele and Freidel 1990:414), or the Jaguar War God, a deity associated with the Moon (Milbrath 1999:123–125). Bassie-Sweet (1991:192–198) specifically associates what she calls the "Twisted Cord Jaguar" with the full Moon and then proposes that the jaguar is the zoomorphic form of GIII.

Although the platform upon which the sun shield rests appears to be a flat bar, it is most likely a throne with four sides (Bassie-Sweet 1991:163). A similar throne, decorated with jaguar masks at either end, is seen in the accession scene depicted on the Palace Tablet. Although the shield above the platform may represent the Sun on the horizon, it may also symbolize the seating of the ruler, K'inich Kan B'ahlam, on the throne.

Like many thrones seen in Classic Maya art, it is supported by two *Pauahtuns* or *Bacabs*. The *Pauahtuns* support the throne with one hand and with the other hand touch the band running across the bottom register of the panel. This band is composed of alternating glyph blocks. One reads *cab* or "earth"; the second glyph block is a profile of God C, who signifies divinity or holiness (Taube 1992:27–31).

The band terminates with the faces of GIII (Milbrath 1999:102). The band of alternating "earth" and God C glyphs, terminating with the faces of GIII, may represent the nightly passage of the Sun through the underworld (Baudez 1996:123). Rather than the west-to-east path of the Sun beneath the earth, it may be more accurate to say that the band represents the horizon that lies between the points of maximum excursion. In any event, the band seems to serve as a symbol of transition. The gesture used by the *Pauahtuns* as they touch the band is reminiscent of the *u-pas-kab* or "hand above earth" glyph, interpreted as "to experience" or "to be born" (Lounsbury 1980:113).

Table 12.2 Dates for Kan B'ahlam's life and their proposed astronomical significance.

L.C. Tzolkin and Haab	Context	GMT+2 Correlation (Gregorian 584285)	Event	Proposed Astronomical Significance
9.10.2.6.6 2 Cimi 19 Zotz	Temple of the Cross	AD May 23, 635	Birth of Kan B'ahlam	Moon at 1 day before new
		AD January 20, 702		Moon at 2.42 days before full Completion of one 18⅔-year cycle of maximum lunar excursion since Kan B'ahlam's Accession
9.13.10.1.5 6 Chichan 3 Pop	Zapata Panel	AD February 20, 702	Death of Kan B'ahlam	Full Moon rises at 1° from zenith azimuth
9.13.13.15.0 9 Ahaw 3 Kankin	Temple XIV Tablet	AD November 6, 705	Apotheosis of Kan B'ahlam	Sun at 4 nights before nadir passage Full Moon rises at zenith azimuth

The underworld is a place of death, transformation, and rebirth. The presence of the underworld deity God L, the *Pauahtun* on the left, ties the scene on the tablet to the underworld. The floating toponyms on either side of the shield, referring to the "7 and 9 place," denote supernatural space (Martin and Grube 2000:194; Miller and Taube 1993:151); perhaps the 7 and 9 refer more specifically to the underworld. On the roof frieze, Kan B'ahlam is identified with God K. On the tablet located in the inner sanctuary—the *pib na*, "steambath" or "underground house" of GIII (Houston 1996; Schele and Freidel 1990:251)—Kan B'ahlam is associated with the Jaguar War God, emblematic of the night Sun, the full Moon, and GIII seated in the underworld (Aldana 2004). Repeating a common theme in Classic Maya art, Kan B'ahlam communes with his deceased father, Janahb Pakal, who reinforces the role of Kan B'ahlam as warrior by offering him the *Tok Pakal*, a miniature "war stack." The tablet thus resonates with metaphorical associations between warfare, the Moon, and the nadir Sun.

Multiple solar positions are implied on the tablet: the rising Sun at summer solstice, the Sun at or below the horizon, and the Sun in the underworld at nadir. The iconography of the tablet, although static, represents the myriad levels of the Maya universe.

This concept is embodied in the crossed spears poised above the throne. Milbrath (1999:272) argues that the Temple of the Sun is oriented to a cross constellation in Sagittarius seen rising in the eastern sky just before Kan B'ahlam's accession and it is this constellation that was represented by the crossed spears in the tablet. If we view the spears from a purely mathematical perspective, we discover yet another layer of interpretation. The angle of the spears is 53°. This angle is repeated in the 53° interior angle of the 3:4:5 proportions of the tablet (Fig. 12.37). The angle of the crossed spears also coincides with the angle between the transverse axis and the diagonal of the temple. In other words, the angle of the spears commemorates the angle between summer solstice and nadir and between summer solstice and the maximum lunar extreme, a deft way of depicting the polarity between Sun and Moon. As it turned out, this angle encompassed Kan B'ahlam's life and death.

CONCLUSIONS

The Temple of the Sun was used to track major stations of the Sun as well as to mark important dates in the reign of Kan B'ahlam. Four new solar hierophanies have been identified within the temple. The morning light observed entering the temple during equinox, summer solstice, and zenith passage is characterized by diagonal rays reaching back to interior corners. The broad beam of light entering the central doorway at nadir passage indicates the transverse axis of the temple.

The diagonal rays of light recorded at equinox, summer solstice, and zenith passage were observable only to a small group of astronomers dedicated to monitoring the passage of the Sun throughout the year. But at summer solstice, the dramatic morning light may have illuminated a noble personage standing in the center of the Temple of the Sun who could have been visible to a larger audience gathered in the

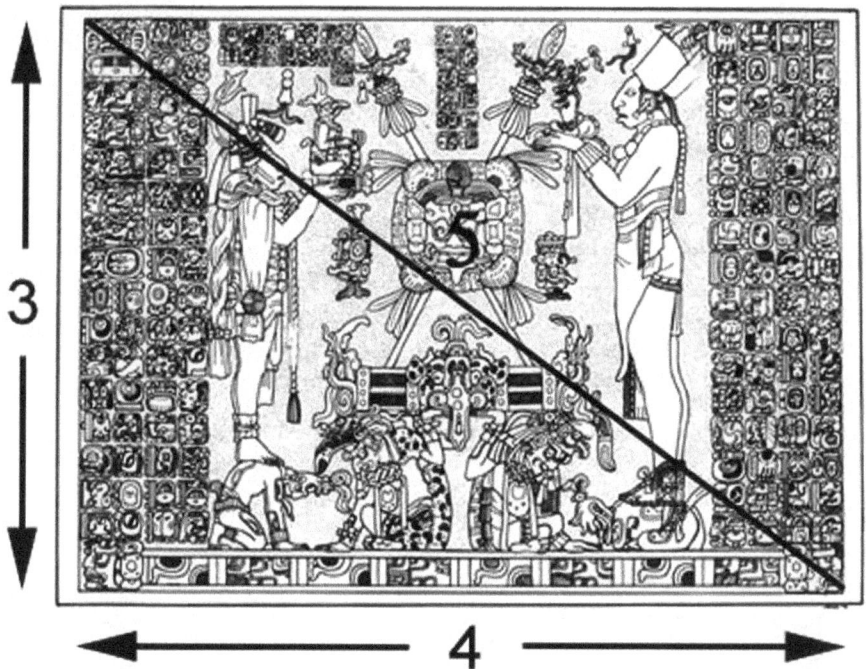

12.37 Principal proportion of the Tablet of the Sun (after Linda Schele).

plaza. The Sun setting directly over the roof comb that evening may also have been witnessed by a group of celebrants. Both events probably played a part in the public pageantry celebrating the anniversary of Kan B'ahlam's heir-designation ceremony. These hierophanies would have strengthened the ties between the earthly and supernatural worlds that the ruler represented.

The private hierophany within the temple would have carried a similar message, however. As we have seen, the precision of the temple's original design and subsequent modifications suggests that the Temple of the Sun functioned as a commemorative structure for ritualized astronomical observations that served to reaffirm the ruler's central place in the cosmic order.

Hierophanies, both public and private, depended on the acute harmony between scientific observations and mathematical knowledge. The precision of the architectural design is apparent in the geometry of the temple, which is founded on the proportions of an integral right triangle. The use of the integral right triangle has been proposed as a significant design element in the layout of building groups at Palenque and Tikal (Grube 2001:230; Harrison 1994:243). The same proportion is repeated in the dimensions of the Tablet of the Sun. As seen in Figures 12.31, 12.33, and 12.37, the geometric proportions of the tablet repeat the proportions of the temple. The geometric proportion of the 3:4:5 integral right triangle contains interior angles that, in turn, relate to the angles of the Sun and Moon.

Whereas the architecture continues to display alignments that apparently influenced the timing and expression of historical ceremonies, the iconography on the tablet is more difficult to read. The problem is that artists were trying to depict three-dimensional space on a two-dimensional field of vision. For example, the earth band and throne define dual terrestrial planes or horizon lines, and the vertical figures and "war stack" stand either on earth or in the underworld below. How space is depicted in two-dimensional art remains a subject for further investigation.

The multiple views of space depicted in the art point to certain overlapping features in the architectural design. The most complex shifts relate to the constant variations between the azimuths of the building's alignment and the visible light, especially at summer solstice and at nadir. The unique setting of the site is responsible for this multidimensional perspective of space. Such a feat would not have been possible on a flat landscape. The mountainous environment of Palenque encouraged, rather than hindered, a profound examination of the horizon and more profound solutions. Just as Palencano artists stretched the possibilities of multiple spatial references in their art, the builders of the Cross Group were cross-referencing horizon events.

Inherent in the brilliant, but problematic, interplay between the "ideal" and real horizons is the Maya concept of duality. Astronomically, that duality can best be appreciated in the synchronization of the lunar and solar cycles: the 120° moonrise every 19 years and the biannual nadir sunrise that mark the transverse axis of the Temple of the Sun; the full Moon setting at the solar nadir position on AD July 26, 690, three days after Kan B'ahlam's bloodletting rite; the full Moon rising at the zenith passage position on AD January 10, 692, when Kan B'ahlam dedicated the sanctuary; and finally, the full Moon setting at 300°, directly opposite the nadir sunrise position on AD January 10, 684, when Kan B'ahlam acceded to the throne. Figure 12.38 shows the full Moon setting at its maximum northern excursion at an azimuth of 299°, less than one degree off the transverse axis of the Temple of the Sun. The maximum excursion, which occurred on December 15, 2005, replicates the lunar hierophany of January 10, 684.

The importance of nadir passage is now obvious in the alignment of the Temple of the Sun. Although no date is associated with the ceremony depicted on the tablet, it is possible that the central image represents the concept of nadir. At nadir, the morning Sun shone directly along the central axis of the temple; that night it illuminated the center of the underworld. This may have been an auspicious moment for communication with the other world through rituals associated with sacrifice, death, and renewal.

Just as the ruler acted as a conduit between worlds, the Sun's position at zenith and nadir passages served as a portal through which the Moon and outer planets passed. Viewed from the Temple of the Sun, the Moon and Sun appeared as complementary opposites whose interwoven cycles were the basis for keeping time and giving measure to space.

The alignment between the Temples of the Sun and Cross at zenith passages may lend support to theories that identify the gods of those temples with the twin pro-

12.38 Full Moon at maximum lunar excursion setting behind the roof comb of the Temple of the Sun, as seen from the Cross Group Plaza (photograph by Alonso Mendez, December 15, 2005).

tagonists of the *Popol Vuh*; that is, GI may represent the older Hero Twin, Hunahpu, the Sun; whereas GIII may represent the younger twin, Xbalanque, the Moon. In view of the astronomical complexities described in this article and to be further explored in subsequent studies, we prefer to consider the complementary relationship empirically. The solstices, zenith and nadir passages, and maximum lunar excursions represent peak transitions in the courses of the Sun and the Moon. Solstices are the extreme extensions of the Sun on the horizon, whereas zenith and nadir are the vertical extremes. These spatial positions demarcated the boundaries of the ecliptic and defined sky, earth, and underworld. As principal actors in the creation myth recorded in the Cross Group, GI and GIII established those cosmological boundaries.

In the text and iconography of the Temple of the Sun, Kan B'ahlam aligned himself with GI and GIII. His affinities were marked by hierophanies in the heavens and on earth. The mathematical and astronomical precision seen in the architecture replicated the "cosmic principles of hierarchical order" that formed the basis of religious thought (Aveni and Hartung 1986:8). Linking the cosmos to earthly events was a powerful affirmation of the divine in a place and time.

ACKNOWLEDGMENTS

We wish to express our appreciation to Moises Morales Marquez and Alfonso Morales Cleveland for generously sharing their knowledge of archaeoastronomy; to Susan Mendez Prins, Xun Mendez, and Catherine B. Kahn for their participation in the observations; and to Thor Anderson, Susan Milbrath, David Stuart, Julia Miller, and Chip Morris for their invaluable comments and suggestions. Finally, we wish to thank Arnoldo González Cruz, Director of the Proyecto Arqueologico Palenque; Juan Antonio Ferrer Aguilar, Director of the Archaeological Zone of Palenque, INAH; and the helpful staff of the Archaeological Zone of Palenque.

REFERENCES

Aldana, Gerardo. 2001. Oracular Science: Uncertainty in the History of Maya Astronomy, 500–1600. Unpublished Ph.D. dissertation, Harvard University.

Aldana, Gerardo. 2004. El trabajo del alma de Janahb Pakal: La cuenta de 819 días y la política de Kan Balam. In *Culto funerario en la Maya Clasico: IV. Mesa Redonda de Palenque*, ed. Rafael Cobos, 283–307. Mexico, INAH.

Andersen, Tom, Peter Hanson, and Ted Leckie. 1997. Starry Night Deluxe. Sienna Software, Toronto, Canada.

Anderson, Neal, Alfonso Morales, and Moises Morales. 1981. A Solar Alignment of the Palace Tower at Palenque. *Archaeoastronomy: The Bulletin of the Center for Archaeoastronomy* 4 (3):34–36.

Anderson, Neal, and Moises Morales. 1981. Solstitial Alignments of the Temple of the Inscriptions at Palenque. *Archaeoastronomy Bulletin* 4 (3):30–33.

Aveni, Anthony F. 1980. *Skywatchers of Ancient Mexico*. Austin, University of Texas Press.

Aveni, Anthony F. 2001. *Skywatchers*. Austin, University of Texas Press.

Aveni, Anthony F., and Horst Hartung. 1978. Some Suggestions about the Arrangement of Buildings at Palenque. In *Proceedings of the Third Palenque Round Table*, ed. Merle Greene Robertson, 173–177. San Francisco, Pre-Columbian Research Institute.

Aveni, Anthony F., and Horst Hartung. 1986. Maya City Planning and the Calendar. *Transactions of the American Philosophical Society* 76:1–84. Philadelphia.

Aveni, Anthony F., and Lorren D. Hotaling. 1996. Monumental Inscriptions and the Observational Basis of Mayan Planetary Astronomy. In *Eighth Palenque Round Table, 1993*, ed. Merle Greene Robertson, 357–367. San Francisco, Pre-Columbian Research Institute.

Barnhart, Edwin L. 2000. The Palenque Mapping Project: Settlement and Urbanism at an Ancient Maya City. Paper presented to the faculty of the Graduate School of the University of Texas at Austin.

Bassett, Leigh. 1999. Maya Date Calculator. P.O. Box 2509, Laurel, Maryland 20709.

Bassie-Sweet, Karen. 1991. *From the Mouth of the Dark Cave*. Norman, University of Oklahoma Press.

Baudez, Claude-François. 1996. The Cross Group at Palenque. In *Eighth Palenque Round Table, 1993*, ed. Merle Greene Robertson, 121–128. San Francisco, Pre-Columbian Research Institute.

Carlson, John B. 1976. Astronomical Investigations and Site Orientation Influences at Palenque. In *The Art, Iconography and Dynastic History of Palenque*, Part 3, ed. Merle Greene Robertson, 107–117. The Robert Louis Stevenson School, Pebble Beach, California.

Chan Chi, Felipe, and Daniel E. Ayala Garza. 2003. Observaciones de un arqueoastronomo en Dzibilchaltun: Tras los pasos del arqueólogo Victor Segovia Pinto. Paper presented at the 13th Encuentro Internacional, Los Investigadores de la Cultura Maya, Universidad Autonomía de Campeche, Campeche, Mexico.

Christenson, Allen J. 2003. *Popol Vuh: The Sacred Book of the Maya*. Winchester, United Kingdom, O Books.

Coggins, Clemency. 1996. Creation Religion and the Numbers at Teotihuacan and Izapa. *Res* 29/30:16–38.

Eliade, Mircea. 1958. *Patterns in Comparative Religion*. New York, Sheed and Ward.

Fernández, Miguel Angel. 1991. *Palenque, 1926–1945*. Comp. Roberto García Mol. Mexico D.F., Instituto Nacional de Antropología e Historia.

Freidel, David, Linda Schele, and Joy Parker. 1993. *Maya Cosmos: Three Thousand Years on the Shaman's Path*. New York, William Morrow.

Gossen, Gary H. 1974. *Chamulas in the World of the Sun: Time and Space in the Maya Oral Tradition*. Prospect Heights, IL, Waveland Press.

Grube, Nikolai. 2001. *Mayas, una civilización milenaria*. Cologne, Konemann Verlagsgesellschaft mbH.

Hanna, William F., Claude E. Petrone, Roger L. Helmandollar, Alfonso Morales C., and Leon Langan. 1996. *Geophysical Surveys at Mayan Archaeological Sites: Palenque, Chiapas, Mexico*. Hanna GPR Report, San Francisco. Precolumbian Art Research Institute.

Harrison, Peter. 1994. Spatial Geometry and Logic. In *Proceedings of the Seventh Palenque Round Table*, ed. Merle Greene Robertson, 243–252. San Francisco, Pre-Columbian Art Research Institute.

Hartung, Horst. 1980. Certain Visual Relations in the Palace at Palenque. In *Third Palenque Roundtable, 1978*, ed. Merle Greene Robertson, 74–80. Austin, University of Texas Press.

Houston, Stephen. 1996. Symbolic Sweatbaths of the Maya: Architectural Meaning in the Cross Group at Palenque, Mexico. *Latin American Antiquity* 7 (2):132–151.

Karasik, Carol, ed. 1996. *Mayan Tales from Zinacantán: Dreams and Stories from the People of the Bat*. Collected and trans. Robert M. Laughlin. Washington, DC, Smithsonian Institution Press.

Kelley, David H. 1976. *Deciphering the Maya Script*. Austin, University of Texas Press.

Kelley, David H. 1980. Astronomical Identities of Mesoamerican Gods. *Archaeoastronomy* 2 (Supplement to the *Journal for the History of Astronomy*):S1–S54.

Krupp, Edwin C. 1983. *Echoes of the Ancient Skies: The Astronomy of Lost Civilizations*. New York, Harper and Row.

Lounsbury, Floyd. 1978. Maya Numeration, Computation and Calendrical Astronomy. In *Dictionary of Scientific Biography* 15, Supplement 1, ed. Charles Coulston-Gillispie. New York, Charles Scribner's Sons.

Lounsbury, Floyd. 1980. Some Problems in the Interpretation of the Mythological Portion of the Hieroglyphic Text of the Temple of the Cross at Palenque. In *Third Palenque Roundtable, 1978*, Part 2, ed. Merle Greene Robertson, 99–115. Austin, University of Texas Press.

Lounsbury, Floyd. 1985. The Identities of the Mythological Figures in the Cross Group Inscriptions of Palenque. In *Fourth Palenque Round Table, 1980*, ed. Merle Greene Robertson, 45–58. San Francisco, Pre-Columbian Art Research Institute.

Lounsbury, Floyd. 1989. A Palenque King and the Planet Jupiter. In *World Archaeoastronomy: Selected Papers from the Second Oxford International Conference on Archaeoastronomy*, ed. Anthony F. Aveni, 246–259. Cambridge, Cambridge University Press.

Malmström, Vincent H. 1997. *Cycles of the Sun, Mysteries of the Moon: The Calendar in Mesoamerican Civilization*. Austin, University of Texas Press.

Martin, Simon, and Nikolai Grube. 2000. *Chronicle of the Maya Kings and Queens*. London, Thames and Hudson.

Milbrath, Susan. 1988. Representación y orientación astronómica en la arquitectura de Chichén Itzá. *Boletín de las Escuela de Ciencias Antropológicas de la Universidad de Yucatán* 89: 25–37.

Milbrath, Susan. 1999. *Star Gods of the Maya: Astronomy in Art, Folklore, and Calendars*. Austin, University of Texas Press.

Miller, Mary, and Karl Taube. 1993. *An Illustrated Dictionary of the Gods and Symbols of Ancient Mexico and the Maya*. London, Thames and Hudson.

Montgomery, John. 2002. *Dictionary of Maya Hieroglyphs*. New York, Hippocrene Books.

Powell, Christopher. 1996. A New View of Maya Astronomy. Unpublished master's thesis. University of Texas, Austin.

Robertson, Merle Greene. 1991. *The Sculptures of Palenque: IV. The Cross Group, The North Group, The Olvidado, and Other Pieces*. Princeton, NJ, Princeton University Press.

Schele, Linda, and David Freidel. 1990. *A Forest of Kings: The Untold Story of the Ancient Maya*. New York, William Morrow.

Schlak, Arthur. 1996. Venus, Mercury, and the Sun: GI, GII, and GIII of the Palenque Triad. *Res* 29/30:180–202.

Stuart, David. 2006. *Sourcebook for the 30th Maya Meetings*. Mesoamerica Center, Department of Art and Art History, University of Texas at Austin.

Taube, Karl Andreas. 1992. *The Major Gods of Ancient Yucatan*. Washington, DC, Dumbarton Oaks Research Library and Collection.

Tedlock, Barbara. 1992. The Road of Light: Theory and Practice of Mayan Skywatching. In *The Sky in Mayan Literature*, ed. Anthony F. Aveni, 18–42. New York, Oxford University Press.

Tedlock, Dennis. 1985. *Popol Vuh*. New York, Simon and Schuster.

Tedlock, Dennis. 1992. Myth, Math, and the Problem of Correlation in Mayan Books. In *The Sky in Mayan Literature*, ed. Anthony F. Aveni, 247–273. New York, Oxford University Press.

Vogt, Evon Z. 1990. *The Zinacantecos of Mexico: A Modern Way of Life*. Fort Worth, TX, Holt, Rinehart and Winston.

Wald, Robert F. 1999. *A Palenque Triad*, ed. Peter Keeler. Austin, TX, Maya Workshop Foundation.

Zeilik, Michael. 1985. Ethnoastronomy of the Historic Pueblos: I. Calendrical Sun Watching. *Archaeoastronomy* 8 (Supplement to the *Journal for the History of Astronomy*):S1–S24.

CHAPTER THIRTEEN

Astronomy, Ritual, and the Interpretation of Maya "E-Group" Architectural Assemblages

James J. Aimers and Prudence M. Rice

In 1924, the archaeologist Frans Blom described a distinctive cluster of structures in the northeastern portion of the Lowland Maya site of Uaxactun, Guatemala (Fig. 13.1; Blom 1924; Ricketson 1928a; Ricketson and Ricketson 1937; Ruppert 1940). Labeled "Group E," this configuration consisted of a western pyramid (Structure E-VII) opposite three north-south aligned structures (Structures E-I, E-II, and E-III) on a low platform defining the eastern edge of the plaza (Blom 1924). Blom recognized that the assemblage was precisely oriented on an east-west axis, and further investigation convinced him that the three eastern structures marked the position of the sun at sunrise on the equinoxes and solstices when viewed from the western pyramid. Therefore, Blom concluded that the E-group complex was a solar observatory. These considerations, combined with the remarkable preservation of Structure E-VII-sub—at that time the earliest known structure in the Maya Lowlands—made

---

Reproduced by permission from *Ancient Mesoamerica* 17:1 (2006):79–96.

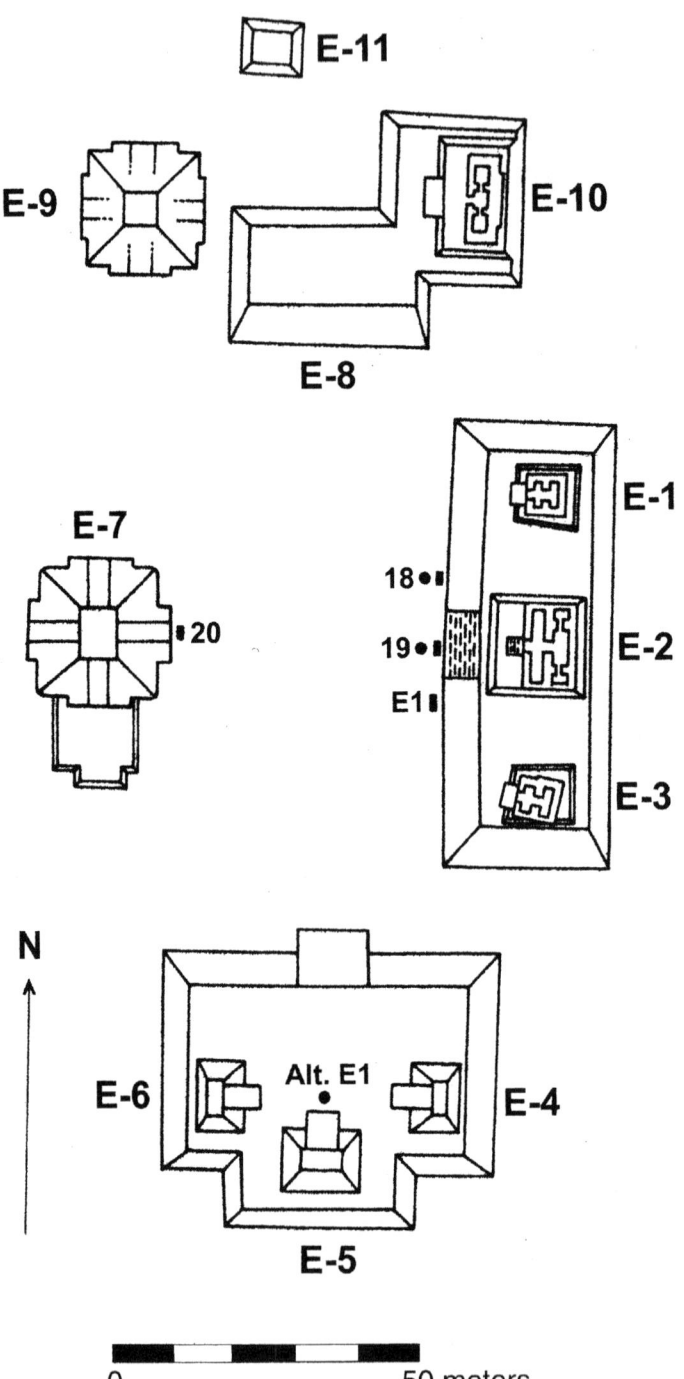

13.1 Group E, Uaxactun, Guatemala (after Rice 2004:Figure 4.3).

Group E at Uaxactun one of the earliest and best-known architectural assemblages in the Maya world.

Since Blom's work, E-group arrangements have been identified, often in variant forms, throughout the lowlands and elsewhere in Mesoamerica (Table 13.1), and these date from the Middle Preclassic (ca. 700–400 BC) through the Terminal Classic (ca. AD 800–950) period. Most generally, the Lowland Maya E-group arrangement consists of a small platform on the west side of a plaza opposite the central of three temples on a north-south-oriented platform to the east. The western structure is often "radial" (Cohodas 1980): a tiered platform, square in plan, with stairways on all four sides, usually without a masonry superstructure.

As more E-group assemblages have been investigated, their astronomical meaning and function have been questioned, and substantial evidence from excavation and epigraphy now can be incorporated into their interpretation. Here we review available data on E-groups, including their history, construction variants, and theories about their functions. We conclude that E-groups were not precise timekeeping instruments in stone but functioned more symbolically as settings for large-scale ritual concerning the solar cycle, the sociopolitical and religious role of which may have varied through time and space.

HISTORY AND GEOGRAPHICAL DISTRIBUTION

Temporal Distribution

The earliest known Lowland Maya E-groups are at Uaxactun, Tikal, El Mirador, Nakbe, and Güiro/Wakna and date to the Preclassic period (see Chase 1983:1245, 1985:36; Coe 1965:23). Middle Preclassic structures in the East Plaza of Tikal's Mundo Perdido, or "Lost World," complex (Fig. 13.2) might be the earliest known examples of this presumed solar observatory configuration in the lowlands (Fialko 1988; Laporte and Fialko 1990, 1995). This arrangement consists of a radial platform, Structure 5C-54-1st, on the west side of a plaza facing a north-south elongated platform with the three temples of Structure 5D-84/88-1st to the east. The first (ca. 700–600 BC) building episode of these structures was simple: a radial structure situated opposite but off-center a long, narrow, north-south mound with stairs on the center line of each side. These were later overbuilt during the late Middle Preclassic period (Tzek ceramic complex and phase, 500–400 BC), and remodeled in the early Late Preclassic (Chuen, 400–200 BC).

Although archaeologists associate E-groups with the Lowland Maya, they are not uncommon elsewhere in Mesoamerica, particularly in the isthmian area (Clark and Hansen 2001). In highland Chiapas, in the lower Central Depression of the Río Grijalva, Escalera-phase (ca. 600–450 BC) civic-ceremonial architecture commonly includes a north-south linear mound, sometimes in cruciform shape, with a centered platform to the west. These structural pairs have been found at "San Isidro, Mirador, Tzutzuculi, La Libertad, Chiapa de Corzo, Finca Acapulco, Ocozocoautla,

Table 13.1. Lowland Maya sites with "E-group" complexes and variants

Central Petén
1. Cenote (Chase 1983:1236–1254)
2. Cerro Ortiz
3. Chachaclun (Chase 1983)
4. Chalpate (Lou 1997) Bajo La Justa (Holtun, La Tractorada; see Grazioso et al. 2001)
5. Dos Aguadas (Aimers 1993; moved from Western Belize list)
6. Naranjo
7. Paxcaman (Chase 1983:1236–1254)
8. Tayasal (Chase 1983:1236–1254)
9. Tikal (see Laporte and Fialko 1990, 1995)
10. Uaxactun (Blom 1924; Ricketson 1928a)

Northeastern Petén
11. El Venado (Aimers 1993)
12. Holmul (Jason Gonzáles, personal communication 2001)
13. Ixtinto (Acevedo et al. 1996:238)
14. Nakum (Ruppert 1977 [1940]; Tozzer 1913)
15. Río Azul (Aimers 1993)
16. Xultun (Ruppert 1977 [1940])
17. Yaxha (N = 2; Ruppert 1977 [1940])

Southeastern Petén (Mopan-Dolores region; Corzo et al. 1998:193–194; Laporte 1996:255; Laporte and Mejía 2002; Mejía et al. 1998; Ruppert 1977 [1940])
18. Buenos Aires
19. Dos Hermanas
20. El Camalote/Melchor
21. El Chal
22. El Naranjal
23. Ixkun
24. Ixtonton
25. Ixtutz
26. La Providencia 1
27. Mopan 3-East
28. Sacul 3
29. Ucanal

Southern Petén
30. Machaquila (Graham 1967)
31. Seibal

Calakmul/El Mirador Region
32. Balakbal (Ruppert 1977 [1940])
33. Calakmul (Ruppert 1977 [1940])
34. El Mirador (Hansen 1992:84)
35. Güiro/Wakna (Hansen 1992:84)
36. La Muñeca (Ruppert 1977 [1940])
37. Naachtun (Ruppert 1977 [1940])
38. Nakbe (Hansen 1992:84)
39. Oxpemul (Ruppert 1977 [1940])
40. Río Bec II (Ruppert 1977 [1940])
41. Tintal (Hansen 1992:84)
42. Uxul (Ruppert 1977 [1940])

Western and Northern Belize
43. Actuncan (Laporte and Mejía 2002:7)
44. Arenal (Aimers 1993)
45. Baking Pot (Aimers 1993)
46. Barton Ramie (Aimers 1993)
47. Blackman Eddy (Garber et al. 2001)
48. Cahal Pech (Clark and Hansen 2001:43)
49. Cahal Pichik (Ruppert 1977 [1940])
50. Caracol (Chase and Chase 1995)
51. Colha (Aimers 1993)
52. Cuello (Aimers 1993)
53. El Pilar (Aimers 1993)
54. Hatzcap Ceel (Ruppert 1977 [1940])
55. Nohmul (Aimers 1993)
56. Pacbitun (Healy 1990)
57. San José (Ruppert 1977 [1940])
58. Xunantunich (Ruppert 1977 [1940])

Yucatan
59. Acanceh (Coggins 1983:3–38)
60. Chel (Anthony Andrews, personal communication 2002)
61. Dzibilchaltun (Andrews and Andrews 1980)
62. Kabah (Andrews 1975)
63. Santa Rosa Xtampak (Coggins 1983:37–38)
64. Yaxuna (David Freidel, personal communication 1999)

More Distant Sites
Comalcalco, Tabasco, Mexico (Aimers 1993; Andrews 1975)
La Florida or El Naranjo Frontera, Petén (Morales 1998)
Uaxac Canal, Baja Verapaz, Guatemala (Ruppert 1977 [1940])
Valley of Guatemala (Kaminaljuyu, Rincon, Rosario-Naranjo, Cruz de Cotio, San Isidro II, Las Charcas; Valdés 1997:183)
Chiapas, Mexico (San Isidro, Mirador, Tzutzuculi, La Libertad, Chiapa de Corzo, Finca Acapulco, Ocozocoautla, Vistahermosa, and five or more other sites; Lee 1989:207, 225; Lowe 1977:224)

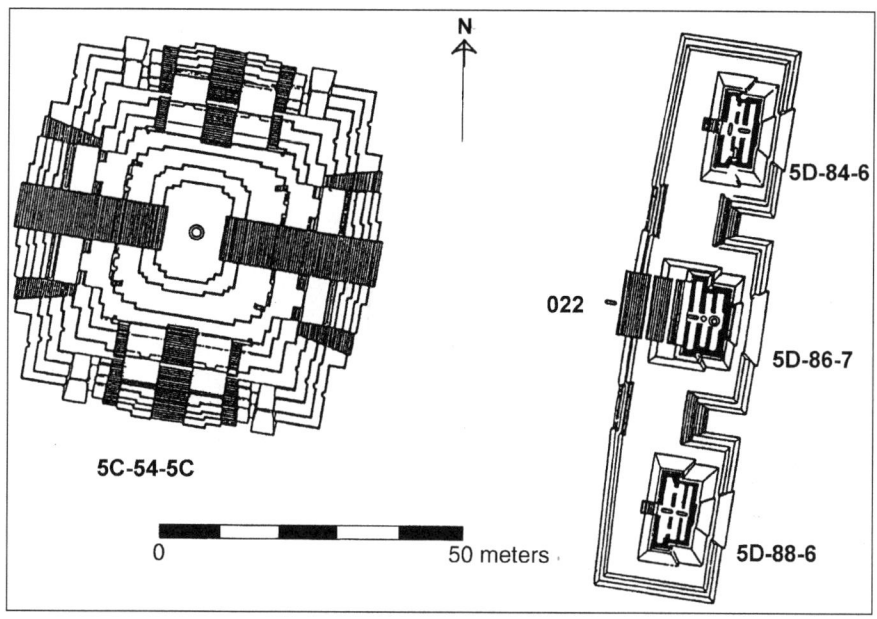

13.2 Mundo Perdido, Tikal, Guatemala (after Rice 2004: Figure 4.4).

Vistahermosa and five or more other sites" in Chiapas (Lee 1989:207, 225; Lowe 1977:224).

The linear-mound-plus-western-platform combination is unusual in the Gulf coastal region, but a similar arrangement can be found at the southern end of the Main Plaza at La Venta, Tabasco, Mexico (Fig. 13.3; Drucker et al. 1959; Reilly 1999): Structure D-8, formerly referred to as the "Long Mound" in Group B, is a long, narrow mound with a north-south axis. A low truncated conical structure (D-1) sits to the west of Mound D-8 on the central axis. A basalt column (Monument 49) was found set into the southern end of the north-south linear Structure D-8, and Philip Drucker (1952:9; see also Lee 1989:fig. 4.10) suggested that three such columns might originally have been set in a line on top of this mound. If so, the three columns might have marked sight lines from the western platform (Structure D-1) to sunrise on the solstices and equinoxes. The linear-mound-plus-platform arrangements found in Middle Preclassic–period (or Formative-period) Chiapas and elsewhere, including the early one at Tikal, originally might have supported similar markers—perhaps smaller stone columns or perishable wooden poles—placed to identify sunrise sight lines or the solar zeniths. Such arrangements could have functioned as early observatory complexes.

Elsewhere, a linear-mound-with-western-platform arrangement appears in the eastern part of the site of Tlalancaleca, in Puebla in the Mexican highlands (García Cook 1981:251). The date of this complex is not certain, although it is likely to

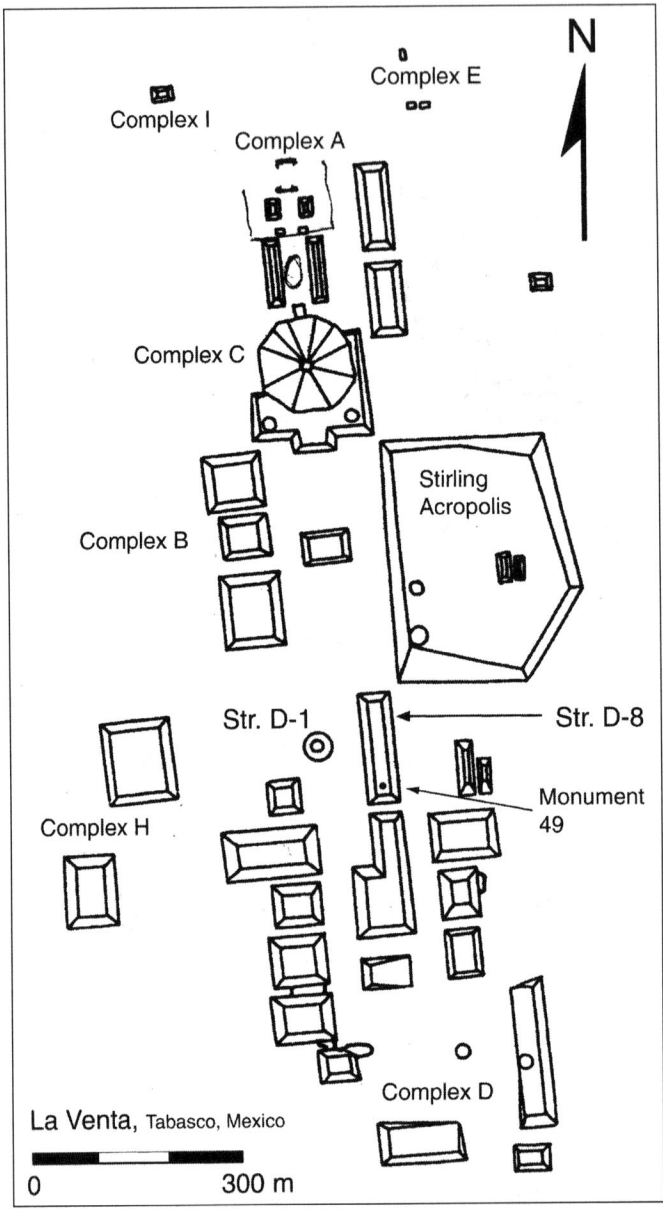

13.3 La Venta, Tabasco, Mexico (after Reilly 1999:Figure 1.1b).

date to the period 800–400/300 BC. E-group arrangements also have been found in the Late Preclassic/Formative period at several Highland Maya sites in the Valley of Guatemala, including Kaminaljuyu, Rincon, Rosario-Naranjo, Cruz de Cotió, San Isidro II, and Las Charcas (Valdés 1997:83, citing Carson Murdy).

In the lowlands, the site of Nakbe has an E-group complex that may date to the late Middle Preclassic period, contemporary with that at Tikal (Hansen 2000). The Tigre Complex E-group at El Mirador seems to date from the Late Preclassic, as do the E-groups at Cenote and Paxcaman (Chase 1985:37), Colha (Hester and Eaton 1982; Hester et al. 1980), Pacbitun (Healy 1990), and a possible E-group variant at Cahal Pech (Awe and Campbell 1988). The latest E-group, according to available evidence, appears to be that at Nohmul, with a construction period sometime in the Terminal Classic or Early Postclassic period (Hammond 1985:47). Thus, the construction of E-groups in general appears to have spanned most of Preclassic and Classic Maya history.

Spatial Patterning

The currently known spatial distribution of E-groups (Fig. 13.4) is wide and probably incomplete, and it provides little information about the significance of this complex, aside from an indication that it was predominantly a Lowland Maya phenomenon (Ruppert 1940:224) and not restricted to large sites. Typically, there is only one E-group at a site, although the medium-size site of Yaxha has E-groups in both Plaza F and Plaza C (Hellmuth 1971; Heyden and Gendrop 1980:52, 93, 137). Clemency Coggins (1983; also Coggins and Drucker 1988) has suggested that there are also two E-group assemblages at Dzibilchaltun.

The similarity in shape between the massive, centrally located Ciudadela at Teotihuacan and the Maya E-group complex has been noted by Juan Pedro Laporte and Vilma Fialko (1990:59; Laporte 2003:215) and others (Cabrera Castro 2000; Morante López 1996). We, however, are not yet persuaded that the resemblance was intentional and agree with George Cowgill (2003:323), who comments, "Seeing a resemblance requires one to ignore the North and South Platforms and everything else in the Ciudadela. If the Ciudadela had been intended as a place to enact the practices connected with Maya astronomical groups, I would expect it to have looked far more like Maya examples." Although evidence beyond form has not yet been offered, and processes that might explain the resemblance are unclear, the argument warrants consideration because interaction between the Maya area and Teotihuacan is well established (see papers in Braswell 2003), and the origins of Teotihuacan's architectural style increasingly appear eclectic (Demarest and Foias 1993).

E-Group Variants

Arlen Chase (1983:1301; see also Chase and Chase 1995) categorized known E-groups into three types on the basis of their formal characteristics. In Chase's first type, exemplified by the assemblage at Uaxactun (Fig. 13.1), the three eastern structures are of approximately equal size and sit on a single platform. In the second type (the "Cenote Style"), this portion of the grouping is less consistent from north to south, and the central section tends to be larger or to have an easterly extension in

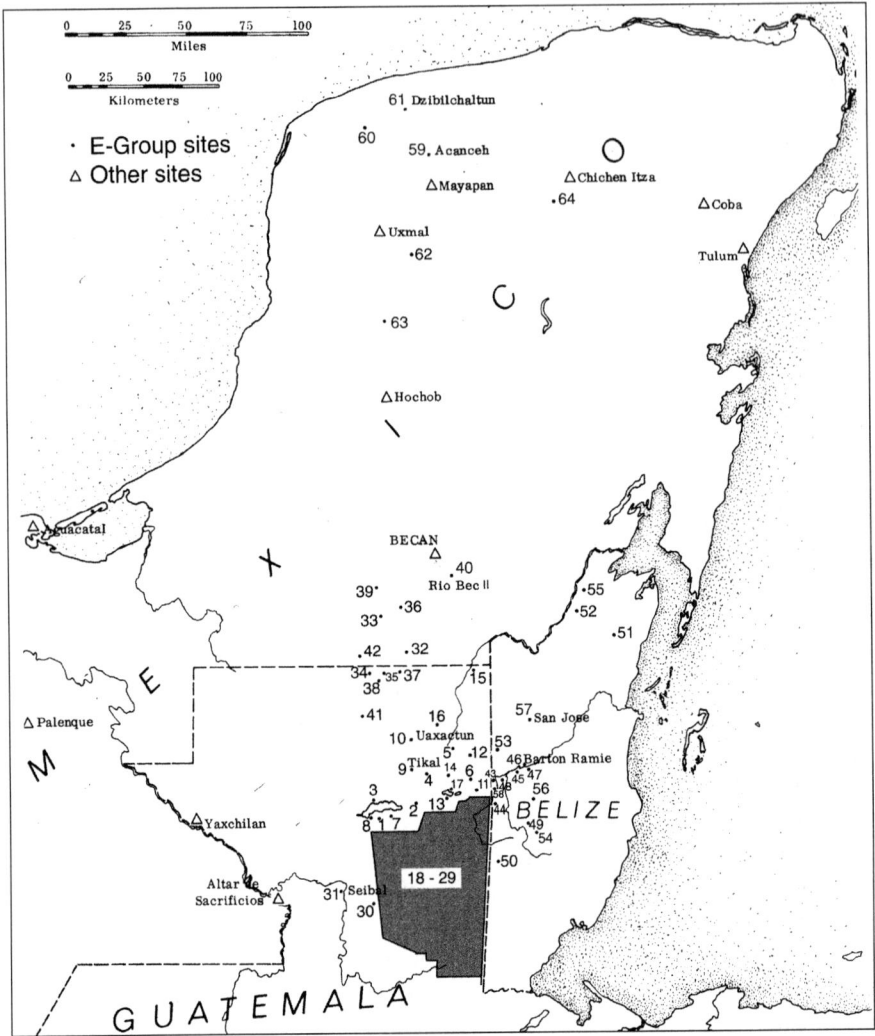

13.4 Map of the Maya area showing E-group sites.

comparison with relatively small southern and northern sections (Fig. 13.5). Members of Chase's third type (the "Cenote Variant") seem to be defined primarily as anomalies, having significant morphological differences from the two more easily recognizable types described earlier.

In an earlier study, James Aimers (1993:figs. 13–15) plotted the spatial distribution of 45 Lowland Maya E-groups classified according to Chase's types, but no obvious geographical patterning was apparent. None of the types defined by Chase appear to be more common at larger sites; nor did specific environments appear to have distinctive E-group types (e.g., Belize Valley, riverine locations, Peten). In addi-

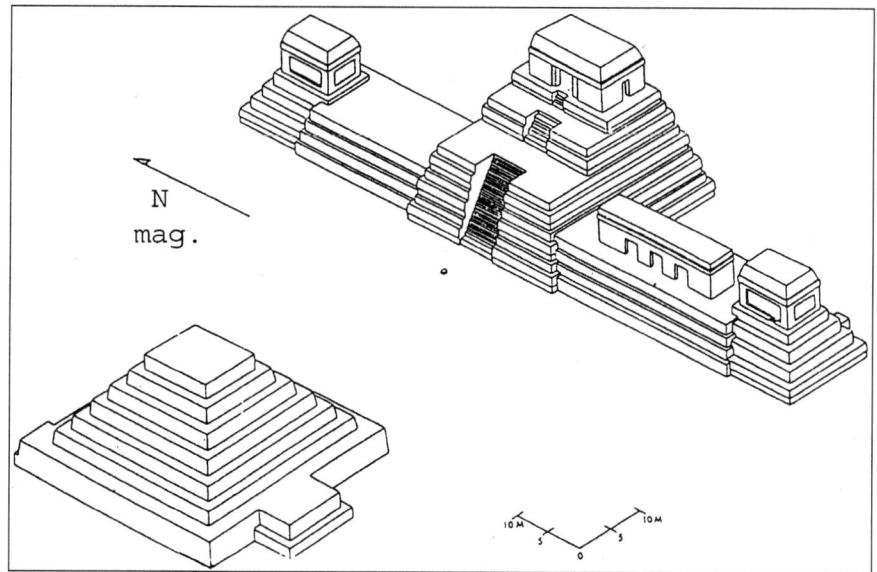

13.5 Cenote, Guatemala (after Chase 1985:39).

tion, examination of site and structure plans provided little indication of strict rules regarding the morphology of the E-group. Each site seemed to be a variation on a recognizable theme. Despite this variability, however, the wide distribution of E-groups as a distinctive civic-ceremonial architectural complex suggests a network of shared beliefs and ritual in Mesoamerica from at least the Middle Preclassic period or earlier (see also Aveni et al. 2003; Culbert 1991).

ASTRONOMY AND RITUAL IN THE INTERPRETATION OF THE MAYA E-GROUP

The Solar Observatory Hypothesis

Blom (1924) suggested that the Late Preclassic E-group configuration at Uaxactun functioned as a solar-seasonal observatory complex marking the dates of the solstices: From a viewing position on the western radial structure, sunrise at the summer solstice occurs over the northern temple, and sunrise at the winter solstice occurs over the southern structure. Later, Oliver Ricketson (1928a, 1928b) suggested that on the equinoxes the sun would rise behind the central eastern structure, but he expressed some doubt as to the astronomical significance of its orientation:

> One question ... can be appropriately raised at this time, and that is whether this complex of buildings is a true observatory, or planned to mark the already known directions of the four significant annual positions of the sun? The writer is strongly in favor of the latter theory.... The writer believes that these buildings were erected

in their respective positions as temples dedicated to the four seasons, or the four most significant positions of the sun in the course of the solar year, and that their erection is to be more closely associated with geomancy than with astronomy. (Ricketson 1928b:439–440)

Despite this caveat, Ricketson compared Group E to the Caracol at Chichen Itza, a round structure typically characterized as an observatory (see Aveni 2001:273–282 for a full discussion; see also Aveni et al. 1975). As a result of this ambiguity, in only four years Group E at Uaxactun was established in the literature as a bona fide astronomical observatory rather than one that merely marked the position of already known celestial phenomena.

Here it is appropriate to clarify what we mean by the term *observatory* in attributing function to E-groups. Anthony Aveni, Anne Dowd, and Benjamin Vining (2003:172) recently argued that, "if seeing the sun [rising points] can be shown to have been a part of the [building] scheme, then regardless of whether the Maya were watching it scientifically or ceremonially, the associated architectural complex may be regarded as an observatory." We, however, prefer a definition aligned more closely with Western scientific usage, which postulates an observatory function as the primary purpose of the structure or complex. By this thinking, if E-group structural arrangements as originally constructed provided accurate sightlines to solstitial sunrises, and then were modified over time in ways that compromised this accuracy, the complex ceased to function as an observatory sensu stricto.

The confusion engendered by Ricketson's 1928 articles is evident in Karl Ruppert's key discussion of E-groups, published more than a decade later. Ruppert (1977 [1940]:222) reported 19 examples in a relatively circumscribed area of northeastern Peten, southern Quintana Roo, and Belize; 13 "in almost pure form" like that of Uaxactun; and six that were "less clear." He concluded that the Uaxactun observational complex developed first, and the other assemblages, which he assumed to be later, functioned merely as settings for derived ritual.

This temporal and functional dichotomy between astronomical and ritual complexes has influenced every subsequent discussion of E-groups and is evident in alternative terms for the E-group arrangement. These include *Complejo Conmemorativo Astronómico* (Fialko 1988) and *Complejo de Ritual Público* (Laporte 1996). Here we maintain the original, functionally neutral term *E-group* to refer to this architectural complex.

Astronomy and Scheduling

Astronomical accuracy is emphasized most strongly in interpretations that suggest E-groups were "calendars in stone" constructed to assist farmers in scheduling agricultural activities (e.g., Rathje 1972, 1978). This is something of a straw man, set up and knocked down by most scholars dealing with the practical implications of the Maya calendar. Ethnographic observation shows that Maya farmers would not need to be told when to carry out the various activities related to the agricultural cycle; weather and visible growth cycles clearly indicate when it is time to perform certain tasks (Redfield and Villa Rojas 1934:44; Redfield and Warner 1940). The assemblages may still commemorate important agricultural dates, however.

William Rathje (1972:233, 1978) also proposed that E-groups developed as precise timekeeping mechanisms to schedule trade in the resource-poor lowlands. Rathje's spatial-distribution model was based on Ruppert's work of 1940, and his hypothesis resulted from the application of a theoretical model derived from economic geography to a body of data to which it may simply be inappropriate. It is now evident, for example, that the E-group has a much wider occurrence than earlier thought. The most significant problem with Rathje's hypothesis was his assumption that E-groups were accurate timekeeping devices, which has never been demonstrated.

A related, but more powerful, consideration comes from recent studies by Aveni and colleagues (2003) who, following previous investigations (Aveni and Hartung 1986), suggest that the E-group alignments were targeted toward observations of the solar zeniths. In particular, they note intervals of multiples of 20 days leading to the first solar zenith, occurring on May 10 at the central Peten latitude of circa 17°31'N. These would have been marked, they argue, in architectural complexes where rituals might have been carried out as "anticipatory sun sightings during the interval leading up to the planting season" (Aveni et al. 2003:162–163).

Ritual and Geomancy

At the opposite side of the interpretive spectrum, John Carlson (1981) suggested that a system of geomancy, not unlike that practiced by the ancient Chinese (and still in use), might explain some of the irregularities of Maya sites, and of the E-group complex in particular. Just as ancient astronomers tried to make sense of the night sky and the movement of the sun through observation and attendant ritual, a system of geomancy "explains" physical geography and provides a system of rules for manipulating the perceived power of the landscape. In fact, geomancy can be considered a terrestrial version of astrology (Wheatley 1971). A tendency to see the landscape as alive and sacred has been documented in many societies (see, for example, Howard 1986:350; Hugh-Jones 1979:235; Thomas 1990:169), including the Maya (see Scully 1991:2–17; Townsend 1982; Vogt 1981). Carlson (1981:188) suggested that, aside from Uaxactun, "other Group E–type structures may align to topographical or other celestial features," an idea that otherwise had been overlooked since it was raised by Ricketson.

TESTING THE SOLAR-OBSERVATORY HYPOTHESIS

Equinoxes

Although many archaeoastronomical investigations are complicated by the changing positions of celestial bodies over time, the position of the sun at particular times of the year remains relatively constant over thousands of years. This means that the solar-observatory hypothesis can be tested. Because the sun rises due east on the equinoxes, if the Maya were interested in accurately "monumentalizing" the position of the sun on these days, the assemblages should have a fairly precise east-west alignment.

Although many older site plans are conventionalized, and some do not distinguish true from magnetic north, examination of site plans suggests that the rising of the sun on the equinoxes would not coincide with the precise center of the eastern structures in the great majority of 45 cases examined (Aimers 1993:table 4), even allowing for a large margin of error in the accuracy of the maps. Eliminating five conventionalized maps from the calculations, fewer than 25 percent (9 of 40) of the assemblages are within 5° of due east. Important for these conclusions is that the sun on the horizon appears to have a width of only half a degree; thus, a very slight skew will be significant (Carlson 1974:108; Flower 1990:14). Even at Uaxactun, the equinoctial alignment is not precise, both because of a deviation of the axis from due east and because in the tropics, the sun, after rising, appears to travel diagonally at an angle of approximately 15° south of vertical.

The sun on the horizon appears to have a diameter of approximately one half of a degree, and it takes only a few days for the sun to move 2° (or about four times its apparent width) along the horizon near the equinoxes. This means that a truly accurate alignment for an observation of the equinoctial rising of the sun would have the architectural configuration aligned within 2° north or south of due east (which is to say, a greater deviation from due east would lead to a large discrepancy in determining the date of the equinox). Of the E-groups examined, only El Venado, Uaxactun, and Ucanal have this characteristic, and the map for El Venado is conventionalized. This is a significant indication that the majority of these structures were not created rigorously to mark equinoctial sunrise positions. Of course, this does not rule out alignments with other celestial phenomena, including the solar zeniths (see Aveni et al. 2003).

Solstices

Near the solstices, the sun moves a distance equal to its own diameter in about 10 days (Aveni and Hartung 1989:459). Thus, architectural alignments must be extremely accurate to serve as precise solstitial markers. Furthermore, at the latitude at which these configurations are located, the angle formed between true east and a vector connecting a given point on the true east line to the position of the sun at the solstice sunrise is approximately 24.5° (Broda 1982:81). Examination of site maps revealed that none of the assemblages has the precise angle required (Aimers 1993).

If the viewer of the sunrise changes vertical position in the E-group, the apparent position of the sun on the horizon also changes, and this could compensate for otherwise inaccurate alignments: "By varying the height of the observer relative to the foresight, whether it be a building or a portion of the horizon, it may be possible to account for many of the orientations skewed 0 degrees to 10 degrees eastward from the cardinal points by sunrise observations at the equinox" (Aveni 1980:249). Nevertheless, the orientations of several of the E-group structures appear to be too inaccurate to be accounted for by this hypothesis (e.g., Pacbitun). Although the limitations of site plans have been noted, many of the plans indicate a skew in the alignments of more than 10° (Aimers 1993:table 4).

Horizontal changes in the position of an observer on the west (radial) building will also change the apparent positions of the southeastern and northeastern buildings necessary for alignments with the solstitial sunrises. Several assemblages can be eliminated as possible solstitial observatories (on the basis of maps) because it is not possible to observe the rising of the sun over the eastern structures: Baking Pot, Cahal Pech, Colha, Comalcalco, El Venado, Kabah, Lamanai, La Muñeca, Naachtun, Río Bec, and Xultun. San José has the correct alignment for the summer solstice, but not for the winter solstice.

In 1991, Aimers and colleagues conducted field observations of sunrise in E-groups at six sites on and near the summer solstice. At the site of Baking Pot, the sun appeared in an appropriate place when viewed from the center of the western structure at 5:55 AM on June 22. The Blackman Eddy assemblage "worked" satisfactorily, with the sun appearing in the appropriate position at approximately 6 AM on June 22. Cahal Pech is unusual in that it does not have a western structure conforming to the typical radial structure associated with other E-groups. There, sunrise was observed from the stairs of the range structure delimiting the western edge of the plaza. At 6:45 AM, the sun appeared somewhat to the south of the center of the southeastern structure. At Pacbitun, the sun appeared at 5:50 AM in a position approximately halfway up the slope of the northern side of the central eastern structure on June 22. Observers at Xunantunich estimated the sun to have first appeared shortly after 6 AM on June 22, close to the southern edge of the northeastern structure. At Yaxha Plaza C, the sun appeared at 6:45 AM on June 20 near the southern edge of the northeastern structure. In sum, only two of the six E-group assemblages accurately marked the position of the sun at sunrise on the summer solstice: Baking Pot and Blackman Eddy. Further, Cahal Pech should probably be eliminated, as it does not have a western structure. These observations do not support the persistent belief that E-groups were astronomically accurate markers of the solstices and equinoxes.

Accuracy

It is possible that these E-group assemblages were planned incorrectly. For example, there is evidence that the Templo Mayor at Tenochtitlan was planned incorrectly and did not accurately mark the sunrise on the equinoxes (Aveni 2003:159; Aveni et al. 1975:985). However, it is difficult to accept the proposition that Maya astronomers, who had "succeeded in tabulating the motion of Venus to .08 part of a day in 481 years" (Aveni 1980:191), would have been unable to produce accurate alignments, and it is particularly hard to imagine that Maya architectural planners *repeatedly* would have produced incorrect alignments. Furthermore, although large masonry observatories are known in the Maya area (e.g., the Caracol at Chichen Itza), massive architectural assemblages would be an unnecessarily time-consuming and labor-intensive instrument for astronomical observation: "Four perpendicular sticks a few inches high, correctly set up on a properly oriented board, would have served the same purpose" (Ricketson (1928b:440). The use of sticks for solar observation is widespread in the New World (particularly in Mesoamerica), and there may be iconographic evidence of such an instrument (Aveni 1980:62–66, 286–311, 2001:20–21; see also Coggins 1983; Digby 1974). In other words, it seems more likely that precise architectural indicators of solstice and equinox positions were less important to the ancient Maya than they have been to archaeologists.

Less functionally utilitarian factors probably influenced the design of these large, centrally located complexes. For example, E-group architects intentionally might have incorporated adjustments and asymmetries into the eastern structures to create hierophanies (interplays of light and shadow that bespeak sacred mysteries; Aveni et al. 2003:173). In addition, Harold Turner (1979:29) has shown that earthly replicas of divine models are often intentionally imperfect. The imprecision of E-groups as solar markers may be due to planning in which astronomical accuracy could be subordinated to other cognitive schema, as it is in contemporary Mesoamerican communities:

> [P]easants have neither the equipment (for instance, accurate compasses) nor the knowledge of the night-time sky necessary to produce accurate alignments. Furthermore, natural features are set there "by the hands of the gods" and not by human design, so that the architectural model is only a vision, an approximation of abstracted armatures, a compromise between the real position on the earth's surface of natural markers such as mountains, solar or astronomical markers, and man made or imagined markers. (Hunt 1977:204)

This explanation for the imprecision of architectural orientations is also valid cross-culturally (see, for example, Guidoni 1975:92, 154, 161, for similar cases among the Indonesian Nias islanders, the Dogon of Mali, and the West African Hausa).

The siting of an E-group could have been planned through a process in which various types of "numerically patterned phenomena, such as geographic directions, seasons, and celestial events" were linked (Hunt 1977:211; see also Brotherston 1982:110–112; Hunt 1977:212–213 provides examples from the Codex Borgia). Aside from many known systems of geomancy (e.g., Chinese, Japanese, West African [Ife], early medieval British, and Southeast Asian), astronomical numerology in European art and architecture fits this pattern (Hersey 1976). In Western astronomy, Copernicus and Kepler addressed astronomical problems with Neo-Platonic and Pythagorean preconceptions of harmony and order that made their models impossibly inaccurate. Johanna Broda (1989: 494) has suggested that "ancient astronomy should be broadened to include the observation of the natural environment in more general terms," and similar geomantic architectural orientation systems have been suggested for Teotihuacan (Heyden 1981) and Monte Alban (Hartung 1981). In the Maya case, an affinity for this sort of transformational structure is well demonstrated by the complex associations among colors, animals, plants, directions, gods, and seasons, as evident from the archaeological record, ethnohistorical accounts, and ethnography. We do not know which of these myriad factors may have influenced E-group design, but the sun's annual movements along the horizon were apparently not the only factors.

Conclusions about the Solar-Observatory Hypothesis

Anthony Aveni and Horst Hartung (1989) made accurate astronomical measurements at Uaxactun in 1988 and concluded that the E-group there accurately would have marked the position of the sun on the solstices but would have been slightly inaccurate in marking the equinoxes. The solstice sun rises not over the center of the southeastern and northeastern structures but over their southern and northern edges, respectively, and the equinoctial sunrise would be slightly off-center. Due to the slow movement of the sun around the solstices, Aveni noted, "[W]e must caution against concluding that . . . the Group E complex in any sense offered a precise means for determining the solstitial dates" (Aveni and Hartung 1989:445).

Aveni and Hartung (1989:451) further suggested that at Uaxactun the originally functioning horizon-based "observatory" went through a series of architectural elaborations that eventually negated its utility as a precise marker of solar movement. Importantly, due to vertical elaboration of the western structure (Structure E-VII) in its final stages,

> all of the sunrise events would have taken place along a natural horizon that lay well above the level of the platform and its three buildings. It is likely that by this time the complex could not have functioned as a solar observatory in any sense. (Aveni and Hartung 1989:447)

This raises the prospect that originally accurate solar observatories became increasingly inaccurate through modifications undertaken in later eras, perhaps after their functional characteristics had been forgotten or otherwise neglected. The authors suggested that "most E-Group complexes might have been non-functioning copies of the astronomically operational archetype at Uaxactun" (Aveni and Hartung 1989:452; but cf. Aveni et al. 2003 for a rethinking of this idea) and were likely the focus for "ritual and ceremony." This idea returns the interpretation of Maya E-groups to Ricketson's suggestion half a century earlier.

Although the accuracy of the E-groups examined thus far would seem to argue against a strict solar-observatory hypothesis as initially argued by Blom and Ricketson in the 1920s, the orientations of the buildings are also too consistent to ignore. Measurements of various angles in the assemblages (Fig. 13.6) were taken to determine whether there is any consistency in the proportion of the structures (Table 13.2). Notably, there is a significant variation in angles a and b, which supports the idea that these were not functioning solar observatories, yet the angles E and F are

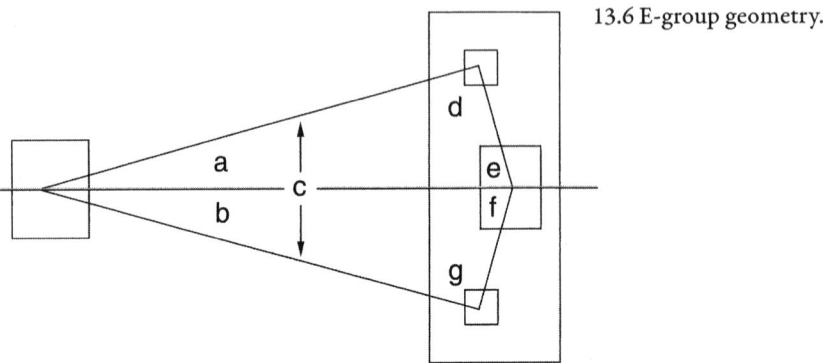

13.6 E-group geometry.

much more consistent in being slightly less than 90° (for approximately 75% of the assemblages). Generally, these measurements suggest that although the geometry and proportion of the overall configuration were maintained rather consistently, the orientation—and, therefore, the possible alignments—vary greatly. The overall geometry and proportion of these structures make them archaeologically recognizable as much as their orientation and may have been more important in their design than accurate solar alignment.

In sum, astronomical measurements and site-plan investigations reveal that none of these sites' E-groups, including that at Uaxactun, are oriented with sufficient precision to have functioned as true observatories for the cyclical positions of the sun. Although it is not possible to fully support Carlson's (1981:188) suggestion that none of the E-groups in the Maya area "have the correct orientation to be functional observatories," it appears that most E-groups were not "correctly" oriented in a functional (i.e., astronomically accurate) sense. The reasons for their construction and use probably differ from the Western notion of "observatory" as a scientific instrument for the precise measurement of celestial movements.

RITUAL CYCLES

Although the reasons for the varied yet consistent orientations of E-groups are still not understood, there are indications of the ritual role of these monumental structures. We suggest that E-groups emerged initially in association with celebrations of annual solar cycling. At some time thereafter, they came to be constructed to celebrate longer calendrical cycles known as katuns and, possibly, longer cycles of 13 katuns, or approximately 256 years, called the *may* (see Rice 2004).

"The Shape of Time": Radial Structures

Monumental architecture is always planned, and the design of large buildings often incorporates symbolic references at various scales, including the overall shape

James J. Aimers and Prudence M. Rice

Table 13.2 E-group geometry

Site	A	B	C	D	E	F	G
1. Baking Pot	11°			86°			
2. Balakbal	23°	24°	47°	87°	78°30'	80°30'	83°30'
3. Barton Ramie	17°	15°	32°	95°	68°	58°	107°
4. Cahal Pichik	20°	20°	40°	84°	76°	40°	120°
5. Calakmul	23°30'	20°	43°30'	60°	96°30'	90°	70°
6. Caracol	12°20'	13°	15°20'	74°30'	93°30'	100°30'	67°
7. Cenote	37°	29°	66°	65°	77°	87°30'	64°
8. Colha	8°30'	10°	18°30'	60°	111°30'	82°	88°
9. Comalcalco	8°	7°	15°	82°	90°	90°	83°
10. Cuello	16°	24°	40°	74°	89°	91°	51°
11. Dos Aguadas	28°30'	29°	57°30'	63°	88°30'	88°30'	62.5°
12. Dzibilchaltun	14°	14°	28°	76°	90°	90°	76°
13. El Mirador	11°	14°	25°	121°30'	47°30'	62°30'	103°30'
14. El Pilar	13°	13°30'	26°30'	76°	91°	82°	84°30'
15. El Venado	10°	10°30'	20°30'	82°	88°	87°30'	72°
16. Hatcap Ceel	24°	25°30'	29°30'	87°	69°	89°30'	65°
17. Ixkun	17°	17°30'	34°30'	104°	59°	57°	105°30'
18. Ixtutz	16°	16°30'	32°30'	71°	85°	78°	84°30'
19. Kabah	12°30'	12°	14°30'	75°	93°	87°	79°30'
20. La Muñeca	15°	16°30'	31°30'	70°	95°	85°30'	78°
21. Naachtun	11°	10°	21°	78°	91°	90°	80°
22. Nakum	15°30'	14°30'	30°	82°30'	82°	94°	71°30'
23. Naranjo	25°	25°	50°	67°	88°30'	91°30'	61°30'
24. Nohmul	17°30'	20°30'	38°	78°	84°30'	80°	79°30'
25. Oxpemul	19°	20°	39°	71°30'	89°30'	86°30'	73°30'
26. Pacbitun	21°	18°30'	39°30'	83°	76°	73°30'	88°
27. Paxcaman	21°30'	25°	46°30'	90°	68°	70°	84°
28. Río Azul	14°	9°	23°	125°	41°	25°30'	145°30'
29. Río Bec	15°	9°	24°	88°	77°	66°	105°
30. San José	32°	22°	54°	103°30'	44°	93°30'	64°
31. Seibal	19°30'	20°	39°30'	90°30'	80°	86°	74°
32. Tayasal	33°30'	29°	62°30'	54°	94°	86°	64°30'
33. Teotihuacan	21°	23°	44°	70°	90°	90°	66°
34. Tikal	18°	17°	35°	74°	88°	89°	74°
35. Uaxactun	17°30'	18°30'	36°	73°	89°30'	90°	71°30'
36. Ucanal	21°30'	20°30'	42°	68°	90°30'	90°	69°30'
37. Uxul	9°30'	9°	18°30'	94°	76°30'	76°30'	94°30'
38. Xultun	11°	11°	22°	77°30'	91°30'	90°	80°20'
39. Xunantunich	15°	13°30'	28°30'	72°	93°	87°	79°30'
40. Yaxha Group C	14°	14°	28°	105°	61°	61°30'	104°30'
41. Yaxha Group F	12°	12°30'	24°30'	76°	92°	90°	77°30'

Source: Aimers 1993:Table 5.

and applied ornamentation. One clue that E-groups were constructed to symbolize cyclical time comes from the radial structures typically situated as the western structure of this group. Because of their four projecting stairways, these structures have the shape of a cross (+) in plan view. Interpretations of Mesoamerican directional

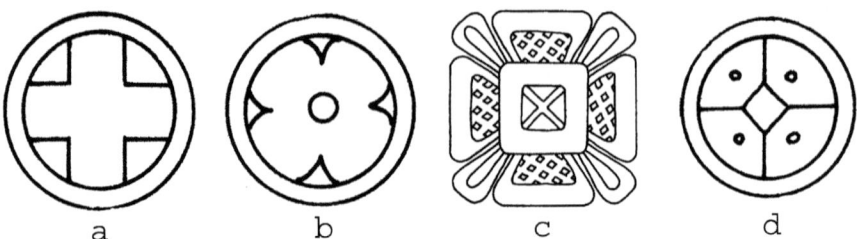

13.7 Quadripartite signs. (a) *Kan*; (b) *K'in*; (c) Maya completion sign; (d) *Lamat* (after Coggins 1980:Figure 2).

symbolism (see Kubler 1962; León-Portilla 1988:Appendix B; Pasztory 1978:110) reveal that quadripartite shapes (Fig. 13.7) play an important role in Maya thought. The Maya word for *day* is *k'in*, which also means "sun." The glyph for *day/sun* is a cartouche containing what might be interpreted as a four-petaled flower, the "petals" defined by four marks at the top, bottom, and sides of the interior. According to Coggins (1980:731),

> The shape of time may ... in one way, be conceptualized as a vertical four-point diagram within the ecliptic band (including a fourth point below). These points or places are: where the sun rises; where it reaches the top; where the sun sets; and where it reaches bottom. This is the equivalent of one day, which the Maya denote with the four-point Kin sign ... a two-dimensional figure that is equal to the completion of a cycle. The steps involved in this quadrangular journey vary somewhat according to Postclassic sources, but it is known that the sun was believed to ascend by steps or levels into the highest heaven, then to descend, and finally to trace a similar pattern in reverse in its journey through the underworld.

Coggins suggested that Uaxactun's Structure E-VII-sub was an intentional architectural representation of this symbol. Such stepped radial pyramids are ideal cosmograms, or symbols of a quartered, stepped universe in which the movement of the ruler up the structure and back down could represent the path of the sun (similar to David Freidel and Linda Schele's [1988] interpretation of the non-radial 5C-2d at Cerros, Belize). Marvin Cohodas (1980:208) agreed with George Kubler and Coggins that radial structures were giant *k'in* signs and were, furthermore, "designed for public participation in rituals regulated by the solar or agricultural calendar."

Thus, the radial structure—like the quincunx; *k'in* day glyph; *k'an* cross; and *lamat*, or Venus glyph—is among the many Maya and Mesoamerican quadripartite figures signifying calendric and cosmic cycle completion (Coggins 1980:fig. 2). These radial structures sometimes displayed stucco reliefs ("masks") with astrological-cosmological themes. For example, the masks on Structure E-VII-sub at Uaxactun refer to the watery underworld, the sun god, and long-lipped gods representing the earth and sky (Freidel 1979:46; Ricketson and Ricketson 1937:84). Freidel (1981) argues that together they represent the sun cycle surmounted by Venus. Importantly, because radial structures typically appear in the middle of open plazas they also represent, in Maya cosmovision, the center of the universe and the joining of the four world quarters.

Ritual and Agriculture

It seems likely that E-group architectural complexes had their inception in rituals of annual solar/agricultural cycling (Cohodas 1980). One explanation for the existence of the 260-day calendar postulates a "ritual or canonical agricultural cycle" Šprajc 2000: 409; see also Aveni 2003:157–158; Tichy 1981) based on the 260 days of the "agrarian year" from roughly February (working the milpa) through October (harvest). Such a cycle would explain why eastern orientations and observations—equinoxes, solstices, quarter years, and so forth, as observed by sunrise positions on the horizon—were important and commemorated architectonically by the Maya and their isthmian neighbors. In the Early Preclassic these observation lines might have been established with perishable or temporary markers, while later permanent architectural markers were constructed to provide an appropriate ritual frame for these activities. Still later, the annual cycles would have been extended to commemorate the conjoined cycling of 365-day solar and 260-day "ritual" calendars.

Evidence for Maya agricultural ritual is extensive (see Aveni and Hartung 1986:56; Pasztory 1978:132); "the agricultural fertility cult was the most widespread, and its myths and deities were probably the most ancient" (Pasztory 1978:130). Contemporary K'iche' Maya conceive of a congruence among plant life, human life, and sunrise (Tedlock 1985:252; Tedlock 1992) and among the Chamula the sun itself is conceived to be composed of maize (Gossen 1965:143). As Joyce Marcus (1987:131) notes, there seems to have been a similar metaphorical relationship among pyramids, mountains, and agricultural fertility in Maya cosmology: "[I]t has been suggested that the lowland Maya, lacking mountains in their natural environment, used temple pyramids as homologous counterparts" (see also Schele and Freidel 1990:427). Maize may grow from the Maya *kawak* glyph, associated with stone, stone buildings, and mountains (Miller and Taube 1993:120; Schele and Miller 1986:45). In K'iche' mythology, mountains contained the maize and water used to make the bodies of the first real humans, and there is a similar Nahua myth (Tedlock 1985:328).

Thus, large, accessible plazas defined by monumental stone buildings may have been appropriate settings for agricultural ritual (see also Chase and Chase 1983:3, 10). Table 13.3 categorizes 41 E-group plazas on a simple scale of restricted, semirestricted, and open based on plaza size and degree of enclosure by buildings. E-groups are associated most clearly with the large, open plazas thought to have been used in large-scale Maya ritual. Only two of the 41 plazas (about 5%) are of a restricted nature. With respect to E-groups, Aveni (2003:162) points out that 12 of these complexes have alignments that "match dates that fall in the midst of the dry season, the most logical points in time to conduct rituals pertaining to the anticipation of the forthcoming crop."

Table 13.3. Plaza characteristics and ballcourt associations.

Site	Restricted Plaza	Semi-Restricted Plaza	Open Plaza	Ballcourt Attached	Ballcourt within 100 m
1. Baking Pot	—	X	—	—	X
2. Balakbal	—	—	X	—	X
3. Barton Ramie	—	X	—	—	—
4. Blackman Eddy	—	—	X	X	—
5. Cabal Pech	—	—	X	X	—
6. Cahal Pichik	—	—	X	—	X
7. Calakmul	—	—	X	—	—
8. Caracol	—	—	X	—	X
9. Cenote	—	—	X	—	—
10. Colha	—	X	—	—	—
11. Comalcalco	—	X	—	—	X
12. Cuello	X	—	—	—	—
13. Dos Aguadas	—	X	—	—	X
14. Dzibilchaltun	—	X	—	—	—
15. El Mirador	—	—	X	—	—
16. El Pilar	—	X	—	X	—
17. El Venado	—	X	—	—	—
18. Hatzcap Ceel	—	—	X	—	—
19. Ixkun	—	X	—	—	X
20. Ixtutz	—	—	—	—	—
21. Kabah	—	—	X	—	—
22. La Muñeca	—	X	—	—	—
23. Naachtun	—	—	X	—	—
24. Nakum	—	—	—	X?	—
25. Naranjo	—	—	X	—	X
26. Nohmul	X	—	—	—	X
27. Oxpemul	—	—	X	—	—
28. Pacbitun	—	—	X	—	X
29. Río Azul	—	—	X	—	—
30. Río Bec	—	X	—	X	—
31. San José	—	—	X	—	X
32. Seibal	—	—	X	X	—
33. Teotihuacan	—	X	—	—	—
34. Tikal	—	X	—	—	—
35. Uaxactun	—	X	—	—	—
36. Ucanal	—	—	X	X	—
37. Uxul	—	—	X	X	—
38. Xultun	—	—	—	X	—
39. Xunantunich	—	X	—	X	—
40. Yaxha Group C	—	X	—	—	—
41. Yaxha Group F	—	—	X	—	X
Total	2	16	21	10	12

Source: Aimers 1993:Table 3.

Ballgame Associations

The association of E-groups with ballcourts, ballgame imagery, and sacrifice (Table 13.3) is notable given that Esther Pasztory (1972, 1978:130) and others have suggested that the major significance of the Maya ballgame concerned the sun's entrance into the underworld and the related renewal of crops and fertility. Cohodas (1975:110) also included the ballgame in a complex of ideas incorporating astronomy, agriculture, and sacrifice:

> [T]he mythological event of the summer solstice, the mating of the sun and moon to conceive the maize, is recorded in the ballgame myth of the *Popol Vuh*.... At the height of its popularity as a cult, the ballgame was probably played on the equinoxes to represent the battle of celestial and terrestrial forces. The sacrifices which culminated the game were employed as sympathetic magic to bring about the two crucial events in the yearly cycle of the sun and of agricultural activity.

This ritual complex incorporates a tension between "celestial and terrestrial forces" and may help explain evidence of ritual sacrifice at E-group locales. The most common form of sacrifice in Classic Maya art is decapitation (Schele 1984:8), and this practice was associated with the ballgame (see Soustelle 1984:3). In the *Popol Vuh,* for example, decapitation of the moon goddess occurs in the ballcourt (Cohodas 1975:109).

At Tikal, the Mundo Perdido E-group might have some links to ballgame ritual, given the presence of Teotihuacan imagery there and in the residential group 6C-XVI (Laporte and Fialko 1990), as well as an unusual configuration of three ballcourts on the north end of the Plaza of the Seven Temples adjacent to Mundo Perdido. A mass grave with the remains of 17 individuals dating to Manik 1 (AD 250–300) was found in front of the central structure of the eastern group in Mundo Perdido, although this sacrifice is more likely to have been dedicatory than ballgame-related (Laporte and Fialko 1995:56).

At Seibal, the stairway of Structure A-10, the eastern structure of an E-group, is flanked by "Stelae" 5 and 7, which display ballgame imagery. The ballgame may have been played against that stairway, perhaps accompanied by the sacrifice of captive opponents on the stairs (Miller and Houston 1987:46, 55) and/or with the sacrificial victims used as balls (Christie 1995:171; see also hieroglyphic steps of Yaxchilan Structure 33). The northern edge of this E-group plaza is marked by radial Structure A-13, which aligns with the A-19 ballcourt; a mass burial in Structure A-13 included parts of 11 people. The individuals in this burial, which is radiocarbon-dated to AD 930, included two women and a child, suggesting that, as at Tikal, the burial was dedicatory rather than that of a defeated ball team (Wright 1994:161, cited in Tourtellot and González 2004:63).

Besides Tikal and Seibal, E-groups at other sites also held burials of sacrificed individuals. The headless skeleton of a 25-year-old woman with no grave goods in Structure E-VII at Uaxactun, the skull caches at San José, and the headless skeletons at Tayasal (Guthe 1921–1922, cited in Ricketson 1928a:69) are also relevant here. In addition, Uaxactun Stela 19, celebrating the 8.16.0.0.0 katun ending (AD 357) in the

E-group, shows a kneeling captive. The presence of kneeling or prone bound prisoners is a common theme on katun-ending monuments in both the Early and Late Classic periods (Rice 2004).

Perhaps there was a Maya ritual complex involving competition in ballcourts, sacrifice on E-group stairways, and burial in nearby plazas. It does appear that Maya sacrifices lasted several days and involved both ritual torture and eventual death, usually by decapitation (Schele 1984:43). Sacrifice after the ballgame is consistent with agricultural ritual on a number of levels (see Monaghan 1990). A central symbol for most human groups, blood was considered by the Maya to be the food of the gods (see, for example, Coe 1988:277). At the Great Ballcourt at Chichen Itza, vegetative matter streams from the decapitated figure, and in what appears to be a reference to ballcourt design at Copan, three carved ballcourt marker-like stones in front of the Jaguar Staircase "show fresh young maize foliage, fertility generated through sacrifice" (Miller and Houston 1987:59).

POLITICAL RITUAL: E-GROUPS AND KATUN CYCLES

Numerous lines of evidence suggest that, out of an early context of celebrating annual solar cycling, E-groups came to be constructed specifically to celebrate longer calendrical cycles of 20 years known as katuns. This also might have been extended to 13-katun cycles of roughly 256 years known as the *may* (Rice 2004). To understand the significance of katuns, a brief background in Maya and Mesoamerican calendrics is needed.

The fundamental unit of time for all Mesoamerican peoples was the day (or "sun"). Days were given numbers and names within two simultaneously running calendars, one consisting of 260 days and the other of 365 days. These two calendars cycled continuously, such that any given day was identified by a number and name in both (e.g., 1 Ajaw 8 Kumk'u), and it took a total of 18,980 days, or 52 years, for the same day names and numbers to recur. The Classic Lowland Maya, however, used the more detailed "Long Count," in which the time was measured from an arbitrary starting date of August 13, 3114 BC. To maintain such records, they registered units of time by multiples of days, mostly but not entirely in units of 20, as follows:

1 day	=	1 *k'in*
20 *k'inob*	=	1 *winal*
18 *winalob*	=	1 *tun* (360 days)
20 *tunob*	=	1 katun (7,200 days, ca. 20 Gregorian years)
20 *k'atunob*	=	1 *b'ak'tun* (144,000 days, ca. 394 years)

Within this system, the Maya were particularly careful to record and celebrate the *completion* (rather than the start) of larger units of time, so-called period endings. These include the completion of katuns and their five-year subdivisions, and the endings of *b'ak'tuns*, by erecting carved, dated stelae. Furthermore, katuns were named by their ending day, always a day Ajaw—as, for example, *Katun* 8 Ajaw.

Many of the earliest katun-celebrating stela known in the Maya Lowlands were found in E-group complexes, particularly in front of or otherwise associated with the eastern structure. For example, in front of the east building of the E-group at Uaxactun, Stelae 18 and 19 commemorate the end of the 16th katun in *b'ak'tun* 8 (or 8.16.0.0.0 in standard notation), a Katun 3 Ajaw ending in AD 357 (Valdés and Fahsen 1995:204). At Tikal, the basal portion of Stela 39 celebrating the completion of 8.17.0.0.0 Katun 1 Ajaw (AD 376; Grube and Martin 1998:81) was found in Mundo Perdido, redeposited in the back room of Structure 5D-86-7, the central temple of the eastern side of the E-group complex. At Nakbe (Hansen 1992:84; Velásquez 1992), smashed and broken Stela 1 was found beside Structure 52, a small, low mound centered on the eastern linear mound (Structure 51) of that site's E-group. The stela, which shows two figures who could be ballplayers, might be related to an enormous altar found sealed beneath a Mamom floor (Hansen 1992:85), suggesting an extremely early date. Laporte (1993:314) proposed a date of circa AD 41, or 8.0.0.0.0, while others suggest an even earlier date around 500–200 BC (Hansen 2000:56). At El Mirador, 13 stelae (1 carved; 12 plain) were all found in the site's Great Acropolis, which is an E-group (Hansen 1991:20; Matheny 1987:332).

Unfortunately, because carved and dated period-ending stelae were not erected in the lowlands until the Late Preclassic period, this katun-ending function cannot be recognized archaeologically in earlier periods (Rice 2004). However, the practice continued through the Early Classic, as Yaxha Stelae 5 (8.16.0.0.0) and 2, stylistically dated to the Early Classic, were erected in front of the eastern structure of that site's Group F E-group, and Stela 6 (undated but stylistically early) was placed in the Group C E-group. At Caracol (Chase and Chase 1987), construction of the E-group of Group A was begun circa 300 BC, and 19 monuments were found in the complex. Stelae 12 (apparently undated) and 20 (AD 487) stood in front of the central structure of the eastern group, and a cache of Early Classic monuments (Stelae 13, 14, 15, and 16 and Altar 7) was found in association with Structure A-5, the northern structure of the eastern group. Three of the latter monuments celebrate the katun endings of 9.4.0.0.0 (514), 9.5.0.0.0 (534), and 9.6.0.0.0 (554).

E-groups continued to be locations for the erection of katun-ending stelae through the Late Classic and early Terminal Classic periods. Ucanal's Plaza A, a Late Classic E-group, was the location of most of Ucanal's monuments, particularly in front of the central-eastern structure; they included two stelae, one altar, and six "monuments" (Laporte and Mejía 2002:8). At Machaquila, 18 of the site's 19 stelae and all of its six altars were found in a "stelae plaza" that appears to represent a structurally altered E-group (Graham 1967). Seven stelae and three altars stood in front of the central and southern structures on the east side; the four stelae with dates commemorate the Late Classic–period endings 9.14.0.0.0 (AD 711), 9.15.0.0.0 (AD 731), 9.15.10.0.0 (AD 741), and 9.16.10.0.0 (AD 761) (Graham 1967). Seibal "Stelae" 5 and 7 are actually paired relief panels celebrating the completion of the half-katun of 9.18.10.0.0 (AD 800); they were set flanking the stairway of Structure A-10, the eastern structure of an E-group. And at Calakmul, the Central Plaza is a large E-group complex where some

40 Late Classic stelae were found, 18 in front (west) of the eastern structure (Folan et al. 2001). Stela 114, dating to AD 435, and Stela 43, dating to the termination of a Katun 13 Ajaw in 514, were reset, probably in Terminal Classic times, into niches constructed at the base of the huge Preclassic Structure II platform south of the E-group complex (Pincemin et al. 1998).

The political significance of katun cycles becomes evident through analogy with Late Postclassic–period and early Colonial-period northern Yucatan. There, as one katun ended and a new one began every 20 years, the new katun was ritually "seated" in a particular town, which controlled tribute rights, land titles, and appointments to public office for the 20-year duration (Edmonson 1979:11, 1982:xvii). As the katun seat changed, so did the administrative lordship or priestly oversight of the period. The chief priest of each katun, who held office for the full 20 years, was referred to as *b'alam* ("jaguar") and had a spokesman or speaker *(chilam),* who was the official prophet of the katun (Edmonson 1982:31).

In addition, the turning or cycling of katuns was accompanied by great ceremonies that Munro Edmonson (1986:21–29, Chapters 12, 29) describes as ritually structured, historico-mythological "dramas" of multiple "acts" and lasting several days. These began with a procession around the town that ended with a drinking ceremony and continued with the official "seating" of the katun as the new *b'alam,* or jaguar priest, of the katun took his place on the "mat of the katun." Following this, the four Yearbearers *(b'akab's,* or calendar priests), who represented not only the years but also various quadripartite entities in nature, were also "seated." Then another ceremonial procession took place in which seven priests "measured" the land, confirmed land titles, and took tribute payments. Next, the mats—symbolic seats of authority—of the lords were "counted" and ranked, and the lords declared their candidacy for various political positions. Sacrifices were carried out, accompanied by music and dancing, which was followed by a ceremonial feast and ritual "interrogation of the chiefs" on the last day of the katun. At this point, the new *chilam,* speaker of the jaguar priest, proclaimed the prophecy for the upcoming katun, and then the assembled lords and priests performed acts of autosacrifice.

From an archaeological viewpoint, the next "act" of the early Colonial-period katun ceremony is most interesting: The ancestors were commemorated by the erection of a cross. In earlier times, however, such commemoration was carried out by the erection of a stela (Morley and Brainerd 1956:212; Roys 1967 [1933]:161, note 1). Bishop Landa (Tozzer 1941:38–39) noted carved stones at Mayapan and reported that the Maya living nearby told him they "were accustomed to erect one of these stones every twenty years." After this event, the celebrations continued with a recitation of the place of the current katun in mythic time and a comic "play," concluding with a "sermon."

Thus, in Postclassic Yucatan, stelae erection was a part of katun-ending celebrations, and 20-year katun cycles played important roles in structuring regional political affairs. The fact that katun-ending stelae were erected regularly in E-groups in the lowlands—perhaps as early as the Late Preclassic and throughout the Classic period—and in twin-pyramid complexes in Late Classic Tikal prompts two observations (see Rice 2004). One is that katun endings were clearly important occasions for ritual celebrations for more than a millennium in the Maya Lowlands, and even early on they might have structured geopolitical activity in the same way they did in early Colonial-period Yucatan. The other is that during the Late Preclassic and Classic periods, the Maya built distinctive architectural complexes that provided an appropriate frame for rituals related to calendrical cycling and their associated stone monuments.

An analysis of burials and caches in E-groups led Arlen Chase and Diane Chase (1995:100) to suggest that these complexes were associated with performance of ancestor rituals. In a related hypothesis, Laporte (1993:314) suggested that the presence of E-groups might be useful for "the definition of territorial units among analogous polities," based in part on his work in the Valley of Dolores, in southeastern Peten, where he has identified more than a dozen E-group assemblages dating to the Late Classic period. According to Laporte (1993:316), "[T]he association of the astronomical complexes with ancestors and lineages would have permitted the sharing of power manifest in the societies of the southeastern Peten." He notes that "the association of the earlier carved stelae known in the central Maya lowlands with this type of architectural compound agrees well with this idea" (Laporte 1993:314).

DISCUSSION

During the late Middle Preclassic or Formative period in Mesoamerica, E-groups—large architectural complexes consisting of a north-south linear mound facing a platform to the west—were constructed in the ceremonial cores of many sites. These widespread Middle Preclassic constructions likely were settings for astronomical observations, possibly using instruments such as poles or stones, to identify—and *predict*—critical transitions in the annual cycle of the sun's points of emergence from the underworld. These astronomical and religious activities probably existed in the Early Preclassic or before, but by the Middle Preclassic, circa 600–500 BC, with the beginnings of formal civic-ceremonial architecture in emerging centers of politico-ritual power, these large, centrally located, and morphologically distinctive architectural complexes became increasingly grandiose commemorations of the sun's journey—and of time itself, in a more metaphorical sense.

A useful distinction can be made between public architecture such as the radial pyramid, portal arch, ballcourt, round structure, and causeway, and elite architectural forms such as the acropolis, range structure, and temple pyramid (Cohodas 1985). Types in the first category are "all specialized forms to serve distinct ritual functions" (Cohodas 1985:51), while dynastic architecture is considered "a propaganda tool designed to validate the ruler's authority" (Cohodas 1985:62). This typology is related to both social and ideological differences between elite and commoner. Following J. Eric S. Thompson (1973), Cohodas (1985:66) argues that while non-elite people were "concerned primarily with deities of nature directly associated with agriculture, the Maya elite adopted more abstract deities and symbols to proclaim their elevated status." For farmers, the seasonal agricultural cycle was of central importance and invariant "no matter who ruled from the great capital cities," while for the Maya elite time came to be "progressive" and linear (Cohodas 1985:66).

A similar distinction about the nature of time—philosophical attitudes toward, or values associated with, or meaning of—has been made by Arthur Miller from a historical perspective. In the Early Classic, Miller (1986:38) argues, successive rulership from among different families was metaphorically associated with the cyclical movements of the sun and the regeneration of gods, while subsequent Late Classic Maya elites began to emphasize linear time to validate the principles of genealogical descent and lineage-based rulership. Architecturally, twin-pyramid groups are the best-known Late Classic representation of the latter emphasis, celebrating the ends of individual katuns as components of longer cycles.

By the end of the Preclassic and beginning of the Early Classic, E-group assemblages, after a long history of development, were widespread throughout the southern lowlands and often dominated the ceremonial architecture at major centers. More than 60 E-group complexes are now known archaeologically in and around the Maya area. However, it is likely that with Maya astronomers' advances in predictive astronomy, any utilitarian observational functions of the E-group assemblage probably had long since become obsolescent, and the architecture of most E-groups probably never accurately tracked the sun's yearly movement, even though the Maya certainly did. The construction of radial platforms in the centers of plazas was later repeated throughout the Late and Terminal Classic lowlands, as exemplified by Copan Structure 4, Seibal Structure A-3, Dzibilchaltun's Temple of the Dolls, and, in the Postclassic period, by numerous radial structures at Chichen Itza and Mayapan.

The large number of katun monuments in E-group plazas convinces us that E-groups were early settings for valedictory ceremonies similar to those associated with the Late Classic twin-pyramid complexes at Tikal and related sites. In the Middle Preclassic period, a small western structure opposite a low north-south-oriented eastern platform may originally have supported markers to create sight lines for observing the sun's annual movements and the yearly agricultural cycle. Later, by the Late Preclassic period, these cycles were commemorated by large monumental structures and associated with celebration of katun endings.

Many interpretations (e.g., Chase 1983, 1985; Coggins 1983; Cohodas 1975; Miller 1986) have emphasized cyclical time and its associated meanings in explaining the emergence of the E-group form. Aveni (2003:162) argues that some E-groups may have been aligned to serve as settings for agricultural ritual coordinated by a dry-season calendar of approximately seven 20-day months, culminating with the rainy season. We believe the cyclical symbolism of the E-group was related originally to a central practice of Maya life, the regeneration of the natural world through public ritual, "a pattern so old that its origins are lost in the very beginning of settled life in Mesoamerica" (Miller 1986:86), and this symbolism was increasingly used as a more overtly political metaphor (Aimers 1993).

CONCLUSIONS

Although their astronomical alignments are inconsistent and continually debated, E-groups have attained the status of legend, glossed by archaeology and astronomy texts alike as an example of the remarkable astronomical achievements of the ancient Maya (e.g., Flower 1990:76; Hammond 1982:294). Although they appear to have been oriented with reference to the yearly movement of the sun, the labeling of E-groups as "observatories" has substituted a portion of their significance for the whole. Solar alignment appears to be only one aspect of some larger set of considerations that may have incorporated concepts of sacred geography, ritual performance in reference to the yearly solar and agricultural cycles, and longer cycles of time, especially katuns. As Aveni (2003:163) notes, the Maya E-group was "performative rather than practical, a theater rather than a laboratory, a planetarium rather than an observatory."

We suggest that E-groups played the same role in the Preclassic-period and Classic-period lowlands as did twin-pyramid groups in the Late Classic in the Tikal region of central Peten: They served as theaters in which calendrical rituals—especially katun celebrations—were enacted, as well as dramatic displays of rulers' agency within a divinely directed cosmos. Their presence throughout the Maya area testifies to the power of the ideology shared by the participating communities. Sites constructing these complexes and erecting period-ending stelae had formal politico-ritual roles as seats of the katun for 20-year periods.

The central role of cyclicity in Maya thought explains why the E-group was one of the most recognizable and enduring monumental forms in Maya architectural history. Although the overall form of these assemblages changed little, their significance may have varied as the meaning attached to cyclical time varied through more than a millennium of sociopolitical and religious development. E-groups were most widely constructed as Maya society was becoming increasingly stratified, an indication that the ritual they framed ensured both cosmic and political order. By expressing a fundamental cosmological concept on a monumental scale, and as settings for religious and political ritual, E-groups provided an experientially powerful and symbolically meaningful condensation of Maya reality. Although limited evidence leaves many questions about the E-group unanswered, we hope that this discussion contributes to more inclusive, less narrowly functionalist interpretations that more closely consider the symbolic richness and complexity of Maya ritual and architectural expression.

REFERENCES

Acevedo, Renaldo, Bernard Hermes, and Zoila Calderón. 1996. Ixtinto: Rescate arqueológico. In *IX simposio de investigaciones arqueológicas en Guatemala, 1995,* ed. Juan Pedro Laporte and Héctor L. Escobedo, 233–251. Guatemala City, Museo Nacional de Arqueología y Etnología and Asociación Tikal.

Aimers, James John. 1993. *An Hermeneutic Analysis of the Maya E- Group Complex.* M.A. thesis, Department of Anthropology, Trent University, Peterborough, Canada.

Andrews, E. W., IV, and E. W. Andrews V. 1980. *Excavations at Dzibilchaltun, Yucatan.* Publication 48. Middle American Research Institute, New Orleans.

Andrews, George F. 1975. *Maya Cities: Placemaking and Urbanization.* Norman, University of Oklahoma Press.

Aveni, Anthony. 1980. *Skywatchers of Ancient Mexico.* Austin, University of Texas Press.

Aveni, Anthony. 2001. *Skywatchers* (rev. ed. of *Skywatchers of Ancient Mexico*). Austin, University of Texas Press.

Aveni, Anthony. 2003. Archaeoastronomy in the Ancient Americas. *Journal of Archaeological Research* 11 (2):149–191.

Aveni, Anthony F., Anne S. Dowd, and Benjamin Vining. 2003. Maya Calendar Reform? Evidence from Orientations of Specialized Architectural Assemblages. *Latin American Antiquity* 14 (2):159–178.

Aveni, Anthony F., Sharon L. Gibbs, and Horst Hartung. 1975. The Caracol Tower at Chichen Itza: An Ancient Astronomical Observatory? *Science* 188 (4192):977–985.

Aveni, Anthony F., and Horst Hartung. 1986. Maya City Planning and the Calendar. *Transactions of the American Philosophical Society* 76 (1).

Aveni, Anthony F., and Horst Hartung. 1989. Uaxactun, Guatemala, Group E and Similar Assemblages: An Archaeoastronomical Reconsideration. In *World Archaeoastronomy,* ed. Anthony F. Aveni, 441–461. Cambridge, Cambridge University Press.

Awe, Jaime, and Mark Campbell. 1988. *Site Core Investigations at Cahal Pech, Cayo District Belize: Preliminary Report of the 1988 Season.* Unpublished manuscript, Trent University, Peterborough.

Blom, Frans. 1924. Report on the Preliminary Work at Uaxactun, Guatemala. *Carnegie Institution of Washington Yearbook* 23:217–219.

Braswell, Geoffrey E., ed. 2003. *The Maya and Teotihuacan: Reinterpreting Early Classic Interaction.* Austin, University of Texas Press.

Broda, Johanna. 1982. Astronomy, Cosmovision, and Ideology in Pre-Hispanic Mesoamerica. In *Ethnoastronomy and Archaeoastronomy in the American Tropics,* ed. Gary Urton and Anthony F. Aveni, 81–110. Annals of the New York Academy of Sciences, Publication 385.

Broda, Johanna. 1989. Significant Dates of the Mesoamerican Agricultural Calendar and Archaeoastronomy (conference paper abstract). In *World Archaeoastronomy,* ed. Anthony F. Aveni, 494. Cambridge, Cambridge University Press.

Brotherston, Gordon. 1982. Astronomical Norms in Mesoamerican Ritual and Time-Reckoning. In *Archaeoastronomy in the New World,* ed. Anthony F. Aveni, 109–143. Cambridge, Cambridge University Press.

Cabrera Castro, Rubén. 2000. Teotihuacan Cultural Traditions Transmitted into the Postclassic, According to Recent Excavations, trans. Scott Sessions. In *Mesoamerica's Classic Heritage: From Teotihuacan to the Aztecs,* ed. Davíd Carrasco, Lindsay Jones, and Scott Sessions, 195–218. Boulder, University Press of Colorado.

Carlson, John B. 1974. Astronomical Investigations and Site Orientation Influences at Palenque. In *Second Palenque Round Table,* Part 3, ed. Merle Greene Robertson, 107–117. Pebble Beach, CA, Robert Louis Stevenson School.

Carlson, John B. 1981. A Geomantic Model for the Interpretation of Mesoamerican Sites: An Essay in Cross-Cultural Comparison. In *Mesoamerican Sites and World-Views,* ed. Elizabeth P. Benson, 143–215. Washington, DC, Dumbarton Oaks.

Chase, Arlen F. 1983. *A Contextual Consideration of the Tayasal-Paxcaman Zone, El Peten, Guatemala.* Ph.D. dissertation, Department of Anthropology, University of Pennsylvania, Philadelphia.

Chase, Arlen F. 1985. Archaeology in the Maya Heartland. *Archaeology* 38 (1):32–39.

Chase, Arlen F., and Diane Z. Chase. 1983. Intensive Gardening among the Late Classic Maya: A Possible Example at Ixtutz, Guatemala. *Expedition* 25 (3):2–11.

Chase, Arlen F., and Diane Z. Chase. 1987. *Investigations at the Classic Maya City of Caracol, Belize: 1985.* Pre-Columbian Art Research Institute Monograph 3. San Francisco, Pre-Columbian Art Research Institute.

Chase, Arlen F., and Diane Z. Chase. 1995. External Impetus, Internal Synthesis, and Standardization: E-Group Assemblages and the Crystallization of Classic Maya Society in the Southern Lowlands. In *The Emergence of Maya Civilization: The Transition from the Preclassic to the Early Classic,* ed. Nikolai Grube, 87–101. Acta Mesoamericana, 8. Verlag Anton Saurwein.

Christie, Jessica Joyce. 1995. *Maya Period Ending Ceremonies: Restarting Time and Rebuilding the Cosmos to Assure Survival of the Maya World.* Ph.D. dissertation, Department of Anthropology, University of Texas, Austin.

Clark, John E., and Richard D. Hansen. 2001. The Architecture of Early Kingship: Comparative Perspectives on the Origins of the Maya Royal Court. In *Royal Courts of the Ancient Maya,* Vol. 2: *Data and Case Studies,* ed. Takeshi Inomata and Stephen D. Houston, 1–45. Boulder, CO, Westview Press.

Coe, Michael D. 1988. Ideology of the Maya Tomb. In *Maya Iconography,* ed. Elizabeth R. Benson and Gillett G. Griffin, 222–235. Princeton, NJ, Princeton University Press.

Coe, William R. 1965. Tikal: Ten Years of Study of a Maya Ruin in the Lowlands of Guatemala. *Expedition* 8 (1):3–56.

Coggins, Clemency C. 1980. The Shape of Time: Some Political Implications of a Four-Part Figure. *American Antiquity* 45 (4):727–739.

Coggins, Clemency C. 1983. *The Stucco Decoration and Architectural Assemblage of Structure 1-sub, Dzibilchaltún, Yucatan, Mexico.* Middle American Research Institute Publication No. 49. Tulane University, New Orleans.

Coggins, Clemency C., and R. David Drucker. 1988. The Observatory at Dzibilchaltún. In *New Directions in American Archaeoastronomy,* ed. Anthony F. Aveni, 17–56. Proceedings of the 46th International Congress of Americanists. BAR International Series 454. British Archaeological Reports, Oxford.

Cohodas, Marvin. 1975. The Symbolism and Ritual Function of the Middle Classic Ball Game in Mesoamerica. *American Indian Quarterly* 11 (2):99–130.

Cohodas, Marvin. 1980. Radial Pyramids and Radial Associated Assemblages of the Central Maya Area. *Journal of the Society of Architectural Historians* 39 (3):208–223.

Cohodas, Marvin. 1985. Public Architecture of the Maya Lowlands. *Cuadernos de Aquitectura Mesoamericana* 6:51–68.

Corzo, Lilian A., Marco Tulio Alvarado, and Juan Pedro Laporte. 1998. Ucanal: Un sitio asociado a la cuenca media del Río Mopan. In *XI simposio de investigaciones arqueológicas en Guatemala, 1997,* ed. Juan Pedro Laporte and Héctor L. Escobedo, 491–214. Guatemala City, Museo Nacional de Arqueología y Etnología and Asociación Tikal.

Cowgill, George L. 2003. Teotihuacan and Early Classic Interaction: A Perspective from outside the Maya Region. In *The Maya and Teotihuacan: Reinterpreting Early Classic Interaction,* ed. Geoffrey E. Braswell, 315–335. Austin, University of Texas Press.

Culbert, T. Patrick. 1991. Polities in the Northeast Peten, Guatemala. In *Classic Maya Political History*, ed. T. Patrick Culbert, 128–145. Cambridge, Cambridge University Press.

Demarest, Arthur A., and Antonia E. Foias. 1993. Mesoamerican Horizons and the Cultural Transformations of Maya Civilization. In *Latin American Horizons*, ed. Don S. Rice, 147–191. Washington, DC, Dumbarton Oaks.

Digby, Adrian. 1974. Crossed Trapezes: A Pre-Columbian Astronomical Instrument. In *Mesoamerican Archaeology, New Approaches*, ed. Norman Hammond, 271–283. Austin, University of Texas Press.

Drucker, Philip. 1952. *La Venta, Tabasco: A Study of Olmec Ceramics and Art*. Bureau of American Ethnology Bulletin 153. Smithsonian Institution, Washington, DC.

Drucker, Philip, Robert F. Heizer, and Robert J. Squier. 1959. *Excavations at La Venta, Tabasco, 1955*. Bureau of American Ethnology Bulletin 170. Smithsonian Institution, Washington, DC.

Edmonson, Munro S. 1979. Some Postclassic Questions about the Classic Maya. In *Tercera Mesa Redonda de Palenque*, vol. 4, ed. Merle Green Robertson and Donnan Call Jeffers, 9–18. Palenque, Chiapas, Mexico, Pre-Columbian Art Research Center.

Edmonson, Munro S. 1982. *The Ancient Future of the Itza: The Book of Chilam Balam of Tizimin*. Austin, University of Texas Press.

Edmonson, Munro S. 1986. *Heaven Born Merida and Its Destiny: The Book of Chilam Balam of Chumayel*. Austin, University of Texas Press.

Fialko, Vilma. 1988. Mundo Perdido, Tikal: Un ejemplo de complejos de conmemoración astronómica. *Mayab* 4:13–21.

Flower, Philip. 1990. *Understanding the Universe*. New York, West Publishing.

Folan, William J., Laraine A. Fletcher, Jacinto May Hau, and Lynda Florey Folan, coords. 2001. Las ruinas de Calakmul, Campeche, México: Un lugar central y su paisaje cultural, 37–42. Universidad Autónoma de Campeche, Mexico.

Freidel, David A. 1979. Culture Areas and Interaction Spheres: Contrasting Approaches to the Emergence of Civilization in the Maya Lowlands. *American Antiquity* 44 (1):36–54.

Freidel, David A. 1981. Civilization as a State of Mind: The Cultural Evolution of the Lowland Maya. In *The Transition to Statehood in the New World*, ed. R. R. Kautz, 188–227. Cambridge, Cambridge University Press.

Freidel, David A., and Linda Schele. 1988. Kingship in the Late Preclassic Lowlands: The Instruments and Places of Ritual Power. *American Anthropologist* 90 (3):547–567.

Garber, James F., M. Kathryn Brown, and Christopher J. Hartman. 2001. The Early/Middle Formative Kanocha Phase (1200–850 BC) at Blackman Eddy, Belize. Available online at: www.famsi.org/reports/00090.

García Cook, Angel. 1981. The Historical Importance of Tlaxcala in the Cultural Development of the Central Highlands. In *Archaeology*, ed. Jeremy A. Sabloff, 244–276. Handbook of Middle American Indians, Supplement 1. Austin, University of Texas Press.

Gossen, Gary. 1965. Temporal and Spatial Equivalents in Chamula Ritual Symbolism. In *Reader in Comparative Religion*, ed. William R. Lessa and Evon Z. Vogt, 135–149. New York, Harper and Row.

Graham, Ian. 1967. Machaquila. In *Archaeological Explorations in El Peten, Guatemala*, 51–99. Middle American Research Institute Publication No. 33. Tulane University, New Orleans.

Grazioso Sierra, Liwy, T. Patrick Culbert, Vilma Fialko, et al. 2001. Arqueología en el Bajo La Justa, El Petén, Guatemala. In *XV Simposio de investigaciones arqueológicas en Guatemala*, ed. Juan Pedro Laporte and Héctor L. Escobedo, 205–209. Guatemala City, Instituto de Antropología e Historia, and Asociación Tikal.

Grube, Nikolai, and Simon Martin. 1998. *Deciphering Maya Politics: The Proceedings of the Maya Hieroglyphic Workshop,* March 14–15, 1998, ed. Phil Wanyerka. Department of Art, University of Texas, Austin.

Guidoni, Enrico. 1975. *Primitive Architecture,* trans. R. E. Wolf. New York, Electra/Rizzoli.

Guthe, C. E. 1921–1922. Report on the Excavations at Tayasal. *Carnegie Institution of Washington Yearbook* 20:364–368.

Hammond, Norman. 1982. *Ancient Maya Civilization.* New Brunswick, NJ, Rutgers University Press.

Hammond, Norman. 1985. *Nohmul: A Prehistoric Maya Community in Belize. Excavations 1973–1983,* 2 vols. BAR International Series 250(i). British Archaeological Reports, Oxford.

Hansen, Richard D. 1991. The Maya Rediscovered: The Road to Nakbe. *Natural History* 91 (5):8–14.

Hansen, Richard D. 1992. El proceso cultural de Nakbe y el área del Petén nor-central: Las épocas tempranas. In *V simposio de investigaciones arqueológicas en Guatemala, 1991,* ed. Juan Pedro Laporte, Héctor L. Escobedo, and Sandra Villagrán de Brady, 81–96. Instituto de Antropología e Historia and Asociación Tikal.

Hansen, Richard D. 2000. The First Cities—The Beginnings of Urbanization and State Formation in the Maya Lowlands. In *Maya: Divine Kings of the Rain Forest,* ed. Nikolai Grube, 51–65. Cologne, Könemann.

Hartung, Horst. 1981. Monte Alban in the Valley of Oaxaca. In *Mesoamerican Sites and World Views,* ed. Elizabeth P. Benson, 41–69. Washington, DC, Dumbarton Oaks.

Healy, Paul F. 1990. Excavations at Pacbitun, Belize: Preliminary Report on the 1986 and 1987 Investigations. *Journal of Field Archaeology* 17:247–262.

Hellmuth, Nicholas M. 1971. *Report on the First Season, Explorations and Excavations at Yaxha, El Peten, Guatemala, 1970.* Unpublished manuscript, Department of Anthropology, Yale University, New Haven.

Hersey, George. 1976. *Pythagorean Palaces: Magic and Architecture in the Italian Renaissance.* Ithaca, NY, Cornell University Press.

Hester, Thomas R., and J. D. Eaton. 1982. *Archaeology at Colha, Belize.* Center for Archaeological Research, University of Texas, San Antonio, and Centro Studi e Recherche Ligabue, Venice.

Hester, Thomas R., J. D. Eaton, and H. J. Shafer. 1980. *The Colha Project, Second Season 1980 Interim Report.* Center for Archaeological Research, University of Texas, and Centro Studi e Recherche Ligabue, Venezia.

Heyden, Doris. 1981. Caves, Gods, and Myths. In *Mesoamerican Sites and World Views,* ed. Elizabeth P. Benson, 1–19. Washington, DC, Dumbarton Oaks.

Heyden, Doris, and Paul Gendrop. 1980. *Pre-Columbian Architecture of Mesoamerica.* New York, Rizzoli.

Howard, Michael C. 1986. *Contemporary Cultural Anthropology,* 2nd ed. Boston, Little, Brown.

Hugh-Jones, C. 1979. *From the Milk River.* Cambridge, Cambridge University Press.

Hunt, Eva. 1977. *The Transformation of the Hummingbird.* Ithaca, NY, Cornell University Press.

Kubler, George. 1962. *The Art and Architecture of Ancient America.* Harmondsworth, Penguin Books.

Laporte, Juan Pedro. 1993. Architecture and Social Change in Late Classic Maya Society: The Evidence from Mundo Perdido, Tikal. In *Lowland Maya Civilization in the Eighth Century AD,* ed. Jeremy A. Sabloff and John S. Henderson, 299–320. Washington, DC, Dumbarton Oaks.

Laporte, Juan Pedro. 1996. La cuenca del Río Mopán-Belice: Una sub-región cultural de las tierras bajas maya central. In *IX simposio de investigaciones arqueológicas en Guatemala, 1995,* ed. Juan Pedro Laporte and Héctor L. Escobedo, 253–279. Guatemala City, Museo Nacional de Arqueología y Etnología and Asociación Tikal.

Laporte, Juan Pedro. 2003. Architectural Aspects of Interaction between Tikal and Teotihuacan during the Early Classic Period. In *The Maya and Teotihuacan: Reinterpreting Early Classic Interaction,* ed. Geoffrey E. Braswell, 199–216. Austin, University of Texas Press.

Laporte, Juan Pedro, and Vilma Fialko. 1990. New Perspectives on Old Problems: Dynastic References for the Early Classic at Tikal. In *Vision and Revision in Maya Studies,* ed. Flora S. Clancy and Peter D. Harrison, 33–66. Albuquerque, University of New Mexico Press.

Laporte, Juan Pedro, and Vilma Fialko. 1995. Un reëncuentro con Mundo Perdido, Tikal, Guatemala. *Ancient Mesoamerica* 6:41–94.

Laporte, Juan Pedro, and Héctor E. Mejía. 2002. Ucanal: Una ciudad del Río Mopán en Petén, Guatemala. *U tz'ib* 1 (2):1–71.

Lee, Thomas A., Jr. 1989. Chiapas and the Olmec. In *Regional Perspectives on the Olmec,* ed. Robert J. Sharer and David C. Grove, 198–226. New York, Cambridge University Press.

León-Portilla, Miguel. 1988. *Time and Reality in the Thought of the Maya,* 2nd ed. Norman, University of Oklahoma Press.

Lou P., Brenda. 1997. Chalpate, análisis del asentamiento y orientación de un centro satélite de Tikal. In *X simposio de investigaciones arqueológicas en Guatemala, 1996,* ed. Juan Pedro Laporte and Héctor L. Escobedo, 373–379. Guatemala City, Museo Nacional de Arqueología y Etnología and Asociación Tikal.

Lowe, Gareth. 1977. The Mixe-Zoque as Competing Neighbors of the Early Lowland Maya. In *The Origins of Maya Civilization,* ed. Richard E.W. Adams, 197–248. Albuquerque, University of New Mexico Press.

Marcus, Joyce. 1987. *The Inscriptions of Calakmul: Royal Marriage at a Maya City in Campeche, Mexico.* Museum of Anthropology, Technical Report 21. Ann Arbor, University of Michigan.

Matheny, Raymond T. 1987. El Mirador: An Early Maya Metropolis Uncovered. *National Geographic* (September):317–339.

Mejía Amaya, Héctor E., Heidy Quezada, and Jorge E. Chocón. 1998. Un límite político territorial en el sureste de Petén. In *XI simposio de investigaciones arqueológicas en Guatemala, 1997,* ed. Juan Pedro Laporte and Héctor L. Escobedo, 171–190. Guatemala City, Museo Nacional de Arqueología y Etnología and Asociación Tikal.

Miller, Arthur G. 1986. *Maya Rulers of Time/Los soberanos mayas del tiempo.* University Museum, University of Pennsylvania, Philadelphia.

Miller, Mary Ellen, and Stephen D. Houston. 1987. Stairways and Ballcourt Glyphs: New Perspectives on the Classic Maya Ballgame. *RES* 14:47–66.

Miller, Mary Ellen, and Karl Taube. 1993. *An Illustrated Dictionary of the Gods and Symbols of Ancient Mexico and the Maya.* New York, Thames and Hudson.

Monaghan, John. 1990. Sacrifice, Death, and the Origins of Agriculture in the Codex Vienna. *American Antiquity* 55 (3):559–569.

Morales, Paulino I. 1998. Asentamiento prehispánico en el Naranjo-Frontera, La Libertad, Petén. In *XI simposio de investigaciones arqueológicas en Guatemala, 1997,* ed. Juan Pedro

Laporte and Héctor L. Escobedo, 123–134. Guatemala City, Museo Nacional de Arqueología y Etnología and Asociación Tikal.

Morante López, Rubén B. 1996. Los observatorios astronómicos subterráneos: ¿Un invento Teotihuacano? *Revista mexicana de estudios antropológicos* 42:159–172.

Morley, Sylvanus G., and George Brainerd. 1956 *The Ancient Maya*. Stanford, CA, Stanford University Press.

Pasztory, Esther. 1972. The Historical and Religious Significance of the Middle Classic Ballgame. In *Religión en Mesoamerica, XII Mesa Redonda of Sociedad Mexicana de Antropología,* ed. Jaime Lituak King and Noemi Castillo Tejero, 441–455. Mexico City.

Pasztory, Esther. 1978. Artistic Traditions of the Middle Classic Period. In *Middle Classic Mesoamerica AD 400–700*, ed. Esther Pasztory, 108–142. New York, Columbia University Press.

Pincemin, Sonia, Joyce Marcus, Lynda Florey Folan, William J. Folan, Maria del Rosario Domínguez Carrasco, and Abel Moralez López. 1998. Extending the Calakmul Dynasty Back in Time: A New Stela from a Maya Capital in Campeche, Mexico. *Latin American Antiquity* 9 (4):310–327.

Rathje, William. 1972. Trade Models and Archaeological Problems: The Classic Maya and Their E-Group Complex. In *XL Congresso Internazionale degli Americanisti* 4:223–235. Rome.

Rathje, William. 1978. Trade Models and Archaeological Problems: Classic Maya Examples. In *Mesoamerican Communication Routes and Cultural Contacts,* ed. Thomas A. Lee Jr. and Carlos Navarette, 147–175. Papers of the New World Archaeological Foundation, No. 40. Provo, UT.

Redfield, Robert, and Antonio Villa Rojas. 1934. *Chan Kom: A Maya Village*. Carnegie Institution of Washington Publication 448. Carnegie Institution of Washington, Washington, DC.

Redfield, Robert, and W. Lloyd Warner. 1940. Cultural Anthropology and Modern Agriculture. In *Farmers in a Changing World*. Washington, DC, U.S. Government Printing Office.

Reilly, F. Kent, III. 1999. Mountains of Creation and Underworld Portals: The Ritual Function of Olmec Architecture at La Venta, Tabasco. In *Mesoamerican Architecture as a Cultural Symbol,* ed. Jeff Karl Kowalski, 14–39. New York, Oxford University Press.

Rice, Prudence M. 2004. *Maya Political Science: Time, Astronomy, and the Cosmos*. Austin, University of Texas Press.

Ricketson, Oliver, Jr. 1928a. Astronomical Observatories in the Maya Area. *Geographical Review* 18:215–225.

Ricketson, Oliver, Jr. 1928b. Notes on Two Maya Astronomic Observatories. *American Anthropologist* 30:434–444.

Ricketson, Oliver G., and Edith B. Ricketson. 1937. *Uaxactun, Guatemala, Group E, 1926–1931*. Part 1: *The Excavations;* Part 2: *The Artifacts*. Carnegie Institution of Washington, Publication No. 477. Washington, DC, Carnegie Institution of Washington.

Roys, Ralph L. 1967 (1933). *The Book of Chilam Balam of Chumayel*. Norman, University of Oklahoma Press.

Ruppert, Karl J. 1977 (1940). A Special Assemblage of Maya Structures. In *The Maya and Their Neighbors: Essays on Middle American Anthropology and Archaeology,* ed. Clarence L. Hay, Ralph L. Linton, Samuel K. Lothrop, Harry L. Shapiro, and George C. Valliant, 222–231. New York, Dover Publications.

Schele, Linda. 1984. Human Sacrifice among the Classic Maya. In *Ritual Human Sacrifice in Mesoamerica,* ed. Elizabeth H. Boone, 7–48. Washington, DC, Dumbarton Oaks.

Schele, Linda, and David Freidel. 1990. *A Forest of Kings: The Untold Story of the Ancient Maya.* New York, William Morrow.

Schele, Linda, and Mary Miller. 1986. *The Blood of Kings: Dynasty and Ritual in Maya Art.* Fort Worth, TX, Kimbell Art Museum.

Scully, Vincent. 1991. *Architecture: The Natural and the Manmade.* New York, St. Martin's Press.

Soustelle, Jacques. 1984. Ritual Human Sacrifice in Mesoamerica: An Introduction. In *Ritual Human Sacrifice in Mesoamerica,* ed. Elizabeth H. Boone, 1–5. Washington, DC, Dumbarton Oaks.

Šprajc, Ivan. 2000. Astronomical Alignments at Teotihuacan. *Latin American Antiquity* 11 (4): 403–415.

Tedlock, Barbara. 1992. *Time and the Highland Maya,* rev. ed. Albuquerque, University of New Mexico Press.

Tedlock, Dennis. 1985. *Popol Vuh: The Definitive Edition of the Mayan Book of the Dawn of Life and the Glories of Gods and Kings.* New York, Simon and Schuster.

Thomas, Julian. 1990. Monuments from the Inside: The Case of the Irish Megalithic Tombs. *World Archaeology* 22 (2):168–178.

Thompson, J. Eric S. 1973. Maya Rulers of the Classic Period and the Divine Right of Kings. In *The Iconography of Middle American Sculpture,* ed. Gordon R. Willey. 52–71. New York, Metropolitan Museum of Art.

Tichy, Franz. 1981. Order and Relationship of Space and Time in Mesoamerica: Myth or Reality? In *Mesoamerican Sites and World-Views,* ed. Elizabeth P. Benson, 217–245. Washington, DC, Dumbarton Oaks.

Tourtellot, Gair, and Jason J. González. 2004. The Last Hurrah: Continuity and Transformation at Seibal. In *The Terminal Classic in the Maya Lowlands: Collapse, Transition, and Transformation,* ed. Arthur A. Demarest, Prudence M. Rice, and Don S. Rice, 60–82. Boulder, University Press of Colorado.

Townsend, Richard Fraser. 1982. Pyramid and Sacred Mountain. In *Ethnoastronomy and Archaeoastronomy in the American Tropics,* ed. Anthony F. Aveni and Gary Urton, 37–62. Annals of the New York Academy of Sciences, Publication No. 385. New York Academy of Sciences, New York.

Tozzer, Alfred M. 1913. *A Preliminary Study of the Ruins of Nakum, Guatemala.* Memoirs of the Peabody Museum of Archaeology and Ethnology, Vol. 5, No. 3. Harvard University, Cambridge, MA.

Tozzer, Alfred M. 1941. *Landa's* Relación de las cosas de Yucatan: *A Translation.* Papers of the Peabody Museum of Archaeology and Ethnology, No. 28. Harvard University, Cambridge, MA.

Turner, Harold W. 1979. *From Temple to Meeting House: The Phenomenology and Theology of Places of Worship.* The Hague, Mouton.

Valdés, Juan Antonio. 1997. El Proyecto Miraflores II dentro del marco preclásico en Kaminaljuyu. In *X simposio de investigaciones arqueológicas en Guatemala, 1996,* ed. Juan Pedro Laporte and Héctor L. Escobedo, pp 81–91. Guatemala City, Museo Nacional de Arqueología y Etnología and Asociación Tikal.

Valdés, Juan Antonio, and Federico Fahsen. 1995. The Reigning Dynasty of Uaxactun during the Early Classic. *Ancient Mesoamerica* 6 (2):197–219.

Velásquez, Juan Luis. 1992. Excavaciones en el Complejo 75 de Nakbe. In *V simposio de investigaciones arqueológicas en Guatemala, 1991,* ed. Juan Pedro Laporte, Héctor L. Escobedo

A., and Sandra Villagrán de Brady, 97–102. Guatemala City, Instituto de Antropología e Historia and Asociación Tikal.

Vogt, Evon Z. 1981. Some Aspects of the Sacred Geography of Highland Chiapas. In *Mesoamerican Sites and World Views,* ed. Elizabeth P. Benson, 119–142. Washington, DC, Dumbarton Oaks.

Wheatley, P. 1971. *The Pivot of the Four Quarters: A Preliminary Enquiry into the Origins and Character of the Ancient Chinese City.* Chicago, Aldine.

Wright, Lori E. 1994. The Sacrifice of the Earth? Diet, Health, and Inequality in the Pasión Maya Lowlands. Ph.D. dissertation, Department of Anthropology, University of Chicago.

CHAPTER FOURTEEN

A Reflection on the Ancient Mesoamerican Ethos

Miguel León-Portilla

The organizers of the Second Oxford International Conference on Archaeoastronomy asked me to reflect on those traits that can be described as essential to the old Mesoamerican ethos. I have dared to accept the invitation in full awareness of the complexity of the subject.

To begin, there are two concepts included in the title I have given to this paper that require at least an introductory comment. One is the idea of *ethos*. The other is related to the possibility of applying that concept to the ancient Mesoamericans as if they had lived in basically one universe of spiritual traditions and forms of behavior.

A number of years ago a so-called "ethos anthropology" developed and exerted various kinds of influence on the disciplines related to the study of culture. While not necessarily adhering to the conceptual implications of the ethos anthropology, I consider it adequate to guide this reflection from the perspective of the concept of *ethos*.

From *World Archaeoastronomy*, ed. A. Aveni (New York: Cambridge University Press, 1989).

WHAT DO WE MEAN BY ETHOS AND HOW CAN WE APPLY THIS CONCEPT TO THE ANCIENT MESOAMERICANS?

Let us turn to the two best dictionaries of the English language you all so often consult. The *Webster New Twentieth Century Dictionary* (the *Webster*) tells us that ethos is "the characteristic and distinguishing attitudes, habits, etcetera, of a racial, political, occupational or other group" *(Webster,* 1978, p. 678). Members of the "Oxford Conference on Archaeoastronomy" will find it necessary also to consult with the classical *Oxford English Dictionary* (the *OED*). Its answer is that ethos is "the characteristic spirit, prevalent tone of sentiment of a people or community, the genius of an institution or system" (*OED*, 1979, p. 314).

In sum, it being our aim to reflect on the ancient Mesoamerican ethos, we have to take into account the available sources, archaeological and documentary, to search for what they can offer about the characteristic attitudes, habits, world view, prevalent tone of sentiment, characteristic spirit, forms of behavior, institutions and, employing the *OED*'s term, "the genius" of the ancient Mesoamericans.

In terms of this description of the concept of ethos, we can ask ourselves if the ancient Mesoamericans, since the classic period and more intensely in the postclassic, shared attitudes and key concepts so basically similar that their world views, institutions and main forms of behavior will be correctly considered as variations derived from one and the same source.

I know this question has been posed many times. In my search for some key traits of the Mesoamerican ethos, I will not go back to the old arguments. Instead, I will present to you some converging evidence derived from the testimonies mainly of the Nahuatl speaking peoples of Central Mexico and of several members of the great Maya family. The evidence derived from testimonies of these peoples living so far away, as well as from other groups like the Mixtec of Oaxaca, will help us to appreciate whether we can speak of an ancient Mesoamerican ethos.

A SENSE OF BELONGING

What we know from the texts of the pre-Hispanic tradition of the Nahua, Yucatec Maya, Quiché and other groups, and also from the familiarity with their contemporary descendants, reveals that among the Mesoamericans there is a deeply rooted sense of belonging. This encompasses the feeling of belonging to a family, nuclear and extended, often enlarged by new spiritual forms of relationship. The members of a family are named in Nahuatl *cencaltin,* "those of the same house," and also those who share a same essence and form of existence, *cenyeliztli.* Each human being is essentially linked to a community that lives in a land, ancestral to all its members, and rich in a variety of symbols. The many levels of belonging include, in addition to the primary one of the family, those of the *calpulli* or ward, the town, province and chiefdom. To a Mesoamerican, whether Maya, Mixtec, Otomi or Nahua, it would have been unthinkable to consider himself a loose entity, kinless or in any form isolated.

The cycles of the feasts and religious ceremonies during the solar year helped to provide the Mesoamericans with an understanding and vital feeling of one's belonging to a sacred space and time. According to the ancient world view, about which there are texts in Nahua, Yucatec Maya, Quiché and Mixtec, man exists in a universe horizontally distributed into four cosmic quadrants, each one full of symbols. This universe encompasses also a series of superior and inferior levels, and is the scene where the gods perform their actions. Those divine performances, although mysterious, determine man's existence and fate on the earth. Everything happens in accordance with the rhythms of time, and of divine forces, those of the Sun in its daily and nocturnal paths, and of the many other celestial bodies which at determined moments enter and abandon the scene of man's visible universe.

Each family, *calpulli*, town, province and chiefdom exists in this universe, distributed into four quadrants. The gods had established the primeval measurements of the four quadrants. Also Huitzilopochtli, the tutelary god of the Mexica, gave the command to his people to distribute their city, Mexico-Tenochtitlan, into four great *calpullis* or wards (Alvarado Tezozomoc, 1975, pp. 74–5).

I will add here a reference I dedicate to those who recently have proposed the idea that, at least to the Maya, the four-fold cosmic horizontal distribution was of little or no meaning. I will quote from the account of archaeologist Richard E.W. Adams about his recent findings in the zone of Río Azul in northeastern Petén of Guatemala (Adams, 1986, pp. 441–2):

> In 1985 we cleared another looted burial, Tomb 12, and found additional beautiful hieroglyphic paintings. On the smooth plaster of the tomb's four walls, in thick red line, are the signs for east, south, west and north. Although the meanings of these hieroglyphs have been known for more than a century, our discovery marked the first time they were found in actual natural context, for their appearance in Tomb 12 correctly matches the real directions.

In Río Azul, evidence was unearthed of the actual and "natural" meaning of the hieroglyphs of the four cosmic quadrants. There, their corresponding spatial references are associated not only with man's existence in this quatripartite universe but also to the remains of those who have already departed to "the Region of Mystery." It is of interest to quote once more from the account of Adams, who points to another association of the hieroglyphs of the cosmic quadrants with the symbols of some of the celestial bodies, those who establish the divisions of time and carry the meanings and "burdens" of it (Adams, 1986, p. 442):

> Moreover each directional glyph was supplemented by another representing its mythical cosmic association with, respectively, the Sun, Venus, darkness and the Moon.

The presence there of the hieroglyphs of these deities–celestial bodies is indeed significant. Their paths in the heavens had to be observed all year round, because as a function of them the destinies *(k'in* in Maya, *pije* in Zapotec and *Tonalli* in Nahuatl) had to be discerned and propitiated.

To man's consubstantial adherence to his family, kinship, town, province and chiefdom, as well as his essential relation to his own sacred space, one has to add his appurtenance to the realm of time, at once cyclic and divine, endowed with a large number of faces, those of the portentous carriers of the universe's and man's destinies. Yes, indeed, Mesoamerican man, far from considering himself in any form loose, insulated or deprived of ties, human and divine, had a deep sense of belonging. Two brief texts from the Nahua tradition give eloquent testimony to this. One, a part of a *Huehuehtlahtolli*, "The Ancient Word," will help us to understand from the inside the meaning the Mesoamericans attached to the arrival of a new human being to the family. Among other things these words of great tenderness—in the mood of the Mesoamerican tone of sentiment—were addressed to the recently born baby *(Florentine Codex,* 1969, Book VI, p. 196):

> The Lord of the Near and by *[Tloqueh Nahuaqueh]*, the Giver of Life, has inclined his heart. Here a precious necklace is flaked off, here a precious feather is spread out. Here we dream, we see in dreams. In your hand [the parents' hands], on your neck, He placed a precious necklace, a precious feather, a jade, precious bracelet. Here we see your face. You have been formed; you have been born....

Since the day of his birth the new human being *belonged* to a family. He was, to his parents and to the rest of his relatives, a precious necklace, a precious feather, a jade, a bracelet....

From a different perspective, the other text describes the intrinsic link of man, from the moment of his conception, to the realm of the divine, from which his destiny descends *(Florentine Codex,* 1969, Book VI, ch. XXII):

> It was said that in the thirteen heaven
> our destiny [*tonalli*] was determined.
> When the child is conceived,
> when he is placed in the womb,
> his *tonalli*, destiny, comes to him there;
> it is sent to him by [the supreme god] the Lord of Duality.

The overall Mesoamerican belief in the *tonalli*, with its rich complexity of meanings and interrelations within the ancient world view, is another element that intrinsically colored the ethos of these peoples.

THE *TONALLI*, DESTINIES OF MAN ON THE EARTH

The universe in which the Mesoamericans lived had been established and re-established several times, four according to the Maya, five in the Nahua tradition. There are more than 20 testimonies, in the indigenous texts and the codices, that speak of those "Suns" or ages which have existed. They came into being and were later destroyed, not by any kind of chance but by a predetermined divine act, the result of a destiny, *tonalli*, *k'in* or *pije*).

An identical and very complex constellation of meanings was attached to these three words, from three different Mesoamerican languages: Nahua, Maya and Zapotec. "Sun, day, feastday, time and destiny" are their basic connotations. That of destiny permeated the others so deeply that for the Mesoamericans the reckoning of any period of time led them always to investigate the *tonalli* or destinies associated with it. Thus each of the "Suns" or ages had been re-established as determined by its corresponding destiny. That is why each "Sun" had its calendric name, precisely the name of the *day-tonalli* in which it began to exist: 4-Water, 4-Ocelot, 4-Rain-of-Fire, 4-Wind and 4-Movement. Those *day-tonalli* names were not mere words, they announced the destiny of each one of the "Suns."

Everything happening on the earth and in man's life from his birth to his death, is the outcome of a *destiny-tonalli*. And every one of the gods also has an intrinsic relation to the realm of the *tonalli-s* of time. Thus each god has likewise a calendric name, expression of the *destiny-tonalli* that belongs to him. Examples of this are 5-Flower, *Macuilxochitl*, He-She protector of the singers, dancers, and other artists, and 7-Serpent, *Chicome-coatl*, She (and also He), the deity of maize and of our maintenance in general.

Time penetrates the cosmic quadrants, endowed with preordained rhythms, as the *Book of Chilam Balam* declares it (Solís Alcalá, 1949, pp. 340–1):

> When [the year] has gone round
> the four quadrants,
> east, north, west, south,
> it is said to be *1-Katun*.
> The same thing is said
> when *1-Cauac* begins,
> and each of the years
> goes through this....

A very similar Nahuatl statement is included in *Madrid Codex* (vol. 8, fo. 269r) "Time penetrating space, the quadrants of the earth's surface, the heavens above us and the underworld, disseminates the *tonalli-destinies*, bringing about the realization everywhere of what has been determined by the gods and, above all, by Him-Her who is like the Night and the Wind."

A distinguishing attitude of the Mesoamericans, an extremely significant element of their ethos, was their concern for time as a conveyor of destinies. Several texts from the pre-Hispanic tradition in Maya and Nahuatl proclaim that the *tonalli*, daytime-destiny of a given person, is precisely *i-macehual*, "that which has been granted to him, what he deserves."

The knowledge about the *tonalli-destinies*, still alive in some isolated spots within contemporary Mesoamerica (Villa Rojas, 1968, pp. 151–9), is an attribute of some of the priests and sages. In the old days these experts were called *ah k'inob* by the Yucatec Maya and *tonalpouhqueh* by the Nahua. *Ah k'inob* means "those of the days-destinies" and *tonalpouhqueh*, "the ones who read and tell the days-destinies."

The knowledge of the *tonal-pohualli* or *tzol-kin*, "count of the destinies," implied very complex calculations and a great familiarity with the sacred beliefs and rites. The *tonalpouhqueh, ah k'inob* and other experts in different places within Mesoamerica, were in possession of various calendric counts whose remote origins must be traced back to the days of the Olmecs, to at least the middle of the first millennium BC (Coe, 1977, pp. 70, 79).

All periods of time, the 13 hours of the day, and the nine of the night, the days, the 13-day groups, the spatial orientations of them, the 20-day "months," the years, and their different positions in the various cycles of 52, 104 or in the Maya Long Count, including their interrelations with the cycles of the "Great Star" (Venus), all this and much more had to be taken into account in order "to read and tell" a *tonalli-destiny*.

The *tonalli-day*, the carrier par excellence of the *tonalli-destiny*, belonged to the 260 days *tonalpohualli*, the count of the days-destinies. This count was basically formed by four divisions of 65 days, each one broken into five "weeks" of 13 days each. The names of the 260 days-destinies were structured by means of 20 day-signs, preceded by a numeral which ran from 1 to 13.

The 260 possible names of the day-destinies were incorporated in the *xihuitl* or *haab*, the solar-year-count of the Nahua and Maya, and colored it with the richness of their sacred connotations, like the mystic presence of one or several gods with their influences, good or adverse, the revelation of a time propitious for the performance of a determined form of action, a religious ceremony, the initiation of a war, the enthroning of a ruler.

Nothing in man's life would be set apart as being unrelated to the realm of the destinies of time. The religious celebrations with their many rites, prayers, hymns and dances; the sacrifices and re-enactments of the primeval happenings—everything had to be attuned to the rhythms of the days and destinies.

The gods brought down the destinies. They were the carriers of those *tonalli-s* that belong to them and which the *tonalpouhqueh* or *ah k'inob* could anticipate by knowing, among many other things, the calendric name of the "arriving" deity.

There are some extant *tonal-amatl*, "books of the days-destinies," such as *Codex Borbonicus* and *Tonalamatl of Aubin*, as well as some parts of the Maya *Dresden*, *Madrid* and *Paris* codices. By means of them one can try to penetrate into the subtleties of the computations of the day-destinies and their constellations of interrelations within the universe of divine realities.

Those books were consulted by the priests asked to advise about the *tonalli-s* which had to do with the birth of a baby, the giving of the most adequate name to the infant, the choosing of the most favorable date for his/her dedication to a school, the date for a wedding or for a declaration of the sexual transgressions to the Goddess who will cleanse them.

Those concerned with agriculture, the arts and crafts, commerce, war and the collection of tributes, the lords and high rulers, i.e. all the Mesoamericans, constantly had to consult with the experts about the meaning of their *tonalli-s* and about the remedies to escape as best one could the bad consequences of an ominous destiny.

Miguel León-Portilla

GODS, DESTINIES, AND ACTS OF DESERVING PERFORMED BY MESOAMERICAN MAN

When there was still night, in a primeval time, the gods had deserved, by their own sacrifice and death, the restoration of the Sun, Moon, earth and man. These gods, who so often appear acting by pairs, are the same who everywhere, in space and time, make themselves present in what seems to be a preordained sequence. He-She, *Ometeotl,* the Dual God in the religion of the Nahua, the Grand Father–Grand Mother of the *Popol Vuh,* the divine Pair who resides in the uppermost of the heavens, as depicted in the Mixtec codices known as the *Selden Roll* and *Gómez Orozco,* is the Ultimate Reality, Begetter-Conceiver of all that exists.

He-She has a variety of names often expressed by a pair of interrelated words. Examples of this are *Tloqueh, Nahuaqueh,* "The One that Is Near," "The One that Is Close"; *Ometeuctli, Omechihuatl,* "Lord of Duality," "Lady of Duality"; *Yohualli, Ehecatl,* "Night," "Wind"; *Tezcatlanextia, Tezcatlipoca,* "Mirror of Day," "Mirror of Night"; *Citlallatonac, Citlalinicueh,* "Star Which Illumines Things," "Skirt of Stars."

An easily noticeable trait is the apparently androgynous character of many gods. Thus, for instance, *Tlalteuctli* is at once Lord and Lady of the Earth, and *Cinteotl* is He-She God of Maize. There are, besides, other gods who appear and act by pairs: Our Lord and our Lady of the Place of the Dead (*Mictlanteuctli* and *Mictlancihuatl*); The Lord of Rain and the Goddess of the Terrestrial Waters (*Tlaloc, Chalchiuhtlicueh*); and, in a very especial form, *Quetzalcoatl,* and *Cihuacoatl,* understood as "The Precious Twin" and the "Feminine Twin"—the word *coatl* meaning both "serpent" and "twin."

On other occasions there are two pairs of divine beings who present themselves acting in what seems to be a preordained sequence. This is the case of the four *Tezcatlipocas,* "Smoking Mirrors," who preside over the four quadrants of the world, or rule, in succession, the four previous cosmic ages *(Borgia Codex,* 1980, vol. I, p. 21). In cases like this, it is often said that "other gods" can play the role of those who integrated the original quartet.

A number of ancient texts also register that one or another of the gods, or a pair of them, is invoked as "Our Mother, Our Father" *(in Tonantzin, in Totahtzin).* Such is the case, for instance, in the following prayer from the *Florentine Codex,* 1969, book VI, ch. IX):

> Mother of the gods, Father of the gods, God
> of Fire, the Old God, *Xiuhtecuhtli, Huehuehteotl,*
> reclined on the navel of the earth, within the
> circle of turquoise....

Several testimonies that pertain to "the ancient word," the *Huehuehtlahtolli,* and some sacred hymns and other chants unveil for us what seems to be at the core of this Nahuatl pantheon, so rich in divine pairs and quartets. Indeed the Ultimate Reality, the divine source referred to as "Giver of Life" *(Ipalnemoani:* literally "Thanks To Whom One Lives"), is thought of as a dual entity.

Quetzalcoatl, the Feathered-Serpent god, symbol of the divine wisdom, was asked by the other gods to take care of restoring the human beings. *Quetzalcoatl* went to *Mictlan*, "Place of the Dead," in search of the precious bones of men who had lived in previous ages. *(Leyenda de los Soles,* 1975, p. 120). In this manner he would restore the human beings to inhabit the re-established earth. In the Region of the Dead, *Quetzalcoatl* had to overcome many obstacles. Once he could gather the precious bones, he took them to *Tamoanchan*, the place of origin, the abode of the supreme Dual God. There the Mother Goddess "took them to grind and put them in a precious vessel."

Quetzalcoatl had to transmit new life to the bones. "He bled his virile member. He and the other gods at once did penance, deserved it *(tlamacehuayah)*." And they said: the human beings have been born, i.e. the *macehualtin*, "the deserved ones, because for our sake the gods did penance, deserved it" *(topan otlamaceuhqueh)* *(Florentine Codex,* 1953, Book VII, ch. 11). In fact, the word *macehualtin*, "the deserved by the gods' pennance," became synonymous with human beings not only in Nahuatl but also, as a loan, in several other Mesoamerican languages. It is true that later a differentiation was introduced between *macehualtin* "human beings" (understood as the common people), and *pipiltin*, "those of lineage," members of nobility, the ruling class. But this did not alter the idea that all, men and women, whether of lineage or commoners, essentially were *macehualtin*, "deserved by the gods' penance."

The key concept of *tlamacehua* denotes the primary and essential relation human beings have with their gods. These, with their own penance and sacrifice, deserved—brought into existence—the human beings. The gods did it because they were in need of some beings who would be their worshippers, the providers of sustenance to foster life on earth. Man also had to perform *tlamacehualiztli,* "penance, the act of deserving through sacrifice," including the bloody one of human beings. If the gods *topan otlamaceuhqueh,* "for us did penance," we ought to follow their example, to deserve our own being on the earth with our blood and life.

The often described as so "utterly detestable human sacrifices," the consuming of small pieces of flesh of the human victim, the blood smeared effigies of the gods, were elements the Mesoamericans thought essential to act and respond in terms of their *tlamacehualiztli*. If the gods had sacrificed themselves when it was still night there in Teotihuacan, and if only thus, with their blood, they had deserved our being, to re-enact that primeval action was indeed to give in return, to pay and also restore. The victims were thus named *teomicqueh,* "the divine dead." With these *teomicqueh* man repayed and did his part in maintaining the flow of life on the earth, in the heavens and in the shadows of the underworld. There is a discourse of the *tlahtoani*, "high ruler," in which he advises his sons to take firm hold upon whatever is related to this *(Florentine Codex,* 1969, Book VI, ch. XVII):

> In this manner there is entry near and close into *Tloqueh, Nahuaqueh,* "The One Who Is Near," "The One Who Is Close," where there is removing of the secrets from His lap, from His bosom, and where He recognized one, shows his mercy to one, takes pity upon one, causes one to deserve things [*macehualtia*] ... Perhaps He

causes one to merit, to deserve [*quitemacehualtia*] virility, the eagle warriorhood, the tiger warriorhood. There He takes, he recognizes as His friend the one who addresses Him well, the one who prays well to Him... In his hands He places the eagle vessel, the eagle tube [instruments for the sacrifice].

This one becomes father and mother of the Sun. He provides drink, He makes offerings to those who are above us [*Topan*] and in the Region of the Dead [*Mictlan*]. And the eagle warriors, the tiger warriors revere Him.

With these words to his sons, the high ruler unveiled the meaning of the forms of action they should take, including that of "providing drink and sustenance" to the gods through human sacrifice. To keep near and close to *Tloqueh, Nahuaqueh,* the Dual God, was difficult but if rulership, government, were to be alive, were to be deserved, *tlamacehualiztli,* the act of deserving it should be reenacted.

Several codices, e.g. the *Borbonicus* and the *Tellerianus,* and some texts written in Nahuatl with the Roman alphabet, describe the great variety of performances that in perfect order were rendered in the public celebrations among the 18 groups of 20 days within the solar calendar. There, and also in private life at home, many rites were carried out to win the divine benevolence, to deserve a good *tonalli.*

The *tonalli,* destiny, depended on what Our Mother, Our Father had deserved and conceded to a person. Because of this, whatever possibility might exist to modify the *tonalli* had to be sought in the same book where one could decipher the mysteries inherent in the divine predestinations. There one would find the most adequate action, in a determined date, to give in return, so as to foster in the best possible way the divine flow of life. This, of course, presupposed the *tlamacehualiztli,* the act of meriting and deserving what is "appropriate and righteous," what Our Mother, Our Father had determined and disposed for us when we were engendered and placed in the womb of our own terrestrial mother.

Perhaps as a projection of the deeply rooted belief in the divine Duality, one can trace a tendency in the Mesoamericans to conceive things and speak about them in pairs. Such a psychological leaning appears in a large number of concepts and expressions. Here are a few examples: the universe is composed by the earth's surface as the visible reality, and "that which is above us, and the Land of the Dead," as an invisible and mysterious realm; the concept of person as 'the owner of a face and of a heart'; the basic two social strata the *pipiltin,* "those of lineage," and the *macehualtin,* a term understood here not so much in its primary connotation but as meaning "the ordinary people, the commoners."

To concepts like these one must add a different kind of dualism, present in a great number of expressions in the Nahuatl and Maya languages. It consists of uniting two words which complement each other, either because they are almost synonyms or because they evoke a third idea, usually a metaphor. This linguistic stratagem can be compared to the kennings, compound metaphorical expressions used in ancient Anglo-Saxon. A few examples from Nahuatl are the following: "flower, song" (meaning poetry, art); "water, fire" (war); "seat, mat" (authority), "blouse, skirt" (womanhood).... From Yucatec Maya one can quote: "eye, ear" (that which permits to know);

"father, mother" (protection); "quetzal and blue bird" (a precious reality); "stick, stone" (punishment)....

To feel and cherish the belonging to a family, clan, town, chiefdom ...; to believe and act in a world where time as an atmosphere permeates everything and introduces the *tonalli-destinies* ultimately determined by Our Mother, Our Father; to think and speak with a dualistic orientation, are traits essentially related to the Mesoamerican ethos.

TIME, *TONALLI*, AND STARGAZING

To live attuned to the rhythms of time, performing adequate *tlamacehualiztli,* "acts of meriting and deserving," thus propitiating the good realization of a *tonalli,* one had always to consult the *tonalpohualli* (the 260-day system), and the *xihuitl* (365-day count). But there were other realities, related also to the cycles of time, which presupposed more complex forms of correlation and adjustment based on astronomical observations. Stargazing was an occupation of some of the priests and other members of the ruling group. Centuries of observation of the movements of the celestial bodies had actually led to the structuring of the basic calendrical systems. And stargazing as well had made it possible to introduce the required corrections in the calendar. As a Nahuatl text (*Coloquios . . . apud,* León-Portilla, 1979, pp. 19–20) expresses it, one had often to consult with:

> those who guide us ... and instruct us how our gods must be worshipped.... Those who see, who dedicate themselves to observing and measuring with their hand the running and the crossing of the stars in the sky.

Thanks to those observers and measurers of the stars' movements a number of other cycles of primary importance to the Mesoamericans were charted. Mention can be made of the cycles of the Great Star (Venus), of the Moon and its eclipses, of the Pleiades (so closely related to the 52-year cycle), and of several other constellations and celestial phenomena.

This, of course, helps us to appreciate the importance attached to astronomical observations as an activity so intrinsic to the accuracy of the calendar and to whatever was ruled by it: the cult of the gods, the wisdom of the destinies, the basic duties of man in his belonging to his family, group, town, chiefdom, his activities as a farmer, warrior, artist, merchant or in any other profession. In brief, to exist for the Mesoamericans one had to observe the sky. Without skywatchers the ethos of this people, its distinguishing spirit, its own genius would not have developed.

At this point, I believe it is time for a warning. This paper is addressed to my friends, the archaeoastronomers. The *tonalli, k'in* or destiny of our moment and concern is bringing about the need for the warning. At least this is the way I see it. If skywatching—"the observing and measuring of the running of the stars"—was an essential part of Mesoamerican culture, modern "discoveries" of any of its ancient achievements will have always to do with celestial phenomena intrinsically related to the world view, religion and practices of the Mesoamericans. "Discoveries" of archaeoastronomers dealing with celestial bodies or cosmic cycles, about whose meaning in the culture itself nothing is known, have to be held suspicious and put in parentheses. Those who claim that the cycles of planets like Mars or Neptune were well known to the Mesoamericans will have to tell us where in the Mesoamerican sources, archaeological or documentary, reference is made to the meanings of such cycles and the ultimate nature of those celestial bodies. There are many accounts about Venus and its cycles, a good number related to Quetzalcoatl-Kukulcan, but have we parallel texts or other references to gods associated to the cycles of Mars or Neptune?

This is just an example of what I want to declare as a "reading" of the *tonalli* which belongs to my friends, the archaeoastronomers. A last word: if you are persuaded, as I am, that stargazing, or if you prefer astronomy, was an essential ingredient of Mesoamerican culture, do your best to study the achievements of its skywatchers, not from the perspective and previous knowledge of western astronomy, but from the point of view of those who, contemplating the sky with their naked eyes, unveiled many of its secrets, discovering at once a universe of celestial meanings they could understand only in terms of their own culture.

REFERENCES

Adams, R.E.W. 1986. Archaeologists Explore Guatemala's Lost City of the Maya, Río Azul. *National Geographical Magazine* 169 (4):420–460.

Alvarado Tezozomoc, F. 1975. *Crónica Mexicayotl.* Trans. from the Nahuatl by A. León. Mexico, Instituto de Investigaciones Históricas, Universidad Nacional.

Borgia Codex (Códice Borgia). 1980. Comentarios de Eduard Seler, 3 vols. Mexico, Fondo de Cultura Económica. First reprint of the 1963 ed.

Coe, D. 1977. *Mexico,* 2nd ed. New York, Praeger Publishers.

Florentine Codex, General History of the Things of New Spain [by] Bernardino de Sahagún, in 13 Parts, 12 vols. Book VII, 1953; Book VI, 1969. Trans. from Aztec into English, with notes and illustrations, by Arthur J.O. Anderson and Charles E. Dibble. Santa Fe, NM, School of American Research and the University of Utah.

León-Portilla, M. 1979. *Aztec Thought and Culture.* Norman, University of Oklahoma Press.

Leyenda de los Soles. 1975. In *Códice Chimalpopoca,* ed. and trans. P. F. Velázquez. Mexico, Instituto de Investigaciones Históricas, UNAM.

Madrid Codex (Tro-Cortesianus Troano). Museo de América, Madrid. Reprinted, 1967, Codices Selecti, vol. 8. Graz, Akademische Druck und Verlagsanstalt.

Oxford English Dictionary. 1979. 2 vols. Oxford University Press.

Solís Alcalá, E. 1949. *Códice Pérez.* Mérida, Yucatán.

Villa Rojas, A. 1968. In *Time and Reality in the Thought of the Maya,* ed. M. León-Portilla. Boston, Beacon Press.

Webster New Twentieth Century Dictionary. 1978. Unabridged, 2nd ed. New York, Collins World.

PART III

The Role of Ethnoastronomy

Imagine Greek and Roman religion, politics, calendar, and astronomy still being practiced on some undiscovered remote island in the Mediterranean. Imagine further what a clearer picture of ancient Athenian and Roman society we would acquire by living and learning among these hypothetical contemporaries. Of course, such a scenario in the Old World seems quite impossible, but as Michael Coe and Miguel León-Portilla have suggested in the previous section, many aspects of pre-Columbian culture remain alive in the New World. Specially trained anthropologists known as ethnologists learn the language, live with the families, eat the food, and talk to the people. Their goal is to seek an accurate description of contemporary culture and their reward is the acquisition of a deeper understanding of what is important in the lives of indigenous people. But probing indigenous astronomy is not an easy task, as anthropologist Stephen Fabian's (2001) text on method in ethnoastronomy details. This is especially true in North America, where native ideologies had for so long been excluded from the white-dominated population. In the Hispanic and

Portuguese Americas, larger segments of indigenous astronomical practice have survived.

Another challenge to investigating ancient astronomy based on its current practice has to do with the question of cultural continuity, especially after Hispanic contact. Can we really be sure that a Zuni *pekwin* (calendar priest) or a Maya *h'meen* (day keeper) knows what his ancient predecessors knew and that he utilizes knowledge the way they did? After all, Western astronomy today bears little relation to astrology, which dominated that science as recently as 500 years ago.

The pieces on Mesoamerican ethnoastronomy selected for this section include Judith Remington's early work on the highland Maya. It details a series of Western-based categories of astronomical phenomena and offers a perspective on just how different each is from those shared by our contemporary culture. In contrast, Barbara Tedlock's later piece "Quichean Time Philosophy" proceeds from a discussion of dialectics couched in a firm linguistic context. It constitutes a more anthropologically based inquiry into indigenous highland Maya astronomy. Her study of the calendar shows that other astronomies often recognize natural cycles that are entirely unknown in the West, for example an 82-day lunar cycle (for more on such cycles, see Aveni, Bricker, and Bricker 2003). Her work has encouraged other investigators to get away from the biased notion of setting up a standard checklist of Western phenomena as a guide to understanding astronomy in other cultures.

Working among the native tribes of Colombia, anthropologist Gerardo Reichel-Dolmatoff introduces us to the concept of modeling one's house after the cosmos, a habit we find repeated in cultures as diverse as the Amazonian Bororo (Fabian 1992) and Barasana (Hugh-Jones 1982), the Warao and Yekuana of Venezuela (Wilbert 1981), the Pawnee (Chamberlain 1982), and the Navajo (Chapter 18) in the Americas; the Batammaliba of Togo and Benin (Blier 1987); the cultures of Oceania (Makemson 1938); and elsewhere. In Trudy Griffin-Pierce's article on the *hooghan*, we encounter the nature-culture connection yet again in the powerful symbolism of the house. Urton's "Quechua Animals and Cosmology" introduces the uniqueness of the Andean constellations so prominent in the summer Milky Way. A parade of indigenous animals depicted by dark clouds rather than by connected stars marches across the sky to the beat of a set of rhythms whose notes sound key points in the life cycle of each animal (shall we call it "astrozoology"?).

To close this section I chose two pieces that raise larger issues. I picked anthropologist Billie Jean Isbell's short commentary (one of a number at the conference on ethonoastronomy and archaeoastronomy in the American Tropics held at the New York Academy of Sciences in 1981; Aveni and Urton 1982) because it underlines the significance of the solar zenith passage, a sky event that can be viewed only in tropically based cultures (see Aveni 1981a [in introduction] for a more detailed discussion of this idea). The contrasting sky orientation makes for a quite different astronomy from what one finds in cultures that developed in temperate climates. Last, folklorist Mary Jane Young's article "Ethnoastronomy and the Problem of Interpretation" deals with the delicate issue of how to reconcile emic evidence rooted in contemporary native religious belief systems regarding their ancestors with that acquired in the field by the etic approach of the outside investigator. It may be read in conjunction with Chapters 27 and 28.

THINGS TO THINK ABOUT

1. What aspects of Maya astronomy and calendar discussed in the pieces by Remington and Tedlock do you think most clearly reflect outside influence from Western astronomy?

2. Compare the cross constellations discussed in Remington's article on contemporary Maya culture with those mentioned in Coe's review of the enthohistoric evidence. In Urton's 1981 book, several Andean crosses are also mentioned. What do they symbolize?

3. Examine the constellations sketched in the articles in this section and find the areas where they are located on a modern star map found in any elementary textbook of astronomy. Show on a copy of your star map how the connector lines of the indigenous constellations are different (or the same). Then see whether you can locate any of them in the sky.

4. In what sense are (were) native skywatchers natural philosophers, as we think of their Greek counterparts? Or was their knowledge of the sky purely practical?

5. This is a question about "homemaking": What aspects of the natural world can you think of that motivate people in modern cultures to design the place they choose to live in? Contrast how we "make our home" with that of the Navajo and the Desana.

SUGGESTIONS FOR FURTHER READING

Aveni, A., H. Bricker, and V. Bricker. 2003. "Seeking the Sidereal: Observable Planetary Stations and the Ancient Maya Record." *JHA* 34:145–161.

Blier, S. 1987. *The Anatomy of Architecture: Ontology and Metaphor in Batammaliba Architectural Expression.* New York, Cambridge University Press.

Chamberlain, V. del. 1982. *When Stars Come down to Earth: Cosmology of the Skidi Pawnee Indians of North America.* Los Altos, CA, Ballena and College Park, MD, Center for Archaeoastronomy.

Chamberlain, V. del, J. Carlson, and M. J. Young, eds. 2005. *Songs from the Sky: Indigenous Astronomical and Cosmological Traditions of the World.* Austin and Leicester, University of Texas Press and Ocarina.

Fabian, S. 1992. *Space Time of the Bororo of Brazil.* Gainesville, University of Florida Press.

Fabian, S. 2001. *Patterns in the Sky: An Introduction to Ethnoastronomy.* Boulder, CO, Westview.

Gossen, G. 1974a. *Chamulas in the World of the Sun: Time and Space in a Maya Oral Tradition.* Cambridge, MA, Harvard University Press.

Gossen, G. 1974b. "A Chamula Solar Calendar Board from Chiapas, Mexico." In *Mesoamerican Archaeology: New Approaches,* ed. N. Hammond, 217–253. Austin: University of Texas Press.

Griffin-Pierce, T. 1992. *Earth Is My Mother, Sky Is My Father: Space, Time and Astronomy in Navajo Sandpainting.* Albuquerque, University of New Mexico Press.

PART III: THE ROLE OF ETHNOASTRONOMY

Hugh-Jones, S. 1982. "The Pleiades and Scorpius in Barasana Cosmology." In *Ethnoastronomy and Astronomy in the American Tropics,* ed. A. Aveni and G. Urton, 183–202. Annals of New York Academy of Sciences 385.

Magana, E. 1988. *Orion y Mujer Pleyades: Simbolismo Astronómico de los Indios Kaliñat Surinam.* Dordrecht, Foris.

Makemson, M. 1938. "Hawaiian Astronomical Concepts." *American Antiquity* 40:370–383.

McCleary, T. 1997. *The Stars We Know: Crow Indian Astronomy and Lifeways.* Prospect Heights, IL, Waveland.

McDonald, J. 1998. *The Arctic Sky: Inuit Astronomy, Star Lore and Legend.* Toronto, Royal Ontario Museum.

McGee, R., and K. Reilly III. 1997. "Ancient Maya Astronomy and Cosmology in Lacandon Maya Life." *Journal Latin American Lore* 20:125–142.

Reichel, E., and J. Arias. 1987. *Etnoastronomias Americanas.* Bogota, Universidad Nacional de Colombia.

Sosa, J. 1989. "Cosmological Symbolic and Cultural Complexity among the Contemporary Maya of Yucatan." In *World Archaeoastronomy,* ed. A. Aveni, 130–142. Cambridge, Cambridge University Press.

Tedlock, B. 1992a. *Time and the Highland Maya,* rev. ed. Albuquerque, University of New Mexico Press.

Tedlock, B. 1992b. "The Road of Light: Theory and Practice of Maya Skywatching." In *The Sky in Mayan Literature,* ed. A. Aveni, 18–42. New York, Oxford University Press.

Urton, G. 1981a. *At the Crossroads of the Earth and the Sky: An Andean Cosmology.* Austin, University of Texas Press.

Urton, G. 1981b. "The Uses of Native Cosmology in Archaeoastronomical Studies: The View from South America." In *Archaeoastronomy in the Americas,* ed. R. Williamson, 285–304. Ballena Press Anthropology Papers 22.

Vogt, E. 1997. "Zinacanteco Astronomy." *Mexicon* 19(6):110–116.

Wilbert, J. 1981. "Warao Cosmology and Yekuana Roundhouse Symbolism." *Journal of Latin American Lore* 7(1):37–72.

Williamson, R., and C. R. Farrer, eds. 1992. *Earth and Sky: Visions of the Cosmos in Native American Folklore.* Albuquerque, University of New Mexico Press.

Zeilik, M. 1985a. "The Ethnoastronomy of the Historic Pueblos, I: Calendrical Skywatching." *Archaeoastronomy* 8 (Supplement to *JHA* 16):S1–S24.

Zeilik, M. 1985b. "The Ethnoastronomy of the Historic Pueblos, II: Moonwatching." *Archaeoastronomy* 10 (Supplement to *JHA* 17):S1–S22.

CHAPTER FIFTEEN

Current Astronomical Practices among the Maya

Judith A. Remington

INTRODUCTION

The importance of divinatory time-counting in ancient Mesoamerica has long been recognized (Thompson 1950, pp. 66–103). The fact that ten of the seventeen relatively complete surviving pre-Hispanic codices are calendrical astronomical (Peterson 1959, pp. 237–239) indicates that the cultural importance of this and associated concepts may not have been sufficiently stressed by modern investigators. Indeed, a world view based on a set of calendrical-astrological-astronomical premises seems indicated for Mesoamerica. Spanish chroniclers at the time of the conquest, such as Sahagun (1938) and Landa (Tozzer 1941), can be used to help explain these premises, but only if the assumptions upon which they were based are known.

First published as "Current Astronomical Practices among the Maya," by Judith A. Remington, in *Native American Astronomy*, ed. A. Aveni. © 1977 by the University of Texas Press. All rights reserved.

At present the data are inadequate to deduce the basic assumptions of the astronomical approaches of the ancient Mesoamericans. Without this information, it is difficult to determine whether or not seemingly related events are, in fact, coincidences. Thus, individuals presently studying the astronomical-calendrical implications of the codices, the artifacts, the architecture, and even the social organization of the ancient Mesoamericans are severely handicapped.

Few systematic efforts to recover astronomical data from the descendants of the ancients have been made. In an attempt to discover if survivals of non-Western cosmological assumptions, practices, and beliefs could be found, I spent three months during the late spring and early summer of 1974 in highland Guatemala among Cakchiquel and Quiche speakers. My particular point of departure was cosmology, but in Mesoamerican culture this is an extremely inclusive term. It includes astronomy, astrology, spirits, the gods, animals, humans, and the works of all of these. I had intended to limit myself to astronomy, but I found that the interest implied a form of shamanism; this required that it all be taken as a whole.

My approach was essentially exploratory. Preliminary work had exhausted the available sources written soon after the conquest; ethnohistoric sources written much later than AD 1600 were found to be inconsistent, bad copies of earlier misconceptions. The alternative was to go ask somebody.

Several considerations led to the choice of the Quiche language group as the focus. In the first place, Guatemala has placed less emphasis than Mexico on absorbing the Indians into Western culture; this seemed to increase the possibility of finding non-Western concepts. In the second place, the Cakchiquel and Quiche written histories demonstrated their membership in the high culture. Their histories state that they were participants in the Early Postclassic migrations; this in turn suggests that they may be the inheritors of Classic ideas. Third, there are a great many Quiche speakers; this increases the probability of preserving a cultural heritage. Finally, the Cakchiquel and Quiche towns near Lake Atitlan, Quezaltenango, and Antigua are relatively accessible.

The central highlands of Guatemala are populated by Maya speakers of Quiche, Cakchiquel, Tzutuhil, Rabinal, and Uspantec. As of 1961, a total of 537,434 Quichean speakers were in the midwest highlands (Whetten 1961, p. 23). Large numbers of Quiche and Cakchiquel speakers, as well as the total population of Rabinal and Uspantec speakers, are found to the east and north of this region. The two villages where I received most of my information were at the extremes of these areas. The Cakchiquel community, in the department of Guatemala, is at the northeasternmost extension of the region of Cakchiquel speakers. The Quiche community, in the department of Quezaltenango, is at the westernmost extension of the region of Quiche speakers.

Although I was looking for astronomical-astrological-cosmological data in both towns, the bulk of the astronomical-astrological data came from the Quiche community and the bulk of the astronomical-cosmological data came from the Cakchiquel community. The data presented in this paper are limited to those which fall into the Western cognitive category of astronomy.

Judith A. Remington

TIME

260-Day Calendar

The 260-day calendar, *cholquih* according to Recinos and Goetz (1953, p. 28), is the *tzolkin* of the Yucatec Maya, the *tonalpohualli* of the Mexicans. A repeating sequence of twenty named days is combined with a repeating sequence of the numbers from one to thirteen. The calendar is used for divination. Only the Quiche informants gave information about the divining calendar. Although the names of the days are general information, the order of the days and the name of the current day are not. Specific knowledge of the calendar is the property of the shaman.

The days given correspond with one exception to those listed by Thompson for the Quiche (1950, p. 68), with two exceptions to those given by Recinos and Goetz for the Cakchiquel (1953, p. 29), and with four exceptions to those given by Caso for Quichean (1967, pp. 8–15). The meanings of the names of the days have been discussed by Thompson (1950, pp. 68–88) and Caso (1967, pp. 8–15) in great detail. Recinos and Goetz (1953, p. 29) give glosses which are close to what I was given. In most cases, however, the meaning of the name is not directly relevant to the divinatory meaning of the day even though it may have been in the past.

Each day has two divinatory meanings. The first meaning is more or less astrological, in that it refers to the characteristics of individuals born on that day. The second meaning is more properly divinatory, in that it refers to the kinds of things which are appropriate to that day.

365-Day Calendar

Neither the Cakchiquel nor the Quiche informants gave any information about the old 365-day calendar (Yucatec *haab*) or the old 400-day calendar of the Quicheans. The Gregorian calendar is used exclusively.

Seasons

Maya winter (rainy season) changes to Maya summer (dry season) when the nights begin to be long and the days begin to be short. Conversely, Maya summer changes to Maya winter when the nights begin to be short and the days begin to be long. This is the reverse of the Western definition of winter and summer. The two times of the year when the sun is directly overhead are also important; this event is perceived as lasting a day and a night. The midpoint of the year is said to be August 20, the end of the general growing season.

ORIENTATION IN SPACE

Directions

The Cakchiquel informants gave the directions in Castellano (Latin American Spanish), but the Quiche informants had indigenous terms; south and north are both

Current Astronomical Practices among the Maya

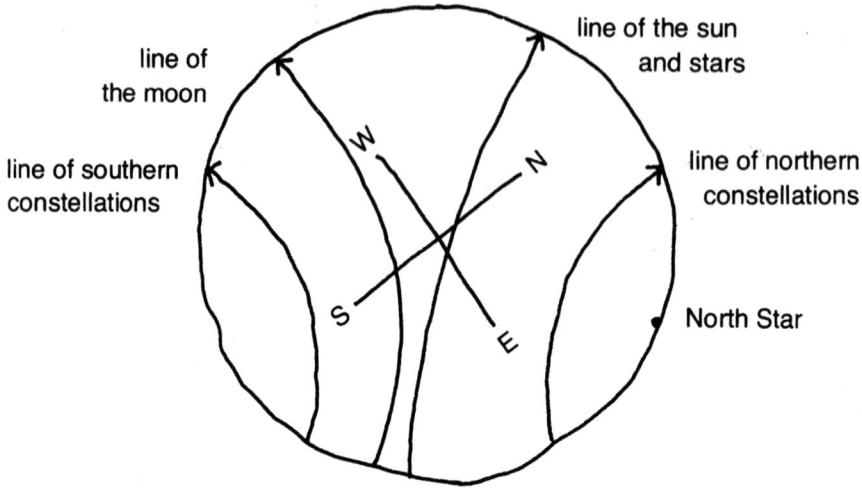

15.1 Rainy-season celestial paths (March 22–September 22).

known as *k'e lik* ("to the side"), east is called *axix* ("above"), and west is called *ikem* ("below").

Celestial Paths

Two maps (Figs. 15.1 and 15.2) were traced from free-hand drawings made by a Cakchiquel informant; other informants checked the drawings and agreed to the accuracy. It can be seen that east is drawn as the base of the maps; astronomic north and south seem to be skewed toward the east.

Certain relationships were suggested by the figures and were confirmed in conversation. The ideal path of the sun and the moon intersect at the equinoxes. In this ideal picture, the moon should be to the south when the sun is to the north, and conversely.

Three classes of stellar paths are recognized. The first is the path of the northern circumpolar or near-circumpolar stars. These are seen as short arcs. Given the irregularity of the countryside, the low latitude, and the haziness of the countryside due to dust and smoke in the dry season and humidity in the rainy season, this is an understandable simplification. Into this category falls Ursa Major. The second category consists of stars which make short arcs to the south. The Southern Cross falls into this category. The third category consists of stars which pass overhead; these are perceived as those which make near half-circles or which are crossed by the paths of the sun and the moon. Orion falls into this category.

The sun was said to rise to the north when the nights are short (rainy season) and to the south when the nights are long (dry season). The length of the night seemed to be the basic reference point, as the length of the day was mentioned secondarily, if

Judith A. Remington

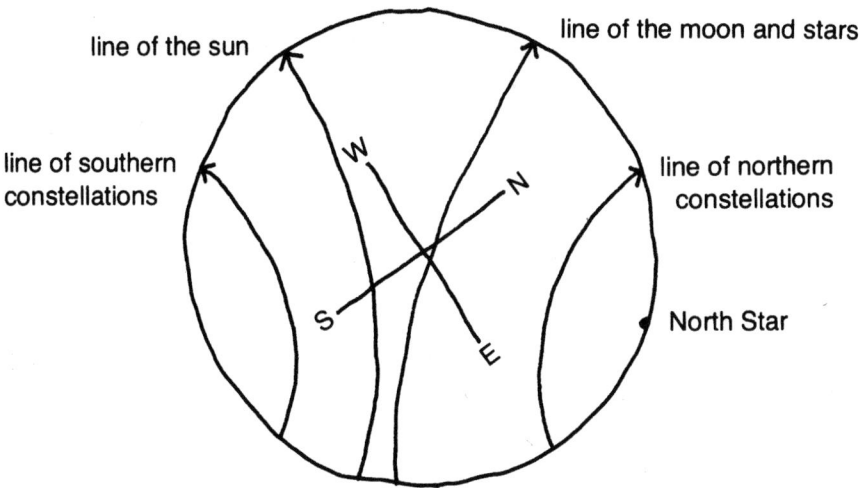

15.2 Dry-season celestial paths (September 22–March 22).

at all. I have reproduced the maps exactly as drawn, although the statement that the sun rises to the north in the rainy season would seem to be contradicted by them. The sun appears to rise to the southeast in both maps, although it rises more to the north in the rainy season. Lacking further information, it therefore seems unwise to draw inferences, however tempting, from the maps.

When working with informants, I used traditional star maps and tables I had worked out which gave the approximate time of day (± one hour) and time of year (± ten days) on which a given hour circle of right ascension was rising, culminating, or setting. One morning session, in order to determine whether or not the informant and I were discussing the same star, I asked where it was now. The question was vague, as I meant to ask where it would be when the sun set. The informant, however, thought for a moment, then pointed downward and eastward at an angle of about thirty degrees. He then pointed upward to where it would be at sunset: "Va por alla." My table checked, and thereafter, when I asked other informants similar questions, I received similar responses.

ASTRONOMY

Eclipses

An eclipse of the sun or moon is described by a Cakchiquel phrase which glosses as "the sun (moon) carries sickness." Cakchiquel women, according to the male informants, are more frightened and say, "The sun (moon) is dying today (tonight)." (For comparison, see Tozzer 1941, p. 138, n. 639.) Not only is the sun (moon) sick itself, but in fact it "carries sickness." Children born near the time of an eclipse will have impediments, such as blindness, muteness, or deafness. Women who leave their houses must be thoroughly covered lest their future children be endangered. In addition, machines will not function correctly. (For comparison, see Thompson 1970, p. 243.)

For the Cakchiquel, a solar eclipse is much more dangerous than a lunar eclipse because, in the event of a solar eclipse, evil spirits of all types come out of the depths of the earth to seize the people. In both types of eclipses, however, the danger to the people is shared by the eclipsed body. The first obligation of the people (including the women who, although covered, are taking a serious risk) is to help the sun or moon. This is done by going to the tops of the hills with every form of noisemaker available, from drums to flutes to bowls beaten with sticks. This noise helps the sun or moon to evade the sickness or death which threatens it (see Thompson 1970, p. 235).

The Quiche use the phrase *skame k'x (k^x)*, or "the sun (moon) died." When an eclipse is about to occur, the people are warned; the church bell may even be rung. People close themselves in their houses and no one is allowed to leave. If it is a solar eclipse, they are afraid that looking at the eclipsed sun will freeze their eyes and cause blindness. Looking at an eclipse of the moon, however, brings all kinds of sickness as with the Cakchiquel. The Quiche also use various kinds of noisemakers. Once these are ready, the whole family kneels, uses the noisemakers, and laments the eclipse. In the case of the moon they say, "I hope she doesn't die. I hope the moon goes on a straight path; perhaps she is in serious trouble because of what is happening." The people may cry in their grief. As one informant said, "Half of the people here call the moon 'lady,' old moon (*mama k^x*). The moon is a god; the sun is a god. He of the church who is god is only god of the earth."

Sun

The word for sun is also the word for day. The sun therefore not only defines the day, it is the day. As mentioned above, the sun is a god; it is occasionally called *ʔx k'aumaʔ* by the Cakchiquel. As *aumaʔ* means "grandfather" and *k'a* is a possessive prefix, the phrase glosses to "the sun, our grandfather." The sun is also the point of reference for the stars. A star's position is given by the hour at which it is in a certain place and by its position in front of (rising before) or behind (rising after) the sun.

Moon

Three words are given for the moon: *ik^x* in both Quiche and Cakchiquel, *nan* ("old woman") in Quiche, and *k'autiʔ* ("our grandmother") in Cakchiquel. In general conversation the Quiche used the *ik^x* form. The Cakchiquel used *k'autiʔ* generally; the *ik^x* form was only used in formal phrases describing the phases of the moon. As mentioned above, the moon is a god.

The moon is described, beginning with new moon, by its appearance on the first evening, the second evening, and so on. The new moon is called by phrases which gloss to "night of the moon" and "the moon is born." The second evening is described by phrases which gloss to "now the moon is seen" and "first day of the moon." From the third evening to first quarter, the gloss is "second (third, etc.) day of the moon." At first quarter, two glosses are used, "the moon is filling" and "half time of the moon." From then until full moon, the evening count continues.

At full moon, the descriptions change to "first evening of the full moon" and "round moon." From then until last quarter, waning, the gloss becomes "the moon rises at seven (eight, etc.)." The waning quarter moon is known again as "half time of the moon" or "midnight moon." The count of rising time continues until the last night before new moon, known as "the end of the moon (month) today." The word ik^x is used here in the sense of month. This was also noted by Lothrop (Recinos and Goetz 1953, p. 28).

The phases of the moon are categorized in yet another way. It is tender while it is waxing, seasoned while it is waning. The moon is dangerously tender, however, when it is new (from 12 to 72 hours, depending on the informant) and when it is full (for 12 hours). When the moon is dangerously tender, there are many prohibitions. If one chops wood, one cuts oneself. If one succeeds in cutting the wood, it cannot be seasoned properly; it twists and bends. If one makes charcoal, it will not burn. If one plants, there will be much growth but no harvest. If one harvests, the crop rots. If one picks fruit, the fruit is juicy but bitter. If one cuts oneself, one bleeds excessively. People are lethargic and work poorly. They are particularly susceptible to sickness. During the rainy season, it rains more when the moon is tender than it does when the moon is seasoned.

The Quiche plant by the calendar, but harvest by the moon. They do not distinguish between dangerously tender and tender for harvesting. They harvest only after the full moon, when the moon is seasoned. This rule is vigorously applied only to milpa crops, specifically corn and dry beans, which are planted around 20 March "to await the rains." Crops planted on irrigated lands and late crops, such as greens, sown between September and December, may be harvested when ready.

The Cakchiquel both plant and harvest by the moon. Their harvesting rules are the same as those of the Quiche. Planting, however, is bound to both the calendar and the moon. The third day of the moon (fourth from new), when the moon is on "this side," is the ideal day for planting. Crops sown on this day grow more rapidly and mature earlier. I was told that a lime tree planted on the third day matures in seven years, but that eleven years are required if it is planted on any other day. Crops can be sown from the third day waxing to the quarter moon waning, at which time the moon is too seasoned and crops will not grow.

The middle of April, during the allowable lunar phases, is the ideal planting time in the Cakchiquel village. During this period, anything can be grown. August is the second best month; 20 August, the middle of the year, marks the end of the general growing season. After this date, beans are the only milpa crop which can be sown, from the eighth to the twenty-ninth of September, when the moon is favorable. In late December and early January, when the three stars of Orion's belt are on the eastern horizon at sunset, chayote can be sown. Ideally, the sowing takes place on 6 January, the Day of the Kings in Roman Catholicism. The three stars of the belt are commonly known as Tres Reyes (Three Kings). Lunar age also takes precedence in this case.

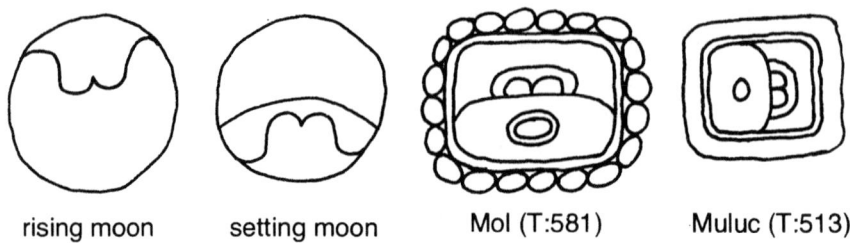

| rising moon | setting moon | Mol (T:581) | Muluc (T:513) |

15.3 Informant's moon drawings with comparison glyphs.

The moon is considered to be yellow. It is inhabited by a rabbit, *umul*. I asked for a drawing of the moon and received two (Fig. 15.3). When the moon is rising, the mouth is down (*boca abajo*).

Planets

Venus is the planet of most importance. The Cakchiquel call it *locera macamil* from Castellano *lucero* or "morning star" and Cakchiquel *macamil* or "very bright star." The Quiche call Venus "Santiago" when it is the morning star and *raskap*, "thing of night" or "something late," when it is the evening star.

According to the Quiche informants, Venus passed from behind the sun (when it was called *raskap*) to in front of the sun (when it was called "Santiago") in January of 1974. In February 1974, Santiago rose at 4:30 AM July is the month of Santiago; 3:00 AM is the hour of Santiago in the month of July. The informants said that Venus would not appear again as the evening star until January of 1975. All this is accurate. I was unable to determine if the positions of Venus were calculated over eight-year intervals (as in the Dresden Codex) or if only running observations were kept.

The Cakchiquel informants gave very little information about Venus. They did, however, mention a "comet" which appeared at 5:00 AM in February of 1974, together with the *macamil*. This may have been a reference to Comet Kohoutek or it may have referred to the appearance of Jupiter as a morning star, approximately a month after the heliacal rising of Venus. Incidentally, this was taken as a sign that the world would end in the year 2000.

I was unable to elicit specific information about planets other than Venus. The Cakchiquel informants did say that there were four *macamil: locera macamil, macamil, macamil g'vn* ("yellow"), and *macamil kvku* ("red"); these seem to correspond to Venus, Jupiter, Saturn, and Mars. This is tentative, however, as the Cakchiquel used *macamil* freely when speaking of any bright star.

Meteors

Meteors are known as "excrement of the stars." No other information was given.

Judith A. Remington

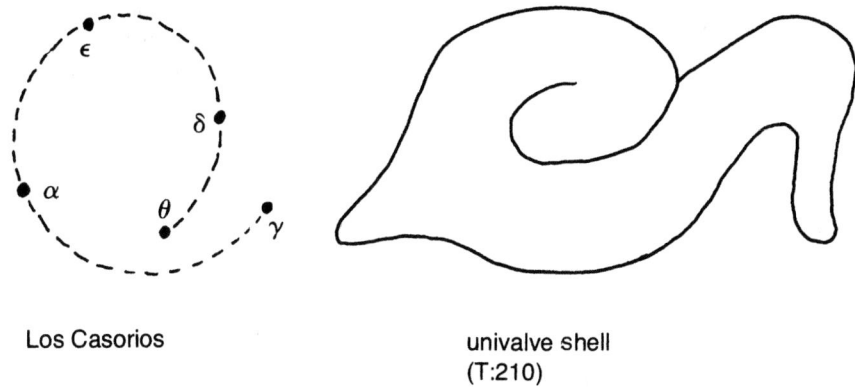

Los Casorios

univalve shell
(T:210)

15.4 Visualization of Maya constellation Los Casorios in Taurus with comparison glyph.

Stars

The stars, *camil* or *c'mil* in Cakchiquel and *c'umil* in Quiche, are used primarily to tell the time of night or to herald the time of year. I have attempted to identify them on the basis of the information given by the informants. Unfortunately, by the time I had learned enough to elicit the kind of information available, the rainy season had begun and viewing was bad to terrible.

The first groups of stars mentioned (and often the only group known) by Maya speakers in Guatemala is the Pleiades. The name *moʔots* (occasionally *muʔuts*) is general, from Chuj to Quiche. In Chuj the name is said to mean "those who travel together." In Quiche it is said to mean "a handful." The Cakchiquel say it is the name of the Seven Kids, Siete Cabritos in Castellano. In late June, *moʔots* is said to rise at 3:00 AM It goes from behind the sun to in front of the sun when the rains begin. This occurs roughly in mid-May. Finally, it is considered the sign of the month of November, when it is called *Fetal akap,* "signal of night," because it rises at sunset. It is said to be visible from 7:00 at night until 6:00 in the morning.

Only the Cakchiquel informants mentioned the constellation of the Wedding Party (Los Casorios), *aʔakolaʔ.* This is the head of Taurus, in which Aldebaran is the godfather, ε Taurus is the godmother, θ and δ are the couple, and γ is the priest or the onlookers. The fact that the terms used come from Catholicism suggests that this constellation may be post-Hispanic. The possibility of pre-Hispanic origin should not be ignored, however, as the group is visualized as a spiral, somewhat similar to that of the conch spiral (Fig. 15.4).

The second most frequently mentioned constellation is that of Orion's Belt and Sword (Fig. 15.5). It is known by the Cakchiquel informants as the Three Kings (Tres Reyes) and was mentioned above in connection with the January planting of chayote. A child born at this time is given the name of one of the kings. It is known in Cakchiquel as *ausiʔ camil,* "three stars." The one Chuj informant gave its name as *os t'ilan* "three in line." To the Quiche it is the Three Marys (Tres Marias); in Quiche it is

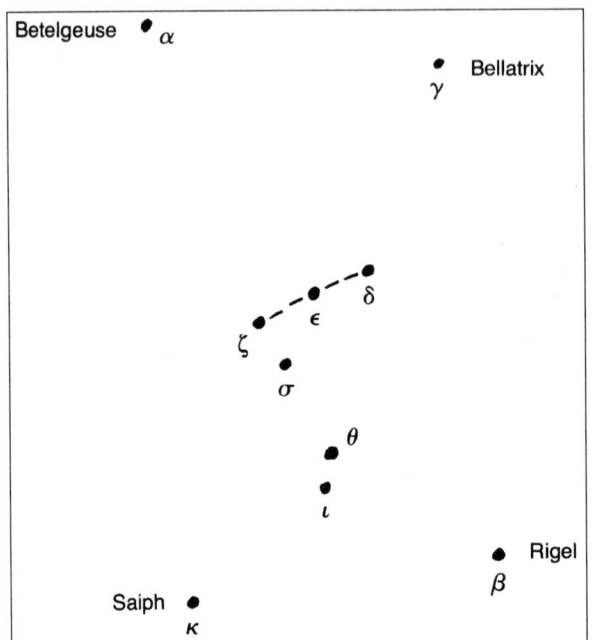

15.5 Visualization of Maya constellation Tres Reyes or Tres Marias in Orion.

c'ok, "the right angle," and comprises both the belt and the sword. For the Quiche the constellation is the sign of the month of December, when it is also called *retal akap*, "signal of night," because it rises at sunset. It is said to be visible from 7:00 PM to 6:00 AM.

The Milky Way is called the Course of Santiago (Corrida Santiago) by the Cakchiquel; my informants did not know of a Cakchiquel name. it is used to predict the coming of the dry season. The Quiche informants called it *raskap*, "thing of night" or "something late," the same term which was used for Venus as evening star. It rises in the evening around Christmas and is used to predict the coming of cold weather.

Only the Quiche informants mentioned the Gemini, known as *kiep c'umil*, "two stars." They are described as two stars which go together, the first of which is "half cloudy" and the second of which is a very large star which follows by about half an hour. There exists the possibility that the two stars referred to are Procyon and β Canis Minor, but ethnohistoric sources indicate that Gemini formed a pre-Hispanic constellation (Tozzer 1941, pp. 131, 192 n. 1015, 220). The constellation is the sign of the month of January, when it rises at sunset. It is said to be visible from 7:00 at night until 6:00 in the morning.

The star Regulus is known by the Quiche as *hun c'umil*, "one star." It is the sign of the month of February. It rises at sunset and is visible from 7:00 in the evening until 6:00 in the morning. The Cakchiquel informants mentioned a star called *la gran lucero* or *macamil* (both of which mean "bright star") which rises at 6:00 in the afternoon on February 9. This is probably Regulus, as 6:00 or 7:00 in the evening (or

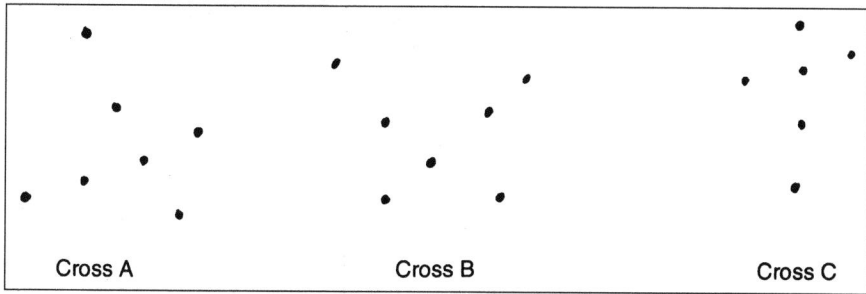

15.6 Copies of drawings of Thieves' Dagger constellations, in order of time but without regard to orientation.

morning) seems to refer to sunset (or sunrise) rather than to the precise hour. If this is the case, the Cakchiquel use the appearance of Regulus as a signal to begin preparing their milpas. This star is then used as an alarm clock until the middle of April when it sets at about 3:00 AM.

Another large star is of importance in the months of January and February. it is known to the Quiche as $sk^xek'ap$, "dark" or "giving notice of the night." It is the opposite of Venus, which announces the day. It is in the "middle of the sky" shortly before sunset in November and disappears in the west at sunset in the latter part of January. Its disappearance means that the dry season (Maya summer) is drawing to an end. It seems likely that $sk^xek'ap$ is Altair.

The next group of interest is the Thieves' Dagger or Cross to the Cakchiquel, the Thieves' Cross to the Quiche. The Cakchiquel call it *krus aloʔomaʔ*; the Quiche call it *ki krusil eloʔomap*, "thieves' cross" in both cases. I have had considerable difficulty in identifying this constellation because it seems that there are at least three similar constellations with the same name. One Cakchiquel informant specifically stated that in mid-May there are three thieves' daggers; all are seen in the south, one at midnight, one at 2 to 3 AM and one at 4 to 5 AM. In early June, one was said to be at its midpoint at 6:00 AM (or sunrise). Figure 15.6 gives copies of informants' drawings of the constellations without regard to orientation. The second cross, in time, is said to be more to the north than the third. Under the assumption that culmination times were given in all three cases, the best estimates of the positions thus seemed to be Right Ascension $21–22^h$, $18–19^h$, and $16–17^h$. Further discussion elicited the fact that there are very many stars in the region of the first two crosses to appear. Even though this identification was verified only with the star maps, the second cross appears to be a group in Sagittarius, consisting of σ, φ, δ, and γ horizontally, and λ, δ, ε, and η vertically (Fig. 15.7).

The first cross, in time, is probably located in Scorpio, as it seems that there was a pre-Hispanic constellation in or near Scorpio (Tozzer 1941, p. 193). Also, when doing linguistic work with Valley Zapotec informants, I was told of a cross consisting of many stars which was found in the Alacran (scorpion). This seemed more likely to be

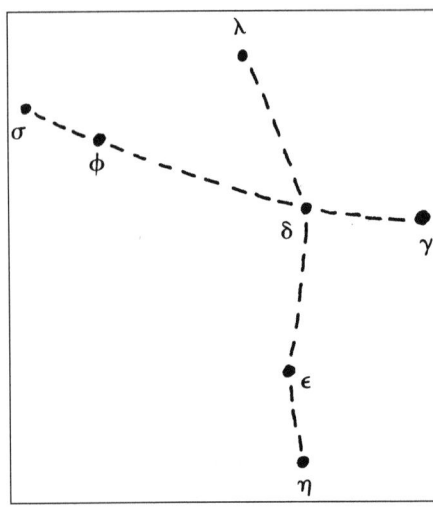

15.7 Possible visualization of Cross B in Sagittarius.

in the Milky Way (therefore Sagittarius?) than in Scorpio, however.

The third cross, in time, seems to fall in Piscis Austrinus and Grus, although this is also speculative. A possible construction which is consistent with Figure 15.6, Cross C, consists of Fomalhaut, α, β, and γ Grus, plus small connecting stars (Fig. 15.8).

The Quiche informants also spoke of a Thieves' Cross that could be seen above from midnight to 3 AM in mid-June. This implies culmination at approximately 1:30 AM and tends to verify the identification of the Sagittarius cross. One informant said that the cross was not straight but rather was inclined to its left. This informant also mentioned a cross which could be seen from January through March, in March at midnight. The Southern Cross fits this description. Obviously a great deal more work needs to be done with these constellations.

A Cakchiquel informant told me of a custom followed by the people of Chinautla. There people receive a "secret" from the Thieves' Dagger. When this constellation culminates at 4 AM, the boys of six or seven years of age are sent out with sticks of cane to stage a mock battle in defense of these stars. This is done so that the boys will use the machete well when they are adult.

The Quiche informants have a different use for the constellation. It is believed that these stars watch over thieves, who also walk at night. Before committing a crime, specifically robbery, it is necessary to get their protection; it brings luck. A potential thief must obtain two large, fully ripe bananas as offerings. Thus, when someone walks past a house at night carrying two bananas, one knows he is considering robbery. The putative thief goes out in the country at midnight, offers the bananas, and makes a speech to the constellation. He uses magical words which attract money to him. This can also be done with *c'ok* when it is visible at midnight.

The last constellation mentioned by the informants was the Big Dipper. The Cakchiquel call it *carro k'o reh*, "car with a tail," and the Quiche call it the Siete Cabreras, "the seven goatherds." No Quiche name was given. The only information given was that the constellation could not now be seen (because of the rainy season?).

I was told by the Cakchiquel informants that at all times, during all seasons of the year, four large southern stars (star groups?) can be seen in the course of the night. In the middle to end of May, one rises at 4:00 PM, one rises at sunset, one rises at 2:00 AM, and one rises at 5:00 AM. If one makes the assumption that they culminate 3–4

15.8 Possible visualization of Cross C, with Fomalhaut and stars in Grus.

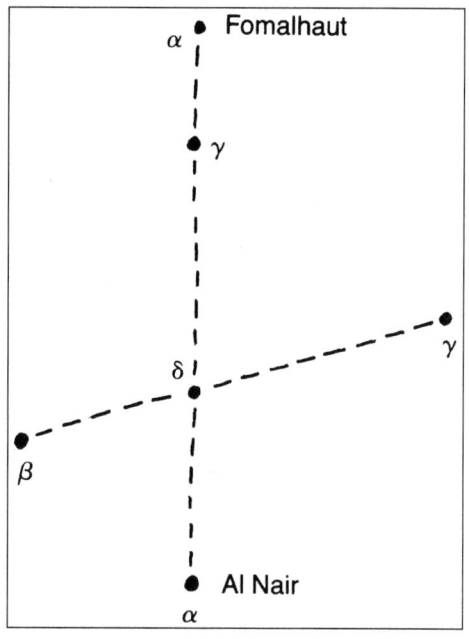

hours after rising because they are far to the south, we have Right Ascension 12, 14, 22, and 1. This suggests α, β Crux, α, β Centaurus, Fomalhaut, and Achernar. The latter star may also have been mentioned by the Chuj informant, who noted a star called *yecl pasco*, "sign of Christmas," which he said was culminating at sunset on Christmas day.

One final note seems indicated. When I was working with the Quiche, I heard that there had been a flying-saucer incident two years before. I asked the shaman who was helping me that day about it. He thought it was most probable the objects were experimental craft of either the Russians or the Americans. I suggested that the technology seemed rather more advanced. He replied that any people capable of putting men on the moon could also develop advanced aircraft. He seemed a little indignant at my lack of faith in technology.

CONCLUSIONS

The three preceding sections have presented current data which indicate that present-day Maya cosmology contains many elements which seem to be survivals of pre-Hispanic times. Among these elements are the 260-day calendar, a non-European spatial orientation, and the names of several stars and constellations. It seems probable that some of the beliefs relating to solar and lunar eclipses, as well as some of the planting and harvesting practices, are also of pre-Hispanic origin.

The purpose of the study from which these data were obtained was to determine if pre-Hispanic cosmological survivals indeed existed, and, if so, if such survivals existed in sufficient quantity to warrant further investigation. I believe that the pilot study showed that both questions can be answered with an unqualified affirmative.

If we can establish a good body of pre-Hispanic data, many of the problems currently facing investigators in associated fields of study, from archaeology to iconography, will disappear. At worst, the variables with which we all must deal will be better defined. It is my hope that more investigators will find the problem of survivals as challenging and that some of us will be meeting again soon, in the field.

ACKNOWLEDGMENT

I am indebted to Anthony G. Ketterling for drawing the figures.

REFERENCES

Caso, A. 1967. *Los calendarios prehispanicos.* Instituto de Investigaciones Historicas. Mexico City, Universidad Nacional Autonoma de Mexico.

Recinos, A., and D. Goetz. 1953. *The Annals of the Cakchiquels.* Norman, University of Oklahoma Press.

Sahagun, B. de. 1938. *Historia general de las cosas de la Nueva Espana,* ed. P. Robredo, 5 vols. Mexico City.

Thompson, J.E.S. 1950. *Maya Hieroglyphic Writing: An Introduction,* 3rd ed. Norman, University of Oklahoma Press.

Thompson, J.E.S. 1970. *Maya History and Religion.* Norman, University of Oklahoma Press.

Tozzer, A. M. 1941. Landa's relacion de las cosas de Yucatan. *Papers of the Peabody Museum, Harvard University,* no. 18. Cambridge, MA.

Whetten, N. L. 1961. *Guatemala: The Land and the People.* New Haven, CT, Yale University Press.

CHAPTER SIXTEEN

Quichean Time Philosophy

Barbara Tedlock

Beginning with the important work of Heinrich Berlin (1958) and Tatiana Proskouriakoff (1960), there has been a revolution in our thinking about the supposed Mayan glorification of or obsession with time. Previously it had been assumed that all Mayan hieroglyphic texts dealt solely with recurrent cyclical astronomical matters (Thompson 1954). Then, with the sudden breakthrough in hieroglyphic textual analysis it became clear that many of the texts on Classic Maya stelae recorded births, dynastic successions, alliances, marriages and deaths of individuals. Nonetheless, not all Mayan texts, even on stone monuments, concern such historical matters. Thus far, the extant Precolumbian codices (Madrid, Paris, Dresden and Grolier) are believed to contain primarily cyclical, ritual and astronomical matters. We have a difficult

---

From *Calendars in Mesoamerica and Peru: Native American Computations of Time*, ed. A. Aveni and G. Brotherston. British Archaeological Reports International Series 174 (1983). Reproduced by permission from the author.

theoretical problem here because in Western thought, beginning with the Classical Greeks, temporal modes are described as either cyclical and recurrent *or* as linear and non-repetitive. This dichotomy ultimately rests on the contradiction between Plato's view of time as the measure of the motion of a body and Aristotle's view of time as the motion of bodies. Since Classic Mayan cyclical and lineal time, models are clearly present archeologically in the same sites at the same time horizon, it would appear that the Mayas had somehow resolved this apparent contradiction, if indeed they even took these models to be in conflict or contradiction. Our problem, then, is to examine this supposed contradiction in order to construct a new model of Mayan time philosophy.

As a first step, let me assume that the Maya may have had different times for separate provinces of their reality—biological, astronomical, mechanical, psychological, historical, religious, social—and that these various time perspectives underwent a process of "totalisation" (as Sartre would call it) recorded and rationalized within the intermeshing cycles of their calendars. I come to this intuition from the study of modern Mayan multimetrical temporal concepts and rituals involving dialectical thought patterns which go far beyond the dialectics of polarization (thesis, antithesis, synthesis), as historically exemplified in Hegelian and Marxist circles, to include the dialectics of complementarity, overlapping or mutual involvement, and reciprocity.[1] These insights concerning the dialectics of Mayan temporal concepts come from reflections upon data gathered during twenty months of anthropological fieldwork, including a formal apprenticeship to a calendar priest in Momostenango, Guatemala.[2]

The community of Momostenango is located in the Midwestern Highlands of Guatemala at 15.04° north of the Equator. Although this community is located in the tropics, the climate is quite temperate because it lies from one to three thousand meters above sea level. The population of this municipality is approximately 45,000 persons, of whom the overwhelming majority (98%) are Quiché speakers. Momostenango is perhaps best known to travelers for its fine woolen blankets and to ethnographers for its celebration of 8 Batz', the largest ongoing calendrical ritual anywhere in Mesoamerica today that is scheduled according to the 260-day (*rajilabal k'ij*[3] "counting of suns and days") sacred almanac. Here reside nearly 10,000 initiated female and male "burners" (*poronel*) and "daykeepers" (*ajk'ij*), who are formally trained and initiated in calendrical rituals (in accordance with both the solar 365-day cycle and the sacred 260-day cycle) by a formal three-tiered hierarchy of male priest-shamans known as "mother-fathers" (*chuchkajawib*).

Unlike many other Guatemalan communities, where calendar experts have been persecuted and peripheralized by Roman Catholic priests and indigenous lay *catequistas,* the Momostecan hierarchy of calendar experts has prevented such religious domination by carefully controlling access to the coveted positions of eldership within the community. As elsewhere in Mesoamerica, one may speak of a "civil-religious hierarchy" in Momostenango, but here the bridge between religious and governmental duties is provided by this group of calendricists rather than by the leaders of the local Catholic confraternities. Thus, men who are actively working toward the respected position of elder within the community tend to avoid serious involvement in the confraternities and instead undergo further training in calendrical ritual, so as to ascend in the local hierarchy.

The first level of religious training is that of a simple "burner," a man or woman who may approach the outdoor community altars in order to burn incense and make offerings to the ancestors and deities. Novices undergoing training for this office are known as "burdens" (*ëkomal*) during the 65-day period known as "washing for the work service" (*ch'ajbal chac patan*), which begins shortly after noon on 1 Cawuk. On this date the teacher arrives at Paja', "In the Water," a shrine which is located in a *barranca*, on the edge of a stream just west of the town center. There the teacher begins the "back part of the path" (*rij ubinibal*, literally "rear or back of the instrument-for-walking or traveling"), which is the current expression in Quiché for indicating the beginning of a set of calendrical rituals. This spatio-temporal concept places the speaker in the position of starting off *behind* the final or "big" day of the initiation rather than as standing *before,* as we express it in phrases such as "the night *before* Christmas."

Beginning the 65-day ritual series slightly after high noon, when the sun is directly overhead, raises a question as to when the Quiché begin and end their days. This issue has been much debated in the literature on other Mayan groups, with noon and sunset being the favourite candidates for the boundary line between two successive days (La Farge 1930: 657; Redfield and Villa Rojas 1934: 184; Lincoln 1942: 110; Villa Rojas 1945: 143–44; Thompson 1950: 102). At first glance the case of the 1 Cawuk would seem to support the noon theory, and so would the way in which the Momostecans sometimes take notice of the fact that the bells of the parish church are striking twelve. Having stopped what they are doing long enough to glance up at the sun, they may then say a brief prayer with their faces still uplifted slightly, addressing the sun by invoking the day—for example, *Sa'j la Ajaw Wajxakib K'anil*, "Come here, Lord 8 K'anil." But then, if several people happen to be together at this time, they will greet one another (as if they had just met) by using the expression that is appropriate for the rest of the afternoon: *xbe k'ij*, "the sun (day) went," in which *x-* indicates completed action—the implication being that the sun has just already done something, rather than that something has just begun. This is quite a disorientating experience to the outsider, especially if that outsider already knows that high noon is ordinarily referred to conversationally in Quiché as *nic'aj k'ij*, "middle of the sun (or day or time)." How can the "middle of the day" also be the time when the day is greeted and asked to enter or come here, and simultaneously be the time when the day "went"? In what sense is noon the beginning, middle, and end of the day, sun, and time? This case is partly resolved when one learns that the path of the sun is referred to in Quiché as *oxib uxucut* "three corners," consisting of sunrise (*relebal k'ij*, "the coming-out place of the sun"), noon (*nic'aj k'ij*, "middle of the sun, day or time"), and sunset (*ukajibal k'ij*, "the going-down place of the sun"). These three "corners" are considered the three main turning points or transitions in time, in which the influences of day (*k'ij*) and night (*ak'äb*) overlap in a dialectics of mutual involvement.

QUICHEAN TIME PHILOSOPHY

At the 15° latitude where Momostenango is located, night comes on rapidly at sunset, which occurs around 6 PM, with only very slight seasonal fluctuations. At this time in the evening one greets people on the path in Momostenango with *xoc ak'äb* "night entered." Night continues on into early dawn. At this time there are black streaks low on the eastern horizon, with a yellow background that slowly intensifies and deepens into orangeness (*ya' k'ak' chuwi xe kaj*, literally "fire is given there at the bottom edge of the sky"). Then all of the colour slowly disappears and the rising of the sun itself brings about a process called *sakir uwäch ulew* ("the face of the earth whitens or lightens"). If one meets a person on the path at this time the proper greeting is *sakiric* "it is getting white or light," and this continues to be the proper greeting until noon. However, even though the sun has now appeared, this time of day is known as *nimak'äb* "big night," which indicates that the night has grown very large or long and is nearing its end. This seems strange until one realizes that the period of time between dawn and noon is considered an especially delicate, cool time when the sun is slowly climbing up the sky. Then when it reaches noon it is quite strong for a time, but it is already spoken of as if it had completed itself.

In the attempt to decide beginning and ending points for a day, one encounters the same sort of problems I have discussed elsewhere in connection with the misdirected search for a beginning or ending point for the 260-day cycle (Tedlock 1982: 93–97). Just as I discovered that Momostecans were more interested in offering a particular day as the middle of the cycle rather than in speculating on the question of its "beginning," so I must suggest here that the important issue at the scale of a single day is the midpoint. At noon, the day is maximally 8 K'anil (or whatever the Day Lord is), and the influence of 8 K'anil rises before that moment and declines after it. A dream occurring on the previous night, at whatever hour of the darkness, will be spoken of as having been handed from 7 Can to 8 K'anil; a dream of the following night is handed over from 8 K'anil to 9 Toj, and so forth. From the fact that a day has a sharply marked middle it does not follow—at least not in Quichean dialectics—that it has sharply marked beginning and ending points. The moments of sunrise and sunset are certainly used to mark time, but they do not provide absolute boundaries for the influence of successive Day Lords. As for "midnight," that is a moment that is reckoned in accordance with events that belong to the night itself rather than a boundary between two "days," as we shall see later on.

Now, returning to the rites performed on the afternoon of 1 Cawuk, we can see more clearly why this time of day is chosen. Once noon has past, the influence of 1 Cawuk is strong but is already declining, pointing forward, as it were, to the rest of the series of rituals which it begins. Starting with the second day in the series (1 E) and from then on, a teacher may well prefer the early morning hours, before the capacity of a Day Lord to listen to prayers has been worn out. He returns on 1 E to the same shrine he visited the first time (Paja', In the Water), where he again asks permission of the Day Lord—and of his own direct ancestors—to train the novice. After this second ritual he begins to give intensive instruction in time reckoning, cosmology, observation of the sun (*ki'j*), moon (*ic'*), and stars (*ch'umil*), herbal and shamanic curing and

WORK SERVICE	MIXING POINTING
(Burner)	(Daykeeper)
. .	1 Quej
. .	. .
. .	1 Junajpu
. .	8 Quej
. .	1 Aj
. .	8 Junajpu
. .	1 Came
. .	8 Aj
1 Cawuk	1 Cawuk
. .	8 Came
1 E	1 E
8 Cawuk	. .
1 Can	. .
8 E	. .
1 Tijax	. .
8 Can	. .
1 Batz'	. .
1 C'at	. .

16.1

proper ritual behaviour to be followed at the outdoor shrines. Now, the rhythm of his visits to the shrines speeds up to 7- and 6-day intervals in order to intercalate or insert the 8-day series (*wajxakibal*) into the 1-day series (*junabal*) that was started first. With the first 8-day, 8 Cawuk, there is also a shift of locale to Ch'uti Sabal, "Little Declaration Place," which is a shrine located on a hilltop half a kilometer due west of Paja'. Six days later on 1 Can the teacher returns to the low shrine at Paja'.

This series of 1-days and 8-days (see Fig. 16.1) rotates back and forth at 7- and 6-day intervals to produce the following meter: 1 Cawuk + 13 = 1 E + 7 = 8 Cawuk + 6 days = 1 Can + 7 days = 8 E + 6 days = 1 Batz' + 7 days = 8 Tijax + 6 = 1 C'at. Expressed in the language of Western music, this 65-day "back of the path" time period opens with a 13/65 time signature, in which the right-hand figure (65) indicates the unit of measurement (1/65 of the total time period under consideration), and the left-hand figure (13) indicates the number of such units in each measure. In this ritual series, however, after only one measure the meter speeds up and alternates back and forth four times from 7/65 to 6/65 through eight measures, producing an irregular multimeter. This multimeter resolves itself and achieves an exciting asymmetrical balance through the principle of dialectical complementarity, in which the distinctions (7 and 6) are simultaneously in an alternating relationship to each other—7, 6, 7, 6, 7, 6, 7, 6—and in dialectical completion of each other—the original 13 is matched with 7 + 6 = 13, repeated four times. A third type of dialectical complementarity, known as direct opposition, is present in the spatial dimension of the rituals, in the shift from low to high place (Paja'/Ch'uti Sabal); low to high number (1/8); east to west; and wet to dry.

After a person has been initiated as a "burner," she or he may decide to go on to be a "daykeeper," by completing another 65-day "back of the path" permission period known as "washing for the mixing pointing" (*ch'ajbal baraj punto*). This period begins at the Paja' shrine with 1 Quej (see Fig. 16.1), which is then followed 13 days later by 1 Junajpu, also at Paja'. At this point the cycle begins the alternating 7- and 6-day multimeter pattern: 1 Quej + 13 = 1 Junajpu + 7 = 8 Quej + 6 = 1 Aj + 7 = 8 Junajpu + 6 = 1 Came + 7 = 8 Aj + 6 = 1 Cawuk + 7 = 8 Came + 6 = 1 E. The first thing to notice about these two 65-day time periods is that the second one, for the more advanced initiation as an *ajk'ij* "daykeeper," actually falls earlier within the 260-day cycle than the first 65-day period for the training as a "burner" (see Fig. 16.1). Secondly, these two 65-day time periods are overlapping, so that the first and second days of the "washing for the work service" are also, respectively, the third-to-last and final days of the "washing for mixing and pointing." Thus, the teacher completes the chronologically later 65-day period (work service) first and then waits nearly 250 days to begin the earlier (mixing and pointing) 65-day period.

The initiation as either a "burner" or a "daykeeper" is celebrated in a two-step procedure which takes place outside of the 65-day (*rij ubinal* "back part of the path") ritual series. In the case of a "burner" these rituals are performed on 7 Tz'i' and 8 Batz', while in the case of a "daykeeper" the days are 8 Batz' and 9 E. The rituals on 7 Tz'i' take place at either the home of the novice or of the teacher beginning at sunset (*ukajibal k'ij*, "the going down of the sun," which is simultaneously the term for the western direction), culminating with fireworks late in the evening. Then, on 8 Batz', the rituals take place at the 8-day shrine, Ch'uti Sabal, beginning very early in the day, after Venus as morning star (*junajpu*) has risen, or if Venus happens to be an evening star (*rask'äb*) or simply is not visible at all, when the earliest rays of light have appeared among the black clouds at the horizon (*xe kaj* "end of the sky"). If the person initiated will become a burner, then this is the final day of his/her initiation. Later, if the person is being initiated as a daykeeper, s/he will celebrate 8 Batz' once more and then go, on the next day (9 E), to a higher shrine known as Nima Sabal, "Large Declaration Place." The first day of these two sets of initiatory days (7 Tz'i' and 8 Batz', respectively) is known as the "broom" (*mesebal*), while the second day (8 Batz' and 9 E, respectively) is known as the "big day" (*nima k'ij*).

The ideal pattern in Momostenango is to speed up the initiatory process by completing these two overlapping 65-day cycles during just one 260-day period, so that the novice is simultaneously initiated as both a burner and a daykeeper (see Fig. 16.1). In order to "double-time" these two initiations the "permissions" must begin with the second, or more advanced "mixing and pointing" series and then, before the completion of this cycle, the first day (1 Cawuk) of the first set of "work service" must be intercalated. This means that the next day (8 Came) of the "mixing and pointing" set must be picked up after the first day (1 Cawuk) of the "work service" set has occurred. This is followed by 1 E, which occurs as the last day in the "mixing and pointing" and the second day in "work service." The double-time series then continues straight on with the remainder of the days of the work service series. Initiation in this double-timed system is celebrated during the three-day consecutive period of 7 Tz'i', 8 Batz' and 9 E. When the two levels of initiation are combined (through a dialectics of mutual involvement) within a single 260-day cycle, 7 Tz'i' is known as the "broom" (*mesebal*), 8 Batz' is the "eve" (*mixprix*) and 9 E is the "big day" (*nima k'ij*).

The pattern whereby rites of passage are marked by three consecutive days of ritual activities at the end of a double-timed cycle is replicated on the higher levels of the religious hierarchy of Momostenango, ranging upward through the mother-fathers of the patrilineages, the cantons and the town as a whole. Thus, for example, in the case of either of the two mother-fathers of the town, the "back part of the days" before the initiation into this office can theoretically be done in two separate time periods. The first one, the "backpack" or *ëka'bal* (literally "pack frame" or "yoke"), stretching from 1 Quej to 1 Toj (see Fig. 16.2), is divided into 65 + 65 + 52 days = 182 days (or one half of the solar year). This involves the burning of great piles of *copal* incense, praying and making of other offerings for the benefit of Momostenango, and even for the entire world. The second is the "washing of the shrine" or *ch'ajbal rech awas* (literally "washing of the taboo"), which stretches from 1 Toj to 1 Batz' and is likewise divided into 65 + 65 + 52 = 182 days. The two parts can be done in two separate solar years, but in 1976 when the new mother-father for Pueblo Viejo was initiated they were double-timed. The "back part of the days" began with the "later" series, the "washing of the shrine," and ran as follows: 1 Toj + 13 = 1 Ik' + 7 = 8 Toj + 6 = 1 Tz'iquin + 7 = 8 Ik' + 6 = 1 K'anil + 7 = 8 Tz'iquin + 6 = 1 Imöx + 7 = 8 K'anil + 6 = 1 Ix = 65 days. What would have been the "earlier" series, the "backpack," began 13 days after 1 Ix on 1 Quej, and was thereafter intercalated with the "washing of the shrine" series. The combined backpack and washing rituals of 1 Quej were followed by a return, 7 days later, to the washing rituals on 8 Ix, a day which is skipped in the backpack series when it is done alone. With the exception of 8 Ix, the two sets of rituals were done together from 1 Quej until 1 Batz', combining the washing of the town shrines (i.e. removal of previously burned copal) and the backpacking of the town. Next the backpacking continued on, alone, but only until 1 C'at. Thus, in actual practice what could have been 2 × (65 + 65 + 52 = 182) = 364 days were overlapped (through the dialectics of mutual involvement) and reduced to 65 + 13 + 65 + 52 = 195.

As in the case of the combined burner and daykeeper initiation discussed earlier, the initiation of mother-fathers of the town took place on three consecutive days. The first or "broom" day was 9 E, the second or "eve" was 10 Aj and the third or "big" day was 11 Ix. If the backpack and washing series had been done over two separate years, they would have each been culminated by a two-day ritual. I could detail many more examples of contrasting two-day and three-day rituals, but the general rule is this: wherever such rituals are the culmination of two overlapping counts of days, the observances will span three days. Otherwise, they will span only two.

Reflecting on the so-called "*tzolkin* triad" found in the eclipse tables of the Dresden codex (pp. 53a–58b), one wonders if this might also have marked off overlapped cycles. Various hypotheses concerning the possible function of this triad have been advanced. For example, Satterthwaite (1965: 623) thought that it might have

QUICHEAN TIME PHILOSOPHY

BACKPACK	WASHING THE SHRINE
	1 Toj
	1 Ik'
	8 Toj
	1 Tz'iquin
	8 Ik'
	1 K'anil
	8 Tz'iquin
	1 Imox
	8 K'anil
	1 Ix
	8 Imox
1 Quej	1 Quej
	8 Ix
1 Junajpu	1 Junajpu
8 Quej	8 Quej
1 Aj	1 Aj
8 Junajpu	8 Junajpu
1 Came	1 Came
8 Aj	8 Aj
1 Cawuk	1 Cawuk
8 Came	8 Came
1 E	1 E
8 Cawuk	8 Cawuk
1 Can	1 Can
8 E	8 E
1 Tijax	1 Tijax
8 Can	8 Can
1 Batz'	1 Batz'
8 Tijax	
1 C'at	
8 Batz'	
1 No'j	
8 C'at	
1 Tz'i'	
8 No'j	
1 Ak'abal	
8 Tz'i'	
1 Ajmac	
8 Ak'abal	
1 Toj	

16.2

been used as ± 1-day allowance (e.g. a three-day error range) in lunar correction or variation. An alternative interpretation is that it was used to shift from one line to the next above it at certain periodic intervals (Andrews 1940: 156; Lounsbury 1978: 796). Most recently, it has been suggested that it functioned as a one-day recession of the window-defining base dates necessitated by the recycling of the table (Bricker and Bricker 1983: 12). Alternatively, I would suggest that the triads in this table indicate that double or even multiple cycles are overlapped, quite aside from the question of variability or recession in the astronomical phenomena that may correlate with these cycles. The Brickers focus on the solar aspect of this table, but following my sense of the Mayan (or at least Quichean) preference for overlapping dialectics and for multi-metrical time reckoning, I follow Kelley in calling these tables "lunar-solar" (1976: 42–43). The ethnographic research needed to pursue these issues further has barely begun (see Remington 1977; Neuenswander 1981). What is specifically needed is more information on contemporary highland Guatemalan conceptualization and observation of the moon, and of the nocturnal sky in general. The richness of what remains to be learned is only hinted at in what I will be able to sketch out here.

There is a general and widespread interest in the night sky in Momostenango. This is particularly true during the cold, dry season months of November, December, January and February, when people rise after midnight and go visiting, consult diviners and attend seances. At this season of the year, on nights without a bright moon, the winter Milky Way or *ube tew*, "ice road," clearly reveals its rift; that part of the Milky Way is called *xibalba be*, "road to the underworld." Night rituals which take place during this season are marked in terms of the rising of certain stars and constellations. These time markers include the rise of Regulus (*jun ch'umil*, "one star"), Orion's Belt (*oxib ch'umil*, "three stars") and the Big Dipper (*pac'*, "cupped hands"). During this same season, on the night of the full moon (*rija ic'*), when the Milky Way is barely visible, people rise when the moon is at its peak near midnight and walk to the various hot mineral springs in the community in order to bathe. These trips to the springs involve courtship and sexual liaisons, which are timed in accordance with the phases of the moon.

When the moon is "small" (*alaj*) it is a time when all of the world is considered tender—animals, plants, trees and people. During the fifteen days (7 + 8) of the waxing moon, butchering, harvesting, woodcutting and sexual relations are avoided. Then, on the night of the full moon and for the following fifteen days until the dark of the moon, when she is "buried" (*mukulic*), all of these activities become propitious, since the moon, and all humanity with her, are now hard or mature (*rij*). These idealized fifteen-day intervals recall the similar intervals that appear seven times in the lunar-solar table of the Dresden codex.

The night of the full moon (*jun ak'ab ube* "one night her road") is a particularly important night each month in Momostenango. Only on this night does the moon's path—called (like the sun's path) *oxib uxucut*, "three corners"—cross the sky from east to west in a single night. The three corners consist of moonrise in the east (*relebal ic'*, "the coming-out place of the moon"), midnight of the full moon (*pa nic'aj* "at the halfway" or "middle") and moonset in the west (*ukajibal ic'* "the going-down place of the moon"). In order to contrast them, the solar and lunar triangles are referred to respectively as *chupam sakil* "in the light" and *chupam k'ekum* "in the darkness." It was not until I understood these triangles that I could understand a remark that a

Momostecan layman once made while we were walking down a path just after sunset. On seeing the full moon that had just come up, he said: "The sun has risen." Here I would suggest that at least some of the *kin* or sun glyphs in the Dresden lunar-solar tables (see 56a, 52b, 54b, 56b, 57b and 58b), superimposed on boundaries between light and dark backgrounds, might be metaphors for the full moon rather than literal indications of the sun.

In the more esoteric world of priest-shamans, there is an even greater and more serious interest in the paths of the heavenly bodies. Those who have been initiated as mother-fathers at any level of the three-tiered hierarchy visit the highest hills within the community on regular schedules, and the two who are at the highest level visit the sacred four-directional mountains as well. Some of these sacred places have good views of the sky and of large stretches of the horizon. The following information on ritual cycles was gathered without knowledge of the possible importance of 82-day cycles ($3 \times 27^{1}/_{3} = 82$) in charting the motion of the moon among the constellations (see Aveni 1980: 67–82).

Momostecan patrilineage leaders visit Nima Sabal, one of the highest shrines, on the following sequence of days: 9 Quej + 13 = 9 Junajpu + 13 = 9 Aj + 13 = 9 Came + 13 = 9 Cawuk + 13 = 9 E + 13 = 9 Can + 4 = 13 Toj = 82 days. On the first day of this series they open their particular patrilineage's shrine at Nima Sabal at sunset and remain for some time, burning incense, praying to their dead predecessors in office by name and observing the night sky. In these prayers they mention the phases of the moon and its position in the night sky. They are particularly interested in the seasonal variations in the moon's path through the summer Milky Way (*saki be*, "white road"), which they compare with its passage through the bifurcated winter Milky Way. They discuss these variations among themselves and occasionally make notes on their Gregorian calendars at home. Certain men are known as experts in predicting rain according to the phases of the moon in its seasonal voyage through the Milky Way. These men carefully observe the night sky on all seven days of this series of 9-days (*belejebal*), but this is not considered necessary by the others, who observe the night sky with any seriousness only on the opening day (9 Quej) and 82 days later on the closing (13 Toj), when they once again arrive at sunset to pray, burn incense and observe the night sky before they close the shrine for a 22-day period. They will repeat this pattern once again on 9 Batz' + 13 = 9 C'at + 13 = 9 No'j + 13 = 9 Tz'i' + 13 = 9 Ak'abal + 13 = 9 Ajmac + 13 = 9 Toj + 4 = 13 Ajmac, at which point another 82-day cycle has passed. Now the shrines remain closed for 74 days before the first cycle, from 9 Quej to 13 Toj, begins again. These 82-day periods are referred to in Quiché as *chac'alic* "to be staked, suspended, stabilized or set," which was explained to us as referring both to the firm placement of a table on four legs and to the forked poles which are planted to support the roof beams of a new house.

The point of particular astronomical interest here is that wherever the moon (if visible) might have been located among the constellations on a given 9 Quej or 9 Batz', it would be in the same position 82 days later on the following 13 Toj or 13 Ajmac, respectively. It requires further fieldwork to confirm the present evidence that this cycle is precisely what the mother-fathers are observing when they make their nocturnal visits to Nima Sabal. I should add that I also have evidence of four different 82-day periods in the rituals performed by one of the two mother-fathers who serve the entire town of Momostenango, but those rituals, which involve overlapping 82-day cycles with others of 40 and 65 days, require a separate paper to themselves.

The reckoning of 82-day periods in Momostenango, and their connection with ritual mountaintop visits that are known to include observation and discussion of the night sky, make it necessary to re-open the question as to whether Mayan astronomy included not only synodic moon-reckoning, but sidereal reckoning as well. On the synodic side, the conceptual identity of the moon with the sun, on just the one night of the full moon, may give us new ways of reading and interpreting the lunar-solar pages in the Dresden codex. And the "*tzolkin* triads" in those same tables may signal the overlapping of two or more cycles, as do three-day rituals at Momostenango, rather than providing for errors of measurement.

The conception of the *oxib uxucut* or "three corners" of the paths of the sun and the full moon may call for a re-opening of the question as to whether ancient Mayan astronomers were interested in angles, though of course the corners in question here are angles attributed to the celestial movements themselves rather than to the geometry of observation.

On the issue of Mayan time philosophy, it is apparent that contemporary Quiche thinking about time frequently follows a dialectics of overlapping or mutual involvement. There are moments of dialectal complementarity, as when time intervals both alternate (as between 7 and 6 days) and complete one another (to add up to the meaningful pattern number of 13), or when two separate shrines (one for 1-days and the other for 8-days) are in direct and unmediated opposition in a particular ritual cycle. On the other hand, a dialectics of mutual involvement is seen when burner (or work service) rituals overlap with daykeeper rituals, when washing rituals overlap with backpacking rituals, and when night overlaps with day.

NOTES

1. For excellent discussions of various forms of dialectics see Gurvitch (1964) and Sartre (1976).

2. This fieldwork was made possible by a Research Fellowship from the State University of New York at Albany and by a summer Faculty Fellowship from Tufts University. Tony Aveni, Dennis Tedlock and Mary Jane Cramer have all made useful suggestions on the first draft of this paper, which the author has incorporated. Any remaining ambiguities are her own. I am most grateful for the support provided by these individuals and institutions.

3. The orthography used for Quiché in this paper is the practical one suggested by the Instituto Indigenista Nacional de Guatemala (see David G. Fox, *Lecciones elementales en Quiché*, pp. 15–18). Vowels are pronounced as in Spanish, ë is like the vowel in English "met," and ö is like the vowel in English "foot." Consonants are also as in Spanish, with an equivalence between c (used before a, o and u) and qu (before e, i), except for k, which is articulated with the tongue farther back than for c or qu; tz, which is like the German Zeit; w, which is like

English w; x, which is like the English sh; and b, which is a glottalized p. Other glottalizations are indicated by '.

REFERENCES

Andrews, E. Wyllys. 1940. Chronology and Astronomy in the Maya Area. In *The Maya and Their Neighbors,* ed. L. Hay et al., 150–161. New York, Appleton-Century.

Aveni, Anthony F. 1980. *Skywatchers of Ancient Mexico.* Austin, University of Texas Press.

Berlin, Heinrich. 1958. El glifo "emblema" en las inscripciones mayas. *Journal de la Société des Américanistes* 47:111–119.

Bricker, Harvey M., and Victoria R. Bricker. 1983. Classic Maya Prediction of Solar Eclipses. *Current Anthropology* 24:1–24.

Craine, Eugene R., and Reginald Reindorp. 1979. *The Codex Pérez and the Book of Chilam Balam of Maní.* Norman, University of Oklahoma Press.

Fox, David G. 1973. *Lecciones elementales en Quiché.* Guatemala, Instituto Linguistico de Verano.

Gurvitch, Georges. 1964. *The Spectrum of Social Time.* Dordrecht, Holland, D. Reidal.

Kelley, David Humiston. 1976. *Deciphering the Maya Script.* Austin, University of Texas Press.

La Farge, Oliver. 1930. The Ceremonial Year at Jacaltenango. *Twenty-third International Congress of Americanists in New York,* 1928. New York, pp. 656–660.

Lincoln, J. Steward. 1942. The Maya Calendar of the Ixil of Guatemala. *Contributions to American Anthropology and History* 38:99–128.

Lounsbury, Floyd G. 1978. Maya Numeration, Computation, and Calendrical Astronomy. In *Dictionary of Scientific Biography,* 15, suppl. 1, ed. C. C. Gillispie, 759–818. New York, Scribner.

Neuenswander, Helen. 1981. Glyphic Implications of Current Time Concepts of the Cubulco Achi (Maya). Manuscript prepared for Centro de Estudios Mayas, Universidad Nacional Autónoma de México.

Prouskouriakoff, Tatiana. 1960. Historical Implications of a Pattern of Dates at Piedras Negras, Guatemala. *American Antiquity* 25:454–475.

Redfield, R., and A. Villa Rojas. 1934. *Chan Kom: A Maya Village.* Carnegie Institution of Washington Publication 448. Washington, DC.

Remington, Judith A. 1977. Current Astronomical Practices among the Maya. In *Native American Astronomy,* ed. Anthony F. Aveni, 75–88. Austin, University of Texas Press.

Sartre, Jean-Paul. 1976. *Critique of Dialectical Reason.* Trans. Alan Sheridan Smith. London, NLB.

Satterthwaite, Linton. 1965. Calendrics of the Maya Lowlands. In *Handbook of Middle American Indians,* 3. Archaeology of Southern Mesoamerica, ed. Robert Wauchope and Gordon R. Willey, 603–631. Austin, University of Texas Press.

Tedlock, Barbara. 1982. *Time and the Highland Maya.* Albuquerque, University of New Mexico Press.

Thompson, J. Eric S. 1950. *Maya Hieroglyphic Writing: An Introduction.* Norman, University of Oklahoma Press.

Thompson, J. Eric S. 1954. *The Rise and Fall of Maya Civilization.* Norman, University of Oklahoma Press.

Villa Rojas, Alfonso. 1945. *The Maya of East Central Quintana Roo.* Carnegie Institution of Washington Publication 559. Washington, DC.

CHAPTER SEVENTEEN

Astronomical Models of Social Behavior among Some Indians of Colombia

Gerardo Reichel-Dolmatoff

Among many native tribes of Colombia, the tropical night sky and the cyclic motions of celestial bodies constitute models for certain forms of human behavior. The sky is seen as an enormous blueprint of everything that did, does, or will happen on the earth; an enormous map replete with information on every aspect of biological and cultural behavior, time, space, evolution, and psychological phenomena; in sum, an encyclopedic body of what one might call "survival information," the knowledge of which alone can give Man a measure of security.

The most obvious aspect is time, cyclic time. In Colombia, the yearly seasonal round is divided into four ninety-day periods that coincide with the beginnings and the ends of the two rainy and two dry seasons. Observation of the sun therefore becomes important for the management of natural resources. By this I refer not only

From *Ethnoastronomy and Archaeoastronomy in the American Tropics*, ed. A. Aveni and G. Urton, Annals of the New York Academy of Sciences 385 (1982).

to agriculture, but also to seasonal events, such as fish runs, bird migrations, the rutting and breeding seasons of different game animals, the various fruiting seasons of trees, and the cyclic availability of edible insects or mollusks, of honey, and of many other food resources. The precise prediction of the onset, progress, and end of each season is important not only to the horticulturalists, but also to the hunter and gatherer, to the prospective traveler, to the canoe builder, and to any group of people who intend to build a house.[1] Cyclic time, then, is observed by the natives on a scale that extends from hourly changes of physiological and psychological functioning and environmental changes in light, temperature, humidity, and so on, to changes in circadian rhythms, to monthly, seasonal, and yearly cycles, and, beyond that, even to the observation of equinoctical precession. Upon these cyclic phenomena of the heavens and of nature, the Indians project cycles of specific cultural relevance, such as the menstruation cycle, the cycle of embryonic development, the human life cycle, psychological developments, plant growth, and any number of other recurring events, as recognized by the natives.

Another aspect is space. Those facets of heavenly space that are of immediate importance to the Indians refer either to stable relationships, such as the outlines of certain constellations, or to dynamic relationships, such as those that exist between celestial bodies.[2] This refers mainly to the changing relationships between the sun, the moon, and the larger planets and to the changing position of the Milky Way. These fixed spaces and fixed orbits are very important to the Indians, who see in them a set of principles of order, of organization. For the same reason, any dissonance in this heavenly harmony is thought to be harmful. Eclipses, comets, meteorites, shooting stars, tektites, and planetary conjunctions are greatly feared, because they are thought to mirror calamitous conditions that exist somewhere on this earth. Dissonances do not predict coming events; instead, they point to malfunctionings that are actually taking place in human society or in nature. The observation of these celestial dysfunctions is thus a diagnostic procedure, not a prognostic one. Native astronomy is not much concerned with astrological prediction, but with learning to read the sky, which mirrors this world; the sky must be scrutinized in every detail because it is a map and a mirror of nature. What counts is the correct reading, because the sky is not only an ecological blueprint for Man's tenancy of this earth, but also a guide to spiritual development and moral integration.

In the following I shall refer to three aboriginal cultures of Colombia: first, the Desana, a tribe of Tukanoan Indians of the equatorial rain forests of the Vaupés Territory in the Northwest Amazon;[3] second, the Kogi Indians of the Sierra Nevada de Santa Marta, in northern Colombia, at about 11° north;[4] and, third, the Muiska Indians of the Andean Highlands, at about 5° north.[5] The Desana and Kogi still number several thousands; the Muiska have disappeared and have become assimilated into the rural mestizo population, so that information on them is derived from archaeological remains and from the accounts of the early Spanish chroniclers.

I shall speak mainly of the equatorial Desana and their neighbors, and only briefly refer to the other tribes, to point out some similarities or differences. In Desana origin

Gerardo Reichel-Dolmatoff

myths a frequent motif is the "search for the center," the "Center of the Day," as the Indians call it. In brief, the story is this: A supernatural hero who carries a staff goes in search of a spot where his staff, when standing upright, will not cast a shadow. He eventually locates this spot on the equatorial line, and it is there where he subsequently establishes his people. In one of many shamanic images, the staff is said to be a shaft of sunlight, which, falling vertically into a womb-like lake, fertilizes the earth. This idea of the "central point" is all-important to the Indians; it is, in essence, a spot where a cosmic sexual contact takes place, a meeting between Sky and Earth, and life on this earth subsequently develops in a bounded space that extends around this center.

The model for this bounded space is perceived in the sky. It consists of the huge hexagon formed by the stars Pollux, Procyon, Canopus, Achernar, T3 Eridani, and Capella. The center of this hexagon is said to be Epsilon Orionis, that is, the central star in Orion's belt.

Now, among the Desana and other Colombian Indians, hexagonal shapes and outlines constitute fundamental ordering principles.[6] These recurring shapes are observed, for example, in the hexagonal structure of rock crystals, which are common shamanistic power objects; they can be seen in honeycombs and wasp's nests, and in the hexagonal plates on the back of a tortoise shell. In the shamanist world view, all these hexagonal shapes are said to be imbued with transformative energies, and in this manner, all places, spots, and objects where a transformation is said to take place are imagined as hexagons, or as hexagonal containers. Thus, in shamanist imagery, the female womb is seen as a hexagonal body; the human brain is seen as a hexagon divided into innumerable hexagonal ventricles; and the structure of a house is imagined as a hexagon. All these hexagonal shapes symbolize continuity by transformation; they symbolize an eternally recurrent natural model that, by its unchanging persistence, expresses a sense of world order.

I have said that the model of bounded terrestrial space is seen in the sky, in the great hexagon of a number of bright stars centered upon Epsilon Orionis. Now, this celestial hexagon is projected upon the earth, where it delimits the tribal territories of the Tukanoan Indians. The image is that of an enormous transparent rock crystal standing upright, the six corners of which are the six stars I have mentioned, while, on earth, the corners are formed by six major waterfalls located on certain rivers (Fig. 17.1). The center, the axis of this crystal tower, is a vertical line between Epsilon Orionis and a large boulder covered with petroglyphs, located approximately at the spot where the equatorial line crosses a north-to-south flowing river.[7] In other words, the crystal axis is the phallic staff that joins the male sky to the female earth. Within this terrestrial hexagon all major celestial phenomena have their counterparts and mirror images. For example, the Vaupés river corresponds to the Milky Way; certain landmarks, such as isolated hills, lakes, or large rocks, are associated with stars or constellations, and Orion's belt is centered upon the equator, in an east-west direction.

According to the Desana and their neighbors, the twenty or so Tukanoan tribes are, ideally, grouped into six phratries, each one consisting of three exogamous tribes. Since each tribe consists of males and their virilocal by residing spouses, a phratry can

427

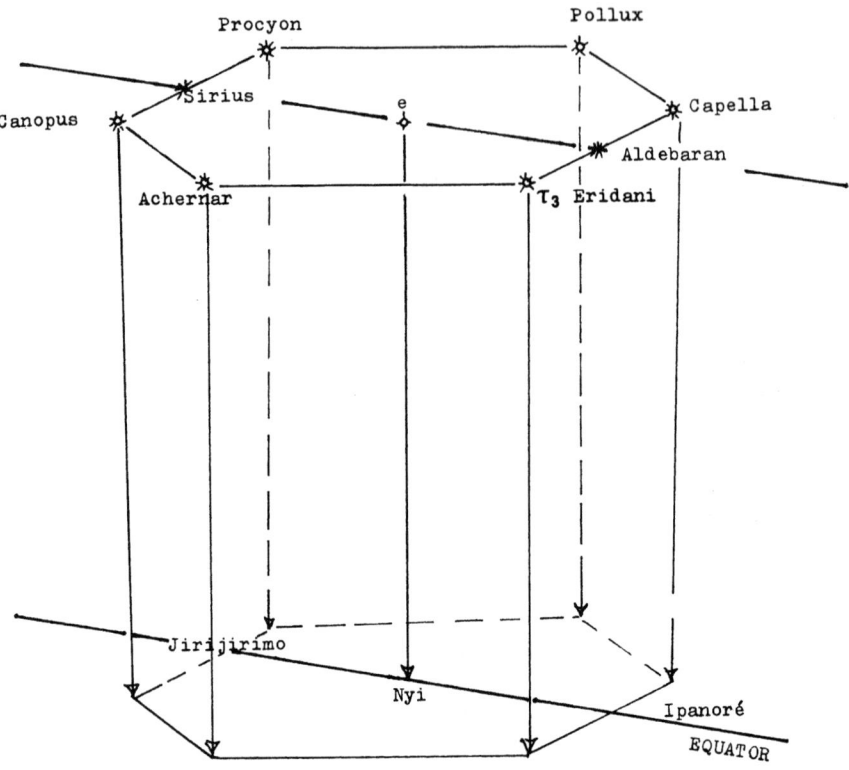

17.1 The hexagonal prism between sky and earth. The falls of Ipanoré are the point of origin of the tribe; the Jirijirimo falls are the largest in the entire territory; the other points correspond to minor falls. The center is the rock of Nyí, a large boulder covered with petroglyphs, located at the spot where the Pira-paraná crosses the equatorial line.

be said to consist of six units. This phratric, tribal, and sexual division is imagined as a series of interrelated, adjoining hexagons. The great celestial hexagon that is projected upon the land is divided into six hexagons representing phratries; each phratry is divided into six triangles, that is, three so-called "male" and three "female" ones (Fig. 17.2). Although, at present, this territorial division does not correspond to a social and geographical reality, it continues to be an important shamanistic model, the true origin of which is seen in the sky, and the dynamics of which constitute a body of complex esoteric lore.

A basic principle of Tukanoan social organization is exogamy. Incest laws are very strict and their origins are elaborated in many myths and shamanistic texts. As is often the case in these tribal societies, there is some overlapping of different situations, such as father/daughter incest, mother/son incest, sibling incest, or plain adultery. The human actors in these dramatic situations are personified in the sky, where Sun is either Moon's husband, father, or brother, while Venus is a daughter of Sun or a son of

Gerardo Reichel-Dolmatoff

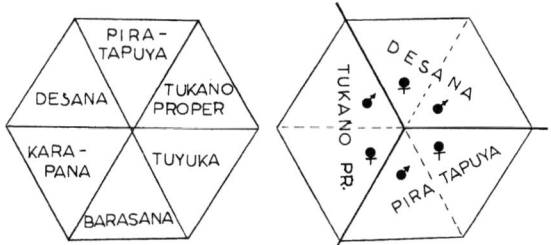

17.2 The hexagon as territorial unit and social model. Left: The Vaupés territory and the ideal distribution of the six original Tukanoan tribes. Right: The ideal distribution of a phratry of three intermarrying tribes.

Moon. The precise identification may vary from tribe to tribe, but the essential interpretation is the same: The incestuous (or adulterous) male is castrated and his penis is thrown up into the sky where it turns into Orion's belt. Sometimes the castrated man is a father figure, at other times he is the son who is being punished by the father; in any case, the victim acquires the proportions of a hero.[8]

In Desana shamanist imagery, Orion is the Master of Animals, a supernatural gamekeeper. He is a mighty hunter who can be seen walking over the sky, over the Milky Way, which is his trail, carrying a game animal, a string of fish, or a basket of fruits, thus announcing the different harvesting seasons.

In what I have said so far, I have referred to a number of symbolic equivalences, and before I can continue I must clarify some of these concepts. In Tukanoan ideology, a symbolic or metaphorical relationship is never limited to a one-to-one comparison; symbolic images are always seen as chains of analogies. For example, the Milky Way can be conceptualized as a river, as a trail in the forest, as an immense cortège of people, a cast-off snake skin, a fertilizing stream of semen, and so on. In the same way, the celestial hexagon can be seen as a rock crystal, a tortoise shell, a honeycomb, a womb, a brain, a ritual enclosure, a tribal territory, a fish trap, and so forth. The entire concept of transformation is thus based upon likeness, not identity. Astronomical interpretations are thus always based on multiple chains of analogies, never on one single image.

To continue—I have spoken here of the astronomical models of tribal territories and social organization and I have touched upon the problem of marriage rules. I now must add some observations on native concepts of fertility and growth in relation to astronomy.

According to the Indians, the energy of the universe is generated by the sun. The sun, too, is imagined as a rock crystal that is said to contain an unlimited amount of what the Indians call "color energies." The principal color energies have a male fertilizing power. The moon, on the other hand, contains a female principle of color energies related to plant growth and the human menstrual cycle. Now, solstices and equinoxes are of relatively little importance on the equatorial line, so the sun's cycle is not measured through them, but is taken to be a continuous, perennial force. However, lunar phases and positions in relation to earth and sun are closely observed and are thought to be correlated with female fertility and, in general, with the growth cycles of animals and plants. Most interestingly, the different lunar phases constitute a calendar for birth control, for game protection, and for the restriction of the production of many materials. In fact, every month, during the sixteen or so days between the first and third quarters, sexual intercourse is prohibited, as are hunting, fishing, and the gathering of raw materials for such things as basketry, wood carving, and pottery making. Shamans recommend instead agricultural labor and gathering activities, mainly of insects and larvae. Of course, these prohibitions are not too severe but, since they are reinforced by shamanist threats of impending illness, they do constitute an effective mechanism of control. The lunar calendar, then, is essentially a guide to the protection of natural resources and a means of population control.

In order to explain the reasons why the first and third quarters of the moon's phases occupy this important position, I must refer to another cyclic phenomenon. The Milky Way is imagined as two huge snakes; the starry, luminous part is a rainbow boa, a male principle, and the dark part an anaconda, a female principle. The cycle of fertilizing forces emanating from the sky is punctuated by the shifting of the Milky Way, which is seen as a swinging motion made by the snakes. Now, in ordinary nature, anacondas and boas are said to copulate at two periods of the year, approximately at the vernal and autumnal equinoxes. Late in March and again in September anacondas swim upriver at night and, now and then, lift about one-third of their bodies out of the water and then slap down with a loud splashing sound. This is part of their mating behavior, but the Indians say that, when the snakes rise out of the water, they watch the stars in order to ascertain the proper time. At these times of the year, and during the nightly shifting of the Milky Way, the Indians transfer the image of the two snakes to that of the intersecting of the ecliptic and the celestial equator, and to the intersections of the path of the moon and the path of the earth, at the first and third quarters. In everyday nature the two equinoctial dates are associated with periods of fertility; by the end of March and the end of September the spawning seasons of fish are beginning and these fish runs provide the model for sib distribution along the rivers, for patterns of reciprocal food exchanges, and for ritual dances in which spawning behavior is equated with human procreation. This, too, is the harvesting season for many wild fruits, and it is the proper time for male initiation rituals.

The image of the intertwining snakes, that is, of two bent, snake-like bodies that intersect at two points—◊—is an important shamanistic icon.[9] It is thought that two snakes lie in the fissure between the two hemispheres of the human brain, and that their rhythmically shifting motion determines the relationship between the unconscious and the conscious in what concerns sex, food, and aggression. Now, this image of the human brain is patterned after the shamanist image of the entire celestial vault as one gigantic brain divided by the great fissure of the Milky Way. The Desana believe that both brains, the cosmic and the human, pulsate in synchrony with the rhythm of the human heartbeat, linking Man inextricably to the Cosmos.

I must return once more to the hexagonal pattern. The great hexagon in the sky is also an architectural model, and traditional longhouses are built according to this celestial plan. The houses are large structures contained within six points of reference

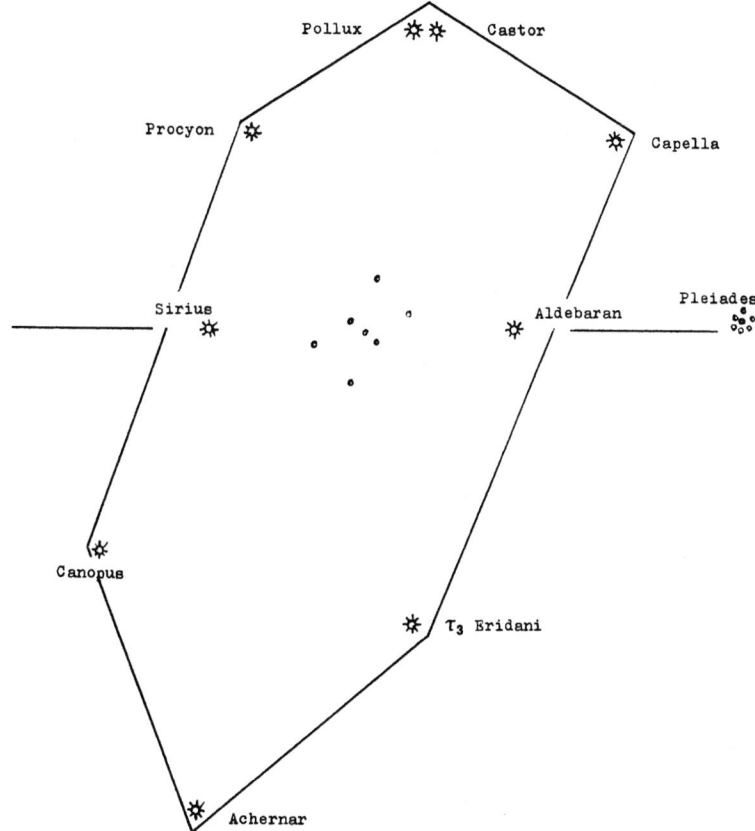

17.3 The longhouse as an astronomical model; Achernar marks the men's door and Gemini mark the women's door.

marked by strong houseposts that are identified with stars (Fig. 17.3). The middle section of the roof is supported by another set of six strong vertical posts that delimit a hexagonal central part that has ritual functions.[10] An imaginary line drawn at a right angle to the longitudinal axis of the house divides both the outer and the inner hexagon into two halves and represents the equatorial line. At the same time, it represents the Orion's belt, Zeta and Delta being the middle houseposts, while Epsilon is not visibly marked but coincides with the true center of the hexagon and the house. But this is not all: The basic outline of the structure delimited by the inner hexagon can be perceived, in a very schematic way, as a longitudinal ridgepole and three parallel cross-beams.[11] The Desana see in this the constellation of Orion and, in one shamanistic image, it is a rack upon which the hero is crucified. In this outline, Betelgeuse and Bellatrix are the summer solstice points, Saiph and Rigel are the winter solstice points, and the belt is called the "Path of the Sun." It is at the center, on Epsilon Orionis, where, on ritual occasions—at the two equinoctial dates—heaps of palm fruits are

431

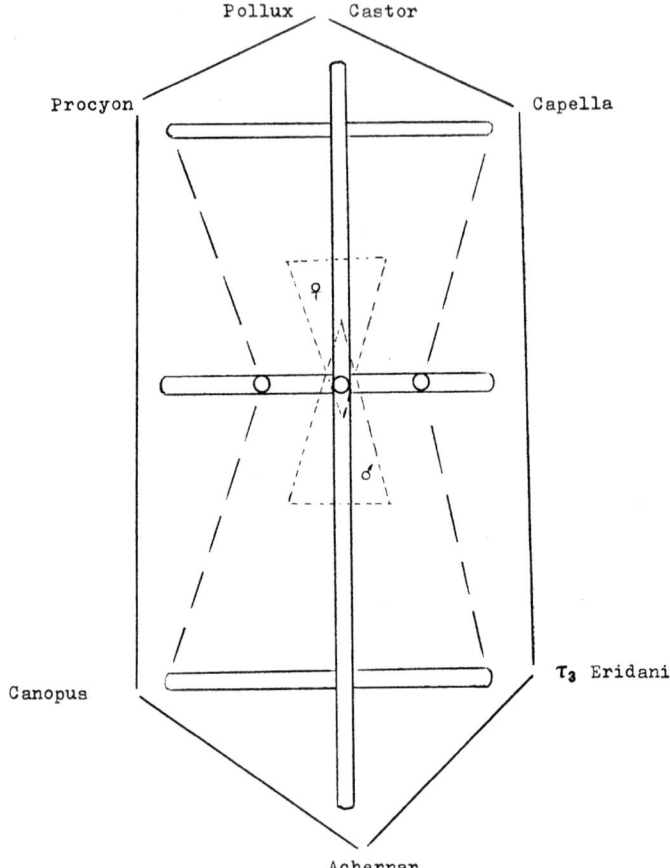

17.4 The longhouse identified with Orion. The inner hourglass-shaped outline marks the overlapping male/female dance pattern. The diamond centered upon Epsilon Orionis represents a vagina.

deposited. In this case, the clusters of fruits are identified with the Pleiades, which rise after sunset in the east late in September and set before sunrise late in March, in both cases announcing the onset of the principal fruiting seasons, the fish runs, and the proper time for initiation rituals.

Within the inner hexagon of a longhouse the Indians perform ritual dances related to the multiple symbolism of Orion. Men and women dance in the center, between the six main posts. They dance back and forth, each group representing a "triangle" and this interpenetration of male and female triangles will trace the hourglass-shaped outline of Orion (Fig. 17.4). The dancers move back and forth over the dividing line, which is Orion's belt. Now, the front part of a longhouse represents an abstract dimension where shamans and chiefs have their seats, while the back part represents ordinary reality, the domain of women, children, servants, and food preparation. The pattern of Orion thus forms an area of articulation between male and female, light and darkness, fertility and restraint.

Obviously, the most important constellation is Orion. Orion is Man: Man the Progenitor, the Hero, the Hunter, the Sinner and, finally, the Victim. But every year he rises anew and, when people see him resurrected in the sky, shamans remind them of everything this constellation stands for. In some images, Orion is a rack upon which the adulterer has been put; in other images, Orion is a tapir hide stretched out to dry between six stakes; or a potstand, or a fish trap. Orion's belt is the severed penis of the incestuous youth, or the castrated father; in another image, the belt is the adze used to extract starch from split palm trunks, a phallic instrument that is the main attribute of the Master of Animals. In still another image, the constellation is seen as an evil shaman turned jaguar who had his penis (or tongue) cut off and was hung from a tree and burned; he disappeared in a cloud of smoke and then reappeared in the sky as Orion. Some shamans see in Orion a jaguar copulating with his sister, a huge snake. And, according to the imagery involved, the Great Orion Nebula becomes a stream of semen, a heap of palm starch, or a cluster of fruits.[12]

What really matters is that this ambivalence of Orion does not offer a simplistic, polarized choice between conventional concepts of Good and Evil but that it presents the intelligent and searching beholder with a guide to self-knowledge and subsequent adaptive behavior. I shall attempt to reduce a large body of related shamanist concepts to its basic propositions in order to elaborate upon this point.

Among the Desana, all mental and psychological processes are believed to arise in the left cerebral hemisphere, which is said to harbor the unconscious, and where latent ideas must be activated by shamanistic procedures; the induction of altered states of consciousness is essential in this case. Once identified and isolated, an idea must be transformed into an image, an icon. This process of increasing awareness is achieved in the right cerebral hemisphere, while the act of transformation takes place in the great fissure, which, as I mentioned before, contains the two rhythmically twisting snakes. Once the idea has become transformed into an icon, that is, into a culturally coded sign, it can be expressed by action. Now, the three steps of this mental process are represented by Orion: The left cerebral hemisphere is Zeta Orionis, the, point of transformation is Epsilon, and the right hemisphere is Delta Orionis.

I have gone into these shamanistic details because I want to make this point: It would be very mistaken to think that the astronomical thinking of these Amazonian Indians refers only to practical affairs of resource management and to the modeling of social organization. Far from it. Shamans are quite explicit about the importance of the philosophical dimensions that capture the mind when contemplating the sky, and will insist that the celestial vault is the one and only model for Man's material and spiritual well-being. Let me give you one more example: The six straight lines of the celestial hexagon that encompass sacred space, both in the sky and on the earth, represent an important metaphysical model, a moral proposition called "the Path, the Way," which the individual must travel in his or her life. In Desana theology the idea of the Quest is very elaborate; it corresponds to a process of individuation, under shamanistic guidance. The starting point for the Quest, for both sexes, is Aldebaran (and, by extension, the Pleiades); this is the place of birth. Thence, men proceed in a counterclockwise direction to Capella, which marks the ritual of naming; from Capella he proceeds to Pollux, where initiation takes place; from there he goes on to Sirius, the star of marriage, and then he turns sharply toward the east and arrives at the center.

This is Epsilon Orionis, the star of procreation, of multiplication, of fatherhood. The last stretch continues toward the east and ends at Aldebaran, now designated as the point of death, of rebirth, and of return. Women travel first in the opposite direction and then, after their marriage at Sirius, they proceed jointly with their husbands. The entire trajectory, a spiritual pilgrimage, divides the human life span into three periods: youth, from Aldebaran to Sirius; maturity, from Sirius to Epsilon Orionis; and old age, from there back to Aldebaran. In shamanistic imagery these three parts of the life span are marked off by the two intersections of the snakes.

Before turning briefly to other tribal cultures of Colombia, I should like to add one more point. In Northwest Amazonian shamanistic thinking there exists a close relationship between astronomical observations, cosmological speculations, and drug-induced trance states. This gives a very peculiar stamp to native astronomy. Under the influence of a narcotic drug an Indian would imagine himself to be flying through space, ever deeper into unknown dimensions. He would believe himself to be moving among dazzling stars, turning sun wheels, and flashing beams of many-colored lights. At the same time, shamans will point out that the narcotic experience is not as one-directional ecstasy, but an inner voyage that also leads into the many-layered dimensions of the human mind. The human brain, shamans say, is modeled after the celestial vault and the human mind functions according to the stars, which are the ventricles and sensoria of the cosmic brain. Drug-induced trance states develop in a dimension where ordinary concepts of space and time are canceled out; in fact, narcotic drugs are expressly said to constitute a mechanism for modifying space and time and to be a means for exploring other cosmic dimensions where hitherto unknown models for human faculties might be discovered. In using drug experiences, together with the behavioral models encoded in them, as a key to interpret celestial phenomena, shamans can operate with a very convincing mechanism, because all Indians who consume drugs under shamanistic guidance are likely to have similar sensations and experiences.

I shall briefly refer to the Kogi Indians of the Sierra Nevada. Kogi culture is very different from that of the Amazonian rain forest Indians, and Kogi religion, philosophy, and historical traditions are of a complexity that can be compared only to that of the High Cultures of Mesoamerica. Although, at present, the Kogi form a scattered peasant population, there still exist lordly and priestly lineages with a strong sense of privilege and rank. The Kogi are agricultualists; hunting and fishing are practically nonexistent.

Kogi astronomical knowledge is based upon the observation of rising and setting points on the horizon and on zenithal observation. A small hole in the conical temple roof admits a ray of sun or moonlight that, in the case of the sun, traces the outline of solstices and equinoxes in the dark interior. Stone alignments, horizon markers, fixed observation points in the form of priestly stone seats, stone circles, gnomons, and similar stone settings, can be found over much of the Sierra Nevada. Kogi villages, ceremonial centers, isolated temples, shrines, and other structures are always sited according to astronomical principles.

The entire Sierra Nevada is imagined to have a hexagonal plan and to constitute one huge rock crystal, very much like the Desana image of their world. The corners of this crystal correspond to six sacred sites, while, in the sky, they correspond to six first-magnitude stars; the celestial center, again, is Epsilon Orionis.[13] In fact, Orion's belt corresponds to the three principal ceremonial centers, which lie in a line, although many kilometers apart, the middle one of which is designated "the only one."

Since the Sierra Nevada has an approximately conical shape, with rivers radiating in all directions and valleys opening toward the lowlands and the sea, the entire mountain massif constitutes one huge sundial with which the priest-shamans watch the horizons of their particular valleys. All constructions—shrines, roads, and bridges—conform to celestial sightlines; there is not a single spot of any importance in the Sierra Nevada of which the native priests will not say that it has some astronomical implication.

The sky over the Sierra Nevada is, again, conceived as a map of the land with all its topographical details and its mythical geography, being peopled by divine personifications, animals, artifacts, and all kinds of personified forces of nature. In one very telling image, a Kogi priest is described as sitting in the center of a dark temple, holding in his hand a mirror facing upward. A vertical ray of sunlight penetrates the roof and falls upon the mirror surface; but this ray of light proceeds from the disk that is the sun's face, and this disk, too, is a mirror. The reflection is, thus, endless, and expresses the Kogi concept of eternity and the dimension of cosmic space.

The astronomical division of the Sierra Nevada implies a division into clan territories and into a number of lineages that are associated with particular celestial phenomena. The most important priestly lineage is the one called the "Keepers of the Rock Crystal" and the sacred number six is repeated in many contexts. Once again, Sun, Moon, and Venus represent a triangle that holds the potential for drama. Incest and adultery on earth can be seen in the sky in eclipses, conjunctions, or in the unusually close proximity between two or more celestial bodies. Since all these phenomena are models of behavior on earth that must be followed in detail, ritual incest is practiced on certain occasions.

The scheduling of ceremonies is entirely geared to astronomical happenings; the main ritual dances take place at the solstices and equinoxes, each in a different village. The rituals of the life cycle have their specific shrines and temple mountains, identified with constellations or individual stars. Marriages are celebrated at equinoxes, while death rituals take place on solstices. Temple architecture conforms to cosmological and astronomical principles. Although, when seen from the outside, a temple has a circular shape, the interior structure shows a combination of circle, square, and hexagon, which corresponds to sacred spaces that are occupied by particular individuals during ceremonies.

Let me finally turn to the highland Muiska, the most advanced culture of prehistoric Colombia. In the early sixteenth century the Muiska formed two incipient states; the ceremonial center of one was the Temple of the Sun, while the center of the other was the Temple of the Moon. Both temples were associated with the principal priestly and lordly lineages. The heart-land of the Muiska was formed by a chain of old Pleistocene lake beds that provided fertile soil.[14] It so happens that this chain of flat valley bottoms extends for more than 200 kilometers in a southwest-northeast direction; that is, in the direction of summer solstice sunrise. In other words, by chance of nature, the entire Muiska territory is oriented in this manner, the chiefdom of the Moon Temple being located in the southwestern section, the chiefdom of the Sun Temple occupying the northeast.

Two highland valleys in this territory are known to present a clear, dark, cloudless sky during most of the year.[15] In both valleys one still can observe huge stone columns, alignments, and other stone settings. As observed from one of these structures, the sun rises on summer solstice exactly over a sacred lagoon whence, according to tradition, the Creator Goddess (Bachué) of the Muiska emerged in mythological times.[16] From the same observer's position an alignment of large cylindrical columns runs in an east-west direction.

A constellation of six sacred lakes[17] is located in the central area and is combined with many mountaintops that constitute astronomical sightlines. In the same region, a pattern of Catholic shrines and pilgrimages continues ancient aboriginal models described by the early chroniclers.

Rock crystals have been found in many shamans' graves, and emeralds, also of hexagonal structure, were important as offerings and in other ritual contexts. The overall similarity of these symbolic interpretations is best illustrated by a Muiska myth recorded at the time of the Spanish conquest. The myth tells that the daughter of a local chieftain was impregnated by the rays of the divine sun at a certain date. She eventually gave birth to a huge emerald. The stone burst open and from it emerged a child who grew up to become a great chief.[18]

Among the equatorial Tukanoans there is little interest in a horizon calendar. Zenithal observation is limited to the mythic motif of locating the "center," and all other practical astronomical knowledge refers to cyclic seasonal phenomena, the interpretation of constellations, the nature and motions of the Milky Way, and some aspects of astral proxemics. The great model is the hexagonal rock crystal. All these aspects of calendrics and astronomy exist among the Kogi who, in addition, have the following elaborations: a very detailed horizon calendar, zenithal observations, temples as astronomical observatories, associations between astronomy and weaving, record keeping with notched sticks, and pilgrimages and dances according to astronomical patterns. In both cultures, the sky is a map. Among the Kogi there is less emphasis on the relationship between astronomy and hallucinogenic drugs, and more weight is given to precise observation, to sightlines, record keeping, and the details of ritual structures. Many of these features seem to have been present among the ancient Muiska Indians, among whom we find astronomically-oriented architecture, a division into sun- and moon-associated states, a pattern of pilgrimages, the Sun/Moon incest theme, the hexagonal symbolism of mineral structures, and other details.

All three cultures have, thus, a truly fundamental astronomical substructure. This substructure consists not only of a body of practical knowledge used in timekeeping, but contains complex intellectual elaborations concerning time/space relationships, the importance of biological cycles, and many philosophical formulations. In trying to explain forms of cultural behavior through native astronomy, it is obvious that we must go beyond the utilitarian level of calendars, architecture, astrology, and so forth, and take into account the intellectual and spiritual aspects as conceived by the Indians, which can be found in, e.g., their concepts of cosmogony and the time/space continuum, and their notion of a participatory universe.

NOTES

1. I would suggest that certain ring-shaped shell mounds or midden sites of early prehistoric cultures might have been used as horizon calendars.

2. In the shamanistic world view, the universe is layered. The Desana refer to a sublunary dimension, followed by the spheres of Sun, Moon, and Venus; then comes the Milky Way and beyond that, another, deeper dimension.

3. On the Desana, see the following works by G. Reichel-Dolmatoff: *Amazonian Cosmos: The Sexual and Religious Symbolism of the Tukano Indians* (Chicago: University of Chicago Press, 1971); *The Shaman and the Jaguar: A Study of Narcotic Drugs among the Indians* of *Colombia* (Philadelphia: Temple University Press, 1976); *Beyond the Milky Way: Hallucinatory Imagery of the Tukano Indians* (Los Angeles: UCLA Latin American Center Publications, 1978); "Desana Animal Categories, Food Restrictions, and the Concept of Color Energies," *Journal of Latin American Lore* 4 (1978):243–291; "Desana Shamans' Rock Crystals and the Hexagonal Universe," *Journal of Latin American Lore* 5 (1979):117–128.

4. On the Kogi, see the following works by G. Reichel-Dolmatoff: *Los Kogi: Una Tribu Indigena de la Sierra Nevada de Santa Marta, Colombia* (Bogotá: 1950–1951); "Notas sobre el Simbolismo Religioso de los Indios de la Sierra Nevada de Santa Marta," *Razón y Fábula* 1 (Bogotá: Revista de la Universidad de los Andes, 1967), 55–72; "Templos Kogi: Introducción al Simbolismo y la Astronomia del Espacio Sagrado," *Revista Colombiana de Antropologia* 19 (1975):199–246; "The Loom of Life: A Kogi Principle of Integration," *Journal of Latin American Lore* 4 (1978):5–27.

5. On the Muiska, see, among others, José Pérez de Barradas, *Los Muiscas antes de la Conquista,* 2 vols. (Madrid: Consejo Superior de Investigaciones, 1950–1951).

6. See G. Reichel-Dolmatoff, "Desana Animal Categories," 265–271, and "Desana Shamans' Rock Crystals."

7. The spot is said to be the rock of Nyi on the Pira-paraná river. See G. Reichel-Dolmatoff, *The Shaman and the Jaguar,* 155–156, and *Beyond the Milky Way,* 138–141.

8. On this particular motif, see R. Lehmann-Nitsche, "Las Constelaciones del Orión y de las Hiadas," *Revista del Museo de la Plata* 26 (1921):17–69.

9. Mesoamericanists might recognize in it the Ollin motif.

10. In reality, the six posts form a rectangle, but the space enclosed by them is said to be "like a hexagon." There are some indications that, in former times, houses had a circular ground plan that included six main posts forming a hexagon.

11. The three crossbeams are designated "jaguars." See G. Reichel-Dolmatoff, *Amazonian Cosmos*, 104–110.

12. Obviously, Orion symbolism is closely related to the *yurupari* initiation ritual, but that is beyond the scope of this paper. On the *yurupari,* see Stephen Hugh-Jones, *The Palm and the Pleiades: Initiation and Cosmology in Northwest Amazonia* (London: Cambridge University Press, 1979).

13. On the earth, the center is located at the highest snow peak. This world axis is imagined as a spindle; see G. Reichel-Dolmatoff, "The Loom of Life."

14. Most of these lakes dried up some 30,000 years ago, but the ancient lake bed and the surrounding land provided the best agricultural soils for the Muiska.

15. These are the valleys of Villa de Leyva and of Ramiriquí, both in the Boyacá district.

16. This observation was made at Saquenzipa, near Villa de Leyva. According to myth, the goddess Bachué rose from the Laguna de Iguaque.

17. These lakes are still in existence.

18. Reference is made to the myth of Goranchacha, the Son of the Sun, as recorded by the Spanish chronicler Fray Pedro Simón in *Noticias Historiales de las Conquistas de Tierra Firme en las Indias Occidentales,* 5 vols. (1623; reprint Bogota: 1882–1892).

CHAPTER EIGHTEEN

The *Hooghan* and the Stars

Trudy Griffin-Pierce

The Navajo homeland is located in the Four Corners area—the Colorado Plateau country where Arizona, New Mexico, Utah, and Colorado come together. The Navajo belong to the Athabascan language family, which they share with Apacheans in the Southwest as well as with Native Americans in Oregon, California, Canada, and Alaska. Although towns exist today on the Navajo Reservation, traditionally the Navajo lived in scattered, small family groups.

A concern for order in human life guides Navajo behavior. The Navajo word *hózhǫ́*, which has no precise English equivalent, expresses concepts of beauty, harmony, blessedness, and satisfaction. The significance and pervasiveness of this idea are reflected in the statement that "abusing means disorderly treatment" (Pinxten 1983:29). All phenomena in the Navajo universe are interrelated and interdependent; all living beings

---

Reproduced by permission from *Earth and Sky: Visions of the Cosmos in Native American Folklore*, edited by Williamson and Farrer. © 1992 University of New Mexico Press.

serve one another to some extent. Consequently, everything has a place and a function in a mutually dependent chain that connects each feature of mind, body, earth, and celestial phenomena and ultimately includes the whole universe. When an individual abuses one element, he or she disrupts the whole system and is made ill by the forces thus unleashed.

This interrelatedness is reflected in the Navajo dwelling, which is called a *hooghan,* and in stories about the *hooghan.* As we will see, the *hooghan* embodies the concept of *hózhǫ́.* The *hooghan* and the sky can also be conceptualized as reciprocals because the *hooghan* reflects cosmological order while the sky can be seen as a sort of *hooghan.*

These concepts are also exemplified in sandpaintings, which are created as a part of ceremonies that take place in the *hooghan.* Sandpaintings are an important part of Navajo healing rituals and play an essential role in the restoration of the patient to a state of health and harmony. These depictions, therefore, and the stories that accompany them reflect the Navajo emphasis on order and harmony, circularity and reciprocity. Sandpaintings synthesize the aesthetic, the sacred, and the medicinal by creating a visual model of the natural-supernatural worlds.[1]

This chapter focuses on the visual expression of one aspect of worldview—the transmission of normative standards of Navajo ethical behavior through stellar patterns. Specifically, it examines star patterns and moral teachings. I examine two constellations—the two *náhookǫs* (the Big Dipper and Cassiopeia)—which are conceptualized by some Navajo medicine men, or chanters, as a *hooghan.* As we explore the significance of the *hooghan* and its celestial counterpart, we will see how the two *náhookǫs* symbolize the concept of *hózhǫ́.*

STAR PATTERNS AND MORAL TEACHINGS

As Keith Basso (1983: 45) has demonstrated with the Western Apache, geographical features of the physical landscape serve as "indispensable mnemonic pegs on which to hang moral teachings of their history." When a Western Apache sees a particular mountain, the name of the mountain evokes a particular historical tale that has moral significance. In this way, "the land makes people live right" (Mrs. Annie Peaches, in Basso 1983: 2).

For the Navajo, celestial phenomena serve a similar function as visual reminders of key values. The stories associated with constellations provide moral guidance for the Earth Surface People, as the Navajo call themselves. Only by adhering to the right values can the Earth Surface People establish and maintain harmony in their lives and in the universe. The constellations serve as powerful symbols because they are universally visible. First Woman, in Newcomb's (1967: 83) version of Creation, refers to this when she says,

> When all the stars were ready to be placed in the sky First Woman said, "I will use these to write the laws that are to govern mankind for all time. These laws cannot be

written on the water as that is always changing its form, nor can they be written in the sand as the wind would soon erase them, but if they are written in the stars they can be read and remembered forever."

To the Navajo, the constellations, sun, and moon are *diyin dine'é*, supernatural beings or Holy People. Even though the Holy People left the Earth Surface People (and are no longer visible to humans) at the time of Creation (Slim Curley, in Wyman 1970: 324), these sacred beings remain nearby and are omnipresent. They are described as experiencing human emotions and are tied genealogically through the clan organization to the Earth Surface People (Reichard 1950: 58–59). Thus, for the traditional Navajo, what Westerners call "celestial bodies" are really a class of living beings sharing emotional, genealogical, and physical proximity with humans.

THE *HOOGHAN*

Traditionally, the Navajo *hooghan* is the place in which the Navajo family lives its day-to-day life. Children are born in the *hooghan;* the family as well as motherless lambs sleep there; meals are cooked over the fire in the center of this dwelling; wool, which later will be woven into rugs, is dyed in pots on the *hooghan* floor; articles of clothing, as well as sacred paraphernalia, are stored within this structure. The *hooghan* is a place of instruction, where grandfathers tell stories during the long winter nights and daughters learn to cook and to spin during the day. As with all homes, it is a refuge of order and peace from the outside world.

The *hooghan* is also a place of healing; any Navajo home may be dedicated for sacred use when prayers identify it with the homes of powerful Navajo supernaturals. The ceremonial and the particular sandpainting for that ceremonial determine the sacred dwelling into which it is transformed. Through this transformation, the *hooghan* may become the first house built on earth in which the *diyin dine'é* planned the creation of the world; or it might be Changing Woman's home when she was still a girl at Huerfano Mesa in New Mexico; it could also become the Sun's magnificent celestial home with its many rooms full of jewels, livestock, singing birds, and beautiful fabrics; or the *hooghan* may be transformed into the luxurious home that the Sun built for Changing Woman in the Western Ocean where the Navajo clans of today were created (McAllester and McAllester 1980: 13).

The primordial and cosmic houses celebrated in Navajo prayers and stories are unlike those in any other published literature; house beams listen and fall into place at the command of supernaturals as these beings create houses with cosmic ground plans (McAllester and McAllester 1980: 13–15). And the houses created by these *diyin dine'é* are made of dawn, with rooms of turquoise and ladders of white shell with rainbows that extend into the interior.

In the beginning, when the first people emerged from the four underground worlds to appear on the earth's surface, they had nothing. Before they could form the earth into a habitable place, they needed to gather together and plan. But in order to establish the order of the world, they needed to build the first *hooghan* where they could meet and discuss things for the future.

It was essential for the orderly unfolding of creation that the structure in which this planning occurred was constructed according to sacred specifications based on the four directions. The gods selected the four main support poles of the *hooghan* following the directional order that is proper in most ceremonies—the sunwise circuit—which begins with the East, moves to the South, then turns to the West, and finally, moves to the North.[2] In keeping with Navajo world view, the builder of the first *hooghan* thought about the suitable poles, discussed them with friends whose aid he would need, and then, with their assistance, brought the appropriate support poles to the site, where they were put in the proper cardinal positions, beginning with the East.

The following stanzas of the Chief Hooghan Songs demonstrate how fundamental the four directions and the *diyin dine'é* of each of these four directions are to the construction of the *hooghan*. By following these sacred specifications, *hózhǫ́*, expressed as a "long life" and "happiness," is built into the very foundation of the *hooghan*.

> Along below the east, Earth's pole I first lean into position. As I plan
> for it it drops, as I speak to it it drops, now it listens to me as it
> drops, it yields to my wish as it drops,
> Long life drops, happiness drops into position, *ni yo o*.
>
> Along below the south, Mountain Woman's pole I next lean in position.
> As I plan for it it drops, as I speak to it it drops, now it listens to me as it drops, it
> yields to my wish as it drops,
> Long life drops, happiness drops into position, *ni yo o*.
>
> Along below the west Water Woman's pole I lean between in position
> As I plan for it it drops, as I speak to it it drops, now it listens to me as it drops, it
> yields to my wish as it drops,
> Long life drops, happiness drops into position, *ni yo o*.
>
> Along below the north, Corn Woman's pole I lean my last in position
> As I plan for it it drops, as I speak to it it drops, now it listens to me as it drops, it
> yields to my wish as it drops,
> Long life drops, happiness drops into position, *ni yo o*.
>
> (Wyman 1970: 115)

Navajo chanter Frank Mitchell (Frisbie and McAllester 1978: 245) explains that after the gods built the first *hooghan* and established leadership within this structure, "That [was] the start of the human race on earth." Because the sacred specifications surrounding its construction were followed, the first *hooghan* became the proper site for the creation of life which could then unfold *nizhónígo*, or "in an orderly and proper way." Today's *hooghan* thus embodies the concept of *hózhǫ́*, both in its planning and in the creative process of events which began within its walls. This is why Father Berard Haile (1947) used the *hooghan* as a conceptual introduction to the Navajo universe and why Farella (1984: 87) refers to the *hooghan* as "one of those master encodings… an economical starting point for understanding the whole of the Navajo world view."

SANDPAINTINGS

Navajo ritual both cures and prevents illness. According to Navajo thought, illness is caused by natural phenomena *(diyin dine'é)*, some species of animals, the misuse of ceremonial paraphernalia or activities, and ghosts (Wyman and Kluckhohn 1938: 13–14). To cure the patient, the practitioner must invoke the cause of illness and bring it under control. The particular cause of the illness, rather than the physical symptoms, dictates the particular ceremonial needed to cure the patient. The use of ceremonial knowledge and the proper performance of orderly procedures in a ritually controlled context restore harmony, beauty, balance, and order. Thus, the actions and states of the individual are clearly linked with the rest of the universe.

Navajo ritual both restores and creates balance. Navajo ritual is organized around chants, which are ceremonials composed of a series of ritual procedures and conducted by a medicine man or chanter for a patient in order to cure illness. Each ceremonial has an associated origin myth that provides the sanction and rationale for the chant ritual by giving an account of how the chant's ritual procedures were acquired by some of the supernaturals and were given to the Earth Surface People. Thus, the chantway myth mediates between the natural and supernatural worlds.

The rites that compose a ceremonial can be combined into a two-night, five-night, or nine-night version. A sandpainting is but one of these component rites within a ceremonial; other rites include the consecration of the *hooghan,* the setting-out of prayer sticks, and the sweat bath. Only a few of the sandpaintings still being used today depict constellations.

The sandpainting assists healing in four ways: it attracts the supernaturals and their healing power; it identifies the patient with their healing power; it absorbs the sickness from, and imparts immunity to, the patient seated on it; and it creates a ritual reality in which the patient and supernatural interact dramatically. The supernatural beings are thought to be irresistibly attracted by seeing their portraits painted in sand; when they arrive, they become one with their images depicted on the *hooghan* floor. After the patient sits on the images of the supernaturals and the chanter has pressed sand from the various parts of the figures' bodies to the corresponding parts of the patient's body, the patient, through identification, becomes like the sandpainted supernaturals, powerful and immune to further harm. Gladys Reichard (1950: 112) described the sandpainting rite as a "spiritual osmosis in which the evil in man and the good of deity penetrate the ceremonial membrane [the sandpainting] in both directions, the former being neutralized by the latter, but only if the exact conditions for interpenetration are fulfilled." It is the final purpose of the sandpainting that most concerns us in this chapter.

The cyclical nature of Native American time (rather than the linear, "progress"-oriented nature of modern Western time) makes it possible for events of the past to

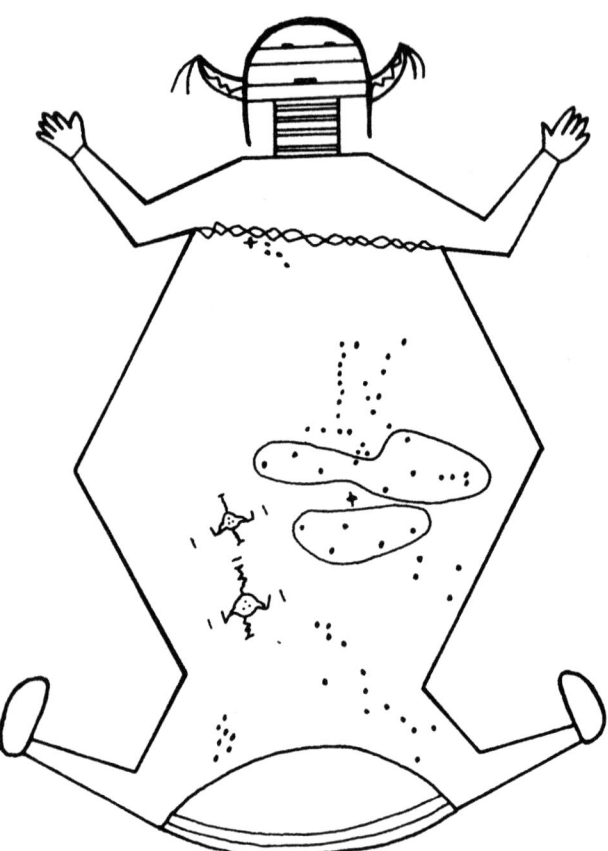

18.1 Chanter A's Father Sky with the Big Dipper and Cassiopeia circled.

occur again in the present.[3] Because the past coexists with the present, the past is accessible to Native Americans in a way that it is not accessible to Westerners. The recitation of the chantway myth through prayers, songs, and ritual procedures is understood as an actual—not a symbolic—reenactment. (See Brown [1989] for a discussion of the Native American cyclical perspective of time.) It is this concept of the re-creation of past events that makes a sandpainting so powerful and results in the patient's transformation from observer to actor as he or she is taken from the context of a family *hooghan* into a mythic world where miraculous events are commonplace.

CONSTELLATIONS IN SANDPAINTINGS

Figure 18.1 depicts the night-sky portion of a sandpainting of the sky and the earth by a Navajo chanter to whom I will refer as Chanter A.[4] *Náhookǫs bikạ'ii* translates as "Male Revolver" and corresponds to the part of Ursa Major that includes the Big Dipper (Griffin-Pierce 1986). (This particular chanter reversed Ursa Major so that the

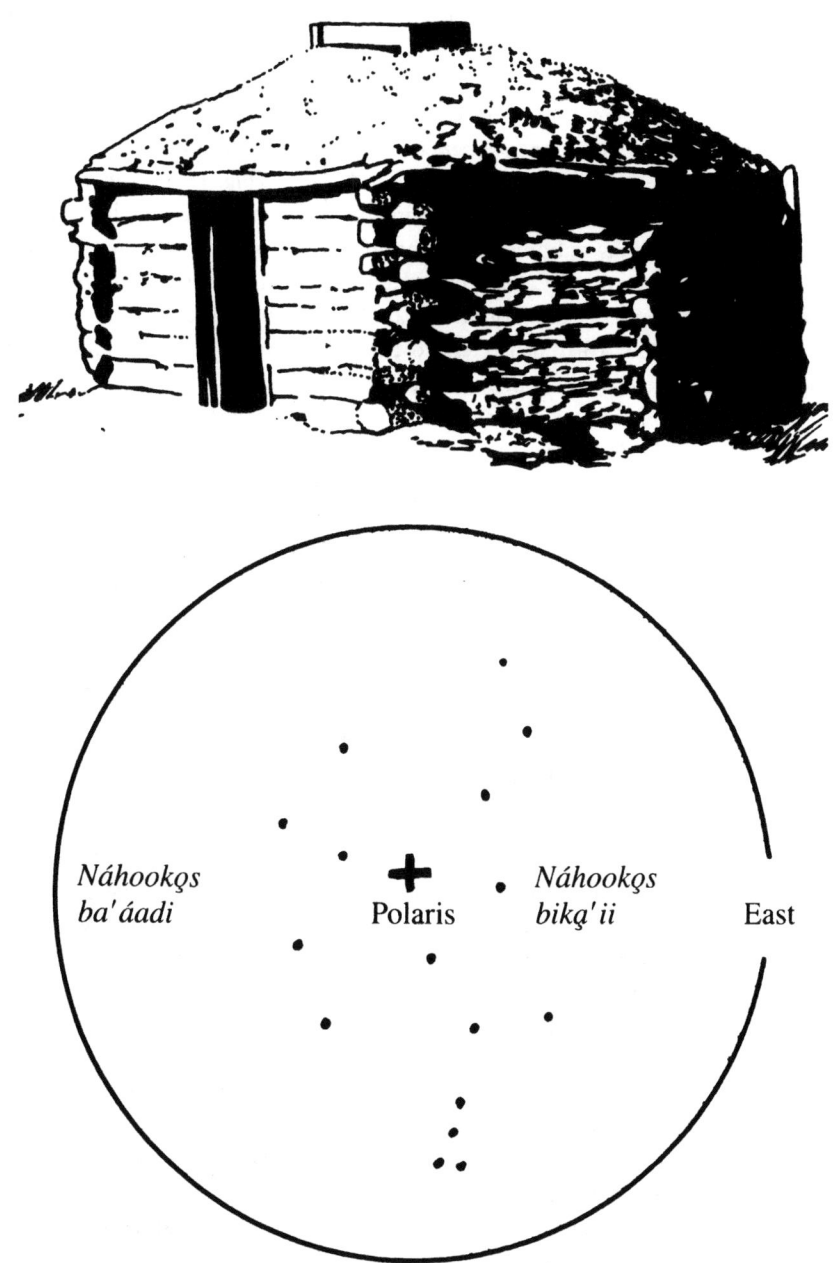

18.2 Upper: Navajo *hooghan*. Lower: the two *Náhookǫs* (from a sandpainting) revolving around Polaris, which represents the central fire in the *hooghan*. Navajo sandpaintings do not reflect the relative brightness of stars within constellations.

two pointer stars point away from Polaris.) *Náhookǫs baʼáadii* (2) is "Female Revolver" and is the same as Cassiopeia. The Navajo names for these constellations describe their circular motion around Polaris which the Navajo refer to as their igniter, the star that illuminates them and provides their fire and light.

Chanter A interprets these constellations (the Big Dipper and Cassiopeia) as a symbol of the Navajo home or *hooghan* as these two star groupings revolve around Polaris, which represents the central fire in the *hooghan* (Fig. 18.2). Together, these constellations represent "old people" or "women folks," he said. "They tell us [by their positive example] to stay at home, to stay around your fire." Here, the implication is that these constellations set a moral example for the Earth Surface People to be at home with their families to carry out their familial responsibilities.

Newcomb (1966: 156) also refers to Polaris as the *hooghan* fire and to Ursa Major and Cassiopeia as "the married couple who circled this fire but never left it to find some other." She also says that two laws—the law against two couples living in the same *hooghan* or doing their cooking over the same fire and the law that forbids a young man to look at the face of his mother-in-law—were "written in the stars" (Newcomb 1966: 156).

Chanter A offered a related but slightly different interpretation when he referred to the two *náhookǫs* as leaders, as sources of wisdom and knowledge always available to the Earth Surface People. These constellations are also visual reminders to leaders on earth that they must always be willing and ready to help their people.

THE *HOOGHAN* AND THE TWO *NÁHOOKǪS* (THE BIG DIPPER AND CASSIOPEIA)

Náhookǫs bikąʼii (The Big Dipper) and *Náhookǫs baʼáadii* (Cassiopeia) are thus a visual metaphor for the Navajo, home or *hooghan,* a central focus for life, as these two beings revolve around Polaris—symbolic of the fire in the center of the *hooghan*—and the other stars revolve around the Big Dipper and Cassiopeia (Chanter A). When I recounted this metaphor to Chanter B, he said, "That chanter gave you the *Hózhǫ́ǫ́ji* [Blessingway] version [of the two *náhookǫs*—the Big Dipper and Cassiopeia]," referring to the ceremonial which is considered to be the cornerstone and backbone of Navajo ritual.

Hózhǫ́ǫ́ji differs from the curing ceremonials not only in form (for example, Blessingway drypaintings differ in materials, designs, and use from the Holyway sandpaintings) but also in intent. The purpose of *Hózhǫ́ǫ́ji* is preventive in nature: Blessingway is held to ensure peace, harmony, success, and good fortune for the Navajo, their relatives, flocks, and other possessions, and, by extension, the whole tribe (Frisbie 1980: 161).

What is of primary interest here is that the central concept of Blessingway is *sąʼa naghái bikʼe hózhó*—related to the name of the rite. *Hózhǫ́ǫ́ji*—reflected on the intent of this ceremonial, which is "to secure a fine result in any phase of the life cycle, from birth to old age" (Haile, in Wyman 1970:7–8). *Sąʼa naghái bikʼe hózhǫ́* has been discussed by many scholars of the Navajo, especially Farella (1984); this complex phrase can be translated as "according-to-the-ideal-may-restoration-be-achieved" (Reichard 1950:47). Although this interpretation is too brief to convey the full essence of such a complex concept, it remains the best gloss of this phrase.

Father Berard Haile (in Wyman 1970: 10) explains the connection between Blessingway and the *hooghan*:

> Blessingway is vastly concerned with the hogan, a term which has been anglicized from the Navajo *hooghan,* the place home. This place home is the center of every blessing in life: happy births, the home of one's children, the center of weddings, the center where good health, property, increase in crops and livestock originate, where old age, the goal in life, will visit regularly. In a word, the hogan spells a long life of happiness.

A derived subceremony of the Blessingway focuses on the *hooghan* as a central focus for the maintenance of the order, harmony, balance, and peace necessary for the continuation and orderly functioning of the universe. This House Blessing Ceremony commemorates the building and blessing of the first mythological *hooghan* and implements instructions from the supernaturals for future generations to continue these orderly procedures, which are the basis for establishing *są'a naghái bik'e hózhǫ́* in the universe.

According to Frisbie (1980: 165–66), the *hooghan* is a home and place of security and is equated with maturity and a willingness to settle down and to plan for the future.

> Without a hogan you cannot plan. You can't just go out and plan other things for your future; you have to build a hogan first. Within that you sit down and begin to plan. (Frisbie and McAllester 1978: 244)

However, *hooghans* are much more than just dwelling places: they are important mythologically and are personified both as deities and as living entities. As we have seen, the *hooghan* is associated with Water Woman, Mountain Woman, Wood Woman, Changing Woman, and the Sun. The individual components of the *hooghan*, such as beams, earth, and fire, as well as its completeness is addressed in song and prayer (Frisbie 1980: 166).

When Chanter A said that the two *Náhookǫs* (the Big Dipper and Cassiopeia) represented a *hooghan* that served as a reminder for "women folks ... to stay at home," he was implicitly referring to the strength of the mother-child bond in Navajo society; this bond has been called perhaps the strongest (Aberle 1961: 166; Lamphere 1977: 70) and most important (Witherspoon 1970:59) tie in Navajo society. A child is born into the mother's clan and preferred residence is in the residence group of the wife's parents. The husband was often absent: in earlier times his absence was a result of hunting and raiding expeditions as well as kinship responsibilities to his sister's children, while today his absence results from wage work. Both in traditional and contemporary times, men's roles serve to reinforce the mother-child bond.

One of the chief *hooghan* songs of Blessingway, quoted below, directs the woman of the family to keep the *hooghan* neat and orderly both inside and outside. Wyman (1970: 118) explains that

> In the song, "away from a woman" means that the hogan is assigned to the woman of the family and it is her duty to make it more attractive by the orderly arrangement of property inside and cleaning rubbish away from the outside premises.

The song to which Wyman refers contains these words:

> ʹe neya ... away from a woman ... away from a woman.
> It is my hogan where, from the back corners beauty radiates, it radiates from a woman.
> It is my hogan where, from the rear center beauty radiates, it radiates from a woman.
> It is my hogan where, from the fireside beauty radiates, it radiates from a woman.
> It is my hogan where, from its side corners beauty radiates, it radiates from a woman.
> It is my hogan where, from the doorway on and on beauty radiates, it radiates from a woman, it increases the radius of beauty, *golghane*.
> (Wyman 1970: 118)[5]

This song also mentions the hearth, or "fireside." The fire in the center of the *hooghan*, represented by Polaris, is also of particular significance. As Ruth Roessel, a Navajo writer and teacher (1981: 72), puts it,

> Without the assistance of the fire and the poker there would be no Navajos today. During the beginning period [after the Emergence], the fire and the poker were capable of speech, and there was communication between the poker, the fire and the early Navajos.

The fire in the center of the *hooghan* is deeply intertwined with the concept of the *hooghan* itself. This is reflected linguistically: "a no fireplace home," or "a home where a fireplace is no more," designates a home in which the fireplace has been abandoned—because the death of a younger person has occurred in that home (Wyman 1970: 10). In such a case, the relative may burn the *hooghan*, or make an opening through the North side to permit the passage of a corpse, or block the entrance and smokehole with timbers to warn that a burial had been made in it (Wyman 1970: 10). No Navajo would approach such a site because of the association between witchcraft and the dead.

The metaphor of the two *náhookǫs* (the Big Dipper and Cassiopeia) as leaders also relates both to the concept of *sąʹa naghái bikʹe hózhǫ́* and the *hooghan* itself. The willingness and the maturity to take responsibility for one's actions are qualities that ensure the maintenance of *sąʹa naghái bikʹe hózhǫ́* in the universe. The chief *hooghan* songs of Blessingway have this title because they refer not only to their contents—a plan for the first *hooghan*—but also to the person who will see that these plans are carried out.

Slim Curley, a Navajo chanter (in Wyman 1970: 112), explains that during the process of Creation, those involved said, "This much is clear, that from the beginning of time this leading chief of ours had full knowledge of things, no doubt about it." This is a reference to the leading headman, or *naat'áanii*, who knows

> which materials are suitable, where to obtain them, how to set them. His "full knowledge" also includes planning the hogan, asking friends for help, directing construction, talking matters over, etc., and is the keyword of one song. (Wyman 1970: 112)

The depth of the leader's knowledge is described in the following chief *hooghan* song:

Of origins I have full knowledge... *holaghai*.
Of Earth's origin I have full knowledge.
Of plant origins I have full knowledge.
Of various fabrics' origins I have full knowledge.
Now of long life's, now of happiness origin I have full knowledge, *holaghai*.
Of Mountain Woman's origin I have full knowledge.
Of various jewels' origins I have full knowledge...

(Wyman 1970: 113)

The song goes on to describe a gradual progression of action that comes to a climax in the last song: first, the headman is credited with possession of full knowledge; then, he has the thought of putting this knowledge into practice; and, finally, he speaks of his purpose to others. This points to another quality of leadership: a strong leader possesses valuable knowledge that he draws upon as he carefully considers how he will put his plans into action before speaking of these plans to others.

The chief *hooghan* songs include a set of "planning songs." The words of the chorus reflect the emphasis on strong leadership:

'e ne ya ... he gives orders and with it he gives orders as he passes by, *ni yo o*.
Now with Earth he gives orders as he passes by,
Now with vegetation he gives orders as he passes by,
With fabrics of all kinds he gives orders as he passes by,
With long life he gives orders as he passes by, with happiness he gives orders as he
  passes by, he gives orders as he passes by, *ni yo o* ...

(Wyman 1970: 121)

This song set is used at ceremonies for the induction of leaders because "headmen are supposed to be dependable and reliable planners in important public matters" (Wyman 1970: 121). Wyman adds that the *hooghan* "is the logical place for planning of any kind" (Wyman 1970: 121). This reiterates Frisbie's (1980: 244) description of the *hooghan* as a symbol of maturity because it is there that one plans for his or her future.

In her analysis of Navajo chantway myths, Spencer (1957: 40, 58–60) observes that the assumption of responsibility for the welfare of the family group and for the larger group are common themes in these myths, which are based on the conditions of Navajo life. The relatively isolated position of the family group imposes the necessity for familial cooperation, with the assumption of personal responsibility on the part of each family member. In chantway myths, indolence and irresponsibility lead to physical disaster, while industry and responsibility bring success.

Only through proper leadership can generational continuity be ensured. The two *náhookǫs* (the big Dipper and Cassiopeia) as leaders serve as a reminder to those on earth that we must provide strong leadership and accept responsibility for both present and future generations.

Finally, the enduring nature of the constellations as well as the image of the *hooghan* they represent is reflected in the songs of Talking God[6] from Blessingway:

> *'e ne ya* ... the same [hogan] will continue, the same will continue, on its surface it passes by.
> Exactly on Earth's surface it passes by,
> On its surface vegetation passes by, on its surface fabrics of all kinds pass by,
> Chief long life, chief happiness, old age one says continues to pass by on its surface,
> the same will continue, *ni yo o*.
>
> (Wyman 1970: 120)

Talking God is saying that the inhabitants of the earth's surface pass on, and even the timbers of this particular *hooghan* may decay and collapse with age, but the concept of the *hooghan* continues on.

To symbolize that the *hooghan* type and its songs (in Blessingway) will continue on indefinitely, two stone slabs which are imbedded in the ground next to the eastern poles (which frame the East-facing doorway) are "set for" the *hooghan*. This means that wherever they are found, these stone slabs indicate that the mandate contained in Blessingway songs to continue the *hooghan* type of dwelling has been followed (Wyman 1970: 14).

Just as the stone slabs symbolize the continuation of the *hooghan* and, thus, of the Navajo way of life, the daily occurrence of the dawn as the sun returns symbolizes the continuation of time and of life itself. Dawn (associated with the white and the East) is one of the four cardinal light phenomena, along with the blue of day-sky (associated with the South), the yellow of evening twilight (West), and the black of darkness (North). Each of these four light phenomena serves as a guide to people's movements and activities (Griffin-Pierce 1988). Dawn causes people to awaken: "he (or she) who has walked in it [the dawn] will enjoy every possession," says chanter Frank Mitchell (in Wyman 1970: 370). Thus, by rising early, by "walking in the dawn," one is assured of abundance. Prayers and offerings at dawn outside the East-facing doorway of the *hooghan* ensure this prosperity.

The Milky Way (*yikáísdáhí*) symbolizes the white corn meal sprinkled by First Woman as she said her morning prayers (Chanter C). Ceremonials accompany the dawn prayers as the sun rises; by visualizing the Milky Way as the corn meal used in these prayers, *yikáísdáhí* serves as a visual reminder to pray to the dawn as a life-giving source.

Prayers and offerings are of particular importance in showing respect and appreciation to the *diyin dine'é*. Harry Walters, director of the Ned Hatathli Museum at Navajo Community College, Tsaile, explains:

> If you see something out of the ordinary, it is a warning to remind you to restore the balance, to act like a Holy Person [*diyin dine'é*], to show respect to the mountains, animals, and people [including yourself]. Every chance we get we should acknowledge our gifts from the Holy People. If we forget the Holy People [including the Stars and Constellations] remind us. This is what it means when you see a Holy Person. You show respect and acknowledge their gifts to us by leaving offerings.

Chanter A explained that the purpose of the stars is to help the Earth Surface People to live the right way: "Before the Twins [the Sacred Twins who slew the Monsters which once inhabited the earth] and the Monsters were created, the *diyin dine'é* said, 'Someday there will be all kinds of misfortune in the world. We are creating the Stars to help the Earth Surface People to find their way, so that they can regain their faith and reestablish their balance and their direction.'"

Blessingway is designed to maintain and reinforce *hózhǫ́*. An important aspect of achieving and maintaining this ideal, valued state is to view life with the proper, reverent frame of mind. Reichard (1944, 1950) and Witherspoon (1977) have demonstrated how powerful thought is in Navajo culture. To think something is to cause it to be. Witherspoon (1977) explicates *są'ah naagháí* as thought and *bi'keh hózhǫ́* as speech to show how intimately related these two processes are considered to be: speech is the outer form of thought, and thought is the inner form of speech. An unbreakable bond exists between thought and action, speech and event.

In Figure 18.3 we see another example of the Navajo emphasis on concentration and clear thought processes. The reason Spider Woman taught the Navajo to make figures in string is to help the Navajo learn how to concentrate. "You learn to think when you make these," a Navajo girl told folklorist Barre Toelken (1979: 95). The girl's father elaborated on the link between clear thought and living a good life, as well as on the link between the string figures and celestial constellations.

> It's too easy to become sick, because there are always things happening to confuse our minds. We need to have ways of thinking, of keeping things stable, healthy, beautiful. We try for a long life, but lots of things can happen to us. So we keep our thinking in order by these figures and we keep our lives in order with the stories. We have to relate our lives to the stars and the sun, the animals, and to all of nature or else we will go crazy, or get sick. (Toelken 1979: 96)

The depiction of the Pleiades *(dilyéhé)* in string represents the importance of clear, unclouded thought in order to receive the guidance necessary to "keep things stable and beautiful" so that one can live a long life. Only when one learns how to concentrate, how to use one's mind in proper ways does one grow to understand and to respect the place and function of all living things in the universe. It is through this understanding of the interrelatedness of all things and through adherence to proper values that one contributes to the order of the universe.

18.3 Navajo woman making a string figure of *dilyéhé* (the Pleiades). (Line drawing after an original serigraph by the author.)

As a cultural map, sandpaintings order and make sense of the natural-supernatural world around us. The patient, ill in spirit as well as in body, cannot help but respond to ceremonial efforts to reestablish order, efforts that occur at several levels of meaning.

First, there is the powerful image upon which the patient sits, the depiction of the *diyin dine'é*, drawn at the direction of a respected and trusted chanter. The chanter has guided the production of a ritually correct sandpainting. This sandpainting, which has previously healed many with similar illnesses, orders both the natural and supernatural worlds by correctly depicting particular images. The patient is aware that this sandpainting has been selected from the many sandpainting images known by the chanter because it is particularly suited for treating the patient's illness. The order reflected in the painting's visual representation of the world assists in reestablishing order in the patient's inner world of thought and feeling.

Just as all sandpaintings order the natural-supernatural world, sandpaintings of the heavens order a particular part of that world by emphasizing those constellations with particular significance for the Navajo. By filtering out constellations that are less culturally significant from the seemingly infinite number of stars strewn across the celestial sphere, the sandpainting depictions of constellations help the knowledgeable individual to focus on a select set of celestial entities.

These particular constellations evoke allegorical stories that help the Earth Surface People to live in the right way. Thus, depictions of constellations serve as mnemonic devices through which to remember moral stories. These visual and verbal images not only tell us what the Navajos find significant about the sky but also how the Navajos conceive of themselves and the right way to conduct their lives.

The visual depiction of constellations in sandpaintings transmits and reiterates the Navajo emphasis on order, balance, circularity, and reciprocity. Not only does the ceremonial itself work actively to restore this universal balance and order, but constellation depictions and the allegorical stories they evoke also remind the individual at a deeper level of how interrelated and interdependent his thoughts and feelings are with those of the rest of the universe.

The Navajo *hooghan* embodies the concept of *hózhǫ́* not only in its planning but also in the sequence of creation begun within its walls. The two *náhookǫs* (Cassiopeia and the Big Dipper), the constellations which symbolize the *hooghan,* are created in sandpaintings on the *hooghan* floor in a ceremonial context and demonstrate the interrelatedness of the universe as their depiction plays its role in the restoration of universal balance and order.

ACKNOWLEDGMENTS

I appreciate the patient and kind assistance of Harry Walters, director of the Ned Hatathli Museum at Navajo Community College in Tsaile, Arizona, as well as other Navajo consultants who must remain anonymous. I also appreciate the assistance of Rain Parrish, Dr. Edson Way, and Steve Rogers, formerly at the Wheelwright Museum of the American Indian, Santa Fe, who made possible photography and documentation of five hundred sandpaintings at this museum. My thanks also go to Dorothy House, librarian at the Museum of Northern Arizona, and to Jan Bell, curator at the Arizona State Museum, who provided similar assistance at their institutions. I also appreciate the suggestions of Ray Williamson and Claire Farrer. I would also like to express my appreciation to Chanter A for allowing me to use his drawings as the basis for my drawings in Figures 18.1 and 18.2. Figure 18.3 is my own original line drawing.

NOTES

1. Most Navajo ceremonials are used to treat the actual or anticipated illness of a patient or patients. Sandpaintings—which, more accurately, should be called "drypaintings" because they are made of sand *and* other materials—are also used when the ceremonial is not concerned specifically with curing. Even when the primary intent is not healing, there is usually a "patient," or "one-sung-over." This chapter deals with sacred sandpaintings, which are created and destroyed in a ritual context. These are not to be confused with commercial sandpaintings—permanent paintings of pulverized dry materials glued onto a sand-covered wood backing—which are secular objects made by Navajo laymen for sale. For a discussion of commercial sandpaintings, see Parezo (1983).

Fieldwork for this chapter was conducted in the Tsaile and Piñon, Arizona, areas in spring 1984, spring and summer 1985, and winter 1985–86.

2. There are actually five poles because the two eastern poles which stand on either side of the entrance are known collectively as the East pole, so that one pole at each cardinal point may be mentioned in the songs. The intercardinal points are not significant to Navajo cosmology, in contrast to Pueblo belief and practice.

3. The concept of cyclical time is not exclusively Native American but is characteristic of many cultures which emphasize the spiritual over the technological. The annual celebration of Christmas and Easter, as well as the concept of the calendar itself, are inherited from a tradition with this kind of cyclical orientation. However, in modern Western culture, the pervasive concept of time tends to be linear and "progress" oriented. In the modern Western conceptualization of time, the past can be celebrated and commemorated but it cannot be reentered in any other than a purely symbolic manner. (See Eliade [1954] for a discussion of conceptions of time in the ancient cultures of Asia, Europe, and America.)

4. I have omitted the names of the Navajo individuals with whom I worked because Navajo opinions about sharing information about sandpaintings and astronomy vary considerably. Thus, I feel it is important to maintain the privacy of those with whom I worked. Although chanters agree on the identification of the major Navajo constellations, no two chanters depict the constellations in an identical manner.

5. This song reflects the extent to which the Navajo emphasize directional concepts related to the construction of the *hooghan* described earlier in this chapter. The "back corners," or rear base corner below the West pole of the *hooghan,* are first mentioned; then the rear center, or the inner center between the fireplace and the West pole; then the "fireside," which is directly beneath the interlocked point of the *hooghan* poles; then the "side corners," which are on either side of the doorway (which always faces East); and, finally, the exit trail out the "doorway. "

6. Talking God, generally addressed as "maternal grandfather" and known as the grandfather of the gods, is one of the great gods. He acts as a mentor, often supplying mythical characters with the solutions to questions put to them by other *diyin dine'é*. Reichard (1950: 476) refers to Talking God as "the only god I have found with a sense of compassion."

REFERENCES

Aberle, David F. 1961. The Navajo. In *Matrilineal Kinship,* ed. David Schneider and Kathleen Gough, 96–201. Berkeley: University of California Press.

Albert, Ethel M. 1956. The Classification of Values: A Method and Illustration. *American Anthropologist* 58:221–248.

Basso, Keith H. 1983. "Stalking with Stories": Names, Places, and Moral Narratives among the Western Apache. In *Text, Play, and Story: The Construction and Reconstruction of Self and Society,* ed. E. Bruner, 19–55. Washington, DC, American Ethnological Society.

Brown, Joseph E. 1989. *The Spiritual Legacy of the American Indian*. New York, Crossroad Publishing Company.

Eliade, Mircea. 1954. *The Myth of the Eternal Return*. New York: Pantheon Books.

Farella, John R. 1984. *The Main Stalk: A Synthesis of Navajo Philosophy*. Tucson, University of Arizona Press.

Frisbie, Charlotte J. 1980. Ritual Drama in the Navajo House Blessing Ceremony. In *Southwestern Indian Ritual Drama,* ed. Charlotte Frisbie, 161–198. Albuquerque, University of New Mexico Press.

Frisbie, Charlotte J., and David P. McAllester, eds. 1978. *Navajo Blessingway Singer: The Autobiography of Frank Mitchell, 1881–1967.* Tucson, University of Arizona Press.

Griffin-Pierce, Trudy. 1986. Ethnoastronomy in Navajo Sandpaintings of the Heavens. *Archaeoastronomy* 9:62–69.

Griffin-Pierce, Trudy. 1988. Cosmological Order as a Model for Navajo Philosophy. Paper presented at the American Anthropological Meeting, Phoenix, Arizona, November 1988.

Haile, Father Berard. 1947. *Starlore among the Navaho.* Santa Fe, Museum of Navajo Ceremonial Art.

Lamphere, Louise. 1977. *To Run after Them.* Tucson, University of Arizona Press.

McAllester, David P., and Susan McAllester. 1980. *Hogans: Navajo Houses and House Songs.* Middletown, CT, Wesleyan University Press.

Newcomb, Franc Johnson. 1966. *Navaho Neighbors.* Norman, University of Oklahoma Press.

Newcomb, Franc Johnson. 1967. *Navaho Folk Tales.* Santa Fe, Museum of Navajo Ceremonial Art. Reprint: Albuquerque, University of New Mexico Press, 1990.

Newcomb, Franc Johnson. 1980 [1964]. *Hosteen Klah: Navajo Medicine Man and Sand Painter.* Norman, University of Oklahoma Press.

Parezo, Nancy J. 1983. *Navajo Sandpaintings: From Religious Act to Commercial Art.* Tucson, University of Arizona Press.

Pinxten, Rik, Ingrid van Dooren, and Frank Harvey. 1983. *The Anthropology of Space: Explorations into the Natural Philosophy and Semantics of the Navajo.* Philadelphia, University of Pennsylvania Press.

Reichard, Gladys A. 1944. Prayer: The Compulsive Word. *Monographs of the American Ethnological Society,* no. 7. New York, J. J. Augustin.

Reichard, Gladys A. 1950. *Navajo Religion: A Study of Symbolism.* Princeton, NJ: Princeton University Press.

Roessel, Ruth. 1981. *Women in Navajo Society.* Rough Rock, AZ, Navajo Resource Center, Rough Rock Demonstration School.

Spencer, Katherine. 1957. Mythology and Values: An Analysis of Navaho Chantway Myths. *Memoirs of the American Folklore Society,* vol. 48. Philadelphia.

Toelken, Barre. 1979. *The Dynamics of Folklore.* Boston, Houghton Mifflin.

Witherspoon, Gary J. 1970. A New Look at Navajo Social Organization. *American Anthropologst* 72:55–65.

Witherspoon, Gary J. 1977. *Language and Art in the Navajo Universe.* Ann Arbor, University of Michigan Press.

Wyman, Leland C. 1952. The Sandpaintings of the Kayenta Navaho. *University of New Mexico Publications in Anthropology,* vol. 7. Albuquerque, University of New Mexico Press.

Wyman, Leland C. 1970. *Blessingway.* Tucson, University of Arizona Press.

Wyman, Leland C. 1983. *Southwest Indian Drypainting.* Albuquerque, University of New Mexico Press.

Wyman, Leland C., and Clyde Kluckhohn. 1938. Navaho Classification of Their Song Ceremonials. *Memoirs of the American Anthropological Association,* no. 50. Menasha, WI.

CHAPTER NINETEEN

Animals and Astronomy in the Quechua Universe

Gary Urton

At the beginning of his treatise on the "errors and superstitions of the Indians" written in 1571, Polo de Ondegardo gives us one of our longest accounts of the constellations recognized by the Incas. Among the constellations are several animals and birds including llamas, a feline, and a serpent. By reference to other Spanish and indigenous chroniclers of Inca culture, the list of animal constellations can be expanded to include the tinamou, the condor, and the falcon.[1]

In addition to the list of animal constellations, Polo gives us the following explicit statement concerning the relationship between celestial and terrestrial animals:

> ... in general, [the Incas] believed that all the animals and birds on the earth had their likeness in the sky in whose responsibility was their procreation and augmentation. (Polo, [1571] 1916: cap. 1; my translation)

Reproduced by permission from *Proceedings of the American Philosophical Society* 125:2 (1981):110–127.

The identification of the Incaic animal and bird constellations has eluded us for some time. In fact, it is not surprising that this has been the case because apparently their recognition even by the Incas was sometimes something of a problem. The chronicler Garcilaso de la Vega gives us the following confession concerning his early astronomical training:

> They fancied they saw the figure of an ewe [llama] with the body complete suckling a lamb [uñallamacha], in some dark patches spread over what the astrologers call the Milky Way. They tried to point it out to me saying: "Don't you see the head of the ewe? There is the lamb's head sucking; there are their bodies and their legs." But I could see nothing but the spots, which must have been for want of imagination on my part. (Garcilaso, [1609] 1966, Bk. II, cap. XXIII)

What was apparently the same llama, and its suckling baby, is also described in the chronicle of Francisco de Avila (1608) in the central Andean community of Huarochirí. The llama, says Avila, was "blacker than the night sky" (Avila, [1608] 1966: cap. 29).

From the testimony of Garcilaso and Avila alone it should have been clear that in order to identify the animal constellations of the Incas, we should look first to the "dark spots." However, the literature on Incaic ethnoastronomy is full of attempts to explain away, or to dismiss entirely without comment, the testimony on the dark cloud animal constellations in the Milky Way (e.g. Lehmann-Nitsche, 1928: p. 36).

Constellations of the type we are discussing have only rarely been reported in the ethnoastronomical literature,[2] and it will be one of our goals here to present as complete a description as possible in light of the data collected thus far in the field. Our major objective is to arrive at an understanding of the general principles of Quechua astronomy and cosmology incorporated in this one category of celestial phenomena. It should be emphasized at the beginning that we would expect to find a close correspondence between Quechua celestial and terrestrial classifications and symbolism. For example, we would expect that the principles of opposition, mediation, and unification which have been found to operate among natural and supernatural forces (e.g. Earls, 1973; Isbell, 1974 and 1979; Mayer, 1974; Ossio, 1976; Urbano, 1974 and 1976; Wagner, 1978; and Zuidema, 1973 and 1977), will be fundamental to the conceptualization and mechanics of Quechua astronomy.

Through ethnoastronomical fieldwork carried out in a number of communities around the city of Cuzco, Peru, it has been possible to identify several animal constellations which conform to the general descriptions given in the Spanish chronicles (cf. Urton 1978a and b, 1979 and 1980; Zuidema and Urton, 1976). In addition to the identifications of the celestial locations of these "dark cloud" animal constellations, information was obtained relating to their role in the life of the various communities, especially in the community of Misminay where the most extensive fieldwork was carried out (Urton, 1979).

The *runa* of Misminay possess an extremely rich and complex body of astronomical lore. At least 48 different single stars, constellations and "dark clouds" are named and observed for purposes ranging from the reckoning of time at night to the timing of the planting of the crops. As has been discussed elsewhere (Urton 1978b and 1979), the Quechua recognize two major types of constellations: a) star-to-star constellations, western-type stellar groupings which represent *inanimate* architectural and geometric forms, and b) the *yana phuyu* ("dark cloud") constellations, all of which represent *animate* forms. The dark cloud constellations are located in the southern portion of the Milky Way where one sees the densest clustering of stars and the greatest surface brightness, and where therefore, the fixed clouds of interstellar dust which cut through the center of the Milky Way (the dark cloud constellations) appear in sharper contrast.

In recent fieldwork in Misminay (June, 1979), it was found that in addition to these two primary classes *of constellations,* the celestial *animals* are apparently also divided into three color categories: *yana* ("dark"), *muru* ("spotted" or "multi-colored") and *rojo* ("red"). We will discuss here the "dark" and "spotted" animal forms, since they appear to represent two types of *yana phuyu,* and leave for another time a description of the "red" animals which are perhaps related to single red stars such as Antares and Betelgeuse.

CELESTIAL LOCATIONS AND PERIODICITIES OF THE *YANA PHUYU*

The dark cloud constellations which have been identified are as follows, listed in the order in which they rise along the south-eastern horizon (see Fig. 19.1):

mach'acuay	—serpent
hanp'atu	—toad
yutu	—tinamou
llama	—llama
uñallamacha	—baby llama (/or llama's umbilicus/or serpent)
atoq	—fox
yuthu	—tinamou

From the head of the Serpent in the west to the tail of the Tinamou in the east, the dark cloud constellations stretch in a line through about 150° of celestial space, straight along the central course of the Milky Way (= *Mayu*—"River"). Figure 19.1 is a drawing which shows all of the dark cloud constellations in the sky at the same time. The view is toward the south from a hypothetical location in the southern Andes near the city of Cuzco (i.e., Cuzco's latitude is −13°30'; thus, as shown in the drawing, the unmarked south celestial pole stands 13°30' above the southern horizon).

As the sky appears to revolve at night throughout the course of a year, it will be unusual to actually see the entire line of dark cloud constellations in the sky on the same night. The time of greatest visibility occurs when the center of the line (i.e., the area around the Southern Cross and the *Yutu)* stands along the north/south meridian at midnight; this occurs around the 23rd of March, the day of the autumnal equinox (in the southern hemisphere). If we rotate the sky so that the entire line of dark cloud constellations is underground at midnight (which will therefore be the night

19.1

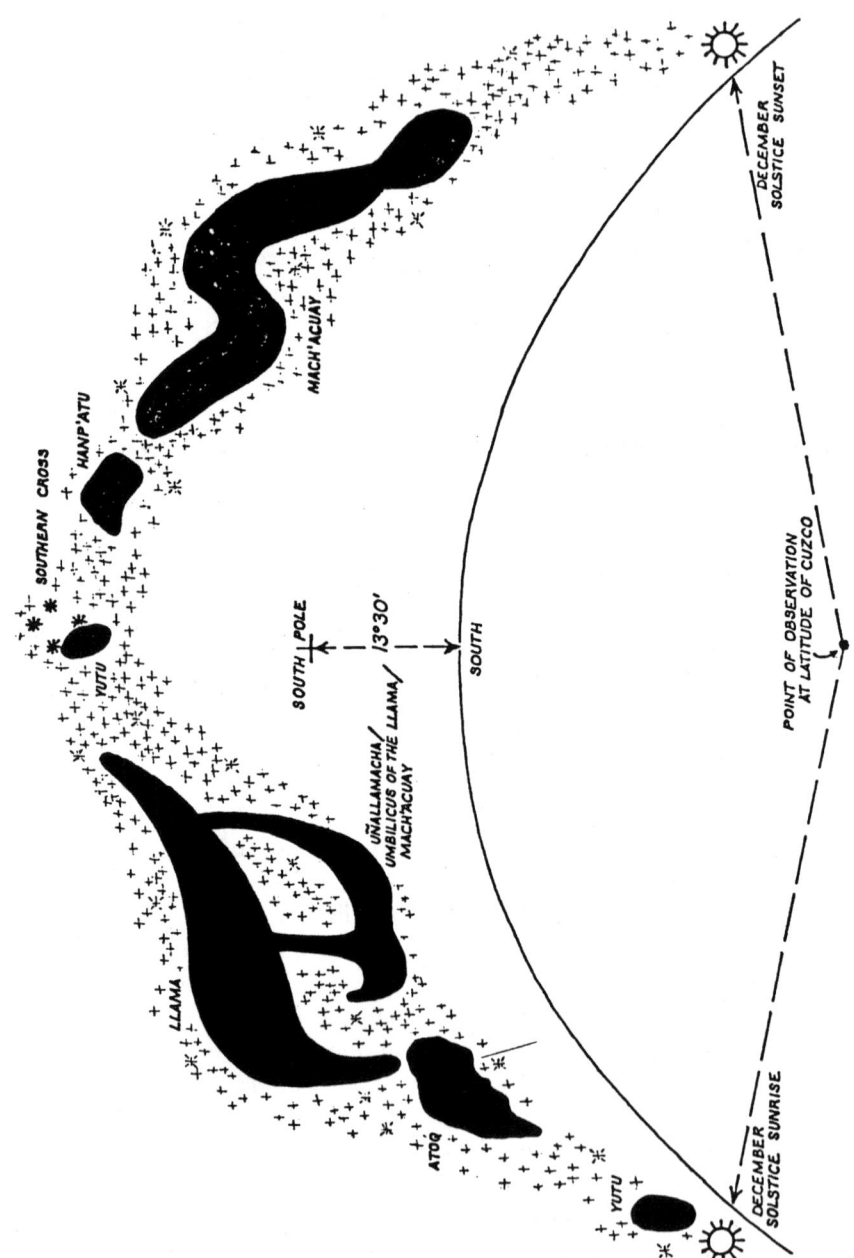

on which the *fewest* dark clouds will be seen), we arrive at a date of September 26, very near the vernal equinox.

In addition to this temporal relationship to the two equinoxes, we see in Figure 19.1 that there is an orientational relationship to the December solstice; that is, the most extreme *northerly* rising and setting points of the dark clouds coincides with the most *southerly* rising and setting points of the sun. Thus, at midnight on the night of the March equinox, the line of dark cloud constellations stretches in an arc through the southern skies from the rising point in the southeast to the setting point in the southwest of the December solstice sun (see the discussion of the relation between the dark cloud constellation of the Llama and the solstices and equinoces in Zuidema and Urton, 1976: pp. 90–94).

Another important observational and temporal characteristic of the dark cloud constellations is their relationship to the rainy season. In the southern Andes, the rainy season begins in October and ends in early April with the heaviest period of rains lasting from December through February. While discussing one of the dark cloud constellations (the Serpent) with informants in Misminay, it was said that the dark cloud Serpent is seen in the sky at night during the rainy season while, during the dry season (May–July), it is below ground at night. A similar description of the relationship between the rainy season and the celestial Llama (*Yacana*) was recorded in the early seventeenth century in Huarochirí by Avila (1966: cap. 29).

In addition to their relationship to the rainy season, the dark cloud constellations are associated in another direct way to water; as mentioned before, they are located along the center of the Milky Way which is called *Mayu*, the "River." While we cannot describe in great detail here the Quechua conception of the Milky Way (cf. Urton, 1979), it is important that we establish well the cosmological significance of these fixed clouds of interstellar dust, in the form of animals, which float through the sky along the celestial River.

In Misminay, the designation of the Milky Way as a celestial river goes beyond the mere metaphorical equation of a linear stream of water on the earth with a stream of stars in the sky. One informant in Misminay explained the relationship between the two rivers by etching a line on the floor of his hut; the line, he said, represented the Vilcanota River which is like a mirror (*espejo*) reflecting the *Mayu*, the River in the sky. Therefore, the Milky Way is equated *directly* with the Vilcanota River which flows from the southeast to the northwest through the Dept. of Cuzco.

Second, the celestial River is believed to carry into the sky the actual water which flows through the Vilcanota River. As the Vilcanota River flows from the southeast to the northwest, it carries terrestrial water to the edge of the earth. The water then flows into the *mar*, the cosmic sea, which completely encircles the earth. As the Milky Way revolves around the earth, it dips into the cosmic sea in the west, takes in the terrestrial water, passes underground and rises again in the east. As the Milky Way moves slowly through the sky above the earth, it deposits water throughout the celestial sphere. The water then returns to the earth in the form of rain where, in its continuous cosmic cycle, it again flows along the tributaries which feed into the Vilcanota River. In this way, the celestial and terrestrial rivers act in concert to continuously recycle water, the source of fertilization, throughout the universe.

It is clear from this description of the cycling of the Milky Way that in order to understand the full significance of the dark cloud animal constellations within it, we must consider the three elements, or parts of the universe, with which they are therefore associated: sky, water, and also earth (since they are subterranean for at least half of every day). There are a number of additional factors which must be considered such as the specific types of animals represented, their biological cycles and behavior patterns, and the question of the correlation between astronomical and biological cycles as suggested in the quotation from Polo de Ondegardo (above). These factors will be discussed later in this article.

Of the three factors mentioned above—sky, water and earth—water occupies a position of mediation since it is the element which is cycled through the other two. It is therefore essential to study the dark cloud constellations by analyzing the connections between sky and water and earth and water.

SKY AND WATER

In his monumental studies of the natural sciences in South America, Alexander von Humboldt made the following observation:

> I have endeavoured to describe the approach of the rainy season, and the signs by which it is announced.... The dark spot in the constellation of the Southern Cross becomes indistinct in proportion as the transparency of the atmosphere decreases and this change announces the approach of rain. (von Humboldt, [1850] 1975: pp. 138–139)

In my own fieldwork in Misminay, I was also told that the dark cloud constellations are observed in the prediction of rain. While informants in Misminay did not describe the specific methods used, nor the times when the observations are made, we have explicit testimony from elsewhere in the southern Andes that such predictions are made during the month of August at the beginning of the planting season and at the time of transition from the dry season to the rainy season. Padre Jorge Lira has recorded the following meteorological and crop predictions in use today in communities around Cuzco:

> If the stars of the sky appear bright and beautiful, everything will be good, materially and spiritually. If, in the Milky Way, there is an *accentuation of the dark areas*, or the "sacks of carbon," it will be a year of pestilence and death. (Lira, 1946: pp. 18–19; my translation and emphasis)

If we combine the accounts of von Humboldt and Lira, we can conclude that since the *obscuring* of the dark clouds indicates the approach of rain, their *accentuation*, as described by Padre Lira, indicates the absence of rain. Therefore, we find a curious kind of inversion in the relationship between water and the dark cloud constellations;

their *appearance* in the night sky is associated with the period of the rainy season, but their gradual fading out, or *"disappearance,"* as a result of increasing atmospheric moisture, announces the actual approach of rain. Since the dark cloud constellations are located in the celestial River, which spreads water from the cosmic sea throughout the celestial sphere, the weather predictions described by Padre Lira reflect a coherent and logical explanation of the operation, and inter-relation, of certain natural phenomena. But beyond the "convenience" of a consistent explanation of the natural universe, the data and processes we have outlined provide a system of prognostication, an essential element in the survival of communities whose livelihood depends upon the success of the crops which, in turn, depends upon the amount of rainfall.

It should also be pointed out here that the basic system we have described above was also understood by the Incas. In order to demonstrate this knowledge, we have only to combine the following statements from Bernabé Cobo and Francisco de Avila. First, with regard to our description of the cosmic circulation of water in modern Quechua cosmology, Cobo tells us the following concerning Inca cosmology:

> They say, in addition, that through the center of the sky there crosses a great river which they take to be that white band which we see from here below and call the Milky Way.... Of this river, they believe that it takes up the water which flows beyond the earth. (Cobo, [1635] 1956, 11: pp. 160–161; my translation)

And concerning the relation between the dark cloud constellations and the transport of water to the celestial sphere, Avila says:

> This Yacana [in the sky] ... is like the shadow of the llama. They say that this Yacana comes down to earth at midnight when it cannot be noticed or seen and drinks all the water from the sea. They say that if she did not drink this water, the entire world would be drowned. (Avila, [1608] 1966: cap. 29; my translation)

Therefore, the dark cloud animal constellations were, and continue to be, important elements in the cosmological relationship of water and sky, and as a category of celestial phenomena, they are important observable indicators of continuity and change in the physical universe.

EARTH AND WATER

Since the water within the celestial River has a terrestrial origin, it is therefore not surprising to find that the animals of the Milky Way also originate from the earth. According to one informant in Misminay, the *yana phuyu* ("dark clouds") are actual pieces of earth which are taken up into the sky by the Milky Way. The informant was uncertain whether the animals are taken up under the earth, during the subterranean passage of the Milky Way, or if they enter it from the mountain-tops where, he said, there are a lot of wild animals.

The terrestrial origin of these celestial animals is further indicated by the fact that even though they are located in the sky, they are classified as *pachatira (pachatierra)*, a name which combines the Quechua and Spanish words for "earth." The name *pachatira* was obtained in a situation which throws additional light on the symbolic significance of the term as used in relation to celestial phenomena. In a long conversation which I had with a group of men and women in Misminay, I asked about the sexual association of various astronomical bodies. It was generally agreed that single stars, as well as the star-to-star constellations, are masculine. When asked about the dark cloud constellations, one man answered immediately that they are female. Later, however, he pointedly returned to the question and said that he had been wrong earlier in calling them female; they are, he said, *pachatira*. Thus, while the dark cloud animals may be thought of as more female than male in opposition to the stars which are male, they are actually neither—they are *pachatira*.

Pachatira is an important concept in Quechua cosmological thought. In the community of Kuyo Grande, Casaverde found that *Pachatierra* is classified as female and is considered to be the malevolent twin sister of *Pachamama*, "Earth Mother" (Casaverde, 1970: p. 150). Oscar Nuñez del Prado gives the following description of the malevolent nature of *Pachatierra* as found in Kuyo Chico:

> She is wicked and eats the hearts of men, who then die spitting blood. She is generally found by cliffs and precipices, and her preferred victims are children or adults who stay asleep in bad weather. (O. Nuñez del Prado, 1973: p. 36)

The femaleness of *Pachatira*, and the relationship of *Pachatira* to the Earth Mother *(Pachamama)*, is also found in the area of Ocongate where three feminine forms of earth *(pacha)* combine to express the total concept of *Pachamama*; the three are *Pacha Tierra, Pacha Ñusta,* and *Pacha Virgen* (Gow and Condori, 1976: p. 6).

> Since the dawn of the universe, Pachamama has said: I am Santa Tierra. I am the one who nurtures and gives suckle. I am Pacha Tierra, Pacha Ñusta, Pacha Virgen. (Gow and Condori, 1976: p. 10; my translation)

In the community of Songo, when blowing a coca *k'intu*, one often calls upon "Mother Earth" which includes *Santa Tira, Pachamama* and *Pachatira Mama* (Wagner, 1976: p. 200). There is no clear-cut distinction made in Songo between *Pachamama* and *Pachatierra;* they are both female and both related to *hallp'a* and *pampa* ("soil, ground"; Catherine A. Wagner, personal communication, 1978).

From these accounts, we find that *Pachatira* refers primarily to the earth and to its powers of fecundity. It is also often associated with a concept of femaleness in relation to *Pachamama*, but the latter does not appear to be a necessary or invariable characteristic (e.g., J. V. Nuñez del Prado, 1970: pp. 75–76). This slight ambiguity in the sexual classification of *pachatira* is well illustrated by the man in Misminay who first called the dark cloud constellations female, but then insisted that they are more properly classified as *pachatira*.

That *pachatira* refers to a concept of earthly or subterranean fecundity is important to our discussion of celestial animals which originate from, and are actually composed of, the earth. We find in the Andes a general belief in the subterranean origin of all animals (cf. Aranguren Paz, 1975: p. 108; and Duviols, 1976: p. 283). As stated by an informant of the Gows in the area of Ocongate:

> It was a very long time before alpacas existed. When it first dawned, they were hidden under the earth where there are springs. Then, when the sun rose again, all the animals came out of a spring. For this reason, we make an offering to a spring and the lakes at the foot of Ausangate. If there had been no subterranean spring, we would not have had animals. The spring and the lakes are the owners of the animals. (Gow and Gow, 1975: p. 142; my translation)

With this quotation, we can begin to understand the relation not only between animals and the earth (*pachatira*), but also between animals and earthly, or subterranean, *water*. Since the animals of the earth, those actually used by humans as food, clothing, and transport, originate from subterranean springs, it is not surprising to find that the animals which inhabit the waters of the celestial River are also related to the concept of the earth as a fecund force (i.e., to *pachatira*). As we saw earlier in our description of the diurnal rotation of the Milky Way, the celestial river passes beneath the earth after first entering the cosmic sea in the west. We can well imagine the tremendous mixing and crossing of subterranean water, earth, and animals which occurs as the Milky Way passes beneath the earth and how, therefore, the animals in the sky are intimately connected with the animals of the earth.

With these observations on the general nature of the category of dark cloud constellations as a background, we can now examine the specific characteristics of each of the terrestrial and celestial animals in order to determine what relationship exists between the behavior of the animals and the behavior of the constellations. Our investigation is prompted by the statement of Polo de Ondegardo to the effect that the animal constellations are responsible for the "procreation and augmentation" of their animal counterparts on the earth. We will begin with the Serpent, at the head of the line of dark cloud constellations, and then proceed eastward along the line of constellations as they rise after the serpent.

MACH'ACUAY ("SERPENT")

Snakes are a relatively rare part of the southern Andean fauna, especially when considered in light of the fact that one of *the* herpetaria of the world, the Upper Amazonian rainforest, lies just to the east and north. The only naturally occurring species of snake above 12,000 feet in the Andes is the mildly poisonous Colubrid, *Tachymenis peruviana*. *T. peruviana* (Fig. 19.2) ranges in size up to about one-half meter and its coloring is a yellowish or pale brown with dark spots and longitudinal streaks running along the upper part of its body. An oblique dark streak runs from the eye to the angle of the mouth. A pair of grooved fangs are positioned below the back of the eye (cf. Boulenger, 1896: pp. 117–119). *T. peruviana* gives birth to live young and its birthing period is between September–October (Fitch, 1970: p. 156). Its altitudinal range is between 6,000–15,000 ft.

19.2

The term *Mach'acuay* is commonly found in the literature on the reptilian fauna of Cuzco, and I would suggest that it refers primarily to the indigenous snake *T. peruviana*. Cayón (1971) gives the term *mach'acuay* for snakes in general and he also mentions three other terms that are in current use:

machato	—"drunken"
uru	—"worm"
maq ta uru	—"mozo (young man) worm"

The following two reptiles have also been described for the Cuzco area and may be considered as varieties of the *mach'acuay* group: *yana-muroq* ("dark spotted"), a small green and white snake; and *oje-muroq* ("grayish spotted"), a small brown to dark brown snake (Cayón, 1971: p. 144; also see Roca, 1966: p. 61). The more impressive South American reptiles, such as the anaconda and the boas, do not occur naturally in the area we are discussing. However, Garcilaso de la Vega tells us that the inhabitants of the Inca empire who lived in the jungle to the east of Cuzco brought huge reptiles (*amarus*) to the Inca as tribute (Garcilaso, Bk. 5, cap. 10). In addition, I would suggest that the present distribution of the medium-sized *Constrictor constrictor ortonii* along the Upper Marañon, and the *Boa hortulana hortulana* of the Upper Madre de Dios east of the Dept of Cuzco (cf. Schmidt and Walker, 1943a: p. 280; and 1943b: p. 305) is a good indication that the large Amazonian reptiles are known today by the inhabitants of the high Andes.

19.3 Bushmaster; Surucucú (from Ditmars, 1937: plate 70).

In the community in which I carried out fieldwork (altitude *ca.* 13,000 ft.), it is common to travel far down the Urubamba Valley for work and to consult the shamans of the lowlands for purposes of divination. Since a similar pattern of lowland travel was common among Pre-Columbian Andean populations, it is entirely feasible to suppose that the medium to large reptiles of the Upper Amazon have been a part of the faunal knowledge of Andean peoples for some time.

In addition to the Serpent constellation called *Mach'acuay,* I have elsewhere (Urton, 1978a) described a constellation, *Sulluullucu,* which was mentioned by an informant immediately after a discussion of *Mach'acuay*. I now think that the word *sulluullucu* may be related to the name Surucucú, the deadly Bushmaster (*Lachesis mutus*—Fig. 19.3) of the tropical forest (cf. Ditmars, 1937: p. 137; Tastevin, 1925: p. 172). If this derivation or relationship of the Quechua term is correct, it is all the more impressive since the constellation *Sulluullucu* was mentioned in the community of Songo, at an altitude of *ca.* 14,500 ft. A further suggestion that large Amazonian serpents are known in the Andes today is found in the Quechua dictionary of Padre Jorge Lira. Lira gives the term *mach'acuay* for serpent, but he also gives a name for the boa, *A'ti mach'akkway*. Lira relates the name *A'ti mach'akkway* to *amaru,* the monstrous serpent of Andean mythology and also, as mentioned earlier, the large serpents which were given as tribute to the Incas.

Amarus are important for our study because the name is applied to rainbows which are believed to be giant serpents. The body of the Rainbow Serpent rises up out of one spring, arches through the sky, and buries the opposite end of its body in

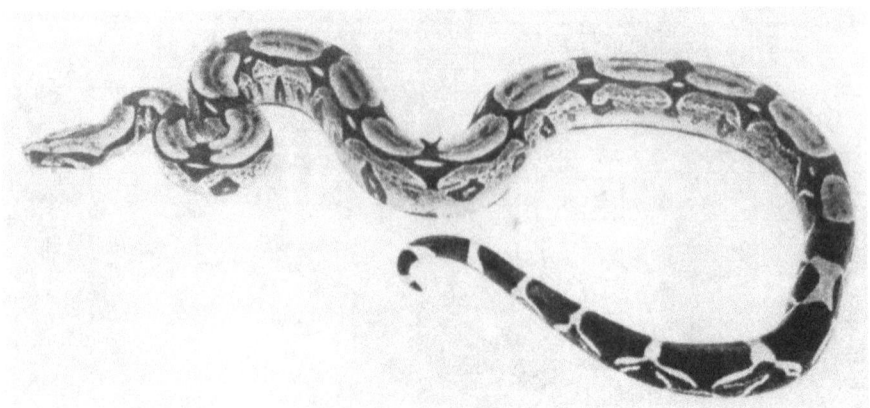

19.4 South American Boa (from Ditmars, 1937: plate 6).

another spring. *Amarus* are thought of as double-headed; one head is buried in each spring.

From these data, we find a relationship between the dark cloud constellation serpent *Mach'acuay* and the rainbow serpent *Amaru*; this is a relation of dark and multi-colored reptiles. Precisely the same relationship or opposition of coloring is found in the ethnoastronomy of tropical forest reptiles. As described in Tastevin's study of the Amazonian legend of Bóyusú (=*surucucú*, the Bushmaster): "...the celestial Bóyusú appears during the day in the form of the rainbow, and at night in the form of a dark spot..."(Tastevin, 1925: p. 183; my translation). Tastevin further shows that the dark spot serpent is a dark cloud in the Milky Way which winds itself round the constellation of Scorpio. Scorpio is important in the Amazonian calendar because when it stands in the zenith in November, it signals the beginning of the rainy season (Tastevin, 1925: p. 173).

In the tropical forest, the relationship of large dark/multi-colored serpents is not without an observable basis in the fauna since the Ringed or Rainbow Boa (*Epicrates cenchris*—Fig. 19.4) exhibits just this opposition or union of colors. The Rainbow Boa has a brownish hue with blackish rings; however:

> Gliding into the sun the reptile is transformed. As the light catches the upper surface at certain slants, patches of iridescence glow in green and blue, like the wings of the morpho butterfly... (Ditmars, 1937: caption to Plate 8)

In addition, the coloration and habits of the anaconda (*Eunectes murinus*—Fig. 19.5) are similar to the descriptions of the celestial and aquatic (rainbow) serpents of Amazonian tradition. Anacondas, which are olive green with large, round black spots, are the most aquatic of the boas. They are nocturnal and spend their days in swamps or sunbathing in low branches over the water. Significantly, the few observations that have been recorded of the anaconda breeding cycle report that they give birth (of anywhere from 20–100 young) "...in the early part of the year" (Burton, 1975: p. 206).

19.5 Anaconda (from Ditmars, 1937: plate 4).

Thus, at the same time that the dark cloud serpent of the Amazon, Bóyusú, stands in the zenith to signal the beginning of the rains, the large aquatic boas of the earth (anacondas) have recently given birth to their young.

Celestial serpents in the Amazon are therefore related to rainbows (water), dark clouds in the Milky Way, and to the beginning of the rainy season. This is the same complex of associations found in Andean astronomy:

mach'acuay —dark cloud in the Milky Way observed at the beginning of the rainy season
amaru —the double-headed, multi-colored rainbow serpent

The emergence of "meteorological" serpents (*amarus*) from the ground immediately following a rain shower, and their reentry as the atmosphere becomes less moist, is an important clue to understanding the relation between terrestrial and celestial reptiles in the Andes. The *amaru,* which rises out of a spring after rain, exhibits a climatological behavior pattern similar to terrestrial serpents which, at the *end* of the cold/dry season and at the *beginning* of the warm/rainy season, emerge from subterranean hibernation. The Andean dry/cold season (May–July) is a period not only of reduced activity among reptilian fauna but also among the fauna upon which reptiles prey. Therefore, terrestrial reptiles in the Andes are variably active and inactive in direct relation to pronounced alternations between dry/cold and warm/rainy seasonal changes (cf. Schoener, 1977: pp. 115–116). Since meteorological serpents (rainbows/*amarus*) only appear during the rainy part of the year, they exhibit a seasonal activity cycle similar to that of terrestrial reptiles.

The principal identification of the dark cloud Serpent (*Mach'acuay*, see Fig. 19.1) is a large zig-zag-like streak of interstellar dust which stretches from a point near the Southern Cross to Adhara (in the Western constellation of Canis Majoris). The head of the Serpent precedes the tail in rising; thus, the movement of the constellation through the night sky can be likened to terrestrial serpents and rainbow serpents which rise out of the earth headfirst and reenter headfirst. Since the dark cloud Serpent stretches over such a large celestial area, we must select a part of its body as a point of departure for our analysis of the cycles of visibility and invisibility of the celestial Serpent. I would suggest for this purpose that we consider the head of the Serpent since the head is crucial in determining the time and place of the emergence and reentry of *Mach'acuay*. The heliacal rise of the head of the Serpent occurs during the first week of August; its heliacal set occurs during the first week of February. Related to the latter date is the observation that *Mach'acuay* stands along the north/south meridian (i.e., at its zenith point) at midnight on February 1st. When we recall that the most intense period of rain in the Andes occurs between the months of December and February and that the planting period begins at the beginning of the change from dry to rainy in August, it is apparent that the periodicity of the celestial Serpent's rising out of the earth and its reentry into the earth during the night brackets the rainy season. In effect, the celestial Serpent, like meteorological serpents, emerges from the earth with the warm/rainy season and reenters the earth at the beginning of the dry/cold season. In addition, we have found that the principal serpent of the Dept. of Cuzco (above 12,000 ft.), *T. peruviana,* gives birth to its young in September–October, just after the onset of the warm/rainy season.

In this first analysis, then, we can suggest that Polo de Ondegardo's statement concerning the responsibility of celestial animals for their terrestrial counterparts refers to the easily observable and cosmologically important correspondence between the periodicity of the presence and absence of terrestrial, celestial, and meteorological reptilian fauna in the universe.

HANP'ATU ("TOAD")

That the celestial Toad appears to "pursue" the Serpent across the sky is ironic in light of the fact that snakes are the greatest predators upon toads and frogs. While certain toads are occasionally known to get the better part in combats with snakes (cf. Noble, 1931: p. 383), this is by no means the usual outcome. In Quechua, a distinction is made (as in our own classification of the amphibia) between the more aquatic frogs (= *ococo, k'ayra* and *ch'eqlla*) and the terrestrial toads *(Hanp'atu)*. We will be concerned here primarily with toads rather than frogs since every reference I have collected myself, or have encountered in the literature, regarding a celestial amphibian has been in relation to *hanp'atu,* the toad (cf. Roca 1966: p. 43; Cobo, 1964, v. 1: p. 352; and Gonzalez Holguin, 1952).

The principal Andean toad is *Bufo spinulosus. B. spinulosus,* which is very resistant to dryness and altitude, breeds principally at the onset of the rainy season in

19.6 *B. spinulosus* (from Blair, 1972: plate 11).

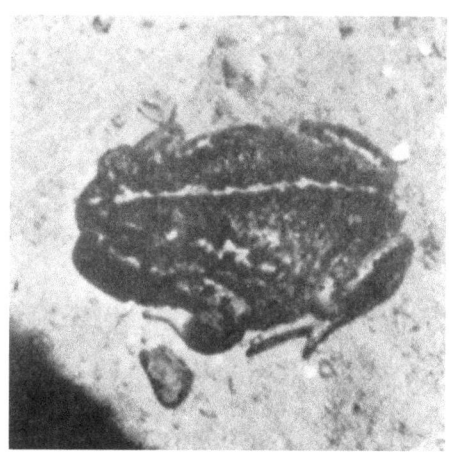

permanent bodies of water (Fig. 19.6). The range of *B. spinulosus* in South America extends along the cordilleras southward to −43° latitude; the altitudinal range to which they are adapted varies between 1,000–5,000 meters (Cei 1972: p. 83).

In a study of the behavior of assorted fauna on the plain of Anta (west of Cuzco), Demetrio Roca recounts that in addition to the name *hanp'atu,* toads are also referred to by the following names: *Pachakuti*—"turning of the earth"; *Saqra*—"devil"; *Pachawawa*—"earth child"; and *Jacinto*—"hyacinth" (Roca, 1966: p. 45).

Toads are called "devils" (*saqra*) because they were created by the devil; because they foretell bad luck when seen (Cobo, 1964, 1: p. 353); and because they are used in the malevolent practices of witches (Roca, 1966: p. 45). The two terms *pachawawa* and *pachakuti* are important for our study because they refer to the common habit of toads to burrow within the earth during the dry/cold season and, like *mach'acuay,* to reemerge with the warm/rainy season (cf. Noble, 1931: p. 421; and Grzimek *et al.,* 1974: pp. 360–367). It is also important to note that amphibia are most active at night when the humidity is much greater than during the day. Therefore, toads are the "children of the earth" (*pachawawa*) in that they hibernate within, and later emerge from, the earth. This cyclical entry and reemergence, coinciding with the cycling of the dry/cold and warm/rainy seasons, is a behavioral pattern well-described by the name *pachakuti* ("turning of the earth").

From his fieldwork and observations of the behavior of toads on the plain of Anta, Roca gives us the following description of this cycle of subterranean hibernation:

> The earth is alive during the month of August, being intensely animated by the toad or *pachakuti* which emerges from the interior of the earth in great numbers. It is noted in the plain (of Anta) that beginning in the month of May, wide deep cracks appear in the earth through which the toads return to the womb of the earth, reappearing in the month of August. (Roca, 1966: p. 42; my translation)

After the initial emergence of the toads in August, their behavior (mating, croaking, etc.) is observed closely for divinatory purposes:

> If in the months of September and October they croak day and night in great numbers, it is an augury that there will be much rain and, as a consequence, the crops will be abundant; but, if during these months they croak only a little and softly,

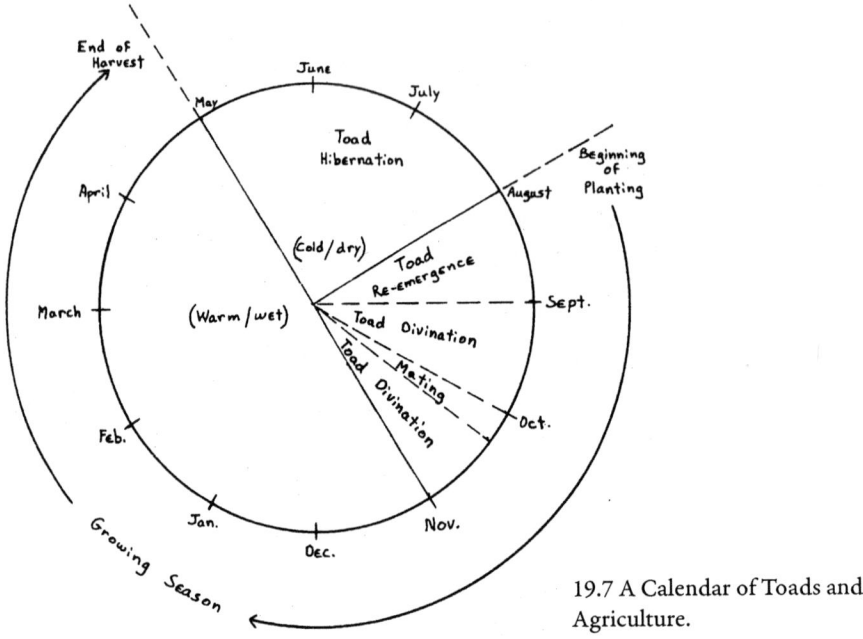

19.7 A Calendar of Toads and Agriculture.

it is a sign that it will not rain and that the frosts will be strong. (Roca, 1966: pp. 58–59; my translation)

In addition to a divinatory connection between toads, weather, and the crops, toads are also related to agriculture because at the beginning of the planting season in August, toads encountered in great numbers indicate that the year's crop will be abundant; if only a few small ones are encountered, the crop will be small (Roca, 1966: p. 59).

If we now summarize the cyclical behavior of toads, we arrive at the following calendar of activities in relation to agriculture and the seasonal cycle (Fig. 19.7). In this "calendar of toads and agriculture," we see a very close coincidence between the cycles of agriculture and amphibian behavior. The points of transition occur along the temporal boundaries defined by the cycling of the cold/dry season and the warm/rainy season. As we have found earlier in our discussion of *Mach'acuay*, a similar correlation is found in the relationship between the seasons and the terrestrial, meteorological and celestial reptilian fauna.

The dark cloud constellation of *Hanp'atu* (cf. Fig. 19.1) is a small patch of inter-stellar dust in the Milky Way which moves between the tail of *Mach'acuay* and the Southern Cross. If we consult a celestial globe, we find that during the first days of the month of October (i.e., at the time of the mating period of terrestrial toads), *Hanp'atu* rises about one and one-half hours before the sun; thereafter, *Hanp'atu* will rise progressively earlier than the sun each morning. In effect, the celestial toad rises

Gary Urton

    a. Puna Tinamou   b. Andean Tinamou   c. Highland Tinamou

19.8 Species of Tinamous (from Blake, 1977: Plate 1). Painting by Guy Tudor.

into the sky in the early morning just after terrestrial toads have emerged from their long period of subterranean hibernation and just at the time of their most intense croaking and mating period.

YUTU (TINAMOU)

Near the center of the line of Dark Cloud constellations, at the point where the Milky Way flows nearest to the south celestial pole, sits the Tinamou (*Yutu*). The *Yutu* of Quechua astronomy is equivalent to the Western constellation of the Coalsack, one of the few dark spots recognized, and named, in our own system of astronomy.

    The partridge-like birds called *Yutu* (Quechua) and *Tinamou* (Carib) resemble game-birds with their short legs, compact bodies and small heads with slender necks (Fig. 19.8). There are some 9 genera and 43–45 species of Tinamous distributed throughout South America and northward to the Tropic of Cancer (cf. Grzimek, 1972: p. 82; and Lancaster, 1964a and b).

    The range of vertical habitat of Tinamous is from the tropical rainforests up to the high, cold puna-land of the Andes (cf. de Schauensee, 1970: pp. 3–9; Roe and Rees, 1979: pp. 475–476; and Traylor, 1952). Tinamous eat mostly seeds and fruits, although they are occasionally known to swallow small animals whole. Lancaster, for instance, observed Tinamous eating frogs and lizards on several occasions (Lancaster, 1964a: p. 171). This observation is especially interesting in the present context since the Dark Cloud constellation of the Tinamou (*Yutu*) "pursues" the celestial Toad through the sky. In a somewhat less antagonistic characterization, one Quechua informant described to me the nightly race between the Toad and the Tinamou. The Toad, said the young woman, always wins the race and therefore, her husband likes to characterize himself as a Toad in opposition to others who are Tinamous.

The characterization of the *Yutu* as a slow animal in relation to other celestial animals is based perhaps not only on these nightly celestial races, but also on the terrestrial behavior of Tinamous. Tinamous are notoriously slow, "stupid" birds (cf. Cobo, 19, vol. 1: p. 321). Not only are they disinclined to flight, but when flushed, they fly low and poorly. Long ago, W. H. Hudson made the following observations on the flight of Tinamous:

> The Tinamou starts forward with such amazing energy, until this is expended and the moment of gliding comes, that the flight is just as ungovernable to the bird as the motion of a brakeless engine, rushing along at full speed, would be to the driver. ... In the course of a short ride of ten miles, I have seen some of these Tinamous dash themselves to death against a fence close to the path, the height of which they had evidently misjudged. I have also seen a bird fly blindly against the wall of a house, killing itself. (quoted in Knowlton 1909: 78–79)

In fact, the most common reaction of the Boucard Tinamou when startled is not to fly but rather to do nothing, to simply "freeze" (Lancaster, 1964a: p. 171). Barring escape by freezing or flying, a Tinamou hard-pressed in open country will often crawl into a hole dug by another animal.

Aside from the above "un-bird-like" features which no doubt make the Tinamou a noticeable part of the bird population in Andean communities, there are a number of additional characteristics which may contribute to its celestial projection in the form of a Dark Cloud constellation. First, Tinamous are distinctive in their solitary, unsociable nature. They are rarely found in coveys, either with members of their own or other species. This unsociable or solitary behavior is extended to breeding and incubation habits. Males begin calling in the early mornings and late evenings at the beginning of the mating season. This pattern increases until the height of the breeding period after which time calling begins to subside. During the breeding period, males attract a number of females by their persistent calling. The females all lay their eggs in the male's nest and depart, wooed to the nest of another male by the flute-like whistle of the male Tinamou. After the several (2–5) females have deposited their eggs in the nest, the eggs are incubated by the *male* (Grzimek, 1972: p. 84).

Thus, unlike most birds, Tinamous are solitary and polygynous, and the typical male/female roles in incubation are reversed. The Tinamou is therefore a model not only of bad social behavior, but in their breeding habits, they exemplify what would be extremely undesirable reproduction habits if practiced by humans, i.e., inconstancy in mating and the abandonment of the children by the mother.

Before leaving the Tinamous of the earth to describe those of the sky, we should mention one other unusual feature, the eggs of the Tinamou: "They [the eggs] may be green, turquoise-blue, purple, wine-red, slate-gray or a chocolate color, and they often have a purple or violet lustre" (Grzimek, 1972: p. 85).

Thus, Tinamou eggs are like segments of a rainbow, cast in an oval and placed in a nest. In the mythology of the Desana Indians, Tinamous are believed to have been the sole survivors of a world fire and were responsible for preserving, through their eggs, all the colors of the rainbow (Reichel-Dolmatoff, 1978: p. 280). One might say, then, that Tinamous lay rainbows, and the Dark Cloud consellation of the Tinamou is located at the center of the arc of the Milky Way which, in Quechua thought, is considered equivalent to a nocturnal rainbow (cf. Urton, 1979: pp. 142–143). Therefore, we find in the example of the Tinamou the elaboration of an idea which appeared with the celestial Serpent (*Mach'acuay*); that is, the equation of a celestial dark spot with the rainbow, an equivalence of black and multi-colored.

As mentioned earlier, the Dark Cloud constellation of the *Yutu* (cf. Fig. 19.1) is located at the foot of the group of stars which, in Western astronomy, is known as the Southern Cross (*Crux*). Therefore, the astronomical periodicities of the Southern Cross will be virtually identical to those of the *Yutu*. The heliacal rise and set dates (September 3 and April 22, respectively) of the principal star of the Southern Cross, α Crucis, give a very close approximation of the agricultural season in the Andes. In addition, α Crucis and the *Yutu* transit the upper meridian on the morning of the December solstice sunrise, and on the morning of the June solstice they are transiting the lower meridian. Phrased another way, the *Yutu* is at its zenith point on the morning of the December solstice, and it is at its nadir point on the morning of the June solstice. The periodicities of the *Yutu* therefore relate the celestial animal both to agriculture and to the solstices.

The mating season of Tinamous in the Cuzco area is not reported in the literature. However, the breeding season of several varieties of Tinamous in the northern hemisphere extends from February through April (cf. Pearson, 1955; and Friedmann, 1950). This period, in the northern hemisphere, is related to the onset of longer days, warmer weather, and the approach of the rainy season following the December solstice. Skutch has shown that "birds transported across the equator seem to adapt their nesting to the season appropriate to their new environment" (Skutch, 1976: p. 72). Thus, when translating the factors related to the breeding season of northern hemisphere Tinamous to the high Andes of the southern hemisphere (that is, lengthening of the day and the onset of the warm/rainy season) we arrive at a period of a couple of months after the *June* solstice, i.e., July through early September. Another factor related to the beginning of breeding in birds is the increased availability of food, a condition which is met for the seed-eating Tinamous not only by the beginning of the rainy season, but also by the August–September planting of seed-crops such as maize.

In his description of the various Incaic agricultural duties of the year, Guaman Poma records the times when the crops must be guarded from birds and animals; the guard duties begin with the planting in late August and end with the harvest in early May. If we combine, in a single calendar, the times when Guaman Poma mentions explicitly the need to guard the crops (Guaman Poma, 1936: pp. 1130ff.) with the times of the heliacal rise and set dates of the celestial Tinamou, we arrive at the close correlation shown in Figure 19.9.

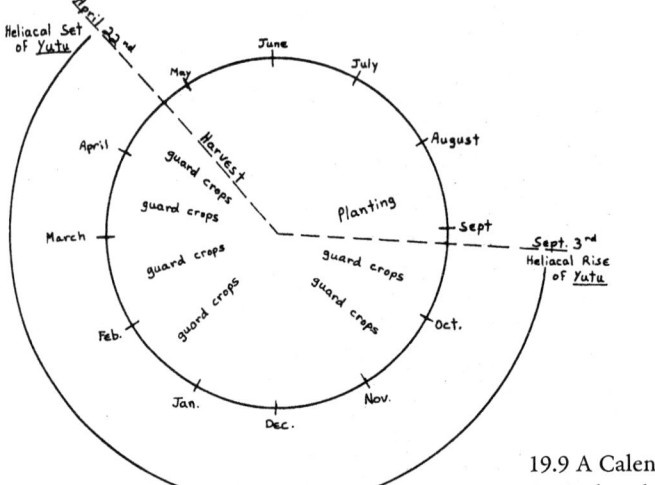

19.9 A Calendar of Tinamous and Agricultural Guard Duties.

When we compare this figure to the earlier calendar of toads and agriculture, and to the data which we discussed earlier concerning the relationship of rainbow/Serpents and the rainy season, we can understand the larger set of associations between agriculture, rainbows and the celestial Tinamou. Rain, and therefore rainbows, occur in the southern Andes primarily during the period from September through April. As we see in Figure 19.9, this is also the period of time when the celestial Tinamou is in the sky. The beginning of the calendrical correlation between the celestial Tinamou and the crops is also related to the breeding period of terrestrial Tinamous, and the total span of the calendar (i.e., from September through late April) is the total period when terrestrial Tinamous represent a threat to agriculture.

LLAMA

Perhaps the most conspicuous Dark Cloud constellation, since it virtually fills the sky overhead during the rainy season, is the Llama. We are in a considerably stronger position when we discuss the symbolic and ritual associations of the Dark Cloud Llama since the Spanish chroniclers were much more explicit in their descriptions of this animal and its relation to Inca rituals (the reader is referred to the article by Zuidema and Urton, 1976, for a more complete discussion of the ethnohistorical material). As our concern here is primarily with the relationship between constellations and the "procreation and augmentation" of animals on the earth, we will confine our discussion to a comparison of the biological and astronomical cycles of terrestrial and celestial llamas.

As Professor John Murra has demonstrated so well through the ethnohistorical documentation (Murra, 1965), the llama has for some time been an essential animal

19.10 Llamas (*L. peruana*) (from Walker, 1964, 2: p. 1375).

in Andean life. In Inca times, the llama was used primarily as a beast of burden but it also provided meat for food, wool for clothing, dung for fertilizer, and it was, and still is, considered an appropriate gift to the gods in the form of a sacrifice (Fig. 19.10).

The breeding period of llamas begins in late December and the gestation period lasts 11 months. Llamas begin to give birth from late November to early December with the birthing period ending in March. Llamas, and the closely related camellid alpaca, give birth between 6:00 AM and 12:00 noon (Dr. Jorge Velasco N., personal communication, 1977). Thus, for Andean pastoralists, the early morning hours of the rainy season in December are important times for caring for newly born calves and for attempting to ward off predators such as the fox.

For the herder who rises early in the morning during the birthing period, an especially fortuitous sight will be the appearance, over the southeastern horizon, of the two bright stars α and β Centaurii. In Quechua astronomy, α and β Centaurii are referred to as *llamacñawin* ("the eyes of the llama"); they are the first part of the huge body of the llama to appear over the horizon. After the heliacal rising of the eyes of the llama in late November, the eyes and the body rise progressively higher in the sky each morning until, in late April, at the end of the birthing season of llamas, the Dark Cloud constellation of the Llama stands along the north/south meridian at midnight.

Llamas were also incorporated in the Inca calendar system in the form of sacrifices which were made at fixed intervals during the agricultural season. Brown and brownish-red llamas were sacrificed from August to September at the beginning of the planting season. White llamas were sacrificed and black ones were tied to a post and starved to death to induce rain and growth of the crops, and multi-colored llamas were sacrificed at the time of the harvest in late April, early May (Polo de Ondegardo, 1916, chapter 6).

The black llamas which were starved in October and the multi-colored ones sacrificed in April–May are of special interest in our study because the dates which they mark in the calendar define an important solar axis in the Inca calendar. It has been shown (Zuidema and Urton, 1976: p. 86) that the inferior culmination of α and β Centaurii at midnight on October 30 was coupled with the zenith sun (which occurs on the same day) as a way of fixing the solar dates for the initiation of young Inca nobles one month later in late November. The superior culmination at midnight of α and β Centauri in April occurred on the same night that the sun stood in the nadir at midnight. Thus, the alternation of the superior and inferior culmination at midnight of the eyes of the Llama was seen in relation to the alternation of the zenith and nadir sun in October and April and these dates were marked in the ritual cycle by the sacrifice of black and multi-colored llamas. Francisco de Avila, who wrote an account of the beliefs and customs of the Indians of the community of Huarochirí in the early seventeenth century, has given us a description of the Dark Cloud constellation of the Llama in which we find explicit references of the Llama descent to the earth at midnight. The Llama drinks the waters of the swollen rivers at the beginning of the rainy season, she is then shorn of her wool which turns out to be multi-colored despite the fact that Avila describes her as having the color of a shadow, and finally she returns to her position in the celestial river (Avila, 1966: cap. 29). Therefore, we find in the case of the llama the same theme of dark and multi-colored symbolism which has appeared earlier and we also find the correlation of a Dark Cloud constellation with rain and the sun.

ATOQ (FOX)

The South American fox *(Dusicyon culpaeus)* inhabits wooded, hilly country ranging up to 4,000 meters in the Andes (Fig. 19.11). Most of the six South American fox species are nocturnal hunters which may contribute to their projection in the dark clouds of the night sky. The diet of the fox is omnivorous and includes birds, rabbits, frogs, toads and an occasional sheep. In addition, Franklin, in a study of the "Social Behavior of the Vicuña," reports that foxes prey both on adult and baby vicuña. The vicuña defense against attacks by foxes is "group mobbing" (Franklin, 1974: 486). In fact, we have an excellent description of the group mob of vicuñas on foxes in the seventeenth-century chronicle of Bernabé Cobo:

> There are usually a large number of foxes where vicuñas live; and the foxes chase and eat the young of the vicuñas. The vicuñas defend their children in the following way. Many vicuñas rush together to attack the fox, striking it until it falls to the ground. They then run over it many times without giving it a chance to get up until they kill it by their blows. The cries of the miserable fox are useless as he succumbs to the feet of the vicuñas. (Cobo, 1964, 1: p. 368)

When we note the position of the celestial fox in relation to the baby llama and the hind legs of the mother llama (Fig. 19.1), we would appear to have fixed in the

19.11 South American Fox (from Walker, 1964, 2: p. 1160).

clouds of interstellar dust in the Milky Way this well-recorded motif of pursuit and trampling. Like the Tinamou, the fox has a tendency to freeze when endangered. There is one account of such a "frozen" animal being approached by a man and remaining motionless even when struck with a whip-handle (Walker, 1964, 2: p. 1160).

The mating season of foxes falls in mid-winter; in South America, the season extends from late June through September (cf. Ewer, 1973: p. 309). With a gestation period of around 10 weeks, baby foxes generally appear from October through December. In the community of Misminay, it is commonly believed that foxes give birth principally on one day of the year: December 25, four days after the solstice. In addition, the *runa* of Misminay pinpoint exactly the spot where baby foxes are born every year. It occurs, they say, on the side of a nearby mountain called Wañumarka ("Storehouse of the Dead"), at a point which is precisely the setting point of the *June* solstice sun as viewed from the community. However, as we will see below, the solsticial relation of foxes goes beyond their birth near the time of the December solstice sunrise at the place of the June solstice sunset.

The Dark Cloud constellation of the *atoq* (Fox) is a rather amorphous dark spot which stretches at a right angle from the tail of Scorpio crossing the ecliptic between the Western constellations of Scorpio and Sagittarius. The importance of this celestial position is that as the sun travels along the path of the ecliptic throughout the year, it "enters" the constellation of the fox at the time of the December solstice. Therefore, as the sun rises in the southeast with the constellation of the fox around the time of the

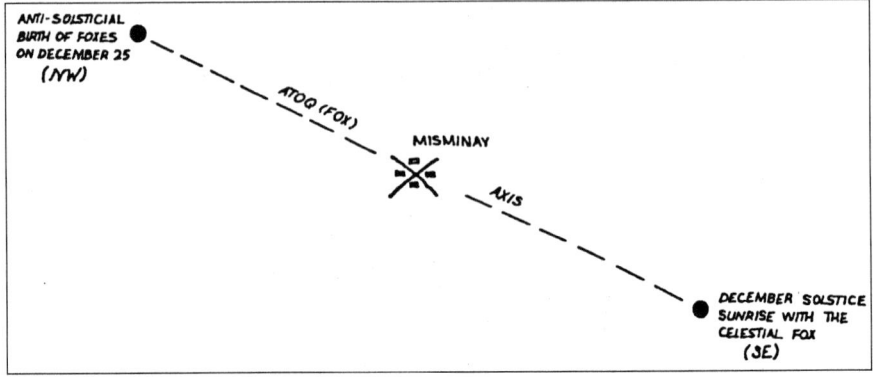

19.12 The Axis of the Fox (from Urton, 1979: p. 108).

December solstice, terrestrial foxes are born on the earth in the anti-solsticial direction (that is, in the direction of the *June* solstice sunset; see Fig. 19.12).

If we extend this solsticial/fox analysis further, we find that the other passage of the sun through the Milky Way occurs at the time of the June solstice, the time when the sun sets on the side of the mountain Wañumarka. Therefore, since we have found earlier that the breeding season of foxes begins in late June and baby foxes are born in December, the life cycle of the fox is directly associated not only with the sun in its two solstice positions, but also with the times and places of the intersection of the sun with the celestial river, the Milky Way.

CONCLUSIONS

Aside from the close correlation which we have found between biological and astronomical phenomena in these data, the Dark Cloud animal constellations serve as the focus for a number of important classificatory and symbolic principles in the Quechua universe. Principal among these are color (dark, light and multi-colored); fertility (the cosmic circulation of water, *pachatira*, animal procreation, etc.); and orientation. The latter is seen primarily in the relationship between rainbows, the Milky Way (as a nocturnal celestial river) and the sun. In the section on the *atoq* (fox), we saw a coincidence in the orientations of the sunrise points of the solstices, as viewed from the community of Misminay, with the orientations of the two axes of the Milky Way during the early evenings at the times of the solstices. These coincidences are diagramed in Figure 19.13.

On the evening of the June solstice, when the sun sets in the northwest, the Milky Way will slowly begin to appear in a line stretched across the sky from the northeast to the southwest (i.e., it will form an arc *opposite* the sun). At the time of the December solstice, the sun sets in the southwest and the Milky Way is seen as a celestial arc running from the northwest to the southeast. Thus, as illustrated in Figure 19.13 on the evenings of the solstices the Milky Way is seen as an arc opposite the setting sun. This,

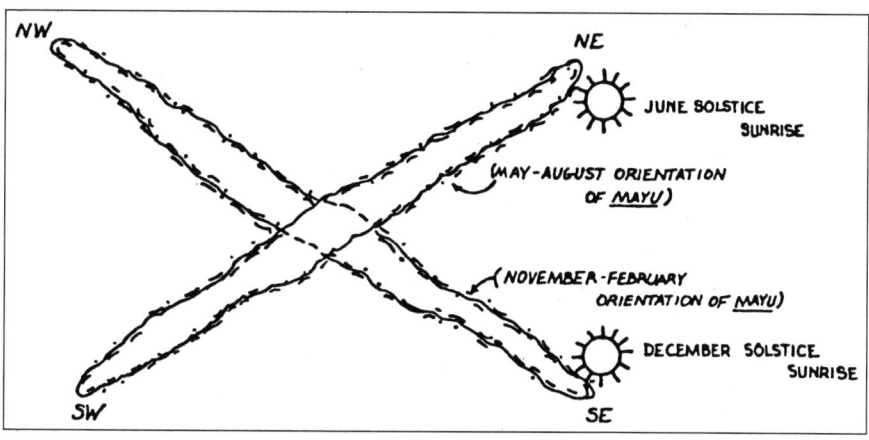

19.13 The Solstices and the Seasonal Axes of the Milky Way (from Urton, 1979: p. 94).

in fact, is exactly the same relationship that exists between rainbows and the sun; rainbows are always seen as arcs (or circles) stretching across the sky directly opposite the sun. Therefore, rainbows and the Milky Way may be equated and opposed to the sun not only because they are continuous arcs passing through the sky, but also because there is a consistently observable relationship between the sun and the celestial arc. In his studies of tropical forest astronomical symbolism, Prof. Lévi-Strauss has described a similar pattern of relationships among the sun, rainbows, the Milky Way and the moon (Lévi-Strauss, 1979: pp. 110–113).

These observations suggest then a synchronized pattern of celestial and meteorological lines and points of orientation which are in constant motion but which retain, throughout the annual cycle, a persistent, internal pattern of oppositions; as the sun moves south, the arcs of rainbows and the Milky Way rotate slowly northward so as to maintain their opposition to the sun. This implies as well that the seasons are not determined by the sun alone but also by celestial arcs of a certain orientation at a certain time of the day, night, or year (Fig. 19.13).

One other major conclusion which we can draw from the material presented in this paper concerns a similar system of correlations, that between the cyclical behavior of animals in the Andes and the astronomical cycles of the Dark Cloud constellations. The universe of the Quechuas is not composed of a series of discrete phenomena and events, but rather there is a powerful synthetic principle underlying the perception and ordering of objects and events in the physical environment. This principle is clearly seen in operation in Quechua astronomy and cosmology, and I would suggest that with a slightly different orientation in our studies, a shift from astronomy to iconography, we could develop a more coherent approach to the study of animal motifs in Andean art from Chavin times onward. At the least, it is hoped that our exegesis of one short sentence from the chronicle of Polo de Ondegardo has contributed toward an understanding of his observation that, "... in general, they believed that all the animals and birds on the earth had their likeness in the sky in whose responsibility was their procreation and augmentation" (Polo, 1916: cap. 1).

NOTES

1. Support for the fieldwork upon which this paper is based was provided in 1975–1976 by the Wenner-Gren Foundation for Anthropological Research, in 1976–1977 by the Organization of American States, and in 1979 by the Research Council of Colgate University. Their support is gratefully acknowledged.

2. Dark cloud-like constellations have been described elsewhere in South America (Nimuendajú 1948:265; Tastevin 1925:182, 191; and Weiss 1969), among Australian aboriginal populations (Maegraith 1932), and in Java (Pannekoek 1929:51–55; and Stein Callenfels 1931).

REFERENCES

Aranguren Paz, Angelica. 1975. Las creencias y ritos magico-religiosos de los pastores puneños. *Allpanchis Phuturinga* 8:103–132.
Avila, Francisco de. 1966. *Dioses y hombres de Huarochirí* (1608). Lima.
Blair, W. Frank, ed. 1972. *Evolution in the Genus Bufo.* Austin, University of Texas Press.
Blake, Emmet R. 1977. *Manual of Neotropical Birds* 1. Chicago, Univeristy of Chicago Press.
Boulenger, George A. 1961. *Catalogue of the Snakes in the British Museum* 1896. 3 London.
Burton, Maurice. 1975. *Encyclopedia of Reptiles, Amphibians and Other Cold-Blooded Animals.* New York.
Casaverde Rojas, Juvenal. 1970. El mundo sobrenatural en una Comunidad. *Allpanchis Phuturinga* 2:121–244.
Cayón, Armelia, Edgardo. 1971. El hombre y los animales en la cultura quechua. *Allpanchis Phuturinga* 3:135–162.
Cei, Jose M. 1972. *Bufo* of South America. In *Evolution in the Genus* Bufo, ed. W. Frank Blair. Austin, University of Texas Press.
Cobo, Bernabé. 1956. *Historia del Nuevo Mundo* (1653), 1 and 2. Madrid, B.A.E.
De Schauensee, Rodolphe Meyer. 1970. *A Guide to the Birds of South America.* Wynnewood, PA, Livingston.
Ditmars, Raymond L. 1937. *Snakes of the World.* New York, Macmillan.
Duvols, P. 1976. *"Un petite chronique retrouvées": Errores, ritos, supersticiones y ceremonias de los yndios de la provincia de Chinchaycocha y otras del Piru.* Edition et commentaire par Pierre Duviols. *Journal de la Société des Américanistes* 63, Paris.
Earls, John. 1973. La organización del poder en la mitología quechua. In *Ideología Mesiánica del Mundo Andino,* ed. Huan Ossio. Lima.
Ewer, R. F. 1973. *The Carnivores.* Ithaca, NY, Cornell University Press.
Fitch, Henry S. 1970. *Reproductive Cycles in Lizards and Snakes.* University of Kansas, Museum of Natural History, Misc. Publications 52.
Franklin, William L. 1974. The Social Behavior of the Vicuña. In *The Behavior of Ungulates and Its Relation to Management* 1, ed. V. Geist and F. Walther. Morges, Switzerland.
Friedmann, Herbert, and Foster D. Smith Jr. 1950. A Contribution to the Ornithology of Northeast Venezuela. *Smithsonian Institution, Proceedings of the United States National Museum* 100:3268. Washington, DC.

Garcilaso de la Vega. 1966. El Inca. *Royal Commentaries of the Incas* (1609), part 1. Trans. and with an intro. by Harold V. Livermore. Austin, University of Texas Press.

Gonzalez Holguín, Diego. 1952. *Vocabulario de la Lengua General de todo el Peru llamada Quichua o del Inca* (1608). Lima.

Gow, David, and Rosalind Gow. 1975. La alpaca en el mito y el ritual. *Allpanchis Phuturinga* 8:141–165.

Gow, Rosalind, and Bernabe Condori. 1976. *Kay Pacha*. Cuzco.

Grzimek, Bernhard. 1974. *Animal Life Encyclopedia*. 5. *Fishes II, Amphibians*. New York, Van Nostrand Reinhold.

Isbell, Billie Jean. 1974. Parentesco andino y reciprocidad. Kuyaq: los que nos aman. In *Reciprocidad e intercambio en los Andes peruanos*, ed. Giorgio Alberti and Enrique Mayer, 110–152. Lima, I.E.P.

Isbell, Billie Jean. 1978. *To Defend Ourselves: Ecology and Ritual in an Andean Village*. Latin American Monographs, No. 47, Institute of Latin American Studies. Austin, University of Texas Press.

Knowlton, Frank H. 1909. *Birds of the World*. New York.

Lancaster, D. A. 1964. Life History of the Boucard Tinamou in British Honduras, Parts I and II. *The Condor* 66 (3):165–181; 66 (4):253–276.

Lehmann-Nitsche, R. 1928. Coricancha. *Revista del Museo de la Plata* 31:1–260.

Lévi-Strauss, Claude. 1979. *The Origin of Table Manners*. London: Harper and Row.

Maegraith, B. G. 1932. The Astronomy of the Aranda and Luritja Tribes. *Royal Society of South Australia, Transactions* 56:19–26.

Mayer, Enrique. 1974. Más allá de la familia nuclear. *Revista del Museo Nacional, Lima* 50:303–333.

Murra, John V. 1965. Herds and Herders in the Inca State. In *Man, Culture, and Animals,* ed. Anthony Leeds and Andrew P. Vayda. Pub. No. 78 of the American Association for the Advancement of Science, Washington, DC.

Nimuendajú, Curt. 1948. The Mura and Pirahá. *H.S.A.I.* 3:255–270, B.A.E. Washington, DC.

Noble, G. Kingsley. 1931. *The Biology of the Amphibia*. New York: McGraw-Hill.

Nuñez del Prado, Juan Victor. 1970. El mundo sobrenatural del los Quechuas del Sur del Perú, a través de la Comunidad de Qotobamba. *Allpanchis Phuturinga.*

Nuñez del Prado, Oscar. 1973. *Kuyo Chico: Applied Anthropology in an Indian Community*. Chicago, University of Chicago Press.

Ossio, Juan. 1976. El simbolismo del agua y la representación del tiempo y el espacio en la Fiesta de la Acequia de la comunidad de Andamarca. *Actes du XLIF Congrès International des Américanistes* 4:377–396.

Pannekoek, A. 1929. Een merkwaardig Javaansch Sterrenheeld. *Tijdschrift voor Indische Taal-, Land- en volkenkunde. Bataviaasch Genootschap* 69:51–55.

Pearson, A. K. 1955. Natural History and Breeding Behavior of the Tinamou, *Nothoproeta ornata. The Auk* 72:113–127.

Polo de Ondegardo. 1916. *Los errores y supersticiones de los indios.* (1571). Lima.

Poma de Ayala, Felipe Guaman. 1936. *El primer nueva cronica y buen govierno* (1584–1614). Paris.

Reichel-Dolmatoff, Gerardo. 1978. Desana Animal Categories, Food Restrictions, and the Concept of Color Energies. *Journal of Latin American Lore* 4 (2):243–291.

Roca W., Demetrio. 1966. El Sapo, la culebra y la rana en el folklore actual de la pampa de Anta. *Folklore, Revista de Cultura Tradicional* 1 (1):41–66.

Roe, Nicholas A., and William E. Rees. 1979. Notes on the Puna Avifauna of Azangaro Province, Department of Puno, Southern Peru. *The Auk* 96:475–482.

Schmidt, Karl P., and Warren F. Walker Jr. 1943a. Snakes of the Peruvian Coastal Region. *Zoological Series of the Field Museum of Natural History* 24 (27):297–324.

Schmidt, Karl P., and Warren F. Walker Jr. 1934b. Three New Snakes from the Peruvian Andes. *Zoological Series of the Field Museum of Natural History* 24 (28):325–329.

Schoener, Thomas W. 1977. Competition and the Niche. In *Biology of the Reptilia* 7, ed. Carl Gans and Donald W. Tinkle. London, Academic Press.

Skutch, Alexander F. 1976. *Parent Birds and Their Young.* Austin, University of Texas Press.

Tastevin, P. C. 1925. La légende de Bóyusú en Amazonie. *Revue d'Ethnographie et des Traditions Populares* 6:172–206.

Traylor, Melvin A. 1952. Notes on Birds from the Marcapata Valley, Cuzco, Peru. *Fieldiana. Zoology.* Chicago Natural History Museum. 34 (3):17–23.

Urbano, H. Oswaldo. 1974. La representación andina del tiempo y del espacio en la fiesta. *Allpanchis Phuturinga* 7:948.

Urbano, H. Oswaldo. 1976. Lenguaje y gesto ritual en el Sur andino. *Allpanchis Phuturinga* 9:121–150.

Urton, Gary. 1978a. Beasts and Geometry: Some Constellations of the Peruvian Quechuas. *Anthropos* 73:32–40.

Urton, Gary. 1978b. Orientation in Quechua and Incaic Astronomy. *Ethnology* 17 (2):157–167.

Urton, Gary. 1979. The Astronomical System of a Community in the Peruvian Andes. Ph.D. dissertation, University of Illinois at Urbana-Champaign. Published as *At the Crossroads of the Earth and the Sky: An Andean Cosmology,* Latin American Monographs, Institute of Latin American Studies No. 55, Austin, University of Texas Press, Fall 1980.

Urton, Gary. 1980. Celestial Crosses: The Cruciform in the Astronomy of the Quechuas. *Journal of Latin American Lore* 6 (1):87–110.

Von Humboldt, Alexander. 1975. *Views of Nature* (1850).

Wagner, Catherine A. 1978. Coca, Chicha, Trago: Private and Communal Rituals in a Peruvian Community. Ph.D. dissertation, University of Illinois at Urbana-Champaign.

Walker, Ernest P., et al. 1964. *Mammals of the World.* Baltimore, Johns Hopkins Press.

Weiss, Gerald. 1969. The Cosmology of the Campa Indians of Eastern Peru. Ph.D. dissertation, University of Michigan, Ann Arbor.

Zuidema, R. Tom. 1973. Kinship and Ancestor Cult in Three Peruvian Communities: Hernandez Principe's Account in 1622. *Bulletin Institut Français des Études Andines* 2 (1):16–33.

Zuidema, R. Tom. 1977. Mito e historia en el antiguo Perú. *Allpanchis Phuturinga* 10:15–52.

Zuidema, R. Tom, and Gary Urton. 1976. La constelación de la Llama en los Andes peruanos. *Allpanchis Phuturinga* 9:59–120.

Culture Confronts Nature in the Dialectical World of the Tropics

Billie Jean Isbell

Aveni and Urton are to be congratulated for organizing a conference that, for the first time, has brought together scholars from various disciplines to discuss the ethnoastronomy and archaeoastronomy of the American tropics. We have had to converse across disciplinary boundaries and I hope that the endeavor has clarified some of the concepts, methods, and perspectives of the different specialties represented.

As an anthropologist, I would like to suggest that the tropics provide a perceptual environment that promotes and enhances a particular "science of the concrete,"[1] whereby perceived order in the environment is the basis for systems of classifications, epistemological structures, and cosmologies. In the American tropics, the science of the concrete takes on a particular character that results in epistemologies founded in what I will call dialectical, reversible dualism.

From *Ethnoastronomy and Archaeoastronomy in the American Tropics*, ed. A. Aveni and G. Urton, Annals of the New York Academy of Sciences 385 (1982).

The native philosophers of the indigenous societies under discussion engage in the study of the nature and limits of knowledge. Their epistemological reflections are embedded in religious and ritual practices. Moreover, the native philosophers, who are usually shamans or astronomer-priests, use methods and metaphorical language that are unfamiliar to us. More importantly, the logic that underlies these systems of knowledge is dialectical rather than rationalistic. As Roy Wagner has pointed out in *The Invention of Culture,* the anthropological definition of the concept of dialectic refers to

> a tension or dialogue-like alternation between two conceptions or viewpoints that are simultaneously contradictory and supportive of each other. As a way of thinking, a dialectic operates by exploiting contradictions against a common ground of similarity, rather than by appealing to consistency against a common ground of differences, after the fashion of rationalistic or "linear" logic.[2]

Examples may help to clarify the two types of logic. In Western science, taxonomies and typologies employ the principle of similarity against a common ground of differences. The underlying assumption of rationalistic logic is linear causality. In sharp contrast, dialectical logic focuses upon simultaneous interdependence and contradiction. Linear causality is not assumed. For example, the most widespread dialectical concept in the American tropics is the *Axis Mundi*. While the cultural contents differ, the necessary tension is maintained between the opposed elements of the upperworld and the underworld. The interdependence is such that one cannot be defined without reference to the other.

By reversible dualism, I mean a logical process whereby the definition of one of the polar opposites must be derived from the view or position of the other. For example, the structure of the relationship between the polar opposites of the *Axis Mundi* necessarily means that one, the upperworld, must be defined by viewing it from the underworld. This principle of reversible dualism applies more generally, and I argue that astronomical phenomena in the American tropics are perceived as dual, dialectical pairs. One is necessarily the vantage point for the cultural definition of the other and vice versa. Therefore, time is perceived as a dialectical tension between two interdependent, but contradictory, elements. In part, this is due to the structural relationships that pertain between solar zenith and nadir phenomena and significant seasonal changes in weather, fauna, and flora. In turn, observable celestial periodicities—such as the phases and positions of the moon and the paths of various constellations—are sought that fit the epistemological structures that explain seasonal changes. For example, the path of the Pleiades is important everywhere in the American tropics for announcing seasonal changes. However, because the predictable appearance and disappearance of the constellation occurs with different periods throughout the American tropics, the significance differs from culture to culture. If I am correct about the structure of epistemologies in the American tropics, the Pleiades should always be opposed to some other celestial body in a dialectical, reversible relationship.

I shall discuss the structural similarities of the various native cosmologies as well as the culture-specific differences that have come to light in this conference. But first, I would like to make an observation about the nature of the science of the concrete. Awareness of the regularity and structure of various naturally occurring periodicities provides the schemata[3] for verifying epistemologies in every culture. I suggest, however, that the structural peculiarities of the observable periodicities in the tropics result in a number of shared epistemological features.

Most significant is the fact that, in the tropics, celestial bodies move on straight tracks, rather than around a fixed point in the sky (the north and south celestial poles).[4] The perceptual consequence of this phenomenon is that the sky is divided into two halves. Moreover, as Urton has explained, the point of observational orientation is the movement of celestial bodies in relation to the observer's own fixed locality, rather than a fixed celestial pole.[5] Aveni has shown that all an observer needs are simple devices, such as crossed sticks directed to the horizon or sighting tubes and gnomons oriented to the zenith, to make accurate astronomical and calendrical calculations and predict the approach of the solstices, the equinoxes, the cycles of Venus, and the zenith passages of the sun.[6]

Observed astronomical phenomena in tropical America play a large role in the development of epistemological structures. Specifically, the vertical orientation to the sky of an observer who is the fixed center around which celestial bodies move promotes epistemologies that have dual, symmetric, and reversible structures similar to those found in spatiotemporal structures of the native astronomy. In order to delineate a few of the similarities of the structures of space, time, and cosmologies, I will begin with an examination of zenith and nadir solar phenomena. In Figure 20.1, I have attempted to depict the symmetry and reversibility of zenith and nadir solar passage dates by drawing them in the shape of a figure eight. The zenith dates for each latitude are on the left; the nadir dates are on the right. Note that for each latitude, there are two zenith dates and two nadir dates, except at the lines of the tropics and at the equator. Zenith and nadir converge with the equinox dates at the equator. They diverge as one moves north and south until zenith passage of the sun coincides with the June solstice at the Tropic of Cancer (23° 26') in the north, and with the December solstice at the Tropic of Capricorn (23° 26') in the south. The relationship is reversed for the nadir dates. Moreover, if you trace the zenith passage dates from north to south, you will find that the dates for the nadir are the same (plus or minus two days) for the same latitudes, moving in the opposite direction from south to north. Consequently, the spatiotemporal relationships of the annual passage of the sun are like two interlaced cords, to borrow Klein's Mesoamerican metaphor for the structure of the cosmos. In studying the articulation between cosmologies and native astronomies in the American tropics, we find that culture confronts nature in an attempt to apprehend the dynamic relationship between time and space shown by the zenith and nadir passages.

In reviewing the data presented at this conference, I find three common principles of organization in the structures of tropical American cosmologies. I find, however, one significant difference. All four principles reflect the dynamic relationship

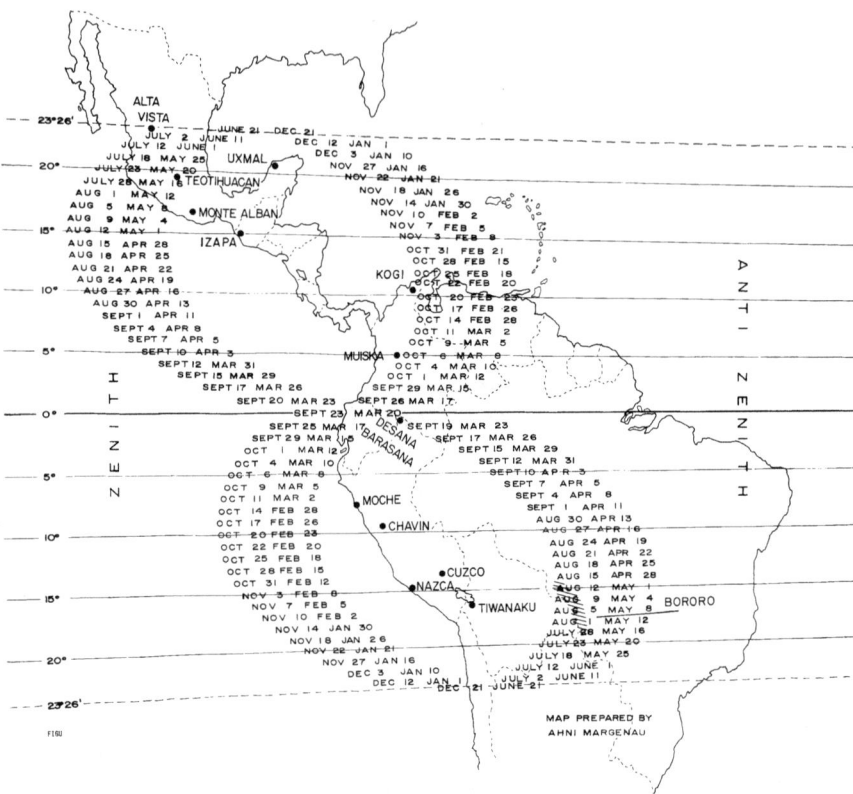

20.1 Zenith and nadir (antizenith) dates taken from the American Ephemeris and Nautical Almanac for 1980, correct for other years ± 2 days.

between time and space in the tropics. The shared organizational principles are as follows: (1) The celestial paths of the sun, moon, and stars are conceived of as cosmic forces whose multiple interactions are responsible for transitions of time, climate, the agricultural cycle, and states of human existence. A corollary of this is that the major metaphors for cosmic order and disorder are based on principles of movement, of transition, and of reversibility, rather than on metaphors of static equilibrium; (2) these dynamic principles, based on observations of interacting cosmic forces, are replicated onto space in settlement plans and architectural and ritual space, as if the built environment of social space were a mirror reflection of the dialectical elements in the heavens; (3) finally, the most prevalent cosmological principle in the American tropics is the *Axis Mundi*, accompanied by what I will call a circulatory cosmology. The *Axis Mundi* is a world axis around which cosmological and celestial forces circulate. It is expressed metaphorically as a world tree, cosmic mountain, or a pyramid encircled or encompassed by a river, gut, an umbilical cord, or woven fabric. This vertical axis and its circulatory cosmic flow form two opposing dynamic principles, which maintain the dialectical tension essential to cosmic order.

The major difference I find is that between the structure of Barasana and Desana cosmologies and that of the cultures to the north and south. The Barasana and the Desana are located near the equator. According to Hugh-Jones and Reichel-Dolmatoff, neither place any importance on horizon observation of the annual movement of the sun. Nor do they make use of heliacal risings and settings to fix events in a calendar. They do, however, observe the zenith passage of the sun, which occurs during the equinoxes, and the annual paths of various constellations. The Pleiades, Scorpio, and Orion are among the important constellations observed as indicators of the changing seasons. Reichel-Dolmatoff makes a point that is true, I think, for both of these equatorial cultures: they are concerned with the diagnosis of the present state of affairs in the universe, not with a prognosis for the future. I would like to suggest that, at the equator, time and space conspire to give the observer the impression of being stationary at the center of the *Axis Mundi,* around which the universe circulates (see Hugh-Jones' excellent description of the structure of the Old and New Path of the Milky Way for details). Additionally, the dual rainy and dry seasons and the absence of a calendar based on tracking the annual movements of the sun on the horizon may intensify the perception that one is living at the center of present time and space. Consequently, the science of the concrete for cultures at the equator should be directed toward knowledge of the variations of the flora and fauna in their environments rather than precise astronomy and prediction of future events. Reichel-Dolmatoff mentions that the Muiska and the Kogi have more complex astronomical systems than do the Desana. Whether they also have a precise calendar is not clear. Fabian hypothesizes that the Bororo, have a complex system of observations and probably a precise calendar. These three cultures are far from the equator and would provide evidence concerning the relationship between the constraints of perceptual realities and the development of epistemologies.

As one moves farther and farther north or south away from the equator, the apparent annual movement of the rising and setting of the sun on the horizon becomes greater.[7] Moreover, the dates of zenith and nadir passages of the sun diverge until, at latitudes 15° north and south, the four zenith and nadir dates divide the year into more or less equal segments. Coggins argues that, at Izapa, 15° north, the zenith passages of the sun define the 260-day ritual calendar. Broda gives an excellent summary of current research on this question for Mesoamerica. Aveni and Urton suggest that zenith and nadir may be important for astronomical alignments and geometric forms among the lines of Nazca.[8] Zuidema argues that, at Tiwanaku, 16° 33' south, zenith and nadir passages divide the year into more or less equal segments[9]—an observation he believes to have been important to cultural developments in the Andes.[10] The site of Alta Vista, located at the Tropic of Cancer, provides sound evidence that the ancient astronomers of the American tropics were aware of the regularity and structure of zenith phenomena. In discussing the location and orientation of this site, Aveni states that the astronomers seemed to have been seeking the actual place where the sun turns around on its northern migration and begins its journey to its southern turn around position.[11] During the two solstices, the sun slows down in its march along the horizon as it approaches its northern and southern extremes. Aveni notes that "the sun will stand perched in the zenith at noon on the longest day, the first day of summer. A shadowless moment occurs as the sun arcs over the zenith and returns to its southern realm."[12] This annual journey of the sun is calculated by watching sunrise and sunset positions on the horizons—hence, horizon astronomy, a name that Aveni has coined.

Aveni and Zuidema's recent research on the astronomy of the Incas of Cuzco, located at 13°30′ south, provides the best evidence for the significance of nadir solar passages.[13] Zuidema argues that the annual cycles of the sun, the moon, and the Pleiades were the three central celestial phenomena that were correlated together in such a way that one observed event announced the other. Urton demonstrated that the same observations are possible for the coast of Peru. In both regions, the series of astronomical observations and the agricultural cycle are conjoined.

The most significant feature of Inca astronomy is that the zenith and nadir were diametrically opposed in space and time. The zenith sunrise point on the eastern horizon and the nadir sunset on the western horizon formed the axis upon which the sun was believed to travel. Zuidema argues that nadir sunset was established by backsighting from the zenith sunrise position to form one of the *ceque* lines. The June solstice sunrise and the December sunset locations were likewise connected by a *ceque* line. The zenith and nadir dates at the latitude of Cuzco are six months apart (see Fig. 20.1). The zenith-nadir dates were used to time the agricultural and ritual cycle. The nadir pair, August and April, are the beginning of planting and the beginning of harvest, respectively. Important rituals were celebrated on both dates. The February and October zenith dates, according to Sherbondy's ethnohistorical research, center upon rituals concerned with the control of water.[14] In February, a ritual was celebrated to signify that the earth was saturated with water and that a transition into the dry season was approaching. In October, a ritual was performed to signal the end of the first irrigation cycle; this period is a transition into the rainy season. Conversely, one pair of opposite dates, the August nadir, the beginning of the agricultural cycle, and the February zenith symbolically form the Andean *Axis Mundi*. The earth is still believed to open up on these dates to receive offerings. These two periods, opposed in time and space, are both propitious and dangerous. It appears to me that Andean people discovered an organizing principle of dialectical, reversible dualism that ordered their time, space, and society. The cultural focus seems to be on determining transitions from one cycle or state to another: from the dry season to the rainy season; from the sun's journey to the north to its journey to the south; from periods of abundance and prosperity to periods of scarcity and poverty. These periods were announced by specific series of celestial events and mapped onto space in axial sight lines (the *ceques*) that were observationally reversible (i.e., zenith sunrise and antizenith sunset) and semantically dualistic. I would like to suggest that a similar structural organization might be discovered for Mesoamerica.

In Mesoamerica, the prevalence of the concept of the *Axis Mundi* (see Coggins' paper) and the importance of the zenith passages for regulating the agricultural cycle suggest that the nadir passages may have more significance than previously believed.

Moreover, both the Andean and Mesoamerican areas of complex cultural development are at latitudes where zenith and nadir passage dates are spaced such that they can be used in similar structural organizations.

Broda points out that, at Tenochititlan, the two zenith dates were commemorated by important rituals in May and July. During the period in between these two dates, the sun entered the abode of the dead, which was believed to be in the north. I assume that this coincided with the sun's journey to the north, where it appeared to slow down before turning around during the June solstice to begin its journey towards its southern extreme. In addition, she discusses the coincidence of the disappearance of the Pleiades with the zenith passage of the sun on 17 May as contrasted with the nadir passage of the sun on 18 November, when the Pleiades were observed at the zenith. Thus, she argues for an "opposite symmetry" between the path of the sun and the path of the Pleiades. Climatologically, these two events were significant. The rains were announced by the disappearance of the Pleiades and the shadowless moment of zenith passage of the sun on 17 May. The dry season was heralded by the appearance of the Pleiades at the zenith and the sun at the nadir on 18 November. Therefore, one can think of the sun and the Pleiades as existing on an axis that announces the changing seasons. They form a dialectical, reversible, dual structure.

Likewise, Zuidema and Urton both argue that the multiple interactions of the zenith/nadir dates, the path of the Pleiades, and the phases and positions of the moon announce the changing seasons in Cuzco and on the coast of Peru. But whereas Broda finds an "opposite symmetry" prevailing between the course of the sun and that of the Pleiades, Zuidema and Urton find convergence. They both argue that the disappearance of the Pleiades at the end of April coincides with the nadir, which announces major transition periods. Zuidema argues that the disappearance of the Pleiades and the sun at the nadir symbolized the death of the sun, the moon, and the Pleiades. Beginning the harvest at this time in April caused Earth Mother to die as well. A period of rebirth was begun with the reappearance of the Pleiades (heliacal rise at dawn) in early June in Cuzco and on the coast of Peru. Notice that the cultural focus is on transitions and transformations of states, rather than on static moments. The logic is dialectical, not rationalistic. If we compare the interpretations of the disappearance of the Pleiades from the information given in Broda's, Zuidema's and Urton's paper, we get Table 20.1.

I have discussed these examples in some detail to illustrate that the same astronomical event (the disappearance of the Pleiades) is taken to be highly significant in both Mesoamerica and the Andes. However, this easily observable event is correlated with other highly significant events: the zenith and nadir of the sun. In order to gain further understanding of the cosmological meanings of this set of multiple observations, we have to examine the cosmological and astronomical context within which they occur. In the Andes, the reappearance of the Pleiades and the approach of the solstice in June announce a period of rebirth for the sun, the moon, the Pleiades, and the earth. What is the next cosmological sequence in Tenochititlan? In addition, do we find "symmetric opposition" in the activities of the mythic figures associated with

Table 20.1 The disappearance of the Pleiades interpreted.

Multiple Astronomical Events Observed	Transition Announced	Location

May 17
  Zenith sun
  Disappearance of the Pleiades
    ("opposite symmetry") or axis
	Onset of rains
	Sun enters world of the dead
		Tenochititlan (Broda)

April 26
  Sun in nadir
  Disappearance of the Pleiades
  New moon (convergence)
	Onset of harvest
	Death of the sun, the moon,
	  the Pleiades, and Earth
		Cuzco (Zuidema)

April 19–22
  Sun in nadir
  Disappearance of the Pleiades
    (heliacal set at dusk)
  Next full moon (convergence)
	Wet to dry season crops
	
	Feast to creator god
		Huarochiri and Coast
		of Peru (Urton)

the sun and the Pleiades in Aztec mythology? In Inca mythology, do the sun and the Pleiades have convergent roles or activities? These are further sets of questions that, once answered, would advance the study of the relationship between native astronomy and cosmology. We need to know the kind of metaphorical language used to express both opposed and convergent astronomical events.

Metaphors relate at least two phenomena together in figurative expression. The multiple interactions of celestial bodies that are observed in order to determine major transitions are better expressed in the language of metaphor, which expresses movement, transition, and reversibility or reflection. For example, the metaphors that come to mind from the papers in this conference include: weaving (Klein), hydraulics (Hugh-Jones, Reichel-Dolmatoff, and Urton), the refraction of light in crystals (Reichel-Dolmatoff and Zuidema), and mirrors (Klein). Even the symbolic forms that at first glance seem to be static, in reality express dialectical concepts: the *ushñu* (Zuidema), the serpent-mountain (Townsend), the caterpillar-jaguar (Hugh-Jones), and volcanoes (Coggins). I am reminded of the session in the Hayden Planetarium when Aveni asked Franklin to speed up the motion of the sky. The effect, as we watched the Milky Way change, was like watching a giant undulating snake move across the sky. We were able to perceive what the language of metaphor attempts to capture when apprehending the spatiotemporal relationships of tropical American cultures.

What we were attempting to capture in the planetarium was a dynamic process that somehow embodies temporal order. The ancient astronomers of tropical America attempted to capture these same processes of structural order in settlement plans and architectural and ritual space. We might label their attempt relational, or dialectical geometry. The paper by Aveni and Hartung illustrates the concern that Mesoamericans had for obligating the perceiver to focus upon the interaction of celestial forces by constructing the environment of ceremonial space around the principle of what Aveni and Hartung have called inter-building relationships. I think that the astronomer-architect-priests were attempting to assure that the dynamic (or dialectical) perspective they had discovered would be maintained from generation to generation. Likewise, the *ceque* lines of Cuzco, which embody the organizational principle of dialectical, reversible dualism, are a means of perpetuating the epistemological discovery that there exists a dynamic relationship between time and space. For example, Dearborn and White's paper confirms Zuidema and Urton's arguments concerning the relationship between the zenith solar passage, the path of the Pleiades (its heliacal rise and set), and the June solstice. The construction of the Torreón obligates the observer to adopt a perspective that focuses upon the relation of one transition of time to another instead of pinpointing the precise date of an astronomical event. I would suggest that such a dynamic perspective is shared by the cultures of the American tropics as culture confronts nature in the reversible world of the tropics. Through the application of dialectical logic, the cosmologies of these cultures construe the contradictions perceived in nature as necessarily interdependent. The contradictions can never be resolved because cosmic order is maintained by the dialectical tension resulting from the reversible relationship pertaining between opposed elements.

ACKNOWLEDGMENTS

I would like to give special thanks to my husband, W. H. Isbell, and Lauris McKee for reading numerous versions of this paper and making helpful comments.

NOTES

1. C. Levi-Strauss, *The Savage Mind* (Chicago: University of Chicago Press, 1966), chap. 1, "The Science of the Concrete."

2. Roy Wagner, *The Invention of Culture* (Chicago: University of Chicago Press, 1975), chap. 3, p. 52.

3. I am using the term *schemata* in the sense of U. Neisser, *Cognition and Reality: Principles and Implications of Cognitive Psychology* (San Francisco: W. H. Freeman, 1976).

4. A. F. Aveni, *Skywatchers of Ancient Mexico* (Austin: University of Texas Press, 1980), "Astronomy with the Naked Eye," chap. 3, pp. 48–132.

5. G. Urton, "The Use of Native Cosmologies in Archaeoastronomical Studies: The View from South America," in *Archaeoastronomy in the Americas,* ed. R. Williamson, pp. 285–304 (Santa Barbara, CA: Ballena Press, 1981).

6. Aveni, *Skywatchers.*

7. Aveni, *Skywatchers,* 63.

8. A. F. Aveni and G. Urton, "A Preliminary Investigation of Geometrical and Astronomical Order in the Nazca Lines," unpublished ms.

9. R. T. Zuidema, "Organizing Space for Computing the Calendar," unpublished ms.

10. R. T. Zuidema, "The Inca Observations in Cuzco of the Solar and Lunar Passages through Zenith and Antizenith," in *Archaeoastronomy in the Americas,* ed. R. Williamson, pp. 319–342 (Santa Barbara, CA: Ballena Press, 1981).

11. A. F. Aveni, "Tropical Archaeoastronomy," *Science* 213 (1981):161–171.

12. Aveni, "Tropical Archaeoastronomy."

13. Aveni, *Skywatchers of Ancient Mexico*, 294–311, and Zuidema, "Inca Observations in Cuzco."

14. J. Sherbondy, "Water in the Ritual Calendar of Cuzco," paper presented at the symposium entitled Myth and Ritual in Andean Societies at the annual American Anthropological Association meeting, Washington, DC, 1980.

CHAPTER TWENTY-ONE

Ethnoastronomy and the Problem of Interpretation
A Zuni Example

M. Jane Young

INTRODUCTION

One way to understand archaeoastronomical sites, whether rock art sites or prehistoric structures, is to elicit interpretations from the Native American groups who inhabit the areas in which such sites are located. Nevertheless, although this is a valid procedure for groups such as the Puebloans of New Mexico who trace their ancestry back through many years of residency in the same geographical area, there are still numerous problems in using the method of ethnographic analogy, especially when it is applied in a simplistic and non-discriminating manner. Despite evidence for cultural continuity, for instance, one cannot project the present onto the past as if cultural concepts were totally static over periods of hundreds of years. Using my own

---

Reproduced by permission from *Songs from the Sky: Indigenous Astronomical Cosmological Traditions of the World*, edited by V. del Chamberlin, J. Carlson, and M. J. Young (West Sussex: Ocarina Books, 2006).

fieldwork recording rock art and interpretations of that rock art at the Pueblo of Zuni to provide examples, I will discuss the use and abuse of ethnographic analogy, emphasizing the problems inherent in the attempt to apply contemporary interpretations of indigenous peoples to the past, particularly when such interpretations have been influenced by the interpretations of non-indigenous peoples.

Questions of meaning and function can best be addressed in a culture, such as that of the Zunis, in which rock art is an ongoing part of the tradition. In such cases, particularly where there is a high degree of cultural continuity, there are greater possibilities for historical reconstruction, based on the evidence of today. For the same reason, there is greater possibility for reliance on ethnographic materials, particularly since most works on Zuni have been written in the last one hundred years or so. In this regard, contemporary perceptions of rock art may be able to tell us something about its historical context, yet there are recognized limitations to such an approach. It is likely, for instance, that rock art is produced for different reasons today than during prehistoric times. Perhaps, too, the meanings of specific design elements have changed through time. For these reasons, discussions with tribal members concerning the meanings of rock art depictions are most informative with respect to those images of recent date. Their interpretations of older figures are certainly an important component of contemporary perceptions, but might or might not reflect the "original meanings" of such images or the reasons for their production. Furthermore, contemporary interpretations might be subject to the influence of "distorting feedback" from sources outside the culture. Those who have studied Zuni culture have influenced it, not only by their presence in the field, but also by their creation of written records of interpretations by certain individuals that give these greater status and broader distribution than they might otherwise have had. For these reasons, contemporary interpretations must be projected onto prehistory, or even recent history, only with the greatest caution and with a full recognition of the dynamic aspects of tradition—the change as well as the continuity.

This is not, of course, meant to suggest that we should despair that Zuni culture is changing or that it "no longer echoes clearly the strains of the past" (Dell Hymes, personal communication, 1982). It is, instead, a cautionary statement, perhaps appearing obvious to many, that we cannot, in our attempt to reconstruct and understand the distant past, treat modern-day interpretations as absolutely accurate statements about that past.

ZUNI INTERPRETATIONS OF ROCK ART

Many Zunis are familiar with the published books, articles and monographs about various aspects of their culture. Despite the fact that they often describe such works as "containing many lies," the Zunis seem to incorporate much of this material into their own oral tradition. For example, in the summer of 1979, a Zuni man interpreted a number of rock art panels at a site called the Village of the Great Kivas in words that were almost identical to the interpretations recorded by Frank H.H. Roberts, Jr., from

21.1 Pecked insect figures, 50 cm × 120 cm, Village of the Great Kivas (ZRAS Site 1), Pueblo of Zuni. (Photograph by Nancy L. Bartman, 1979)

his Zuni workmen while excavating the site in 1930 (Roberts 1932:140–152; Young 1979). According to both accounts, Figure 21.1 is a depiction of poisonous insects that the war chief carved on the rocks as he sang a song requesting the insects to sting the Navajo, the traditional enemies of the Zuni. Figure 21.2 was said to consist of a humpbacked flute-player (referred to as a rain priest), a horned toad and an insect, all of which were "pictured on the rocks for the purpose of attracting clouds and moisture" to this area (Roberts 1932:150; Young 1979). The spirals in Figure 21.3 were described as representing the period in the Zuni emergence myth when the people were traveling about looking for the Center, where they would build their permanent home. The deer with elongated antlers was "a record ... of an unusually successful hunt ... placed there in order to propitiate the spirits of the slaughtered animals and to attract others to the region" (Roberts 1932:150; Young 1979). The turtle, also identified as a "horny toad," is an important water creature; its appearance here was said to be for the purpose of bringing rain to the Zuni area.

Although the Zuni man I talked with in 1979 had read Roberts' written account and thought it to be "correct," it is unclear whether his interpretations were based on accounts in Roberts, or whether they flowed directly from a stable and widespread oral tradition about this site. If the latter was the case, his explanations of the meanings of these images to me would constitute one performance of a traditional tale and the telling by Roberts' workmen would constitute another.

A second, perhaps clearer, example of the adoption by Zunis of interpretations made and promulgated by outsiders concerns another panel at the same site (Fig.

21.2 Pecked flute-player, horned toad, animal or insect, 1 m × 60 cm, Village of the Great Kivas, Pueblo of Zuni. (Photograph by M. Jane Young, 1979)

21.3 Pecked spirals, deer, horned toad, 1 m × 80 cm, Village of the Great Kivas, Pueblo of Zuni. (Photograph by M. Jane Young, 1979)

21.4). In the summer of 1979, my Zuni colleague explained the meaning of some of the rock art elements on this panel, as had Roberts' workmen, by referring to a common Zuni folktale:

> The zigzag from the Moon and star to the owl is the owl's flight during the night, when it would spy on the Navajo and then return to Zuni to report the number and location of the enemy to the Zuni war chief. (Roberts 1932: 151; Young 1979)

A year later, the same Zuni colleague suggested that the same petroglyphs represented the "supernova explosion a long time ago" (Young 1980). In the summer of 1981, the tribal historian also gave this interpretation of the panel, saying that someone who had visited the pueblo had told him about the "Crab Nebula" and the suggestion of some astronomers that this panel and several other similar ones in the Southwest recorded the supernova explosion of AD 1054, which resulted in the formation of the Crab Nebula (Young 1981).

The 1054 supernova explosion was a striking event (likely visible in the Southwestern United States), which would have initially appeared as an extremely bright star close to the crescent Moon in the morning sky. This bright star would have been

21.4 Pecked elements, 2 m × 180 cm, Village of the Great Kivas, Pueblo of Zuni. (Photograph by M. Jane Young, 1979)

visible, first in the morning sky and then in the night sky, for approximately 650 days. Some archaeoastronomers have consequently argued that, because of "the propensity in men to notice and record unusual sky events, particularly if they correspond with a terrestrial event of great importance" (Brandt and Williamson 1979:33), an event of this magnitude and visibility might well have been recorded in the rock art of the area.

Rock art configurations such as the one at Zuni (Fig. 21.4), which depict a star in close association with the crescent Moon, are used to support this hypothesis (Brandt and Williamson 1979:1–38). Whether or not this might actually be the "original meaning" of these images may be argued, but the comment of the tribal historian suggests that the idea did not begin with the Zunis. Rather, he and other Zunis learned of and adopted the recent suggestion of these astronomers that this panel is a record of the supernova explosion. Apparently then, a story about this site, told by non-Zunis, has influenced the interpretations of some Zunis. Interestingly, although the identification of individual images at this site was generally straightforward—owl, star, crescent Moon, owl's path at night—the meaning conveyed by the cluster of images as a whole is somewhat ambiguous. In some cases, the meaning of all of the images taken together adds up to a whole that is more than just "the sum of its parts." Thus, in addition to referring to the Zuni folktale described above, the cluster can take on a completely different meaning, under the influence of suggestions from non-Zunis.

An interesting final irony to this particular example of the process of reinterpretation is that some Zunis seem to have associated this entire panel of rock art elements, rather than just the star and crescent, with the supernova, so that subsequently, they could associate it with the zigzag alone. I was surprised during my initial period of

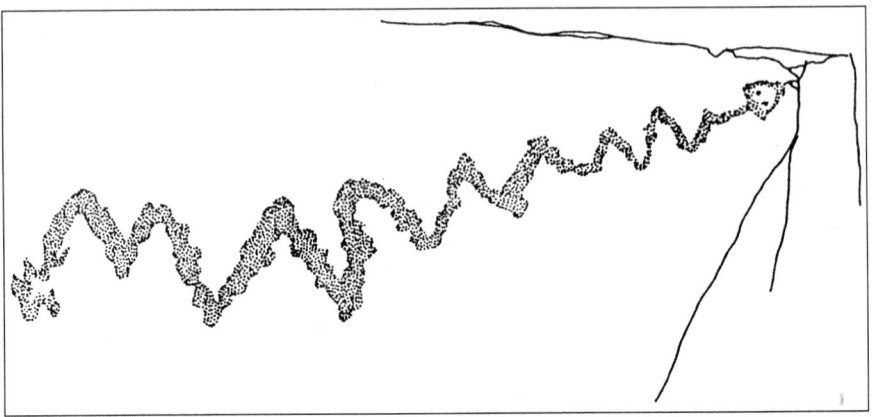

21.5 Pecked snake figure 10 cm × 60 cm, ZRAS Site 16, Pueblo of Zuni. (Drawing from slide by Murray Callahan)

fieldwork to find that several Zunis identified Figure 21.5 as the "Crab Nebula" while others described it only as "a snake with a zigzag body" (Young 1979–81).

The discoveries at Fajada Butte in Chaco Canyon (Frazier 1979; Sofaer, Zinser and Sinclair 1979:283–291) and their ensuing wide distribution in the media have led to another similar instance of reinterpretation at Zuni. Because at the Fajada Butte site so-called "daggers of light" interacted with spiral petroglyphs, several individuals went to Zuni to inquire about the meaning and use of spiral symbols for contemporary Zunis. As a result, however, one of my Zuni colleagues now believes that the spirals at the Village of the Great Kivas (Fig. 21.3) are bisected by streaks of light at the winter solstice (Young 1981)—an impossibility since the necessary special configuration of nearby boulders, or any other physical feature that might produce such phenomena, is completely missing. Thus, this man, the same one who reinterpreted another panel at this site as "Crab Nebula," once again changed his interpretation of a panel since his original commentary in 1979; at that time he had told me that the spirals represented "the journey of the ancestors in search of the Center." This change in meaning seems to have been sparked directly by a visitor who told him about the site at Chaco Canyon and suggested that he might find something similar in the rock art near Zuni. The visitor must have been quite convincing, for my colleague was prepared to go to some lengths to observe this phenomenon. He said: "In the winter near the solstice I want to spend the night up there at the Village of the Great Kivas. I'll look at the spirals just before the Sun comes up. I bet the same thing might happen" (Young 1981). Furthermore, because it has been implied that the spirals at Fajada Butte represent "time" (i.e., the motion of the Sun during the year), this Zuni colleague wonders if the same might be true for the spirals at the Village of the Great Kivas. When I visited Zuni in 1984 he said: "Maybe those spirals are supposed to be the travels of the Sun during the year. There are two of them—maybe they're the two halves of the year" (Young 1984). Yet, when I reminded him that three years earlier he had said that these same spirals depicted the "journey in search of the Center," he replied, "They mean that too."

CONCLUSION

I suggest that at least part of the reason some Zunis are so willing to accept and incorporate these "astronomical interpretations" of rock art into their contemporary tradition is that it provides a way of emphasizing that science is not the sole property of the Euro-American. In a sense, they are saying, "we knew about that back then," thus validating the knowledge of their ancestors and the credibility of traditional accounts. It is certainly the belief of many ethnoastronomers, myself among them, that the ancestors of the contemporary Pueblo Indians had an extremely sophisticated knowledge of naked-eye astronomy that had been transmitted orally for hundreds of years, and it is not unlikely that some of it was recorded in material forms, such as rock art. Because of this depth of tradition, however, one cannot simply "decode" astronomical motifs in rock art solely on the basis of present-day interpretations, or the "well, it looks that way to me" stance of the outside observer.

REFERENCES

Brandt, John C., and Ray A. Williamson. 1979. The 1054 Supernova and Native American Rock Art. *Archaeoastronomy*, Supplement to *Journal for the History of Astronomy* 1:1–38.

Frazier, Kendrick. 1979. The Anasazi Sun Dagger. *Science 80* 1(1):56–67.

Roberts, Frank H.H., Jr. 1932. The Village of the Great Kivas on the Zuñi Reservation, New Mexico. *Bureau of American Ethnology Bulletin* 111:149–152. Washington, DC, Government Printing Office.

Sofaer, Anna, Volker Zinser, and Rolf Sinclair. 1979. A Unique Solar Marking Construct. *Science* 206:283–291.

Young, M. Jane. 1979–1984. Tapes and notes from fieldwork at the Pueblo of Zuni. Tapes and notes in Young's possession.

PART IV

The Classic Maya
A Testing Ground for Precise Astronomy in the Written Record

The Templo Mayor, or "Great Temple," of the Aztecs is oriented 7½° south of east. If you use all the available archaeoastronomical software you can find, you will discover that nothing of any apparent significance happens in the sky along that line of sight (except maybe sunrises in late February and mid-October). But a statement written by a well-respected Spanish chronicler tells us that King Moctezuma deliberately aligned ancient Mexico's most sacred place of human sacrifice so that the equinox sun would appear in the notch between the twin temples that crown its summit. He even had to tear the building down and rebuild it to get it right! Apparently, worshippers looked up (at an approximately 20° angle) from the plaza below to witness the event that scheduled the ritual. The need to skew the building 7½° toward the north from true east, where the equinox sun crosses the horizon, can be attributed to the sun's steep angular ascent toward the southeast in tropical latitudes after it rises in the morning sky.

I cite this example for two reasons. First, it illustrates how easy it is to misinterpret the archaeological record; and second, it demonstrates the enormous advantage of studying ancient astronomy in cultures that offer us a *written record*. Although the information about the Great Temple comes from Spanish historians who lived in the capital after the conquest, and consequently whose word must be treated with caution, the ancient Maya offer us a pre-Columbian writing system exclusively their own, the only such system in the New World that has been deciphered.

We use the term "classical," a label usually reserved for the ancient Greco-Roman civilizations, to characterize the zenith of ascendancy of Maya culture (ca. AD 200–900). This was a time when great art, architecture, sculpture, and written manuscripts (many of which pertained to astronomy) abounded in the great cities that dotted the Yucatan peninsula. Although most Maya documents were destroyed by the Spanish priests who came after the conquistadors to Christianize the natives, and who suspected the content of the ancient writing had been inspired by the devil, what remains of these documents, and of the hieroglyphs carved on monuments, has yielded an astonishing number of astronomical references. Maya writing also affords us an excellent opportunity to learn about the role of astronomy in a civilization that can be characterized as a state, that is, a body of people who permanently occupy a territory and are politically organized under a government free from outside control and possessing the coercive power to maintain order.

In the West, states and empires gave rise to the quantitative scientific practice that would develop into the highly intricate astronomy and calendar we know today. For all of these reasons, I decided to devote a section of *Foundations* to articles about Maya astronomy that might serve as a test case. Do complex state-level societies acquire and express astronomical knowledge in the same way? Are their astronomies applied to the same ends as those in the ancient West? What specific advantages does writing offer a culture for storing and disseminating astronomical knowledge? And how is this sort of astronomy different from what we find in tribal and hunter-gatherer societies? (My book *People and the Sky: Our Ancestors and the Cosmos* [2008] explores this subject further.) These are the sorts of questions addressed by the four pieces reproduced in the present section.

I open with John Justeson's thorough overview of the Maya written record. For want of space, it has been necessary to amend this rather lengthy piece. And, as it is quite detailed, less-initiated readers may require some preparatory reading (see, e.g., my *Stairways to the Stars* [1997] for a primer in Maya astronomy); but those who persist will be rewarded with an understanding of the rather exotic Maya view of time, as well as of the sky and its denizens conceived by a culture dedicated to precision and exactitude, yet driven largely by religious and dynastic concerns. (For an update on the Maya Venus table, see Bricker and Bricker [2007].) Reading Justeson's overview, you may well wonder what could have been going on in the minds of the astronomer-scribes who painted these extraordinarily complex hieroglyphs, calculated time cycles bordering on millions of years, and predicted eclipses and planetary stations with uncanny accuracy—all of it in a "low-tech" culture? What was their ontology of astronomy? Did they, too, contemplate the universe "for itself" the way Western astronomers do? Were they really like us?

Part IV: The Classic Maya

The first of two contributions by linguist Floyd Lounsbury is about Maya calendar dates that accompany a marvelous mural painting at the ruins of Bonampak in the rainforest of Chiapas, Mexico. It shows a scene of captivity and sacrifice in the aftermath of war, above which is positioned a portion of a zodiac accompanied by hieroglyphs that may represent the planet Venus. Lounsbury's investigation of the relevant dates, which has since been critiqued (cf. Aldana 2005), was the first study to suggest that the schedule of raids, the presentation of captives, and the accession of a new king who has just proven his military prowess to a waiting public were all events fixed by key points in the cycle of the planet Venus—Maya star wars!

The second piece by Lounsbury shows how astronomy was employed to strengthen the dynasties that ruled the Maya states. A Palenque king adopts Jupiter and uses its retrograde twists and turns to symbolize—perhaps even to schedule—major turning points in his life. He proudly discloses the cosmic connection as he displays his life story in hieroglyphs on major monuments at the great Classic Maya city of Palenque that he ruled for so long. (For more on Jupiter retrograde and the Maya, see Milbrath 2000.)

Prior to publication of Victoria Bricker and Harvey Bricker's article on the so-called seasonal tables, most aficionados of the Maya codices believed that, except for an eclipse, a Venus (and possibly a Mars) ephemeris (a table giving precise astronomical data), the remaining portions of the indigenous written documents were simple almanacs that dealt only with endless cycles of time used as a device to schedule rituals to be performed on lucky and unlucky named days. But Bricker's work, reproduced here, broke new ground by demonstrating that many of the almanacs in the codices actually refer to "real-time" events; that is, they are historically grounded and, important for us, frequently adjusted in accordance with changing astronomical phenomena—the way we update our modern almanacs. As a result, real-time studies are now an active component of research on Maya astronomy (cf., e.g., Vail and Aveni 2004).

THINGS TO THINK ABOUT

1. Compare the Maya zodiac with other zodiacs you have heard about from around the world. Do you think zodiacs are universal? Why do skywatchers invent them? Why do you think animals are such a prominent feature of zodiacs? In addition to researching the outside literature, use some of the articles in *Foundations* to assist you in thinking about these questions.

2. How do astronomical alignments discussed in other articles in *Foundations* relate to the complex issues about calendar keeping raised in the articles in this section? Write a short essay titled "Maya Cultural Astronomy: It Isn't Just About Alignments."

3. Consult the zodiac in the Paris Codex. You can find it online at famsi.org. How many animals can you identify? Which ones are the most difficult to make out?

4. Timing wars by the stars certainly does not seem like a winning strategy, especially if the enemy watches the sky as well. Why then do you think the Maya adopted it?

PART IV: THE CLASSIC MAYA

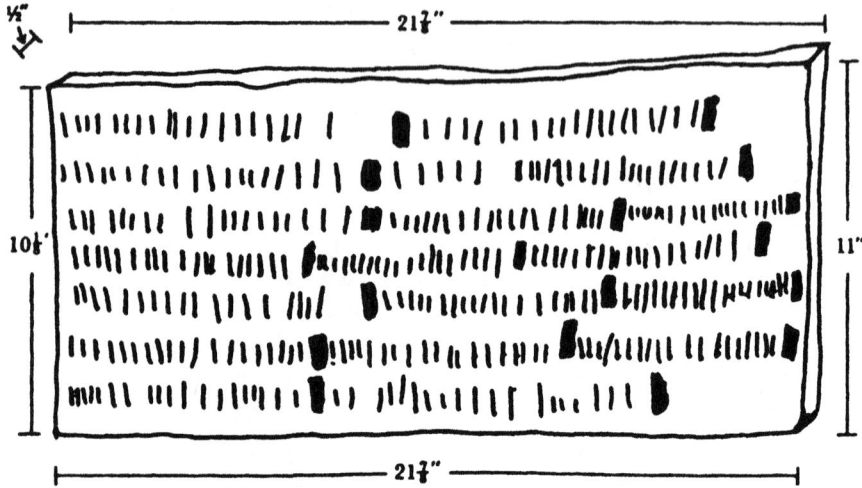

IV.1 A contemporary Maya "calendar board."

5. Lounsbury's method for charting Jupiter retrogrades, although accurate, is now obsolete. To do the same, you can use any of the modern software packages (e.g., Voyager, EZ Cosmos, or simply enter "Astronomical Software" online to find a program that suits you). Try figuring out when the next retrograde cycle of Jupiter is scheduled to commence. See also whether you can identify the dates of "first and second stationary" that mark the period of retrograde motion. Also find when the next morning heliacal rise of the planet Venus is expected and compare the number of days that have elapsed since the previous one. Finally (for avid skywatchers), how accurately can you detect these changes with the naked eye? (For some hints on how to proceed with the last question, see Aveni and Hotaling 1994.)

6. Above is a reproduction of a Chamula ʔotalkʼakʼal ("Counter of Days," or solar calendar). Brought to light by anthropologist Gary Gossen in 1969, it measures approximately 22 inches by 10 inches and records a calendar. Rendered in charcoal on wood, each mark stands for a day. Note that the last mark in each cycle is the most heavily emphasized part of the Maya habit of stressing the day on which cycles are completed. What cycles do you think are being marked out and why has the calendar keeper arranged them in this manner? What, if anything, about this contemporary calendar resonates with what you know about the ancient Maya calendar and counting system?

SUGGESTIONS FOR FURTHER READING

Aldana, G. 2005. "Agency and the Star War Glyph: A Historical Reassessment of Classic Maya Astrology and Warfare." *Ancient Mesoamerica* 16:305–320.

Aveni, A. 1997. *Stairways to the Stars*. New York, Wiley.

Aveni, A. 2008. *People and the Sky: Our Ancestors and the Cosmos*. London, Thames and Hudson.

Aveni, A., ed. 1992. *The Sky in Mayan Literature*. New York, Oxford University Press.

Aveni, A., and L. Hotaling. 1994. "Monumental Inscriptions and the Observational Basis of Maya Astronomy." *Archaeoastronomy* 19 (supplement to *JHA* 25):S1–S54.

Bricker, H., and V. Bricker. 1997. "More on the Mars Table in the Dresden Codex." *Latin American Antiquity* 8:384–397.

Bricker, H., and V. Bricker. 2007. "When Was the Dresden Codex Venus Efficacious?" In *Skywatching in the Ancient World: New Perspectives in Cultural Astronomy*, ed. C. Ruggles and G. Urton, 95–120. Boulder, University Press of Colorado.

Bricker, V., and H. Bricker. 1986. "The Mars Table in the Dresden Codex." *Middle American Research Institute* 57:51–80.

Love, B. 1995. "A Dresden Codex Mars Table?" *Latin American Antiquity* 6:350–361.

Milbrath, S. 2000. *Star Gods of the Maya*. Austin, University of Texas Press.

Vail, G., and A. Aveni, eds. 2004. *The Madrid Codex: New Approaches to Understanding an Ancient Maya Manuscript*. Boulder, University Press of Colorado.

CHAPTER TWENTY-TWO

Ancient Maya Ethnoastronomy
An Overview of Hieroglyphic Sources

John S. Justeson

Dedicated to Floyd Lounsbury

What remains of Lowland Maya astronomical theory and observation has come down to us in ethnohistoric records, mostly ethnographic accounts compiled by their Spanish conquerors; in the alignments of their surviving "permanent" structures; and in native accounts, mainly in their hieroglyphic records. This chapter is an overview of data from sources of the third kind. The chief advantage of these sources is the precise temporal information they explicitly provide; patterns of dates in and between many hieroglyphic texts demonstrate their astronomical significance. These provide much material for investigating the formal means by which the Maya analyzed and anticipated celestial events.

Astronomical content of Maya hieroglyphic manuscripts was demonstrated almost from the beginning of serious research on Maya writing, in extensive computing tables. The extant copies date to within a few centuries of the Spanish Conquest,

---

From *World Archaeoastronomy*, ed. A. Aveni (New York: Cambridge University Press, 1989).

between AD 1200 and 1500 (M. D. Paxton. Stylistic and iconographic analysis of a Maya manuscript. Unpublished PhD dissertation, University of New Mexico, 1986) but internal evidence in the table indicates that they were adapted over the years from prototypes going back to ca. AD 755.

Successes with the manuscript tables were soon extended to the carved inscriptions, almost all of which date to the Classic period of AD 300–900. Apart from the recognition of records in a lunar calendar, the results were generally not as reliable as those in the manuscripts; the reasons are sketched in the section "Explicitly Astronomical References." Since the discovery that most hieroglyphic texts record the activities and pedigrees of rulers, attention has focused on the construction of dynastic sequences and a skeleton of historical events involving the persons named in them. But, as Maya astronomy concerned the behavior of the sky gods, "deities whose activities vitally influenced human affairs" (Kelley and Kerr, 1973, p. 180), astronomical correlates of historical events have begun to be recognized in the essentially historical narratives. Classic texts almost never mention these correlates; seldom do they make any explicitly astrological statements, referring instead to associated human events. As in the interpretation of structure alignments, these unstated correlates must be inferred from distinctive patterns, and demonstrated by statistical argumentation. While initially impeding recognition of celestial observations, this pattern of reference provides a means for recognizing the cultural significance of celestial phenomena in the events the Maya say occurred on these dates.

This chapter surveys hieroglyphic evidence for the means by which Maya astronomers attempted to develop formal models for the activities of the sky gods, and analyzes the intrinsic potential and consequences of these methods for the models they could develop. It sketches the relations of celestial to human affairs, so far as hieroglyphic sources reflect them; unfortunately, we are allowed only isolated glimpses into these relations, since the evidence comes from elite sources that treat relatively few topics, and these mainly as they involve the ruler himself.

MESOAMERICAN CALENDARS AND ASTRONOMY

The Maya Calendar System

Although Maya predictive astronomy had both spatial and temporal dimensions, what survives in hieroglyphic texts is concerned almost entirely with predictions for the timing of events. The standard units of time measurement were not simply the tools by which the dates of celestial events were recorded, but also tools for recognizing the periodicities in those events via periodic repetition at or near the same point in one or more of these standardized measures of time. As the celestial objects were sacred deities, influencing the fates of human beings, the measure of sacred time in the ritual calendar was primary among these tools; but all the basic calendars were involved. Maya temporal astronomy must therefore be surveyed, as it developed, first of all in terms of Maya calendars.

John S. Justeson

Structure. There were three basic Maya calendars. One was a ritual calendar of 260 days, used throughout Mesoamerica for prognostication; it is often called the sacred round (SR). There was also a so-called vague year of 365 days, and a civil year or tun of 360 days within a cycle called the long count; periods of 20 360-day years (the katun) and 400 360-day years (the baktun) were also commonly employed; higher powers of 20 were more rarely used to form multiples of the 360-day year. Most historical dates were specified in both the 260- and 365-day calendars, and in that order; the two commensurate in 73 sacred rounds, or 52 vague years, forming an 18,980-day cycle called the calendar round.

The names above are those used by scholars today. There is confusion in the literature over the referents of the Mayan words attached to these cycles. Names applied to the ritual calendar meant "count / recitation of days" or "series of days" (e.g., Cholan *čol k'in,* Yucatec *¢olkin,* Quichean *alah q'ix*), but these may refer instead or in addition to the almanacs for prognostication in terms of the ritual calendar days. A glyphic spelling that designates 260-day spans (but not the ritual calendar itself has been identified by Mathews (Toniña dates 1. A glyph for the period of 260 days? *Glyph Notes* 8, privately circulated, 1979); it can be transliterated ^{13}SAK-HAB, perhaps reading *sak haʔb'* (lit., "white year," "white" having connotations of "diminished substance" in some Mayan languages, "resplendent, magnificent" in others); I suspect that the numeral 13 is a semantic determiner.

The pan-Mayan word for "year" approximates *haʔb'* in almost all Mayan languages; this term was applied at Spanish contact to both the 365-day and 360-day years. A word apparently unique to the Lowland Mayan languages was *tu·n,* meaning the end of a year of either 360 or 365 days, e.g. a station in an absolute time count in the civil calendar or from another anchoring event whose anniversary was celebrated especially installation into rulership (Fox and Justeson, 1984, p. 53, n.32). This word is conventionally used by Mayanists for the period of 360 days. This misuse is minimized but not entirely avoided here; when it occurs, it is distinguished by Roman typeface from the Mayan usage of *tu·n.* The 360-day years of the civil calendar were grouped into 20s; this grouping is referred to as the katun, after the Yucatec name for the period (**k'atun* < earlier **k'al tu·n* "20 year [endings]"), but it was evidently also referred to as **may* in Quichean languages, hieroglyphic texts, and colonial historical texts (Justeson and Campbell. The linguistic background of Maya hieroglyphic writing: arguments against a Highland Mayan role, unpublished manuscript, 1982). No Mayan name for the 400-year baktun is known (baktun is a scholarly term based on Yucatecan *b'á·k* "400" and *tù·n* "year," on analogy with *k'atun*), or for higher multiples of the civil year. No Mayan term for the calendar round is known.

Although these were the principal calendars, and the only ones appearing in the Postclassic hieroglyphic manuscripts, other calendars or temporal cycles were also recorded in the inscriptions. The most common was a nine-day cycle (designated Glyph G, and indexed G_1 through G_9). The hieroglyphs corresponding to the days in this cycle evidently named deities rather than days, deities generally equated with the nine Lords of the Night (a group of nine Aztec deities who ruled the fortunes of each night in succession); Kelley (1972), Kelley and Kerr (1973, p. 201) and Schlak (The gods of Palenque, unpublished manuscript, 1985) have suggested planetary identities for these deities. The glyphs referring to them seem to state that their authority has ended, suggesting that they are daytime records. Because 9 and 18 980 are relatively prime, a calendar round date referencing this nine-day cycle was thereby fixed historically in a cycle of nine calendar rounds or 468 years.

Next most commonly attested are records in a lunar calendar (see the section "Lunar Models"), about 170 cases being known. Rarely recorded are two other cycles. One is of 819 days, the least common multiple of three calendrically and ritually important numbers (9, 13 and 7); texts in which it occurs enumerate the days elapsed since the last station in the cycle, stations being 819 days apart. A final unknown cycle is Glyph Y, with coefficients 1–6; no current proposal successfully accounts for its coefficients. Formerly a Glyph Z was distinguished, always followed immediately by Glyph Y, but it is simply a numeral coefficient (5 + the grammatical suffix –*b'is*) of Glyph Y.

The position of a date in the above cycles was presented almost invariably in the following order: long count—ritual calendar—nine-day cycle—Glyph Y—lunar series—vague year. The 819-day calendar usually occurs between ritual calendar and vague year dates, the intervening data being suppressed.

History. Calendars are essentially systems of cyclical numeration, specialized in content for application to days. The structures of the basic Mesoamerican calendars directly reflect the general Mesoamerican system of numeration: in all Mesoamerican languages the numeral systems were in base-20, and subdivision into cycles of 20 days is a common feature of the 260-, 360- and 365-day calendars. Thereby, they parallel the subdivision of lunar months into groups of ten days in the calendar sticks of the Cheyenne, Osage, Winnebago and Zuni, numeral systems in the languages of these North American Indian groups being base-10. This is evidently a formal, numerical pattern, unrelated to natural cycles.

The calendar periods have been examined for astronomical bases. Interestingly, the Maya made no attempt to keep their 365-day vague year in line with the seasons or the sun, in spite of a clear seasonal basis for at least two of the Yucatec month names—**yá?š=k'i•n*, the time of new growth, and **k'an=k'i•n*, the time of ripening or maturity. It was not subdivided into "natural" units, e.g. approximating lunar months, but only into 18 named groups of 20 days each, followed by a year-ending period of five "nameless," unlucky days.

The 260-day period has been taken as a deliberate approximation to various natural intervals, most significantly the time to birth since missed menses (e.g. Brotherston, 1983) and the time of zenith passage at c. 15° north latitude[2] (most recently, by Malmström, 1973, 1978; Coggins, 1982). Undoubtedly, such correlates were noted and used once the system was in place (cf. Kelley and Kerr, 1973, p. 180), but it seems unlikely that the system intentionally approximated *any* interval. Structurally, the ritual calendar is a permutation of two cycles, one of 20 named days and one of 13 numerals. Such a structure is unlikely to arise in a calendar whose essential rationale was its overall length; subdivision in such instances is usually into sequential units. Rather, it parallels the structure of the calendar round: there, two separate, coexisting cycles together formed a 52-year cycle; they came to be cited together since their permutation was useful for fixing dates in historical time, but no one doubts that the constituent cycles were independent. Most likely, the 260-day period was also the effect of combining two pre-existing ritual cycles, one of 20 named days and one of 13 numerals.

The separate origins of the constituent cycles are lost. The rationale for the 13 numerals designating days in the ritual calendar is uncertain (13 was an important mythic and ritual number, but these may be results rather than sources of its position in the structure of the ritual calendar). The cycle of 20 named days presumably reflects the base-20 system of numeration that was universal in Mesoamerica; it recurs in the subdivisions of the vague and civil years. This 20-day cycle can be reconstructed linguistically to a time before ritual calendar records are attested archaeologically (see below).

The period of 360 days was a later approximation to the 365-day period. It was used in a system of positional notation for recording arbitrary spans. One of its chief applications was to fix the cyclic dates of the ritual calendar and the vague year in a time count whose cycle was so long that it was effectively linear during the historical era; it is known among epigraphers as the long count. The long count was an enumeration of days from a mythological base some 3000–4500 years prior to the Maya historical era, expressed as a sum of years and days via multiples of 20; multiples of the higher powers were named before multiples of the lower powers in the long count and in everyday Mayan speech. Thus, a Maya date would be given effectively as, e.g., "9 baktuns, 15 katuns, 1 year, 6 uinals (a 20-day unit), and 3 days" (abbreviated 9.15.1.6.3), this sum being counted from an unstated but fixed historical date assignable to late August or early September of 3114 BC. Arbitrary spans of time counted from other, expressed dates were represented in the same system but, in the inscriptions, with the baktuns, katuns, years, uinals and days presented in reverse order.

The 360-day year clearly developed to accommodate the 365-day period to some other exigency; not only are these spans of almost equal length, but the same words, *haʔbʼ* and *tu·n*, were applied to both. Most likely, the accommodation was to the base-20 system of positional notation, for recording long count dates or other temporal spans: the 360-day unit is found only among those Mesoamerican groups that recorded the months of the vague year; and the place values in the base-20 system of positional notation were in series, not as $1, 20, 20^2, 20^3, 20^4, \ldots, 20^n$ (as in Mayan languages), but rather as $1, 20, 360, 20 \times 360, 20^2 \times 360, \ldots, 20^n \times 360$. However, it could have become the base for positional notation after having been devised for another purpose.

The only calendrical cycles shared with other Mesoamerican systems were the ritual calendar and the vague year, along with their permutation as the calendar round. Both cycles are documented by c. 500 BC in Oaxaca, the ritual calendar (Marcus,

1976) evidently a century or two earlier; and ritual calendar dates may be recorded by c. 900–700 BC on Olmec-style artifacts and murals. Based on their seasonal correlates, the months of the vague year seem to have been named, among the Maya, by c. 500–400 BC, but the earliest documentation for the Maya or their neighbors is from the era 100 BC–AD 200. The named days of the Mayan ritual calendar existed by c. 600 BC, judging from linguistic data (by 1000 BC, if glottochronological dates are relied upon). The lunar calendar and nine-day cycle are documented on a Preclassic Mayan text dated AD 199. The Glyph Y cycle is documented by AD 498, the other cycles not until after AD 600.

The Astronomical Content of the Codices

Tables of numerals in the hieroglyphic manuscripts are invariably for counts of days between recorded positions in the ritual calendar; in some, the recorded time intervals had clear astronomical significance. The astronomical tables are now understood in fair detail, due especially to the trailblazing work of Förstemann (1904, 1906), Meinshausen (1913), Willson (1924) and Teeple (1925, 1930), but advances concerning essential features continue to the present lay. Paxton (unpublished dissertation, 1986, pp. 35–50) provides a balanced assessment of controversial aspects, although neglecting Kelley (1983).

Planetary tables. The Venus table spans five pages of the *Dresden Codex* (Table 22.1), introduced by a one-page preface. 584 days accumulate on each of these five pages, via smaller intervals of 236, 90, 250 and 8 days. The full-page interval of 584 days approximates the 583.92-day average synodic period of Venus. The subintervals approximate the average times of visibility of Venus as morning star, invisibility around superior conjunction, visibility as evening star, and invisibility around inferior conjunction, respectively; but, as Aveni (1980, p. 187) comments, "it is puzzling that the 90-day interval in the table is so different from the true disappearance interval (about 50 days) and that the morning and evening star intervals are represented as being unequal." Another puzzle has been the use of a consistent value, 584, for the length of the synodic period. The synodic period varies annually, but has the same integer approximation every fifth year; the five-year cycle of Venus years repeats in a symmetrical sequence, 587, 583, 580, 583, 587. Thus, each of the pages could have been assigned a Venus year of 580, 583 or 587 days and thereby accurately reflected the variations in the Venus cycle. Likely solutions are noted in the section "Predicting Planetary Motions."

The chief controversy concerning the Venus table surrounds the Maya use of a set of four numerals in the preface to the table. They are generally seen as means of generating a new, viable starting point for the table from stations in the current cycle, on a day 1 Ahau (sacred to Venus) at or near heliacal rising, once calculations from the current 1 Ahau base began to accumulate observational error. Closs (1977) demonstrates that the numerals can be used instead as devices for locating the recorded

Table 22.1 Scheme of the Venus cycle on *Dresden*, pp. 46–50 (restored and corrected). (From Thompson, 1972.)

Line	Page 46				Page 47				Page 48				Page 49				Page 50			
	Cib	Cimi	Cib	Kan	Ahau	Oc	Ahau	Lamat	Kan	Ix	Kan	Eb	Lamat	Etz'nab	Lamat	Cib	Eb	Ik	Eb	Ahau
1	3	2	5	13	2	1	4	12	1	13	3	11	13	12	2	10	12	11	1	9
2	11	10	13	8	10	9	12	7	9	8	11	6	8	7	10	5	7	6	9	4
3	6	5	8	3	5	4	7	2	4	3	6	1	3	2	5	13	2	1	4	12
4	1	13	3	11	13	12	2	10	12	11	1	9	11	10	13	8	10	9	12	7
5	9	8	11	6	8	7	10	5	7	6	9	4	6	5	8	3	5	4	7	2
6	4	3	6	1	3	2	5	13	2	1	4	12	1	13	3	11	13	12	2	10
7	12	11	1	9	11	10	13	8	10	9	12	7	9	8	11	6	8	7	10	5
8	7	6	9	4	6	5	8	3	5	4	7	2	4	3	6	1	3	2	5	13
9	2	1	4	12	1	13	3	11	13	12	2	10	12	11	1	9	11	10	13	8
10	10	9	12	7	9	8	11	6	8	7	10	5	7	6	9	4	6	5	8	3
11	5	4	7	2	4	3	6	1	3	2	5	13	2	1	4	12	1	13	3	11
12	13	12	2	10	12	11	1	9	11	10	13	8	10	9	12	7	9	8	11	6
13	8	7	10	5	7	6	9	4	6	5	8	3	5	4	7	2	4	3	6	1
14	4	14	19	7	3	8	18	6	17	7	12	0	11	1	6	14	10	0	5	13
	Yaxkin	Zac	Zec	Xul	Cumku	Zotz'	Pax	Kayab	Yax	Muan	Ch'en	Yax	Zip	Mol	Uo	Uo	Kankin	Uayeb	Mac	Mac
16	N	W	S	E	N	W	S	E	N	W	S	E	N	W	S	E	N	W	S	E
17	A	B	C	D	E	F	G	H	I	J	K	L	M	N	O	P	Q	R	S	T
18	Red‡	Red‡	Red‡	Red‡	Red	Red	Red	Red	Red	Red	Red	Red	Red	Red	Red	Red	Red	Red	Red	Red
19	Venus	Venus	Venus	Venus	Venus	Venus	Venus	Venus	Venus	Venus	Venus	Venus	Venus	Venus	Venus	Venus	Venus	Venus	Venus	Venus
	236	326	576	584	820	910	1160	1168	1404	1494	1744	1752	1988	2078	2328	2336	2572	2662	2912	2920
20	9	19	4	12	3	13	18	6	2	7	17	5	16	6	11	19	15	0	10	18
21	Zac	Muan	Yax	Yax	Zotz'	Mol	Uo	Zip	Muan	Pop	Mac	Kankin	Yaxkin	Ceh	Xul	Xul	Cumku	Zec	Kayab	Kayab
	T	A	B	C	D	E	F	G	Winged	Winged	Winged	Winged	Winged	Winged	Winged	Winged	Winged	Winged	Winged	Winged
22	Winged	Winged	Winged	Winged					Chuen	Chuen	Chuen	Chuen	Chuen	Chuen	Chuen	Chuen	Chuen	Chuen	Chuen	Chuen
	Chuen	Chuen	Chuen	Chuen					H	I	J	K	L	M	N	O	P	Q	R	S
23	Red	Red	Red	Red	Red	Red	Red	Red	:	:	:	:	Red	Red	Red	Red	Red	Red	Red	Red
	Venus	Venus	Venus	Venus	Venus	Venus	Venus	Venus					Venus	Venus	Venus	Venus	Venus	Venus	Venus	Venus
24	E	N	W	S	E	N	W	S	E	N	W	S	E	N	W	S	E	N	W	S
25	19	4	14	2	13	3	8	16	7	17	2	10	6	16	1	9	0	10	15	3
	Kayab	Zotz'	Pax	Kayab	Yax	Muan	Ch'en	Ch'en	Zip	Yaxkin	Uo	Uo	Kankin	Cumku	Mac	Mac	Yaxkin	Zac	Zec	Xul
26	236	90	250	8	236	90	250	8	236	90	250	8	236	90	250	8	236	90	250	8

bases, in the long count from the ostensible base in the table's preface at 9.9.9.16.0. Regardless of the nature of this mechanism, the location of the implied historical sequence of basedates in the long count is secure from 10.10.11.12.0 to 11.5.2.0.0 (c. AD 1038–1324), Teeple's full sequence, under the usual correlation of Maya with European chronology. Lounsbury (1983b) relates the 9.9.9.16.0 date to a historical base one cycle earlier than Teeple, at 10.5.6.4.0; Thompson (1950) interposed a full series of bases beginning 9.10.15.16.0. A clear presentation of the alternatives, and the basic arguments for each, are given by Paxton. Paxton and Closs prefer Teeple's sequence, while Aveni (1992) and I favor Lounsbury's; Thompson's sequence appears to lack current support.

The synodic periods of Mars, Jupiter, Mercury and Saturn have been identified in other tables in the manuscripts. Since the tables are structured as normal multiplication tables for time spans relevant to the major calendars, intervals recalling planetary periods or their subdivisions could be coincidental; because none shows deviations conforming more to planetary motions than to multiplication tables, no proposed table of planetary "mean motion . . . has ever survived the stage of simple suggestion" (Aveni, 1980, p. 199). Thompson (1972, pp. 22–3, 107–8) provides a brief, scathing critique of these hypotheses.

In spite of Thompson's critique, Willson's (1924) view that the table on pp. 43b–45b of the *Dresden Codex* relates to Mars continues to be entertained (Aveni, 1980, pp. 195–8; Kelley, 1980, p. 30S, 1983, pp. 178–9). Based on Bricker and Bricker's (1986) results, the 3 Lamat base of the table in the ritual calendar can be interpreted as the canonical date for the heliacal rise of Mars, assuming a correlation in the Goodman family; and this parallels the use of canonical heliacal rise as the ritual calendar base of the Venus table. The table is 78 days long (subdivided into three rough quarters of 19 days and one of 21 days), re-entered ten times for a total of 40 stations spanning 780 days (= 3 × 260), just a couple of hours longer than the average synodic period of Mars. A table of multiples of 780 in the preface contains four numbers deviating by 260 from multiples of 780, recalling the corrective numerals in the preface to the Venus table; Bricker and Bricker (1986) and the section "Predicting Planetary Motions" show that they recover Martian heliacal rise at the 3 Lamat base of the table. The 78-day span evidently relates to the roughly 75-day period of retrograde motion. What I consider least securely established about this table is the function of the subdivisions (prior to the Brickers' work, it was whether the table related to Mars at all); I also differ with them on the placement of tabular retrograde motion in the long count, although their argument is in some respects more straightforward than my own.

Lunar tables. The Dresden eclipse table is a sequence of 70 stations spanning 46 sacred rounds; the stations, occurring at intervals of six and, more rarely, five lunar months (177–8 or 148 days), are those on which solar eclipses might be visible. Details of the use of the tables are controversial, but their essential identity was realized on the basis of the six- and occasional five-month groups of lunar months. The positions of the nodes can be determined, to within a few days, from two lines of evidence. (1) A five-month period between eclipse dates joins an eclipse near the postnodal extreme to one near the prenodal extreme; the nodes fall about 14 days later. (2) The range of variation of the recorded dates from the 173.31 days between nodes can be calculated, from an arbitrary base; the position of the node will lie near the midpoint. The results of the two types of calculation are in agreement, placing a node at, or within a couple of days, of the first and last stations of the table. Given these placements of the nodes, the number and internal arrangement of lunar groups is within a day of a true eclipse-possible date of new moon. Pictures are placed after the five-month groups, evidently referring to lunar eclipse possibilities or to actual lunar eclipses. Means of updating the table are provided, as in the Venus tables.

Calendrical Commensuration of Synodic Cycles

The Venus table, Mars table, eclipse table, and prognostication tables are all entered via positions in the ritual calendar; the registered accumulation of time reaches specified stations in that calendar; and the full length of the tables is always an integral multiple of 260 days. The 405 lunations of the eclipse table amount on the average to 11,959.89 days, less than three hours short of the 11,960 days spanned by 46 ritual calendar periods. In the Venus table, five canonical Venus years (one pass through five pages) commensurate the vague year: $5 \times 584 = 2920 = 8 \times 365$; this commensuration was also exact or only a day short for the true Venus year, any five adding to 2919.6 ($= 5 \times 583.92$) days. 65 canonical Venus years commensurate the ritual calendar in two calendar rounds, and the Venus tables do cover this 104-year period; they move ahead of 65 true Venus years by only 5.09 days. The true Venus year commensurates 137 sacred rounds to within 0.78 days, four Venus years before the end of the table; provision is made in the preface to the Venus tables for a shift of tabular base at the point of this commensuration. The accumulation of 0.78 days' positive error every 61 Venus years is balanced by provision for commensuration of 128 sacred rounds to 57 Venus years every fifth or sixth basedate shift, yielding an error of ±0.4 days; the next (unrecorded) base would have been the occasion for such an adjustment.

In less than 16 years, the 260-day calendar commensurates the synodic period of the moon and of all the visible planets to within 4.31 days (see Tables 22.6 and 22.7 and the section "Predicting Planetary Motions"). Three sacred rounds equal one Mars year, four are three days less than nine Mercury years, five are 16 hours less than 44 lunar months, 16 are a day more than 11 Saturn years, and 23 sacred rounds are three days less than 15 Jupiter years. Allowing spans as long as the Venus table, the synodic periods of Mercury, Venus, Mars and the moon commensurate with the sacred round to less than a day; the other calendars do not improve on the 260-day cycle in regard to Saturn and Jupiter.

Such commensurations were in use not only in the Postclassic manuscripts but also in the inscriptions of the Classic period. In both, the Maya joined simple numerological calendrics with their use of astronomical periodicities. Lounsbury (1976) established a pattern of backward projection of current historical events to a prior event of the same sort—a birth projected back to a birth, an accession to an accession—occurring on a date that fell in the same point in various calendrical cycles. Often, the projections are quite large, a common variant being a projection back to a sacred era, just before the base of the long count; this base was apparently itself a projection from the katun 7.6.0.0.0 backward by two cycles of $365 \times 1440 = 73 \times 7200$ days each, commensurating the vague year and the katun (Justeson et al., 1985, n.32). Other examples, recognized as such since Lounsbury identified the basic pattern, occupy small segments of real historical time, and refer back to genuine historical precursors on dates occupying similar positions in calendrical cycles.

This numerological use of planetary and lunar periodicities is frequent relative to other recognized astronomical references. Long-term cycles commensurating basic calendrical intervals with true astronomical synodic intervals, especially eclipse cycles, are amply attested in the Classic period. In the case of planetary cycles, Classic period usage is often essentially calendrical, involving the span of a canonical planetary year rather than approximate average planetary cycles: for long calculations, the whole-number values for synodic periods were presumably known to be inaccurate in fixing the date of a phenomenon, since the discrepancies become appreciable even during the career of an individual skywatcher; however, the synodic cycles of Mercury, Jupiter and Saturn were difficult to commensurate effectively within the calendrical framework (see the section "Predicting Planetary Motions"). One text, Caracol Stela 3, records several intervals that do not closely approximate an average number of planetary synodic cycles or commensurate planetary with calendrical cycles, but do fall within the range of variation of synodic intervals of two planets at once (Kelley and Kerr, 1973, pp. 197–201; Kelley, 1975, 1977a, 1983, pp. 184–93); this suggests observational rather than canonical or predicted linkages with the cycles of the planets involved. There are many dates of planetary and lunar phenomena, explicitly marked as such in accompanying texts, that reflect the observations on which longer-term generalizations were based.

Maya Model Building

The basic approach to temporal predictive astronomy among the Maya was in long-term, cyclical commensuration of synodic cycles with basic calendrical cycles; such commensurations were codified by Postclassic times, some having Classic traces, with substantial observational records dating at least to the Classic period. By systematically applying this approach ourselves, its potential for developing useful predictive models can be determined; in some respects, the kind of astronomical knowledge codified in the manuscripts can be clarified. The section "Lunar Models" treats these topics for lunar records, the most common celestial records explicitly designated as such in the texts; the section "Predicting Planetary Motions" treats planetary records. In both cases, the predictive knowledge exemplified by the Postclassic tables can be seen as emerging via attempts to commensurate the ritual calendar with other cycles.

John S. Justeson

LUNAR MODELS

Commensuration with the Ritual Calendar

Several multiples of 260 days commensurate the 29.530588 average synodic period of the moon (Table 22.2). All commensurations accurate to within ±2½ days constitute either an eclipse interval or an interval halfway between successive eclipse intervals; and the category alternates with successive commensurations. Each off-phase commensuration is three months longer than one eclipse interval, and three months longer than another (when doubled, of course, each is a full eclipse cycle; half are to within ±2½ days, the rest deviating by ±2½–5 days). Thus, when the Maya attempted to rationalize the lunar month with the ritual calendar, that calendar highlighted eclipse recurrence intervals; the recovery of good eclipse cycles was *imposed* on the Maya astrologer, even if he did not immediately notice the eclipse relationship. The eclipse cycle of 405 lunar months in the *Dresden Codex* emerges as the shortest to rationalize the lunar synodic month with the sacred round to within less than a day; to improve the fit, a 4772-month cycle is required—almost 12 times longer.

These lunar/sacred round cycles could have been known quite early, especially if not only eclipses but also the monthly disappearance of the moon was an occasion for ritual prognostication. As soon as one of these cycles was applied to a lunar eclipse series—consecutive visible eclipses occurring every six months for two or three years—the dates reached would be a series in which lunar eclipses would again be visible, though with gaps in as many as half the positions. Thus, applying the cycles to lunar eclipse series provides immediate success in long-term eclipse prediction. Conversely, Mesoamerican astronomer-priests must soon have noticed that relatively small multiples of 260 days often separated pairs of visible eclipses, and that it is always the same multiples that do so.

Since eclipses were times of ritual danger, Mesoamerican astrologers would surely have used these ritual-eclipse cycles to project from a visible eclipse to its next few recapitulations in ritual time; after adjusting by up to 2½ days to capture the correct date of full moon, these would reliably warn of certain dates as eclipse possibilities. Table 22.3a illustrates the results of such projections, applied to eclipses visible in the Yucatan peninsula in the fifth century AD; the figures in the table represent the distances between projected stations. In addition, it is assumed that the ritual-eclipse cycles were unknown until detected in that century. Accordingly, the first few eclipses cannot be anticipated; 14 viable eclipse stations occur before the shortest of the ritual-eclipse cycles has elapsed, 38 before the next shortest. In fact, eclipses were seen 10, 26, 36 and 46 sacred rounds after the first visible eclipse in fifth-century Yucatan. Thereafter, 37 out of 46 visible eclipses are anticipated by projecting ritual-eclipse cycles from prior visible eclipses, a success rate that actually begins somewhat earlier, after the first 25 years of eclipse observation. Together with the warnings for dates on which eclipses did not occur, 113 of 148 eclipse stations are projected after this point; and the structure is clear enough that the 35 missing stations could be filled in, apart from some uncertainty in the placement of the 148-day spans. Two or three

Table 22.2 Commensurations of ritual calendar with lunar months to within ±2½ days during a span of four calendar rounds.

Eclipse intervals ± three months		Eclipse intervals	
44L =	5·260 − 0.65	88L =	10·260 − 1.31
○ 132L =	15·260 − 1.96	229L =	26·260 + 2.50
273L =	31·260 + 1.85	317L =	36·260 + 1.20
361L =	41·260 + 0.54	● 405L =	46·260 − 0.11
449L =	51·260 − 0.77	493L =	56·260 − 1.42
537L =	61·260 − 2.07	634L =	72·260 + 2.39
678L =	77·260 + 1.74	722L =	82·260 + 1.08
766L =	87·260 + 0.43	810L =	92·260 − 0.22†
854L =	97·260 − 0.88	898L =	102·260 − 1.53
942L =	107·260 − 2.19	1039L =	118·260 + 2.28
1083L =	123·260 + 1.63	○ 1127L =	128·260 + 0.97
1171L =	133·260 + 0.32	1215L =	138·260 − 0.34†
1259L =	143·260 − 0.99	1303L =	148·260 − 1.64
1347L =	153·260 − 2.30	1444L =	164·260 + 2.17
1488L =	169·260 + 1.52	1532L =	174·260 + 0.86
1576L =	179·260 + 0.21	1620L =	184·260 − 0.45†
1664L =	189·260 − 1.10	1708L =	194·260 − 1.76
1752L =	199·260 − 2.41	1849L =	210·260 + 2.06
1893L =	215·260 + 1.40	● 1937L =	220·260 + 0.75
1981L =	225·260 + 0.09	2025L =	230·260 − 0.56†
2069L =	235·260 − 1.21	2113L =	240·260 − 1.87
(2157L =	245·260 − 2.52)	2254L =	256·260 + 1.95
2298L =	261·260 + 1.29	2342L =	266·260 + 0.64
2386L =	271·260 − 0.02	2430L =	276·260 − 0.67†
2474L =	281·260 − 1.33	2518L =	286·260 − 1.98

Filled circles mark commensurations used in the Classic period, the 405-month *Dresden*/Palenque cycle and the 1937-month Copan cycle; unfilled circles mark other commensurations that are likely to have been in use; and daggers follow multiples of the *Dresden*/Palenque cycle. Italics mark commensurations with the 360-day year. None of the cycles commensurate the 365-day year (and thus the calendar round), but a close approach occurs at four calendar rounds: 2571L = 4CR + 3.14d, an eclipse interval. The clean alternation between commensurations falling into an eclipse cycle, and those three months out of phase with an eclipse cycle, was presumably known to the Maya.

decades of eclipse observation and recording are necessary and sufficient to produce a model for the timing of eclipses so complete that a system for anticipating all eclipse-possible dates would be revealed—a model essentially identical in structure to that of the *Dresden Codex*. Since the lunar commensurations not forming eclipse cycles are spans three months larger and smaller than eclipse cycles, these intervals might also have been applied to the same problem; the smallest periods generated thereby, of ±3 months, are the 41- and 47-month divisions of the saros eclipse cycle (47 + 41 + 47 + 41 + 47 months). Use of the eclipse cycle ±3 month groupings would generate an eclipse warning table such as Table 22.3a more quickly and completely.

A similar sequence of eclipse warnings can be projected into any arbitary period by using the smallest few ritual-eclipse cycles, if records have been kept for at least 33 years (46 sacred rounds) before that period. I constructed such models for each katun from 8.13.0.0.0 through 10.0.0.0.0. Any given katun contains 41 or 42 eclipse sta-

Table 22.3a An eclipse warning table based on eclipses visible during the fifth century AD in the Yucatan peninsula.

AD 400							
[1920]	354¹	[326]	177²⁴	177³	[176]	177²	
[502]	178¹	177¹	502¹	177²³⁴	178¹³	177²	
177¹	[679]	177¹	177²³	502¹	177¹²⁴	325¹⁴	
[354]	[176]	[177]	178²³	177²⁴	177¹²	177¹	
[178]	178²	178¹	177¹²³	177³⁴	177²³	177¹²	
[1211]	177¹	177²	177²³⁴	177¹²³⁴	325⁴	177¹⁴	
176¹	177¹	177²³	177³⁴	177¹⁴	177¹³	177²⁴	
502¹	679¹	325¹	325¹	178¹²	177²⁴	177²³⁴	
177¹	[178]	177¹³	177²	502⁴	177¹²	177²	
	[354]	177¹²³⁴	177¹²³⁴	177¹³⁴	178¹⁴	148¹	
	354²	177¹	177³	177¹	177¹	177¹³	
	325¹	177³	178¹²	177²⁴	502¹³	177²⁴	
	177¹²	178³	177²³	178³⁴	177¹	177¹²⁴	
	177¹²³	325¹	[502]	[176]	178¹⁴	177¹²³	
	176¹	177¹	177³⁴	177²	176¹³	177²³	
	177¹²	177²	177¹²³⁴	325³	177¹³	177 ²³	
	177²	177¹²	178²³⁴	178²⁴	178²³⁴	[325]	
		177¹²	177³⁴	177³⁴	[325]	177¹	
		177²³	177¹⁴	[177]	177¹²⁴	177¹²³	
		177²³⁴	502¹³	177¹³	177³	[177]	
		[325]	177²³⁴	[177]	177³	177³	
		177¹⁴	178⁴	325¹	178²	177³⁴	
		177¹²³⁴		177²⁴		177³	
		177²				325⁴	
		177⁴				177²³	
						177²³	
						AD 500	

Stations are situated at the dates of visible lunar eclipses, and at full moons occurring within ± 2½ days of 10, 26, 36 or 46 sacred rounds after a visible lunar eclipse; figures are the distance from the last station to the current station. Brackets indicate that a station was the occasion of a visible lunar eclipse that was not projected; boldface indicates visible eclipses that were projected. Stations projected from eclipses, but on which no eclipse was visible, are in normal type. Superscripts indicate which ritual-eclipse cycles join the station to a prior visible eclipse, 1, 2, 3 and 4 corresponding, respectively, to 10, 26, 36 and 46 sacred rounds. Spans between eclipses are from Aveni (1980, Table 18), with columnar format keyed to his Table 19; selection among 176-, 177- and 178-day spans is also based on Aveni's Table 19, to facilitate comparison; they have not been checked against computed dates of full moons.

tions; for comparability, projections were into a span, not of a katun, but of 42 eclipse stations. Each projection was forward from a visible eclipse, by the four smallest ritual-eclipse intervals (10, 26, 36 and 46 sacred rounds); projection therefore begins with eclipses visible 46 sacred rounds prior to the first day of the katun. About 80 per cent of the eclipse stations were projected by the ritual-eclipse cycles, most of them by more than one. By including also the off-phase commensurations (5, 15, 31 and 41 sacred rounds) ±3 lunar months, about 90 per cent of the stations are projected, most both by off-phase projections and by ritual-eclipse cycles.

Since the pattern of six-month groupings, with rarer five-month groups, emerges clearly in the last two-thirds of Table 22.3a, it could readily be reconstructed for the few cases of 11-, 12-, 17- and 18-month groupings, apart from some leeway in the

Table 22.3b An eclipse warning table projected from lunar eclipses visible in fifth-century Yucatan.

AD 400							
	177			177	177	*176*	177
354	177	326†			177	178	177
177	178	177			*177*	177	
		177	502			177	325
		177†	177	325		177	177
		178	178	177		148	177
	679†	177	177	177	177	177	177
	176†	177	177	177	177	177	177
	178	325	177	177	177	177	177
1211	177	177			178	177	177
177†	177	177	325		*177*	178	148
	502	177	177			177	177
		177	177	325			177
502†			178	177	177		177
177	177			178	177	502	177
	178†	325		177	177	177	177
354†	*177*	177			178	178	177
[178]	177†	177			176†	176	
		177	502†	177		177	[325]
	354	177	177		148	178	177
		177	178		177		177
679	325	177			178	[325]	[177]
177	177	177			177†	177	177
177	177	325†			177	177	177
177†	176	177			177	177	177
176	177	177			177†	178	
	177	177					325
		177			325		177
502					177		177
177							AD 500

The table is a fuller version of Table 22.3a: all spans reach forward to eclipse stations; boldfaced spans reach visible eclipses projected, forward and/or backward, from visible eclipses, with daggers marking those reached only by backward projections; and italicized spans reach stations projected backward from visible lunar eclipses. Gaps are introduced to highlight positions at which eclipse stations are not projected.

placement of the five-month groups within the 11- and 17-month subdivisions. In fact, the century charted in the table is more than twice the length necessary to work out the pattern; forward projections into years 25–40 (column 3 and the lower part of column 2) would be quite adequate. Thus, a good model for predicting all eclipse-possible dates, applicable to both solar and lunar eclipse prediction, could have been generated solely via commensurations of the lunar synodic month with the ritual calendar given records of less than half a century; such tables could well have been known by the beginning of the Early Classic in southeastern Mesoamerica.

Even without extrapolating a complete eclipse warning table, successfully projecting so high a proportion of observed eclipses would have to have structured the astrologers' conceptions of the eclipses they had *not* anticipated. Referring again to our fifth-century Yucatan model, most of the eclipses that were not anticipated could be postdicted from subsequent eclipses, by *backward* projection of ritual-eclipse cycles (Table 22.3b); they were "normal" eclipses, in that visible eclipses follow them at ritually appropriate intervals. In fact, 72 of the 75 lunar eclipses visible in Yucatan during the fifth century are separated from another visible in the same century by an eclipse interval of 10, 26, 36 or 46 sacred rounds, none longer than the Dresden's ritual-eclipse cycle. Given that they were considered normal eclipses, how might they be anticipated? Backward projection would immediately yield the answer: almost every one could be projected forward by a ritual-eclipse cycle from stations that they *had* projected, but on which no eclipse had been seen. In this way, they were indistinguishable from the eclipses projected from visible eclipses, and were captured by the same ritual constructs.

Such a synthesis of projected and unanticipated eclipses must have affected the astrologers' conceptions of the projected stations on which *no* eclipse was visible; since eclipses could be reliably projected forward from them, they were probably considered occasions of some sort of eclipse phenomena—not simply as mistaken predictions. Furthermore, by projecting forward from projected eclipse stations, *every* eclipse station occurring after the establishment of the two smallest ritual-eclipse cycles is captured in the fifth century, developmental model and in the mature, katun models (even without the off-phase projections).

The Lunar Calendar in the Inscriptions

Accompanying many long count dates are records in a Maya lunar calendar. Teeple (1930) showed that lunar months were enumerated, from first to sixth moon numbers. The days within a lunar month were enumerated, roughly from new moon; moon ages from 3 to 29 are attested. Finally, the total number of days in the lunar month was specified as 29 or 30. Other information, of an uncertain character but correlating with the moon number, was given in Glyph X. Occasionally, in place of the moon age a compound appears that was formerly treated as corresponding to a moon number of 0; along with an absence of any moon age record, it corresponds to the period of the moon's invisibility (F. G. Lounsbury, personal communication, 1978; L. D. Schele, personal communication, 1985).

Different records of the same date sometimes differ in moon number and/or moon age. Within a site, only one instance of a given date is normally contemporaneous; because the others are several years before the date of the monument on which they appear, discrepancies here may be due to back-calculation from contemporaneous records via a formal system. Between sites, the duplicate dates are always civil year endings, mostly contemporaneous; these discrepancies reflect different models or different parameters of shared models (for moon ages, also different observational results). The systems can only be worked out in detail on a site-by-site basis.

There is scant opportunity to determine arithmetically whether moon ages specify days elapsed within a lunar month, or the current day (see Lounsbury, 1978, p. 776).

Readings of hieroglyphs, however, suggest that they were elapsed; I read the usual moon age compound as "*n* (days) have passed." The verbal suffix habitually associated with the count of days seems surely to represent either *-i(h)* or *-iš* (Fox and Justeson, 1984, pp. 58–62), either one of which indicates action already completed; the usual verb seems to be "to pass" (Justeson, Norman and Hammond, 1988; cf. Schele, 1982, pp. 83–5; Stuart, 1984).

The starting point of this count is almost always within a day or two of new moon; disappearance, new moon, and reappearance all remain under consideration. Deviations of three to four days in the placement of moon ages relative to each other suggested to Lounsbury (1978, p. 774) that moon ages were calculated from different bases at different sites, and at different times within the same site. The count was from reappearance in texts explicitly stating that the moon age is the number of days since the moon was born (Lounsbury, 1978, p. 774); "the birth of the moon" (e.g., colonial Yucatec *yi'x ù·h*) is a Mayan metaphor for the first appearance of the crescent after new moon. However, this starting point was probably not usual when it was not explicitly stated: moon ages of one and two days are apparently never recorded, evidently falling in the period of invisibility; new moon or the moon's disappearance must frequently have been the starting point. Lounsbury (1978) takes estimated new moon (conjunction) as the alternative to reappearance, while A. Schlak (unpublished manuscript, 1985) argues for disappearance; the differences involve a two-day discrepancy in the correlation constant each prefers. New moon ties better to the strongest candidate for a solar eclipse date; and, in the early system of moon numbering reconstructed below for Tikal, new moon was probably the first day of the lunar month.

That moon numbers range from 1 to 6 indicates that months in the lunar calendar were enumerated in groups of six or less; this, in turn, suggests that that calendar was at least loosely based on eclipse seasons, though not necessarily upon a formal system of eclipse cycles. The earliest evidence for calculations involving the cycle of the Dresden eclipse table is from AD 683 (Lounsbury, 1978, pp. 775, 811). Most of the Maya area was then using the Uniform System of numbering moons in groups of six, with no five-month groupings as in the eclipse cycle (Teeple, 1930); it was a purely formal system. Teeple hypothesized that this system replaced one keyed more faithfully to an eclipse cycle, although this has not been demonstrated in the intervening years.

Reviewing all lunar records prior to the ostensible introduction of the Uniform System, I find support for Teeple's hypothesis in the earliest records at Tikal. Records there, through 9.4.13.0.0, are consistent with an enumeration of elapsed lunar months, with the last month in an eclipse cycle beginning at new moon on the date of a solar eclipse possibility. Tentatively, the early Copan sequence appears to have been in an eclipse cycle, offset by three months so that the date of a solar eclipse would be assigned to the third month; a similar system may have been present after the end of the Uniform period at nearby Quirigua. If this is so, it may relate to the lunar month commensurations with the lunar calendar in a span three months offset from an eclipse cycle.

Further support for Teeple's hypothesis is provided by the deviant moon number in the earliest known lunar calendar record (J. S. Justeson, P. Mathews and F. G. Lounsbury. The chronological portion of the Seattle stela and the early history of the Maya eclipse calendar, unpublished manuscript, 1982). A miniature Preclassic stela in the Seattle Art Museum records a moon number of 17, with no compound for the moon age. The date, 8.8.0.7.0 (AD 199), occurs after 47 + 17 lunar months had just been completed since the last solar eclipse visible in the area. Apparently, the subdivisions of the saros cycle were recognized at the time (hence the cycle was begun after 47 months were completed), though not necessarily the cycle itself; the 47-month group could be a 44 + 3 month period, involving the shortest lunar/sacred round commensuration, with three months added to relate it to the eclipse cycle. This record suggests that the subdivisions of the saros cycle had been recognized, although the further subdivisions had either not been recognized or not incorporated into the lunar calendar, by AD 199 among the Lowland Maya.

Since no moon age is recorded on the Seattle stela, the reference is ostensibly to invisibility; given Thompson's correlation constant, 584283, one day had been completed since new moon, so the moon would not have been visible. If the month count was of elapsed months, then the first month of an eclipse cycle included a solar eclipse station at new moon; otherwise, the last month included one. If the moon age was counted from reappearance and the 584283 correlation is correct, the eclipse station fell near the end of the month, and the moon numbering system was essentially equivalent to Tikal's. If moon age was counted from disappearance or new moon, or if the 584285 correlation is correct, the eclipse station fell near the beginning of the month, and the Seattle system of moon numbering was distinct from Tikal's. Unfortunately, because the monument was recovered by looters its source is unknown.

The Uniform System replaced the eclipse cycle system at Tikal during the Early Classic. Satterthwaite (1958a, pp. 132–3) and Jones and Satterthwaite (1982, p. 57) postulated its adoption between 9.4.0.0.0 and 9.4.13.0.0, since the latter and all subsequent Tikal dates are in the Uniform System, while the former date is not. Subsequently, John Graham (1972, pp. 106–14) established a different Uniform System as having been used at Altar de Sacrificios (it was also at neighboring El Pabellón) until the Uniform Period. This was a system of regular six-month groupings, as in the classical Uniform System, but the moon numbers were smaller by one. Both Satterthwaite's and Graham's observations are overlooked in the recent literature on the subject, which still maintains the late introduction of Uniform moon numbering. The only other sites with many pre-Uniform moon numbers are Pusilha and Piedras Negras; Pusilha also used Uniform moon numbering from the start. At all other sites near Tikal, which have only one or two pertinent texts, the Uniform System is found back to 8.16.0.0.0, and the Altar de Sacrificios variant seems to be found elsewhere along the Usumacinta (and at Uaxactun, which Mathews [1986] suggests had ties with Yaxchilan). The Uniform System was evidently in use throughout the Maya area from the beginning of the Early Classic, a conclusion reached independently by Schele (personal communication, 1985) in her examination of the Early Classic lunar records.

A few sites, with only one or two legible pre-Uniform lunar dates, do not agree with Uniform coefficients; they usually deviate by only one. These discrepancies could reflect different uniform bases, or eclipse cycles, but with only one or two dates at each site the discrepancies cannot be profitably investigated. Piedras Negras is the exception: it fits neither a uniform system nor an eclipse cycle, yet it has seven pre-Uniform moon ages with known long count positions. Although there is still not enough evidence to work out the details, the Piedras Negras system seems to advance about one month per five civil years faster than the Uniform System, as though one five-month period was introduced during this period. There are enough dates at five-year intervals to show that the same moon number occurs after 59 lunar months (60 in the Uniform System); 59-month groups would have to consist of nine six-month periods and one five-month period. 59 months form an eclipse recurrence interval, one that occurs in each 135-month division of the Dresden's 405-month eclipse table, but discrepancies in assignments of moon numbers to true eclipse stations begin to appear late in the second group of 59. The system would normally provide the same moon number to all eclipse stations during a given hotun, and thus may have served as a short-term eclipse calendar; perhaps the Uniform System itself began as an accommodation (12 months long) of an eclipse cycle to the civil year.

Implications of the early data are that eclipse cycles were being tracked, with accurate predictive models, during the Preclassic and in the first half of the Early Classic—at least in the interval from AD 199 to AD 514. From about AD 350 onwards, however, most sites adopted simpler systems of moon numbering. Most common was the Uniform System, in the Tikal orb, with a minority variant along the Usumacinta having moon numbers smaller by one; the majority variant may have enumerated current months, the minority elapsed months. After c. AD 750, some sites are thought to have attempted to adapt the lunar calendar to the eclipse cycle once again. This hypothesis rests mainly on deviations from uniformity; moon numbers have not been shown to conform to stations in an eclipse cycle.

Although moon numbering does not generally fit into an eclipse-cycle system, eclipse cycles commensurating the sacred round were used in long back-calculations to compute moon ages associated with mythological dates. At Palenque, the ratio of 405 lunar months to 46 sacred rounds was the basic cycle, implemented by a ratio of 43 30-day months to 38 29-day months (Lounsbury, 1978, p. 775) that produces a theoretical synodic month of 29.530864 days. The cycle at Copan was different, a ratio of 79 30-day to 70 29-day months corresponding to a theoretical month of 29.530201 days. Although, according to Lounsbury, "It has not been determined how the astronomers of Copan arrived at this ratio or on what theory it was based," the Copan system is simply a different implementation of the same theoretical apparatus as was in use at Palenque: a long eclipse cycle was commensurated with the ritual calendar. The difference was in the commensuration chosen. The Copan cycle was one of 1937 lunar months, nearly 157 years; this span falls on the average just 18 hours short of 220 sacred rounds, and gives precisely the 79:70 ratio of 30-day to 29-day months.

John S. Justeson

Lounsbury (1978) notes that seven ratios involving small numbers of 30- and 29-day months give better results than the Copan system, and that five give better results than the Palenque system. Similarly, the Dresden and Copan ritual-eclipse cycles were not the shortest available. It appears that the lunar cycles used by the Maya were those for which the number of months in the ritual eclipse cycle could be subdivided evenly by one of the short and accurate month ratio systems. Table 22.4a displays the most accurate commensurations from Table 22.2, in order of accuracy, along with the 30- to 29-day month ratios they impose. Only one of the eclipse cycles (a doubled half-cycle) commensurates with the ritual calendar with greater accuracy than the Palenque and Copan systems. The difference was not significant; it would take over 8000 years for its average performance to better Palenque's system by a day, and over 200 to better Copan's. Further, the most accurate short-term behavior of that system requires a 30- to 29-day month cycle about ten times longer than the Palenque and Copan systems. So Copan's and Palenque's systems were not measurably inferior to any other commensuration of eclipse cycles with the ritual calendar; and, among these commensurations, they admitted the simplest systems of month-length alternations. Short and accurate, the Palenque/Dresden system was the best practical system available to the Maya for long-term use.

Table 22.4b shows that the commensurations of the ritual calendar with lunar positions three months (half a cycle) out of phase with an eclipse cycle almost all involve these long 30- to 29-day cycles; the exception, a 32-month system, is detectably less accurate only after about a century. Finally, Table 22.4c displays the many month-length alternation systems less than 200 lunar months long, together with the shortest eclipse cycle they fit into evenly, in order of the accuracy of the synodic period estimates they induce. The only ones that have been recognized in the texts are the Palenque and Copan systems, suggesting that commensuration with the ritual calendar was of prime importance in selecting among the possible alternation systems. Note, however, that such systems are represented so poorly that this restriction does not strongly support a preference for ritual-eclipse cycles over off-phase commensurations. Indeed, the 132-month off-phase commensuration fits the month alternation scheme of the Dresden eclipse table (as the 81-month Palenque system does not), perhaps because the related 132 + 3 month Metonic cycle is a major subdivision of the table (and the only eclipse cycle to subdivide the 405-month ritual-eclipse cycle).

The Copan and Palenque systems represent equally viable solutions to the formalization of eclipse seasons via the same essential model: both were extremely accurate in their arrangement of lunar months over the short term while admitting full accuracy over the testable long term. The difference in the accuracy of the Copan and Palenque commensurations of the ritual calendar with the eclipse cycle amounted to only one day in over 800 years (i.e., it was not noticeable). The Palenque eclipse cycle of 46 sacred rounds has the smallest absolute error, in days, of any single cycle short of 542 sacred rounds; the Copan cycle has the smallest error of any other single cycle of its length or less. Copan's system was inferior to Palenque's: its commensuration took five times as long, and its 30- to 29-day month alternation schema was almost twice as long.

The most serious alternative to the Palenque and Copan systems would be a ritual-eclipse cycle of 128 sacred rounds, about a day less than 1127 lunar months. It is one of the more accurate of these cycles, the only one other than Copan's and Palenque's with a short enough month alternation system to be viable. Although a less accurate ritual-eclipse commensuration than Copan's, it takes almost 200 years for it to accumulate an extra day of error. Its advantages are that it is half the length of Copan's, and the 1127-month cycle subdivides into 23 segments of 49 months; the 49-month month alternation scheme is at once the most accurate and the shortest available. The 1127-month or 128-round cycles are important candidates to watch for as investigations into Classic period commensurations continue. No good evi-

Table 22.4 Ratios of 30-day to 29-day lunar months and corresponding ritual calendar commensurations.

Month ratio				Synodic estimates		Error
30:29		months:rounds		30:29	days in cycle months	SR − true
(a) Eclipse cycles						
621	550	2342	266	29.530316	29.530516	−.00027†
43	38	405	46	29.530864	29.530864	+.00028
79	70	937	220	29.530201	29.530201	−.00039
203	180	1532	174	29.530026	29.530026	−.00056†
668	591	2518	286	29.530580	29.530374	+.00079†
26	23	1127	128	29.530612	29.529725	−.00086
1121	992	2113	240	29.530525	29.531472	+.00088
(b) Eclipse cycles ± 3 months						
633	560	2386	271	29.530595	29.530595	+.00001
150	133	1981	225	29.530540	29.530540	−.00005
209	185	1576	179	29.530457	29.530457	−.00013
656	1581	2474	281	29.530315	29.531124	+.00054
1098	971	2069	235	29.530691	29.531174	+.00059
17	15	1664	189	29.531250	29.531250	+.00066
335	296	1893	215	29.530903	29.529847	−.00074
(c) 30-day:29-day month-length alternation systems						
						30:29 − true
● 26	23	1127	128	29.530612	29.529725	+.00002
87	77	9676	1099	29.530488	29.530798	−.00010
○ 95	84	537	61	29.530726	29.534451	+.00014
61	54	7475	849	29.530435	29.530435	−.00015
69	61	12740	1447	29.530769	29.530612	+.00018
96	85	9412	1069	29.530387	29.530387	−.00020
● 43	38	405	46	29.530864	29.530864	+.00028
○ 35	31	132	15	29.530303	29.545455	−.00029

continued on next page

Table 22.4—*continued*

Month ratio 30:29		months:rounds		Synodic estimates 30:29	days in cycle months	Error SR − true
(c) 30-day:29-day month-length alternation systems—*continued*						
103	91	5626	639	29.530928	29.530750	+.00034
○ 60	53	678	77	29.530973	29.528024	+.00039
● 79	70	1937	220	29.530201	29.530201	−.00039
77	68	4640	527	29.531034	29.530172	+.00045
44	39	7387	839	29.530120	29.530256	−.00047
94	83	5133	583	29.531073	29.530489	−.00049
97	86	17019	1933	29.530055	29.530525	−.00053
53	47	8100	920	29.530000	29.530864	−.00059
○ 17	15	1664	189	29.531250	29.531250	+.00066
62	55	5265	598	29.529915	29.530864	−.00067
71	63	5494	624	29.529851	29.530397	−.00074
80	71	3020	343	29.529801	29.529801	−.00079
89	79	6216	706	29.529762	29.530245	−.00083
93	82	15575	1769	29.531429	29.530658	+.00084
98	87	15170	1723	29.529730	29.530653	−.00086
76	67	8866	1007	29.531469	29.530792	+.00088
59	52	3108	353	29.531532	29.530245	+.00094
101	89	9500	1079	29.531579	29.530526	+.00099

Synodic month estimates differ depending on their derivation from ratios of the two month lengths vs. ritual calendar commensurations. (a) and (b) list all ritual calendar commensurations yielding synodic month estimates accurate to within ±0.001 day in a span of four calendar rounds or less; in (a), daggers mark cycles that are twice a commensuration of months with sacred rounds in an eclipse cycle ± 3 months. (c) lists all month length ratios whose synodic month estimates are accurate to within ± 0.001 day and that involve systems of alternation in a cycle of fewer than 200 lunar months. Six systems appear in both (a) or (b) and (c), simultaneously achieving an accurate month alternation system and a short and accurate commensuration of the ritual calendar with the lunar month; filled circles mark month ratios that correspond to ritual-eclipse cycles, while empty circles mark ratios that correspond to ritual half-eclipse cycles. At least two of these systems, both eclipse cycles, were used in the Classic period. Notations are the same as in Table 22.2.

dence for the 49-month cycle has been published, but Lounsbury (personal communication) has discovered one indication of its use. On Palenque's Tablet XIV, Chan Bahlum's apotheosis on ritual day 9 Ahau is related to the birth of the god G-I of the Palenque Triad on the day 9 Ik 3000 years earlier, which is linked to a primordial event on the same day nearly a million years ago. The distance between the latter two dates is 2 × 11 × 41 × 1447 × 260; in terms of Maya numerology, the equivalence of 1447 days with 49 lunar months (= 1446–999 days) is highly suggestive; 2 × 41 × 260 is also a ritual-eclipse cycle, although not a particularly accurate one.

Some sites may not have used commensurations to the ritual calendar to fix their lunar ratios. At Uaxactun, the magnitude of error in the estimate by which a moon age of three days was projected back to 7.5.0.0.0 suggests that they used a ratio of 9:8, 25:22 or 100:89 30- to 29-day months in their cycles, unless their alternation system was much longer than elsewhere; the two shorter periods also form eclipse intervals, of 17 and 47 months, while the longer is an eclipse half-cycle. The 17-month alternation system fits within the 493-month eclipse cycle (= 17 × 29 months), commensurating 56 sacred rounds. However, the approximation is poor by the standards of Copan and Palenque; a difference of at least four days would have accumulated since Uaxactun's lunar records began, while the 405-month Palenque cycle would still have shown no discrepancy. Similar results are found for the 47-month alternation system. A very long alternation system, using a month ratio of 620:551, also agrees with the Uaxactun data and fits into the most accurate eclipse cycle in Table 22.4a, of 2342 months (though a 621:550 ratio is more accurate); but Uaxactun's astrologers are unlikely to have projected moon ages using this uniquely long and recognizably inaccurate system of month projection in place of the very accurate eclipse cycle system it presupposes. Finally, seven eclipse cycles of 405 months each can be subdivided into 15 groups of 189 months, but the 81-month cycle provides detectably better short-term results while fitting within the eclipse cycle (5 × 81 = 405). Thus, Uaxactun was probably not making use of eclipse cycle commensuration with the ritual calendar.

An alternative in the spirit of the lunar month commensurations is that the projection involved commensuration of an eclipse half-cycle, transformed by adding ±3 lunar months into an eclipse cycle—here, of 44 + 3 months. If so, a 47-month cycle is suggestive of the Preclassic system of the Seattle stela. Otherwise, the projection may have been based solely on a short-term, formal arrangement of 30- and 29-day lunar months (more likely in 17 or 47 than in 189 months).

Finally, Classic inscriptions record another lunar cycle whose nature is undetermined. The Glyph X cycle consists of 12 known forms, half of which are *prima facie* variants of each other; they are designated 1, 1a; 2, 2a; 3, 3a; 4, 4a; 5, 5a, 5b; and 6a. They are not enumerated periods, since they lack numeral coefficients; they appear to function as names. Teeple (1930, pp. 61–2) observed that a given moon number normally admitted only two possible forms of Glyph X, and a given form of Glyph X normally occurs with only two moon numbers. The same correlations of the moon numbers and Glyph X forms is found, whether or not five-month groups are interposed among the more frequent six-month groups. Accordingly, the Glyph X cycle is keyed directly to the moon number cycle. Because there are at least five different systems for moon numbering before 9.12.15.0.0—the Tikal eclipse calendar, two separate Uniform Systems, the Piedras Negras 59-month count, and the more obscure Copan system—Glyph X must also be treated on a site-by-site basis. Linden's (1986) attempt to predict the coefficient of Glyph X in a pan-Mayan system based simply upon the long count date while maintaining a strict correlation of Glyph X and Glyph C variants is therefore internally inconsistent. However, his proposal that Glyph X reflects an 18-lunar month cycle might be accommodated by supposing that it was reduced to a 17-month cycle when a five-month group occurred.

PREDICTING PLANETARY MOTIONS

The success in anticipating visible eclipses, and the accuracy with which lunar phases could be anticipated over long spans of time, must have confirmed for the Mesoamerican astrologer the essential linkage of ritual time with sacred time—with the activities of the deified sun and moon. Its application to the motions of the planets was a logical extension. The much broader range of variation, in days, separating recurrences of the same events in the cycles of the planets would make their short-term motions more difficult to capture via projections of long-term commensurations; the only universally recognized planetary table includes no mechanism for dealing with such variation, nor, in my opinion, does the Mars table. Immediate success would again emerge with application of the ritual cycle to the motions of two of the planets, in fact within a single planetary cycle: three sacred rounds closely approximate the synodic period of Mars, and one approximates the 263 days of continuous visibility of Venus, as morning and evening star. While this augured well for the results to be obtained by the traditional commensuration, longer-term planetary cycles could not emerge so readily as did the lunar eclipse patterning; and, for Saturn and Jupiter, the commensuration would never be very successful. This may have been a serious cosmological problem for the Maya. But, applied to Venus, the brightest of the planets, the success of the standard approach would be impressive: the canonical synodic period of 584 days commensurated the vague year in the same span as it did the ritual calendar, and it did so 13 times during that span.

The Venus Table

Table 22.1 is a schematic of the five pages of Venus table stations. Three lines of vague year dates are given, lines 14, 20 and 25. The last entry of each (13 Mac, 18 Kayab, and 3 Xul) is the starting point (basedate) of a Venus cycle; the three recorded bases correspond to successive historical bases near the heliacal rise of Venus, about a century apart. From a given station, one proceeds forward in time to the next station by the amount specified in line 26 below the station reached (e.g., by 90 days from 19 Kayab to 4 Zotz'). The ritual calendar dates of the stations are given in lines 1–13. The first pass through the five pages of the table uses the first line of ritual calendar dates, and each subsequent pass uses the next line below; the final entry of the 13th line is the basedate in the ritual calendar, 1 Ahau. The span of the table is thus 65 Venus years (= two calendar rounds).

Canonical heliacal rising. The structure of the Venus table is such that its base is located at a date of predicted heliacal rising. Its content is such that this base occurs on a day 1 Ahau in the ritual calendar. 1 Ahau was the name of the god Venus; the day was probably sacred to the god, whenever it occurred, and was presumably of special significance when it occurred on the date of any special event in the life of the planet, such as heliacal risings and settings or stationary points and renewed motion. This combined structure and content imposes much of the remaining detail of form and content of the tables.

To have a regular system for anticipating heliacal rises on 1 Ahau, recourse might be had to records of such risings; but in any long series the pattern is quite erratic owing to variations in the time between successive inferior conjunctions, in periods of invisibility after inferior conjunction, and in viewing conditions (although the Maya could obviously have taken account of the latter in developing a model to handle the first two types of variation). However, it would have been sufficient to look instead at data concerning *all* ritual calendar dates and their recurrences *near* heliacal rising to get a good idea of when a new base could be anticipated.

Such recurrences can take place in very short order: four synodic periods average 4.31 days short of nine sacred rounds, and perfectly commensurate it after the 580-day year in the five–Venus year cycle ($587 + 583 + 583 + 587 = 2340 = 9 \times 260$); depending on the position in the five-year cycle and with an unusually' long disappearance interval, it can return very rarely after eight or 53 Venus years. However, it takes records of about a century to recognize when the recurrence of heliacal risings on roughly the same date in the ritual calendar would *regularly* occur, and another few decades to smooth out the variations into a regular system based on integer values. Apart from the near miss after four Venus years, the same point in the Venus year normally first returns on the same SR only after 57 or 61 Venus years, averaging, respectively, 3.53 days later or 0.78 days earlier. Any earlier pattern is off, on the average, by more than five days. So regular recurrences are noted only after a bit over 91 and 97 solar years—just under two calendar rounds (104 vague years).

These spans define the period during which the table *had* to be used from one 1 Ahau basedate before shifting to a new 1 Ahau basedate for heliacal rising—at least 57 Venus years—and the times (57 or 61 Venus years) at which corrections would have to be instituted. This period was convenient in terms of Maya analytic constructs: the least common multiple of the 584-day Venus year and the 260-day ritual calendar defines the length of the table, with no correction; this period was two calendar rounds, just four to eight Venus years longer than the timing of the base correction. So the period of a "complete" table was just long enough to accommodate the period after which the first basedate substitution would have to be made.

The synodic periods of Venus go through a regular, symmetrical pattern of variation in length that repeats every five Venus years: 583, 587, 580, 587, 583. Thus, from a particular basedate, a given page of the Venus table always corresponds to the Venus year variant. The table, however, consistently registers the *average* figure, 584. Although this failure to register regular variation in the synodic period that so perfectly fits the structure of the table has occasioned comment and various types of explanation, the Maya *could not* incorporate these variations and still be able to recycle the table by recapturing new 1 Ahau basedates. When basedate substitution recovers 1 Ahau as a date of heliacal rising, it also produces a phase shift of one or two Venus years within the cycle of five Venus years. Thus, if the five pages corresponded under the first base to Venus years with synodic periods 583, 587, 580, 587, 583, and the basedate is shifted after 57 Venus years, the pages would subsequently correspond to the series 580, 587, 583, 583, 587; a basedate shift after 61 years would give the series 587, 580, 587, 583, 583. So the regular 584-day intervals are simply the consequence of the reinstallation of 1 Ahau basedates for heliacal rising after two calendar rounds had almost run their course.

This regular interval, in turn, constrains the precise values of the adjustment numbers. Since any pair of computed heliacal risings are separated from 1 Ahau by some multiple of 584 ($= 20 \times 29 + 4$), only five days of the 20-day veintena, separated by intervals of four days each, could serve as dates of any of the four stations of the Venus year. Since the basedate shift must preserve this feature when recapturing a heliacal rising on 1 Ahau, the numbers used to shift basedates have to be less than an even multiple of 584 by a small multiple of four days (any multiple of 260 ($=4 \times 65$) will differ from a multiple of 584 by *some* multiple of four). The table of basedate adjustment factors consists of a series of four numbers; they are read from largest to smallest, as usual for multiplication tables. All are multiples of 260 days, as necessary to lead from one day 1 Ahau to another. The three largest represent the numbers of days necessary to move from one date of heliacal rising on 1 Ahau to another, roughly correcting for accumulated errors; all three adjust by a multiple of four days less than a multiple of 584, as required. They could be used to calculate new bases from each other or from the canonical 1 Ahau 18 Kayab base, and thus are related directly to the structure of the table.

The first and smallest in the series of adjustment numbers is 9100 days, written 1.5.5.0. Like the other three, it is a multiple of 260 and thus leads from one day 1 Ahau to another. However, it disagrees with the other three intrinsic characteristics of the other three numbers in the adjustment table: (1) It does not approximate a multiple of 584 and thus cannot lead from one heliacal rising date to another. Either there is an error here, or the use of the number was markedly different. In moving from one 1 Ahau date to another, at least one of the dates involved must not be heliacal rising but presumably some other Venus station. (2) The rough magnitude of this number is such that about one day's error would have accumulated in the relation of the 584-day count to the average period of 583.92 days; if it is a correction factor of some sort, it corrects for an error of about one day. This difference is related to the first, since the deviations by multiples of four were a consequence of leading from one heliacal rising date to another within the table. (3) 9100 is not a small multiple of four days less than a multiple of 584; it factors as $15 \times 584 + 340$ or $16 \times 584 - 244$.

Partly because 9100 does not lead from one date of heliacal rising to anywhere near another, and because it is unique among the correction factors in this regard, few discussions of the table's structure accommodate it; Thompson (1950, p. 225) posited scribal error and reconstructed another figure (also deviating in feature 3). The recorded figure joins recorded stations of the table (canonical settings of morning and evening star); this is probably relevant, since only two other multiples of 260 can (along with their differences from $146 \times 260 = 65 \times 584$, the span of the table), and suggests the recorded number be accepted as written. Since it is in series with the basedate adjustment factors, its use was most likely related in some way to theirs, although it could not be used itself to effect such adjustments. Its precise function remains enigmatic, but one likely class of hypotheses is developed on pp. 532–3. The Venus cycle was associated with eclipse cycles in a variety of ways (see section "The

Venus-Eclipse Cycle"). It is argued below that the number 1.5.5.0 was used with the 4.18.17.0 and 9.11.7.0 adjustments to locate eclipse intervals within the table, and that it may relate historically to the 1 Ahau 18 Uo basedate.

In summary, the attempt to regularly capture heliacal risings on the day 1 Ahau generates the regular 584-day lengths of the Venus years, and thereby the sequence of heliacal rising dates from a given basedate for two calendar rounds (after which the cycle, with no shift of base, would repeat). It also fixes the timing and magnitude of the correction factors. The epoch of these innovations can be secured via Lounsbury's reconstruction of the rationale for the 1 Ahau 18 Kayab base of the table. Evidently no correction was applied during the first cycle through the table from that base, in AD 934, but a correction using the procedures registered in the preface to the tables was applied to institute the 1 Ahau 18 Uo basedate in AD 1129, and regularly thereafter near the end of the table, each 57 or 61 Venus years. So the uniform 584-day period was instituted with the 18 Uo base, if it was not in place earlier for other reasons.

The other stations. The remaining structure of the table is defined by the internal subdivision of the 584-day periods into four portions. These subdivisions of the cycle are of the same length in each of the five pages of the planet's cycle. The lack of variation of the subdivision probably reflects the consistent 584-day length of the overall period rather than requiring separate explanation in terms of the basedate shift or some other factor; given the uniformity assigned to the Venus year, there is scant rationale for variation in the subdivisions. Because the invariance of the subintervals imposes or presupposes the uniformity of the 584-day period, the subintervals too were presumably defined by AD 1129. It is possible that the subdivisions were reached by different intervals at that time than they were when the next base, 13 Mac, was instituted, c. AD 1227; since that base was evidently the current one when this copy of the table was written down (see Lounsbury, 1983b, p. 7; Kelley 1983, p. 177), the present intervals were established no later than AD 1227.

The final station in a given Venus year, reached by a count of eight days, provides the base for the count to the first position of the next Venus year; the final position in the table gives the 1 Ahau base of the entire table. The eight-day interval is identified with the period of invisibility leading up to heliacal rising as morning star because it matches the average duration of the period; Aztec sources specify this canonical length for the period; and the associated iconography agrees with the ethnohistoric discussions of the activities of the gods on that day. The subsequent intervals approximate the periods of visibility as morning star (c. 263 days, here 236); invisibility around superior conjunction (c. 50 days, here 90); and visibility as evening star (c. 263 days, here 250). Under this system, visibility as morning star is projected to end 17 to 29 days too soon (an average of 23 days early); visibility as evening star is projected to begin three to 14 days too late (an average of nine days late).

Since the approximations to the visibility and invisibility periods fall outside the range actually observed, either these periods are not what the subdivisions are intended to monitor, or they, like the 584-day units, were adopted to effect some kind of accommodation to another cycle.

Gibbs (1977) has proposed that the unusual units reflect ritual implications of certain days in the veintena for the stations selected; the chief evidence is that half of the days of the veintena do not occur in the table, while all would occur if the canonical series of intervals were the average values 263 + 50 + 263 + 8. She notes more specifically that there is a special structure to the placement of these veintena dates with respect to Venus stations: veintena days on which the evening star could set are the same as those on which the morning star could rise or set; and the veintena days on which the evening star could rise are midway between the veintena dates for the other stations. With various additional assumptions, she shows that the intervals recorded are the closest to the canonical values that could be used.

Although the schema developed by Gibbs is possible in principle, and has been favorably received (cf. Aveni, 1980, p. 190; 1992), no positive data strongly supports or even suggests it. External support for the most basic assumption, that the days are ritually selected, is extremely tenuous: the absence of an assignment of certain veintena days to any station. Because the position of each Venus round is offset four days in the veintena from the position of the previous round, five different veintena days out of the 20 will occur with each station, one on each page; this gives at most four sets of names available for any station under any system. Given the eight-day period correctly attributed to invisibility around inferior conjunction, the set of veintena days for heliacal setting as evening star at the end of one Venus round is *forcibly* the same as for heliacal rising as morning star at the beginning of the previous Venus round; this agreement is not imposed by special ritual concerns, except that the specific set involved is imposed by assignment of heliacal rise to a 1 Ahau basedate. Any one of the four sets could occur with either of the other two stations, giving 16 possible set assignments. Six of these assignments place one or the other of the stations at the same veintena sets as the stations around inferior conjunction, while another puts both in this set; and ten out of the 16 assignments restrict the possible veintena days in the table to at most half (one to a quarter) of the 20. So the assignment of sets found in the codex, using only half the days of the veintena, is typical of a chance assignment, i.e. one not based on concerns for days of the veintena. Given the actual assignment of veintena sets to Venus stations, there are 25 possible assignments of specific veintena days to a given Venus year; 20 per cent of these would place the specific day in the set for setting of morning star halfway between the day for the preceding and following stations, and of course this is only one of many "special" formal patterns that are possible. So the assignment of sets is typical of their possible assignments, and the assignment of days within the actual sets is typical of their possible assignments. There is no evidence here for special selection of veintena dates having influenced the structure of the table.

There is positive evidence for another line of explanation. The tabular deviations from the reference points in the Venus cycle result in dates at which the planet was visible near the point of invisibility; the hieroglyphs above each day-count have been interpreted as referring to the visibility of the planet. Further, Teeple (1930) noted that the temporal distances were roughly equal to multiples of lunar months, or a half lunar month greater: $236 = 8 \times L$, $90 = 3 \times L + 1½$; $250 = 8½ - 1L$. Since the last figure is imposed by the other two, given the eight days assigned to invisibility, with 1½ days' leeway around an exact lunar month the likelihood of getting such agreement by chance would be at most 4/29 for the 90-day case (87–90 being within 1½ of 88½) and 3/29 for the 236-day case (235–237 being within 1½ of 236); the likelihood of getting both at lunar month intervals would be no more than 12/841, less than 1.5 per cent. Teeple's observation therefore seems very likely to be relevant.

Putting these observations together, the essential structure of the table is determined: the first station specifies the last date on which the morning star is still visible when the moon is at the same phase as when it rose heliacally; the next specifies the first date on which the evening star is visible when the moon is again at that phase; the next station specifies the expected date of heliacal setting of the evening star, one half lunar phase offset from the previous three dates; and the last station specifies heliacal rising of morning star eight days later.

This system accounts for the difference between the Dresden intervals and the natural periods they approximate, and in particular for the fact that unequal intervals are used to approximate the 263-day morning star period and the 263-day evening period.

The apparent interest in placing the stations of a Venus year at a fixed position in the lunar month is tantamount to an association of the Venus stations with an eclipse cycle. Although not part of astronomical reality, a relationship of Venus to eclipses is verified by the ethnohistoric association of Venus with eclipses, thought to be due to the planet's appearance during solar eclipses before any star or other planet. More to the point, there was a definite *calendrical* association of the Venus cycle with eclipse cycles: the calendric commensuration of the sacred round with the Venus cycle in 65 Venus years effectively commensurated it with eclipse cycles (see section "The Venus-Eclipse Cycle"), and historic basedate corrections were made by means of shorter Venus-eclipse cycle commensurations with the sacred round. Kelley (1977a) has postulated such longer-term Venus-eclipse cycles as constructs in the inscriptions, and Aveni (1992; see also below) has shown that the historical basedates of the Venus table can be linked to a cycle of lunar eclipses visible in Yucatan.

Given the long-term calendrical Venus-eclipse cycles preserved in the Venus table, a shorter-term association of Venus stations with an eclipse cycle is plausible. It is most strongly supported by the specific number of lunar months specified in the Venus table: 8 + 3 lunar months and 3 + 8½ lunar months are spans frequently separating, respectively, pairs of visible lunar eclipses and a lunar-solar or solar-lunar eclipse sequence. The placement of Venus stations at consistent lunar phases would capture short-term repetitions of eclipses within a Venus year, if they happen to occur at the appropriate stations.

The one oddity is that 88 or 89 days might be expected where 90 is consistently recorded, and the 250-day figure correspondingly increased by one or two to complete the sum of 584 days (alternatively, that the 250-day figure is one or two days short of what would be expected, and the 90-day figure being correspondingly increased from 88½). The expectations are based on average lengths of lunar months, of course, and three synodic months can amount to 90 days observationally. This suggests that the model was based on a specific historical point at which the 90-day figure was correct. It could have been either an observed length of three lunar months, or the correct length within a 584-day grouping projected from a basedate. That the point was significant enough to incorporate an unusual lunar interval within a lunar phase recurrence system into the essential structure of the table suggests that it was the date of an eclipse—associated with the corresponding Venus station—one that occurred 326 days after the canonical date of Venus' heliacal rising as morning star with a new or full moon, before the current base of AD 1227. It may be significant, then, that 326 days (=236 + 90 = 178 + 148) form an eclipse recurrence interval.

Eclipses visible in the Maya area rarely occur at the margins of Venus' visibility around superior conjunction. One of them is of special interest. On 11 January, AD 1126, a lunar eclipse occurred that was visible in the Maya area, c. 236 + 90 days after Venus rose heliacally on a full moon (uncertainties concerning viewing conditions make it impossible to know the exact date of past heliacal risings); probably more to the point, it was exactly on the 326th day of the third 584-day Venus year before the 1 Ahau 18 Uo base.

This association is the more significant because Lounsbury found that this date was one on which the correction mechanism was inaugurated. The previous basedate, 1 Ahau 18 Kayab, was one on which Venus and Mars rose heliacally and in conjunction as morning stars. The event led to the backward projection of six calendar rounds, the common multiple of the Venus and Mars canonical rounds, to a prior "historical base," verifying the relevance of the Mars association. Although a new 1 Ahau basedate for heliacal rise can be instituted near the end of two calendar rounds, the Maya apparently passed by the opportunity to do this on 1 Ahau 8 Yax, after 61 Venus years; they permitted the Venus calendar to slip out of phase with the planet's cycle. Lounsbury suggests that the celestial display associated with the prior base was considered too special to permit human tampering.

A further reason may be suggested, based on Aveni's (1992) results: the 8 Yax date was not closely preceded by a lunar eclipse, whereas both the 18 Kayab basedates were preceded by a lunar eclipse a month before the basedate, as were the prior 18 Kayab basedates projected back through six calendar rounds.

The next feasible basedate occurred 57 Venus years after the second 1 Ahau 18 Kayab base, on 1 Ahau 18 Uo, 61 Venus years since the missed 8 Yax base. 18 Uo was instituted as a base, recorded as such in the preface to the tables. Lounsbury argues that it did so because it recapitulated the display of 18 Kayab but with Jupiter in the role of Mars; it was also preceded by a lunar eclipse a month earlier, as the 18 Kayab bases had been. Whether these events provided a celestial mandate for the correction scheme, or simply rendered the original base a bit more prosaic, the corrections were regularly applied thereafter: 61 Venus years later a 13 Mac base was instituted, the current base whose vague year stations are recorded in the uppermost of the vague year sequences; 61 Venus years later, a 3 Xul base would be instituted, provided in the lowermost of the vague year sequences. These had neither the planetary nor eclipse correlates of the previous bases. The next expected base would be 1 Ahau 8 Ch'en, 57 Venus years later.

Accordingly, the lunar eclipse on the 326th day of the Venus round just prior to the one ending on the new base, 18 Uo, heralded both the heliacal rising of Venus as evening star and the inauguration of the new base of the Venus calendar. To complete the parallels between the bases, the 18 Kayab base of AD 934 had itself been preceded by an eclipse, this one a solar eclipse, that occurred in the lunar month just before heliacal setting of morning star (i.e. at the first station of the Venus table); it was on the 238th day of the seventh Venus round before 1 Ahau 18 Kayab (on or near the 236th day since heliacal rising). The following basedate would be heralded by an eclipse at the following Venus station. All of this strengthens the parallelism of the 18 Uo base with the 18 Kayab base, and thereby supports Lounsbury's proposed rationale for the installation of the 18 Uo basedate. Evidently, it was the occasion for the values assigned to the intervals within the canonical Venus year. It was probably also the occasion for their approximation to lunar month groupings in eclipse cycle intervals. Just as the Maya had been observing the occurrence of eclipses during the month just before the last station of the planet in the lunar month preceding the basedate of the Venus cycle (Aveni, 1992), they had also observed eclipses during the month just before heliacal setting of the morning star before the original 18 Kayab base and in the month after heliacal rising of evening star before the 18 Uo base. The four significant stations in the planet's visibility were thereby provided the same kind of formal relation to eclipse recurrence, canonical rising as morning and evening star being separated by an 11-month eclipse interval, canonical setting as morning and evening star by an 11½-month eclipse interval.

In summary, the subintervals of the Venus year, the ritual calendar dates associated with each, and the corrective mechanisms for reaching new basedates were all established with the installation of the 1 Ahau 18 Uo basedate. The essential difference of the form of the table present at that time is that a sequence of vague year positions would have been given that was calculated from the 18 Uo base, and the line of vague year dates calculated from the 3 Xul base would not have been present. The 18 Uo line would have been the uppermost line of vague year dates and the 13 Mac line the lowermost, if the association of the upper line with the current base and the lower line with the coming base was conventional.

The Venus-eclipse cycle. It is not only the short-term eclipse recurrences within a single Venus year that relate the Maya conception of Venus to eclipses. The Maya Venus cycle is intimately linked with an eclipse cycle by virtue of its length of two calendar rounds. This span is 13.19 days longer than an integral number of lunar months, and the position of the nodes advances by a little over five days in that time. Passing the two calendar rounds of the canonical Venus cycle can lead from a solar to a lunar eclipse, or conversely. Further, moving 13.19 days short of the Venus cycle, the nodes fall back about eight days; so, from postnodal eclipses, a Venus cycle less half a lunar month is an eclipse station, and from prenodal eclipses a Venus cycle plus half a lunar month is an eclipse station. The result is that the position of the eclipse station falls back by half a month for two Venus cycles, then moves forward a full month and thus reinstates an eclipse station within three or four days of the original position. Depending on the position of the nodes, there is a possibility of missing a lunar eclipse station every fourth cycle, when the node is roughly centered on new moon; in this case, a solar eclipse station appears about where the lunar eclipse station was two cycles earlier. Because the nodal limits are wider for solar eclipses, their stations are consistently reinstated. Finally, the fit to 65 true Venus years is even closer than to the canonical cycle; the node returns to virtually the same position, while the position in the lunar month advances by eight days.

Aveni (1992) has identified a historical sequence of lunar eclipses that not only fits this pattern, but does so in the lunar month just prior to the historic basedates of the Dresden Venus table. A lunar eclipse during the lunar month completed before the disappearance of Venus as morning star is associated with each of the 18 Kayab basedates; the repetitions are a consequence of the Venus-eclipse cycle just noted. In addition, an eclipse occurred in the lunar month just before the 18 Uo basedate. However, this was not due to the Venus-eclipse cycle, for the distance to the corrected basedate was eight Venus years shorter than that cycle. An association of eclipses proximate to Venus basedates is linked in the previous section to the placement of the stations of the Venus cycle at eclipse recurrence intervals from each other within the Venus year.

Additional evidence of a linkage of eclipse cycles with the Venus cycle is preserved in the Venus table. The factors that serve to link basedates also function to link stations of the table in various eclipse cycles. The most obvious of these is the 4.12.8.0 factor; it joins the last historic 18 Kayab basedate to the 18 Uo basedate via a commensuration of 128 sacred rounds with an eclipse cycle of 1127 lunar months, the second or third best commensuration of the lunar and ritual calendars. It is because of this commensuration that the sequence of lunar eclipses occurring just before canonical disappearance of evening star just prior to each of the 18 Kayab basedates was also reflected in the same eclipse association for the 18 Uo basedate.

This correction is only associated with baseshifts after 57 Venus years. Among basedates registered in the table, only one involved such a shift, and this was a historical anomaly due to the failure to apply a baseshift after the first 18 Kayab base; 18 Uo could also have been reached as the second of two basedate shifts of 61 years, if a shift had applied after the first 18 Kayab base. Corrections occurring after 61 years are exact multiples of the sacred round that do *not* commensurate the lunar month in an eclipse cycle. Thus, when a prior basedate occurred with the true heliacal rise of Venus, 128 sacred rounds overcorrects the accumulated error; its application is accurate only if a canonical basedate is maintained for the better part of two Venus cycles rather than one, or after four 61-year base shifts have been applied. The former is the situation postulated by Lounsbury for the Dresden: the historical 1 Ahau 18 Kayab basedate at 10.5.6.4.0 was retained as the base of the next Venus cycle, at 10.10.11.12.0; the correction instituting the base at 18 Uo was applied to this now discrepant canonical base. Subsequent corrections were near the end of a single Venus cycle, after 61 Venus rounds; this interval does not approximate an eclipse recurrence interval, and the cycle noted by Aveni fails to continue just prior to the next two bases (13 Mac and 3 Xul).

In the table, the (current) basedate 13 Mac is reached by adding 9.11.7.0 to the 18 Kayab base; this factor is the sum of 4.12.8.0 (leading to 18 Uo) and a one-cycle correction of 4.18.17.0. It is here, I believe, that the significance of the problematic 1.5.5.0 correction factor can be discerned. Although the 9.11.7.0 correction is not an eclipse cycle, the 1.5.5.0 adjustment transforms it into one. 9.11.7.0 less 1.5.5.0 is 2025 lunar months—five lengths of the Dresden eclipse table; similarly, a one-cycle correction of 4.18.17.0 less 1.5.5.0 is an eclipse cycle of 898 lunar months commensurating 102 sacred rounds. The shortening links an eclipse station near the last 18 Kayab basedate to one in the 16th Venus year before the 13 Mac basedate. By adding rather than subtracting the interval, the nodal and lunar month positions both advance by ten days; thus, the interval leads from an eclipse station near the 18 Kayab basedate to ten days after an eclipse station in the 16th year of the new 13 Mac basedate.

The relevance of these positions is not only in their characterization as eclipse stations: 1.5.5.0, is also one of only six multiples of 260 that can lead from any station in the table to any other in less than one full Venus cycle. Specifically, it links canonical setting of morning star with a subsequent canonical setting of evening star. Being associated with the basedate correction factors linking the last 1 Ahau 18 Kayab base to the next two basedates of the table, the linkage between these setting dates is most plausibly of setting of evening star just before a basedate with a prior setting of morning star. If the linkage is of actual stations, the only historical position for the linkage would be between canonical setting of morning star on 1 Ahau 3 Zotz' in AD 1105 and canonical setting of evening star on 1 Ahau 18 Uo in AD 1129. The reason is that this was the only time that the new basedate fell on canonical setting of evening star, owing to the failure to institute a new basedate after the first full Venus cycle; normally, the new basedate fell four days after canonical setting of evening star and four days before canonical rise of morning star. However, this date-linking interval may have been more a formal device for indicating an important eclipse recurrence interval imbedded in the structure of the table, giving a longer-term eclipse relationship to the short-term eclipse cycles that defined the positioning of these stations in the first place; it may not be historically anchored at all.

The 4.18.17.0 ± 1.5.5.0 interval links the original historical 1 Ahau 18 Kayab base to stations around superior conjunction during the same Venus cycle, the one ending on the new basedate of 18 Uo. Since the original basedate was preceded by a lunar eclipse a month earlier, they link these stations to prior eclipse stations, but neither was near a visible eclipse of the sun or moon. This suggests a formal relevance of long-term Venus-eclipse cycles to the canonical positions of the stations of the Venus table around superior conjunction in the years leading up to the 18 Uo basedate.

The pattern discovered by Aveni associates the Venus table basedates with visible lunar eclipses, highlighting the significance of the Venus stations as eclipse stations. In particular, the timing of the shift from an 18 Kayab to an 18 Uo basedate is such as to preserve an eclipse association of the basedates, or more precisely of the heliacal setting of evening star about eight days before the basedate, for this association was lost at the next 1 Ahau 18 Kayab; and the rationale for the placement of the Venus stations in the lunar month proximate to true setting of morning star and rising of evening star based on historic eclipses at those positions is clarified by the habitual association of visible lunar eclipses in the period of visibility during the lunar month prior to heliacal setting of evening star just before the basedates of each Venus cycle. If the 1.5.5.0 factor was used to derive longer-term eclipse cycles relating these same stations to heliacal rising basedates, that use connects with these interrelated eclipse recurrences near the significant stations of the Venus cycle during the years leading up to the establishment of the 18 Uo base.

Commensuration Models for the Other Planets

The special structure of the Venus table is partly due to the special relation of its cycle with the 365-day year; its 5:8 ratio permitted recycling of the same vague year positions every fifth Venus year, with an adjustment for slippage by returning to the same position in the ritual calendar once that period had commensurated the true average synodic period. For the other planets, this kind of structure is not feasible.

Mars. Willson (1924) proposed that the almanac on pp. 43–5b of the *Dresden Codex* functioned as a Mars table, and aspects of Bricker and Bricker's (1986) study of the table provide solid support for this interpretation. With good reason, a relationship to Mars has long been disputed. Apart from the table of multiples of 780 days in its preface, the most striking piece of evidence for this function was the ring number of 352 days, a period approximating the interval from conjunction to retrograde motion. Ring numbers are distances prior to the basedate of the long count, always reached by projecting backward from a significant historical date by a common multiple of 260 and another topically pertinent interval. This ring number is no exception, being the smallest multiple of 780 (= 3 × 260) that can reach to before the long count base; if the historical basedate was chosen for its intrinsic relation to the Mars cycle, the fact that the ring number achieved was 352 must be a coincidence. In fact, given that the basedate in the sacred round was 3 Lamat, *any* ring number had to be of the form $780 \times n + 92$, $780 \times n + 352$, or $780 \times n - 78$; the latter two correspond to canonical distances between significant points in the Mars cycle. Either the 3 Lamat base was contrived, or the connection of the ring number to the Mars cycle is coincidental, under Willson's hypothesis.

Willson's evidence that this table concerned Mars could therefore all be construed as coincidental. Thompson (1950, 1972) also argued that the structure of the table

Table 22.5 A hypothetical Mars table, commensurating the vague year and calendar round.

1	14 Pax	6 Zotz'	19 Kayab	11 Zec
2	4 Uayeb	1 Xul	9 Pop	6 Yaxkin
3	14 Uo	11 Mol	19 Zip	...
		...		
72		...11 Kayab	19 Ceh	16 Cumku
73	4 Kankin	16 Pop	9 Muan	1 Zip
	4 Imix	4 Etz'nab	4 Chuen	4 Lamat
	273	117	273	117

The stations are essentially arbitrary.

provides positive evidence *against* a Mars interpretation. The most unnatural aspect of this hypothesis has been the fact that its basic length is 78 days, so that a subdivision into even tenths is being used; and each of these is further subdivided, about equally, into segments of 19, 19, 19 and 21 days in sequence. A table having 40 evenly spaced stations does not correspond to natural subdivisions of a planetary cycle. Secondly, even apart from the issue of the number of stations involved, the table's structure differs radically from that for Venus, the only unmistakable planetary table in the manuscripts, and accords with that of the non-astronomical almanacs.

Following the method of the previous subsection, it can be shown that the structure of the Venus table could not provide an appropriate model for Mars, and, based largely on work by Bricker and Bricker (1986), that the subdivision of the 780-day cycle neatly accommodates a ritual Mars cycle.

The 1:3 ratio of the Martian integral synodic period to the ritual calendar is so close to the true average synodic period that the error is just under one day per calendar round. This means that, once a station in the cycle departs from a given ritual calendar date, it cannot return in historical time. Thus, the standard Mesoamerican association of deities with a ritual calendar date was not feasible for Mars: corrections to a table of positions for the planet, keyed to dates within three calendar rounds, would have to return the same vague year position while shifting the ritual calendar position; and the essential structure of the table would give fixed ritual calendar dates for stations, which would cycle through diverse vague year positions. A fragment of such a model might look something like that of Table 22.5. Such a table would include 73 vague year positions for each station, each being a multiple of five days' distance from every other position for that station; and a position 22 Mars years further along reaches a station exactly five days later in the vague year. A corrective mechanism therefore must balance five days of accumulated error. This requires about 79 (i.e., 5/[780-779.93651]) Mars years, and thus could occur after approximately one pass through the entire table; it would be implemented by beginning the next pass on line 58 rather than line 7 (or perhaps on line 52 rather than line 1). After this length of time, the average position of Mars in the ritual cycle would have shifted five days in the ritual calendar; one line would have to be given for each ritual calendar basedate under this shift, parallel to the vague year lines of the Venus table. However, it would take about 200 years before a single correction would be applied. Adjusting for accumulated average error of the

John S. Justeson

Mars cycle within a reasonable time span therefore requires an approach that does not reuse expressed dates, if the ritual calendar is involved. The manuscripts express all dates in the ritual calendar, and their almanacs are always structured for reuse; this precludes construction of a Mars table adjusting for average accumulating error, the *raison d'être* of the Venus table. In a non-adjusting table of the Mars cycle, then, vague year stations are superfluous; a Mars year almanac would be constructed with its stations located in the ritual calendar only. The failure of Willson's proposed Mars table to accord with the structure of the Venus table is therefore unavoidable.

Because a system of simple ritual calendar stations is intrinsic to representation of the Mars cycle in the ritual calendar, the traditional structure of a divinatory almanac—among which astronomical almanacs must be numbered—was appropriate. One feature of these almanacs is that their stations span a divisor of their total span; the almanac is usually re-entered four, five, ten or 20 times before completion of a full round. The 78-day length of the repeating portion of the table accommodates subdivision into tenths; it also closely approximates the period of Mars' retrograde motion, as Willson noted. These two features are sufficient to make the 78-day structure quite a feasible one for a Mars table. The conformity of the proposed Mars table of pp. 43b–45b with the structure of non-astronomical almanacs is therefore to be expected; it does not argue against the validity of Willson's hypothesis. The exceptional appearance of the Venus table—presenting fixed stations in the vague year—arose because the Venus cycle does not accommodate ritual calendar stations; they would regularly repeat in the 260-day year only after prohibitively long intervals, while repeating reliably within the 365-day year at eight-year intervals for a relatively long time.

The features which have been seen as inimical to Willson's hypothesis are therefore predictable characteristics of a Maya Mars table. Furthermore, certain of Bricker and Bricker's (1986) results concerning pp. 43–45b strongly support the view that it represents the synodic cycle of Mars. The ostensible base of the table is 3 Lamat, for this is the day of the ritual calendar with which the prefatory table of multiples of 780 is associated, as well as that of the ostensible historical base of the table (9.19.7.15.8). The Brickers show that 3 Lamat was a significant anchor in the Mars cycle. In the Terminal Classic and Early Postclassic periods, the date of heliacal rising is restricted to the quarter of the ritual calendar starting three days before 3 Lamat, with the concentration at this end of the range being on 3 Lamat itself. Thus, the ostensible base-date of the table is evidently selected for its relationship to heliacal rise, just as the base of the Venus table was the date of heliacal rise. We can interpret 3 Lamat as a viable canonical date of Mars' heliacal rise during the era to which the manuscript pertains; it provides a *terminus ante quem* for predictions of heliacal rise.

The Brickers argue that the retrograde period of Mars is the emphasis of the almanac as a whole. This is supported by their observation that the historical base of the almanac precedes a retrograde period by only 50 days. This raises a question concerning the relevance of the 3 Lamat base, ostensibly related to heliacal rise, since the historical 3 Lamat base is the one 260 days *after* canonical heliacal rise. The answer, I believe, can be inferred from a consideration of this preceding 3 Lamat. If canonical heliacal rise was the computational base from which the almanac was entered, the beginning of observed retrograde occurred within a day or two of the beginning of the fifth period of 78 days, that running from 3 Ahau to 3 Etz'nab; so perfect an alignment of the retrograde period with the regular tenths of the almanac is quite rare. This, I would argue, is the reason for the selection of this particular retrograde period as the one inaugurating the table, one that has the strength of accommodating both the treatment of heliacal rise as the tabular base and its apparent concern with retrograde motion. Further, just as 3 Lamat is a *terminus ante quem* for heliacal rise, the subsequent 3 Lamat provides a loose *terminus ante quem* for the retrograde period; this is because retrograde follows heliacal rise by an average of around 292 days.

This property suggests a function for "aberrant" multiples of 260, those that are not multiples of the canonical planetary synodic period, that parallels the use of such numbers in the Venus table. In the latter, multiples of 260 that deviate from multiples of 584 are used to recapture appropriate dates of heliacal rise. In the case of Mars, multiples of 260 that deviate from being multiples of 780, by being added to or subtracted from the historical 3 Lamat base, *must* lead to dates of canonical heliacal rising: subtracting $780 \times n + 260$ leads backward, while adding $780 \times n - 260$ leads forward, to canonical heliacal rise on 3 Lamat.

Any such number is appropriate to capture canonical heliacal rising. In the Venus tables, the usage was more precisely to maintain the canonical date of heliacal rise on or near its observed date in the ritual calendar. In the case of Mars, the synodic period is quite variable. However, variations in planetary synodic periods are quite regular, defined by commensurations of the planet's sidereal period with the solar year. The difference $n - m$ between any two factors $780n - 260$ and $780m - 260$ must be such that $(n - m) \times 779.93651$ is within a few days of being a multiple of both the sidereal period of Mars (686.99576) and the solar year (365.2422); the Maya did note sidereal/synodic commensurations. The pattern for synodic Mars years is $(7 + 8 + 7 + 8 + 7) + (7 + 8 + 7) + (7 + 8 + 7 + 8 + 7)$, in which the 8's may be further subdivided as $1 + 7$. The aberrant multiples actually entered in the table are $17 \times 780 - 260$, $40 \times 780 - 260$, $4 \times 780 + 260$ and $93 \times 780 + 260$; interpreting them as reaching canonical heliacal rise from the basedate, they must be written $-4 \times 780 - 260$, $-93 \times 780 - 260$, $17 \times 780 - 260$ and $40 \times 780 - 260$. Their magnitudes do in fact differ by appropriate intervals, the separation of each from the largest $(40 \times 780 - 260)$ being 133×780, 44×780 and 23×780, respectively. Thus, if any one of the aberrant multiples reaches an observed heliacal rise on 3 Lamat, all of them should be within a few days of it. Without detecting the pattern of intervals relating these aberrant multiples of 260, nor the systematic basis for such a relation given the position of the historical base with respect to canonical heliacal rise on 3 Lamat, Bricker and Bricker (1986, p. 74) discovered that forward counts of $17 \times 780 - 260$ and $40 \times 17 - 260$ do lead to dates on or near true heliacal rise; I calculate that $-4 \times 780 + 260$ and $-93 \times 780 + 260$ do as well.

Accordingly, Willson appears to have been correct in proposing that pp. 43–45b of the Dresden treat the synodic cycle of Mars.

The remaining structure of this table is its subdivision into four segments of 19, 19, 19 and 21 days each. The Brickers' conclusions concerning this structure are at the very least controversial. They treat the almanac as a means for anticipating variations in the length of the synodic period, specifically with respect to the timing of retrograde. When observed heliacal rising falls into the first or last quarter of a 78-day span, a correction of 19 or 21 days is applied to the expected beginning of retrograde—otherwise no correction is applied—and addition or subtraction depends on whether heliacal rise is in the first or second half of the range of tzolkin dates on which it can occur. This model does place the widely varying positions of the retrograde period appropriately, but this is not a very compelling result. Details aside, some such model has to work: an extreme synodic interval is the cue, not anything structural in the table; and movement from an extreme quarter to a middle half can seldom fail to keep the retrograde period roughly centered in the 78-day span. The success of the model stems from the division of this span into roughly equal quarters, given that that span approximates the retrograde period. Accordingly, the table can be used in the way the Brickers propose, but there is no evidence that it was in fact so used.

Moreover, such a model is far less accurate than models that would have been available to the Maya. With the amount of data that would have been used in constructing this scheme, they could not have failed to detect the regularity in the patterning of the variations. All visible planets undergo a regular, symmetric pattern of variation in their synodic periods that repeats when the sidereal and solar years commensurate; these commensurations perforce commensurate the average synodic year, and commensuration of this span with the solar year and/or the vague year is established for the Maya. Over a period of several hundred years, from the beginning of the Early Classic to near the end of the Late Postclassic, I find less than one day's variation in the length of a given variant for Mars; relatively short-term patterns of c. 15–22 Mars years give far more useful anticipatory results than the model proposed. Had the Mars table been divided so that seven or eight synodic periods were explicitly displayed, the pattern of variation could have been captured; reusing the same synodic chart for all variants, as in the Venus tables, prevents exploitation of the seasonal patterning of synodic variation.

Rather than providing a crude mechanism for tracking highly predictable variations, the subdivision of the 78-day unit into four segments of 19, 19, 19 and 21 days each—or some other subdivision into approximate 19½-day quarters of the 78-day span—is in fact required to accommodate the 78-day almanac to other stations of the Mars cycle. The points of observational interest are heliacal set and rise around the period of invisibility (lasting about 1½ × 78 days), and the stationary points (about 78 days apart, and separated from the adjacent rise/set date by about 3¾ × 78 days). Thus, the sequence of 19, 19, 19 and 21 days, summing to 78, permit one to capture fairly closely the average positions of heliacal rising and setting and both stationary points. With heliacal rise after invisibility as the base, the sequence is as shown in Table 22.6. They can also accommodate the interposing of opposition between the stationary points, dividing the interval into rough halves of 38 and 40 days (the former more

Table 22.6 Mars table intervals (in days), in relation to the Mars cycle.

Defining intervals	Average	Modeled as	Canonical
Helical rise to sp_1	292½	$3 \times 78 + 19 + 19 + 19$	292
			$10L - 3$
sp_1 to sp_2	75	$78 (21 + 19 + 19 + 19)$	78
sp_2 to helical set	292½	$21 + 78 + 19 + 19$	294
			$10L - 1$
Helical set to helical rise	120	$19 + 21 + 78$	118 $4L$

sp = stationary point; L = lunar month.

closely approximating 37½); and the interposing of conjunction between heliacal set and rise, dividing 118 into 59 + 59 (19 + 21 + 19 + 19 + 19 + 21 in the scheme above). Finally, the period from stationary point 2 to stationary point 1 is roughly 10 + 10 + 4 lunations (295 + 118 + 295 days), a possible eclipse interval. In the four-station table, the essential results are independent of which point in the Mars cycle is chosen as the base, except that for two of the four possible bases the assignments of 292- and 294-day values are interchanged; other variations of two days would be possible in a six-station table.

Mercury, Jupiter, and Saturn. The cycles of the other planets are less tractable, with respect to the calendar round, than are those of Venus and Mars. The pertinent data are summarized in Table 22.7. They show that it is not feasible to structure a table of positions for Mercury, Jupiter or Saturn in terms of the number of calendar rounds that commensurates the integral approximation to the synodic period: in all these cases, the accumulated error by the end of such a table would amount to several synodic periods, vitiating its usefulness; and the table would span a period far beyond historical time, making its construction an unverifiable extrapolation.

The ritual calendar does rationalize each true planetary synodic period within the span of the Venus table to within the accuracy of that table, i.e. to within less than 5.09 days in less than 104 years. The Venus table's corrective capabilities, to within 0.78 days in 137 sacred rounds (61 Venus rounds), can also be met for Mercury (0.70 days in 119 sacred rounds) and Saturn (0.99 days in 16 sacred rounds), and approximated for Jupiter (1 × 45 days in 112 sacred rounds). Finally, five calendar rounds rationalize with the average synodic periods of Saturn and Mercury, quite precisely for Saturn; recorded Saturn dates are ancient enough that at least the third such pentad was elapsing when the *Dresden Codex* was being compiled. The Jupiter period commensurates with 12 calendar rounds; one such had probably elapsed by the time of the *Dresden Codex,* although the Maya should have been able to project the commensuration via the rate of regression of the Jupiter cycle with respect to perhaps five or six calendar rounds, or by projecting forward from the separate commensurations of the vague year and the ritual calendar (571 = 3 × 161 + 88 Jupiter cycles = 3 × 247 + 135 sacred rounds; 571 Jupiter cycles = 2 × 237 + 97 = 2 × 259 + 106 vague years; see Table 22.7).

Table 22.7 Commensuration of ritual calendar and vague year with planetary cycles.

	Commensuration of true synodic period with		
	calendar round	ritual calendar	vague year
Mercury		4 : 9 − 2.90	20 : 63 − 0.28
		29 : 65 + 7.96	116 : 365 + 44.70
		37 : 83 + 2.16	
		41 : 92 − 0.73	
		119 : 267 + 0.70	
		160 : 359 − 0.04	
	5 : 819 − 3.70		
	29 : 4750 + 1.69		
Venus		9 : 4 + 4.31	**8 : 5 + 0.4**
		128 : 57 − 3.53	
		137 : 61 + 0.78	
	2 : 65 + 5.09	**146 : 65 + 5.09**	**104 : 65 + 5.09**
Mars		**3 : 1 + 0.06**	47 : 22 − 3.60
	3 : 73 + 4.63		*156 : 73 + 4.63*
			203 : 95 + 1.03
Jupiter		23 : 15 + 3.26	47 : 43 + 2.98
		112 : 73 − 1.45	106 : 97 − 1.77
		247 : 161 + 0.36	153 : 140 + 1.21
			259 : 237 + 0.56
		399 : 260 + 30.11	*399 : 365 + 42.26*
	12 : 571 − 2.88		
	399 : 18985 + 203.27		
	= 18980 × 399		
Saturn		16 : 11 + 0.99	29 : 28 − 1.58
		189 : 130 + 11.97	231 : 223 + 0.47
	5 : 251 − 1.11	365 : 251 − 1.11	260 : 251 − 1.11
		397 : 273 + 0.86	
	189 : 9487 + 260.44		*378 : 365 − 33.61*
	= 9480 × 378		

Format is $n:m + r$, where n is the number of calendrical cycles, m is the number of average planetary synodic periods, and $n = m$ periods $+ r$ days. Several commensurations are given in some cases, with increasing accuracy as higher numbers of units are involved in the commensurations. Boldface entries represent commensurations occurring in Maya planetary tables. In italicized (and some bold) entries, the specified number of calendar periods exactly commensurates the integral approximation to the planet's synodic period; for the true average synodic period, the accuracy in most of these cases is lower than for prior commensurations.

So the important calendar cycles do commensurate in reasonable spans for tables, except for Jupiter in the case of the calendar round, and sometimes more precisely than did Venus. The kinds of structures that can be based on them are akin to the lunar tables, with a succession of synodic periods laid out end to end, not repeating until the entire sequence has been completed; the interlocking of the planetary year with the ritual calendar and vague year, seen in the Venus table and feasible for a Mars table, was not available. Close commensurations to the 360-day year are also found, illustrated in Table 22.8. These were almost certainly noted and used, since all numerical records were kept in terms of it. At least the 13-tun Mars cycle, the 21-tun Saturn cycle, and the 73-tun Venus cycle would have been significant: a 13-tun period was

Table 22.8 Commensuration of the moon and planets with the 360-day civil year.

Moon	Mercury	Venus	Mars	Jupiter	Saturn
5:61 – 1.37	9:28 – 4.57	73:45 + 3.53	13:6 + 0.38	41:37 + 1.28	21:20 – 1.84
21:256 – 0.17	19:59 + 3.22	= 72 vague years		277:250 – 1.05	209:199 – 0.32
	28:87 – 1.35				
	75:233 + 0.53	133:82 – 1.58			

given special importance by the Maya within the katun; the 21-tun period commensurates the lunar cycle to a fraction of a day; and the 73-tun cycle commensurates the tun with the vague year, while a 73-katun period appears to have been the basis upon which the basedate of the long count was fixed.

Possibly relating to planetary cycles is the compound epigraphers have dubbed Glyph Y, distinguished by its T739 mainsign and by its environment, between the glyphs of the nine-day count and the lunar series. In this environment, it takes a numeral prefix between 2 and 6, or takes no prefix. (In one instance, at Xcalumkin, it is preceded by a skull sign that represents the numeral "10," via a rebus on pCT, pYu *la·xu·n* "ten" vs. pCT, pYu *lax* "to end; to die." In this case, it is evidently the "end" reading that is appropriate, a conclusion reinforced by the death markings on the Glyph Y mainsign.) A Glyph Z, formerly distinguished as a "companion" of Glyph Y, is simply an elaborate spelling of the numeral 5, as 5-bi-IH (= either 5-(bi)-BIŠ or 5-b(i)-IŠ) for *ho-b'-iš* "5 (days) later"; its separation is due to the presence of the *-b'iš* suffix, restricted to use with numerals 5 and 7.

The main sign of Glyph Y also occurs as one of the nominal hieroglyphs in the 819-day count passage, although the position of a date in the 819-day calendar appears to be uncorrelated with the coefficient of Glyph Y on that date. With the numerical prefix 9, it is G_6, indicating the sixth position in the nine-day count (J. Linden, 1981, Glyph Y of the supplementary series, unpublished manuscript, and 1982, The Supplementary Series, unpublished MA thesis, Tulane University; F. G. Lounsbury, 1982, Glyph G_6 of the supplementary series, unpublished manuscript). These two contexts are related in that the basedate of the 819-day count falls on the last G_6 before the basedate of the long count; one usage is presumably a partial explanation for the other. Lounsbury and I have speculated that this count relates to the motions of Saturn and/or Jupiter, because of common factors (21 for Jupiter, 3 × 21 for Saturn) of their synodic periods with 819 and a high frequency of Saturn and Jupiter events associated with 819-day counts, and the factors of 819 (7, 9 and 13) show up in structurally parallel ways in canonical synodic periods of all the planets except Venus. This suggests that the 819-day count may have served as a formal tool for tracking the behavior of the planets, or perhaps the relative behavior of pairs of planets, in the short term, perhaps via seven-, nine-, 13- and/or 21-day counts.

If the 819-day count did have such a function, Glyph Y is both a likely candidate mechanism for tracking planetary motions and the most promising candidate for investigating this hypothesis. Its function has not been successfully identified.

It takes varying numeral prefixes, often occurs with suffixes indicating time counts, and occurs between various counts of days and lunar months; evidently it is a count of units of some sort. The day count suffix for *-biš, specific to numerals 5 and 7 in Mayan languages and writing, is sometimes represented when the numeral prefix to Glyph Y is 5; otherwise, the distance number suffix for *-e·x (or *-xe·y), indicating counts of arbitrary units forward (or backward), is suffixed to the unit. Thus, it is not clear whether days within a unit or units are being counted, although probably it is uniformly one or the other. At this writing I have not successfully fit the data to any formal count of units having fixed length, and am still investigating the hypothesis via formal and naturalistic units of variable lengths reflecting subdivisions of planetary cycles and accommodations among cycles. Independently, Victoria Bricker (1983a) suggests that it is a count of lunar months from or to Jupiter's period of retrograde motion, but this hypothesis fits only about two-thirds of the cases.

Useful constraints on a solution are provided by a series of relatively closely spaced records at Yaxchilan, from 9.16.0.0.0 through 9.16.10.0.0; also of special usefulness is the apparent "end of Y" record at Xcalumkin, suggesting either the end of a unit of more than one day's duration or the possibility of more than six units in the count.

Cultural Associations of Celestial Events in the Inscriptions

Major celestial events were used to schedule human affairs, ostensibly as sacred mandates for elite decision-making. Postclassic astronomical tables provide temporal information concerning the timing of celestial events, showing that the elite could anticipate coming astronomical events; this would aid preparations for activities of their own selection in the guise of submission to the divine will. Although indicating the enhanced capabilities of the Maya elite for implementing their decisions about public action, the manuscripts say little if anything about those decisions. Thus, investigation of the astrological scheduling of activities depends primarily on Classic inscriptions, in which the events taking place on a given day are explicitly noted.

General content. Maya historical texts are typically quite terse in their statement of events. There is usually a single main verb. Topics covered are few: birth, death, marriage, and genealogy; accession to office; performance of various rituals; and raiding and warfare. There are numerological exercises in which past, mythological events are projected backward by significant chronological units, evidently to manufacture "precursors" for the current events, providing anachronistic "precedents" for their timing. The only verbose parts of the texts are those naming the ruler: these may refer to him often by a series of epithets, e.g. a personal name, a series of titles of office, designations as captor of various individuals, ruler of his site, descendant of his royal predecessors, etc. In much of the area, each event was mentioned after the date of its occurrence, but especially in the southern highlands many successive events might be mentioned without reference to the dates on which they occurred (a more typical pattern in the languages). Even in these texts, time anchors are given for the initial and final events, as well as for some intermediate events.

The most common types of event referred to in the hieroglyphic texts were the ceremonies conducted by the ruler at the end of the civil year. Usually, these events were recorded only at the end of the katun, a 20-year cycle, or more rarely at the quarter or half divisions of that cycle, or after 13 civil years had passed in the cycle. Other points, "arbitrary" years in the katun, were seldom the occasion for the erection of monuments, although year-ending ceremonies were most likely performed at the end of every year. In the Postclassic, the year-ending (and year-opening) ceremonies were performed annually, but in relation to the vague rather than civil year.

Celestial events treated. Few intrinsically astronomical signs or compounds are known: these include signs for "sun," "moon," and "star" (including "planet"; specific stars or planets are typically specified linguistically by compounds of the form noun + "star," e.g. "great star" [Venus], "rattle star" [Pleiades]). As a result, the only firm evidence concerning the timing of human activity in relation to celestial events involves the sun, moon and planets. Zodiacal signs or constellations are mentioned in a few texts. Research to date has neither focused upon nor established any pattern of scheduling according to stellar phenomena; however, events timed for fixed dates in the solar year could be connected with the position or visibility of a particular star on that date, and the year-ending rituals recorded on Stela 2 from Bejucal, Guatemala, may commemorate the appearance of a supernova that year. Similarly, dates of the appearances of comets are not securely tied to historical records; Halley's comet may be reflected in the taking of the royal name *k'uk'* by three rulers of Palenque who acceded four to six years after its appearance (A. Schlak, unpublished manuscript, 1985), but this pattern is probably coincidental.

The solar, lunar and planetary events are all, though not always, associated with rituals; apart from bloodletting, they do not involve offerings, but some are of an unknown character. So far as I am aware, eclipse dates are not recorded as dates of *other* activities, apart from such accidental associations as upon the birth of Chan Bahlum. Major stations in the cycles of the planets were occasions for initiating other activities, the prime examples being accessions and military campaigns. Most seem to have been anchored in observation rather than being scheduled according to formal or predictive cycles; this is likely, for instance, in Lounsbury's (1983a) examples of events occurring upon the first perceptible motion of Venus, Jupiter or Saturn from a stationary point. Such activities were presumably planned in advance: the occasions for their enactment would have been anticipated by rough prediction, and watched for; observation of the anticipated celestial event would then provide the signal for initiating human action.

EXPLICITLY ASTRONOMICAL REFERENCES

Almost all explicit lunar references in the inscriptions are records in the lunar calendar. Around 200 of these records are known, and a substantial minority are associated with non-year-ending dates. Although these records are calendric rather than astronomical, it would be useful to examine them for regularities in the time of month of the more frequent activity classes recorded in association with these dates, with particular attention to records near full or new moon. To my knowledge, this has not yet been done, and it is not attempted here; the extent to which phases of the moon may have structured the recorded activities of the Maya elite remains unknown. Explicit astronomical references to the moon are restricted to a few likely solar eclipse dates. One, at Poco Uinic, is marked by a solar eclipse sign quite similar to that of the codices (Teeple, 1930, p. 115); another, at Tres Zapotes, may be accompanied by solar iconography (J. A. Fox, 1986, The eclipse record on Tres Zapotes Stela C, unpublished paper presented at the Second Oxford International Conference on Archaeoastronomy, Merida, Yucatan; see p. 495 for abstract); a few seem to be marked by compounds whose root was a word for "black(ness)" (Kelley, 1977b); and one or two are seemingly designated "hidden lord" (L. Schele, personal communication).

The only recurrent astronomical references are those using the star sign; other events, referred to in text without the star sign, are depicted with it in accompanying scenes. Since the star sign appears repeatedly in the Venus tables, usually in a compound read "great star" (the Maya name of Venus) and rarely as the generic "star," these events have been investigated for Venus correlates (Kelley, 1977a; Closs, 1979, 1981; Lounsbury, 1983a); as a result, stages in the cycle of Venus are the most securely documented planetary events. Other events linked to star events also show Venus correlations; Lounsbury (1983a) argues that many such dates at Bonampak are identifiable as Venus events, though not explicitly marked as such, both because of their relationship to explicitly indicated star events and because the sheer frequency of apparent Venus dates at that site seems too great to be due to chance (I have not tested this, however).

Venus Events in the Inscriptions

The events associated with star signs are quite limited in content. One, at Copan, is an heir designation, ostensibly of Venus, and its verb is the same one used, roughly, to indicate the planet's visibility in the Dresden Venus table. In the inscriptions, this verb is used only on occasions of accession, accession anniversaries, and heir designations. Although such events need not occur on Venus dates, associations of these events with Venus in some instances appear to be significant. At Bonampak and Piedras Negras, accessions and accompanying bloodletting rituals occurred on Venus dates within a few months of prior Venus dates, marked as such by the star sign; this association strengthens the case for the significance of the Venus dates of accession. At Palenque, the ruler dubbed Chan Bahlum acceded when Venus was at or near maximum elongation, and celebrated the fifth Venus-year anniversary of his accession when Venus was again at maximum elongation (A. Schlak, unpublished manuscript, 1985; Schele, 1982, had previously noted that this date was the eighth solar anniversary of his accession).

In almost all other events the star sign aids in the identification of what epigraphers have dubbed the "shell star" and "earth star" compounds. The essential referent of these compounds was first identified by Riese, in 1977 (cf. also Riese, 1982), as warfare; much hieroglyphic evidence has subsequently borne out Riese's conclusion, and the accompanying iconography is transparently war-related. V. Miller (1986, "Star warriors at Chichen Itza," unpublished expanded draft of a paper presented at the 1985 meeting of the Society for American Ethnohistory) shows that when star signs occur in iconographic context, the scenes are related to warfare or to human sacrifice in the ball game.

There are contrasts of content between the rituals and battles associated with Venus events, and those not so associated. Among rituals, only accessions, bloodletting and the ball game are clearly associated with Venus dates; offertory rituals, apart from bloodletting, are not, and bloodletting may be associated only or primarily in cases related to accessions and accession anniversaries. Contrasts among types of battles require more extended discussion.

Soon after Riese pointed out the military implications of the star events, Mathews showed that these events refer to major intersite warfare: the "shell star" events typically specify the site at which or against which war was waged, and at the site so designated hieroglyphic stairways were erected by the victors to commemorate the victory. Mathews is probably correct in positing the T575 "shell" as a grammatical suffix on the verb, rather than contributing referential meaning, a view also held by Schele (1982, p. 101). The "earth star" compounds are simple variations on the same structure, with "earth" appearing where the place name would be expected. "Earth" (comprising "world," "land," "dirt," and "territory" in Lowland Mayan languages) presumably functions as the place designation; it is not clear whether it refers to foreign territory, home territory, or both.

Mathews distinguished these cases of territorial conquest from the more common instances of armed conflict. These refer to the seizing of individual named captives (Proskouriakoff, 1960, 1963–4) by a verb generally read *čuk* "seize." Subsequent references to the ruler responsible would often designate him as "captor of / who captured" these individuals (Proskouriakoff, 1960, 1963–4)—apparently via a word, **kan*, used in many Mayan languages for the hunting of game (Norman, 1984; cf. Houston, 1984)—or as "he of *n* captives" (Stuart, 1985); rulers are not cited as conqueror or ruler of a conquered site. Rulers and their associates are depicted in battle dress for these captures.

One battle or series of battles could be designated in both ways, as at Piedras Negras, Tortuguero and Yaxchilan. In none of these cases is a site name specified; Tortuguero specifies "territory." Other occasions of the taking of captives, though not designated as intersite war, were clearly not simple ambushes of isolated individuals, because several captives were taken and many of these episodes lasted for an extended period; and, in some cases, place names are cited in connection with the names of the captives. The events that are referred to via the ruler's taking of captives may be construed as raiding; and the taking of captives may have been more the propagandistic emphasis than the purpose of such raids, which probably involved the pillaging or despoiling of villages and lands. Especially since some cases are explicitly tied to warfare, and that at Tortuguero seemingly to "earth," some of this raiding probably had a territorial dimension; since intersite warfare is not normally involved, this was probably aimed at wresting control over adjacent hinterlands, but they were not oriented to the actual capture of centres of power. Given the closer similarities to the raiding in chiefdoms than to war in state societies, and the propagandistic emphasis in the repeated references to the ruler as the captor of this or that individual, Webster's (1975) model for the role of territorial war at political boundaries in strengthening and maintaining a stable central authority appears apropos, although here in a primitive state rather than an advanced chiefdom.

Lounsbury (1983b; cf. Table 22.9, column b) showed that star events and, at Bonampak, raids were often scheduled for significant points in the Venus cycle. Now W. Nahm (unpublished, 1988) has demonstrated an even stronger tendency for both types of warfare to be scheduled for limited periods during the visibility of Venus. Mary Miller (1986) relates the Bonampak raids to the taking of captives in preparation for accession to office; since some level of success in warfare may have been prerequisite to the taking of office in this case at least—there are other examples of accession on the heels of captures or war, as at Piedras Negras and (Hassig, 1988) routinely among the Aztec—the scheduling may have been in effect a kind of anticipatory accession comparable to heir designation. However, since raiding is elsewhere connected with war, it seems more straightforward to interpret this as territorial warfare, whether at the boundaries or in an unsuccessful (thus unrecorded) attempt to take over a neighboring centre, or in a defense of their own territory (LaMar Stela 3 evidently records the capture of an individual from Bonampak about this time).

The timing of these raids in the solar year supports this territorial interpretation. Arguing from the records of Aztec warfare, Hassig (1988) has found that intersite war could not be sustained at long distances except when corn was standing in the fields of the intended victim; food carried in on the back would be consumed by its bearers within a relatively short distance from their starting point. In the Maya area the relevant part of the year varies somewhat, but can be considered as concentrated between the autumnal and vernal equinoxes. Examination of all cases designated as war by the Venus/star verb (Table 22.9, column g) shows that they are indeed seasonally restricted and in the expected way. The restriction is on the borderline of statistical significance: six or seven of 23, and seven or eight of 24 dates fall outside the expected region, uncertainties being (a) the categorization of a date falling just four days before the autumn equinox, and (b) whether a Piedras Negras capture date associated with some of the dates where the star event glyph is removed should be allowed in the sample. These uncertainties leave a range of probabilities from a low of 1.73 to a high of 7.58 per cent for a chance concentration of dates in the expected region. The dates are actually more concentrated than this, occurring mainly in a "war season" running from mid-November through mid-February, Gregorian.

Table 22.9 Dates of events marked by the star glyph, and their planetary associations.

a	b Venus	c Event	d Saturn	e Jupiter	f Mars	g Gregorian	h Mercury
War events:							
9.9.18.16.3	rE + 5	war				631 Dec 25	
9.10.17.2.14		war	op	op	1:0	649 Dec 21	me + 1
9.10.19.8.4		war	2: −1			652 Mar 30	
9.11.9.8.12?		war	1: −31	2: −8		662 Feb 14	me − 1
9.11.16.11.6		war			2:+6	669 Mar 3	
9.11.19.4.3		war	op	2: −23		671 Sep 26	me − 11
9.12.0.8.3	Dme	war	2:+17	op + 13		672 Dec 9	me − 11
9.12.5.9.14		war	op			677 Dec 14	
9.13.4.17.14		war		2:+8		697 Feb 12	me − 2
9.13.13.7.2	sM	war			2:+15	705 May 30	
9.14.2.0.14		war	1: −21		op+ *3	713 Dec 6	me − 7
9.15.3.8.8	ic − 2	battle	2: +4		1:−7	735 Jan 19	
9.15.3.8.12	rM − 2	battle	2: +8		1:−3	735 Jan 23	
9.15.4.6.4	rE − 1	war	op			735 Dec 1	me − 12
9.15.9.3.14	rE	battle		2: −18		740 Sep 15	me − 2
9.15.12.2.2		war				743 Jul 30	me
9.15.12.11.13		war		2: +17		744 Feb 6	me + 7
9.15.16.7.17	ic	ballgame				747 Nov 1	
9.16.4.1.1		war	1: −5			755 May 7	
9.17.9.5.11		war	1: +7		2: +20	780 Mar 26	me − 3
9.17.10.6.1	me	capture	1:0	2: +22		781 Mar 31	
9.17.15.3.13	me	faces captive				786 Jan 16	me − 12
9.18.1.15.5	ic − 2	battle	1: −17			792 Aug 4	me − 13
9.18.9.4.4	rE + 4	war	1: −13			799 Nov 15	
9.15.1.6.3		death (in war?)	2:−3	op − 5	1:−6	732 Dec 15	me − 10
Accession events:							
9.12.11.6.8	Dme	accession		1: +10		683 Sep 8	
9.15.15.12.16	rE − 6	heir designation	1:+6			747 Feb 13	
9.17.5.8.9	me	accession, autosacrifice	2: −9	1: −8		776 Jun 13	me + 2
9.17.10.9.4	rM	accession	op			781 Jun 2	
9.18.1.2.0	rE	dynastic ritual (heir designation?)				791 Nov 13	

All planetary dates are given in the 584 283 correlation.

Abbreviations. 1: stationary point preceding inferior conjunction, at which retrograde motion begins; 2: stationary point following inferior conjunction, at which retrograde motion ends; D: departure from; E: Venus as Evening Star; ic: inferior conjunction; M: Venus as Morning Star; me: maximum elongation; op: opposition; r: first appearance (rising), as morning or evening star; s: last visibility (setting), as morning or evening star. Deviations are given in days; − indicates that the recorded date precedes the event, + that it follows. Positive deviations should be reduced by two, negative deviations increased by two, if the 584 285 correlation favored by Lounsbury is used.

The Bonampak dates fall in the war season, and thus pattern with major territorial warfare. Nahm shows that the capture events are just as strongly correlated with this war season; evidently the Bonampak case is typical, raids often being long-term or long-distance campaigns. Accessions are also independent of season, except that some are apparently tied to success in war.

Several positions in the Venus cycle are associated with these events. They include inferior conjunction, as well as heliacal setting of the evening star before it and heliacal rising of the morning star after it; stationary points, or perceptible motion from them; maximum elongations; and heliacal rising as evening star. Not included are maximum brightness, heliacal setting as morning star, or superior conjunction. The number of agreements of star events with points in the Venus cycle is statistically significant: ten or 12 warfare dates out of 25 explicitly marked as star events have precise associations with points in the Venus cycle; given that none occur during invisibility around superior conjunction, there is less than a 0.15 per cent chance of getting as many as ten accidental agreements of this sort. Military iconography is associated with the Venus gods depicted in the Venus tables, paralleling Aztec ethnohistoric accounts (Seler, 1904).

Although the cycle of Venus affected the scheduling of a limited class of events, there is no clear relationship between the position of a date in the Venus cycle and the type of event scheduled for that date. There is a certain asymmetry in that war events seem almost always to have been scheduled in connection with planetary cycles, while accessions were so scheduled in a minority of cases—perhaps only or disproportionately when war events led up to the accession. But by patterning similarly with respect to the Venus cycle, the events so scheduled form a single class that differs from other types of events, including the performance of at least those rituals not directly related to accession. The common denominator could in fact be warfare, if the only accessions so scheduled were those related to war, but this is unlikely since some occur at heliacal rising of evening star, long after any prior Venus event could have occasioned the scheduling of battle. Otherwise, the common denominator linking successful intersite warfare, accession and heir-designation, and what distinguishes accession-related rituals from offertory and other rituals that are not directly related to accession, is the taking or perhaps holding of power/authority. This alternative provides a link between the structure of the Venus table and the positions of the Venus cycle at which battles and accessions were located. In the tables, Venus is the subject of a verbal compound that, in Classic texts, describes only accession-related events. It consists of a hand holding a mirror—a symbol of authority, worn on the forehead (Schele and Miller, 1983). The stations nearest superior conjunction are placed so as to ensure that they never actually fall on a date of the planet's invisibility, whereas the dates of heliacal setting of the evening star before and of heliacal rising of the morning star after inferior conjunction are placed at the average positions, guaranteeing that they frequently fall during invisibility (and increasingly for heliacal setting as the stations move ahead with respect to the true Venus cycle by 0.09 days per Venus round). By interpreting the verb in the codex as referring to the regency or power of the Venus god, the deviation from canonical rising and setting near superior conjunction indicates that Venus was still powerful during invisibility around inferior but not superior conjunction. The latter, then, was a period during which accessions and wars should *not* be undertaken, at least under the auspices of Venus.

Along with the scheduling of events via temporal correlates in the Venus cycle, the position of Venus in the sky might correlate with the location or orientation of the events scheduled for those dates. There is little opportunity to know in the case of rituals, except those few whose architectural context is known. In the case of conquests, it can readily be tested, and there is no obvious directional connection in cases in which the conquered and conquering sites are known. One reason may be the constraint on timing of warfare in relation to the agricultural year. For example, significant points in the cycle of Venus as morning star are distributed through about 70 days of the 584-day cycle; a battle against an eastern opponent could be delayed several years waiting for an eastern position for the planet. If political and military exigencies dictate battle during the proximate war season, the range of possible Venus dates will be restricted. Indeed, if superior conjunction falls in that season, dates near maximum elongations, stationary points, or inferior conjunction are not available, and if the season begins shortly after or ends shortly before invisibility around superior conjunction, maximum elongation is the only choice if even it is available.

The Other Planets

Less data are available concerning the cultural correlates of other planets. Only isolated examples of events involving Jupiter, Saturn and Mars have been noted in Classic inscriptions, almost all at stationary points or first detectable movement from them. The only Mars events in Classic texts that I know of and find plausible are ones coinciding with a stationary point of Saturn (Fox and Justeson, 1978; discussion of the Mars association was omitted from the published version) and one involved in triple conjunctions with Saturn and Jupiter, and possibly conjunctions of Mars on Caracol Stela 3 (Kelley, 1983, pp. 190–1); in addition, Lounsbury (1983b) has pointed out that Venus and Mars rose heliacally and in conjunction on what seems to have been the historical base of the Dresden Venus table. Stationary points of Saturn have also been recognized only in association with other planets, Mars (noted above) and Jupiter (Tate, 1988) being attested, while Caracol Stela 3 seems to relate the motions of Saturn and Jupiter (not necessarily by conjunctions, although this is often assumed).

Although these planetary dates are too few to permit secure conclusions concerning the kinds of events enacted on them, they appear to be of the same type as those occurring at all positions of the Venus cycle. The Saturn-Jupiter stationary points are identified as occasions for accession-related ritual by Lounsbury and Tate. The Mars-Saturn stationary point was probably the occasion for a battle or raid: it is marked as the death date of an aged ruler of El Cayo, and occurs in the middle of the war season.

Indeed, this was a war season in which there may have been no viable Venus position; the rising of the evening star would have been at or before its beginning, and it ended before maximum eastern elongation.

This leads to a hypothesis that significant points in the cycles of Saturn or Jupiter may have been selected as occasions for warfare when no Venus date is appropriate to the war season in a given year. I have reanalyzed the 25 war dates marked as star events in relation to the stationary points of Saturn and Jupiter, and their disappearance around inferior conjunction (Table 22.9, columns d and e). The suspected association was not confirmed, for the same proportion of Saturn and Jupiter events is found for dates with Venus associations as for those without. Unexpectedly, however, dates apparently associated with Saturn or Jupiter outnumber those associated with Venus: while half of the dates related to positions in the Venus cycle, fully three quarters of them occur at the significant points in the cycle of Saturn or Jupiter. Eleven of them lie within ten days of a stationary point of Saturn and/or Jupiter, and five within ten days of opposition; eight of the Jupiter/Saturn dates are among the 13 with no Venus association, and eight are among the 12 having Venus associations. Turning to the five accessions marked as star events, and discussed by Lounsbury (1983a), the same associations appear: two prove to be at significant points in the Saturn cycle, one in the Jupiter cycle, and one in both cycles. These associations are statistically significant, whether opposition dates are excluded or included. The Saturn-Jupiter dates discussed by Tate and Lounsbury are also dates of rituals—non-offertory, except for bloodletting rituals—that are related textually to prior accessions.

With this expanded sample of Saturn and Jupiter dates, it can be seen that the same types of events are being scheduled according to the same essential points of the planetary cycles as for Venus. They also show the same avoidance: there are no examples during Jupiter's or Saturn's disappearance around superior conjunction. However, this apparent avoidance is not statistically significant given the current sample size.

The association of not only Venus but also Jupiter and Saturn with dates of war and other events marked by the star sign suggests a check of Mars as well (Table 22.9, column f), particularly since it is brighter and faster than Saturn. Six of the 25 dates occur within 20 days of a stationary point in the 780-day synodic period of the planet, and another is within two or three days of opposition; these, too, are, statistically significant numbers. The explicitly marked star events therefore relate to planets generally, not specifically to Venus as the star par excellence; correlatively, the prime referent of the star sign is in fact either planet or star, as Kelley has long argued, and not specifically Venus as is generally supposed. The rare use *e·k'* "star" rather than the compound *čak e·k'* "Venus," lit. "great star," to refer specifically to Venus, is simply the use of a generic term in a context in which its specific referent has already been established.

It can be seen in the table that, although a few dates are within a few days of opposition of Saturn, Jupiter or Mars, only one occurs on a date for which no Venus event or stationary point of Jupiter, Mars or Saturn was also proximate. Thus, it is possible that only stationary points were relevant in the case of the superior planets. Alternatively, it may be their retrograde periods that are relevant (recall the axe event of Dresden's

78-day Mars almanac); of the 13 war dates lacking Venus correlates, nine fall during and 12 within ten days of the retrograde period of Saturn, Jupiter or Mars.

Taking the four major planets together, the likelihood of a given date falling within ten or 20 days of a significant point (for the superior planets) or at the more precisely defined points in the Venus cycle, is quite high—between 36.02 per cent and 64.50 per cent. Nonetheless, the large number of positive cases renders the overall pattern statistically significant, with only a fraction of 1 per cent probability of attaining the observed number of dates showing a planetary association with ten- or 20-day leeway, including or excluding dates of opposition, without a genuine association. Given 20 days' leeway, only one date (9.15.12.2.2) fails to have an association. Lounsbury notes that Venus and Mercury were in conjunction on this date; Mercury was also at maximum elongation.

In summary, almost every star date is correlated with one of the seven major parts of the Venus cycle or with stationary points (with or without oppositions) of Jupiter, Mars or Saturn. This need not indicate a rigorous planning of intersite warfare according with planetary behavior, however. Because the data suggest that ten or 20 days may separate an actual celestial event from the date recorded in association with it, it is usually possible to find an appropriate planetary event near a time chosen on the basis of political, economic or military opportunity: in individual cases, the probability of falling within 20 days of a significant date is either 52.67 per cent or 64.50 per cent, depending on the inclusion of opposition dates; even restricting to cases within ten days, the likelihood is 36.02 per cent or 43.55 per cent. Although the Maya were clearly manipulating the dates of their enterprises, or ceremonies associated with them, in relation to the planetary cycles, this evidently imposed little constraint on more mundane considerations governing the strategic timing of events. It may have served more to supernaturally sanction strategic decisions of the elite, or to present them as the decisions of the god planets.

As a caveat, the leeway may seem broader in our data than it was in Maya practice, for it may have a number of different sources; for example, apart from questions of perceptibility of motion and conditions of observation, the activity most precisely associated with the planetary event may not be the one recorded. There is no evidence for any leeway at all with respect to the retrograde period, from one perceived stationary point to the next.

The case of Mercury is not treated here. Its period is so short that, with any deviation from the exact date of a phenomenon, an extremely high proportion of dates must fit to be able to show a significant association. Because the planet's proximity to the sun makes for extreme difficulty in viewing, A. Schlak (unpublished manuscript, 1985) has suggested that proximity to the position of maximum elongation would be the prime point of interest in its cycle; this position was relevant to Venus as well so its pertinence is plausible. In fact, nine of the 24 star dates are within ten days of Mercury's maximum elongation (Table 22.9, column h). This is comparable to the numbers of Venus, Saturn and Mars dates, so an association of Mercury's maximum elongation with star events is feasible; indeed, the one star date not associated with any other planet occurs exactly on the date of Mercury's maximum elongation in Thompson's correlation (at 18.6°). Although the hypothesis is reasonable and may be correct, even this high a proportion of Mercury dates has a 52.4 per cent chance of occurring at random.

Thus, an association of star dates with Mercury's elongation from the sun cannot yet be substantiated. If substantiation is to occur, a substantially increased body of evidence will have to be gathered. Schlak has proposed a correlation of Mercury's maximum elongation with two classes of events, accessions and the holding of a symbol of authority representing God K (Bolon Dzacab), but his examples are selective rather than exhaustive. The selection criteria may be related systematically to maximum elongation, or to some associated characteristic such as a calendric pattern, since the sample appears to be biased in favor of such dates; and in a larger sample, including all accessions listed in Schele (1982), I find no statistically significant association of accession dates with large elongations of Mercury. The association with the holding of the God K scepter may turn out to be genuine, but I have not attempted to test this. If it does, this would provide circumstantial support for its relevance in the case of the star events.

CONCLUSIONS

Cyclic commensuration in the ritual calendar was the basic approach used by ancient Mesoamerican astronomers to arrive at models for predicting the occurrence of celestial events in linear time. This paper has examined the potential and limitations of this method, applied to the moon and the visible planets; related the celestial events explicitly mentioned in texts to the activities scheduled in terms of them; and addressed methodological problems faced by epigraphers attempting to recognize unstated astrological content in hieroglyphic texts. Applied to the lunar cycle, Maya astronomical method was strikingly successful: it would lead to immediate success in long-term eclipse prediction, and to accurate placement of eclipse stations once records of roughly 50 years of eclipse observation were available. It was also successfully applied to Venus, and in a more limited way to Mars. For the other planets it was unsatisfactory. It meshed neatly with the synodic period of Venus and the 365-day vague year to create a long-term Venus calendar, one whose structure also incorporated lunar synodic month stations and eclipse recurrence intervals. The Mars cycle matched a tripled ritual calendar so closely that a single synodic period was presented in tabular form, reduced further to the subdivided ritual calendar spans typical of Maya ritual almanacs, but corrective mechanisms could not accommodate the very slow departure of the planet's motions from their ritual calendar stations. On the other hand, it did not accommodate long-term motions of Mercury, Jupiter or Saturn.

The Maya calendar priest's approach to the anticipation of celestial events did not simply permit the development of useful predictive models; it also constrained the essential form that these models could take. Included in this set of constraints was a failure to accommodate regular, symmetrical variations in the synodic periods of the planets. The Maya regularly made use of solar/sidereal/synodic commensurations that structure the recurrence of synodic variants in dynastic ritual (and of no other multiples of sidereal periods), but the tables' imposed structures required reuse of a given synodic chart with different synodic variants; an average value was the closest that could be used.

Such constraints, however, did not preclude the development of simple formal *calendars* related to the moon and perhaps to Venus that differed in structure from the predictive models of celestial activity. Stations in planetary cycles were occasions for sacred and secular war. The main verb of the Mars table of the *Dresden Codex* refers to warfare and is so used consistently in the inscriptions; the Venus table depicts the god star Venus in warrior's garb. In the Classic period, events explicitly related to the cycles of the visible planets (save perhaps Mercury) were recorded as occasions for territorial warfare and for certain non-offertory rituals. The latter are mainly rituals of accession to supreme political power (perhaps especially when otherwise linked to warfare) and bloodletting rituals that were linked to warfare or accession, recalling references to Venus as the subject of an accession verb in the *Dresden Codex*. The planet involved is evidently not specified in the Classic records.

Demonstrating the pertinence of astronomical correlates of most Maya dates depends upon their recurrent association with iconographically or epigraphically definable complexes; even events explicitly marked as having celestial correlates refer essentially to human activities, and some activities with demonstrably pertinent celestial correlates lack any explicit indication of them. Statistical validation therefore proves as crucial in assessing epigraphic evidence for ancient astronomical practices as it is in architectural and archaeological studies. Such assessments are made here for events explicitly marked as referring to "stars." The dates of these events are verified as taking place at significant points in the cycles of Venus, Mars, Jupiter and Saturn. The test depends, however, on a correlation of Maya to Christian chronology in the Goodman family, which I accept along with most other epigraphers; Kelley (1983) presents a cogent critique on epigraphic and ethnohistoric grounds that has to be rebutted in detail before this hypothesis can be considered secure.

Positional astronomy is reflected only in the increasing evidence for reference to asterisms along the ecliptic ("the Maya zodiac"), but epigraphic evidence for Maya astronomy is otherwise exclusively in temporal patterns; this is a bias of the source rather than a reflection of the full range of Maya astronomical knowledge and inquiry.

ACKNOWLEDGMENTS

Thanks are due to James Fox, Peter Mathews, Linda Schele, and especially Floyd Lounsbury for discussion of various astronomical aspects of Maya hieroglyphic texts. Anthony Aveni, Victoria Bricker, Werner Nahm, Merideth Paxton, Arthur Schlak and Gordon Whittaker provided access to unpublished work; this overview has been substantively improved by reference to these works, both in content and via the issues they have raised. Special thanks are due to Dennis Sinnott, who provided his computer program for computing planetary, solar, and lunar positions, and to Manfred Kudlek, for his 1978 tables of eclipses visible at Tikal.

REFERENCES

Aveni, A. F. 1980. *Skywatchers of Ancient Mexico.* Austin, University of Texas Press.

Aveni, A. F. 1992. The Moon and the Venus Table: An Example of Commensuration in the Maya Calendar. In *The Sky in Mayan Literature,* ed. A. Aveni, pp. 87–101.

Aveni, A. F., ed. 1977. *Native American Astronomy.* Austin, University of Texas Press.

Aveni, A. F., S. L. Gibbs, and H. Hartung. 1975. The Caracol Tower at Chichén Itzá: An Ancient Astronomical Observatory? *Science* 188:977–985.

Bricker, H. M., and V. R. Bricker. 1983. Classic Maya Prediction of Solar Eclipses. *Current Anthropology* 24:1–24.

Bricker, V. R. 1983a. Classic Maya Observations of Planetary Retrograde Motion. Paper presented at the Conference on Ethnoastronomy, Washington, DC, September.

Bricker, V. R. 1983b. Directional Glyphs in Maya Inscriptions and Codices. *American Antiquity* 48:347–353.

Bricker, V. R., and Bricker, H. M. 1986. The Mars Table in the Dresden Codex. In *Research and Reflections in Archaeology and History: Essays in Honor of Doris Stone,* ed. E. W. Andrews V. New Orleans, Middle American Research Institute Publication 57.

Brotherston, G. 1983. The Year 3113 BC and the Fifth Sun of Mesoamerica: An Orthodox Reading of the *Tepexic Annals.* In *Calendars in Mesoamerica and Peru: Native Computations of Time,* ed. A. Aveni and G. Brotherston, 167–220. Oxford, British Archaeological Reports Series S174.

Closs, M. 1977. The Date-reaching Mechanism in the Venus Table of the *Dresden Codex.* In *Native American Astronomy,* ed. A. F. Aveni, 89–99. Austin, University of Texas Press.

Closs, M. 1979. Venus and the Maya World: Glyphs, Gods, and Associated Phenomena. In *Tercera Mesa Redonda de Palenque,* ed. M. G. Robertson, 147–166. Pebble Beach, Robert Louis Stevenson School.

Closs, M. 1981. Venus Dates Revisited. *Archaeoastronomy* 4:38–41.

Closs, M., A. Aveni, and B. Crowley. 1984. The Planet Venus and Temple 22 at Copan. *Indiana* 9:221–247.

Coggins, C. 1980. The Shape of Time: Some Political Implications of a Four-part Figure. *American Antiquity* 45:727–739.

Coggins, C. 1982. The Zenith, the Mountain, the Center, and the Sea. In *Ethnoastronomy and Archaeoastronomy in the American Tropics,* ed. A. F. Aveni and G. Urton, 111–123. Annals of the New York Academy of Sciences, 385.

Edmonson, M. S., ed. and trans. 1982. *The Ancient Future of the Itza: The Book of Chilam Balam of Chumayel.* Austin, University of Texas Press.

Förstemann, E. 1904. Page 24 of the Dresden Maya Manuscript. *BAE Bulletin* 28:431–443. Washington, DC, Smithsonian Institution.

Förstemann, E. 1906. *Commentary on the Maya Manuscript in the Royal Public Library of Dresden.* Peabody Museum of Archaeology and Ethnology, Paper 4.2, Cambridge, MA.

Fox, J. A., and J. S. Justeson. 1978. A Mayan Planetary Observation. *UCARF Contributions* 36:55–59.

Fox, J. A., and J. S. Justeson. 1980. Mayan Hieroglyphs as Linguistic Evidence. In *Palenque Round Table, 1978*, Part 2, ed. M. Greene Robertson. Palenque Round Table Series, vol. 5, 204–216. Austin, University of Texas Press.

Fox, J. A., and J. S. Justeson. 1984. Polyvalance in Mayan Hieroglyphic Writing. In *Phoneticism in Mayan Hieroglyphic Writing*, ed. J. S. Justeson and L. Campbell, 17–76. Albany, IMS Publication 9.

Gibbs, S. L. 1977. Mesoamerican Calendrics as Evidence of Astronomical Activity. In *Native American Astronomy*, ed. A. F. Aveni, 21–35. Austin, University of Texas Press.

Graham, I. 1967. *Archaeological Explorations in El Petén, Guatemala*. New Orleans, Middle American Institute Publication 33.

Graham, J. A. 1972. *The Monumental Art and Hieroglyphic Inscriptions of Altar de Sacrificios, Guatemala*. Peabody Museum of Archaeology and Ethnology, Paper 64.2. Cambridge, MA.

Hassig, R. 1988. *Aztec Warfare: Imperial Expansion and Political Control*. Norman, University of Oklahoma Press.

Houston, S. D. 1984. An Example of Homophony in Maya Script. *American Antiquity* 49: 790–805.

Houston, S. D., and P. Mathews. 1985. *The Dynastic Sequence of Dos Pilas, Guatemala*. Pre-Columbian Art Research Institute, Monograph 1.

Jones, G., and L. Satterthwaite. 1982. *The Monuments and Inscriptions of Tikal: The Carved Monuments*. Tikal Report 33, Part A. Philadelphia, University Museum.

Justeson, J. S., W. M. Norman, and N. Hammond. 1988. The Pomona Flare: A Preclassic Maya Hieroglyphic Text. In *Maya Iconography*, ed. E. P. Benson, pp. 94–151. Princeton, NJ, Princeton University Press.

Justeson, J. S., W. M. Norman, L. Campbell, and T. Kaufman. 1985. *The Foreign Impact on Lowland Mayan Language and Script*. New Orleans, MARI Publication 53.

Justeson, J. S., and L. Campbell, eds. 1984. *Phoneticism in Mayan Hieroglyphic Writing*. Albany, IMS Publication 9.

Kelley, D. H. 1972. The Nine Lords of the Night. In *Studies in the Archaeology of Mexico and Guatemala*, ed. J. A. Graham. *UCARF Contributions* 16:58–68.

Kelley, D. H. 1975. Planetary Data on Caracol Stela 3. In *Archaeoastronomy in Pre-Columbian America*, ed. A. F. Aveni, 257–262. Austin, University of Texas Press.

Kelley, D. H. 1976. *Deciphering the Maya Script*. Norman, University of Oklahoma Press.

Kelley, D. H. 1977a. Maya Astronomical Tables and Inscriptions. In *Native American Astronomy*, ed. A. F. Aveni, 57–73. Austin, University of Texas Press.

Kelley, D. H. 1977b. A Possible Maya Eclipse Record. In *Social Processes in Maya Prehistory*, ed. N. Hammond, 405–408. New York, Academic Press.

Kelley, D. H. 1980. Astronomical Identities of Mesoamerican Gods. *Journal for the History of Astronomy* 9, Suppl. 2, S1–54.

Kelley, D. H. 1983. The Maya Calendar Correlation Problem. In *Civilization in the Ancient Americas: Essays in Honor of Gordon R. Willey*, ed. R. M. Leventhal and A. L. Kolata, 157–208. Albuquerque, University of New Mexico Press.

Kelley, D. H., and A. Kerr. 1973. Maya Astronomy and Astronomical Glyphs. In *Mesoamerican Writing Systems*, ed. E. P. Benson, 179–215. Washington, DC, Dumbarton Oaks.

Linden, J. 1986. Glyph X of the Maya Lunar Series: An Eighteen Month Lunar Synodic Calendar. *American Antiquity* 51:122–136.

Lounsbury, F. G. 1976. A Rationale for the Initial Date of the Temple of the Cross at Palenque. In *Primera Mesa Redonda de Palenque, Part 2*, ed. M. G. Robertson, 5–19. Pebble Beach, Robert Louis Stevenson School.

Lounsbury, F. G. 1978. Maya Numeration, Computation, and Calendrical Astronomy. *Dictionary of Scientific Biography* 15:759–818. New York, Scribner's.

Lounsbury, F. G. 1982. Comment on Bricker and Bricker 1982. *Current Anthropology* 24:22.

Lounsbury, F. G. 1983a. Astronomical Knowledge and Its Uses at Bonampak, Mexico. In *Archaeoastronomy in the New World*, ed. A. F. Aveni, 143–168. Cambridge, Cambridge University Press.

Lounsbury, F. G. 1983b. The Base of the Venus Table of the Dresden Codex, and Its Significance for the Calendar-Correction Problem. In *Calendars in Mesoamerica and Peru*, ed. A. F. Aveni and G. Brotherston, 1–21. London, British Archaeological Reports S 174 (International series).

Lounsbury, F. G. 1984. Glyphic Substitutions: Homophonic and Synonymic. In *Phoneticism in Mayan Hieroglyphic Writing*, ed. J. S. Justeson and L. Campbell, 167–184. Albany, IMS Publication 9.

Malmstrøm, V. H. 1973. Origin of the Mesoamerican 260-day Calendar. *Science* 181:939–941.

Malmstrøm, V. H. 1978. A Reconstruction of the Chronology of Mesoamerican Calendrical Systems. *Journal for the History of Astronomy* 9:105–116.

Marcus, J. 1976. The Origins of Mesoamerican Writing. *Annual Review of Anthropology* 5:35–67.

Martínez Hernández, J. 1929. *Diccionario de Motul, maya español, atribuido a fray Antonio de Ciudad Real y arte de lengua maya por fray Juan Coronel*. Merida, Tipográfica Yucateca.

Mathews, P. 1986. Early Classic Maya Monuments and Inscriptions. In *The Maya Early Classic*, ed. G. R. Willey and P. Mathews. Albany, IMS Publication 10.

McCluskey, S. C. 1983. Maya Observations of Very Long Periods of Venus. *Journal for the History of Astronomy* 14:92–101.

Meinshausen, M. 1913. *Über Sonnen- und Mondfinstennisse in der Dresdener Mayahandschrift.* Zeitschrift für Etnologie 45 (2):221–227.

Miller, M. 1986. *The Murals of Bonampak*. Princeton, NJ, Princeton University Press.

Norman, W. M. 1984. Grammatical Analysis of Mayan Hieroglyphs. Paper presented at the 2nd Annual Workshop on Maya Hieroglyphs, University Museum, University of Pennsylvania, Philadelphia.

Proskouriakoff, T. 1960. Historical Implications of a Pattern of Dates at Piedras Negras, Guatemala. *American Antiquity* 25:454–475.

Proskouriakoff, T. 1963. Historical Data in the Inscriptions of Yaxchilan. *Estudios de Cultura Maya* 3:147–167.

Proskouriakoff, T. 1964. Historical Data in the Inscriptions of Yaxchilan. *Estudios de Cultura Maya* 4:177–201.

Riese, B. 1982. Kriegsberichte der klassischen Maya. *Baessler-Archiv: Beiträge zur Völkerkunde* 30:255–321.

Satterthwaite, L. 1958a. Early "Uniformity" Maya Moon Numbers at Tikal and Elsewhere. *Thirty-third International Congress of Americanists* 2:200–210.

Satterthwaite, L. 1958b. Five Newly Discovered Monuments at Tikal and New Data on Four Others. *Tikal Reports* 4. Philadelphia, University Museum.

Schele, L. D. 1982. *Maya Glyphs: The Verbs*. Austin, University of Texas Press.

Schele, L. D., and J. H. Miller. 1983. *The Mirror, the Rabbit and the Bundle: Accession Expressions in the Classic Maya Inscriptions*. Studies in Pre-Columbian Art and Archaeology 25. Washington, DC, Dumbarton Oaks.

Seler, E. 1904. Venus Period in the Picture Writings of the Borgian Codex Group. *BAE Bulletin* 28:353–391. Washington, DC, Smithsonian Institution.

Spinden, H. J. 1916. The Question of the Zodiac in America. *American Anthropologist* 18:53–80.

Spinden, H. J. 1924. *The Reduction of Maya Dates*. Peabody Museum of American Archaeology and Ethnology, Paper 6.4, Cambridge, MA.

Stahlman, W. D., and O. Gingerich. 1963. *Solar and Planetary Longitudes for Years −2500 to +2000 by 10-day Intervals*. Madison, University of Wisconsin Press.

Stone, A. 1983. Epigraphic Patterns in the Inscriptions of Nah Tunich Cave. In *Contributions to Maya Hieroglyphic Decipherment*, ed. S. D. Houston, 1:88–103. New Haven, CT, HRAF.

Stuart, D. 1984. A Note on the "Hand-scattering" glyph. In *Phoneticism in Mayan Hieroglyphic Writing*, ed. J. S. Justeson and L. Campbell, 307–310. Albany, IMS Publication 9.

Stuart, D. 1985. The "Count of Captives" Epithet in Classic Maya Writing. In *Fifth Palenque Round Table, 1983*, ed. M. Greene Robertson and V. M. Fields, Palenque Round Table Series, 7:97–101. San Francisco, Pre-Columbian Art Research Institute.

Tate, C. 1986. Summer Solstice Ceremonies Performed by Bird Jaguar III of Yaxchilan, Chiapas, Mexico. *Estudios de Cultura Maya* 16:85–112.

Teeple, J. 1925. Maya Inscriptions: Glyphs C, D, and E of the Supplementary Series. *American Anthropologist* 27:108–115.

Teeple, J. 1930. *Maya Astronomy*. Washington, DC, CIW Contributions to American Archaeology 1.2.

Thompson, J.E.S. 1935. *Maya Chronology: The Correlation Question*. Washington, DC, CIW Publication 456, Contribution 14.

Thompson, J.E.S. 1950. *Maya Hieroglyphic Writing: An Introduction*. Washington, DC, CIW Publication 589. (Reprinted 1960, 1971, 1979 by University of Oklahoma Press.)

Thompson, J.E.S. 1972. *A Commentary of the Dresden Codex, a Maya Hieroglyphic Book*. Philadelphia, APS Memoir 93.

Webster, D. 1975. Warfare and the Evolution of the State: A Reconsideration. *American Antiquity* 40:464–470.

Whittaker, G. 1986. The Mexican Names of Three Venus Gods in the Dresden Codex. *Mexikon* 8 (3):56–60.

Willson, R. W. 1924. *Astronomical Notes on the Maya Codices*. Peabody Museum of American Archaeology and Ethnology, Paper 6.3, Cambridge, MA.

CHAPTER TWENTY-THREE

Astronomical Knowledge and Its Uses at Bonampak, Mexico

Floyd G. Lounsbury

Needless to say, an unchallengeable correlation of the two calendars [Mayan and European] would be immensely helpful in identifying astronomical data in the texts, although I myself am far from convinced that planetary observations were recorded on stone monuments, unless favourable phenomena perhaps governed a ruler's accession date.

—J.E.S. THOMPSON (1974: 85)

This paper is concerned with Maya attention to the planet Venus, focusing on information that comes from one site, Bonampak, in the state of Chiapas, southern Mexico, during the sixth through the eighth centuries of our era. References are made to data from a few other sites for purposes of comparison. Six pages in the Maya hieroglyphic book known as the Dresden Codex tell us most of what we have known

---

From *Archaeoastronomy in the New World*, ed. A. Aveni (New York: Cambridge University Press, 1982).

about the Maya knowledge of Venus. Five of these pages detail the subdivisions of the five synodic periods that correspond approximately to eight solar years, the anciently known eight-year cycle of Venus [$5 \times 584 = 8 \times 365, = 2920$]. The schema of canonical periods for morning star, superior conjunction, evening star, and inferior conjunction [$236 + 90 + 250 + 8 = 584$] is repeated thirteen times on each of these five pages, bringing it into concord with the 260-day divinatory almanac, and uniting these into a grand cycle of 65 synodic periods of Venus, equal to 146 of the almanac and to two of the 52-year rounds of the calendar [$65 \times 584 = 146 \times 260, = 2 \times 52 \times 365$]. The sixth page, a preface to the other five, marks the historical institution of a device that ingeniously accommodates this scheme to the need for correcting accumulations of small errors (5.2 days in 65 Venus periods or 104 years), accomplishing this without altering the fit 'of the Venus calendar to the almanac, and thus obviating the otherwise inevitable need for total revision. (Transcriptions and analyses may be found in Teeple 1930, Thompson 1950 and 1972, and Lounsbury 1978.) A substantial sequence of observations and record-keeping must have preceded the discovery of the basic cycles, and even more the invention of the corrective device. The "explanatory hypotheses" that were posited by the Maya to account for these regularities may be surmised from the accompanying figures and hieroglyphs. These depict and name the several presiding deities, the five warrior-guises of the Morning Star, and some of his likeliest victims. It is known from Mexican sources, both historical and pictographic, that the heliacal rising of the planet following inferior conjunction was regarded as a time of especial danger.

From the Codex we learn also the hieroglyph of Venus, which occurs in two principal variants (Fig. 23.1a, b). These appear to be in free variation, since their selection correlates in no consistent way with either astronomical or textual context. The glyph is composite. Its first component, prefixed or superfixed, is the sign that is otherwise attested as signifying the color "red" and as being phonetically *chac* in Yucatec Maya. Whether the Maya saw Venus as red, or whether the sign was employed here as a rebus for chác meaning "giant," is uncertain. Yucatec names for the planet Venus given in the 16th-century Dictionary of Motul are *chac ek* and *noh ek,* respectively "red (or perhaps giant) star" and "great star." In other Mayan languages names that mean "great star" predominate, though there are others, particularly for the Morning Star. If one of these was applicable to the hieroglyph (which in the Codex table stands for all aspects and phases of the planet) one is led to conclude that the second component of the glyph, in either of its principal forms, was literally and most simply "star." In one instance however, where its reference to Venus is certain (Fig. 23.1c), and in five others where its reference is probably the same, the presumed "star" component is employed by itself without the prefix. The point of this—and for occurrences of the glyph in the inscriptions it is a pertinent point—is that the main component by itself can be taken in either of two senses: either in a general sense as "star," or in a specific sense as "THE star" or "Venus." Further, it may be noted that this "star" or "Venus" sign serves occasionally in the inscriptions as an alternate sign for the day Lamat, whose more usual form is derivative.

Floyd G. Lounsbury

23.1 Hieroglyphs denoting Venus.

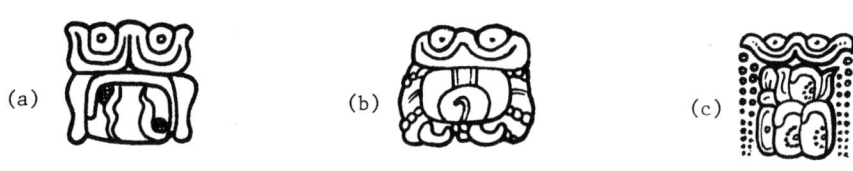

23.2 Hieroglyphs of "star over earth," "star over shell," and "star over Seibal."

There is yet another set of contexts in which this "star" or "Venus" sign appears, in which it has posed a curious puzzle. In these it stands as a superfix over an "earth" sign *(cab, Caban)*, or alternatively over a so-called "shell" sign (of uncertain meaning and reading), or—as yet another alternative—over one or another place-name sign, i.e., over the main component of the "emblem glyph" of a site, with an adjoined locative prefix. And in these it is usually flanked by a pair of like affixes which are diagnostic though of unknown value. These glyphs (Fig. 23.2) had once been suspected of having an astronomical meaning, possibly in relation to Venus, or possibly even to some other "star." But then some occurrences became known that put the matter in a quite different light and led to another, nonastronomical, hypothesis. Two of these occurrences are from a pair of inscriptions that tell the same story, in which the glyph in question designates the first of a sequence of three events, the second and third of which are the capture of an enemy king in battle on the second day and a standard ritual event (possibly a blood offering of some sort) six days after that. On both of these monuments the bound captive is shown under the feet of the victor, and on one of them his name and his emblem glyph (his rank and his home address) are still legible. They identify him as the king of Seibal. But it is precisely the Seibal site name, with a prefixed locative, that stands under the "star" or "Venus" sign in the first of the three event glyphs, the one in which we are interested. This seemed to force a conclusion that the "star-over-Seibal" glyph designates the raid, or some circumstance of the raid, on that enemy capital; and that the associated date is the day of that raid. Accordingly, the glyphs of this category—"star over earth," "star over shell," "star over place-name"—have more recently been interpreted as having to do with warfare; and this has turned out to be a tenable and productive hypothesis. Yet, following the line of the earlier suspicion, David Kelley in exploratory papers (1973, 1977) was able to show that in some cases—of sufficient number to warrant interest—the dates of such events could be sorted into sets, within which they were separated from each other by intervals that tended to approximate some multiple of 584, or they occurred on almanac days that tended to cluster around one or another of the canonical days for a phenomenon of Venus in the Dresden Codex table. This, then, has made it difficult to dispel the suspicion of an astronomical or astronumerological significance for at least some of the "star-over-X" dates.

With this as background, attention can be turned now to Bonampak.

THE BONAMPAK MURALS

The site of Bonampak is in eastern lowland Chiapas, at 16° 44' north latitude and 91° 05' west longitude. Its principal known structure, a three-room building set on a low platform pyramid, with once richly painted interior walls and vaults, was discovered in 1946. The murals were photographed and rendered by artists within the next three years, during brief field trips under difficult circumstances. Further copies, both photographic and painted, were made in 1964. Few of the photographs, from either 1946–49 or 1964, have yet been made available. So it is the artists' renderings on which we must principally depend. The hieroglyphs, unfortunately, received something less than their proper due in these attempts at documentation, for neither with photographers nor with artists could they compete for attention against the rich and fantastic costuming in the ceremonial scenes of Rooms 1 and 3, or against the drama of the battle scene and the humiliation of prisoners in Room 2. And though the artists did an otherwise remarkable job of copying, they were not epigraphers. Nevertheless, the two dates in the hieroglyphic text of Room 1 have been ascertained with confidence; and the reconstruction of a third date, in Room 2, can be accepted I believe with equal assurance. As for what happened on those dates, at least the event associated with the third of them, in Room 2, is clear from the murals: it was a victorious raid on an enemy kingdom, resulting in the capture of prisoners. And as for those associated with the first two dates, it can suffice to refer to them in general terms as dynastic rituals; for the murals of Room 1 guarantee at least that much of a conclusion. If there was a date in Room 3, where the victory celebration is in progress, it has not survived.

Eric Thompson (in Ruppert et al. 1955) read the initial date of the text in Room 1 as 9.18.0.3.4 in the Maya day count (i.e., day no. 1,425,664), which is 10 Kan in the 260-day almanac, and 2 Kayab in the 365-day calendar year. In Thompson's correlation of the Mayan and European calendars, this would be AD 790, December 10 of the Julian calendar. The date was further characterized in the Maya manner for its attributes in the lunar calendar, namely moon age, position in the half-year, month duration (whether 29 or 30 days), and another detail not understood. It was the dark of the moon, precisely a date of lunar-solar conjunction. Three months had elapsed in the half-year, which at that time was reckoned in such a way that eclipse seasons came approximately in the middle of the half-year. It was not, however, an eclipse date; although it was just five lunar months (148 days) after the only known instance of an apparent solar eclipse date specifically designated as such on a Maya monument (cf. Teeple 1930: 115).

The second date of the text is stated to be 16-score and 16 days after the initial date. It is 8 Ahau 13 Muan in the calendar round, of which the 13 Muan is still legible

23.3 Cartouches of the northeast vault of Room 2. After Tejeda (1955), modified in accord with copy by R. Lazo in the National Museum of Anthropology and History, Mexico, and with information from M. Miller (1981: 100) based on examination of photographs.

but the 8 Ahau is not. The redundancies in all this insure the accuracy of the interpretation, even where parts are no longer legible. This date, by the same correlation, would be AD 791, November 11. As already noted, the events on these two dates appear to have been dynastic rituals of some considerable importance.

The third date is in a brief text accompanying the depiction of the victorious warrior-king in Room 2, as he administers a ritual coup to the principal captive at the culmination of that day's battle. The numerical coefficients of the two components of the calendar-round date are crystal clear: the day was 13-something 13-something. But the day sign and the month sign are faded beyond recognition. Since the sequence of the events depicted in the murals of the three rooms pretty surely corresponds to the numbering that has been assigned to these rooms, and since there is no accompanying distance number or other anchor with this date, it can be assumed to be the *next* "13-whatever 13-whatever" after the last date in Room 1. Now this is a problem that allows of a fairly large number of solutions. The real problem is how to choose from among them, in case there should be reason for wanting to know which one it actually was. As it happens, there was someone who had such a reason and did want to know; and that is how I got drawn into all this.

Early in September of 1980 Miss Mary Miller, then a graduate student in Art History at Yale University who was writing a dissertation on the murals of Bonampak, showed me this problem and inquired about the possibility of a solution. She expressed the suspicion, moreover, that this now hidden date might have had some astronomical significance for the Maya. She had some interesting reasons for the suspicion, deriving from her analysis of the murals where the date is found.

At the top of the sloping vault of the north wall of Room 2 (actually northeast), over the scene of the display of prisoners, but over and opposite the scene of the battle, are four large cartouches containing animal and human-like figures (see Fig. 23.3). They are above a "sky band," which position identifies them as representing celestial objects. The leftmost is a pair of peccaries, with "star" signs about them. The rightmost is a turtle, with three "star" signs on its back. The two in between are human-like figures, each with an adjacent "star" sign. One holds a star in his left hand and some other object in his right, while the other partially supports himself with his left while apparently hurling a dart or wielding a baton with his right. All of this is suggestive of zodiacal astrology, for it is known from a thirteenth-century hieroglyphic codex (the Paris Codex) and from the sixteenth-century Motul Dictionary, as well as from twentieth-century native testimony (Thompson 1950: 116), that the Maya have recognized a "turtle" constellation. Further, a band of astral or zodiacal signs inscribed on the Casa de las Monjas at Chichen Itza also confirms this and gives evidence for a "peccary" as well, with both the turtle and the peccary depicted over "star" signs within their respective square cartouches.

It should be apparent now what it was that gave rise to Miss Miller's suspicion and her inquiry about the possibilities for the partially obscured battle date. These "star" figures and signs loom over the scenes of battle and victory. She could not but wonder whether they, or any one of them, might be an architectural and pictorial manifestation of the concept that lay behind the "star-over-earth/shell/placename" glyphs (Fig. 23.2), which in a number of cases can be understood as designating raids or battles, but which in some cases—as Kelley showed—seem yet to have a connection with Venus calendrics. Her question was: Is it possible that this Bonampak raid might have been timed for a Venus phenomenon? Or for any other astronomical phenomenon? A related question was whether the peccary and turtle constellations could be pertinent to the date of that raid. And it was her suggestion that, whatever the calendar-round day might be, it should be the next one of its kind after the dates of Room 1, and probably at not too long an interval after the second of these.

To answer whether it was possible for the raid to have been timed for a Venus or other astronomical phenomenon is of course easier than to answer whether—assuming a positive answer as to possibility—it actually was so timed. It requires only an enumeration of the possibilities and checking them out. But checking them out, with planetary tables etc., requires a calendar correlation that can be applied with confidence. A few years ago, in an article on Maya numeration and astronomy (1978), I stated that the precise correlation between the Mayan and the Julian day counts was still uncertain. That was a concession to the current state of professional opinion on the subject, such as is appropriate in an encyclopaedia article, which that was. As for myself, I had little doubt then, and have even less now, that the Thompson correlation is correct. However, as I admitted, not all are agreed. So a new opportunity to test his or any other hypothetic values should be welcome. Here I shall restrict myself to Thompson's. Three different values of the correlation constant have been associated with his name: his original of 584285, his later revision (to accommodate Highland Guatemalan and Central Mexican data) of 584283, and the in-between value (proposed by Beyer) of 584284. It is clear now that for Classic Maya it is Thompson's original value, 584285, that is appropriate. It is that which I shall use here.

The dates recorded in Room 1, with Julian Calendar equivalents by the 584285 correlation, are as follows:

9.18. 0. 3. 4,   10 Kan 2 Kayab   (AD 790, December 10)
      16.16   (recorded interval)
9.18. 1. 2. 0,   8 Ahau 13 Muan   (AD 791, November 11)

Miss Miller suggested that the 13... 13... date of the battle in Room 2 ought to follow these by a rather short interval. The possibilities for the next few years are the following:

9.18.1.15. 5.	13 Chicchan 13 Yax	(792, August 2)
9.18.2.13.10.	13 Oc 13 Mol	(793, June 23)
9.18.3.11.15.	13 Men 13 Xul	(794, May 14)
9.18.4.10. 0.	13 Ahau 13 Zotz	(795, April 4)
9.18.5. 5. 0.	13 Ahau 13 Kayab	(795, December 20)
9.18.5. 8. 5.	13 Chicchan 13 Uo	(796, February 22)
9.18.7. 1.10.	13 Oc 13 Mac	(797, September 30)

Now it happens that the very first one of these is an astronomically interesting date, offering the possibility of a positive answer to Miss Miller's question, while the remaining dates are of no apparent interest. August 2 (Julian) of the year 792 was exactly the date of an inferior conjunction of Venus.

This is not all. At the latitude of Bonampak (16° 44') this date, which would be August 6 in retroactive Gregorian for that year, was also precisely the date of the sun's zenith passage, returning to the side of the south after having spent 89 or 90 days on the north side of the zenith. And with the ecliptic at 90° to the horizon on this date, it is of some interest that this inferior conjunction of Venus was one of those when the planet was at just about maximum celestial latitude (8° to the south, according to the Tuckerman tables); which, if I am not mistaken, means that the period of invisibility of Venus before and after the conjunction would have been just about minimum, perhaps as little as a day either side.

There is yet more in this date. At this conjunction Venus and the sun were at the celestial longitude of 133° 35'. At the same time (AD 792) the celestial longitude of Regulus was 133° 05', while its latitude was 0° 25'. The three heavenly bodies were in conjunction on that day. It is worth taking a moment to picture what the Bonampak sky watchers would have been seeing (weather permitting) during the period leading up to this date. A month and a half earlier (44 days to be exact) Venus, approaching from the west, had come into conjunction with Regulus. It was a close encounter, for they had only about a quarter of a degree of latitude between them at that time. Venus moved on, reaching its first stationary point 24 days later, having gone about eight and a half degrees beyond Regulus; at which point it went into retrograde and came back to Regulus, reaching it as both went into conjunction with the sun. At their first encounter they were 42½ degrees (nearly three hours) east of the sun; at the time of Venus's turnaround its elongation was 27 degrees (a little less than two hours); and now at the second encounter they went down with the sun, to emerge on the other side into the morning sky. We now that the Maya knew enough to understand (in their terms) where the planet and the star were at this time, even though temporarily lost to view in the brightness of the sun.

Such was the date of August 2 of the year 792, which, if the Thompson correlation is correct, was 13 Chicchan 13 Yax, 9.18.1.15.5. But while the 13's in this Maya date are verified, being clearly legible, the Chicchan and the Yax are hypothetic—merely one possibility out of several that would satisfy the strictly calendrical requirements. On what grounds then can we assume that this actually was the 13... 13... date that was recorded for the battle scene of Room 2?

There are some tests, the outcomes of which will render the hypothesis either more, or less, plausible. One of them is this. The Chicchan-Yax combination was entertained because, in addition to being the first in the set of possibilities, it was the only one that yielded an "astronomically interesting" date. Now if that quality should render the date unique in this respect among those recorded in the murals and in the inscriptions at Bonampak, the date would be exceptional, without precedent, and in no way expectable for the site. It could then be considered as possibly no more than a chance coincidence, however intriguing. If, on the other hand, it should turn out that it is not unique in this respect among the dates recorded at Bonampak, if it should in fact be that the majority of the dates recorded there are astronomically interesting ones, then it would be in conformity with the others, expectable, and rather more likely the correct interpretation of the evidence. And any other of the calendrical possibilities for the 13... 13... date would then be in some measure exceptional and less likely. Mary Miller's suspicion about this problematic date, and the apparent possibility of a positive answer to her question, prompted some testing of other dates at Bonampak. The results came as a surprise. They are of interest in their own right, and together they justify confidence in the 13 Chicchan 13 Yax solution of the battle date, which is at home among them, and which may be considered to have "passed" this test.

The second test relates to the turtle and the peccaries, whether these constellations (if identifiable) were in any way pertinent to the hypothetic date of the battle, and thus whether they could be seen as significant components in its depiction. This matter will be discussed later; but in anticipation it may be noted here that, to the extent identifiable, they are appropriate and significant.

A third test relates to the question posed by Miss Miller, whether the placement of these "star" figures over the battle and victory scenes of the murals might be considered an architectural and artistic rendering of the concept whose hieroglyphic expressions were the "star-over-earth," "star-over-shell," and "star-over-placename" glyphs. This prompted testing of the date associated with the "star-over-Seibal" and some of the other glyphs of this category. Only a beginning has been made and the results so far are mixed. About half of those tested have clearly interpretable and significant results in terms of critical points in the Venus cycle, while the remainder are of uncertain significance, relating possibly to a more general Venus criterion, or in some cases perhaps indicating no more than that there was a Venus-derived linguistic and hieroglyphic idiom for warfare. In terms of probabilities, the basis for the idiom in notions about Venus, particularly in relation to critical points in the planet's movements, is surely borne out by the results so far obtained. Eventually the matter will have to be submitted to a proper statistical evaluation; but before that can be done the examples will have to be collected and studied in relation to their hieroglyphic textual contexts, and their dates will have to be checked for astronomical attributes. A few of those already studied, which furnish comparative evidence pertinent to the Bonampak problem, will be reported here.

APPEARANCES OF THE EVENING STAR

It was the "star-over-Seibal" glyphs on Aguateca Stela 2 and Dos Pilas Stela 16 (Graham 1967) that gave the first clear evidence for the military meaning which glyphs of this category are now known to have had. The two monuments record the same events: a "star" event on 8 Kan 17 Muan, 9.15.4.6.4 (apparently the date of the raid); a decisive battle with capture of the local ruler on the next day; and a follow-up event six days later. The hieroglyph of the first event suggests a Venus significance for that day. Its date was included by Kelley (1977) in the set of those that cluster around the 16.6 positions in the Dresden Codex table (the canonical days for beginning the evening-star period). If such a table had general currency among the Maya of this period, and if it was being followed to the letter, or numerologically, then the dates associated with a particular critical point in the Venus cycle might be expected to conform, right to the day, to those of the table. The fact that they do not, but that they depart in varying small amounts from them, suggests that they were not determined solely by the calendar, but were more likely dependent on observations. The date of the "star-over-Seibal" event, by the 584285 correlation, was November 29 (Julian), AD 735. According to the Tuckerman tables, superior conjunction of Venus was 30 days earlier, on October 30. An interval of 30 days after superior conjunction is a reasonable one in the tropics, where the ecliptic is high and nightfall is quick, for a first sighting of Venus as evening star (seen but briefly after sunset, before it too sets—about 28 minutes after the sun in this case). The tables indicate an elongation of 7.17 degrees for Venus on November 29 of that year. Under good observing conditions in those latitudes the minimum necessary elongation for a first sighting appears to vary from about five to about ten degrees, depending on the time of the year and on the planet's celestial latitude at the time. (Similar limits obtain also for a first sighting of the morning star at its so-called heliacal rising, though the time required for attainment is much less.) Thus, the position of Venus on this date was probably appropriate for a first visibility of the evening star. That the circumstance should have held significance for the Maya is inferrable in part from the glyph that marks the date in these two inscriptions, and in part from the similar situation of Venus on some of the star-marked dates at other sites. Two of the Bonampak dates also catch Venus at this point in its cycle, though they are without glyphic "star" markings.

In Table 23.1 are listed some dates on which the eastern elongation of Venus falls within the indicated limits, and which are marked by one of the "star-over-X" glyphs, or by an unambiguous Venus glyph that includes the distinguishing prefix (as in Fig. 23.1a, b), or by a skull with distinctive markings and teeth that is also a Venus symbol (Fig. 23.1d), or by this in company with one of the regular, "star" signs. Also the two Bonampak dates that fall within this category are included. The dates are listed and

Table 23.1 First of evening star.

	(a)	(b)		(c)	(d)	(e)	(f)	(g)
(1)	9. 9. 18. 16. 3	100v	+5	631 Dec 24	275.69	280.76	5.07	22
(2)	9. 11. 0. 0. 0	87v	−1	652 Oct 11	201.44	209.31	7.87	31
(3)	9. 14. 0. 0. 0	50v	+4	711 Dec 1	252.81	258.61	5.80	24
(4)	9. 15. 4. 6. 4	35v	−1	735 Nov 29	250.94	258.11	7.17	30
(5)	9. 15. 9. 3.14	32v	+1	740 Sep 13	174.20	181.72	7.52	28
(6)	9. 15. 15. 12.16	28v	−6	747 Feb 11	326.33	343.33	7.00	29
(7)	9. 18. 1. 2. 0	*	*	791 Nov 11	233.02	240.22	7.20	29
(8)	9. 18. 9. 4. 4	5v	+4	799 Nov 13	235.10	243.40	8.30	34

numbered in chronological sequence, but will be discussed here in a different order. The columns of the table give the following information: (a) the day number in Maya notation; (b) the interval, in multiples of 583.92 [symbolized by the letter "v"] plus or minus some number of whole days, between any given date and a reference date which is indicated in this column with two asterisks; (c) the equivalent Julian date by the 584285 correlation; (d) the celestial longitude of the sun for the day in question [ca. 10:00 AM local time]; (e) the celestial longitude of Venus at the same time; (f) the eastern elongation of Venus; and (g) the interval in days after superior conjunction.

Date 4 in this list is the "star-over-Seibal" date. It is recorded at two sites in the southern Peten of Guatemala (those noted above) which were under the dominion of some branch of the Tikal dynasty. In the inscription of the Dos Pilas monument the event is denoted by the hieroglyph illustrated here (Fig. 23.1c). In that of Aguateca it is denoted by a similar one, which compounds the "star-over-shell" glyph with the main component of the Seibal emblem glyph and a superfixed locative sign. These glyphs, whose connotation seems to relate to Venus, apparently designated the raid or a distinctive attendant circumstance. The engagement of forces took place the next day, and the king of Seibal was taken prisoner. He is so depicted, deprived of most of his finery and bound with ropes, but identified by his name and emblem glyph, under the feet of his captor on both monuments. Six days later was the follow-up event that has been mentioned above. Its nature is far from clear, although there is quite a bit that can be said about it with fair assurance. What is pertinent here is that it was apparently not necessarily lethal. Though the conqueror took over at Seibal, the captive king was kept alive for nearly another twelve years. He came to his end eventually as a sacrificial victim at a ritual ballgame—timed for an inferior conjunction of Venus! (Details of this will follow under the appropriate heading further on.)

Date 1, its event, and its designating hieroglyphs are an earlier instance of the same sort as those of Date 4 which have just been reviewed. Curiously, the date is just one day earlier in the calendar-round, being on 7 Akbal 16 Muan. (Date 4, it will be recalled, was on 8 Kan 17 Muan.) The interval between them is equal to two calendar-rounds and one day, or to 65 uncorrected Venus periods and a day, or 104 Maya calendar years and a day; which would be just a day more than one complete run of the Dresden Codex table. And like Date 4, this one also is recorded in two places, on Stela 3 at Caracol (Belize) and on the Hieroglyphic Stairway of Naranjo (Peten of Guatemala). The records are of an engagement between Caracol and Naranjo, the outcome of which was the conquest of Naranjo by the ruler of Caracol, who then held power in both places. In the Caracol inscription the event of this date is designated by the "star-over-shell" glyph compounded with the main component of the Naranjo emblem glyph and a locative superfix, while in the Naranjo inscription (presumably promulgated by the victor after taking over) it is designated by a "star-over-Naranjo" glyph without the shell. Even the glyphic variation is parallel to that of the previous case. Astronomically, while the previous case represented a quite typical position of Venus at this point in its cycle, the present one is about as close to the inside limit as is possible, with an eastern elongation of little more than five degrees, and a time interval after superior conjunction of only 22 days. It is an extreme case, and one wonders whether the Caracol war chief in his anticipation of the appearance of the evening star may not have jumped the gun on it.

Date 6 is from the right-hand panel of the east doorway of Temple 11 at Copan (Honduras). It is noteworthy for its association with a complete Venus hieroglyph, which includes the distinctive *chac* prefix as well as the principal "star" component. This renders its primary signification unambiguous; it definitely pertains (in some way) to Venus. The recorded day, which is 5 Cib 9 Pop, was included in Kelley's list (1975) for this category because it approximates to within two days the 3 Ix 7 Pop position for the beginning of one of the evening star periods in the Dresden Codex table. If this is justified (and I believe that it is), then it requires the chronological position assigned to it here. It is one of the three earlier dates recorded in the temple from which it comes, and there are reasons for considering it to be the date of the ruler's designation as successor or acting regent in lieu of his predecessor who was being held captive at Quirigua. His accession to full title did not take place until seventeen years later (very likely because his predecessor, the legitimate holder of the title, was still alive). The event glyph in the passage with which we are concerned here has its two principal elements in common with those that accompany the Venus hieroglyph in each of the columns in the Dresden Codex table, which give the canonical days for the beginnings and endings of the morning and evening star periods. The passage could thus easily be taken for a strictly astronomical record. But the glyph is also similar—even more similar because of a distinctive affix—to one that is employed in inscriptions to designate acts of heir designation, which glyph also employs the same principal elements. Moreover, the Venus hieroglyph, complete with its *chac* prefix, is employed elsewhere at Copan as a component in the string of appellatives and titles of the above-mentioned ruler. This usage, together with other elements in the present context, including its being preceded here by a "lordship" title, suggests a literally applicable double meaning for the inscription: it was a date of the appearance of Venus as evening star, and it was the date of the designation of this ruler as acting regent and eventual successor to the captive king. (At least this seems to be the best hypothesis at the present time.) The Venus symbols and skulls that appear as iconographic elements in some of the monuments for which he was responsible may be indicative of the image that he acquired or had confirmed on this day. Astronomically the date is a typical one for this category, with an eastern elongation of seven degrees and an interval of 29 days after superior conjunction.

Date 8 is from what is probably the last hieroglyphic record produced at the site of Palenque (Chiapas, Mexico), a ceramic piece known as the Initial-Series Vase because of the fully specified date with which its covering text begins. It is the date of the inauguration of the last known ruler at that site. His six-glyph nominal and appellative phrase includes two components that have Venus associations, one of which is the "star-over-earth" glyph, which also has connotations of warfare. Whether he carried this title solely by virtue of his accession date, or whether he had already acquired a reputation which his accession date symbolizes, is not known. The date is slightly on the generous side as a member of this set, with an eastern elongation of 8.3 degrees and a 34-day interval after superior conjunction.

Date 2 is a katun ending, from the middle panel of the Temple of Inscriptions at Palenque. As a katun ending it would have been recorded anyway, together with a declaration of the rituals performed for the occasion. At Palenque, during this period, other items of note were sometimes included. In the passage for this one a reference to Venus is indicated by a distinctively marked skull (Fig. 23.1d) which is also a symbol of Venus. Since katun endings are determined by numerical criteria unrelated to Venus reckoning, the occurrence of a Venus phenomenon on this date is a pure coincidence. But it was one that did not escape the notice of the priests and the ruler of the place. What they did about it, other than include it in the record as a noteworthy feature of the katun ending, will not be known until the remaining hieroglyphs of that passage are better understood. Astronomically it is typical for the set, with an eastern elongation of 7.87 degrees and an interval of 31 days after superior conjunction.

Date 3 is another katun ending, three katuns later. A Venus phenomenon is implied for it in the record of Stela C at Copan (Honduras), and some sort of association with Venus seems to be implied by the iconography of Stela 16 at Tikal (Peten, Guatemala) which also commemorates this date. On Copan Stela C the evidence is indirect. A Venus event is twice imputed—in parallel passages in a couplet arrangement—to a mythological and numerologically reckoned antecedent date more than four and a half millennia earlier. In the first reference it is designated by a compound containing the usual Venus "star" sign (as in Fig. 23.1c), and it is preceded by an appropriate event glyph. In the second it is designated by the Venus skull sign (as in Fig. 23.1d), this time with verbal affixes. The ascription to a mythological antecedent, in Maya practice, implies ascription of a similar or related attribute to the current date with which the mythological one is paired. But the three glyph blocks where this might have been recorded are destroyed, so direct confirmation is lacking. The theme recurs, however, on the other side of the stela, where the same kind of event is ascribed to yet another mythological antecedent. The three-katun interval between Dates 2 and 3 is five days short of an integral multiple of the Venus period [$3 \times 7200 = 37 \times 583.92 - 5.04$]. Date 3 was 24 days after superior conjunction, and the eastern elongation of Venus was 5.8 degrees. It is another one near the inside limit. But the date was determined by a numerical criterion; and the first appearance of the evening star, if not actual, was imminent. On the Tikal stela pertaining to the same katun ending the text is brief. It records the day, that it was the completion of fourteen katuns in the baktun, and that the required ritual was carried out by the ruler, whose name and titles follow. The text is disposed in three framed areas around the larger depiction of the ruler himself, in regal attire, holding the emblem of office, but wearing a headpiece of which the central component is the Venus skull backed by the Venus "star" sign. With the Copan and Palenque precedents in mind, it might be supposed that the headpiece was motivated by the same consideration that is assumed to lie behind the manifestations of Venus significance in those monuments. It must be noted, however, that the same sort of headpiece turns up again a generation later at Tikal, in Lintel 3 of Temple IV. One would look, then, for a Venus attribute in one of the four dates recorded in the text of that lintel. None of them, however, are at any one of the usual critical points in the synodic cycle of Venus. The event glyph of one of them is the "star-over-shell" glyph (as in Fig. 23.2b). The date is 9.15.12.2.2, which by the correlation assumed here was July 28 of the year 743. On that date Venus as morning star (western elongation 23.63 degrees) was in precise conjunction with Mercury, with 0.64 degrees of latitude difference between them. Whether this provides an adequate explanation for the presence of the headpiece is not certain. But the headpiece is at least indicative of an astrological concern with Venus and some preoccupation with the power attributed to it.

Date 5 is from Bonampak, from Lintel 3. The lintel depicts a local ruler in one of the standard poses for administering a coup to a captive. Though carved in stone and differing in details, in its style and general outlines it corresponds to a focal scene in the battle murals of Room 2, which scene depicts the ruler of a generation (or possibly two) later in a similar pose, and contains the brief text with the 13... 13... date that started this investigation. The victorious ruler on the lintel wears a pendant jade skull mask hung from his neck, with the prominent row of even teeth that characterizes the Venus skull, of which it is quite surely a representation. The date of this event is a typical one of the set of first appearances of the evening star, with an eastern elongation of 7.52 degrees and an interval of 28 days after superior conjunction. Since both the date and the pendant mask appear to implicate Venus in the event of this lintel, they offer a supporting precedent for the reconstruction that was proposed for the missing components in the 13... 13... date of the analogous scene in the mural, which also implicated Venus and implied other astrological interests as well.

Date 7, also from Bonampak, is the second of those in the mural text of Room 1. As noted earlier, it is the date of a dynastic ritual of some sort (one suggestion that has been put forward is that it may concern the formal presentation of a young heir to the throne). That it was a very important occasion can be judged from the costuming and activities depicted in the mural scenes. The date is another one typical for this set. The eastern elongation of Venus is 7.2 degrees, and the date is 29 days after superior conjunction. It is a third instance of an apparent astrological concern with that planet in the timing of important undertakings at Bonampak.

MAXIMUM EASTERN ELONGATION

It was noted earlier that if it should happen to be the case that the majority of the dates of planned events at Bonampak were astronomically interesting ones, then that would impart an additional measure of plausibility to the reconstructed battle date; for it would then be one of a kind. The two additional Bonampak dates which have been reviewed above appear to do just that. The case would of course be strengthened if there should be others. A check of the dates on the stelae of Bonampak brings up the possibility of another significant category, that of the maximum eastern elongation of the evening star. It is one that is supported by a few "star" dates from other sites as well. Those that have been looked into carefully are listed in Table 23.2, and are reviewed below. The listings in the table are divided between two categories. The first (category A) contains dates on which Venus is literally at its maximum eastern elongation, having arrived at its goal, so to speak, in its journey away from the sun. For convenience, this will be called the "arrival" category. The second (category B) contains dates on which Venus has made what may perhaps be interpreted as the first perceptible movement toward a departure from that position, initiating the return journey back to the place of the sun. In these the eastern elongations range from a quarter to three-quarters of a degree less than those that were the corresponding maxima. For naked-eye astronomy, with relatively simple sighting instruments, these may perhaps have been just-noticeable differences, enough to indicate that the return trip was beginning. This will be called the "departure" category. The listings of the dates are by chronological order, but separately within the two categories. Their review, however, will again take them up in an order different from that of their listing. The numbering is continuous, proceeding from that of Table 23.1, rather than beginning over again. The columns, except for the last, give the same categories of information as do those of Table 23.1, namely: (a) the Maya day number; (b) the interval to or from the chosen reference date, which is that with the asterisks; (c) the Julian date; (d) the celestial longitude of the sun; (e) the celestial longitude of Venus; and (f) the eastern elongation of Venus; while the last column (g) this time gives the number of days before the next inferior conjunction. In part B of the table, the bracketed figures in columns (f) and (g) give for comparison the corresponding maximum eastern elongations and the intervals from their dates to inferior conjunction. The bracketed figures thus represent positions of the same kind as those of part A of the table, and are those away from which the presumed first perceptible moves of the planet have been made. The amount of the movement may be seen by subtracting any given unbracketed figure in column (f) from the bracketed one just below it; and the number of days from the date of precise maximum to the recorded date can be seen from a similar comparison of the figures in column (g).

Three stelae are known from Bonampak, on which there are recorded a total of five dates. Two of these are what Mayanists call hotun-ending dates, marking the ends of round-numbered five-tun intervals in the day count (the Maya lustra, quarter-katuns, = 5 × 360 days), which were the occasions for erecting commemorative ste-

Floyd G. Lounsbury

Table 23.2 Maximum eastern elongation.

	(a)	(b)	(c)	(d)	(e)	(f)	(g)
A. Arrival:							
(9)	9. 12. 0. 0. 0	* *	672 Jun 28	99.12	144.72	45.60	71
(10)	9. 12. 11. 6. 8	7v +1	683 Sep 7	167.13	213.75	46.62	72
B. Departure:							
(11)	9. 17. 5. 8. 9	* *	776 Jun 11	83.74	128.31	44.57 [45.31]	57 [71]
(12)	9. 17. 10. 6. 1	3v +0	781 Mar 29	12.50	57.52	45.02 [45.68]	58 [70]
(13)	9. 17. 15. 3.13	6v +0	786 Jan 14	298.66	345.10	46.44 [46.99]	60 [73]

lae at many sites. Dates of this category are determined by strictly numerical criteria, and so may be excluded from consideration in connection with astronomical matters, except when such a date happens to coincide with an astronomical phenomenon of interest, of which note is made in the hieroglyphic text. Two examples that did merit such notations, which were of katun endings, were seen in dates 2 and 3 (Table 23.1); and another, yet to be discussed, is date 9 (Table 23.2).

Stela 1 at Bonampak commemorates the hotun ending of 9.17.10.0.0, recording the date and reporting the enactment of the required rites by the then ruler, whose name, parentage, and titles follow, and whose standing full-figure portrait is the focus of the monument. The date is not one of any special astronomical interest, and no such notation is included in the text. The next stela, chronologically, is the one that is archaeologically designated as Stela 3. It commemorates the hotun ending of 9.17.15.0.0, recording the date, the ritual, and the name of the same officiating ruler, and depicting him facing a kneeling bound captive. Again the hotun ending is of no particular astronomical interest, and the inscription contains no such reference. But there is a subsidiary passage in which the record of the event is almost totally obliterated due to erosion, but in which there are clues to the date. The passage begins with a one-block (and therefore two-digit) "distance number" that is mostly lost due to breakage; but it is apparent that the number of uinals (units of 20 days) was either 3 or 4, while there is little secure evidence as to the number of days beyond that. (See Robertson 1980 for a photographic plate, and Mathews 1980 for a drawing.) Following that there is a clear "posterior date indicator," and then a calendar-round day that has been interpreted in the drawing as 12-something 16 Cumku, the identity of the day sign being unclear. A check of all possibilities, however, shows that there is no two-digit distance number that can lead from 9.17.15.0.0, which is 5 Ahau 3 Muan, to a day 12-something 16 Cumku. It is a mathematical impossibility. But a distance number of 3.13 will lead to 13 Ben 16 Cumku. The conclusion that must be drawn is that the coefficient of the day sign (with two dots and an intermediate space filler) has been misinterpreted (the space filler needs to be another dot), and

579

that the originally recorded distance number was 3.13 and the posterior date was 9.17.15.3.13.13 Ben 16 Cumku. Reexamination of the photograph shows that it permits of this interpretation.

This is date 13 in Table 23.2. As indicated there, the eastern elongation of Venus on that day was 46.44 degrees, which is very close to the maximum attainable; and the date was 60 days before the next inferior conjunction. Precise maxima, however, run closer to 72 days before inferior conjunction. This one, as estimated by interpolation from data in the Tuckerman tables, was 73 days before, with 46.99 degrees elongation. The difference of about half of a degree is of the magnitude that I have supposed was noticeable to the Maya observers and was sufficient as an indication to them that the planet was embarking on its return to the sun. Alternatively one might grant them less observational acuity, and suppose that this was within a vague range of positions, any of which might have been judged simply as maximum elongation, admitting to the category any date that fell within say a 13-day range one side or the other of the precise maximum. But the other available candidates for the category, and the intervals between them, give grounds for rejecting a "vague maximum" hypothesis of this sort and for entertaining seriously the more precise "departure" hypothesis. These are dates 11 and 12, to which attention is now turned.

The latest of the three Bonampak stelae is the one known as Stela 2, which records two dates and the corresponding two events. The latter of these was an autosacrificial act, wherein the ruler offers blood to a deity or an ancestor. The preparations are depicted in the sculpture, in which the ruler's mother offers him the stingray-spine perforating instrument and the shallow bowl of folded paper which was the standard receptacle for blood drawn from one of one's bodily members, while his wife, a lady from Yaxchilan, holds another such bowl in reserve. Though it is apparent from the inscription that the occasion was an anniversary of something, it is not clear what that was. That it may have been something specifically related but prior to the ruler's accession is suggested by the fact that the other date on this stela, with which the inscription opens, was that accession date. This is date 11 of Table 23.2. As can be seen from the table, it was 14 days after the date of precise maximum elongation of the evening star, as best this can be judged; and during this interval the elongation had diminished by about three-quarters of a degree. Like date 13, this one also could be a first perceived move away from the extreme. Interestingly, the interval between this accession date on Stela 2 (date 11) and the necessary reconstruction of the second date of Stela 3 (date 13) is 9.13.4, or 3504 days, which is an integral multiple of 584. Since they concern the same ruler, it is difficult not to suspect that this was intentional. One may even imagine what might have been in the obliterated event phrase; a reference, perhaps, to the sixth Venus anniversary of the ruler's accession. The bound captive may have been intended to have a part in that commemoration.

Date 12 of Table 23.2 is from another site, Piedras Negras in the western Peten, on the Guatemalan side of the Usumacinta River. It is one of the dates recorded on Throne 1 of that site, and it has the "star-over-shell" glyph (as in Fig. 23.2b) in the associated event phrase. This suggests a Venus phenomenon. As can be seen in the table, it is another obvious member of the "departure from maximum eastern elongation" set. And it is exactly halfway between the two dates from Bonampak which have just been considered. From date 11 to date 12 is an interval of 3×584 days, and from date 12 to date 13 is another 3×584 days.

In slight, but I think significant, contrast to these of the "departure" category are those that, for the sake of a concise label, I have called the "arrival" category (though "resting" or "waiting" might have been more accurate as a description). So far, they are only two: date 9, from Palenque, and date 10, from Bonampak. Date 9 (like date 2 of Table 23.1) is a katun ending from the middle panel of the Palenque Temple of Inscriptions. As in the case of other katun endings, it would have been recorded anyway. But as mentioned earlier in connection with dates 2 and 3, there are instances of the notation of Venus phenomena in katun-ending records when these happened to coincide. In the passage which contains date 9, the notation is quite explicit. It has the Venus-skull glyph (nominal, as in the date 2 passage from the same inscription—cf. Fig. 23.1d) followed by the "star-over-shell" glyph (verbal), these being followed then by phrases with first the "east" and then the "west" direction glyphs, and then finally by a compound glyph with a sitting figure who has his head buried in his arms, which are clasped around his knees. The east-west sequence would seem to imply reversal. Reversal there is, not of the side of the sun, but of the direction of movement: eastward away from the sun (or toward a horizon landmark for maximum eastern elongation), and westward toward the sun (or away from the extremity landmark). The sleeping figure may very well symbolize the wait at the resting place in between; for that is exactly the situation of Venus on the date of that katun ending. At the 1978 Round Table meeting at Palenque, Michael Closs proposed a similar interpretation of this passage (cf. Closs 1979, 1982); but I was not then of a mind to anticipate this kind of an astronomical record in a Maya inscription, and I remained skeptical. The picture looks somewhat different now.

Date 10, from Bonampak, is recorded on a piece that is known as Sculptured Stone 1. It is an accession monument, as both the glyphic text and the iconography declare. The date, which had been a difficult one to interpret, was successfully deciphered by Peter Mathews (1980). It is another one that has Venus right at the point of maximum eastern elongation. The interval between date 9 and this one is 11.6.8, or 4088 days, which are equal exactly to 7×584, or to 7×583.92 plus a 0.56 fraction of a day. (In Table 23.2 this appears as 7v +1, because of the "nearest whole day" convention for disposing of fractions.

FIRST STATIONARY POINT

With the amount of attention to Venus that is manifest in the dating of undertakings at Bonampak, the lunar interest that it was possible to see in the initial date of the mural text of Room 1 now seems almost an anomaly; and one is led to wonder whether that date too might not have held something of interest in connection with Venus. The date, it will be recalled, was 9.18.0.3.4, 10 Kan 2 Kayab, which by the assumed correlation was December 10 of the year 790. Consulting again the tables, one finds that on this date the planet, as evening star, was exactly at its first stationary point. It was 21 days before inferior conjunction, and the planet's eastward progress across the field of the fixed stars had come to a halt. From here it would reverse itself to begin the retrograde motion that would carry it through inferior conjunction and, as morning star, after yet another 21 or 22 days, on to the second stationary point; whereupon it would stop, change direction again, and resume its direct or orthograde motion. If the selection of this kind of a date for the event that is commemorated in the first part of the mural text was by design rather than accidental, it indicates that the Bonampak Maya were using more than one reference sphere against which to chart the progress of the planet; for the stationary points are in a way analogous to the maximum elongations, with the difference that they are critical points in the planet's path in relation to the equatorial sphere, with the fixed stars as landmarks, whereas the maximum elongations are in relation to the horizontal sphere, with—for primitive astronomy—topographic features of the horizon or constructed stationary sighting frames furnishing reference points for charting the planet's positions. The ancients of the Old World observed the stationary points and noted them in records; so it should not be too surprising to find indications of the same in the New World.

INFERIOR CONJUNCTION AND HELIACAL RISING

In Table 23.3 are listed five dates, of which three are from Bonampak, one from Seibal, and one from Piedras Negras. The dates pertain to two categories: (A) inferior conjunctions, and (B) probable heliacal risings of the morning star. The columns of the table are as before, except for the last two, which pertain only to category B. They are: (f) the western elongation of Venus, and (g) the number of days after inferior conjunction. The numbering of the dates continues from the last of the previous table. They are listed in chronological order within each category separately. The discussion again will follow a different order.

Date 16 is the battle date in the mural of Room 2 at Bonampak, the one whose reconstruction initiated this inquiry, and which I believe can now be regarded as secure. Its attributes were discussed earlier in the paper. It was at an inferior conjunction of Venus, as well as at a zenith passage of the sun and a conjunction of Venus with Regulus, apparently seen as a propitious date for the undertaking, the outcome of which was such as to confirm the opinion.

Date 15 is from the hieroglyphic steps at Seibal. It was the date of a Maya ballgame. Such events were also the occasion for sacrifices, and ballgame iconography often shows a human figure inside of a large rubber ball as it rolls down the flight of stairs that forms the side of the ball court, to be struck in a moment by the yoke-girded player poised below. A hieroglyph based on this stereotypic representation designates sacrificial ballgames of this sort. The event ascribed to date 15 is denoted by this glyph, with an infixed large-toothed skull suggestive of a Venus association, and with the king of Seibal as the apparent victim and his erstwhile captor as officiant. The date coin-

Table 23.3 Inferior conjunction and heliacal rising.

	(a)	(b)	(c)	(d)	(e)	(f)	(g)
A. Inferior conjunction:							
(14)	9.15. 3. 8. 8	36v −4	735 Jan 17	301.07	same	0	0
(15)	9.15.16. 7.17	28v −2	747 Oct 30	220.00	same	0	0
(16)	9.18. 1.15. 5	* *	792 Aug 2	133.62	same	0	0
B. Heliacal rising:							
(17)	9.15. 3. 8.12	29v −2	735 Jan 21	305.12	298.02	7.10	4
(18)	9.17.10. 9. 4	* *	781 May 31	73.14	65.65	8.48	5

cides with an inferior conjunction of Venus. Compared with date 16, the Bonampak battle date, it is prior by an interval equal to within two days of 28 mean Venus periods. The named victim is the same king who was captured nearly twelve years earlier as a result of the raid initiated on date 4 (Table 23.1), the "star-over-Seibal" date that first led to recognition of the "appearance of the evening star" category.

Date 18 is from the same Piedras Negras throne as was date 12 (Table 23.2). It is the next one thereafter, and its relation to the previous one is indicated by a recorded distance number of 3.3, or 63 days. That date was marked by a "star-over-shell" glyph, probably representing a military exploit timed to coincide with the apparent first move of the evening star away from its position of maximum eastern elongation. Date 18 was the occasion of the accession to the throne by the same man who was involved in the previous event, and the accession was apparently timed to coincide with a heliacal rising of Venus as morning star. It was four or five days after inferior conjunction, and the elongation of the planet amounted to 7.10 degrees of longitude and about a degree and a half of latitude. That the protagonist was aware of the Venus position, and that the timing was deliberate, is implied by the context with its "star" marking of the preceding date.

Dates 14 and 17, just four days apart, are from Lintels 2 and 1, respectively, at Bonampak. These lintels, like Lintel 3 from the same place (cf. date 5, Table 23.1), depict a warrior in the act of administering a symbolic coup to a captive. All three figures have hanging from their necks the pendant mosaic jade skull mask with large teeth that has Venus associations. This fact, together with a glyphic phrase which I had considered to be the beginning of their appellatives, had led me to identify the three figures as representing one and the same individual; in which case the recorded calendar-round dates require the chronological placements assigned to them here. Recently, however, Peter Mathews (1980) has published very cogent reasons for considering that the figures may represent three different persons, with identifications for those on Lintels 1 and 2 that would require placement of their dates one calendar round later. The question is not yet resolved. It hinges on a hieroglyph whose meaning may or may not be what has been supposed. If the dates turn out to belong where I have them here, then date 14 is exactly right for an inferior conjunction of

Venus, and date 17 for a heliacal rising of Venus; and the two lintels would then have the Venus association seemingly implied by the pendant masks and by their similarity to Lintel 3 whose date is secure and is clearly associated with Venus. If, on the other hand, they properly belong a calendar round later, then they would have no significant associations with Venus (they would be about three and seven days after superior conjunction, an unlikely time for astrological attributions). The fact that their erroneous placement produced such seemingly significant results as in dates 14 and 17 would then exemplify no more than the sometimes ironic turns of chance. And then one would wonder how many of the others might be merely tokens of the favors of chance.

CONCLUSION

The foregoing data, even at worst, I believe are sufficient to indicate that the Maya had attained some fair precision in their astronomical observations, just as the Dresden Codex gives evidence of their attainments in mathematical reasoning with the data. And I think the cases reviewed here also give an idea of why a people in a "primitive kingdom" stage of social and political development were interested in astronomy, and why it was fostered by those who held power. Finally, I believe it offers strong support for the Thompson correlation, specifically the value 584285 which he first proposed.

POSTSCRIPT: THE TURTLE AND THE PECCARIES

The 16th-century manuscript of the Motul dictionary (John Carter Brown Library, Providence, R.I.) has the following entries:

> *ac*: tortuga, galápago, ycotea.
> *ac*, 1. *ac ek:* las tres estrellas juntas que están en el signo de géminis, las quales con otras hazen forma de tortuga.

These explain the three stars on the back of the Bonampak turtle (Fig. 23.3). A modern-day informant told Thompson that the turtle constellation is Orion (Thompson 1950: 116). The term "signo" is used in both Spanish and Maya sources for calendrical zodiacal divisions, not for constellations. By the sixteenth century, when the Motul dictionary was being assembled, the constellation of Gemini, because of precession, was already mostly in the "sign" of Cancer, and the stars in the "sign" of Gemini included those of Orion and much of Taurus. The Bonampak turtle, then, must represent Orion; and its three stars are there for an obvious reason. So also is the turtle itself there for a reason; it is appropriate to the 8th-century early August date of the battle depicted below it. The peccaries and the other figures are not yet securely identified. (Note that the interior ridge line of the vaulting cannot represent the ecliptic; for that would not be in accord with the orientation of the building, and it would put the turtle on the wrong side.)

REFERENCES

Closs, M. P. 1979. Venus in the Maya World: Glyphs, Gods and Associated Phenomena. In *Tercera Mesa Redonda de Palenque,* Part 1, ed. M. G. Robertson and D. C. Jeffers, 147–165. Palenque, Chiapas, Mexico, Pre-Columbian Art Research Center.

Closs, M. P. 1982. Venus Dates Revisited. *Archaeoastronomy* 4:4. College Park, MD, University of Maryland.

Dictionary of Motul. See Martinez Hernandez, J., ed. (1929).

Graham, I. 1967. Archaeological Explorations in El Peten, Guatemala. MARI Publ. 33. New Orleans, Tulane University, Middle American Research Institute.

Kelley, D. H. 1977. Maya Astronomical Tables and Inscriptions. In *Native American Astronomy*, ed. A. F. Aveni, 57–73. Austin, TX, University of Texas Press.

Kelley, D. H., and A. Kerr. 1973. Mayan Astronomy and Astronomical Glyphs. In *Mesoamerican Writing Systems*, ed. E. P. Benson, 179–215. Washington, DC, Dumbarton Oaks.

Lounsbury, F. G. 1978. Maya Numeration, Computation, and Calendrical Astronomy. In *Dictionary of Scientific Biography*, ed. C. C. Gillispie, vol. 15, Supplement 1, pp. 759–818. New York, Charles Scribner's Sons.

Martinez Hernandez, J., ed. 1929. Diccionario de Motul, Maya-Español. Merida, Yucatan, Mexico.

Mathews, P. 1980. Notes on the Dynastic Sequence of Bonampak, Part 1. In *Third Palenque Round Table*, Part 2, ed. M. G. Robertson, pp. 60–73. Austin, TX, University of Texas Press.

Miller, M. 1981. The Murals of Bonampak, Chiapas, Mexico. Doctoral dissertation, Yale University, New Haven, CT.

Robertson, M. G. 1980. The Giles G. Healey 1946 Bonampak Photographs. In *Third Palenque Round Table*, Part 2, ed. M. G. Robertson, pp. 3–44. Austin, TX, University of Texas Press.

Ruppert, K., J.E.S. Thompson, and T. Proskouriakoff. 1955. *Bonampak, Chiapas, Mexico.* CIW Publ. 602. Washington, DC, Carnegie Institution of Washington.

Teeple, J. E. 1930. Maya Astronomy. *Contributions to American Archaeology* 1 (2):29–116. CIW Publ. 403. Washington, DC, Carnegie Institution of Washington.

Tejeda, A. 1955. *Ancient Maya Paintings of Bonampak.* [Authorship of accompanying text unindicated.] CIW Supplem. Publ. 46. Washington, DC, Carnegie Institution of Washington.

Thompson, J.E.S. 1935. Maya Chronology: The Correlation Question. *Contributions to American Archaeology,* 14. CIW Publ. 456. Washington, DC, Carnegie Institution of Washington.

Thompson, J.E.S. 1950. *Maya Hieroglyphic Writing: An Introduction.* CIW Publ. 589. Washington, DC, Carnegie Institution of Washington. (2nd ed. 1960, Norman, University of Oklahoma Press.)

Thompson, J.E.S. 1972. A Commentary on the Dresden Codex. *APS Memoirs*, vol. 93. Philadelphia, American Philosophical Society.

Thompson, J.E.S. 1974. Maya Astronomy. In *The Place of Astronomy in the Ancient World*, ed. F. R. Hodson. Phil. Trans. R. Soc. Lond., A. 276, pp. 38–98. London, The Royal Society.

Tuckerman, B. 1964. Planetary, Lunar, and Solar Positions, AD 2 to AD 1649, at Five-day and Ten-day Intervals. *APS Memoirs,* vol. 59. Philadelphia, American Philosophical Society.

CHAPTER TWENTY-FOUR

A Palenque King and the Planet Jupiter

Floyd G. Lounsbury

The introductory passages of the inscription of the Temple of the Sun at Palenque[1] are concerned with mythological matters, in particular with the birth of the second of the deities of the "Palenque Triad," an event that is declared to have taken place on a day which was number 1.18.5.3.6 (275466) in the Maya day count and 13 Cimi 19 Ceh in the Maya calendar-round, this, being equivalent, in European chronology, to October 28 (retroactive Gregorian) of the year 2360 BC.[2] The presentation of this date in the "initial series" of the inscription is followed by its further orientation in what are known as "supplementary series," which specify—amongst other things—its position in the lunar calendar (26th day since first visibility of the current moon, in this, a 30-day lunar month) and its position in an 819-day cycle.[3] But the passage which purports to place it in this latter cycle contains some glaring and puzzling discrepancies. These are well known to students of Maya epigraphy but have yet to be offered reasonable explanation.

From *World Archaeoastronomy*, ed. A. Aveni (New York: Cambridge University Press, 1989).

An "819-day passage" is one which orients an initial-series date according to its position within a Mayan cycle of that length, doing so by stating the number of days since the last "station" or epochal day of the cycle, and then—along with other details—naming that station. In the present case, the inscription states that 13 Cimi 19 Ceh, 1.18.5.3.6, the date of the initial series, was 1.2.11 (411 days) after the 819-day station of 1 Ik 10 Tzec. But this is an arithmetical impossibility. The interval from 13 Cimi 19 Ceh back to the most recent 1 Ik 10 Tzec is 10.10.8 (3,808 days); while that between 13 Cimi 19 Ceh and the most recent 1 Ik 10 Tzec that was a station in the 819-day cycle is 11.10.10.4 (83 004 days).[4] This was the 1 Ik 10 Tzec of 1.6.14.11.2. If we translate the record of the inscription, it amounts to saying that the day November 14 of the year 2360 BC was 411 days after August 14 of 2587 BC, which is a bit short of the mark: it was more than two-and-a-quarter centuries. This is one of the problems of the inscription. And, apart from the arithmetical problem of the 1 Ik 10 Tzec date, there is another, namely that the proper 819-day station for the initial date of this inscription, according to all existing precedents, should be 1 Imix 19 Pax (1.18.4.7.1), at a removal of 285 days (14.5) from the initial date; for this is the next prior station of that cycle. A second problem, then, is in the bizarre choice of station for the 819-day passage in this inscription: instead of the next prior (as are all other recorded examples), it is the 101st prior.

After the mythological prologue, the text goes on to record what appears to have been one of the most important events in the life of the then ruler of Palenque, a king to whom we can refer confidently by his proper name, "Serpent-Jaguar" in translation, or "Chan-Bahlum" in a Chol-Mayan rendering, for both the sense and the reading of his nominal hieroglyph are now quite secure. This historical event was a three-day affair, obviously concerned with the gods, and with more than just the Triad. It began on the day 2 Cib 14 Mol, 9.12.18.5.16, which was July 20, AD 690; it continued on to the next day, 3 Caban 15 Mol; and it was concluded with an offering of some sort on the third day. The human protagonist who dealt with the gods on this occasion was Chan-Bahlum. The events are recorded not only in the Temple of the Sun, but also in the Temple of the Cross and in the Temple of the Foliated Cross (these accounts complementing one another in a certain respect); and the occasion was commemorated, along with its 12th year anniversary, also in the Medallion Series that was placed across the piers and the lintels of the face of the Temple of the Inscriptions. The nature of the events and of the occasion that prompted them has been one of the many intriguing questions that are posed by the inscriptions of Palenque. The remaining passages of the inscription from the Temple of the Sun are concerned with other chronological matters relating to Chan-Bahlum, viz the date on which he, as a boy of six, was named as heir-designate and future successor of his father Pacal; the numerical relation of this date to that of a similar event for a predecessor of similar age a century and a half earlier; the five-day proximity of the date to the summer solstice; the date of Chan-Bahlum's birth; his precise age on the day of his heir-designation; and the interval from this event to the half-katun date. These constitute the subject matter of the main text, apart from the passages that are inserted into the iconographic part of the central panel. Of those latter, one of them is again concerned with the date of his heir-designation, and the other with that of his inauguration as ruler 45 years later (his father lived to be 80 years old). The date of this last event will be one of those to concern us later, but at this point we return to the problem of the apparent error in the 819-day passage.

One suspects that an error so gross is unlikely to have been due to any ordinary sort of mistake in arithmetic. There is, in fact, only one other instance in Mayan inscriptions—from any site—of an error in an 819-day passage. That, curiously, is also from Palenque, and from the same "Cross" group of temples, namely, from the Temple of the Foliated Cross. In that passage the interval that is recorded (the "distance number" in Mayanist jargon) was computed correctly, but it was mistakenly added to the initial-series date instead of being subtracted from it as it should have been. It is as if a competent calendar priest had done the initial calculation—the hard part—and then turned the result over to an assistant to take it from there, but to an assistant, unfortunately, who was still too much of a novice to know that in this context (in contradistinction to most others) it should be subtracted rather than added. The error in the Temple of the Foliated Cross is thus explicable; but the discrepancy in the Temple of the Sun is of an order that calls for some other kind of explanation. In any case, its properties merit inquiry before dismissing it as hopeless; for the possibility of numerological manipulation is something which experience has taught us not to overlook.

Calendrical primes, such as are employed in certain other instances of Mayan numerology, fail to yield anything of significance in application to this problem. That being so, planetary constants are the next most plausible to be tried. The number 584, for Venus, also is unproductive; but 399, for Jupiter, offers a more promising result. The actual interval from the 1 Ik 10 Tzec 819-day station (at 1.6.14.11.2) to the initial-series date of 13 Cimi 19 Ceh (1.18.5.3.6), as previously noted, is equal to 83,004 days (11.10.10.4). The recorded interval, as also previously noted, is equal to 411 days (1.2.11). The difference between the true interval and the recorded one—i.e. the amount of the apparent computational "error"—is thus equal to 82,593 days (11.9.7.13). This is an integral multiple of 399, the whole-number approximation to the mean synodic period of Jupiter (398.88 days, when averaged over very long time spans). It has thus the earmark of a Mayan numerological projection, which is typically in terms of integral values without application of a correction formula.[5]

This result, though suggestive of an interest in Jupiter, could yet be due simply to chance. But there is another facet to the puzzle. A day 1 Ik 10 Tzec is also recorded in a pair of medallions over one of the piers of the Temple of the Inscriptions. This is the same calendar-round day as the one recorded for the 819-day station accompanying the mythological initial date of the Temple of the Sun. An analysis of the Medallion Series by Peter Mathews (1980) showed that the role of 1 Ik 10 Tzec in the text of the medallions was that of the 819-day station proper to the initial-series date of the text (the fourth of those commemorating the 2 Cib 14 Mol event of 9.12.18.5.16), and that it was recorded there properly distanced from that initial date (by an interval of 2.3.14, i.e. 794 days prior).

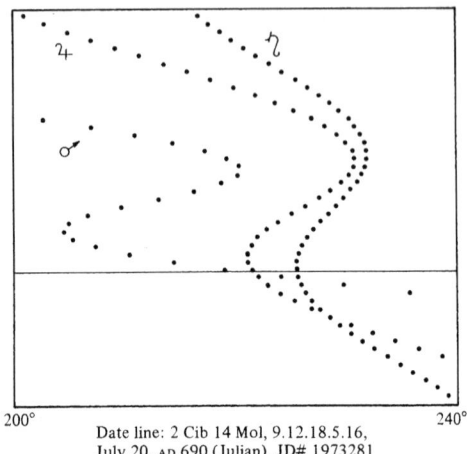

24.1 Celestial longitudes of Jupiter, Saturn and Mars before, on, and after the date of Chan-Bahlum's rites for the gods at Palenque, July 20, AD 690.

Date line: 2 Cib 14 Mol, 9.12.18.5.16, July 20, AD 690 (Julian), JD# 1973281

An 819-day station can occur on the same calendar-round day only once every 63 calendar rounds (3276 years, omitting leap-year days). Or, to put it another way, an appropriate calendar-round day—such as 1 Ik 10 Tzec—can recur again as an 819-day station only after a lapse of 63 × 52 Maya years. That is the interval between the 1 Ik 10 Tzec of the 819-day station of the Temple of the Sun (necessarily 1.6.14.11.2) and the 1 Ik 10 Tzec of the 819-day station of the Medallion Series (9.12.16.2.2). Since the choice of the station to which to relate the initial date of the Temple of the Sun is entirely counter to the normal rule and to all precedents (being the 101st-prior such station rather than the first-prior), one is led to ask whether there could have been some overriding desideratum in this case, giving cause for replication in mythological time of the station that is appropriate in the historical context. And, since the numerical property of the "error" in the Temple of the Sun is suggestive of Jupiter, one is led to inquire into the circumstances of Jupiter on the historical date initiating Chan-Bahlum's important rites for the gods (2 Cib 14 Mol, 9.12.18.5.16) and on the date of the 819-day station prior to those rites (1 Ik 10 Tzec, 9.12.16.2.2).

For the 2 Cib 14 Mol date—equal to AD 690 July 20 (Julian) by the 584285 correlation—data from the Tuckerman planetary tables may be seen as favourable to a hypothesis involving not only Jupiter, but also Saturn and Mars. During a period of some six months prior to that date, Jupiter, Saturn and Mars were in close proximity, Jupiter and Saturn having barely missed one of their rare "triple conjunctions" with each other (see Fig. 24.1).[6] And at intervals following this date there occurred first a conjunction of Mars and Jupiter, then one of Mars and Saturn, and then, somewhat later, one of Jupiter and Saturn (what would have been their third conjunction, had their paths actually crossed and recrossed on their first close encounter). This suggests that it might have been the unusual and impressive gathering of these three celestial figures—conceived to be gods—that occasioned the special Palenque observances.[7]

Floyd G. Lounsbury

24.2 Celestial longitudes of Jupiter and Mars before, on, and after the date of the 819-day station prior to the rites, May 17) AD 688.

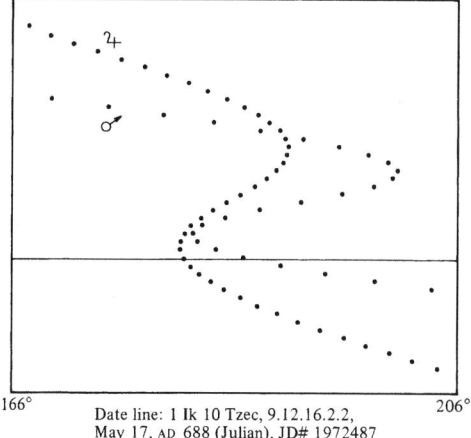

166° 206°

Date line: 1 Ik 10 Tzec, 9.12.16.2.2,
May 17, AD 688 (Julian), JD# 1972487

But this is only a vague description of the situation. When put into precise terms, it is inadequate to account for the choice of the day 2 Cib 14 Mol (July 20, AD 690) for the observances. The closest approach of Jupiter and Saturn to each other, and of Mars to them, prior to this date was some four months earlier. After a trend of separation for a few months, and then of reversal of that trend, they were now closing in on each other again, Mars moving rapidly and the other two more slowly, heading—no doubt obviously—for conjunctions. But the conjunction of Mars and Jupiter was not until several days after the Palenque observances had been concluded, that of Mars and Saturn only after another several days, and that of Jupiter and Saturn not until another month and a half after that. Thus, although suggestive, these seemingly notable aspects of the situation fall short of accounting for the particular date.

On the other problematic date, that of the historical 819-day station 1 Ik 10 Tzec (9.12.16.2.2), May 17, AD 688 (Julian), the situation of Jupiter was in a certain respect similar. Mars was, and had been, in proximity (these two also having barely missed a triple conjunction—see Fig. 24.2), though Saturn was more distant. But there is a more exact similarity in the situations of Jupiter on the two dates of interest: in each case the planet had moved about one-third of a degree (0°.32 and 0°.30, respectively) in departure from dead center of the second stationary point. This—very likely a "just noticeable difference" for naked-eye observation—suggests another hypothesis, viz that it could have been Jupiter's departure from the second stationary point that was the critical detail of interest in these situations. It appears to be the only one that allows specification of a precise similarity on these two dates.

Two instances, however, are insufficient to make a case for the hypothesis suggested by them; their similarity could have been due to chance. But if Chan-Bahlum or his court astrologers had really been interested in Jupiter's departures from the second stationary point, then some other dates might be expected also to give evidence of it. If so, the ruler's accession date is one that should be looked into.

A Palenque King and the Planet Jupiter

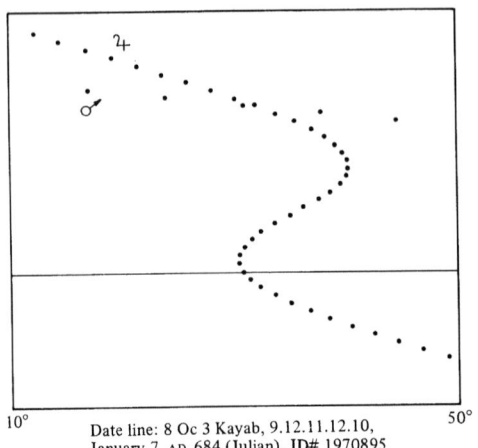

24.3 Celestial longitudes of Jupiter before, on, and after the date of the royal accession of Chan-Bahlum at Palenque, January 7. AD 684. (Also of Mars, top portion of chart.)

Date line: 8 Oc 3 Kayab, 9.12.11.12.10, January 7, AD 684 (Julian), JD# 1970895

The date of Chan-Bahlum's accession to rule is recorded at least seven times at Palenque, in six different inscriptions: it was on 8 Oc 3 Kayab, 9.12.11.12.10, which translates as January 7, AD 684 (Julian). On this date, it turns out, Jupiter was indeed at virtually the same point in its synodic period as it was on the two dates previously reviewed; its position this time was 0°.42 past dead center of the second stationary point (see Fig. 24.3).

The Medallion Series from the Temple of the Inscriptions had already been mentioned, and two of its dates have concerned us. The first was that of its initial series, which was one of the four recordings of 2 Cib 14 Mol, 9.12.18.5.16, the date of Chan-Bahlum's special observances for the gods (first noted here in connection with its record in the Temple of the Sun). The second was the date of its 819-day series, the historical 1 Ik 10 Tzec station of that cycle, which has provided a possible rationale for the exceptional treatment of the 819-day passage in the Temple of the Sun. The concluding text of the Medallion Series made reference to yet a third date, which was the 12-year anniversary of the 2 Cib 14 Mol event. It included the 12-year distance number 12.3.0 (12[360] + 3[20] + 0[1] = 12[365]) and the calendar-round day to which this leads when applied to 2 Cib 14 Mol. That day is 1 Cib 14 Mol (9.13.10.8.16 = July 17, AD 702), for the 12 × 365-day increment is one which has the property of reducing the trecena component of a calendar-round date by an amount of one, while returning the veintena and the haab components unaltered. It must be supposed that this calendrical property provided a prime rationale for the selection of that particular date for an anniversary commemoration of the 2 Cib 14 Mol event. But was it the only criterion? Can there have been others?

The 12-year interval (4380 days) is just seven or eight days short of the 11th multiple of the mean synodic period of Jupiter (11 × 398.88 = 4387.68). The planet would necessarily have been again on point of departure from its second stationary point, though its movement would probably have been still too slight for naked-eye detection. The amount was actually 0°.08 (see Fig. 24.4). The 12-year period is also

24.4 Celestial longitudes of Jupiter before, on, and after the date of the 12-year anniversary of the rites for the gods, July 17, AD 702. (Also of Mars, at top of chart.)

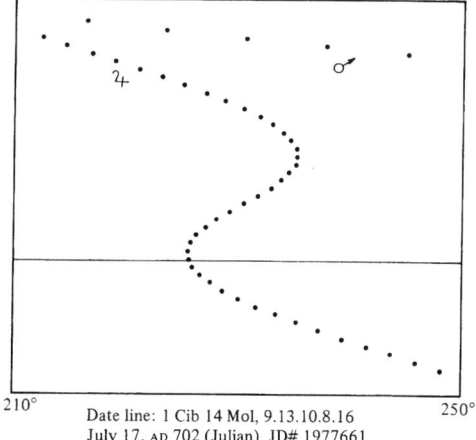

Date line: 1 Cib 14 Mol, 9.13.10.8.16
July 17, AD 702 (Julian), JD# 1977661

one which brings Jupiter back to approximately the same region of the zodiac. The mean sidereal period of the planet is 4332.59 days. Its celestial longitude on 2 Cib 14 Mol was 221°.43; on 1 Cib 14 Mol, 12 years later, it was 225°.73, just 4°.3 beyond its earlier position. This, then, was Jupiter's first departure from a second stationary point in the same constellation of the zodiac as it has been on the 2 Cib 14 Mol date 12 years earlier. On both occasions it was in the region of Beta and Delta Scorpii.[8]

Chan-Bahlum died on 6 Chicchan 3 Pop, 9.13.10.1.5 (February 16, AD 702), only half a year before the anniversary date of 1 Cib 14 Mol; and his next-younger brother, who had already been designated to succeed him, was installed in office just 48 days before this date. Thus, although the planning for the 12-year anniversary observance and for the commemorative Medallion Series could well have been Chan-Bahlum's, their execution must have fallen to his brother and successor (whom we call Kan-Xul, or Hok, though not with the degree of confidence that we have in the names of Chan-Bahlum and Pacal).

Another monument to Chan-Bahlum, with yet another posthumous date, is in Temple XIV at Palenque. Its tablet shows him executing the dance that symbolizes emergence from the underworld, while he is offered the symbol of rulership by his long deceased mother.[9] (The dance step and the arm movement are known from other examples at Palenque and elsewhere, as well as from vase paintings which relate it to its mythological precedents.) The inscription of the panel begins with dates of events in an antiquity of deep cosmological proportions. It moves then from one of these by an interval of 5.18.4.7.8.13.18 (nearly a million years!) to the day 9 Ahau 3 Kankin (9.13.13.15.0 = November 2, AD 705). On this day there occurs an event of which Chan-Bahlum is the protagonist and which involves again some of the deities. The iconography and the posthumous timing implicate it as the date of the ruler's apotheosis, although the hieroglyphs designating the event are not yet sufficiently well understood to secure the interpretation from that line of evidence. The four preceding examples which we have seen of Jovian second-station dates lead us to inquire now

A PALENQUE KING AND THE PLANET JUPITER

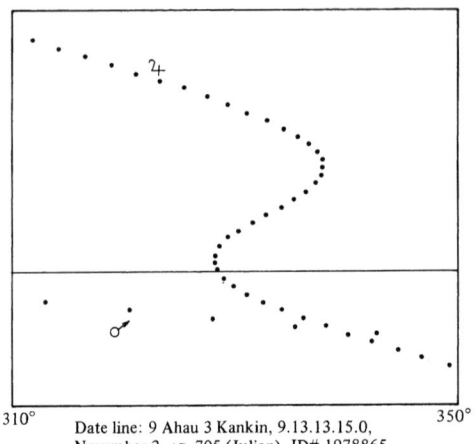

24.5 Celestial longitudes of Jupiter before, on, and after the date of the apotheosis of Chan-Bahlum, November 2, AD 705. (Also of Mars, after this date.)

Date line: 9 Ahau 3 Kankin, 9.13.13.15.0, November 2, AD 705 (Julian), JD# 1978865

into the position of Jupiter on this date. It turns out again to be the same: Jupiter has moved 0°.29 off dead center of the second stationary point (see Fig. 24.5).

We have now, in all, five examples that appear to be indicative of an interest in the movement of the planet Jupiter. The critical datum would seem to be the planet's first perceptible move in departure from the second stationary point—the deity's embarkation, so to speak, on the next leg of his journey in his nearly 12-year circuit of the zodiac. Of these five examples, three are the dates of events whose timings were subject to human control: Chan-Bahlum's inauguration as ruler, his special rites for the gods, and his apotheosis, all of these coinciding with the same critical point in the synodic period of Jupiter. The other two are dates whose determinations followed independent arithmetical or calendrical criteria; but they happened to exemplify the same astronomical phenomenon, and the exploitation of the date in each case was clearly deliberate and in response to the astronomical coincidence. One of these involved the extraordinary projection, employing Jupiter numerology, of a historical 819-day station back 63 calendar-rounds to create a mythic-time precedent; and the other gave recognition to an approximate sidereal, synodic, and calendrical anniversary of a Jupiter date and an event that were climactic in the life of this ruler.

For the sake of perspective on the statistical significance of these coincidences in timing, we should see them in relation to the entire corpus of recorded dates pertaining to this ruler. There are a total of 13 or 14 if the third day of the rites for the gods is included. (Reference is made to the third day in the inscriptions, but it is not listed by name as are the first two days of the rites.) Of the 13 named days, five are those discussed above, associated with the second stationary point of Jupiter, and another of the 13 must be included with these, since it is that of the second day of the rites. This leaves a remainder of seven. Of the seven, two are period-ending dates in the long count: 9.10.10.0.0 (December 3, 642), and 9.13.0.0.0 (March 15, 692), strictly numerical in determination and with no significance in relation to critical points of Jupiter's period. They serve as round-number anchors in the long count for fixing dates given in the calendar-round. Two others are of natural events, Chan-Bahlum's birth, 9.10.2.6.6 (May 20, 635), and his death, 9.13.10.1.5 (February 16, 702), whose timings presumably were not subject to prior astrological calculation. (His birth, however, was the day before a predictable eclipse date,[10] a coincidence exploited in other ways in the inscriptions and mythological inventions of Palenque.) With these four removed from consideration in relation to the Jupiter phenomenon, the remainder is down to three.

One of these is the date of the rite that we call "heir designation," this being the only interpretation that survives comparison of all of its hieroglyphic records and their contexts. For Chan-Bahlum, as for a ruler a century and a half earlier with whose similar rite his is compared, it took place when he was six years old—6.2.17, to be exact. (There is wide variation in the ages of the rulers for whom it is recorded: the youngest were those designated in childhood or youth to succeed lineally, with actual succession coming many years or decades later; the oldest were those designated in adulthood to succeed laterally, with succession following after a relatively short interval.) An heir-designation, surely, is an occasion susceptible to astrological timing.

The heir-designation rite for Chan-Bahlum took place on 9 Akbal 6 Xul, 9.10.8.9.3, which was June 14 (Julian Calendar) or June 17 (retroactive Gregorian) of the year 641. In respect to the situation of Jupiter, this date, strictly speaking, ought not be eligible for inclusion in the same category as the five "second-station" dates that have already been discussed, because the amount of the planet's movement away from the second stationary point was a bit more, really too much to be considered a "just noticeable" difference or a "first" indication of departure from the rest station. It had moved 1°.18 by that date. As compared with the others (0°.32, 0°.30, 0°.42, 0°.29 and 0°.08), this is of a different order, though of course still miniscule in relation to the range of possibilities. In a different respect, however, the situation bore an aspect of similarity to those of the 1 Ik 10 Tzec and 2 Cib 14 Mol dates (of 47 and 49 years later), this being in an involvement of Jupiter with Mars. In fact, it was the most special of all of these, and surely the most striking to observe. The rite for Chan-Bahlum followed upon a remarkable triple conjunction of those two planets, about the most symmetrical and "perfect" example that might ever be witnessed, the conjunctions coinciding, respectively, with (1) Jupiter's first stationary point, (2) the virtually simultaneous oppositions of Mars and Jupiter to the sun, and (3) Jupiter's second stationary point. (See Fig. 24.6.) It was a celestial drama which the astrologers must surely have been monitoring. And having long since personified and supernaturalized these zodiacal travelers, they could hardly have failed to attach significance to their encounters. This is what immediately preceded the rite that was performed before the gods for the boy Chan-Bahlum by his father Pacal, dedicating him and designating him (if our interpretation is correct) as his future heir and successor. But the date of that rite, while similar to the others reviewed here, was 13 or 14 days later (about 0°.75 further) in respect to Jupiter's departure from the second stationary point.

Another attribute of this date, however, deserves mention: it was the date of new moon (lunar-solar conjunction) just prior to the summer solstice—"five days prior," as is specially noted in the inscription of the Temple of the Sun.[11] The fact that attention

A Palenque King and the Planet Jupiter

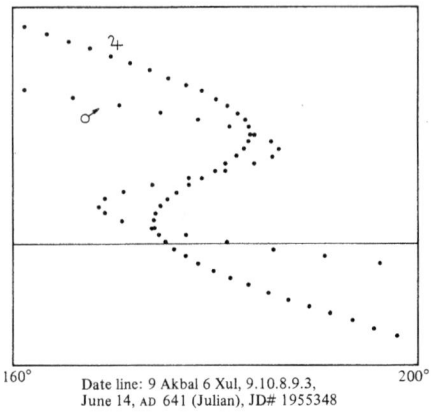

24.6 Celestial longitudes of Jupiter and Mars before, on, and after the date of the rite of heir-designation for Chan-Bahlum, at age six, June 14, AD 641.

Date line: 9 Akbal 6 Xul, 9.10.8.9.3, June 14, AD 641 (Julian), JD# 1955348

is directed to this in the inscription suggests that the lunar and seasonal timings may also have been a desideratum, in which case the slightly greater measure of Jupiter's removal from the second stationary point could be seen as a minor accommodation to the lunar schedule. Also to be considered is that this was the first of these Jupiter timings at Palenque, preceding the others by from 43 to 64 years. And it is only with Chan-Bahlum that they are associated; the dates for no other ruler at Palenque conform to this pattern.

There is one other important date in Chan-Bahlum's official curriculum vitae: it is the eight-year anniversary of his accession to rule. This was observed on 5 Eb 5 Kayab, 9.12.19.14.12, which equates with January 7 (Julian) or 10 (Gregorian), AD 692. Although the accession date (v. supra) is one of those exhibiting the Jupiter timing discussed here, this eight-year anniversary date is not of that category. (One at 12 years would have done better.) For present purposes, i.e. for the hypothesis of deliberate timing of important events to coincide with Jupiter's departure from the second stationary point, this counts as a definite negative instance. It had, however, other attributes of potential astronomical interest. Eight years is the Venus recurrence period—five synodic periods, 13 sidereal periods (to within a day and a half), with completion of these at the same time of the year. The Mayan Venus cycle, as known from the later *Dresden Codex,* was of eight civil years, 2920 days. This anniversary, however, was after eight tropical years, 2922 days; it was 5 Kayab, not 3 Kayab. (Both the accession day and its anniversary equate with January 7 in the Julian calendar, with leap-year intercalations.) On both occasions the morning star Venus was close to its greatest elongation from the Sun (a little over 45°) though still about 1° from the maximum. And for whatever it may be worth in this context, it can be noted that on the morning of this anniversary date, January 7, 692, there was very likely (i.e. as nearly as the writer can determine) a heliacal rising of Jupiter, a first appearance at the eastern horizon after conjunction.[12] Thus the 5 Eb 5 Kayab date appears to have had its own qualifications suitable for such an occasion. But it pertains to an entirely different category from that which has emerged from the other cases here considered.

These are the dates connected with the name of Chan-Bahlum. The relevant data are summarized in Table 24.1. One would like to have had more dates to test; but even with this number it would be difficult to suppose that the degree of association of controllable timings with a particular phenomenon of Jupiter had been a matter of chance, or that the emphasis given to the chance associations had been otherwise motivated. It appears that the ruler had associated his identity and his fortunes in some way with that planet or deity.

There survive no Mayan Jupiter tables, such as we have for Venus and for the eclipse cycle in the Dresden hieroglyphic codex. The skywatchers at Palenque, however, must have been in possession of a sufficient corpus of records of prior observations of Jupiter to have arrived at the correct integral value for the mean of the synodic periods; for this was employed numerologically in the reckonings of the 819-day station which accompanies the initial date in the inscription of the Temple of the Sun. And with such records, and in the process of determining the mean, they would have acquired some knowledge also of the extent and the patterns of deviation from the mean. A comparison of the intervals between the Palenque dates shows that this deviation was taken into account. If we consider the dates of the three events whose timing was entirely free of calendrical, numerical, or other astronomical constraints (those marked with two asterisks in Table 24.1), we have the following:

```
9.12.11.12.10    8 Oc 3 Kayab, accession (January 7, 684)
   +6.11. 6     = 2386 days = 6(398.88) –7.28 days

9.12.18. 5.16    2 Cib 14 Mol, rites to the gods (20 July, 690)
  +15. 9. 4     = 5584 days = 14(398.88) –0.32 days
9.13.13.15. 0    9 Ahau 3 Kankin, apotheosis (November 2, 705).
```

It can be seen that the second interval, the longer one, is as close as possible to an integral multiple of the true mean synodic period of Jupiter (398.88 days), while the first, the lesser of the two, falls more than seven days short of an integral multiple of that period. When first noted, this was disconcerting, but it turns out to be in accord with proper expectations, supporting rather than contradicting the interpretation. The Jupiter period is a little more than a month longer than the solar year. Consequently each succeeding Jupiter period "begins" approximately a month or so later in the calendar year than did the one preceding. If we count "'beginnings'" from the second stationary point—as it appears the Maya at Palenque may have done—then we can say that the Jupiter periods that "begin" in the spring months of the year are the shorter ones, including those that are as short as about 395 days; those that "begin" in the fall months of the year are the longer ones, including those that are as long as about 402 days; and those that "begin" in the summer and in the winter months run closer to the mean. In the above tabulation it can be seen that the six Jupiter periods that intervene between the first and the second dates have winter "beginnings" (more or less average in length) and spring "beginnings" (short); hence the accumulated shortfall from

Table 24.1. Recorded dates pertaining to the ruler Chan-Bahlum, with positions of the planet Jupiter, in degrees of celestial longitude, and in days, relative to a preceding second stationary point.

Date (Julian)	Event		Degrees	Days
May 20, 635	Birth (father's age 32)		37.83	204
Jun 14, 641	Heir-designation (age 6)	?*	1.18	28
Jan 7, 684	Accession to rule (age 49)	**	0.42	15
May 17, 688	"1 Ik 10 Tzec" 819d station	*	0.30	14
Jul 20, 690	"2 Cib 14 Mol" rites	**	0.32	15
Jan 7, 692	Eight-year anniversary of accession		25.14	154
Feb 16, 702	Death (age 67)		40.18	247
Jul 17, 702	"1 Cib 14 Mol" commemoration	*	0.08	8
Nov 2, 705	Apotheosis of deceased ruler	**	0.29	12

(**) marks instances in which Jupiter's departure from the second stationary point appear to have been a criterion for the selection of the date.
(*) marks those instances in which the date was otherwise determined, but where its special observance appears to have depended on this position of Jupiter. The one marked (?*) may have had a similar connotation.
Unstarred items are on dates when Jupiter was not in immediate departure from the second stationary point.

an integral multiple of the mean is to be anticipated. In contrast, the 14 periods that intervene between the second and third dates start with summer "beginnings" (more or less average) and then include beginnings that range over an entire year thereafter (wherein the short balance the long); hence the optimum approximation to an integral multiple of the mean period in their total duration. (This is a drastic oversimplification, though true enough to illustrate the point in question. A more suitable locus for "beginnings" would have been the planet's opposition to the Sun.)

Two hypotheses, related but distinct, have been advanced. The first is that a departure of Jupiter from its second stationary point marked a preferred time for dynastic rituals for a particular ruler at Palenque. Support for it derives from the associations discussed above and summarized in Table 24.1, and further, also from the intervals separating the doubly starred items of Table 24.1, whose deviations from mean-value multiples correspond to the actual deviations at those times. The second is that the coming together of Mars or of Saturn with Jupiter provided a favorable or motivating context in some cases, though it was not in general a necessary condition. This rests primarily on the fact that the close congregation of Jupiter, Saturn and Mars during their simultaneous retrograde periods prior and up to the 2 Cib 14 Mol event furnishes the only discoverable motivation for that event and for the manner in which it is recorded in the inscriptions. The triple conjunction of Mars with Jupiter prior to the 9 Akbal 6 Xul heir-designation event is also suspect as one of the determining components in its timing; and the approximation to a repetition of that relationship prior to the 1 Ik 10 Tzec 819-day station, although a near miss so far as a triple conjunction is concerned, is suspect as one of the reasons for the importance attached to that date. Such assemblages of the planets during simultaneous retrogressions are rare enough to warrant attaching some significance to these, and much too rare to have been made a general requisite for the timing of dynastic rituals.[13]

This chapter can be concluded with a note returning to the matter with which it began, viz the anomalous 819-day passage in the Temple of the Sun. Its chronology can be reconstructed now as follows:

1.18. 5. 3. 6	13 Cimi 19 Ceh, birth of deity G3 (November 14, 2360 BC)
−1. 2.11	= 411 days, recorded interval
1.18. 4. 0.15	5 men 13 Yax, posited Jupiter phenomenon (September 29, 2361 BC)
−11. 9. 7.13	= 82,593 days (=207 × 399 days)
1. 6.14.11. 2	1 Ik 10 Tzec, recorded station of the 819-day cycle and hypothetic Jupiter phenomenon (August 14, 2587 BC).

An 819-day passage normally specifies only a single interval and a single date. The date is that of the next-prior "station" or starting point of the 819-day cycle, and the interval is that which mediates between it and the date of the initial series. The chronology of such a passage, if presented in the manner above, would require only three lines instead of five. And so it is also in the Temple of the Sun, where the format is normal though the content is not; the information contained in the third and fourth lines of the above reconstruction has been omitted, leaving the passage arithmetically absurd if interpreted literally. Whether this was by design, through an oversight, or due to an uncomprehending editor's deletions for preservation of conventional format, remains an unanswered question. The intent of the person who made the original computations appears to have been to set down a distance number representing the interval from the initial-series date back to a next-prior supposed Jupiter station, and then to go from there by successive mean Jupiter periods to a date that was both a hypothetic Jupiter station and an 819-day station with the same calendar-round name as the one that preceded the historical 2 Cib 14 Mol event, thus arriving at an ancient precedent or mythological "charter" for the latter—a manipulation itself with ample precedents in the Mayan legacy. That the 63-calendar-round cycle necessary to this end must doom the result to failure (not being commensurate with the mean Jupiter period) appears not to have been an important consideration. In any case, pertaining to a mythological era, it was exempt from empirical verification.

A survey of the dates from other Mayan sites, with a view to testing for similar observations of this or of other phenomena of Jupiter, has not yet been made. A few instances, however, have been noted. An interesting one is the date of Lintel 25 of Yaxchilan: 5 Imix 4 Mac, 9.12.9.8.1, or October 20, AD 681 (Julian). This is the date of the blood offering and serpent vision that is depicted on that famous lintel (housed in the British Museum),[14] and it is the date of the inauguration of Shield-Jaguar as ruler of Yaxchilan. As of that date, 17 days past the date of dead-center of the second stationary point, Jupiter had moved 0°.18 away from its position at that time. Whether this is indicative of a pattern for that principality, or for that particular ruler, or whether it was an isolated chance occurrence, remains to be seen. It was 809 days before the inauguration of Chan-Bahlum at Palenque, an interval that is equal to two mean synodic periods of Jupiter plus 11 days, or to six days more than the two actual synodic periods that intervened (deviating on the long side from the mean).

Another one of interest is the date 6 Eb 0 Pop, 9.15.12.11.12, which is on Lintel 2 from Temple 4 at Tikal (one of the spectacular wooden lintels in the Ethnographic Museum in Basel).[15] As of that date, February 3, AD 744 (Julian), Jupiter had moved 0°.50, in 18 days, away from dead-center of the second stationary point. Mars also was in the scene on that date. The next day, 7 Ben 1 Pop, is recorded in the same inscription as the day of a "star-timed" war expedition against an enemy principality. "Star"timings of such undertakings in synchrony with critical points in the cycle of Venus have been previously noted (Kelley, 1977; Lounsbury, 1982). This one from Tikal Lintel 2, lacking any such Venus credentials but with a significant one from Jupiter, offers confirmation of a point that Kelley has made about the "star" sign in these hieroglyphs, namely that its reference is not inherently restricted to the planet Venus, unless a specific prefix or an implicating context indicates it as such.

ACKNOWLEDGMENTS

This paper is an abridgment (in its first parts) and an expansion (in its later parts) of one begun but not completed in the spring of 1977 while a Senior Fellow in Pre-Columbian Studies at Dumbarton Oaks. I wish to acknowledge my indebtedness to Dumbarton Oaks, and to Elizabeth P. Benson, then Director of Pre-Columbian Studies, for the opportunities made available to me there to work on problems of Palenque epigraphy.

I owe thanks also to John S. Justeson for helpful criticisms received during the revision and preparation of the final manuscript of this paper, and to Roger W. Sinnott for the use of his computer program, *Planet Positions,* which has facilitated computation, and which, with an appended graphing program, produced the figures which appear here.

The subject matter of this paper was presented publicly on a number of occasions prior to its presentation in Merida at the 2nd Oxford Conference viz at University of California at Los Angeles, Anthropology Graduate Student Association, April 13, 1982; Wellesley College, Henry R. Luce Lecture, April 26, 1983; Yale University, Anthropology Colloquium, September 15, 1983; State University of New York at Albany, Department of Anthropology and Institute of Mesoamerican Studies, October 7, 1983; University of Pennsylvania, University Museum Maya Symposium, April 7, 1984. It was also the topic which I first proposed as a contribution to the then forthcoming International Conference on Archaeoastronomy at Oxford (September, 1981), but which I later changed, when the exciting results concerning Venus phenomena at Bonampak and other sites began to accumulate while pursuing the problem to which Mary Miller introduced me in the fall of 1980 (cf. Lounsbury, 1982).

Floyd G. Lounsbury

NOTES

1. The archaeological site of Palenque is in lowland Chiapas, Mexico, at 92° 5' 20" west longitude and 17° 29' 30" north latitude (Maudslay, 1889–1902, vol. V: "Text [for] vol. IV," p. 8). The inscriptions with which this paper is concerned date from the last two decades of the seventh century and the first of the eighth century of our era. The king to whom reference is made in the title lived from AD 635 to 702, and he reigned from 684 to 702. For reproductions of the panel from the Temple of the Sun the reader is referred to Maudslay (1899–1902, plates 87 and 88) (A. P. Maudslay's photographs and Annie Hunter's line drawings), and to Robertson (1980, p. 105) (Linda Schele's more recent drawing of the text). But for a single exception (note 9), discussion of problems of hieroglyphic and iconographic interpretation of the Palenque tablets is not included in this present article.

2. The Maya day count, or "long count" as it is generally known, was a native chronological system wherein the days, starting from a far-distant prehistoric epoch, were assigned numbers in sequence. In principle it was analogous to the Julian Day count introduced by Joseph Justus Scaliger in 1583 as a scheme for unifying Old World historical chronology. The institution of the day count in Middle America, however, antedated the invention of the Julian Day count in Europe by at least 16 centuries, its earliest surviving examples (carved on stone monuments) dating from the first century BC. Its epoch, or day zero, was August 13 (retroactive Gregorian) of the year 3114 BC (−3113 in astronomical chronology). Like the epoch of the Julian Day count (January 1, retroactive Julian, 4713 BC), this was a backward projection, via cyclical reckonings, from its date of invention sometime prior to the mentioned earliest attestations.

The Mayan numeral system is vigesimal. Its application to the day count, however, entails a departure from base 20 in one position, viz a unit in the third position is equal to 18 (rather than 20) of the second. This can be viewed in either of two ways, depending on whether the "tun" (360 days, unit in third position) or the "kin" (day, unit in first position) is taken to be the fundamental unit of the count.

The Maya calendar-round was a native calendrical system wherein the days were designated by their coordinates in three modular dimensions: a cycle of 13 days, represented by the numerals 1 to 13; a simultaneous cycle of 20 days, represented by day names; and another simultaneous cycle of 365 days, numbered within 18 named 20-day periods and an additional five-day residue period. The first two of these cycles taken together define the 260-day sacred almanac, or "tzolkin," wherein the days receive binomial designations (e.g., 13 Cimi, as in the date just cited in the text). The third cycle is that of the civil year, or "haab," wherein the days are designated by numerals prefixed to the names of the subperiods (e.g., 19 Ceh, of the date cited in the text). The tzolkin and the haab, running simultaneously, together define the cycle of the "calendar-round," 18,980 days in length ($=52 \times 365$). This system is older in Middle America than the chronological system described in the paragraph above, and with even earlier attestations.

The equation of Mayan dates with dates of the European calendar and Christian chronology presumes the correctness of the "584285" correlation (Maya Day Number + 584285 + Julian Day Number). This is the first of the values proposed by J. Eric S. Thompson (1927, 1935) for what is generally known as the Thompson (or Goodman-Martinez-Thompson) correlation. Although Thompson later (1950) proposed a two-day reduction of its value, and although there are a few scholars who are not yet convinced of the correctness of either of Thompson's two values, the present writer has no doubts whatever concerning the validity of the first of these for the Lowland Classic Maya inscriptions (restricting the second to Central Mexican

601

and Highland Mayan dates, and admitting a minor asynchrony within the Mesoamerican area). Proofs will be presented in a future publication.

3. The term "initial series" commonly designates the series of hieroglyphs standing at the beginning of an inscription, which introduce and specify the chronological and the calendrical attributes of its opening date. The term "supplementary series" has been applied to the hieroglyphs which follow immediately after the initial series and specify the attributes of the date in the lunar calendar; and it has been extended by some writers to cover also the further specifications given in the 819-day orientation of a date. The term for this series is a relic of a time prior to the discovery of the significations of its glyphs. Usage has varied, some writers employing the dichotomy between "initial" and "supplementary" strictly on the basis of positions in an inscription, while others have employed it on a conceptual basis, with reference to content. In the positional usage, which was prior, the "initial series" included also the glyphs known as "G" and "F" (placing the date in a nine-day cycle), and it might or might not include those that place the date in the haab (the civil-year component of the calendar-round). In the conceptual usage, on the other hand, it has come to include the day number (long-count number) and the almanac or calendar-round specifications regardless of where they are found, even if not initial to an inscription, and even to the inclusion of long-count dates recorded in the *Dresden Codex*.

4. The interval between any two given calendar-round days (cr) can be determined from the following relationships:

$\Delta tz = 40 \times \Delta tr - 39 \times \Delta v$, mod 260;
$n(H) = \Delta tz - \Delta h$, mod 52;
$\Delta cr = 365 \times n(H) + \Delta h$;

where the independent variables Δtr, Δv and Δh are defined as follows:

Δtr = the interval between the two trecena coordinates (in the cycle of the repeating series of numbers 1 to 13);

Δv = the interval between the two veintena coordinates (in the cycle of the repeating series of 20 day-names: Imix, Ik, Akbal, Kan, Chicchan, Cimi, Manik, Lamat, Muluc, Oc, Chuen, Eb, Ben, Ix, Men, Cib, Caban, Etznab, Cauac, Ahau);

Δh = the interval between the two haab coordinates (in the cycle of 365 days, numbered 0 to 19 in the 18 named 20-day periods "Pop, Uo, Zip, Zotz, Tzec, Xul, Yaxkin, Mol, Chen, Yax, Zac, Ceh, Mac, Kankin, Muan, Pax, Kayab, Cumhu," and 0 to 4 in the five-day residue period "Uayeb");

and where the dependent variables tz, $n(H)$ and cr are as follows:

Δtz = the interval between the two tzolkin days (in the cycle of 260);
$n(H)$ = the number of whole haabs (365-day years) contained in the interval between the two calendar-round days; and
Δcr = the interval between the two calendar-round days.

5. For examples of Mayan numerology involving calendrical, lunar-solar and planetary periods, see Lounsbury (1976, 1978, pp. 804–8; 1983).

6. Note concerning the figures. Celestial longitudes are plotted at ten-day intervals. Longitude increases from left to right, and the passage of time from top to bottom. The date of the noted event in each figure is that of the horizontal axis (designated the "date line") one-third of the distance up from the bottom of the chart. Each figure displays the longitudes of Jupiter at ten-day intervals within a 40° sector of the zodiacal circle which includes its opposition to

the Sun and its stationary points before and after opposition. Also displayed are the longitudes of Saturn and Mars when either or both of these is (or are) within the same 40° sector. Any of these at opposition to the Sun is visible in the sky throughout most of the night. At first stationary point visibility is more in the morning sky, and at second stationary point more in the evening sky. The sequence of the figures represents the order of their consideration in the text. Their chronological sequence is 20.6, 20.3, 20.2, 20.1, 20.4, 20.5.

7. Dieter Dütting has also proposed this as a possible explanation of the significance of the 2 Cib 14 Mol date (Dütting, 1982). I came to it by way of a solution to the problem of the discrepancy in the 819-day passage in the Temple of the Sun, and I take that route again in presenting the arguments here. The original hypothesis has been modified, however, in the light of the additional data here introduced.

The months during the spring and summer of 1981, just prior to the (First) Oxford Conference, presented a rare opportunity for the naked-eye observer to gain an appreciation of the events that led up to the 2 Cib 14 Mol observance at Palenque—a phenomenon still fascinating for a "primitive" astronomer, as the writer can attest. The subject of the frequency of such occurrences received attention in several publications during the year preceding. See especially Meeus (1980).

8. Equatorial coordinates for Jupiter on the two dates of interest, July 20, 690 (2 Cib 14 Mol) and July 17, 702 (1 Cib 14 Mol), both at 6.00 PM CST (90° west), and for Delta, Beta and Alpha Scorpii during that period, were as follows (coordinates in degrees with decimal fractions; data for the stellar coordinates, AD 700, from Hawkins, 1968):

	RA	Decl.
Jupiter, July 20, AD 690 (Julian)	219.23	−14.57
Jupiter, July 17, AD 702 (Julian)	223.47	−15.92
Delta Scorpii, AD 700 (+/−)	221.40	−18.03
Beta Scorpii, AD 700 (+/−)	222.94	−15.36
Alpha Scorpii, AD 700 (+/−)	227.96	−22.55

9. Illustrations of the carved and inscribed panel from Temple XIV may be seen in Robertson (1974, p. 173; photographed by Merle Green Robertson) and Robertson (1979, p. 204; line drawing by Linda Schele). The "dance" iconography and its significance are discussed in Lounsbury (1985, pp. 53–6).

10. The eclipse of May 21, AD 635 (the day after Chan-Bahlum's birth) was not observable in the Americas. The supposition that it would have been anticipated as a possibility by the Maya at Palenque is based on the fact that there are several attestations to knowledge of the 11,960-day eclipse cycle (405 lunations, 69 internodal intervals) at Palenque during Chan-Bahlum's lifetime and after, and that the mythological birth date of the Sun god, recorded with accompanying eclipse symbolism, is related to the birth date of Chan-Bahlum by means of that cycle and one of its subdivisions. This was before the institution of the 12 Lamat base of the table in AD 755 (November 8), and was some three to four centuries before the base of the update of the table which we have in the *Dresden Codex*.

11. That this is the import of the hieroglyphs at Q7 to Q8 in the inscription of the Temple of the Sun is premised on the following considerations: (1) the inscription states that on the fifth day after the heir-designation there was another event, involving [as protagonist(?)] the Sun; (2) that day, by the 584285 correlation, was June 19, AD 641 (Julian), or June 22 (retroactive Gregorian); (3) the summer solstice of that year was exactly on that day, at 14.35 UT, or 8.35 CST; (4) although the event phrase is without other attestation in this value, there

is not much else that can be said about the Sun that would be of relevance on that particular date. (Another interpretation that has been placed on this event phrase would have the Sun as a predicate noun and the heir-designate as the subject, to the effect that the latter [metaphorically] "became the Sun" on the fifth day after his initial designation [Schele, 1984]). Should that turn out to be the correct interpretation, one may ask "Why on the fifth day?"; in which case, there being no other reason or precedent, the answer could only be that the summer solstice was a symbolically appropriate time for such a status transformation.

12. The situation of Jupiter in respect to the Sun on the morning of the eight-year anniversary date of the accession, January 7, 692 (Julian), at 6.00 AM CST (12.00 UT), was as follows:

	Longitude	Latitude	Altitude	Azimuth	Magnitude
Sun	290.20	0.00	−10.8	110.00	−26.8
Jupiter	278.50	−0.18	0.4	114.7	−1.9

In ecliptic coordinates the difference in longitude was 11°.7 and in latitude 0°.18. In horizontal coordinates (at Palenque's geographic longitude and latitude) the difference in altitude (the "arcus visionis") was 11°.2 and the difference in azimuth 4°.7.

13. The frequency of triple conjunctions of various pairs of planets, and of planets with stars, has been discussed by Meeus (1980). He notes that a theoretical frequency of triple conjunctions of Jupiter with Saturn, assuming circular orbits, had been calculated (by another writer) as an average of one every 125 years, and of Mars with Jupiter an average of one every 110 years. Their actual frequencies are much less, however, due to factors such as orbital eccentricities, perturbations, and other irregularities; and their occurrences cannot be predicted from any simple periodicities. He counted 20 of Jupiter and Saturn in the tables computed for the 3100 years from 101 BC to AD 3000, the shortest interval between them being of 40 years and the longest of 377 years. The 2 Cib 14 Mol event at Palenque, however, gives witness to the fact that near misses may also excite interest. Allowing a tolerance slightly greater than that exemplified on that occasion, and assuming circular orbits, it appears that approximations of that degree might be expected every 91 or 109 years, with occasional misses; but the sources of irregularity destroy also that periodicity and lessen the frequency of the phenomena.

14. For excellent photographs, and for an illuminating glimpse into the current state of our understanding of the ancient Maya, see the related publication by Schele and Miller (1986). For excellent drawings, see Graham (1977). Lintel 25 is on Plate 63 in Schele and Miller (1986, p. 199) and on p. 3:55 in Graham (1977).

15. For illustrations of Lintel 2 of Temple 4 at Tikal, and drawings of its inscription, see Maudslay (1889–1902, vol. 3, Plates 72–4).

REFERENCES

Dütting, D. 1982. The 2 Cib 14 Mol Event in the Inscriptions of Palenque, Chiapas, Mexico. *Zeitschrift für Ethnologie* 107:233–258.

Graham, I. 1977. *Corpus of Maya Hieroglyphic Inscriptions*, vol. 3 [Yaxchilan], part 1. Cambridge, MA, Peabody Museum, Harvard University.

Hawkins, G. S. 1968. Astro-archaeology. In *Vistas in Astronomy*, ed. A. Beer, vol. 10, pp. 45–88. New York, Pergamon Press.

Kelley, D. H. 1977. Maya Astronomical Tables and Inscriptions. In *Native American Astronomy*, ed. A. F. Aveni, 57–73. Austin, University of Texas Press.

Lounsbury, F. G. 1976. A Rationale for the Initial Date of the Temple of the Cross at Palenque: The Art, Iconography, and Dynastic History of Palenque, Part 3. In *Proceedings of the Segunda Mesa Redonda de Palenque,* ed. M. Greene Robertson, 211–224. Pebble Beach, CA, Pre-Columbian Art Research, The Robert Louis Stevenson School.

Lounsbury, F. G. 1978. Maya Numeration, Computation, and Calendrical Astronomy. In *Dictionary of Scientific Biography,* ed. Charles Coulston Gillispie, vol. 15 (Supplement 1), pp. 759–818. New York, Charles Scribner's Sons.

Lounsbury, F. G. 1982. Astronomical Knowledge and Its Uses at Bonampak, Mexico. In *Archaeoastronomy in the New World,* ed. A. F. Aveni, 143–68. Cambridge, Cambridge University Press.

Lounsbury, F. G. 1983. The Base of the Venus Table of the Dresden Codex, and Its Significance for the Calendar-Correlation Problem. In *Calendars in Mesoamerica and Peru: Native American Computations of Time,* ed. A. F. Aveni and G. Brotherston, 1–26. Oxford, British Archaeological Reports (Intl Series) no. 174.

Lounsbury, F. G. 1985. The Identities of the Mythological Figures in the Cross Group Inscriptions of Palenque. In *The Palenque Round Table Series,* vol. 6, ed. M. Greene Robertson and E. P. Benson, Fourth Palenque Round Table, 1980, pp. 45–58. San Francisco, Pre-Columbian Art Research Institute.

Mathews, P. 1980. The Stucco Texts above the Piers of the Temple of the Inscriptions at Palenque. *Glyph Notes* 10, 6p. Cambridge, MA.

Maudslay, A. P. 1889–1902. *Archaeology,* vol. 4: *Palenque.* London, R. H. Porter and Dulau. Facsimile reprint 1974, prepared by F. Robicsek. New York, Arte Primitivo.

Meeus, J. 1980. Les conjonctions triples Jupiter-Saturne. *L'Astronomie et Bulletin de la Société Astronomique de France* 94:27–36.

Robertson, M. Greene, ed. 1974. *Primera Mesa Redonda de Palenque [1973, Part 1].* The Palenque Round Table Series, vol. 1. Pebble Beach, CA, Pre-Columbian Art Research, The Robert Louis Stevenson School.

Robertson, M. Greene, ed. 1979. *Tercera Mesa Redonda de Palenque [1978, Part 1].* The Palenque Round Table Series, vol. 4. Monterey, CA, Herald Printers.

Robertson, M. Greene, ed. 1980. *Third Palenque Round Table [1978, Part 2].* The Palenque Round Table Series, vol. 5. Austin, University of Texas Press.

Schele, L. 1984. Some Suggested Readings of the Event and Office of Heir-designate at Palenque. In *Phoneticism in Mayan Hieroglyphic Writing,* ed. J. S. Justeson and L. Campbell, Publication no. 9, Institute for Mesoamerican Studies, pp. 287–305. Albany, State University of New York.

Schele, L., and M. E. Miller. 1986. *The Blood of Kings: Dynasty and Ritual in Maya Art.* Fort Worth, Kimbell Art Museum.

Thompson, J. E. 1927. A Correlation of the Mayan and European Calendars. *Publications of the Field Museum of Natural History, Anthropological Series* 17 (1), 22p. Chicago.

Thompson, J. E. 1935. Maya Chronology: The Correlation Question. *Contributions to American Archaeology* 14, 104p. Washington, DC, Carnegie Institution of Washington.

Thompson, J. E. 1950. *Maya Hieroglyphic Writing: An Introduction.* CIW Publ. 589. Washington, DC, Carnegie Institution of Washington (2nd ed., 1960; 3rd ed., 1972, Norman, University of Oklahoma Press).

Tuckerman, B. 1964. Planetary, Lunar, and Solar Positions, AD 2 to AD 1649, at Five-day and Ten-day Intervals. *Memoirs of the American Philosophical Society* 59.

CHAPTER TWENTY-FIVE

Archaeoastronomical Implications of an Agricultural Almanac in the Dresden Codex

Victoria R. Bricker and Harvey M. Bricker

Ernst Förstemann (1906:159) pointed out that pages 38b to 41b of the Dresden Codex contain a double tzolkin almanac of 520 days, divided into five periods of 104 days each (Fig. 25.1). The eleven pictures on those pages are clearly concerned with agricultural subjects like rainfall, drought, and planting; in most cases, the captions over the pictures also seem to refer to events of interest to the Maya farmer. At the same time, however, the first and the last captions in the almanac contain eclipse glyphs. The appearance of eclipse glyphs in an almanac 520 days in length implies that it is also concerned with the correspondence between a double tzolkin and a triple eclipse half-year, as is the major eclipse table on pages 51 to 58 in the same Codex (Teeple 1931:88–91; H. Bricker and V. Bricker 1983).

The first caption must refer to the beginning of the 520-day almanac, and the last must refer to the end. The intermediate dates follow intervals of 104, 208, 312, and

---

Reproduced by permission from *Mexicon* VIII (2) (1986):29–35.

Implications of an Agricultural Almanac in the Dresden Codex

25.1 The Agricultural Almanac on Pages 38b to 41b of the Dresden Codex (after Villacorta C. and Villacorta 1976:86, 88, 90, 92).

416 days. None of these intervals corresponds to the length of the eclipse half-year, 173.31 days (Teeple 1931:90), or one of its multiples (e.g., 346.62 days). This means that the captions with eclipse glyphs must refer to the extremes of the almanac, to its first and last tzolkin dates.

Table 25.1 shows in abbreviated form the structure of the tzolkin dates and intervals mentioned or implied by the agricultural almanac on pages 38b to 41b. The tzolkin date that immediately precedes the first caption is 6 Cauac. That caption contains two eclipse glyphs, the first representing a darkening of the Sun and the second a darkening of the Moon (Fig. 25.1). Our work with a similar pair of eclipse glyphs in the Mars table on pages 43b to 45b of the same Codex suggests that the 6 Cauac date falls between dates of solar and lunar eclipses (V. Bricker and H. Bricker 1986). The tzolkin date that introduces the last caption in the almanac is 13 Ben (Table 25.1). That caption refers to a single eclipse, a darkening of the Sun (Fig. 25.1), which may occur on the 13 Ben date itself or in the six-day interval that links it to the 6 Cauac date at the beginning of the next run of the almanac.

The question is where within the time period relevant to ancient Maya civilization do such conditions prevail? It can be shown quite simply that they do not occur later in Maya history than AD 919. Eclipse dates cluster in three restricted zones of a double tzolkin during any given period of several decades (Teeple 1931:90; H. Bricker and V. Bricker 1983); in our previous work on the eclipse table in the same Codex we referred to these zones as "eclipse danger windows" (H. Bricker and V. Bricker 1983). Table 25.2 defines the limits of these "windows" for the 33-year period covered by the version of the eclipse table actually shown on pages 51 to 58, the so-called 12 Lamat Original, which began in AD 755. The 6 Cauac date that precedes the first caption in the agricultural almanac falls near the end of the first window (Table 25.2).

The eclipse table can be used five times (ca. 165 years) without correction, but by the beginning of the sixth multiple the astronomical nodes have moved backward through the tzolkin by slightly more than eight days, and the table must be corrected (see H. Bricker and V. Bricker 1983, for a discussion of nodal recession and the correction necessary to counteract it). By that time, the 6 Cauac date falls about two days after the window closes, and it can no longer be used for predicting eclipses, not only in the eclipse table itself, but also in the 520-day agricultural almanac (Table 25.2).

Table 25.1 Structure of the agricultural almanac on D 38b–41b.

Cycle		Tzolkin Date	Distance Number	Cum. Total of Days
1		6 Cauac		1
			+ 16	
	=	9 Men		17
			+ 8	
	=	4 Akbal		25
			+ 11	
	=	2 Ix		36
			+ 10	
	=	12 Kan		46
			+ 1	
	=	13 Chicchan		47
			+ 12	
	=	12 Caban		59
			+ 6	
	=	5 Akbal		65
			+ 12	
	=	4 Men		77
			+ 11	
	=	2 Cimi		88
			+ 11	
	=	13 Caban		99
			+ 6	
2	=	6 Akbal		105
		.		
3	=	6 Manik		209
		.		
4	=	6 Chuen		313
		.		
5	=	6 Men		417
		.		
		2 Ik		504
			+ 11	
		13 Ben		515
			+ 6	
1	=	6 Cauac		521 (=1)

Thus the beginning date of the sixth multiple of the eclipse table, 2 August AD 919, can serve as a terminus ante quem for the agricultural almanac. The 6 Cauac date at the beginning of the almanac could not fall after a solar eclipse and before a lunar eclipse of the same eclipse season after AD 919.

Only one set of dates in the 12 Lamat Original version of the eclipse table conforms to the conditions implied by the first and the last captions in the agricultural almanac. Column 41 predicts a solar eclipse on 6 May AD 775, followed by a lunar

609

Implications of an Agricultural Almanac in the Dresden Codex

Table 25.2 Solar eclipse danger windows (bracketed) for the 12 Lamat Original version of the eclipse table. The number to the left of each date is its sequential position in the 260-day *tzolkin*. Numbers marked with an asterisk refer to dates in the second half of a 520-day double *tzolkin*.

Window 1			Window 2			Window 3		
251*	4	Chuen	164	8	Kan	77*	12	Caban
252*	5	Eb	165	9	Chicchan	78*	13	Edznab
253*	6	Ben	166	10	Cimi	79*	1	Cauac
254*	7	Ix	167	11	Manik	80*	2	Ahau
255*	8	Men	168	12	Lamat	81*	3	Imix
256*	9	Cib	169	13	Muluc	82*	4	Ik
257*	10	Caban	170	1	Oc	83*	5	Akbal
258*	11	Edznab	171	2	Chuen	84*	6	Kan
259*	12	Cauac	172	3	Eb	85*	7	Chicchan
260*	13	Ahau	173	4	Ben	86*	8	Cimi
1	1	Imix	174	5	Ix	87*	9	Manik
2	2	Ik	175	6	Men	88*	10	Lamat
3	3	Akbal	176	7	Cib	89*	11	Muluc
4	4	Kan	177	8	Caban	90*	12	Oc
5	5	Chicchan	178	9	Edznab	91*	13	Chuen
6	6	Cimi	179	10	Cauac	92*	1	Eb
7	7	Manik	180	11	Ahau	93*	2	Ben
8	8	Lamat	181	12	Imix	94*	3	Ix
9	9	Muluc	182	13	Ik	95*	4	Men
10	10	Oc	183	1	Akbal	96*	5	Cib
11	11	Chuen	184	2	Kan	97*	6	Caban
12	12	Eb	185	3	Chicchan	98*	7	Edznab
13	13	Ben	186	4	Cimi	99*	8	Cauac
14	1	Ix	187	5	Manik	100*	9	Ahau
15	2	Men	188	6	Lamat	101*	10	Imix
16	3	Cib	189	7	Muluc	102*	11	Ik
17	4	Caban	190	8	Oc	103*	12	Akbal
18	5	Edznab	191	9	Chuen	104*	13	Kan
19	6	Cauac	192	10	Eb	105*	1	Chicchan
20	7	Ahau	193	11	Ben	106*	2	Cimi
21	8	Imix	194	12	Ix	107*	3	Manik
22	9	Ik	195	13	Men	108*	4	Lamat
23	10	Akbal	196	1	Cib	109*	5	Muluc
24	11	Kan	197	2	Caban.	110*	6	Oc
25	12	Chicchan	198	3	Edznab	111*	7	Chuen
26	13	Cimi	199	4	Cauac	112*	8	Eb
27	1	Manik	200	5	Ahau	113*	9	Ben
28	2	Lamat	201	6	Imix	114*	10	Ix

eclipse on 21 May AD 775. The tzolkin date, 6 Cauac, falls 14 days after the predicted solar eclipse and one day before the predicted lunar eclipse (the eclipses in question occurred two days later than predicted, according to Oppolzer [1887:190, 355]). Column 44 predicts a solar eclipse on 19 October 776, 518 days after 6 Cauac, four days after 13 Ben and two days before the 6 Cauac date that begins the next run of the almanac (this eclipse also occurred two days later than predicted, accord-

Table 25.3 Alternative 6 Cauac dates and their Gregorian equivalents.

12 Lamat 1st (Original)	20 May AD 775
12 Lamat 2nd	16 Feb AD 808
12 Lamat 3rd	14 Nov AD 840
12 Lamat 4th	13 Aug AD 873
12 Lamat 5th	13 May AD 906

ing to Oppolzer [1887:190]). Columns 41 and 44 also predict eclipses with the same relationship to 6 Cauac in each of the next four multiples of the table. Table 25.3 lists the Gregorian equivalents of the relevant 6 Cauac dates for each of the first five runs through the table. For calculation purposes, we have used the Modified Thompson 2 correlation constant of 584,283 (Thompson 1960:305) because it agrees best with ethnohistorical evidence from Yucatan and Central Mexico. The eclipse data have provided us with five possible years for the agricultural almanac: AD 775, 808, 840, 873, and 906. Other information in the almanac, its iconography and what some of the other captions are saying, allows us to reduce the number of alternatives to two.

The first picture on page 38b depicts rain falling from a sky band onto an anthropomorphic vulture (Fig. 25.1). The second picture on the same page shows the rain god, Chac, planting with a digging stick in the rain; Chac appears again with his digging stick in the third picture, this time without rain, walking on two glyphs, one representing the earth. The first picture on page 39b shows the rainbow goddess, Chac Chel, pouring water from a jug. Chac reappears in the second picture on the same page with his digging stick, but without rain; he is seated beneath a sky band from which rain is cascading in the third picture on that page. The first picture on page 40b shows Chac hanging down from a sky band, wielding an axe (Figs. 25.1 and 25.2). The caption above the picture states that Chac is in the sky (Fig. 25.2). The first collocation can be read as *ti caan(na)* or *ti caan* "in the sky" and the second as *chac(ci)* or *chac* "rain" in Classical Yucatec (Pío Pérez 1866–77:37, 64). Both the picture and its caption imply that Chac (and the rain) has left the earth and has returned to heaven. The second picture on that page depicts an anthropomorphized macaw *(moo* in Classical Yucatec [Pío Pérez 1866–77:223]) brandishing two torches below a sky band (Figs. 25.1 and 25.2). The caption above the picture (Fig. 25.2) refers to the macaw (the main sign in the third collocation) setting fires in heaven *(u-kak ti caan)* and ends with a *kin-tun-yabil* collocation, which signifies "drought" (Pío Pérez 1866–77:177; Thompson 1960:269). In the third picture on the same page, a dog *(tzul)* hangs down from a sky band *(caan)* holding a lighted torch (Figs. 25.1 and 25.2). The caption reads *u-toc tzul caan* "the dog burns [in] the sky" (Fig. 25.2). The rains reappear in the two pictures on page 41b (Fig. 25.1), signalling the end of the drought. The pictures and several captions on pages 38b to 41b seem to be describing meteorological events and agricultural activities that take place during the rainy season. This information allows us to narrow down the five alternative dates for the almanac suggested by the eclipse data to two.

The first and the last dates in Table 25.3 fit best with the seasonal information in the pictures and captions on pages 38b to 41b. The rainy season normally opens sometime in May in the Yucatan peninsula and lasts through October (Page 1933:420–421).

IMPLICATIONS OF AN AGRICULTURAL ALMANAC IN THE DRESDEN CODEX

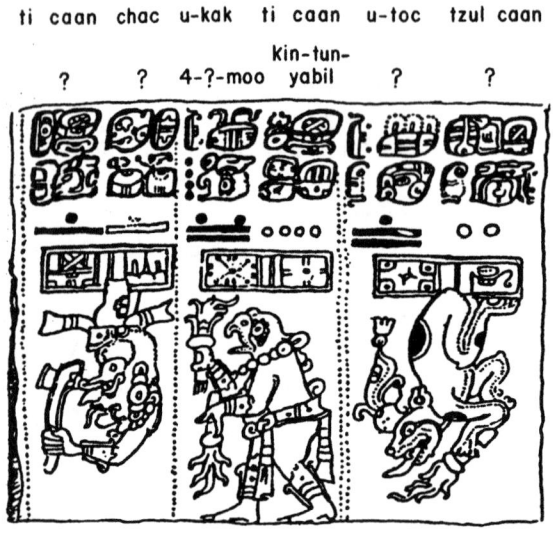

25.2 D 40b (after Villacorta C. and Villacorta 1976:90).

In Ebtun, Yucatan, where I spent the spring and summer of 1979, planting begins as soon as the first heavy rains have fallen. Correlating the 6 Cauac date at the beginning of the almanac with 20 May 775 means that it coincides with the onset of the rainy season. The first drought picture and caption occur 58 days after the beginning of the almanac, on 17 July 775. The dry spell continues for almost a month, finally breaking on 15 August. This unusual dryness is referred to as a drought by the *kin-tun-yabil* collocation at the end of the caption above the second picture on page 40b (Fig. 25.2). The pictures on page 41b imply that once the rains have resumed, they continue for at least 17 more days, until 1 September (Fig. 25.1). The 6 Cauac date also falls at the beginning of the rainy season if it is correlated with 13 May 906. In that case, the drought would commence on 10 July and end on 8 August of the same year.

Although another possible date for the beginning of the almanac, 13 August AD 873, falls within the rainy season, it occurs more than two months after planting can take place and is, therefore, not in agreement with the planting scenes in the middle of pages 38b and 39b (Fig. 25.1). The third date in Table 25.3, 14 November AD 840, falls about two weeks after the close of the rainy season and five or six months after the time for planting. No harvest scenes are shown in the pictures on pages 38b to 41b. Therefore, it cannot be the 6 Cauac date that introduces the almanac on those pages. Similarly, the second date, 16 February AD 808, falls in the heart of the dry season and cannot be correlated with the text and the pictures on pages 38b to 41b.

It is apparent, then, that the seasonal information in the agricultural almanac corresponds perfectly with the first and fifth dates suggested by the eclipse data and very poorly with the other three. The only other astronomical references in the almanac that might permit a choice between the eighth-century and tenth-century alternatives are the "sky bands" in the first, sixth, seventh, eighth, and ninth pictures (Fig. 25.1). Only the last of these is iconographically unambiguous: the band from which the dog is hanging contains the sign for Venus immediately adjacent to the sign for the Moon (Fig. 25.2). The obvious question is, therefore, whether there was anything remarkable or noteworthy about the positions or behaviors of these two celestial bodies on or near the tabulated day, 4 Men, in the eighth or tenth century.

The eighth-century date, AD 4 August 775, was the day before a new moon (Goldstine 1973), and the Moon would almost certainly have been invisible. On the previous day, however, which should have been the last day of visibility for the waning lunar crescent, the Moon and Venus, which was then a "morning star," were in conjunction (Table 25.4), rising almost together shortly before 4:00 AM, approximately 90 minutes before sunrise (Kluepfel 1980). The tenth-century date, 28 July 906, was a new-moon date, and the most probable day of last lunar visibility was two days earlier. On 26 July 906, the Moon and Venus were again very close together, in or near conjunction (Table 25.4), rising about 3:00 AM, approximately two and one-half hours before sunrise.

The data on Venus and the Moon discussed above suggest that the ninth picture in the agricultural almanac may be one example of a Maya iconographic convention for the close propinquity in the night sky of two celestial bodies (a propinquity that in some cases conforms to the concept of "conjunction" as used by western astronomy). We are not aware of any other example in the Maya hieroglyphic corpus that duplicates exactly the iconography of the picture in question, but a somewhat similar picture on another page of the Dresden permits a preliminary, very partial test of this hypothesis.

On page 56b, which is part of the eclipse table, a solar eclipse glyph is shown hanging from a sky band composed of three elements—the glyph for Venus on the left, the glyph for the Moon in the middle, and a glyph of unknown connotation on the right (Fig. 25.3). The picture is associated with column 58 of the table, which contains, according to our understanding of the table's structure (H. Bricker and V. Bricker 1983), three eclipse-prediction dates 15 days apart, 10,040 days after the table's base-dates A (solar eclipse), B (lunar eclipse), and C (solar eclipse), respectively. Each of the eclipse-prediction dates is at the center of an explicitly stated three-day range of uncertainty. The base-date-A increment in column 58, to which the picture most probably refers, corresponds to 7 May AD 783, the most probable last day of lunar visibility before a new moon on 9 May. One day before this, on 6 May 783, within the three-day range allowed for by the table's structure, Venus and the Moon were in close propinquity in the pre-dawn sky (Table 25.4). Although the characteristics of the eclipse table make it unlikely that the pictures refer to the base-date-C increments, the later date in column 58 is also appropriate to the iconography. The base-date-C increment corresponds to 6 June 783, two days before a new moon and a solar eclipse. On the previous day, 5 June 783, Venus and the waning lunar crescent were again in close propinquity in the two hours or so preceding sunrise (Table 25.4). These data, which are obviously consistent with our hypothesis about the meaning of an object

Implications of an Agricultural Almanac in the Dresden Codex

Table 25.4 Dates of close propinquity or conjunction of Venus and the Moon.[1]

Date and Time	Planet	Longitude	Latitude	Δλ	d
2 Aug 775 2,004,336 4:05 AM	Venus Moon	107.55 97.42	0.31 3.83	10.13	10.72
*3 Aug 775 2,004,337 4:00 AM	Venus Moon	108.77 109.24	0.35 4.41	0.47	4.08
4 Aug 775 2,004,338 5:00 AM	Venus Moon	110.05 121.59	0.39 4.81	11.54	12.34
25 Jul 906 2,052,175 3:20 AM	Venus Moon	89.90 77.28	−0.73 3.45	12.62	13.29
*26 Jul 906 2,052,176 3:20 AM	Venus Moon	91.10 89.47	−0.68 4.14	1.62	5.09
27 Jul 906 2,052,177 4:00 AM	Venus Moon	92.33 102.26	−0.63 4.66	9.93	11.24
5 May 783 2,007,169 3:30 AM	Venus Moon	2.60 349.11	−1.80 5.20	13.49	15.19
*6 May 783 2,007,170 3:30 AM	Venus Moon	3.74 1.89	−1.83 5.00	1.85	7.10
7 May 783 2,007,171 3:45 AM	Venus Moon	4.90 14.59	−1.85 4.55	9.69 -	11.61
4 Jun 783 2,007,199 3:25 AM	Venus Moon	37.38 23.90	−1.92 4.13	13.48	14.77
*5 Jun 783 2,007,200 3:25 AM	Venus Moon	38.56 36.05	−1.90 3.34	2.51	5.81
6 Jun 783 2,007,201 3:45 AM	Venus Moon	39.75 48.21	−1.88 2.39	8.46	9.48

suspended from a very limited segment of a sky band, support our interpretation of the ninth picture in the agricultural almanac.

Although the investigation of the sky-band information appears, on present evidence, to have defined a specific iconographic convention for planetary propinquity or conjunction, it has not permitted an unambiguous choice to be made between the eighth- and tenth-century dating alternatives for the agricultural almanac. On bal-

25.3 Column 58, Column 59, and the Intervening Picture on Pages 55b and 56b of the Dresden Codex (after Villacorta C. and Villacorta 1976:120, 122).

ance, we prefer the eighth-century dating for three reasons: a) the cross-tie to the eclipse table for the eighth century is to the version of that table that actually appears in the Codex, not to a recycled multiple that must be calculated; b) the true date of the propinquity of Venus and the Moon is closer to the date predicted by the agricultural almanac for the eighth century (one day before 4 Men) than for the tenth century (two days before 4 Men); and c) Venus and the Moon were closer together on the eighth-century date than on the tenth-century date. None of these reasons can be regarded as conclusive, however, and ambiguity remains.

In summary, the almanac on pages 38b to 41b of the Dresden Codex is concerned primarily with seasonal and agricultural matters, but it also mentions eclipses that refer the almanac to the spans of time covered by the eclipse warning table on pages 51 to 58 and its fourth multiple. Thus we have the very nice case where an almanac without long count dates can be dated to no later than AD 906 and to as early as AD 775 because of its interlocking relationship with another table in the same Codex, and this is confirmed by several kinds of independent data from climatology and archaeoastronomy.

col. 58

col. 59

NOTE

1. Gregorian equivalents and Julian Day Numbers are given for each of the four dates discussed in the text (marked with an asterisk) as well as for the day before and the day after it. The times shown for each date are local times for the Yucatan peninsula (20°30' north latitude, 88°30' west longitude) approximately 15 minutes after Moon-rise or Venus-rise (whichever occurred later) on the day in question. Ecliptic longitudes and latitudes, tabulated in decimal degrees, and other astronomical data were obtained from Kluepfel (1980). The quantity "$\Delta\lambda$" is the absolute difference, in degrees, between the longitudes of Venus and the Moon. The quantity "d," also in degrees, is the overall resultant angle between Venus and the Moon, taking

into account differences in both longitude and latitude. This value, calculated by the algorithm given by Duffett-Smith (1981:52), would seem to be a more relevant measure of how close two celestial bodies appear to be to a terrestrial observer than just the difference in longitudes.

REFERENCES

Bricker, Harvey M., and Victoria R. Bricker. 1983. Classic Maya Prediction of Solar Eclipses. *Current Anthropology* 24:1–23.

Bricker, Victoria R., and Harvey M. Bricker. 1986. The Mars Table in the Dresden Codex. In *Research and Reflections in Archaeology and History: Essays in Honor of Doris Stone*, ed. E. Andrews, pp. 51–80. Tulane University Middle American Research Institute Pub. 57.

Duffett-Smith, Peter. 1981. *Practical Astronomy with Your Calculator*, 2nd ed. Cambridge, Cambridge University Press.

Förstemann, Ernst. 1906. *Commentary on the Maya Manuscript in the Royal Public Library of Dresden*. Papers of the Peabody Museum of American Archaeology and Ethnology, Harvard University, Vol. 4, No. 2. Cambridge, MA.

Goldstine, Herman H. 1973. *New and Full Moons 1001 BC to AD 1651*. American Philosophical Society Vol. 94. Philadelphia.

Kluepfel, Charles. 1980. Planets. (An astronomical software package available from the author.)

Oppolzer, Theodor. 1887. *Canon der Finsternisse*. Denkschriften der Kaiserlichen Akademie der Wissenschaften, Mathematisch-Naturwissenschaftliche Classe 52.

Page, John L. 1933. The Climate of the Yucatan Peninsula. In *The Peninsula of Yucatan: Medical, Biological, Meteorological and Sociological Studies*, ed. George Cheever Shattuck, 409–422. Carnegie Institution of Washington Pub. 431.

Pío Pérez, Juan. 1866–1877. *Diccionario de la lengua maya*. Merida, Imprenta Literaria de Juan F. Molina Solís.

Teeple, John E. 1931. *Maya Astronomy*. Carnegie Institution of Washington, Contributions to American Archaeology Vol. 1, No. 2.

Thompson, J. Eric S. 1960. *Maya Hieroglyphic Writing: An Introduction*. Norman, University of Oklahoma Press.

Villacorta C., J. Antonio, and Carlos A. Villacorta. 1976. *Códices mayas*, 2nd ed. Guatemala, Tipografía Nacional.

PART V

Cultural Astronomy's Greatest Mysteries

One problem with cultural astronomy lies not only in the different ways it is practiced but also in the way it is perceived. I believe the old term "archaeoastronomy" had become a curse for many of those who practiced it, because literature carrying that label too often was taken to be slightly crackpot by large segments of the established disciplines tangent to it. Frequently, archaeoastronomy was not taken seriously because of the media's emphasis on its sensationalism and entertainment value (recall the three examples of pop-archaeoastronomy mentioned in the preface). These factors had a lot to do with changing its name from "archaeoastronomy" to "cultural astronomy."

In most circles, the popular examples constitute just about all that the average person knows about the astronomy of the ancients—fabulous accounts of the creation of ancient artifacts attributed to alien visitors, or bizarre earthbound (yet ungrounded) explanations of ancient constructions made by lost civilizations that entirely miss the culture connection.

PART V: CULTURAL ASTRONOMY'S GREATEST MYSTERIES

One of the world's most famous ancient attractions for bizarre explanations is the Nazca lines, giant geoglyphs etched on the desert coast of southern Peru. Variously hypothesized as an ancient airport that once accommodated intergalactic travelers (the most famous of all Nazca fantasies by a wide margin), a world map, a tidal calculator, and an ancient Olympic-style raceway (see Aveni 1990, 1999, for references and much more), these lines, according to the standard explanation in many textbooks, are a part of the largest astronomy book in the world (Kosok 1965). The foremost proponent of the idea that the Nazca lines are astronomically aligned and that the animal drawings represent constellations was Maria Reiche (1968), whose admirable labor in preserving and analyzing the geoglyphs spanned more than fifty years. But she may have been wrong.

The first piece in this section, coauthored in 1991 by me and archaeologist Helaine Silverman, who has worked on Nasca remains for much of her career (cf. Silverman 1993; Silverman and Proulx 2002), uses ground-based evidence to reassess the archaeological record pertaining to alignments. As readers will discover, the underlying natural phenomenon that emerges as the essence of the lines is not the shining of stars but the flowing of water; however, as perceptive readers will discover, astronomy may still have played a role.

When astronomer William Miller published a short note in a regional journal in 1955 about his discovery of "Two Possible Astronomical Pictographs Found in Northern Arizona" (Miller 1955), little did he know that he would generate one of New World archaeoastronomy's major controversies. The second selected piece, coauthored by a team of interdisciplinary scholars, analyzes petroglyphs showing crescents alongside starlike images similar to those documented by Miller. They expand on Miller's hypothesis that each instance is a record of a cataclysmic event that took place in the constellation of Taurus on the evening of July 4, 1054—the great Crab Nebula supernova explosion.

The little-known counterpoint article that follows, by anthropologist Florence Hawley Ellis, deals with what we know about indigenous Zuni skywatchers since European contact. Based on ethnographic information, she argues that the star and crescent moon can best be explained as records of the far more common phenomenon of the crescent moon passing the morning star—a conjunction. As I have stressed throughout *Foundations*, readers should decide for themselves; but I cannot pass over these pieces without noting our contemporary culture's fascination with cataclysmic events—from hurricanes and tornadoes to the sudden extinctions of species and collapses of whole civilizations, not to mention colossal explosions in general, both terrestrial and extraterrestrial.

Another petroglyph that has raised controversy is the famous Fajada Butte Sun Dagger. It consists of a pair of spirals carved atop an eminence in Chaco Canyon, New Mexico. As explained in the next piece I have chosen for this section, coauthored by an artist and a physicist, the petroglyphs are really components of Native America's first clock. Its armature is a dagger of light that strikes the limits of the petroglyphic clock face at key points in the solar cycle. At night a moon dagger takes the place of

Part V: Cultural Astronomy's Greatest Mysteries

the sun dagger. A film narrated by actor Robert Redford and also more recently an animated computer model show how it works. The companion piece by astronomer John Carlson marshals evidence that argues against this explanation. For other takes on this well-publicized controversy, see Palca's 1989 article and the piece by Jim Judge (Chapter 34). See also the recent compendium of papers on Chaco astronomy (Sofaer 2008). And again, as always, decide the truth for yourself.

Despite the controversy over Chaco Canyon's Fajada Butte site, the possibility that native North Americans marked the lunar standstills was strengthened by the interdisciplinary studies of Malville and colleagues at Chimney Rock in southwestern Colorado. Like Chankillo (see Chapter 21), Chimney Rock's lofty isolated setting, far from any known subsistence base, along with the inclusion of an artificial platform jutting out in the direction of the pair of vertical peaks that mark the lunar standstills between which the moon is positioned at its 18.6-year northern standstill, strengthens the case.

Space did not allow me to add other pieces on controversial studies in cultural astronomy to the collection. Prominent among them is the case of the descending serpent of light on the Castillo (or Temple of Kukulcan) at Chichén Itzá, thought by some to have been designed to celebrate the equinoxes. Readers interested in this topic are invited to consult the references (see, e.g., Carlson 1999; Aveni 2000).

Understanding astronomies that emanate from cultures other than our own is a tricky business. Often we can be way off the mark in our attempt to comprehend what was going on in someone else's mind—particularly when that other mind functioned in a world of experience very different from our own. Recall British historian Jacquetta Hawkes's (1967, cited in Part I) oft-cited suggestion that every age gets the Stonehenge it desires and deserves. As the papers in this section demonstrate, anyone who studies cultural astronomy should always raise two important questions: Are we seeing what we want to see? and Was their message intended for us?

THINGS TO THINK ABOUT

1. Do you see any similarities between the structure of the Nasca lines and what Zuidema's paper tells us about the organization of the Inca capital of Cuzco?

2. Discuss the astronomer's versus anthropologist's view of ancient petroglyphs. Under what conditions might these views be compatible?

3. Look up the many ways different Native American cultures name each of the full moons of their seasonal year (see Chapter 28 or Aveni 1989, ch. 3, for a start). Choose one native culture group and write a short essay on what the names of their moons tell you about the activities of that particular culture.

4. What culture-based evidence is offered to help verify the case of the Chaco sun dagger? What does Carlson consider to be the weak points in the case? What aspects of the case does he concede?

SUGGESTIONS FOR FURTHER READING

Aveni, A. 1989. *Empires of Time: Calendars, Clocks, and Cultures*. New York, Basic. [Revised and reprinted in 2002 by University Press of Colorado, Boulder.]

Aveni, A. 1990. "An Assessment of Previous Studies of the Nazca Geoglyphs." In *The Lines of Nazca*, ed. A. Aveni, 1–40. Philadelphia, American Philosophical Society.

Aveni, A. 1999. *Between the Lines*. Austin, University of Texas Press.

Aveni, A. 2000. *The Book of the Year: A Brief History of Our Seasonal Holidays*. New York, Oxford University Press.

Breunig, G. von. 1980. "Nasca: A Pre-Columbian Olympic Site?" *Interciencia* 5:209–219.

Carlson, J. 1999. "Pilgrimage and the Equinox: 'Serpent of Light and Shadow' Phenomenon at the Castillo, Chichén Itzá, Yucatan." *Archaeoastronomy: The Journal of Astronomy in Culture* 14(1):136–151.

Hawkes, J. 1967. "God in the Machine." *Antiquity* 41:174–180.

Kosok, P. 1965. *Life, Land, and Water in Ancient Peru*. New York, Long Island University Press.

Miller, W. 1955. "Two Possible Astronomical Pictographs Found in Northern Arizona." *Plateau* 27(4):6–12.

Palca, J. 1989. "Sun Dagger Misses Its Mark." *Science* 244:1538.

Reiche, M. 1968. *Mystery on the Desert*. Stuttgart, Eigenverlag.

Silverman, H. 1993. *Cahuachi in the Ancient Nasca World*. Iowa City, University of Iowa Press.

Silverman, H., and D. Proulx. 2002. *The Nasca*. Oxford, Blackwell.

Sofaer, A., ed. 2008. *Chaco Astronomy: An Ancient American Cosmology*. Santa Fe, Ocean Tree Books.

von Daniken, E. 1971. *Chariots of the Gods*. New York, Bantam.

CHAPTER TWENTY-SIX

Between the Lines
Reading the Nazca Markings as Rituals Writ Large

Anthony Aveni and Helaine Silverman

The winter solstice—on or about June 21 in the southern hemisphere—is a day like any other on the plain that stretches from the coastal hills of southern Peru inland to the Andes. Almost always the day dawns clear; here, in the rain shadow of the Andes, years can pass between storms that drop measurable rainfall. As the dew burns off the rocky surface of the desert, purplish with an oxidized patina called desert varnish, the horizon begins to dance in the heat. By afternoon the wind picks up, and a dusty haze hovers near the horizon. Only as the sun slips down through the haze toward the Pacific, toward the west, does the heat finally break.

It was late on such a day, June 21, 1941, that the American geographer-historian Paul Kosok and his wife Rose went for a picnic on the pampa, or plain, of Nazca, a wedge-shaped plateau of desert perhaps fifteen miles across at its widest, bounded on two sides by river valleys and on the third by the foothills of the Andes. Paul Kosok

From *The Sciences* 31:4 (1991):36–42.

26.1 Great Triangle, 1979. Photograph by Marilyn Bridges.

had come to Peru to study the irrigation systems built by ancient agricultural peoples to nourish crops in the coastal desert. While in Lima he had heard rumors of a web of shallow canals crossing the plain of Nazca, and he had gone south to investigate.

The "canals" turned out to be nothing of the sort. What Kosok found instead were broad stripes etched on the desert surface, running arrow-straight across rises and gullies for hundreds of yards, even several miles. Scattered among them were other markings, in the form of tight spirals or plazalike trapezoids as much as half a mile long. Limned on the desert floor among the geometric markings were vast spindly animal and plant figures including a bird measuring 100 feet across. Clearly this was no ancient irrigation network. But Kosok found no other clues on the empty pampa about who had made the lines, or when, or why.

It was on a small rise in the strange inscribed landscape that Paul and Rose Kosok watched the sun set that evening fifty years ago. Accounts differ about whether it was he or she who first noticed that a line radiating from their vantage was aimed directly at the setting sun. But, as Kosok later wrote, they made the intuitive leap together: "With a great thrill we realized at once that we had found the key to the riddle." The pampa and its markings, Kosok soon became convinced, were "the largest astronomy book in the world," a great chart recording significant celestial objects and alignments.

Kosok was far from the first to see the Nazca lines; Peruvian archaeologists in the 1920s knew about the patterned desert, and the markings had become a familiar sight to the pilots of Peru's national airline when it started regular service in the 1930s. But Kosok's epiphany, evoking an ancient class of astronomer-priests who recorded

26.2 Great Spiral and Trapezoids with Bus, 1987. Photograph by Marilyn Bridges.

their knowledge on the face of the desert, had a broad, mystic appeal. Maria Reiche, a German mathematics tutor from Lima, began collaborating with Kosok late in 1941, charting the lines and cataloguing possible matches with the key positions of the moon, the stars and the sun. As a result of her work, which she carried on after Kosok left Peru, the idea that the Nazca lines are an ancient astronomical text has become received wisdom, enshrined in popular and scholarly writing alike.

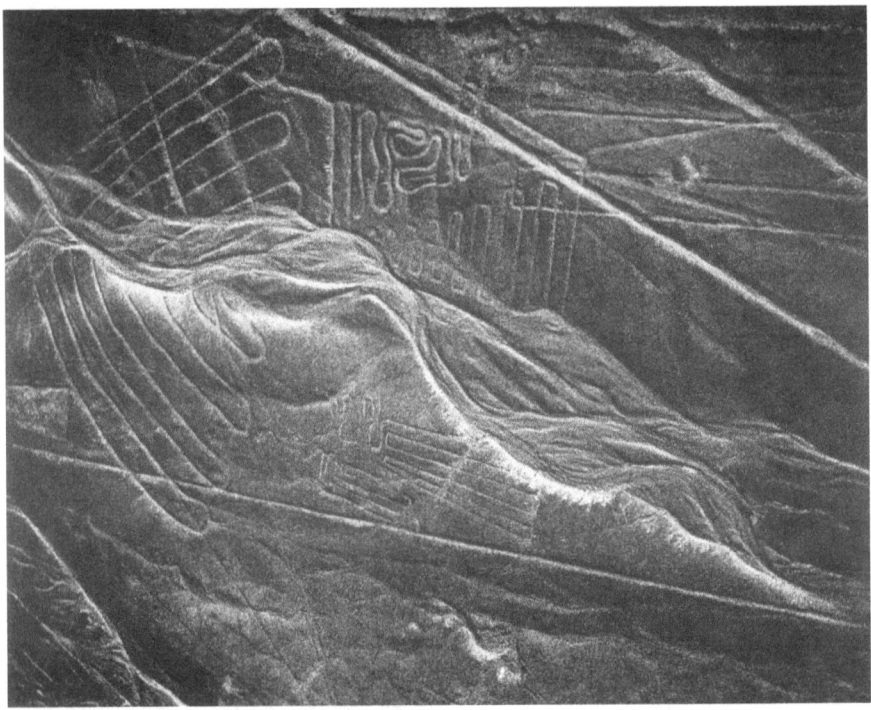

26.3 Feathers, 1979. Photograph by Marilyn Bridges.

But a new look at the geometry of the lines has cast doubt on the astronomical interpretation. The orientation of the lines, on closer scrutiny, suggests not a sky chart but a reflection of the inevitable concern of an agricultural people in a searing desert—water. Not that the lines are turning out to be irrigation ditches after all; their link with water is more likely to have been ceremonial. What kinds of ceremonies might the lines have been part of? How were such rituals organized and carried out?

Any speculations have to be based on a factor that Kosok and Reiche largely ignored in their explorations of geometry and the stars: the people who made the lines. Now, after decades of interest in the lines themselves, the people who drew them are emerging from the shadows, thanks to archaeological work in the valleys bordering the desert plain. We now know the identity of the line makers. And even though they left no written testimony about why they drew patterns on the desert some 2,000 years ago, what the lines meant to them can be glimpsed by analogy with the ritual observances and practices of later, better-known cultures. The Spanish conquerors of Peru, who destroyed so much, did preserve records—distorted and incomplete ones, to be sure—of preconquest Andean beliefs and practices. Some of those records have survived among traditional Indian communities of the Peruvian highlands. By analogy with what is known of later cultures, archaeologists are at last beginning to read the messages inscribed on the desert.

Anthony Aveni and Helaine Silverman

26.4 Trapezoids and Right-Angle Overview, 1979. Photograph by Marilyn Bridges.

Maria Reiche, now in her eighties, still lives on the plain of Nazca, in the landscape that captivated her fifty years ago. Ever since then she has devoted herself to protecting and recording its markings. She has concentrated her surveys on the spirals, the great geometric patterns and the living forms—birds, a spider, a killer whale, a monkey, a plant—that are clustered in one corner of the pampa. Reiche has also shown how the patterns were made. For aeons, strong, sand-bearing winds have scoured the face of the desert, leaving it paved with fist-size rocks brought down from the high Andes in ancient episodes of flooding. Recently it was determined that where the rocks are exposed to the sun, bacteria have deposited a patina of purplish iron and manganese oxides: desert varnish. But the underside of each rock and the soil below are light in color. Strip off the surface rocks, and a bright mark is left on the desert for centuries. The dark rubble from the stripped earth can then become an outline for the figure; at Nazca each line or shape is edged by a rock berm as much as a yard high. The work probably went quickly, as one of us (Aveni) discovered with a group of twelve volunteers from Earthwatch Expeditions. It took Aveni and his volunteers just an hour and a half to make a vivid spiral, thirty feet across.

Reiche herself brightened the color difference between the surface and the subsoil by sweeping many of the lines. Her intended viewers were the tourists who fly over the pampa in small planes, marveling at the cat's cradle of lines and figures. Indeed, Reiche's books about the Nazca markings have made them a popular tourist destination. And with the growth of popular interest, sensational explanations for the lines have multiplied.

Prominent among them are proposals that the lines served as runways for spaceships carrying extraterrestrial visitors or that they marked the courses of prehistoric foot races. Another explanation, inspired by the striking sight the markings present from the air, holds that the people of ancient Nazca made the outsize patterns to be admired by the gods—or by pre-Columbian aeronauts floating over the pampa in hot-air balloons.

Reiche's explanation, an elaboration of Kosok's astronomical hypothesis, is more sober. She identified several of the great animal figures as star maps: the monkey figure, she proposed, maps the positions of key stars in the constellations Leo and Ursa Major; the spider charts the stars in Orion. Each alignment and constellation, Reiche argued, signified a specific time of year. The patterned pampa served as a vast astronomical calendar.

The idea is compelling. After all, as Reiche has pointed out, an agrarian people living in a desert might need an accurate calendar to foretell the seasonal rise and fall of the rivers. But we saw reasons to be skeptical. The hypothesis treats the complex of lines and figures as a single, unified system. Yet the markings bear no resemblance to a single astronomy text, even a complicated one. Seen from the air, the pampa of Nazca looks more like an unerased blackboard at the end of a busy day of class, cluttered with overlapping but unrelated markings.

What is more, Reiche assumes the same stars and constellations that dominate the European picture of the sky were also important to the ancient people of Nazca. Many of the alignments and constellation maps she has described refer to features prominent in Western star lore, and some of them, such as Ursa Major, are scarcely visible above the haze-bound horizon at Nazca. Not surprisingly, as the anthropologist Gary Urton of Colgate University has shown, the natives of the Andes today see the sky quite differently from the way Europeans do (see Chapter 19).

Taken as a whole, Reiche's hypothesis says too little about the people who made the lines and about how the markings may have reflected their customs, rituals and social structures. It treats the desert markings as sterile geometry, uninfluenced by culture; they might just as easily have been made by any ancient culture, given a canvas as inviting as the pampa of Nazca. The lack of a human context does not disprove the astronomical hypothesis, of course, but it strengthened our sense that the hypothesis was incomplete at best. Astronomy did play a part in the design of other ancient American ritual structures—for example, the *ceques*, invisible sacred lines that radiated from the center of the Incan capital Cuzco—but other preoccupations of the society also left their marks.

The first systematic test of the astronomical hypothesis was made in 1968 by Gerald S. Hawkins, an astronomer then at Boston University who was fascinated by ancient observatories. Five years earlier Hawkins had announced that the gray monoliths of Stonehenge were erected as a specialized astronomical observatory for predicting eclipses. He decided to explore the role of astronomy in the Nazca markings by applying the same techniques that had been used at Stonehenge. On the pampa of Nazca, however, the result was quite different. After cataloguing the orientations of all the lines and trapezoids that cross a sample strip of the pampa, then comparing the data with the directions of rise and set for the sun, the moon and certain bright stars at key times of year, Hawkins concluded that, on the whole, the Nazca markings have little to do with astronomy.

Still, it seemed he might have overlooked something. Like Reiche, Hawkins brought to his studies a largely northern hemisphere view of the night sky. The stars he chose as benchmarks are prominent in the northern sky, and he left out features, such as the southern end of the Milky Way, that have made a stronger impression on southern observers. And though he looked for lines that might have marked the solstices, he did not investigate the possibility of other alignments during the solar year, which might have been critical in an agricultural calendar. Finally, even in casting doubt on an astronomical explanation, Hawkins's negative findings made the question of what the lines did mean loom as large as ever.

To put the astronomical theory to a more definitive test and search for other clues about the meaning of the lines, Aveni, together with Urton; the archaeologist Persis B. Clarkson of Athabasca University in Edmonton, Alberta; the computer scientist Clive L. N. Ruggles from the University of Leicester in England; and a varying group of Earthwatch volunteers and Colgate students, spent five field seasons surveying the Nazca markings at ground level. Our search for correlations between the lines and the natural features on earth and in the sky was guided by the internal organization of the markings. As Reiche was the first to note, many straight lines radiate, spoke-like, from points known as line centers. The points—apparently the organizing centers for the markings—quickly became the focus of our research as well.

In addition to the prominence of the line centers, another factor drew our attention to them: the resemblance of their hub-and-spoke pattern to the ceque system of ancient Cuzco. Like the lines emanating from each line center, the Incan ceques radiated from a single point, the great temple that stood at the confluence of Cuzco's two rivers. Both the invisible ceques and the highly visible Nazca lines are quintessentially straight, rarely veering off course as they cross hills or valleys. If the line centers and the ceques did prove to be analogous, we thought, the Spanish chroniclers' account of the ceque system might shed light on the meaning of the Nazca markings.

We began by picking out as many line centers as we could in the tangle of markings on the pampa, in order to learn how they fit into the overall pattern. To our surprise, more than 90 percent of the lines radiate from line centers, most of them set on low hills and marked by one or more piles of boulders. Standing at a line center, we found, one gets a better view of the basic organization of the markings than anywhere else on the Nazca pampa. Pale lines—most of them only a yard or two in width, but others as wide as an avenue—strike out from the center in all directions. Some end after a few yards; others continue across the undulating terrain and vanish over the horizon. A line may peter out, stop abruptly at the foot of a hill or the edge of a dry

26.5 Overview, 1979. Photograph by Marilyn Bridges.

wash, break up into a zigzag or end at another line center. Still other lines ultimately broaden into one of the vast geometric clearings on the pampa, thereby linking these figures to the line centers. (The animal and plant figures lie outside this network, and our analysis ended up saying little about them.)

To test the astronomical hypothesis, we set up a surveyor's transit at each of the sixty-two identified line centers and took sightings along the 700-odd radiating lines. The resultant catalogue of orientations did little to support Kosok and Reiche's hypothesis. The directions of the lines seemed nearly random, with little clumping at any single angle—far from the pattern one would expect if the markings constituted an astronomical calendar. Even within the range of angles traced out each year by the changing positions of sunrise and sunset, the lines did not seem to favor any particular direction—with one possible exception. The number of lines pointing toward the positions of sunrise and sunset in late October and mid-February was about 50 percent higher than one would expect if the lines were oriented at random.

Here, as it turned out, was the first hint of overarching order in the lines. Even now, late October marks a critical point of the agricultural cycle in the valleys that border the pampa of Nazca. At that time, locals say, dark clouds gather over the Andes, heralding the rains that will swell the flow in the rivers and the ancient underground irrigation canals that thread the oases on the desert rim. Perhaps some of the Nazca lines served to mark the position of sunrise and sunset at the time of year when ample water would flow.

Indeed, water emerged as a central theme of the lines when we moved to a new phase of the study, correlating the orientations of the lines with geographic rather than celestial benchmarks. Most of the line centers are crowded toward the edges of the pampa; they lie either at the ends of the rows of hills that extend fingerlike into the pampa from the mountains on the east or along the elevated rim that separates the pampa from the river valleys on the north and south. Lines radiating from the ends of the fingers, we noticed, tended to continue in roughly the same direction as the rows of hills: nearly half the lines lie within thirty degrees of the ridge axes.

The hilly promontories, like the ridges on a drainboard, define the direction in which water flows as it cascades down the mountains and onto the pampa during the rare rainstorms. Perhaps the lines had been laid out to mimic the direction of water flow. It was also possible, though, that the ridge axes themselves, or the bearings of other high points on the far side of the pampa, had guided the ancient surveyors.

Another correlation seemed to bear out the importance of water flow. Many of the large geometric figures are themselves elongate, seeming to define some favored direction on the pampa. In addition to surveying the figures from the ground, we pored over aerial photographs, measuring the difference between the orientation of each figure and of the nearest dry gully. The match was impressive, though a sizable number of the figures also lie precisely at right angles to the dry watercourses. Yet even such perpendicular figures may have signaled the direction of water flow, because most of them are oriented in such a way that "upstream" lies on the left for an observer entering the feature from the direction of a line center.

Thus it was not so much an obsession with the night sky as an inescapable concern about water that drove the Nazca line makers. Hawkins was correct; and we also learned that the resemblance of the lines to the ceques of Cuzco is more than superficial. The Spanish chroniclers tell us that, among many other functions, the ceques served as a guide to water rights and water rituals. These imaginary lines divided the watershed of the Cuzco valley into wedges, each marking out the water rights of a certain *ayllu*, a social subdivision related by descent from a common spiritual ancestor. Yet the ceques were more than mundane property boundaries, since water was a gift of ancestors residing in the underworld. At springs, wells and river bends along the ceques members of each ayllu performed devotions—casting sacrificial objects into the water, for example—to keep the precious liquid flowing.

Even today, some Andean cultures carry out water rituals along real or imaginary lines on the landscape. Johan Reinhard, an American anthropologist working in the high Andes, has described how villagers in Bolivia traditionally gather on a hilltop marked by the end of an actual straight line, singing and dancing to please the mountain gods who bring rain. Turning from contemporary cultures to archaeology, Reinhard has suggested that the Nazca lines may also have served as stage directions for fertility rituals. We agree; the geometric connection we found between the lines and water, together with the analogy of the ceques, suggests the markings may have played some part in ceremonies designed to summon water from its sources underground or high in the mountains.

Compelling as the similarity may be between the Nazca lines and the ceremonial spokes such as the ceques, another resemblance calls for attention. The late Peruvian archaeologist Toribio Mejía Xesspe called the lines ceremonial roads when he described them sixty years ago, and a close look at them reveals features characteristic of actual Andean roads. Like the roads that knitted together the Incan empire a thousand years later, the Nazca lines run straight over hills and gullies between raised borders of rubble, and they are marked with stone cairns at places where they widen or narrow. One line, which ran clear across the pampa of Nazca, has even (to the dismay of archaeologists) been graded and made into a segment of the Pan-American Highway.

Could the lines have been both ritual guidelines and roads of some kind? Perhaps the irrigation ceremonies in which the lines are implicated were enacted as people walked or danced along them. Many of the lines are far wider than a practical road needs to be, but faint fossil paths, perhaps traveled by the lines' makers, run within the lines, paralleling their raised borders. Two thousand years ago, at climatically significant times of year, fantastically costumed votaries may have danced their way along the paths to summon runoff from the Andes.

The image of the lines as ceremonial walkways or dance floors differs fundamentally from earlier conceptions. If the markings were elements of an ancient astronomy text or mystic messages to the gods, they could have been appreciated only from above, by astronomer-priests on hilltop observatories or by the gods on high. Given the dramatic spectacle the etched landscape presents from an airplane flying over the pampa, such images come naturally to twentieth-century observers. But the ancients may have appreciated the patterns quite differently as they laid them out and, later, moved along them. One must bear in mind that although the markings are most impressive from above, they were meant to be seen and used from ground level.

Parallels between the lines and ritual patterns in other Andean cultures, then, hint at the important role of the lines in the religious lives of their makers. Other analogies suggest how the people of Nazca may have organized their rituals and even how they may have shared the task of making the lines. Andean societies often divide communal labor—whether it is the observance of sacred rituals or the more mundane work of farming—among kinship groups, known today as they were in Incan days as ayllus. Each ayllu tends a specific plot of ground known as a *chhiuta*. The ceques of ancient Cuzco, for instance, marked the boundaries of wedge-shaped chhiutas where members of each ayllu performed ritual duties and maintained the irrigation system. Studying an existing village in the Andes, Gary Urton found that the labor of caring for a ceremonial plaza was divided among nine kinship groups, each one responsible for a single strip of terrain.

Perhaps the lines and geometric figures on the pampa each marked an arena of devotional labor. The task of etching each figure on the desert and carrying out water rituals in the cleared space may have fallen to a single ayllu. The idea that the Nazca patterns were the handiwork of many groups, working independently over a long time, could explain why the inscribed pampa looks so much like an unerased blackboard. It also implies that the markings on the pampa may reflect not just what their makers believed but also how they organized their society.

Who were the line makers? Archaeological studies on the pampa itself have been little help in answering the question, mainly because apart from the markings and scattered fragments of pottery, the terrain offers few signs of human activity—no walls, hearths or fields. In fact, until recently the only evidence came from potsherds. Among them are pieces decorated with the bright motifs of the so-called Nasca culture, which built temples and villages and tended fields in the neighboring river valleys between 200 BC and AD 600. Various of the decorative motifs on the sherds match the enormous plant and animal figures on the pampa—good evidence that the Nasca people were responsible for those images at least.

But who made the lines, the zigzags, the spirals and the great trapezoids remained in doubt. After all, the animal and plant figures are clustered in one corner of the pampa, many of them half obscured by an overlay of lines and geometric shapes. The lines and shapes might well have been the handiwork of a different, later culture. Only now, as a result of survey and excavation done by one of us (Silverman) in the adjacent river valleys, can we confirm that the Nasca people made the Nazca lines.

A Nasca site known as Cahuachi, in the Nazca valley on the southern side of the pampa, was the first to yield evidence for the link. At that site, once thought to have been a great city, Silverman's fieldwork in 1984 and 1985 revealed instead a ceremonial center complete with pyramids, plazas, temples, ritual offerings and burial sites.

On a small pampa immediately south of Cahuachi—and clearly a part of the ceremonial center—lies a complex of lines and trapezoids resembling the ones on the main pampa. Cahuachi also seems to be a point of reference for the markings on the pampa: many of the line centers are clustered nearby, and a major line extending all the way across the pampa ends and begins at Cahuachi.

On the other side of the pampa, in the valleys to the north, Silverman's recent survey of archaeological sites bolstered the case for a Nasca origin. Small clusters of lines and geometric etchings lie adjacent to Nasca villages, cemeteries and ceremonial sites. Other line complexes in the valleys could not be linked to Nasca structures, but they could be dated from the decorated potsherds on their surface. Fully 80 percent of the valley markings bore potsherds dating from various phases of the Nasca period.

Thus we conclude that on the pampa, as in the valleys, the great geoglyphs are largely the work of the Nasca people. From the present vantage, 2,000 years later, the Nasca period may look short, but in fact it lasted centuries longer than Europeans have been in the Americas. The persistence of the practice of etching the desert floor, even beyond the Nasca period, is testimony to the deep significance the tradition must have held. For more than 1,000 years, a period that saw the rise and fall of the holy city of Cahuachi, shifts in pottery styles and repeated floods and droughts, the ancient inhabitants of Nazca continued to climb onto the pampa and hillsides overlooking their villages, tracing their beliefs onto the barren surface.

CHAPTER TWENTY-SEVEN

Possible Rock Art Records of the Crab Nebula Supernova in the Western United States

John C. Brandt, Stephen P. Maran, Ray Williamson, Robert S. Harrington, Clarion Cochran, Muriel Kennedy, William J. Kennedy, and Von Del Chamberlain

INTRODUCTION

Dynastic records of the Sung Dynasty, the *Sung-Shih* describe a bright object in the constellation Taurus which appeared on 4 July, AD 1054 (Julian calendar), was visible in daylight for 23 days, and faded from the night time sky after 653 days (Duyvendak 1942, Needham 1959). Association of the Crab Nebula with the object described in the *Sung-Shih* was apparently first suggested by Hubble (1928) when he noted that about 900 years would be required for the Crab Nebula to reach its present size at the measured (Duncan 1921) rate of expansion. Elaborate determination and discus-

---

First published as "Possible Rock Art Records of the Crab Nebula Supernova in the Western United States," by John C. Brandt, Stephen P. Maran, Ray Williamson, Robert S. Harrington, Clarion Cochran, Muriel Kennedy, William J. Kennedy, and Von Del Chamberlain, in *Archaeoastronomy in Pre-Columbian America*, ed. A. Aveni. © 1975 by the University of Texas Press. All rights reserved.

 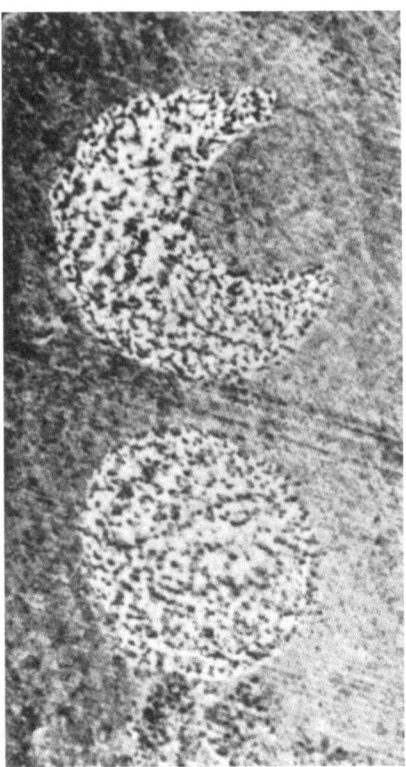

27.1 The northern Arizona sites described by Miller (1955a, b). Left: The White Mesa pictograph (Cave designated NA 5561). The drawing appears to have been made with a lump of red hematite. The diameter of the circle is approximately 10 cm. Right: The Navaho Canyon petroglyph (site NA 5653). The circle is approximately six inches in diameter. (Photographs courtesy of Wm. C. Miller.)

sions of the expansion time of the Crab Nebula have been carried out (Duncan 1939; Baade 1942; Trimble 1968) and the currently accepted date of outburst based on the assumption of unaccelerated motion is AD 1140 ± 10. The difference of 86 years between the oriental records and the purely astronomical extrapolation is a problem to which we will address ourselves later.

Acceptance of the AD 1054 event as the supernova cause of the Crab Nebula came in the 1940's (Duyvendak 1942; Mayall and Oort 1942) and this view prevails today (e.g., Mayall 1962; Scargle 1967; Davies and Smith 1971).

Many authors had noted the absence of written records of the Crab supernova in Europe and the Middle East (e.g., Mayall 1962; Shklovsky 1968, p. 52). Hence there was considerable interest in Miller's (1955a, b) announcement of two possible rock art records (Fig. 27.1) of the event in northern Arizona. Each consists of two design elements, interpreted by Miller as representing the crescent moon and the supernova. Mayall and Oort (1942) estimated the visual magnitude at approximately

John C. Brandt et. al

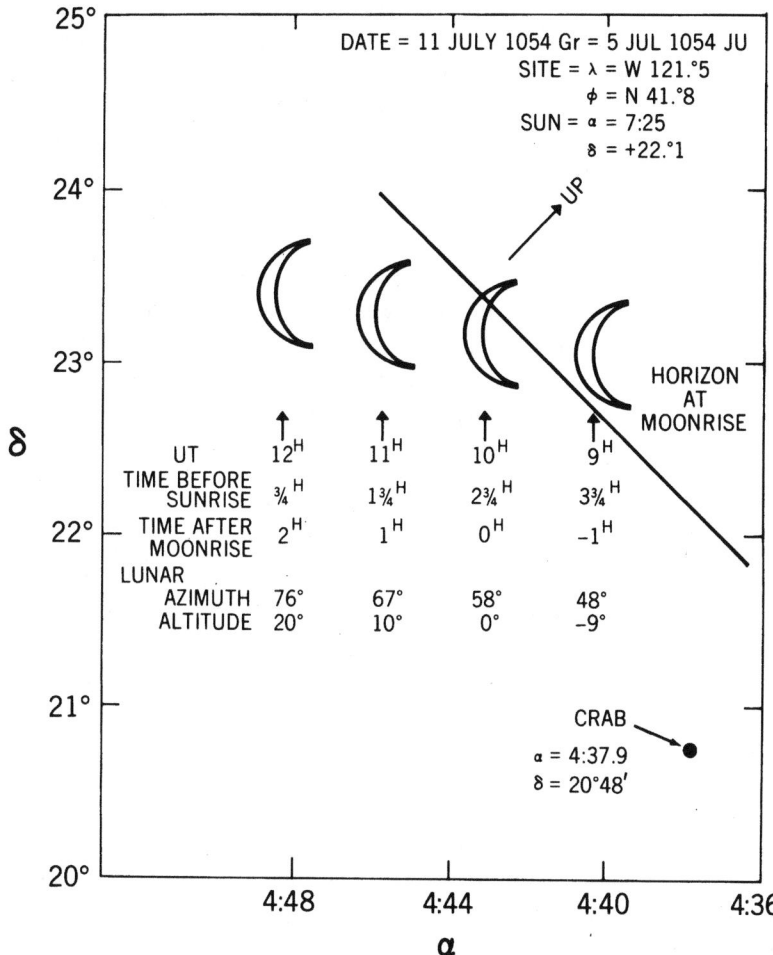

POSITIONS OF MOON AND CRAB 5 JULY 1054 AS SEEN FROM FERN CAVE

27.2 Astronomical circumstances of the conjunction on the morning of July 5, 1054, as seen from northern California. The calculations are based on the U.S. Naval Observatory's data for the "ancient moon."

−5 or a brightness about 6 times that of Venus (but see Minkowski 1971 where the assumed type of supernova is questioned). This is brighter than either Tycho's supernova of AD 1572 (−4) or Kepler's supernova of AD 1604 (−2). Chinese historical records concerning the color of the AD 1054 supernova near maximum are probably not reliable because of the Imperial yellow color (Duyvendak 1942). The astronomical circumstance that made Miller's suggested interpretation quite plausible was the close conjunction of the crescent moon and the position of the Crab Nebula just before dawn on the morning of 5 July 1054. This was pointed out by Miller, and our independent calculations based on the U.S. Naval Observatory's positional data for the "ancient moon" are shown in Figure 27.2. Analysis of pottery fragments by R. C. Euler, Museum of Northern Arizona, indicate that the region of the rock art cited by Miller was inhabited for several centuries including the year of the supernova (Miller 1955a, b). Because the moon moves approximately its own diameter with respect to the celestial background in one hour, the close conjunction of the moon and the supernova was basically only observable in western North America. Clearly "close" is a subjective judgment, but the moon would have moved at least eight times its own diameter to the east of the point of closest conjunction before it would have been observable from China. Positional astronomy was well developed in eleventh-century China (Needham 1959; Needham, Wang Ling, and Price 1960), and the bright object described in the *Sung-Shih* was referred not to the moon, but to a nearby 3rd magnitude star, ζ Tauri. (Note that a discrepancy occurs in the position with respect to ζ Tauri. We discuss the situation below.)

In this paper, we describe several other sites in California and New Mexico where rock art exists that resembles the two cases reported by Miller, and which, therefore, may also represent the Crab Nebula supernova event. Preliminary results have already been reported (Brandt et al. 1972; Brandt et al. 1973). In addition, Miller (1972) has informed us of the existence of another candidate in northern Arizona. In each case, the relationship of the star to the crescent is reasonable, the eastern horizon is visible from the vicinity of the rock art, and the available archaeological evidence is compatible with habitation of the site circa the middle of the eleventh century. We believe that this evidence is consistent with the observation and recording of the conjunction of the Crab Nebula supernova and the crescent moon by several groups of American Indians.

It should be noted that the interpretation of ancient rock art is a very uncertain subject, beset with pitfalls and inherently subjective. We have approached this investigation in the spirit of attempting to see what evidence *might* exist for rock art records of a spectacular astronomical phenomenon. We recognize, however, that it is not possible to prove the conclusions suggested below.

OUR WORK

Our involvement in this work began with an effort to find archaeological evidence for a much brighter supernova in the southern hemisphere (see Brandt et al. 1971b; Alexander et al. 1971; Maran et al. 1973). An appeal was made for archaeological help in locating a dateable record of the Vela supernova (Brandt, Maran, and Stecher 1971a) and a description of this appeal appeared in the 27 March 1972 issue of *Time*. One of the pictures discussed by Miller was used, to illustrate the *Time* article and a response from Muriel Kennedy at Lava Beds National Monument initiated the present study.

The search for and tentative identification of possible rock art records of the supernova are enhanced by the extreme rarity of crescents in rock art as noted by

Miller (1955a, b) for northern Arizona rock art, and by Schaafsma (1973) and Bain (1973) for New Mexico rock art. We have searched the rock art literature and confirm Miller's observation, as discussed below. Details relevant to a particular crescent motif site are given below when they are available.

THE FERN CAVE SITE

This possible record of the supernova is on the west wall of Fern Cave, in the northeast part of Lava Beds National Monument in Northern California. The cave is listed as archaeological site Tlk-2, and it contains a large number of rock paintings. In 1935, test trenches were dug in the floor of the cave, and charcoal, stone awls, mat or basket fragments, arrowheads, and other evidence of past occupancy were found.

Fern Cave is one of many archaeological sites in the Monument area described by Swartz (1964). He found evidence for occupancy at various times dating back to before 1500 BC. Although he noted the radiocarbon date of one timber in a house-pit as AD 803 (± 160 years), he described the relevant period of occupation, called "Component III," as having been probably of short duration sometime between AD 500 and 1869.

Fern Cave was formed in a lava flow in the late Pleistocene or Recent Epoch and is classified geologically as a lava tube. Entrance is possible at only one point, a hole in the ceiling above a mound of rocks. The most impressive array of pictographs is on the east wall of the cave (Grant 1967, p. 102, figure inverted). These paintings have received a fair amount of attention and were photographed by archaeologists more than 30 years ago. At that time, chalk was applied to increase the contrast in making the photographs. This practice, once common, is now recognized as undesirable. However, on the opposite (west) wall of the cave, many less spectacular pictographs were spared this treatment. Among them is the rock painting that includes a crescent (Fig. 27.3) done in a dark pigment that may be charcoal.

The crescent is about 140 centimeters above the cave floor and its diameter is 14 cm. The distance from the left-hand cusp to the circular figure directly below is 17 cm, while the distance from this cusp to the circle located above and to the left is about 21 cm. The crescent and the lower circle are located on a rock panel that slopes down and away from the viewer at an angle of about 35° to the vertical. The upper circle is drawn on a rock surface with a different slope. This photograph was taken with a camera held near the floor of the cave, with its optical axis essentially orthogonal to the surface containing the crescent and the lower circle (which may represent the supernova). This pictograph is recorded in Swartz (1963).

The correct orientation of the moon and the supernova is shown in Figure 27.2. These computations were made at the U.S. Naval Observatory to determine the phase and location of the moon as seen from the Lava Beds region with respect to the precessed location of the Crab, and with respect to the horizon, on the morning of 11 July 1065 (Gregorian) or 5 July 1054 (Julian). Note that the crescent, although resembling the cave drawing in phase, is concave towards the right and upwards. The moon

27.3 The Fern Cave pictograph as described in the text.

reached an altitude of 10° at $1^h\ 45^m$ before sunrise, and an altitude of 20° at 45^m before sunrise.

We must now deal with the incorrect orientation of the crescent (Fig. 27.3) and its reversal (Fig. 27.1). Miller (1955a, b) has suggested that the reversal of one of the Arizona crescents was due to what may be a common error in recording one's recollection of the appearance of the moon. Although there is clear danger in assuming that people of very different cultures will make the same kind of error in recalling the orientation of a celestial body, we tried an experiment. At the University of Maryland, John Carlson arranged for 29 liberal arts undergraduates who were taking the elementary astronomy course to view the moon and Jupiter on 18 July 1972 when the moon was at first quarter. The class assumed that this was an ordinary observing session. Two days later, however, they were asked to draw the moon and Jupiter as they had appeared in the sky. Twenty-three students did a fairly good job of it, but six others reversed the bright and dark halves of the moon, and one of the six even drew Jupiter on the wrong side of the moon. The results are consistent with Miller's suggestion that

John C. Brandt et. al

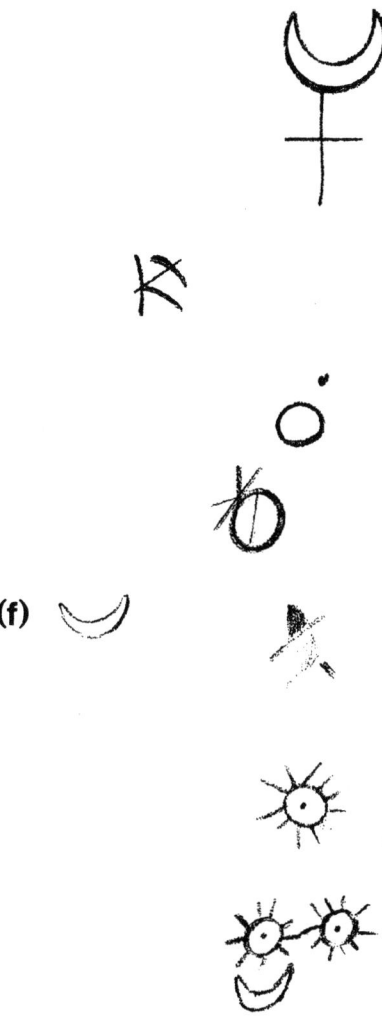

27.4 Charcoal sketch of the Symbol Bridge pictographs based on photographs and direct inspection of the site. Part of the pictograph is weathered and the feature marked (f) is faint but distinct. (Sketch by Dorothy Brandt.)

errors in recording the orientation of the crescent moon are common.

We now return to Miller's point that crescents are rare in Arizona rock art. The rock art of the Lava Beds region is not thought to be related to that found in Arizona. Instead, it is classified with the "Great Basin" region pictographs and petroglyphs of Nevada and eastern California. These have been studied in great detail by Heizer and Baumhoff (1962). Although they identified 58 basic shapes or "design elements," crescents were not included in their list. We examined the sketches of several thousand Great Basin petroglyphs and pictographs in their 1962 book and found only a few crescent-like figures. We also examined several thousand rock art figures from the Klamath Basin (which is a smaller region that includes the Lava Beds) as depicted by Swartz (1963) and again found only a handful of crescent-like figures, one of which is the Fern Cave painting discussed here. Finally, we looked at the hundreds of pictographs and petroglyphs from all over North America that appear in Grant's (1967) book. The only crescent among them is one of the two reported by Miller in northern Arizona.

THE SYMBOL BRIDGE SITE

This site, which may be related to the Fern Cave site, is approximately 5 miles southwest of Fern Cave. Symbol Bridge is on the southeast slope of Schonchin Butte in Lava Beds National Monument and is marked on the National Park Service maps of the Monument. The pictographs are shown in Figure 27.4; they have also been recorded by Steward (1927, fig. 4, p. 61, inverted).

641

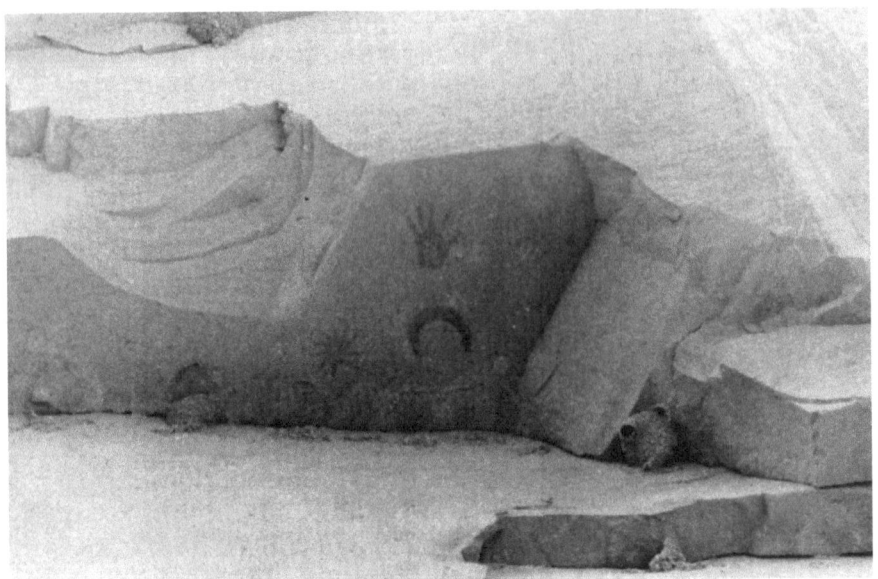

27.5 The Chaco Canyon pictograph as described in the text.

Our interest in this pictograph is in the three crescents shown adjacent to round objects. Refer to the arguments concerning the rarity of crescents in rock art given in the discussion of the Fern Cave site. Here, we have three crescents in a pictograph on the same rock and the overall result is suggestive of the supernova-moon conjunction.

THE CHACO CANYON SITE

This pictograph (Fig. 27.5) was discovered in June 1972 by an archaeological survey team from the University of New Mexico. It is located 6 meters above the canyon floor on the ceiling of a very shallow sandstone cave in the west wall of Chaco Canyon (site no. 29 SJ 427), about 500 meters northeast of Penasco Blanco ruin. Penasco Blanco is one of six large pueblos in the Canyon. Built on a mesa on the southwestern edge of the Canyon, it commands an impressive view of most of the Canyon. The vertical sandstone walls east of Penasco Blanco contain many petroglyphs typical of Chaco Canyon designs; except for a sun symbol on the vertical wall directly below the suspected supernova drawing, no other pictographs are known in the area.

Although habitation of the site at some time in the past is indicated by pottery sherds and viga holes adjacent to the pictograph, no dating is yet available for the sherds. However, tree ring dating of the nearby Penasco Blanco ruin (Bannister 1965, p. 200, fig. 43) establishes that it was inhabited between AD 900 and 1100.

The reddish-brown pigment of the pictograph is probably hematite, a substance commonly found in the canyon. The diameter of the crescent is about 18 cm, and the distance between the center of the crescent and the center of the star is about 30 cm. The image of a hand just above the crescent is about life size. Because the cliff wall is oriented north-south at this point in the canyon and the pictograph is painted on a surface perpendicular to the cliff, the star image appears south of the crescent, and the horns of the moon point westward in Figure 27.2. The hand is therefore east of the crescent.

Just what connection the hand bears to the star-moon image is problematic. Handprints painted on rock are common in places sacred to present day pueblos (Ellis and Hammack 1968), and the presence of a sun symbol on the vertical wall may indicate that the spot was a sacred one. Cushing (1941, p. 128) in his popular account of his first stay at Zuni mentions "a pillar sculptured with the face of the sun, the sacred hand, the morning star, and the new moon," but exactly where this pillar is or what the relative orientation of the design elements is we do not know. If the petroglyph seen by Cushing at Zuni is an ancient one it might be a record of the AD 1054 supernova and could be considered by modern Indians to be a record of Venus in conjunction with the crescent moon.

OTHER SITES

Col. James G. Bain of Albuquerque, New Mexico has kindly supplied us with photographs of two other possible Crab Nebula petroglyph records in New Mexico, one in the Village of the Great Kivas, and the other near Scholle. Both of these examples are associated with other rock art, but no other information is yet available. These sites are presently under study.

DISCUSSION

We can ask whether the ancient native Americans were in fact likely or inclined to record astronomical phenomena in their rock art. We know of no written record or oral tradition going back to the eleventh century that can answer this question. Perhaps the best we can do is to look at Indian customs in historical times. Navajo gourd rattles and sand paintings were often decorated with symbols of the Pleiades and Orion (Haile 1947), and there are likewise Navajo cave paintings of the eighteenth century and perhaps earlier that have been interpreted as representing constellations (Schaafsma 1966, 1973; Britt 1973). Certainly recognizable constellations are common in modern Navajo sand paintings. Stephen (1936) gives many examples of Hopi designs of celestial objects, including unmistakable representations of the Milky Way, the Pleiades, Orion, the crescent moon, the sun and the Morning Star. The kiva murals of Kuaua (Dutton 1963) also contain celestial motifs.

There are pictographs and petroglyphs from various parts of the United States and Mexico that LaPaz (1948) described as records of meteors; most impressive of all are unmistakable paintings of the Leonid shower of 1833, made on a buffalo skin by the Dakota Sioux (Mallery 1893, p. 723). The Dakota Sioux also recorded the total eclipse of the sun which passed through their territory on August 7, 1869 (Mallery 1893, p. 286). Finally, Grant (1966, plate 25) has noted two pictographic representations of comets in the rock art of the Chumash; one of the authors (J.C.B.) has inspected a striking petroglyph of a comet near Los Alamos, New Mexico. Thus astronomical phenomena do appear in Indian art.

We have also investigated the possibility that the rock art in question might be representations of another supernova—the oriental records (e.g., Xi and Po 1966; Shklovsky 1968) indicate a supernova in May of AD 1006. However, during the months of May, June, and July in 1006 the moon and the alternate supernova candidate were never closer than 21° in the sky and the moon was always gibbous at the time of closest approach. As a test of the *Sung-Shih* record, we have also checked the positions of the bright planets for 5 July 1054 and none of them were near the crescent moon or ζ Tauri.

Ho, Paar, and Parsons (1972) have questioned the identification of the Crab Nebula supernova with the Guest Star discovered by the Chinese on 4 July 1054 as described in the *Sung-Shih*. The principal evidence for this contention is (1) the reversal of the directions of the supernova, from the star ζ Tauri; the Crab is actually northwest of ζ Tauri while the *Sung-Shih* states southeast. Also (2) the difference in the extrapolated date (AD 1140 ± 10) from AD 1054 is regarded as significant. We do not regard these objections as convincing. The difference in directions is in fact just a *reversal* of directions—the kind of error one might expect in a report transcribed many times. The difference in dates is probably not serious because the quoted error represents only the internal consistency of the astronomical measurements and *not* the uncertainty in the interpretation (Trimble 1968, 1971).

These considerations notwithstanding, if the identification of the Crab Nebula supernova with the 4 July 1054 Guest Star is correct, then a spectacular conjunction of the crescent moon and the supernova was visible from western North America on the morning of 5 July 1054. We have listed several cases of rock art that appear to resemble such an event, and which are found in Arizona, California and New Mexico. In each case the archaeological evidence is consistent with habitation in the eleventh century AD Thus, these records are consistent with and supportive of the *Sung-Shih* interpretation criticized by Ho, Paar, and Parsons (1972).

As we noted at the outset, we cannot prove that the rock art records discussed here refer, in fact, to the Crab Nebula supernova. All that we can do is to build a circumstantial case. Since the close conjunction would have been visible from Mexico, we would appreciate the assistance of Mexican archaeologists and astronomers in locating additional possible records of this fascinating phenomenon.

ACKNOWLEDGMENTS

Ray A. Williamson, ordinarily at St. John's College, Annapolis, Maryland, acknowledges support from the NASA Goddard Space Flight Center (Grant NGS 21002-033).

REFERENCES

Alexander, J. K., J. C. Brandt, S. P. Maran, and T. P. Stecher. 1971. The Gum Nebula: Further Evidence from Spacecraft and Ground-Based Instruments. *Astrophysical Journal* 167: 487–490.

Baade, W. 1942. The Crab Nebula. *Astrophysical Journal* 96:188–198.

Bannister, B. 1965. Tree-Ring Dating of the Archaeological Sites in the Chaco Canyon Region, New Mexico. *Southwestern Monuments Association, Technical Series* 6 (2):119–214.

Brandt, J. C., S. P. Maran, R. S. Harrington, and M. M. Kennedy. 1972. A Northern California Pictograph That May Be Another Record of the Crab Nebula Supernova Explosion. *Bulletin American Astronomical Society* 3:319.

Brandt, J. C., S. P. Maran, R. S. Harrington, M. M. Kennedy, R. Williamson, C. Cochran, W. J. Kennedy, and V. D. Chamberlain. 1973. Possible Records of the Crab Nebula Supernova in the Western United States. *Bulletin American Astronomical Society* 5:20.

Brandt, J. C., S. P. Maran, and T. P. Stecher. 1971a. Astronomers Ask Archaeologists' Aid. *Archaeology* 21 (4):360.

Brandt, J. C., S. P. Maran, T. P. Stecher, and D. L. Crawford. 1971b. The Gum Nebula: Fossil Stromgren Sphere of the Vela X Supernova. *Astrophys. J. Letters* 163:199.

Britt, C., Jr. 1973. Early Navajo Astronomical Pictographs in Canyon de Chelly, Northeastern Arizona, U.S.A. In *Archaeoastronomy in Pre-Columbian America,* ed. A. Aveni. Austin: University of Texas Press.

Cushing, F. H. 1941. *My Adventure in Zuni.* Santa Fe, Peripatetic Press.

Davies, R. D., and F. G. Smith, eds. 1971. The Crab Nebula. *I.A.U. Symposium* 46, Reidel, Holland.

Duncan, J. C. 1921. Changes Observed in the Crab Nebula in Taurus. *Proceedings National Academy of Sciences U.S.* 7:19–180.

Duncan, J. C. 1939. Second Report on the Expansion of the Crab Nebula. *Astrophysical Journal* 89:482–485.

Dutton, B. P. 1963. *Sun Father's Way: The Kiva Murals of Kuaua.* Albuquerque, University of New Mexico Press.

Duyvendak, J.J.L. 1942. Further Data Bearing on the Identification of the Crab Nebula with the Supernova of 1054 AD. *Pub. Astron. Soc. Pacific* 54:91–94.

Ellis, F. H., and L. Hammack. 1968. The Inner Sanctum of Feather Cave, a Mogollon Sun and Earth Shrine Linking Mexico and the Southwest. *American Antiquity* 33 (1):25–44.

Grant, C. 1966. *The Rock Paintings of the Chumash.* Berkeley, University of California Press.

Grant, C. 1967. *Rock Art of the American Indian.* New York, Thomas Y. Crowell Co.

Haile, B.O.F.M. 1947. *Starlore among the Navajo.* Santa Fe, Museum of Navajo Ceremonial Art.

Heizer, R. F., and M.A. Baumhoff. 1962. *Prehistoric Rock Art of Nevada and Eastern California.* Berkeley, University of California Press.

Ho, Peng-Yoke, F. W. Paar, and P. W. Parsons. 1972. The Chinese Guest Star of AD 1054 and the Crab Nebula. *Vistas in Astronomy* 13:1–13.

LaPaz, L. 1948. Meteoritical Pictographs. *Popular Astronomy* 56:324–330.

Mallery, G. 1893. Picture-Writing of the American Indians. *Tenth Annual Report of the Bureau of Ethnology to the Secretary of the Smithsonian Institution* 1888–1889, by J. W. Powell, director. Washington, DC, Government Printing Office (reprinted in 2 vols. by Dover Publications, New York, 1972).

Maran, S. P., J. C. Brandt, and T. P. Stretcher, eds. 1973. *The Gum Nebula and Related Problems.*

National Technical Information Service NASA SP-332, Springfield, VA.

Mayall, N. U. 1962. The Story of the Crab Nebula. *Science* 137:91–102.

Mayall, N. U., and J. H. Oort. 1942. Further Data Bearing on the Identifications of the Crab Nebula with the Supernova of 1054 AD. *Pub. Astron. Soc. Pacific* 54:95–104.

Miller, W. C. 1955a. Two Possible Astronomical Pictographs Found in Northern Arizona. *Plateau* 27 (4):6–12. Museum of Northern Arizona.

Miller, W. C. 1955b. Two Prehistoric Drawings of Possible Astronomical Significance. *Astronomical Society Pacific Leaflet* 314:1–8.

Needham, J. 1959. Mathematics and the Science of the Heavens and the Earth. *Science and Civilization in China* 3. Cambridge, Cambridge University Press.

Needham, J., Wang Ling, and D. J. de Solla Price. 1960. *Heavenly Clockwork*. Cambridge, Cambridge University Press.

Scargle, J. D. 1967. The Crab Nebula—913 Years after Its Outburst. *Astron. Soc. Pacific Leaflet* 457.

Schaafsma, P. 1966. *Early Navajo Rock Paintings and Carvings*. Santa Fe, Museum of Navajo Ceremonial Art.

Schaafsma, P. 1973. *Rock Art in New Mexico*. Santa Fe, State Planning Office.

Shklovsky, I. S. 1968. *Supernovae*. London, Wiley Press.

Steward, J. H. 1927. Petroglyphs of California and Adjoining States. *University of California Pubs. Am. Arch. and Ethn.* 24:47.

Swartz, B. K., Jr. 1963. Klamath Basin Petroglyphs. *Archives of Archaeology* 21. University of Wisconsin Press.

Swartz, B. K., Jr. 1964. Archaeological Investigations at Lava Beds National Monument, California. Doctoral thesis, Department of Anthropology, University of Arizona.

Trimble, V. 1968. Motions and the Structure of the Filamentary Envelope of the Crab Nebula. *Astronomy Journal* 73:535–547.

Trimble, V. 1971. Dynamics of the Crab Nebula. In *The Crab Nebula*, ed. R. D. Davies and F. G. Smith, 12–21. Holland, Reidel.

Xi, Ze-zong, and Shu-jen Po. 1966. Ancient Oriental Records of Novae and Supernovae. *Science* 154:597–603.

CHAPTER TWENTY-EIGHT

A Thousand Years of the Pueblo Sun-Moon-Star Calendar

Florence Hawley Ellis

In the fall of 1972 two photographs were sent me showing the petroglyph of a hand-print, a horizontal crescent moon upside down, the Great or Morning Star, and the Pueblo sun symbol (two concentric circles with a center dot) in a combination (29SJ427) running down a protected ledge facing east just below Penasco Blanco, the westernmost of the big ruins in Chaco Canyon in northwestern New Mexico (Fig. 28.1). One photograph came from Mr. Tom Windes of the Chaco Center, for which I act as consultant, and which had discovered the petroglyphs during their survey. The other came from Mr. Clarion A. Cochran who had been working as a summer ranger in the Chaco Canyon National Monument and was struck by the parallel between the petroglyphs and a group described by Cushing near Zuni, with which I also was

---

First published as "A Thousand Years of the Pueblo Sun-Moon-Star Calendar," by Florence Hawley Ellis, in *Archaeoastronomy in Pre-Columbian America*, ed. A. Aveni. © 1975 by the University of Texas Press. All rights reserved.

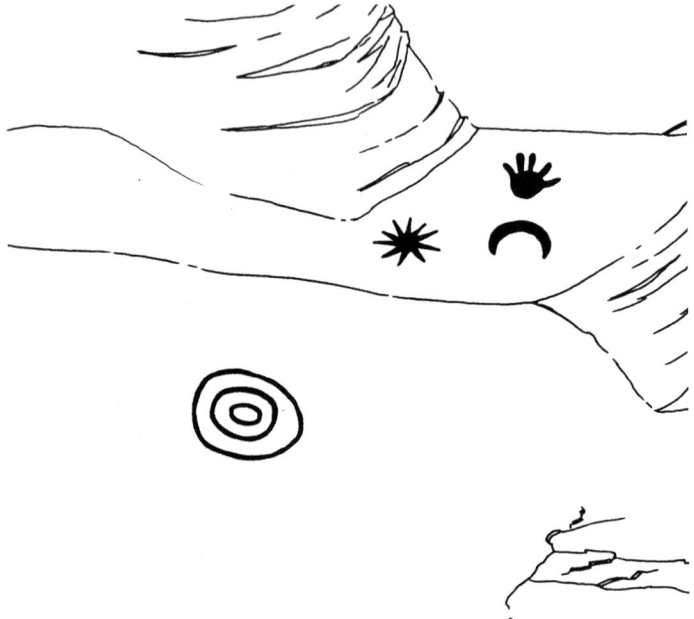

28.1 The group of petroglyphs comprising what we believe to be an 11th-century sun-watcher's station and shrine below Penasco Blanco, Chaco Canyon National Monument, New Mexico.

familiar. The find was, indeed, something over which to become excited, not because it was spectacular but because it permits an unusual glimpse into the agricultural-ceremonial-calendrical system used by the Pueblo Indians of Chaco Canyon a thousand years ago.

Tree ring dates prove that the great Chaco sites were occupied from the 10th into the 12th century, and the suggestion that this combination of symbols recorded the advent of a supernova in the sky in AD 1054 has been made by astronomers. This concept, however, might be questioned by anyone who knew Southwestern ethnology in some detail. If the supernova was visible in the Southwest, as is said to have been the case, the prehistoric Pueblo people no doubt saw it and were impressed. This is no reason to suppose that they recorded it. In his article on the Crab Nebula which is believed to represent the debris of the 1054 explosion, Oort (1957) notes that "Strangely, although the event must have been witnessed by practically everybody in Europe, not a single mention of it has been found in any European chronicle," though in China and Japan the advent of this new star was duly set down. Although at first it could be seen even in the daytime, after two years it no longer was visible to the naked eye. The literature on Southwestern Indians of the historic period and my own questioning in various villages indicates that the Pueblo people characteristically did not record exciting events. For example, although we are positive that the volcanic eruption of Sunset Crater in the San Francisco Mountains of northern Arizona in AD 1067 must have been visible to the Hopi on their mesas only 80 miles away and that peoples living near the crater were forced to abandon their ash-filled houses, their ash-buried fields, and even the region as a whole for some years, no petroglyphs recording this event are known. The Hopi themselves state that they do not even have legends dealing with this traumatic experience in which some part of their ancestors quite certainly were observers if not participants. We are equally without legends pertaining to some of the painful and exotic punishment of Pueblos in the Spanish conquest of New Mexico. The Pimas and Papagos of southern Arizona had calendar sticks, the closest thing to an actual record of events in the Southwest, but although one or more men of a village might choose to keep such a stick, it could be "read" only by the owner because his symbols were no more than mnemonic devices serving to recall selected events to his own mind. He used what symbols he liked to remind himself of what events he thought worth remembering, just as we tie a string around one finger to remind a man to buy a bottle of whiskey, his wife to go to the hairdresser, and his son to put gas into the car. Only one symbol was consistently employed by all the Pima calendar keepers and always for the same event: a dot which specifically referred to their major ceremony. This "Prayer stick" ceremony paralleled the Pueblo solstice celebration in some respects, though given earlier in the fall and only every four years north of the present international border and every two years in northern Sonora (at least in the historic period), and one may wonder whether the dot could have been an abbreviation for the sun symbol.

Other standardized symbols were used in the historic Southwest, but for a different type of purpose. Sun, moon, four pointed stars, stepped elements, and triangles representing clouds which may have a fringe of falling rain, katcinas, the flute player, the horned serpent, mountain sheep, corn, snakes, the four or six lines crossed within a circle which Parsons (1939, p. 359) refers to as glyphs of the directions, the friendship marks of interlocking half circles, and the squares within squares which symbolize a sipapu (all identified by reliable Pueblo informants involved in their use) have been found on cliffs, in caves, on kiva walls, on boundary stones, as well as on masks, pottery, and other items. We know that when the Hopi went on one of their ritually embellished salt-gathering trips to the Grand Canyon, each man pecked his clan symbol into a certain great boulder as his signature, and when Eagle clansmen went out together for some duty they left their clan mark on a rock (Parsons 1939, pp. 358–360). The people of New Mexico similarly claim to have used clan symbols for signatures at places of prayer and repeated group activities; clanless Pueblos such as those of the Tewa presumably employed symbols to represent the names of their bilateral extended families. The most widespread sun symbol of the Pueblos (other than their sun mask used for decorative events) is the Maya glyph for sun (Ellis and Hammack 1968), obviously an import from the south as apparently was true of the water symbol and probably of some other glyphs. The main question of concern to us is why they were placed upon the rocks or elsewhere. Except for occasional scenes of hunting or warfare, the symbols could be said to be nouns with no implied verbs other than "was here" or "occurred" or "pertains to this spot." As discussed by Pueblo Indians, they relate only to the usual, the repeated. A major katcina associated with the Hopi winter solstice ceremony, for instance, carries a crescent moon and the Big or Morning Star on his mask, designs associated with the solstice. Cushing (1967) wrote about Zuni sun-watching in 1893 done at the nearby ancestral pueblo of Matsakya; the pillar marking the spot carried exactly the complex of symbols found on the Chaco ledge. I have in my notes the account of a Rio Grande sun and new moon–watch from a conical white stone outcrop marked with a big star symbol, this watch coming in connection with the winter solstice, as we shall later describe. With Alexander Stephen's considerable body of data on the process of dating ceremonies and work periods and the symbols involved, collected largely between 1891 and 1894 when he was living at Keams Canyon and on the First Hopi Mesa, we need not guess about the Chaco group of symbols.

One way of knowing that the Chaco find of the four associated symbols does not pertain to the supernova is having the data to prove that this complex does pertain to observations pertinent to dating the solstice and the agricultural round. But basic to our understanding of this is some knowledge of the old native Pueblo ceremonial and agricultural calendar.

THE PUEBLO SUN, MOON, AND STAR CALENDAR

At the peak of the Pueblo was a Sky deity, sometimes said to have been sexless but commonly equated with the Christian God by the Rio Grande Indians today and possibly related to the Old Fire God. Of more immediate concern to the Pueblo people was Father Sun and Mother Earth who provided fertility, food, and warmth. Sun's apparent movement from the north to the south was understood to create the seasons. Since it appeared that Sun hesitated a few days in his southern "house," the Pueblos deemed it wise to put on a ceremony which would spur his decision to again take up his march northward. As part of the ritual at that time, the ceremonialists sought omens relating to the length of the growing season and hence to its success in the months ahead. The succession of horizon points over which the sun rose or set became calendar markers. The periodic reappearance of a new moon provided another reference for measuring time, and the consideration of moon and sun together was thought to strengthen chances for good omens because Moon could aid in influencing Sun. Some say that Morning Star, symbol of the beloved Twin War gods and also of the Star of Sky god already mentioned, was watched because it followed Moon so closely, but this is not all of the story. Prayers were made to Sun, Moon, the Morning Star, Orion, and the Pleiades at this time and on some other occasions. It is interesting to note that these heavenly bodies are pricked into the surface of gourd rattles used by Navajo ceremonialists, most of whose lore was borrowed from the Pueblos.

It is unfortunate that in generalizing, some students of native Southwestern calendar systems have rather overlooked the lunar and certainly the Venus features, though in other places the same persons may have described these matters. We read, for instance, that the Hopi of northern Arizona establish dates for all their winter ceremonies by the position of the sun setting over the western horizon and the dates of their summer ceremonies by its position in rising above the eastern horizon (Fewkes 1897; Nequatewa 1931, p. 386; Parsons 1939, pp. 496–497). The statement also has been made that the Eastern Pueblos of the northern Rio Grande put less emphasis on sun-watching than the Western Pueblos, but we believe that this should be understood only in the present tense. The use of our own calendar was forced onto the Rio Grande peoples centuries ago by Spaniards who made dates to be met and imposed saints' days to be observed.

The lunar calendar which nature has neatly divided could be used alone even though lunar periods vary somewhat, but that would mean taking no account of the seasons. Moreover, lunar periods can be rather easily lost. Each so-called Pueblo "moon" begins with the appearance in the sky of the new crescent moon. The two preceding days in which no moon is visible are not counted; they fall between lunar periods. Named intervals in a lunar period at Hopi are spoken of as "new moon" or "crescent first seen," "moon vertical," "crescent moon horizontal," "moon half gone," and "last of moon" (Parsons 1939, p. 1040). The lunar calendar is not purely ceremonial, for both Stephen and Titiev report the Hopi planting according to the lunar as well as the solar sequence. Major ceremonies such as those which lead up to the winter solstice and those others which comprise the very important spring pre-planting complex are dated by the positions of both sun and moon, as if one were reading time by the two hands of a clock. In Zuni, prayer stick offerings still are made in each lunar period by the curing societies, the katcina groups of the Shalako complex, and the impersonators of Sun Youth. On some years we impatiently wait for a late announcement of the big Zuni Shalako celebration and finally hear that the Pekwin has found difficulties in reconciling the position of the Sun, which should be near his southern "house," and Moon, which should be full, a type of problem familiar to the priests of Mexico. Acoma and Laguna explain that Moon, to all the Keres except Zia, a female (as to the Zuni and the Maya), though male to the other Pueblos (as to the prehistoric people of the Valley of Mexico), must be in correct position so that she can bring her influence to bear on Sun for the good of mankind at the winter solstice.

The Pueblo calendar of thirteen lunar months (paralleling the old Maya and Mexican religious calendar), like that of the solar divisions, is best preserved today at Hopi, where every ceremony is associated with a moon and every moon with a ceremony. When a sequence of retreats (the private portion of a ceremony) is to be held by the religious societies, the first is dated by solar and/or lunar observation and the others by the count of days elapsed since that observation or since the preceding society's retreat. Tally cords with knots to be untied, sticks into which notches are cut, and marks on a floor or wall, which could be erased as the days passed, customarily were used in cutting the calendar into shorter intervals when necessary. If an official lost a lunar month, as Titiev (1972, pp. 166–167) noted for Hotevilla, the mistake was covered as long as possible by patient and loyal followers of the watcher but eventually must be rectified by a public announcement that a later month must be skipped.

The data collected by Titiev (1938, fn. 2; 1944, pp. 174–176) in the 20th century and by Stephen in the late 19th (Parsons 1939, pp. 1036–1037) give us the names of successive moons, the month of our calendar in which each supposedly should fall, and the major activities reserved for it. To make the scheme more graphic we have added a translation of the names, the date on which the new moon initiating each lunar period actually appeared in 1892 and 1893, and also the calendar date for those moons 80 years later in 1972 and 1973.

WINTER

Kel-muya—November moon. Crescent on 20 Oct., 1892; 6 Nov., 1972. Known as "the initiate or sparrow-hawk moon" when youths receive their second or tribal initiation and thereafter (in the past) became warriors, of which various hawks are symbols. The making of new fire and the first appearance of katcinas also mark this ceremony, which leads up the Soyal or winter solstice ceremony as the Shalako sequence does in Zuni. Wuwuchim is dated by the chief of one of the three important societies involved in watching the setting sun. At the right point he informs the Chief of Wuwuchim who calls his men that evening to smoke over prayer plumes and decide that the ceremony must be announced next day at dawn by the Crier Chief, but the date, 16 days hence, is that of the closing of the ceremony when offerings are deposited on shrines, a customary type of announcement and confusing to whites.

Kya-muya—December moon. Crescent on 19 Nov., 1892; 5 Dec., 1972. Known as "the sacred but dangerous moon" because it includes the 5 days of imminent disaster from witchcraft, falling through the now fragile covering of the earth, and other frightening possibilities, a complex borrowed from prehistoric Mexico's feared 5 days at the end of their Vague year. The winter solstice ceremony, dated by observation of Sun's reaching his "house," the most southern sunrise point on the horizon, is elaborate and lengthy with all religious societies participating. It opens about 12 Dec., the date depending on being in certain relative positions, as elsewhere in the Pueblo area.

Pa-muya—January moon. 18 Dec., 1893; 3 Jan., 1973. The "play moon"; when gaming and some minor dances comprise the entertainment because weather is too cold for outside activities. During full moon, beans and some corn are secretly planted in containers carried into Chief kiva which is kept heated to produce early sprouting and growth.

Powa-muya—February moon. 17 Jan., 1893; 3 Feb., 1973. "The purification moon." This is the second lunar period after the winter solstice. The Mayan people of Yucatan held a festival at this time to open their agricultural season. The Hopi term for the period refers to the first initiation of young children in which the katcinas, who have come for the big Powamu ceremony to insure fertility and precipitation for the fields, administer a purificatory whipping to initiate and disclose the secret of masking to represent gods. The Powamu pre-planting ceremony clearly is based on the elaborate masked month-long pre-planting ceremony of post-Classic Mexico (Spence 1912, pp. 48, 56–58), which began early in May at the opening of the rainy season.

At "horizontal crescent moon," the Hopi pull up the bean and corn sprouts to be distributed by Sun Youth katcina to women during his visit to springs, clan houses, and katcina clan kivas. More corn and bean seed is planted in containers in kivas, the height reached being interpreted as an omen foretelling success of the coming agricultural season. Uninitiated children receive gifts of katcina doll representations, rattles, etc. from the masses of katcinas who appear. At the Pamurti ceremony held at First Mesa in late January, Sun katcina arrives accompanied by the family of Chakwena, whose mask is glossy black with a horizontal crescent moon on one cheek and the many spiked Great Star on the other (duplicates of the Chaco petroglyph symbols). The Chakwenas are involved with warfare but also with childbirth.

Isu-muya—March moon. 16 Feb., 1893; 4 Mar., 1973. The "moon of moderating winds." A second term, "cactus moon," refers to the considerable dependence on cactus for food at this time when stores from the previous summer's harvest were running short. During this moon the sun is watched, for when it reaches "the first point" a little early corn is planted in protected spots so that the katcinas will have it to bring in during their final Niman or Home-going ceremony, which falls before the main harvest.

Kwya-muya—April moon. 17 Mar., 1893; 2 Apr., 1973. "The moon of setting up windbreaks for small plants." When the second solar point is reached, people get their seeds ready for planting.

Hakiton-muya—May moon. 16 Apr., 1893; 2 May, 1973. "The waiting moon," when major planting cannot yet be done. At the 3rd solar point (mid-April), the leading families are told to plant in succession and others as they please, the first crop put in being corn. At the 4th solar point the first string beans, melons, and squash are planted and at the 5th the main planting of these is done. Any planted after the 6th point would not mature. At the 7th, lima beans are planted. At the 8th the general bean crop is put in.

SUMMER

Kel-muya—June moon. 15 May, 1893; 31 May, 1973. "The planting moon." The 7th and 8th points mentioned for the preceding lunar period may fall into this period if we take the actual dates given. Men are called for communal work in planting the fields of the Town Chief.

Kya-muya—July moon. 13 June, 1893; 30 June, 1973. At the 10th solar point, fields to be irrigated with run-off water are planted and at the 11th, sandy areas are planted, for winds have died down. The summer solstice rites at the 12th point, Sun's "summer house," mark the end of all planting for the plants could not mature before frost.

Pa-muya—August moon. 13 July, 1893; 29 July, 1973. "The moisture moon" when rains should begin. The big Niman katcina ceremony which opens 25 days after summer solstice ends the growing season, and katcinas go home to rest in the San Francisco Mountains until November. Chief of the Snake society watches the sun for

16 days after this moon opens and then starts the Snake-Antelope or, alternately, the Flute ceremony.

Powa-muya—September moon. 11 Aug., 1893; 27 Aug., 1973. "The big feast moon." Four days after the Snake ceremony comes the smoke meeting of one of the three Women's societies, usually followed by a dance.

Angok-muya—October moon. 9 Sept., 1893; 26 Sept., 1973. "The harvest moon." One of the four names which Stephen gives for this month is that which Titiev gives for the 13th month. The Oa'qul harvest basket dance by one of the Women's societies is followed by unimportant dances such as the buffalo and the butterfly.

Isu-muya—October moon. 25 Oct., 1973. Not given by Stephen, but listed by Fewkes and Titiev.

The problem in length of growing season and in actual dating of moons is well illustrated in the difference of sixteen days between the record of the same new moon in Stephen's time and our own, but even in using Stephen's dates alone, one sees that the moon expected to fall in a certain one of our months often actually appears within the preceding month.

The Hopi think of the moons of their calendar as being in a continuous succession which we can best picture as a moving chain encircling a wheel, the upper half of which represents our world and the lower half the underworld. With a strong feeling for duality they sometimes speak of two divisions of the year because they believe, theoretically at least, that when we are experiencing any given winter moon, the underworld moon is at an equal position in a summer season. Major ceremonies given in one season characteristically have a short counterpart performed in the opposite season when the dead who reside in the underworld are believed to be performing their ritual, for everything pertaining to the dead is opposite to that which pertains to the living. This explains why the summer solstice ceremonies are minimal in comparison to the big celebrations of the winter solstice. Stephen could obtain no names for the last five months of his year, the group being lumped as moons without names. In the 1930's Titiev found five of the names of the first seven successive moons, all referring to seasons, agricultural conditions, or ceremony, repeated in the names of the next six lunar periods, but even some of the chiefs had difficulty remembering the names of the lunar periods in order between September and the first of November. The shift in dates of successive new moons and of the same moon in successive years is one cause for confusion of the natives and even of some anthropologists. If we figure our 13 moons as averaging 29 days from crescent to crescent, we have 377 days, some 17 more than those of the solar year. By accurate reckoning, the named moons thus would move farther and farther from the seasonal breakdown of the solar year for which the ceremonies were primarily intended as an aid to growth. Moreover, crops would suffer as lunar dates for planting and other agricultural activities shifted in relation to the solar calendar. Something obviously would have had to give, possibly by the priest-chiefs occasionally skipping one of the nameless moons to bring the calendar back to reality.

Stephen found that in his day (late 19th century) the two portions of the native year, taken together, could be referred to either as one year or as two years (Parsons 1939, p. 1039). Apparently the Hopi, and presumably the other pueblos, were thinking in terms of two chains encircling their wheel at once, one chain made up of alternate groups of seven and six successive moons and the other of two divisions marked off by the solstices in the 365 day solar year. In one of Stephen's notebooks (1883–1884) from ten years earlier, however, he remarked that the katcina or ceremonial year was reckoned in 13 moons but the work year was computed in two successive periods of eight moons, each of these 240 day periods being known as a "lesser year." The first lesser year began with the Initiates' moon in November which opened the winter solstice sequence. It ended in June. The second opened in July and ended in February. The third began in March and ended in October. The fourth, like the first, began in November. Each bundle of four lesser years was known as "the Year of the Great Moon." Stephen does not refer to this eight month work division again and may have been in error, though this seems unlikely if we think of the detail in this scheme. Parsons thought it quite possible that this was the old Hopi system, already dying out in the 1880's, and being replaced, with attendant confusion, by the seven and six month divisions of the 13 moon katcina calendar, which brought the native system closer to that of the whites.

The old Zuni system duplicated that of the Hopi, with a two part year of seven named moons followed by six moons sometimes said to be nameless, though Cushing states that they took the color names of the six directions. Stevenson (1904, p. 108) speaks of the two six month divisions of the year at Zuni, her informants evidently modernizing their system at that time. Each of these months then was broken into three parts, a division not mentioned elsewhere. Like the Hopi, the 20th century Hopi-Tewa recognize a 5 moon summer and a 7 moon winter season, 5 of the moon-period names being repeated in each season (Parsons 1925, p. 1041), a capitulation to the total of 12 in our calendar.

OFFICIAL WATCHERS OF THE CELESTIAL BODIES

Who does or did the sun and moon-watching and made adjustments when necessary? Today some pueblos such as Santa Ana are without sun-watchers and merely note when the gold of turning aspens creeps down the western slopes of a near-by mountain. Other pueblos first observe the date on one of our wall calendars and then send out a representative to observe the sun and/or moon shortly before a specified ceremony, but in the past and even today in some of the really conservative villages several men have been involved at once in sun- and moon-watching and in discussion and interpretation of the results.

Answering the question of sun-watchers necessitates taking a moment to meet the array of Pueblo officials in the old native system who represent supernatural beings of the pantheon from whom they take their authority. The secular officers of the New Mexico pueblos (governor, lieutenant governor, etc.) were imposed by Spain so that dealings with the religious hierarchy which controlled each village could be avoided. For present purposes we can forget the secular slate. Although the native political systems and officials differed somewhat between villages, the two most important religious officers in each were the Town Chief, now generally known as the cacique in Rio Grande villages, and the War priest or Outside Chief. The Town Chief (two in each Tewa town of the northern or Upper Rio Grande who alternated between the long summer and the short winter seasons but only one in each of the other pueblos) was chief priest and generally said to represent Mother Earth. He spent much time in prayer for rain and fertility but he also was a major sun-watcher and after the observations were made he scheduled the principal ceremonies of the year for his own village. Pertinent data from the New Mexico pueblos are meager except for Zuni and one village of the Middle Rio Grande, to be discussed later. At Hopi in northern Arizona where the Town Chief is said to be the earthly representative of Eototo, leader of the katcinas (Titiev, p. 114), the system differs somewhat. On First Mesa a Sun chief, as such, watches the sun after the summer solstice and through the winter solstice. The leader of the Flute Society then takes up the duty, to run through the summer solstice. Oraibi, the Third Mesa Hopi village which was so disrupted in the split of 1906, shows an approximation of this system (Parsons 1939, pp. 502–504 and fn.; Titiev 1944, p. 17). The Hopi have one site for summer sun-watching and another for winter observations, which seems to have been common in the other pueblos. The division of sun-watching periods does not correspond to the two sections of the Hopi ceremonial year in which ceremonies of the agricultural period, from February (Powamu) to late July (Niman), which heavily feature the katcina rain-fertility spirits, are controlled by the Katcina clan chief and the shorter period from Niman ceremony (July) to Powamu (February) by the head of the Badger Clan, claimed to have been first to emerge from the underworld and hence a man of power. Here we see again the ceremonial division into 13 parts and the agricultural division into two, though the attempt to tie plant growth, heavenly observations, and the katcina rain-fertility spirits into a threesome is obvious.

Among supernatural beings Sun, as the embodiment of male qualities, is a warrior and patron of warriors, though his two sons, the Twin War gods, especially the older and leader of the two, were even more specifically the patrons of warriors. In the pueblos of New Mexico these sacred twins are thought of as youths but the Hopi tales depict the twins as children and give the War Star God, the Sky God already mentioned, the qualities of the older of the Twin gods as known in New Mexico. Both personages are symbolized by the Morning Star. In the Southwest, this was the most important of stars, even in Mexico where it symbolized Quetzalcoatl, who was more or less the prototype of our Sky or Star god. As far as we can make out, in early times each pueblo had a War Priest or Outside Chief who served for life as representative of Sun. Working with him were two men who represented the Twin War gods. As the War Priest was required to have taken a scalp, that office has largely disappeared, though it still is found, without scalp requirement, in Hopi, Zuni, Acoma, and Jemez. In Zuni we still have the Pekwin or War Priest and the two Priests of the

Bow represent the War gods. The representatives of these two War Gods in the Rio Grande once held lifetime offices, it is said, but since Spanish days, and by Spanish decree, they have been re-appointed or re-elected annually and are known as captains, a position the Spaniards could understand. They have taken over some of the duties of the old War Priest and the Town Chief has had to accept others. The Star God is represented in Hopi by a Star Chief who emphatically twirls the sun symbol in one of the winter solstice rituals as a hint that Sun should hurry his start northward once more.

The association between stars and war was universal throughout the pueblos. A five pointed star (the Evening or Big Star could have four or more points) was drawn on the wall of the War Chief's chamber at Hopi and on interior walls of a number of other religious chambers and effigies of Orion and the Pleiades were used on his altar (Parsons 1939, pp. 9, 84, 87, fn. 1; figs. 2, 5, 64a). Bourke reported that for Hopi in 1883 "upon the walls of the estufas (kivas) are rudely etched and painted the symbols of sun, moon, and Morning and Evening Star, and Pleiades" (Bourke 1884). In Hano, also known as Hopi-Tewa (the First Mesa village established in the early 1700's by Tewas from the northern Rio Grande who chose to avoid the Spaniards by moving to Hopi) the sun-watcher for the winter solstice was keeper of the sacred images of the War gods, whose symbols were the Morning and Evening stars. The Town Chief is the watcher for the summer solstice, and the six stars of Orion, which make a backwards figure seven, are watched in May to be interpreted as an omen relating to a long and hence good summer. These stars, which are painted on the wall of the Chief Kiva on the Third Hopi Mesa, are shown with four points, and in this case Morning Star also has but four.

In the Rio Grande, the eastern edge of Pueblo country, the oldest religious society of which we have record is the Koshare, whose members are known as "sons of the sun." A bowl decorated with a Koshare figure found in Sapawe, one of the ancestral Tewa pueblos, proves that this society is at least 600 years old. The chief of the Koshare, as might be expected, is one of the sun-watchers as well as being of utmost importance in village councils. Both the caciques or Town Chiefs do sun-watching in the northern Tewa pueblos, the place of the sun's setting being watched for the winter solstice and the place of its rising for the summer solstice and other dates. The duty of watching the Evening and Morning Stars and the constellations fell to heads of the religious societies. The movements of the Great Dipper as well as of the six stars of Orion and of the Pleiades were used for timing during the night and the shadow of a small stick set upright into soft soil or the position of the sun in its arc served for timing during daylight. In Nambe, solstice dating was handled by the caciques noting where sunlight coming through an east window struck a buckskin hanging on the west wall. The rising stars of Orion's belt are watched in relation to both solstices; the statement is that if these appear on 4 and 6 May there will be neither early nor late freezes (Parsons 1939, pp. 175–176, fns.). The Morning Star and the constellations of Orion and the Pleiades are clearly depicted on one of the ritual vessels which held corn meal for morning offerings to the sun at Sapawe, the Tewa ruin mentioned above (excavated under the direction of F. H. Ellis for the University of New Mexico, report in preparation).

We can summarize by saying that a religious officer associated with Mother Earth customarily functioned as sun and moon-watcher and possibly also as star-watcher during the long growing or summer period, which included the summer solstice, and that an officer associated with Father Sun and winter handled this watching from that time through the very important winter solstice, at least while such an officer was available. Working with them were representatives of the Sky or Star God and/or the older of the Twin War gods.

SYMBOLS AND CEREMONIES PERTAINING TO THE MOST IMPORTANT CELESTIAL EVENTS

Let us now come back to the problem of the group of associated petroglyphs found in the Chaco, our approach to be through known ethnographic data. The simplest form of sun symbol throughout the Pueblo area consists of two concentric circles with a dot in the center, the outer symbol symbolizing his rays, the second circle symbolizing his body, and the dot being his umbilicus from which all good things are supposed to come for mankind (Ellis, Field notes pertaining to Rio Grande and Hopi). Rays sometimes were added all around the outside circle or in projecting groups, usually four in number. A very fine example of a large sun symbol with rays and lesser representations of several in smaller scale, with and without rays, are to be seen on boulders just below the village of Shabik'eschee toward the eastern end of Chaco Canyon seven miles up from Pueblo Bonito where they unquestionably mark a sun-watcher's shrine. On the date of the fall equinox, 21 Sept. 1973, the sun rose over a very distinctive projection on the eastern horizon. Its first rays settled on the big Shabik'eschee sun symbol (C. M. McLeod, observer). On a direct line between the rising sun and the stone, a 94° azimuth, at a point ⅔ mile out from the stone on a small point below the cliff top, is a collapsed cairn. This, one may infer, was a shrine where prayer plumes and perhaps other offerings would have been placed, but the "home of the sun" in the pueblo would have been at the big stone with the sun design where the rays of the rising sun first struck, if we may judge by the beliefs of today. The moon, important in dating and also because it supposedly exerts an influence upon sun, usually was represented by a crescent because time periods, each known as "the moon," began with the first appearance of the crescent moon, but half and full moons also were depicted. A crescent lying on its back was considered to be a bad omen because it was dry, but one with tips pointing downward or in vertical position was good because it was a wet moon. Parsons wondered if these interpretations could have been borrowed from white neighbors, but the concept extends from the eastern to the western pueblos and appears deeply ingrained. Sun and moon could be used together in dating and in foretelling the future by omens. Whether the 584-day Venus calendar which the Mixtec-Cholula people of Mexico thought important was of concern to the Pueblo priests we do not know, but Venus as the Morning Star certainly was very important. Although the most common Pueblo representation of a star was and still is an equal armed cross, the Great or Morning Star could be shown with five points or surrounded with spikes.

The hand design which accompanied the sun, moon and Great Star symbols in the Chaco group is rather frequently found in groups of pictographs over the Southwest. This is a common Pueblo signature mark but with religious rather than secular connotations. Girls who have completed the replastering of a kiva finish by drawing cloud and lightning designs on a beam and add handprint, whether in Hopi or in the Rio Grande. Stephen (Parsons 1939, pp. 198, 202) was told that this symbolized the girls' prayer for rain. I have been told that at least in some pueblos the leader of a religious society which had completed a ceremony dipped his hand into wet clay and signed off by pressing it onto a wall to show that he had completed his duty. A footprint also could be used similarly by the Pueblos. Shrines in the open could be marked with a handprint so that the supernatural beings would know who was responsible for the prayers and offerings made there and hence could make accurate reciprocation (Ellis and Hammack 1968). Further, we have the intriguing note (Vaillant 1944, p. 197) mentioning the "imprinting of hand impressions" by captives who were to be sacrificed at the Aztec winter solstice ceremony (a southern use of the hand at this date possibly related to "the sacred hand" at Zuni) and the note that at a feast shortly preceding this and referred to as "the arrival of the Gods" (something of a prototype in concept to the Zuni Shalako, known by the same term) the announcement that one or more supernaturals had arrived depended upon finding the print of a foot in a small cake of cornmeal set out for this purpose (Anderson and Dibble 1951, p. 21), again something of a signature mark.

We have spoken so far of crescent moons and solar horizon observations, but that which Cushing describes for opening spring agricultural work at Zuni was of a somewhat different sort. In the 1500's the Zuni ancestors were living in several pueblos in the Zuni Valley, all of which were abandoned in the Pueblo Revolt of 1680. Survivors constructed the Zuni of today across the river from the site of one of those old towns. Matsakya (Matsaki: Mat'saka) when first seen by the Spaniards in 1581 was a village of a hundred houses four and five stories high, arranged in compounds, but when Cushing was at Zuni in the 1880's Matsakya had been only a ruin on a hill at the northwestern base of To'wayal'lanne, Thunder Mountain, for two centuries. Yet it was not forgotten. When Cushing (1967) was living in Zuni in 1893, each morning at dawn in late February or early March he saw the Zuni Pekwin, their Sun Chief, walk three miles up the valley to Matsakya followed by the head Bow Priest, representative of the older War God (Morning Star). While the Bow Priest waited at a distance, the Pekwin entered what Cushing calls a small tower and sat down before a pillar marked with the symbols found in our Chaco complex and waited, praying and chanting, for the sunrise. His concern was not simply the point at which the sun rose over the horizon but on which morning "the shadows of the solar monolith, the monument of Thunder Mountain, and the pillar of the gardens of Zuni" lay in a single line. When this occurred, the Pekwin thanked the sun, the Bow Priest cut a notch "in his pine

28.2 A stone plug for a small window of the type found in prehistoric Pueblo houses.

wood calendar," and both hurried back to announce that the time to begin farm work had arrived. To many this meant moving from the village out to field houses near the plots they would cultivate.

Cushing adds, "Nor may the Sun Priest err in his watch of Time's flight; for many are the houses in Zuni with scores on their walls or ancient plates imbedded therein, while opposite, a convenient window or small port-hole lets in the light of the rising sun, which shines but two mornings in the three hundred and sixty-five on the same place [see Fig. 28.2]. Wonderfully reliable and ingenious are these rude systems of orientation, by which the religion, the labors, and even the pastimes of the Zunis are regulated." We may recall that on 7 March and six months later, but on no other days of this year, a beam of sunlight entered the dark inner room of the 15th century great house at Casa Grande in southern Arizona by passing through a narrow tubular hole in an inside and an outside wall of the building and probably a matching hole in the surrounding compound wall (Pinkley and Pinkley 1931).

For dating the summer solstice, which like the winter solstice is referred to as "the middle" (of time), the Zuni Pekwin, who is solely responsible for solstice dates and who may be impeached for what is believed to be an error in observation or a "bad heart" which will affect crops adversely, watches sunsets from the pillar at the sun shrine of Matsakya. The sun has been moving northward and passes the moon at a specific named spot. At about the end of May, when the sun descends behind a certain point on the mesa to the northwest, the Pekwin calls out for all Zunis to prepare prayer sticks for Sun, Moon, the dead, and the katcinas. General prayer stick planting is on June 22, after which the summer retreats for rain begin. The solstice is considered to fall on the last of the four days during which Sun sets at the same spot behind Great Mountain, northwest of Zuni. Winter observations are taken from a petrified stump (petrified wood which resembles bones being symbolic of the War Gods who slew the early monsters which preyed upon man) on the east side of the pueblo where sunrise points over Thunder Mountain are noted. When the sunrise reaches a certain point, he notifies the elder brother or head Bow Priest, his lieutenant, and the latter in turn notifies the first group of religious societies to hold a retreat. Two prayer sticks for Sun Father and two for Moon Mother, his sister, are placed in the sun shrine on Thunder Mountain and four for former Sun priests are deposited in a cultivated field. Four days later four more are placed in the sun shrine and after another four days four more are placed in the field. On the morning after the final planting the Pekwin announces that in ten days the sun will rise from "the Middle Place," his southern house, the solstice marker. During that period the Pekwin continues his sunrise visits to the stump for observations and prayers. In 1891 the actual ceremony began on 22 December, after general prayer stick planting on the 21st, and continued for eleven days (Stevenson 1904, pp. 108–141; Bunzel 1934, pp. 534–540).

Between the first and the fourth days, prayer stick offerings are made by members of the religious societies for the sun, the moon, the War Gods, the lightning makers, the ancestral dead (rain makers: katcinas) of the six cardinal points, and the Beast Gods who give aid to curing. We know that the Priestess of Fecundity (today no longer existent), their one female officer and representative of Mother Earth, a parallel to the Town Chief or cacique in the other pueblos, prepared "a number of cotton cord loops" which made up or represented a small sacred white blanket as a gift for the sun. All prayer sticks except those made as personal offerings of the makers or their families are tied together, later to be deposited at the base of the knoll on which the sun shrine observatory stands in Matsakya. Most of the religious societies go into retreat on the second day after the Pekwin's announcement. For eight days there are ceremonies, including those intended to cure the sick, and visiting between the fraternities is prevalent. New War God images are carved from cottonwood, "all varieties of seeds" being placed in a hole representing the umbilicus in the body of the older god and covered by a serrated horizontal projection which is inserted and then decorated by attaching a miniature bow and arrow, shield, and war club, and plumes to symbolize clouds and lightning. An abalone shell pendant is hung around his neck, a belt of raw cotton is added at the waistline, and a netted shield is made for each image to stand upon. The figures are surrounded with prayer plumes, and gaming items for the older god are prepared. After the images have been honored by a kiva display and rituals they are placed in their respective shrines on Thunder Mountain. Among the many activities of this period is the plucking of one of the live captive eagles to obtain the needed soft eagle breast feathers, the selection of a man to make the new fire, the ceremonies held by members of every kiva, and further depositing of prayer plumes in other shrines and in holes dug in family fields. The Shalako, which celebrates the annual visit of the Council of the gods to Zuni and of the six great bird images who will carry the prayers of the people to the cardinal directions, is the most famous in the sequence of the many specific ceremonies (Stevenson 1904, pp. 108–141; Bunzel 1934, pp. 534–540).

Mrs. Stevenson (1904, fig. 3) has provided us with a photograph of the little "tower" at Matsakya sun shrine, which she describes as made up of a "stone wall, semicircular in form, about 3 feet high, the inner space being 3 feet wide and opening to the east. A sandstone slab about 2 feet high and 14 inches wide, with a symbol of the sun 4 inches in diameter etched (pecked) upon it stands against the apex of the wall. A smooth-surfaced stone, on which are cut a number of lines, is inserted in each side of the wall about 8 inches above the base. Some of the priests declare that the lines on the south side of the wall indicate the number of years the previous Sun Priest held the office and the one on the north side the number of years ... (his successor) served. Nine concretions form a square on the ground before the etching of the sun, and there are three smaller ones in line in front of these. Concretion fetishes, valued as bringing fructification to the earth, are found in all the fields. A small flat stone rests on two of the larger concretions." That flat stone may have been the seat on which Cushing saw the Sun Priest place himself. The slab evidently was erected in front of the open side where observations were made. In the photograph one can make out the sun symbol on the face of the slab, four groups of projecting lines coming out from the circle to represent rays as in the Zia sun symbol copied by the state of New Mexico on its flag. Two eyes and a mouth provide features for the face. The other glyphs described by Cushing unfortunately are not visible in this dim photograph.

Very little has been written concerning calendric observations on the Rio Grande, but the word picture we have obtained from one Tanoan pueblo, which must remain nameless, adds appreciably to our understanding of the Pueblo concept of Sun, coordinated celestial observations, and interpretation of shadow in relation to another "solar monolith." In the mythological system here, Sun's southern house is said to be Eagle Pueblo, the eagle being symbolic of Sun throughout the Southwest. There the Sun clan developed and there the arrangements for Sun's movements are made. In his manlike form, Sun is personified as Patiabu, Sun Man who was created in the underworld, the womb of Mother Earth, whence came mankind, plants, animals, and even the supernatural beings. Sun's body was covered with bright paint in all colors, and when he came up into our world his first gesture was to rub off some of this paint onto the plants so that man would have flowers. Flowers thus came to be symbols of Sun (probably a concept borrowed from Mexico). Before any man was permitted to emerge from the underworld, each was taken to an altar to be painted red, black, yellow, or some other flower color, like Sun Man. This was their initiation into the Sun cult and the hard and exacting life of ceremonial activities and precepts of old Pueblo pattern. Today young men may protest at the requirements imposed upon ceremonialists, but the elders believe that as long as one does not abandon the responsibilities imposed on his ancestors by the example of Sun Man and first accepted by Badger Man, he can be assured of good fortune through life.

Within a pueblo, the shrine or home of Sun Man is where the rays first strike in the morning. This also is the home of Moon, here a male spirit. The winter solstice ceremony is thought of as centering in offerings, which are spoken of as the "results of the ceremonies" such as prayer plumes and the new clothing for Sun, Moon, and Evening Star, all to be placed at the sun shrine in concluding the ceremony.

The top man in the little body of sun-watchers in this pueblo is the leader of the Koshare society. He, as Sun Chief, cares for the sun fetish, and wields more authority than any other man of the village. As representative of Sun, his presence in the local Catholic church is requisite to performance of the mass. It is he who gives orders for all the religious societies to perform the necessary ceremonies, including recognition of Sun, for the incoming secular officers just after their installation in January. The other sun-watchers are the Town chief or cacique, the War Priest (still an office of power here), the head War Captain who represents the older War God, and the leader of the important priestly group which represents Mother Earth and directs one of the two men's societies into which all youths are taken. It is said that at one time in the past this group of observers included a Star-watcher, presumably a representative of Big Star–Sky god as at Hopi.

When the time of the June solstice is known to be approaching, the sun-watcher scatters a corn meal offering and walks a short distance north of the pueblo to a stone slab on which he stands while noting the point of sunrise. Sun's "summer house" is reached when that body rises over a narrow peak known as "the heart," across a little arroyo from the watcher. When the cacique receives the watcher's report, he orders the sequence of religious society retreats due at this season. But there is yet another observation to be made. Late in the afternoon on 24 June, celebrated as San Juan's Day by all the Eastern Pueblos since Spanish times and close enough to 21 June to substitute for the summer solstice in today's calendar, the head War Captain and the War Priest go to a solar monolith at the western edge of the pueblo. This is a white outcrop some two feet high and six inches thick, with an equal-armed cross representing a star cut into its east face. The shadows of the mountain range to the west are watched as they fall across this stone. If the shadow completely covers the stone, the omen is interpreted as foretelling a long summer; if the shadow is short the summer also will be short and early frosts will ruin some late crops. The watchers at once hurry to announce the result of their observation to the religious society which is in retreat on this day and its members rejoice or lament according to the shadow's length.

The watchers for the winter solstice are the War Priest and the head War Captain who, each evening, note the place of sunset and keep track of the appearance of the moon from the crest of the ridge on the east side of town. The moon is believed to travel between the north and the south just as the sun does but at opposite seasons so that their paths cross at one point. The matter of prime importance to these calendar priests, during the watch which precedes the date of the winter solstice, is to observe Sun and full Moon exactly when they most closely approach each other, a problem duplicating that of the Zuni Pekwin in trying to properly place the Shalako celebration of the same period. The meeting of Sun and Moon is said to be guarded by the Cacique's society. Should this conjunction not be seen, Moon, which competes with Sun, might steal a little of his time and the year thus be shortened by frosts before harvest. Another problem in some years is Sun's tendency to dally for three or four days in his "summer house," rising at the same point on the horizon repeatedly before starting southward. This also shortens the agricultural year and permits a freeze before Sun and Moon meet. The summer ceremonial cycle, which consists of two series of four day retreats for each religious society, now begins on the calendar date of 20 June and is intended to end about 15 August. If frost occurs the retreats are dropped, for their function is to provide constant aid for growth and maturation of crops during the agricultural season.

The winter solstice ceremony begins about 12 December, the exact date being announced by the Town Chief after the observation of Sun and Moon's meeting. He opens the ritual with the request that the ceremonialists provide the Evening Star deity, the Morning Star deity, the Pleiades, the seven stars of Orion, the Sun and the Moon with clothing, food, and "medicine" so that they may have power to sustain themselves and bring about a long and fruitful year. All of the societies hold a four day retreat in their separate houses, which is followed by a night of dancing in each other's houses, much as in the Hopi Powamu and after the Zuni Shalako. One of the features of this celebration always is the making of new fire with a fire drill and a block of dry wood, another being held in reserve and a loop of cord at hand to make the twirling of the stick more effective in case a spark is not obtained at first. This new fire is guarded in the house of one of the women's societies for four days. At the end of the solstice ceremony, the deities are presented with miniature clothes. During the retreat, each society leader has set up a loom and begun weaving a small feather blanket (2 ft. × 15 in.), the feathers being laid across tightly twisted warp cords of home-grown cotton and held in place by a loosely twisted weft cord. The top strip, with eagle feathers, later will be clipped off to serve as a wrap for Sun. The second strip, with pale yellow feathers from the small "wild canary," will be clipped off for Moon. The third strip, with blue-jay feathers, is explained as being for Fire and the stars. We already have mentioned the small white blanket prepared for Sun at the winter solstice in Zuni. Clothing also is offered to Sun in Acoma, Isleta, and probably through all the pueblos.

SUMMARY

In closing our hurried resume of what at present is known of the use of celestial observations in the Pueblo Southwest, we find the immediate concern to have been primarily the calendar and the foretelling of omens pertaining to the important economic matter of crops, both within the all-encompassing mantle of native religion. Basic concepts and even much of the detail originated in Mexican and Maya originals. The handling, like Pueblo culture itself, was simpler than that in Mexico, though it may have been appreciably more elaborate in late pre-Hispanic times than after its integrity and integration were appreciably broken by a long combination of Spanish and then of Anglo force and persuasive influence. Sun, the Morning Star, Moon, and a Sky God sometimes confused with Sun and sometimes with the Great or Morning Star were the chief recipients of attention in their own specific movements and in the inter-relationships of those movements and, understandably, in receiving compensatory rituals and offerings in return for their cooperation and services to man. The officers of the native religious system, who were responsible for keeping track of the movements of these heavenly bodies and in calculating the native calendar, were the local representatives of those bodies, especially in winter, and of Earth Mother, especially in summer.

The year was cut into two equal divisions by the observed solstices for which the associated ceremonial activities were intended to keep Sun functioning properly, for he and Mother Earth were responsible for fecundity. That equal dual division was less used, however, than one into two unequal divisions comprising thirteen lunar periods which started with that of the winter solstice and was intended to separate the agricultural period from the non-agricultural period in which concentration was upon warfare, hunting, and curing. A more elaborate calendar system of the past is hinted in Stephen's early notes, but we have no evidence of interest in such matters as lunar and solar eclipses other than a brief note that such events were feared as possibly causing the death of children and even of unborn infants (Parsons 1939, pp. 86, 181 fn.).

Like the Aztecs, the Pueblos distinguished divisions of the night by the position of the moon and of Orion and the Pleiades, and to a lesser extent, of the Great Dipper. Their rising or their passage over the uncovered hatchway, which served as a ladder opening and smoke hole directly above the kiva fireplace, commonly was watched for timing rituals. The position of Morning Star, a patron of hunters and warriors, the recipient of prayers for propagation of domestic animals today, and most revered of all heavenly bodies other than Sun and Moon, also was considered in timing. Beyond this we really know little about man's relationship to the stars. Stephen (Parsons 1936, pp. 857–863) reports that Aldebaran, the Broad Star, received prayers in a Hopi curing ritual. The Galaxy or Milky Way is revered by all the Pueblos. On Zuni and Acoma masks and altars it is represented by a white band or a band of black and white squares or a ladder, for like the rainbow it was believed to provide a bridge from earth into the heavens, but the occasional addition of eyes and mouth (Parsons 1939, pp. 182, 340 fn.) indicates that it was personified like the stars as a whole which are spoken of as Night People, "our fathers and mothers," and "little priests." There are statements about stars of cardinal directions, but Parsons found no recognition of the North Star, perhaps because it was too dim to have served as a northern marker in the time of their prehistoric ancestors. The orientation of prehistoric pueblos and of kivas is a subject which may provide some insight into Pueblo beliefs. If we examine the orientation of existent Pueblo towns (Stubbs 1950) we find that all but Sandia, Zia, Zuni, and all the Hopi villages other than Mishongnovi are oriented with marked accuracy to the cardinal points. This undoubtedly has to do with intent of the leaders who originally laid out the sites and properly buried a shrine offering in the plaza and established a directional shrine on each periphery. Rio Grande kivas characteristically were oriented to the east in historic and prehistoric sites but in the Mesa Verde, Chaco, and Kayenta areas, the preferred orientation was to or toward the south. Why the difference?

Could the much discussed "towers" of the prehistoric Southwest have been observatories from which designated calendar priests kept track of the movements of heavenly bodies? This was my proposal at the meeting of the Society for American Archaeology in Mexico City in 1970, and I still believe that their structure, which raised them somewhat, if not above the roofs and trees within a village area, together with their common association with kivas in the Mesa Verde and Chaco areas, makes this a very possible explanation, though they probably also were used for smoke and fire signals. The fact that structures dedicated to Quetzalcoatl especially in his guise as Ehecatl, the Wind God, in Mexico were circular and that this deity eventually arose to become the Morning Star would fit into such thinking, and the little "tower" at Matsakya still used for at least solar observations by Zuni suggests a residue from something more important in the past. This certainly does not, however, proffer a hypothesis that all observation of heavenly bodies formerly was done from towers, the distribution of which was relatively limited, and which may have been used primarily for watching stars even within the limits of that geographic distribution. The presence of towers certainly points to a greater sophistication of culture in some prehistoric Southwestern districts than in others, but with our present knowledge we hardly can say more. We reiterate our statement that the group of petroglyphs near Penasco Blanco in the Chaco quite certainly was a sun-watcher's station, pre-dating that at Matsakya by 300 years. About 15 miles northeast of Zuni is the site known as the Village of the Great Kivas, contemporary with and very closely related to the big Chaco pueblos, and on the cliff face just above the talus which rises behind that site are petroglyphs. One pair consists of an equal-armed cross or star and a horizontal crescent moon with tips turned down. Near it is the simplified drawing of an owl and a long zigzag line which Roberts (1932, p. 151, plate 62b) was told by Zuni workmen referred to a tale told to children by the War Chief, the owl coming at night to lead the chief to the camp of their enemy, the Navajo. Roberts carefully explains that he cannot say whether the explanations he received were or were not those of the prehistoric people who made the designs. Nor can we, but we are dubious of the moon and star symbols having pertained to the story given. That association of crescent moon and star, we would say, is more likely to have marked a sun and moon–watching station. We have another example in the Chama drainage of the Upper Rio Grande where, in the center of an open cave, the hand, the sun symbol of concentric circles, a Great Star and a lesser star (Morning and Evening Stars?) have been painted onto the natural rock wall. In 1955 Miller suggested that two northern Arizona pictographs showing a vertical crescent moon and a circle on a rock surface, one at White Mesa and the other in a tributary of Navajo Canyon, depicted some special celestial event, possibly the supernova of AD 1054. I would be inclined to interpret these as probably markers for sun-watcher's stations, the sun in this case being a circular plaque. These complexes lack the hand and the star, and the Chama drainage example lacks the crescent moon, but from our knowledge of the white stone outcrop merely marked with the cross-star symbol but used as one of the sites for calendric divination and dating by members of a Rio Grande pueblo today, we can be sure that the complex of four features need not be found together on a station. Miller has illustrated the possibility of his hypothesis by records of the crescent moon having been in the sky on 4 and 5 July of that year, but we would quote the Pueblo people again in saying that although the supernova may have been seen and discussed, there is no reason to think it would have been recorded. The long and very serious concern of Pueblo people with the solstices and the lunar periods, however, is amply documented.

REFERENCES

Bourke, J. 1884. *The Snake-Dance of the Moqui of Arizona*. New York, Charles Scribner's Sons.

Bunzel, R. 1932. Introduction to Zuni Ceremonialism. *Bureau of American Ethnology Annual Report* 47:467–544.

Cushing, F. H. 1967. My Adventures in Zuni. Reprint (Filter Press, Palmer Lake, CO) from original article in *The Century Magazine* 25 and 26, 1882–1883.

Ellis, F. H., and L. Hammack. 1968. The Inner Sanctum of Feather Cave, a Mogollon Sun and Earth Shrine Linking Mexico and the Southwest. *American Antiquity* 33 (1):25–44.

Fewkes, J. W. 1897. Tusayan Katcinas. *Bureau of American Ethnology Annual Report* 15. pp. 267–312. Washington, DC.

Nequatewa, E. 1931. The Place of Corn and Feathers in Hopi Ceremonies. *Museum Notes* 3:9. Reprint, in *Hopi Customs, Folklore, and Ceremonies, Museum of Northern Arizona Reprint Series*, no. 4, 1954.

Oort, J. H. 1957. The Crab Nebula. *Scientific American* 196 (3):53–60.

Parsons, E. C. 1936. *Hopi Journal of Alexander A. Stephen*. New York, Columbia University Press.

Parsons, E. C. 1939. *Pueblo Indian Religion*. 2 vols. Chicago, University of Chicago Press.

Stevenson, M. C. 1904. The Zuni Indians. *Bureau of American Ethnology, Annual Report* 23, pp. 1–634, Washington, DC.

Stubbs, S. A. 1950. *Bird's Eye View of the Pueblos*. Norman, University of Oklahoma Press.

Titiev, M. 1944. Old Oraibi: A Study of the Hopi Indians of Third Mesa. *Papers of the Peabody Museum of American Archaeology and Ethnology, Harvard University* 22:1. Cambridge.

Titiev, M. 1972. *The Hopi Indians of Old Oraibi: Change and Continuity*. Ann Arbor, University of Michigan Press.

Vaillant, G. C. 1944. *Aztecs of Mexico*. Garden City, NY, Doubleday.

CHAPTER TWENTY-NINE

Astronomical Markings at Three Sites on Fajada Butte

Anna P. Sofaer and Rolf M. Sinclair

A seven-year study of prehistoric pueblo sites of the Chaco culture has revealed that these people possessed a sophisticated astronomy. Evidence of this prehistoric astronomy consists of multiple light markings on petroglyphs and of several alignments of major structures. This paper presents the results of a recent study of 13 markings at three sites on Fajada Butte in Chaco Canyon, New Mexico; these results are presented in the context of earlier research in this area. Each marking is a distinctive pattern of shadow and light that appears on a petroglyph at a key point in the solar or lunar cycle, i.e., at an extreme or midposition of these cycles, including the meridian passage of the sun at solar noon. [Note: times quoted throughout this chapter are in apparent solar time, in which noon occurs each day when the sun is due south on the meridian. The basic astronomical concepts used here are explained in Aveni (1980) and Krupp

---

Reproduced by permission from *Astronomy and Ceremony in the Prehistoric Southwest*, ed. J. B. Carlson and W. J. Judge. Papers of the Maxwell Museum of Anthropology, No. 2, 1987.

(1977).] Many of the markings simultaneously record two key points in different cycles, such as noon and solstice or noon and equinox. A minimum of 17 key points are indicated by the markings at these three sites.

The markings define the outer boundaries and midpoints of the recurring cycles of the two most basic celestial bodies: the sun and the moon. The accuracy and redundancy of the markings of these symmetric points, as well as the strong visual effect of the markings themselves, indicate their intentional quality. Several of the markings record meridian passage of the sun within a few minutes, and one marks equinox to within a day. An accuracy equivalent to that of these markings is found in the cardinal alignments of major constructions in the area.

Fajada Butte stands 135 m high at the south entrance of Chaco Canyon. The remains of various structures, including a small kiva, and the presence of potsherds and rock art on and near the butte summit show that it was frequented by prehistoric Pueblo people and by Navajo people. The content and style of the glyphs at the three sites reported here indicate that these markings are of prehistoric Pueblo origin and were probably made between AD 900 and 1300.

From about AD 900 to 1150 Chaco Canyon was the center of a complex prehistoric Pueblo society that thrived in the arid environment of northwestern New Mexico. The Chacoan people planned and constructed large multistory structures, ceremonial centers, and extensive roads throughout the 70,000 sq km San Juan Basin. These achievements indicate sophisticated engineering and surveying skills. Ethnographic reports concerning the historical Pueblos (some of whom are descendents of the Chacoan people) reveal their keen interest and skills in observing the sun and moon for ritual and agrarian purposes (see references cited in Sofaer et al. 1979).

The Fajada markings, although similar to others recently reported in the Southwest, are unique among known archaeoastronomical sites in several respects: (a) they use the changing altitude of late morning and midday sun to indicate the solstices and equinoxes, (b) they record solar noon on rock art, and (c) most especially, they combine recording of both the season and noon in the same markings. The markings at one site are the only known ones in the New World that combine recordings of the extremes and midpositions of the moon and the sun. The use of the features and topography of a prominent land mass to create these complex and varied markings on carved petroglyphs is also unique in our current knowledge of prehistoric astronomy.

Following a brief summary of earlier findings at one of the three sites near the top of Fajada Butte (see also Sofaer et al. 1979, 1982a), this chapter describes in detail the recently discovered markings that combine noon and seasonal recordings at two additional sites on the east and west sides of the butte (Sofaer et al. 1982b). New observations and analyses regarding the earlier findings are then presented. The sites are referred to in this paper as the three-slab site, the east site, and the west site. The markings of these sites were first observed and studied by Sofaer between 1978 and 1983. The markings are summarized in Table 29.1. Ethnographic and archaeological correspondences with the markings are also discussed.

Table 29.1 Markings of astronomical cycles on Fajada Butte.

Petroglyph	Solar					Lunar	
	Midday				Sunrise	Moonrise	
	Noon	Summer Solstice	Equinox	Winter Solstice	Equinox	Northern Major Standstill	Northern Minor Standstill
Large spiral (three-slab site)	—	X	—	X	X	X	X
Small spiral (three-slab site)	—	*	X	—	—	—	—
East spiral	X	X	X	X	—	—	—
East snake	X	—	X	—	—	—	—
East rectangle	X	X	—	—	—	—	—
West double spiral	X	—	X	—	—	—	—
West rectangle	X	—	—	—	—	—	—

* The *absence* of light at this location can be taken as a marking (see text).

SUMMARY OF EARLIER FINDINGS

At the three-slab site, six markings record solar and lunar positions (Figs. 29.1, 29.2, 29.3). An unusual configuration of three large stone slabs, each about 2 m high, collimates sunlight each day in the late morning and near midday onto two spiral petroglyphs pecked on a cliff face. The streaks of light so formed change noticeably with small changes in the sun's declination. The vertical light/shadow patterns on the petroglyphs thus go through an annual cycle in which the solstices and equinoxes are marked by the intersection of these patterns with the primary features of the spiral forms (Sofaer et al. 1979). This site also marks the northern minor and major extremes of the 18.6-year lunar standstill cycle by a separate pattern of light and shadow at moonrise (Sofaer et al. 1982a). This light and shadow pattern marks events that are unrelated to those of the solar cycle, yet it is also formed on the primary features of the larger spiral by another edge of one of the three slabs that form the midday patterns.

Since the moon can appear throughout the declination range of the sun, all the solar markings listed in Table 29.1 can also be formed at certain times by the moon. We do not consider most of these to be possible lunar markings owing to a lack of further evidence. The northern minor standstill marking can be formed at certain sunrises away from the solstices and equinoxes; we term it lunar because it occurs as a pair with the northern major standstill marking (which cannot be solar since the moon then exceeds the maximum solar declination). The equinox sunrise marking (Fig. 29.1) also indicates the middle of the moon's monthly or 18.6-year declination cycles, in a manner analogous to the markings of the standstill extremes, and is thus possibly also a lunar marking.

While the solar markings of the three-slab site use the seasonal altitude changes of the late-morning and midday sun to indicate the specific time of the solstices and equinoxes, they do not accurately mark noon. New evidence reveals that two other

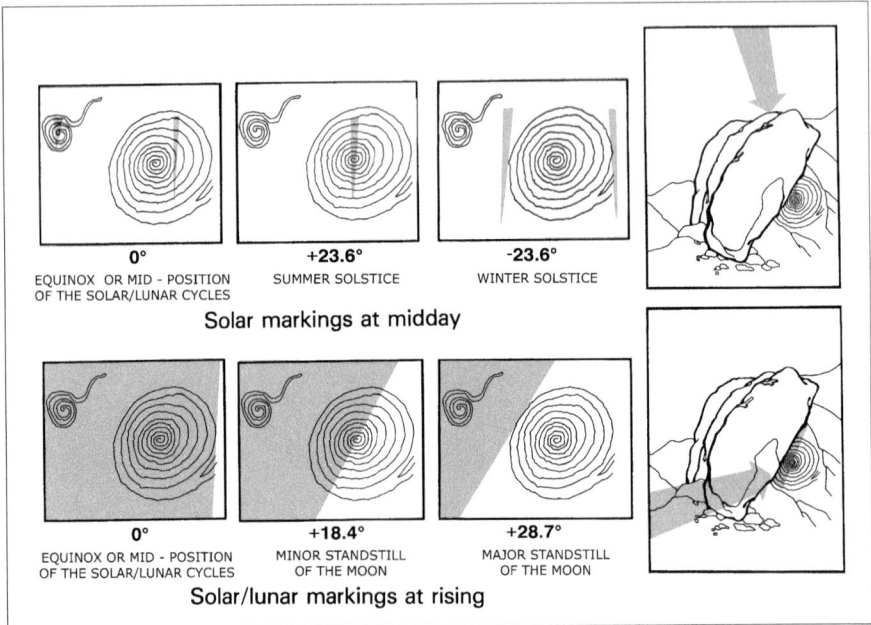

29.1 The three-slab site. Right: Formation of solar and lunar shadow/light patterns by the three slabs near meridian passage (upper) and at rising (lower). Left: Schematic of the resulting patterns on the spirals at the indicated declinations and seasons (see Sofaer et al. 1979, 1982a for detailed photographs). (Note: an error in an earlier publication [Sofaer et al. 1982a:Fig. 4] of "rising 0°" is corrected here.) Illustration by Pat Kenny, © The Solstice Project.

sites on the butte mark the specific time of noon at the seasonal points. Several features of the three-slab site are shared by the two other sites: spirals, dagger-shaped light patterns, and the repeated use of the same glyph or pair of glyphs at quarter points of the seasons.

RECENT FINDINGS OF NOON/SEASONAL MARKINGS

At two sites located a short distance below the three-slab site (Fig. 29.4), five petroglyphs are crossed by visually compelling patterns of shadow and light at a time close to solar noon. The imagery of these patterns distinguishes the solstices and equinoxes in most instances. These shadow patterns form seven markings indicating 11 key points in the daily and seasonal cycles of the sun—i.e., midpositions and extremes (Table 29.1). The markings occur in combinations, most of them in striking conjunctions: pairs that are visible at the same site and that vary with the seasons.

The shadows that form the noon markings are cast by rock edges of the butte, which is locally very irregular. Since the sun's elevation at meridian passage changes

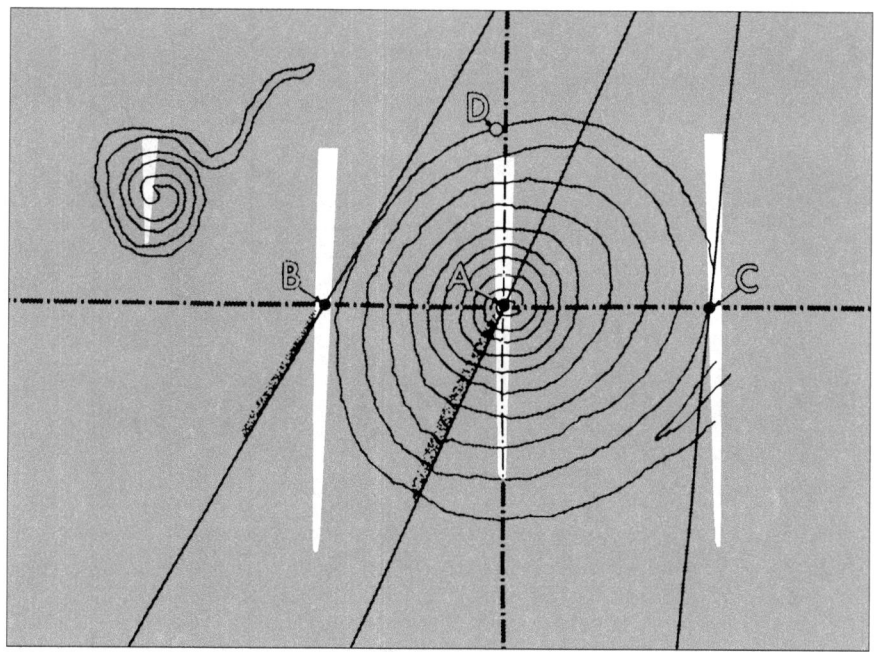

29.2 The three-slab site. Superposition of the main elements of the six markings (see Figure 29.1) showing the multiple uses of certain key features of the large spiral (A), points of tangency at left and right (B and C), the top (D), and the horizontal and vertical axes. Also shown are the two pecked grooves (see text). Illustration by Pat Kenny, © The Solstice Project

annually from 31 degrees at winter solstice to 78 degrees at summer solstice, the particular edges that cast shadows on the glyphs through the year differ greatly in distance and bearing from them. Some edges are less than a meter away, while others are up to 30 m distant. During the sun's daily meridian passage, the glyphs at the east site change from fully lit in morning sun to fully shadowed in the afternoon, and those at the west site go through the reverse transition of shadow to light.

At the east site (Figs. 29.5 and 29.6), which is located about 25 m below the butte summit, three adjacent glyphs—a nearly vertical rattlesnake (22 cm long), a rectangular figure (14 cm wide), and a spiral (15 cm wide)—occur 2.0–2.5 m above the current cliff base. The rattlesnake and the rectangular figure are particularly deeply incised. The spiral is pecked within a rectangular area that appears to have been worked. A shadow edge crosses the spiral glyph within 10 minutes of noon throughout the year, forming a seasonally changing pattern. This pattern is momentarily symmetric about the center of the spiral within a few minutes of noon, forming a wedge at summer solstice, a quartering at equinox, and a bisecting at winter solstice.

The seasonal variation is reinforced by simultaneous markings that occur on the two nearby glyphs only within a few weeks of summer solstice and equinox, close to

29.3 Summer solstice light marking on the larger spiral of the three-slab site. The arrows indicate the two pecked grooves with which the lunar shadow patterns are aligned. Photograph by David Brill, © 1982, The National Geographic Society.

Anna P. Sofaer and Rolf M. Sinclair

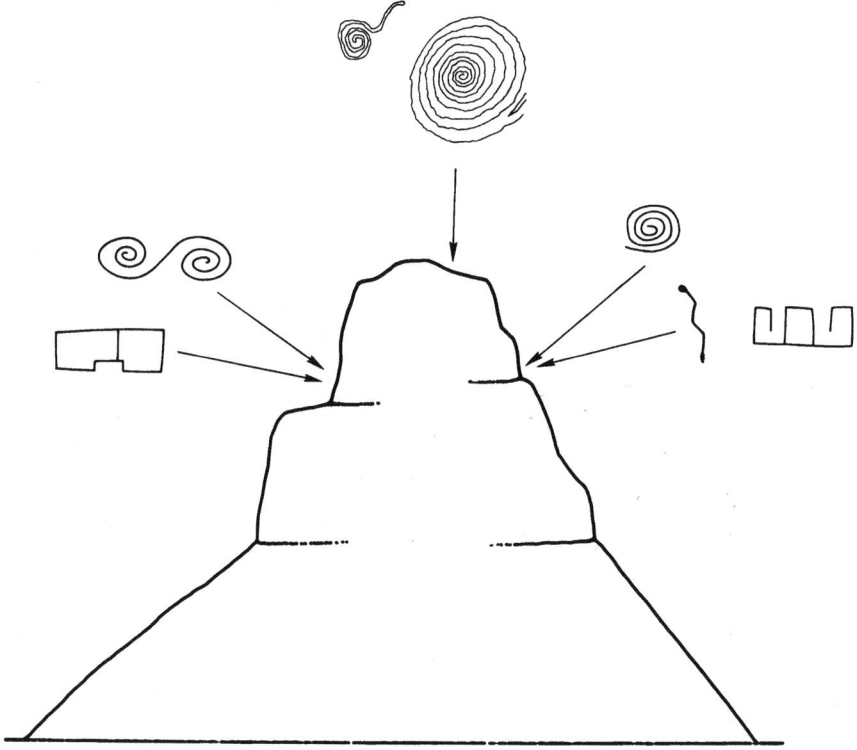

29.4 Schematic of Fajada Butte (looking north) showing the locations of the petroglyphs at the east, west, and three-slab sites (which are not intervisible). Illustration by Pat Kenny, © The Solstice Project

noon. At equinox, at the same time that the spiral is quartered, an unrelated shadow edge crosses the snake, touching all parts at once: the head, the body, and the rattles. This shadow pattern deviates noticeably from alignment with the full body of the snake two weeks earlier or later (Fig. 29.6). At summer solstice, at the time that the wedge marks the center of the spiral, the shadow edge is aligned with the right edge of the rectangular figure (Fig. 29.5). This combination is noticeably different two weeks before and after summer solstice. Thus, the seasonal points of the year are indicated both by the differing patterns on the individual glyphs and, at summer solstice and equinox, by the unique paired combinations of the markings.

At the west site (Figs. 29.7–29.9), which is located about 30 m below the summit, a double spiral (55 cm across) and a rectangular figure 17 cm wide) are about 6 m and 4 m, respectively, above the current cliff base. On days near equinox, close to noon, a narrow dagger-like pattern of light moves in an upward diagonal course through the double spiral in about 22 minutes. An unrelated vertical shadow edge moves across the rectangle in nearly the same interval. At equinox the light pattern moves through

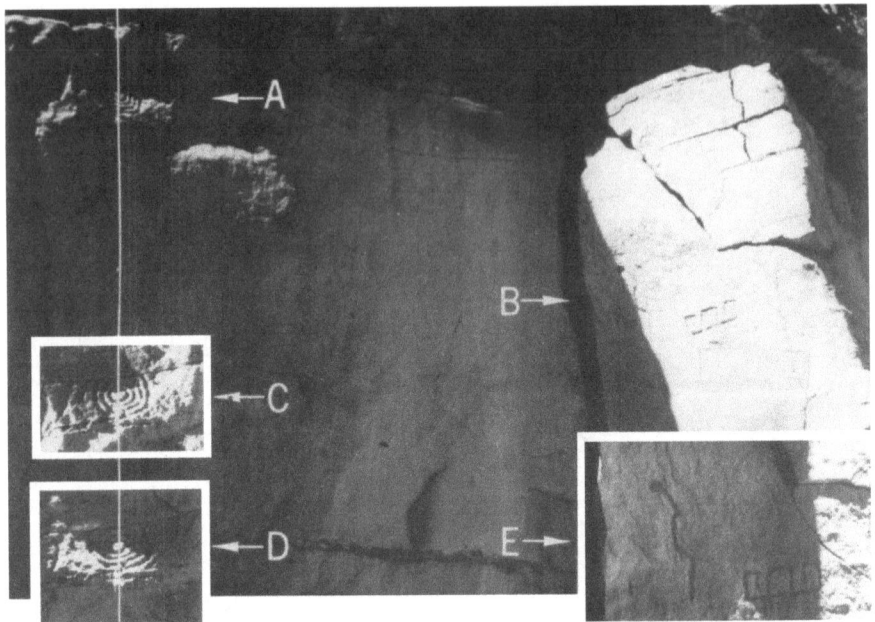

29.5 East site. Simultaneous markings at equinox on (A) spiral and (B) snake (September 20, 1979, 12:04 PM). Insets: (C) winter solstice marking on spiral (December 20, 1980, 12:00.5 PM), (D and E) summer solstice simultaneous markings on spiral and rectangular figure (June 22, 1978, 12:03 PM). The line through ACD points out the change in the shape of the shadows that centrally mark the spirals at each season around the time of noon. Photograph by A. Sofaer and K. Kernberger, © The Solstice Project

the center of the right whorl of the double spiral, with the whorl and the rectangle being bisected simultaneously about 9 minutes before noon. The moving light patterns reach the far right edges of both glyphs within a few minutes of noon.

Movement of light across the double spiral near equinox is a sensitive indicator of the sun's declination (which is then changing most rapidly). The track of the pattern shifts 4.5 cm each day, left to right in the spring and right to left in the autumn, so that the pattern of light movement for each day is quite distinct during this period. Figure 29.9 shows the light pattern on several dates near equinox. In the photographs of the double spiral near equinox, the light form is shown at the same height in its upward diagonal course for each day. The "event" of equinox can occur through the years at different times on the nominal "day" of equinox. Figures 29.7 and 29.8 show that a shift in successive years of only 12 hours in the time of equinox can be seen. By noting the exact position of the light pattern with respect to the center of the right whorl, one can pinpoint equinox to within at least a half day.

The shift of the track of this light pattern from left to right on the double spiral could be used to anticipate the approach of spring equinox, since even to a casual observer a one- or two-day shift from equinox would be evident. While no external

29.6 East site. Markings on the snake showing the alignment of the shadow edge with the petroglyph at equinox. (A) 16 days after fall equinox (October 9, 1982, 12:00 m), (B) 3 days before fall equinox (September 20, 1979, 2:03.75 PM). [Note: data taken a certain time after (or before) spring equinox are equivalent to data taken the same time before (or after) fall equinox.] Photographs by A. Sofaer, © The Solstice Project

evidence is available to determine whether this site was used to mark the time of equinox so accurately, it was clearly possible.

The markings on the rectangle are of special interest in that they indicate noon accurately throughout the year. Each day a vertical shadow edge crosses the rectangle within 5–20 minutes of noon and aligns momentarily with the right edge within 3 minutes of noon throughout the year.

Figure 29.10 illustrates the intervals around noon between first and last touchings by a shadow edge on each glyph at both the east and west sites. This figure also

29.7 West site, 1.5 hours before fall equinox (September 22, 1984, 11:50.5 am). The shadow edge moves across the rectangular glyph (lower right) from left to right, crossing the right edge within three minutes of noon throughout the year. (The outline of this glyph is artificially emphasized in this illustration.) Photograph by Colin Franklin, © The Solstice Project

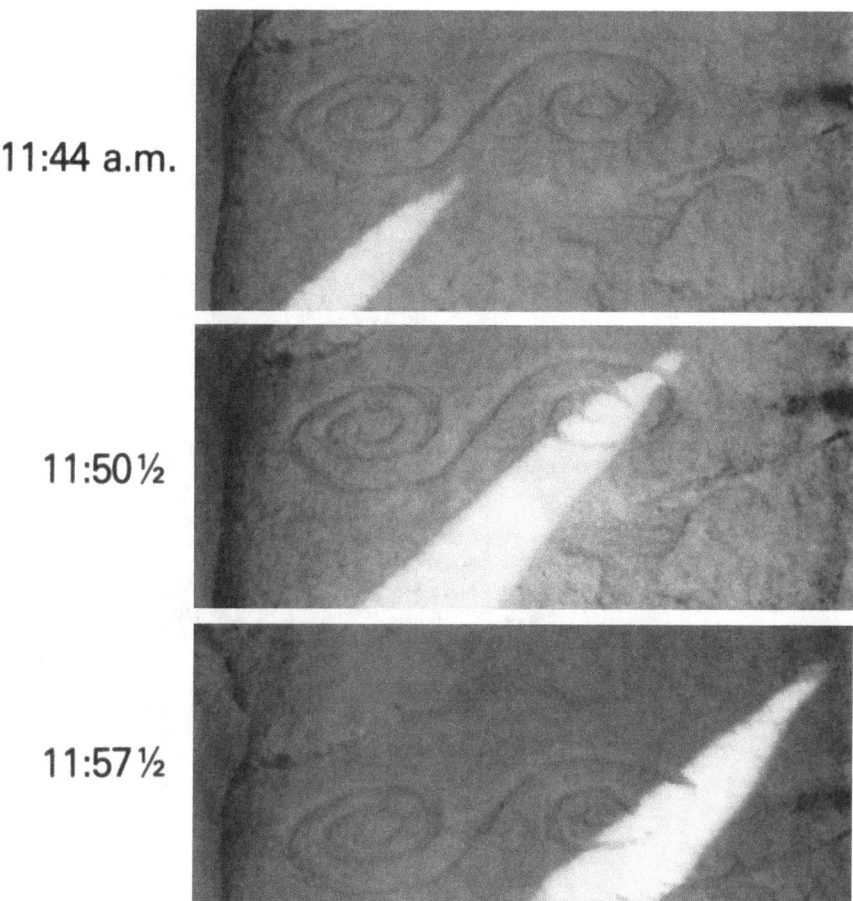

29.8 West site. Marking on double spiral 21 hours before fall equinox (September 22, 1983, top to bottom: 11:44 am, 11:50.5 am, 11:57.5 am). Photographs by Peggy Wier, © The Solstice Project

shows certain times within these intervals when several of the glyphs are marked by distinctive and centrally aligned patterns; these are illustrated in Figures 29.5–29.9 as well. Several of these patterns occur within a few minutes of noon.

Several factors underscore the intentional quality of the markings described here. First, in all instances the light patterns bisect or align with the edges of the glyphs at key points in the solar cycles. Second, six of the seven markings record two such events (noon and a solstice or equinox). Third, the visual effect is striking, most especially when two symmetric shadow alignments occur simultaneously on two nearby glyphs. Fourth, six of the seven markings occur in such simultaneous pairs and are formed by unrelated shadows cast by different parts of the butte. Fifth, three markings use the same glyph in different patterns at different seasons. And sixth, four of the markings

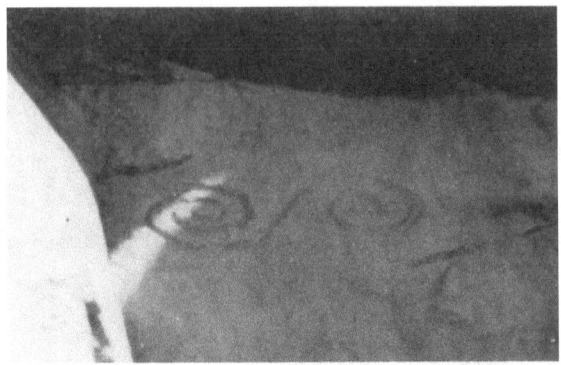

8 days before spring equinox

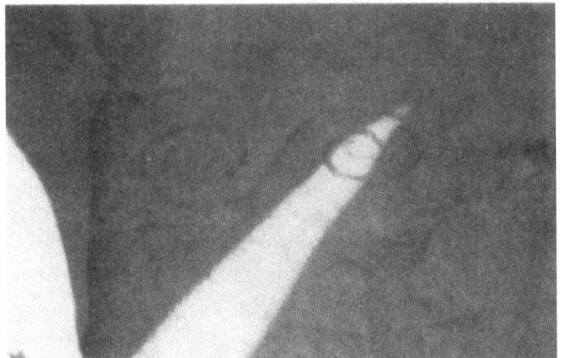

the day of equinox (9 hours after)

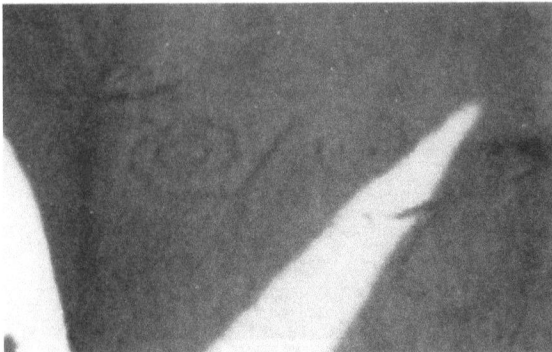

3 days after spring equinox

29.9 West site. Marking on double spiral near equinox. Top: September 30, 1984, 11:42.5 am (8 days after fall equinox). Middle: March 20, 1984, 11:48.5 am (9 hours after spring equinox). Bottom: March 23, 1984, 11:53.5 am (3 days after spring equinox). Photographs by Rolf Sinclair and Michael Marshall, © The Solstice Project

Anna P. Sofaer and Rolf M. Sinclair

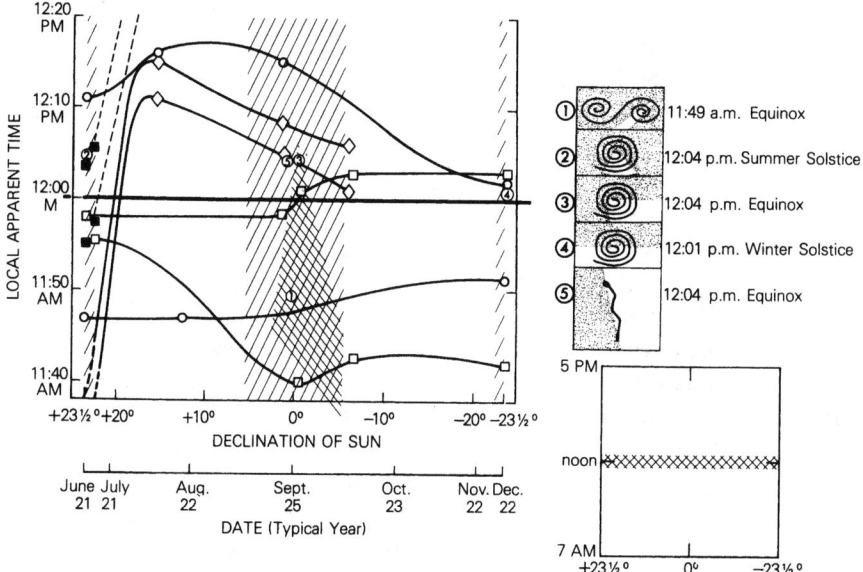

29.10 Left: Time that shadow first touches a glyph and then just covers it (bottom and top curves of each pair). East site: ■ rectangle, ◊ snake, ○ spiral. West site: □ rectangle, \\\\ range of marking on double spiral. Most points represent multiple observations on adjacent days. ///// are the intervals of one month, centered on solstices and equinox. Right: Moments near noon when shadow edges are centrally aligned on certain glyphs. These cases are located by numbers on the graph (left). (1) West site double spiral, March 20, 1984. East site spiral: (2) June 22, 1978; (3) September 20, 1979; (4) December 20, 1978. (5) Snake, September 20, 1979. Inset: Replotting of graph on left to show the fraction of daylight hours involved in the markings. Illustration by Pat Kenny, © The Solstice Project

bisect or involve the centers of spiral glyphs (in a manner similar to markings at the three-slab site).

The rock edges casting the pattern on the rectangular figure at the west site in winter are vertical and about 3 m away; in summer they are horizontal and about 30 m distant. It is remarkable that a point was found where the shadow patterns cast by such a wide range of the butte's irregular surfaces would remain vertical throughout the year and cross the glyph consistently at noon.

The glyphs are so placed that no other shadow crossings occur near noon except those that form the markings; in general the glyphs are either steadily illuminated or shadowed at other times. The markings occur at the universally recognized extremes or midpoints in the annual solar cycle, as do those at the three-slab site. There are no equivalent markings at these sites that occur at times other than the key points in the solar cycle.

The astronomical markings involve a significant fraction of the known rock art on Fajada Butte. There are about 20 clearly formed glyphs at various locations on the butte, primarily scattered over its uppermost one-third (Sofaer and Crotty 1977), and some 20 other amorphous peckings and scratches and historical graffiti. When the five glyphs that are involved in the east and west site noon/seasonal markings are added to the two glyphs at the three-slab site, the marked glyphs constitute about one-third of the Fajada Butte rock art. (The other two-thirds appear not to be astronomically marked.) At each site the marked glyphs are dominant in clarity, size, and number over the few other nearby ones. Four of the seven spiral glyphs on the butte are used in astronomical markings, as are the only two rectangular figures.

Although the markings are referred to here as occurring near noon, it is likely that they were designed to commemorate both a spatial and a temporal halving of the day by recording when the sun is due south. (Note that it is more accurate to determine meridian passage or noon by observing the sun's changing azimuth around a north-south line than by monitoring changes in altitude.)

Central to these accurate noon/seasonal markings are several important astronomical concepts and their implementation. First, the creators of the markings singled out for emphasis midday and the year's quarter points. Then they had to establish the north-south meridian to the necessary accuracy, note the sun's transit, and identify the specific times of the equinoxes and the solstices. With this information they examined the shadow transitions on the Fajada cliffs throughout the year for possible sites. The weathering of the sandstone edges that form the markings has made it difficult to determine whether the developers of the markings worked these edges or used them in their natural state.

We conclude that the three Fajada sites were made to exhibit the extremes and midpoints of basic astronomical cycles considered significant by the Chacoan people. The markings occur only in a narrow band of times of day and year that is centered on astronomically recognizable events. An argument could be that the petroglyphs were originally placed with no regard for illumination and that shadows randomly cross some of them at times (particularly when the sun and shadows are moving fastest). This argument, however, cannot easily explain the pairings of markings, the multiple use of certain carvings, the repetitions of the alignment of two simple shadow patterns with the basic geometry of the petroglyphs at just the times of significance, or the consistent use of the spiral motif.

FURTHER OBSERVATIONS AND ANALYSIS OF THE THREE-SLAB SITE

Lunar Markings

New evidence and further analyses of this site support earlier findings that the northern minor and major lunar standstills are marked at moon rising and refute a speculation that the moon's meridian passage is marked.

The large spiral glyph at the three-slab site contains a total of nine and one-half turns, making it unique among the numerous recorded rock art figures in Chaco Canyon. No other spirals recorded in the canyon have so many turns; most have more than half as many. This fact strengthens the suggestion made earlier (Sofaer et al. 1982a:176) that this number was chosen for the spiral to represent the number of years in the lunar standstill cycle. During the period of 9–10 years between the minor and major standstills, the position of the shadow cast by the most northerly rising moon each year shifts gradually across the nine and one-half turns of the spiral.

Further examination of the site has disclosed a pecked groove tangential to the large spiral and aligned with the major standstill shadow marking. This groove is similar to the pecked groove aligned with the minor standstill shadow marking (Fig. 29.3; Sofaer et al. 1982a:173); these alignments draw attention to the shadow patterns as markings (Figs. 29.1, 29.2).

It has been reported (Sofaer et al. 1982a:178) that a standstill could potentially be marked at four moonrises in the years of the extremes. Further analysis shows that the criteria can be met by as many as 13 risings in some of these years (Michael Zeilik, personal communication 1981). This increased frequency would make the marking of the standstills that much easier to conceive of and achieve.

It was speculated earlier that a possible lunar marking could occur at the moon's meridian passage at the declination of the major standstill, when the moon achieves an altitude of about 82 degrees (Sofaer et al. 1979:290). Recent measurements at the site show that an overhanging cliff edge about 10 m above the slabs would block the rays of the moon at this altitude so that no such marking could occur.

It is likely that the ancient Chacoans noticed the 18.6-year lunar cycle in the course of making horizon observations. During the 9–10 years between the standstills there is at Chaco a 13-degree shift in the extreme northerly and southerly positions of moon risings and settings; this shift is centered around the winter and summer solstice positions. The lunar extreme positions remain close to each standstill limit for a year or more. (The lunar standstill cycle would also be evident as the moon stood higher or lower in the sky in successive years and in the correspondingly changing nocturnal shadow and light patterns in the canyon and around and in the buildings.) In Chaco, the open, flat horizons and the clarity of the desert air of a thousand years ago would have made such observations far from difficult, especially for a people skilled in surveying. These rhythms of lunar cycles would become quite evident after some years of observation. The creators of the Fajada lunar markings did not necessarily know the standstill cycle to any better accuracy than approximately a year, which is what can be perceived in the movements of the shadows on the large spiral of the three-slab site.

These lunar markings do not necessarily signify that the Chaco culture was predicting lunar eclipses. The 18.6-year cycle is of limited use for this purpose. It is worth noting, in this regard, that the modern Pueblo cultures have shown far more interest in cycles than in unusual events (Ellis 1975:60–63). The extremes and midpositions of recurring cycles are themselves significant because they define the limits and order of the cosmos. Recognition of these cycles provides reason enough to commemorate them.

Interrelationship of the Markings and the Assembly of the Three-Slab Site

The three-slab site combines a number of markings in a unique manner. New analyses presented here provide insights into its design and development.

Because of the extensive seasonal altitude changes of the midday and morning sun and the angles of the slabs relative to the cliff face, many different portions of the slabs create different parts of the complex set of markings. Nine separate surfaces are involved in casting the shadows that form six markings of five different declinations. (Note a possible seventh marking: Light is almost entirely blocked by the left and middle slabs at summer solstice [Sofaer et al. 1979:286]; only 10 days later a streak of light is much more evident and then grows through the months to become the bisecting equinox pattern. The effect at summer solstice may have been intended to focus attention solely on the bisecting pattern on the large spiral [Table 29.1].)

Each of the five solar and lunar markings on the large spiral interlocks with the positions of the other markings, in that each of these markings falls on the points and lines of symmetry that define the large spiral (Fig. 29.2). There are five points that define the spiral: center, left edge, right edge, top, and bottom. The first four of these points are clearly part of the astronomical patterns. At the extremes and midpositions of the solar and lunar cycles a pattern of shadow and light strikes one and sometimes two of them. Conversely, most of these four points are involved in two markings. To a lesser extent the fifth point, the bottom edge of the spiral, is also defined by the markings.

At summer solstice the bisecting dagger-shaped pattern is centered simultaneously on both the vertical and horizontal axes of the spiral (Fig. 29.1). The minor lunar standstill marking, which is formed by a different rock edge, also crosses the center of the large spiral.

Three shadow markings are tangent to the right and left edges of the large spiral. These are the winter solstice pattern, the lunar major standstill marking, and the lunar/solar 0-degree declination marking (Fig. 29.1). These markings emphasize the right and left edges of the spiral.

The top edge of the large spiral is defined by another aspect of the markings. At summer solstice the dagger-shaped pattern starts as a dot of light on the top turn of the spiral. Only 10 days earlier or later this effect is lost: the first light on the cliff face is 10 cm above the spiral. The bottom edge of the spiral is related to, and hence perhaps defined by, the momentary symmetrical positioning (mentioned above) of the summer solstice light dagger.

It is possible to arrive at some estimates of the degree of sensitivity in the relationships among the rock slabs and the markings. Because of the oblique angle between the slabs and the cliff face, moving the rock edge that casts the moonrise shadows 1 cm to the north or south would create twice this displacement in the positions of the lunar shadows on the spiral and thus discernibly shift the markings. A movement of either the eastern or middle slab by 2–3 cm to the east or west would displace the shadows that form the midday light patterns by that much. The summer solstice pattern is only 2 cm wide, so such a change could block the light entirely. Because each of the slabs forms a part of up to five markings, any such shift would change and probably destroy more than one marking.

A recent report (Newman et al. 1982) presents a scenario in which the rocks could have fallen naturally close to their present positions. The evidence presented in that report does not, however, exclude the possibility of later deliberate movement of the rocks to create the markings. Indications of possible shaping of the slabs and the cliff face where the spirals were pecked are discussed in an earlier publication (Sofaer et al. 1979:289).

Both the interlocking nature of the markings and their sensitivity to the exact positions and shapes of the rock slabs indicate that moving and shaping of the slabs very likely took place. A few people could have moved and adjusted the slabs in small increments by simple techniques. Such manipulation would have made it easier to attain the interrelated markings. The complexity of the markings suggests that extensive planning, observation, and experimentation were required to achieve them.

The technology for shaping, moving, and using large slabs was well established in Chaco Canyon (Hewett 1936:87–88; Mindeleff 1891:148) and at other places in this cultural area, and many of these slabs are significantly larger than those at the three-slab site (Hayden 1878:429; Mindeleff 1891:58, 147–148; Newcomb 1966:137). Some isolated and implanted slabs are also known to have been used in historical times for astronomical purposes (Mindeleff 1891:86, 148).

Caution is appropriate in assessing the degree of artificiality in structures of the Pueblo culture. The natural appearance of many historical and prehistoric sites, especially shrines, conceals elements of construction and may be intended for the purposes of protective secrecy and integration with nature (Alfonso Ortiz, personal communication 1983). Stevenson's description of a shrine at Zia Pueblo illustrates the subtlety of construction at these sites:

> [A] stone slab rested so naturally on the hillside that it had every appearance of having been placed there by other than human agency. The removal of the slab exposed two vases side by side in a shallow cave. (Stevenson 1894:90)

PREHISTORIC PARALLELS TO THE FAJADA MARKINGS

Shadow-light formations mark critical times in the solar cycle at many prehistoric sites in the Chaco cultural region and elsewhere in the world. Many share the motifs of the Fajada markings: daggerlike shapes of light, spirals, snakes, and vertical shafts channeling light patterns at noon.

It is reported that winter solstice is marked when the rising sun's rays are collimated by a window in Pueblo Bonito, a major structure in Chaco Canyon (Reyman 1976). At Hovenweep, in southeastern Utah, an ancient Pueblo community lying just to the northwest of the Chaco region, several light markings are reported (Williamson 1981:68–70). At the solstices and equinoxes light is collimated by portholes to fall in the corners and midpoints of tower structures. At a nearby rock art site two light daggers bisect two concentric circles and a spiral at summer solstice sunrise. Shadow/light formations mark these glyphs again at equinox sunrise, at which time they also mark a snakelike form. Recent reports of solstice and equinox markings at an eastern Arizona site near the Chaco region include a great number of bisecting dagger shapes of light on rock art, often on spirals. Similar observations are reported in California (Krupp 1983:129–137).

A number of the sites of monumental architecture in Mesoamerica have been noted for their symbolic display of shadow and light (Aveni 1980:284–286). For example, a snake-shaped shadow is formed at equinox along the edge of a Mayan pyramid at Chichen Itza. While this marking is several hundredfold larger than the small snake marked by a shadow on Fajada's east site, the parallel is curious: both sites record equinox with shadow alignments on nearly vertical snakes. At two other sites in Mesoamerica the sun's rays at zenith passage are channeled through vertical shafts to fall as discs of light on the subterranean floors of the structures (Aveni 1980:253–256).

Despite these parallels the Fajada markings remain distinctive in at least three respects: their combining of noon and season, their combining of sun and moon, and their use of rock art and rock slabs in numerous markings clustered on a single, prominent butte. Some of the particular characteristics of the Fajada markings can be understood in the context of historical Pueblo culture.

ETHNOGRAPHIC BACKGROUND

Ethnographic accounts of the historical Pueblo cultures describe concepts fundamental to the Pueblo people's world view. These concepts often have significant parallels among findings about prehistoric Pueblo communities.

Some discontinuity between prehistoric and historical Pueblo culture over the intervening 800–1000 years should be expected as a result of environmental and cultural change and migration, including the Spanish entrada in the 1500s (Berry 1982; Cordell 1984; Upham 1984). Certain gaps exist in the ethnographic record of astronomical practices because the Pueblo traditions of secrecy protect all ceremonial activities, including solar and lunar observations.

> Everyone who has worked among Pueblo Indians realizes only too well how averse they are to revealing the details of this manner of life. This attitude on the part of native informants makes it virtually impossible to secure a complete record of any Pueblo tribe. (Titiev 1944:4)

Fewer than 20 of the 80 or more historical pueblos extant when the Spaniards first came survived to be included in ethnographic studies. Thus, a great deal of information about Pueblo culture is lost or not available. The information that does exist can be used to provide general insights into the significance of prehistoric findings rather than to verify or refute interpretations of specific phenomena.

The historical Pueblos conceived of complementary roles of sun and moon (see references cited in Sofaer et al. 1982a) and used both of them intensively in timing of planting and ceremonial activity. These activities are consistent with the overlapping systems of marking of the solar and lunar cycles at the three-slab site. A recent study (Tedlock 1984:6) describes clearly the complementary roles of sun and moon at Zuni. In timing of the ceremonies for the solstices, the "weak light" of the winter solstice sun is matched with the "bright light" of the full moon, and the "bright light" of the summer solstice sun is matched with the "weak light" of the new moon. The location of the astronomical markings high on the seemingly remote and inaccessible Fajada Butte is in keeping with ethnographic reports of the historical cultures. Buttes, mesas, and mountains are often regarded as sacred places, and shrines placed on their tops (Boas 1925–1928:39–40; Dumarest 1919:206–207; Ortiz 1969). Some shrines on high sites are used for solstice ceremonies (Kallestewa et al. 1984; Stevenson 1904:109, 149).

Specific features of the complexly organized communities of the prehistoric Pueblos are not found or reported among the historical Pueblos. For example, the elaborately planned multistory architecture and engineered road system of the Chaco culture are not present among the historical Pueblos. Precise equinox markings and cardinal alignments are also not evident. The general period of the equinoxes is of major significance among the Tewa Pueblos, however (Ortiz 1969). Similarly, there are no reports in the ethnographic record of knowledge or markings of the 18.6-year lunar cycle among the historical Pueblo cultures, although there is a possible hint of previous knowledge of this cycle in the cosmogony of one pueblo (Stevenson 1894:71).

There are numerous reports of Pueblo people observing shadow and light patterns cast by upright slabs or by windows and doorways on walls and floors of houses, kivas, and chief's houses to time ceremonial and agrarian activities (Cushing 1979:117; Lange 1959:56, 249; Mindeleff 1891:86, 148). On ceremonial occasions light might fall on a quartz crystal, a bowl of water, a deer skin, or a kiva bench (Lowie n.d.; Parsons 1929:176; Titiev 1944:105).

Meridian passage is known to be significant in the cosmology of historical Pueblo people and is an integral part of many myths and ceremonies, often involving light markings. One such instance involves the winter solstice noon.

> In the roof of the ceremonial room there is a hole through which at noon the sun shines on a spot on the floor near where the chief stands.... All sing the song of "pulling down the sun."... This is noon time when for a little while the Sun stands still. (Parsons 1932:292–293)

A similar ritual is practiced at summer solstice (Parsons 1932:297). In keeping with the complementary roles of sun and moon, a report states, "The ritual of bringing down the moon seems to be much the same as that of bringing down the sun," and in this ritual the moon is said to stay "until noon" (Parsons 1932:330). Another study reports that the moon's meridian passage is noted in the timing of Pueblo ceremony (Tedlock 1984:94, 108).

Several other ritual practices and traditions of the Pueblo culture convey the significance of noon as the time when the sun stands still or rests for a short time in the middle of its course through the day (Boas 1925-1928: 284; Dumarest 1919:222) and as the time when "he stops for dinner" (Curtis 1926:104; Dumarest 1919:222). The cacique conducts ceremonies in which the sun as a disc is moved across a screen in the arc of its day's course; he orders the sun at the middle of its course to stop for a short while at its noon position (Dumarest 1919:198; Lange 1959:267).

In one pan-Puebloan tradition, the Sun impregnates a virgin and she gives birth to the son(s) of the Sun (Parmentier 1979:609). The impregnation and the birth often occur at noon and sometimes near summer solstice (Benedict 1931:31, 1935:46; Cushing 1931:429, 431, 436; Dumarest 1919:217; Gunn 1917:129; Parsons 1932:393, 1940:55-56; Stephen 1929:11-14). In several versions of this tradition the sun's rays penetrate a window or hatchway of a dwelling and fall on the lap of the maiden. In one version the virgin (yellow woman of the north) stops to rest on her journey to the center of the earth, and the Sun when it is "over the middle of the world" at the "middle of the day" embraces her and she becomes pregnant (Stevenson 1894:44-45). This theme of the impregnating power of the sun's rays is also evident in the tale of the Sun's son who, as a baby, identifies his father by crawling to the rays of the sun on the kiva floor (Alfonso Ortiz, personal communication 1981; Parsons 1940:56-57).

The use of the rattlesnake as a marked glyph is consistent with the snake's wide ceremonial use and symbolic meaning in historical Pueblo culture, including a close association with the sun. A rattlesnake effigy and a ceremonial staff with the rattles of a rattlesnake indicated on it were found at Pueblo Bonito, which suggest a possible ritual involvement with the rattlesnake as part of the Chaco culture (Pepper 1920:147). Among its many roles, the snake connects the below and above worlds (Tyler 1964:222, 234). At one pueblo it is connected with the zenith and the sun: "Huwaka (Serpent of the Heavens) has a body like a crystal, and it is so brilliant that one's eyes cannot rest upon him; he is very closely allied to the sun" (Stevenson 1894:69). Similarly, at another pueblo a snake with glistening scales is said to fly up to the sun each day and brilliantly reflect the sun's rays (Charles Loloma, a Hopi religious leader, personal communication 1983). Some reports on Pueblo cultures also connect the snake with fertility and equinox (Tyler 1964:228-229, 245-247).

While there is no conclusive information concerning interpretations of the spiral among the historical Pueblos, some Pueblo people have indicated that it conveys the movement of people and clans. In one report, the spiral was described as the movement of people in their search for the center of the earth in the origin story (Roberts 32:151). The spiral has also been reported to represent the annual movement of the sun (Charles Loloma, personal communication 1983).

ASSOCIATIONS WITH THE PREHISTORIC PUEBLO CULTURE AND CHACO SOCIETY

The Fajada markings were probably developed when Pueblo people settled and flourished in the canyon between about AD 900 and 1300. The spiral petroglyph is identified with prehistoric Pueblo people in this region during this period (Schaafsma 1980:135–136; Polly Schaafsma, personal communication 1983).

The achievements of the Chaco culture—elaborate roads, irrigation works, and multistory architecture, all constructed between AD 1000 and 1150—involved highly developed skills of planning, engineering, and surveying (Kincaid 1983; Lekson 1984; Marshall et al. 1979; Powers et al. 1983; Vivian 1974). During this period the Chaco society used these skills to align accurately several major constructions to the cardinal directions. Primary elements in the symmetric designs of the isolated great kiva at Casa Rinconada and of one of the central pueblos, Pueblo Bonito, and the initial portion of the North Road are oriented to within 0.25–0.5 degrees of the cardinal points (Sofaer et al. 1986; Stein 1983:8-1; Williamson et al. 1977:208–212; the canyon alignments were confirmed by surveys carried out by the Solstice Project).

The north/south alignments are particularly significant to the phenomenon of noon markings. A likely method of identifying noon is to watch shadows from a vertical object cross the north/south line as the sun crosses the meridian. Knowledge of true north within 0.25 degrees permits knowledge of solar noon at the latitude of Chaco to within less than one minute throughout the year. Similarly, knowledge of east and west to within less than 0.5 degrees allows the determination of the day of equinox. The architectural alignments indicate the Chaco culture's interest in and capability of achieving such accuracy. These alignments may also represent divisions of space that correspond to the markings on Fajada that divide the day and the year.

CONCLUSION

Although the astronomical markings and alignments of Chaco incorporate utilitarian calendric information, they do so with a redundancy and accuracy far beyond the practical requirements of time-keeping devices. For example, precise noon markings high on a steep butte serve no apparent useful purpose, nor do the markings of the lunar standstill cycle. The accurate alignments that define the major axes of the central ceremonial structures of the canyon are similarly abstract; for example, the east-west alignments could not have been used to determine (or have been determined by) equinox sunrise/sunset because of the locally elevated horizons. Similarly, the North Road was built elaborately and accurately in a direction that serves no apparent utilitarian purpose (Sofaer et al. 1986; Stein 1983:8-1). Rather, what may be seen in the alignments and markings is the geometric expression of astronomical concepts and of the culture's cosmology.

Chaco Canyon was the center of an extensive road network and outlier system. Recent analysis indicates that it may have been a center for pilgrimage and ritual. Fajada Butte may have been a center for the culture's ritual activity related to the sun and the moon.

The clustering of markings found there is so far unique; surveys by the Solstice Project of all the prominent landforms in the Chaco cultural region have disclosed no further astronomical marking sites. When the sun in "the middle of the day" is over "the middle of the earth," the butte's glyphs commemorate this special moment in time and space. The extremes and midpoints of the solar and lunar cycles are integrated on Fajada. The daily and seasonal passages of the sun are united, as are the sun and moon and the earth and sky in the play of shadow and light on the rock carvings atop Fajada Butte.

ACKNOWLEDGMENTS

We are indebted again to Walter Herriman, former superintendent, and the staff of the Chaco Culture National Historic Park for their continued help and cooperation. Peggy Wier, Michael Marshall, Karl Kernberger, James Grant, Colin Franklin, Kenneth Butterfield, and Sheila Rotner helped collect the data, often under adverse conditions. We are grateful to Alfonso Ortiz, Gerald Hawkins, Fred Eggan, Cesare Marino, Rob Blair, Evelyn Newman, Robert Mark, Harold Malde, Fred Nials, and particularly LeRoy Doggett and Stephen McCluskey for a number of helpful conversations. Pat Kenny is to be especially thanked for preparing the illustrations. We again thank our families and friends for their assistance and understanding.

REFERENCES

Aveni, Anthony. 1980. *Skywatchers of Ancient Mexico*. Austin, University of Texas Press.

Benedict, Ruth Fulton. 1931. *Tales of the Cochiti Indians*. Bureau of American Ethnology Bulletin 98. Washington, DC.

Benedict, Ruth Fulton. 1935. *Zuni Mythology*. Columbia University Contributions to Anthropology 21. New York.

Berry, Michael S. 1982. *Time, Space and Transition in Anasazi Prehistory*. Salt Lake City, University of Utah Press.

Boas, Franz. 1925–1928. *Keresan Texts* (2 parts). Publications of the American Ethnological Society 8. New York.

Cordell, Linda S. 1984. *Prehistory of the Southwest*. New York, Academic Press.

Curtis, Edward S. 1926. *The North American Indian* 17. Norwood, MA, Plimpton Press.

Cushing, Frank M., comp. and trans. 1931. *Zuni Folk Tales* (reprint). New York, Alfred A. Knopf. Originally published in 1901 by G. P. Putnam's Sons, New York.

Cushing, Frank M. 1979. *Zuni* (reprint), ed. Jesse Green. Lincoln, University of Nebraska Press. Originally published in 1882 by the Peripatetic Press, Santa Fe.

Dumarest, Noel. 1919. *Notes on Cochiti Mexico*. Memoirs of the American Anthropological Association 6 (3), ed. Elsie C. Parsons. Lancaster, PA.

Ellis, Florence H. 1975. A Thousand Years of the Pueblo Sun-Moon-Star Calendar. In *Archaeoastronomy in Precolumbian America*, ed. A. F. Aveni, 58–87. Austin, University of Texas Press.

Gunn, John M. 1917. *Schat-chen: History, Traditions and Narratives of the Queres Indians of Laguna and Acoma*. Albuquerque, Albright and Anderson.

Hayden, F. V., ed. 1878. *Tenth Annual Report—Geological and Geographical Survey of the Territories*. Washington, DC.

Hewett, Edgar L. 1936. *The Chaco Canyon and Its Monuments*. Albuquerque, University of New Mexico Press.

Kallestewa, B., J. Niiha, A. Peywa, A. Pinto, R. Quam, A. Nastacio, and W. Eriacho. 1984. Statements by Zuni Religious Leaders on Kolhu wala:wa. Testimony presented to the Select Committee on Indian Affairs, U.S. Senate, April 3, 1984. Washington, DC.

Kincaid, Chris, ed. 1983. *Chaco Roads Project Phase I.* Bureau of Land Management, Albuquerque.

Krupp, E. C. 1977 *In Search of Ancient Astronomies.* Garden City, NY, Doubleday.

Krupp, E. C. 1983. *Echoes of the Ancient Skies.* New York, Harper and Row.

Lange, Charles M. 1959. *Cochiti: A New Mexico Pueblo, Past and Present.* Austin, University of Texas Press.

Lekson, Stephen. 1984. *Great Pueblo Architecture of Chaco Canyon, New Mexico.* National Park Service Publications in Archaeology 18B. Albuquerque.

Lowie, R. H. N.d. Hopi Indian inside a Kiva. Photo No. 283545. New York, American Museum of Natural History.

Marshall, Michael P., John Stein, Richard Loose, and Judith E. Novotny. 1979. *Anasazi Communities of the San Juan Basin.* Albuquerque, Public Service Company of New Mexico, and Santa Fe, New Mexico State Historic Preservation Bureau.

Mindeleff, Victor. 1891. A Study of Pueblo Architecture. In *Eighth Annual Report of the Bureau of American Ethnology, 1886–1887,* pp. 3–228. Washington, DC.

Newcomb, F. J. 1966. *Navajo Neighbors.* Norman, University of Oklahoma Press.

Newman, E. B., R. K. Mark, and R. G. Vivian. 1982. Anasazi Solar Marker: The Use of a Natural Rockfall. *Science* 217:1036–1038.

Ortiz, Alfonso. 1969. *The Tewa World: Space, Time, Being, and Becoming in a Pueblo Society.* Chicago, University of Chicago Press.

Parmentier, Richard J. 1979. The Pueblo Mythological Triangle: Poseyemu, Montezuma, and Jesus in the Pueblos. In *Southwest,* ed. Alfonso Ortiz, 609–622. Handbook of North American Indians, Vol. 9. Washington, DC, Smithsonian Institution.

Parsons, Elsie Clews. 1929. *The Social Organization of the Tewa of New Mexico.* Memoirs of the American Anthropological Association 36. Menasha, WI.

Parsons, Elsie Clews. 1932 Isleta, New Mexico. In *Forty-Seventh Annual Report of the Bureau of American Ethnology for the Years 1929–1930,* pp. 193–466. Washington, DC.

Parsons, Elsie Clews. 1940. *Taos Tales.* Memoirs of the American Folklore Society 34. New York, G. F. Stechert.

Pepper, George M. 1920. *Pueblo Bonito.* Anthropological Papers of the American Museum of Natural History 27. New York.

Powers, R. P., W. B. Gillespie, and S. H. Lekson. 1983. *The Outlier Survey.* Reports of the Chaco Center 3. Albuquerque, National Park Service, Division of Cultural Research.

Reyman, Jonathan E. 1976. Astronomy, Architecture, and Adaptation at Pueblo Bonito. *Science* 193:957–962.

Roberts, F.H.H., Jr. 1932. *The Village of the Great Kivas on the Zuni Reservation, New Mexico.* Bureau of American Ethnology Bulletin 111. Washington, DC.

Schaafsma, Polly. 1980. *Indian Rock Art of the Southwest.* Albuquerque, University of New Mexico Press.

Sofaer, A., and J. Crotty. 1977. Survey of Rock Art on Fajada Butte. Ms. on file, Chaco Center, University of New Mexico, Albuquerque.

Sofaer, A., M. P. Marshall, and R. M. Sinclair. 1986. Cosmographic Expression in the Road System of the Chaco Culture of Northwestern New Mexico. Paper presented at the Second Oxford Conference on Archaeoastronomy, Merida, Yucatan, Mexico.

Sofaer, A., R. M. Sinclair, and L. E. Doggett. 1982a. Lunar Markings on Fajada Butte. In *Archaeoastronomy in the New World,* ed. A. Aveni, 169–181. Cambridge, Cambridge University Press.

Sofaer, A., R. M. Sinclair, and L. E. Doggett. 1982b. Noon Markings on Fajada Butte, Chaco Canyon, New Mexico. *Bulletin of the American Astronomical Society* 14:872.

Sofaer, A., V. Zinser, and R. M. Sinclair. 1979. A Unique Solar Marking Construct. *Science* 206:283–291.

Stein, John R. 1983. Road Corridor Descriptions. In *Chaco Roads Project Phase I,* ed. C. Kincaid, pp. 8-1–8-15. Albuquerque, Bureau of Land Management.

Stephen, A. M. 1929. Hopi Tales. *Journal of American Folklore* 42:1–72.

Stevenson, Matilda Coxe. 1894. The Sia. In *Eleventh Annual Report of the Bureau of American Ethnology for the Years 1889–1890,* pp. 3–157. Washington, DC.

Stevenson, Matilda Coxe. 1904. The Zuni Indians: Their Mythology, Esoteric Fraternities, and Ceremonies. In *Twenty-Third Annual Report of the Bureau of American Ethnology for the Years 1901–1902,* pp. 3–634. Washington, DC.

Tedlock, Barbara. 1984. Zuni Sacred Theater. *American Indian Quarterly* 7 (1):93–110.

Titiev, Mischa. 1944. *Old Oraibi: A Study of the Hopi Indians of Third Mesa.* Papers of the Peabody Museum of American Archaeology and Ethnology 22 (1). Cambridge, MA, Harvard University.

Tyler, Hamilton A. 1964. *Pueblo Gods and Myths.* Norman, University of Oklahoma Press.

Upham, Steadman. 1984. *Polities and Power: An Economic and Political History of the Western Pueblo.* New York, Academic Press.

Vivian, R. Gwinn. 1974. Conservation and Diversion: Water-Control Systems in the Anasazi Southwest. In *Irrigation's Impact on Society,* ed. Theodore Dawning and McGuire Gibson, 95–112. Anthropological Papers of the University of Arizona 25. Tucson, AZ.

Williamson, Ray A. 1981. North America: A Multiplicity of Astronomies. In *Archaeoastronomy in the Americas,* ed. R. A. Williamson, 61–80. Los Altos, CA, Ballena Press.

Williamson, R. A., H. J. Fisher, and D. O'Flynn. 1977. Anasazi Solar Observatories. In *Native American Astronomy,* ed. A. F. Aveni, 203–217. Austin, University of Texas Press.

CHAPTER THIRTY

Romancing the Stone, or Moonshine on the Sun Dagger

John B. Carlson

In June of 1977, artist Anna Sofaer was a volunteer participant in a rock art recording field school in Chaco Canyon, New Mexico. She was given the famous 135 m high Fajada Butte at the south entrance of the canyon to explore as part of the summer survey. According to published accounts (e.g., Frazier 1978, 1979; Marshall 1982; Sofaer et al. 1979a, 1979b), it was just before midday on June 29 at an obscure petroglyph site near the top of the butte that Sofaer serendipitously first observed the light-and-shadow phenomenon that has come to be known as the "Sun Dagger" (Sofaer and Ihde 1982).

The petroglyph site consists of two spiral-type petroglyphs pecked into the cliff face of the butte and situated behind three near-vertical 2–3 m high sandstone slabs balanced precariously on soil and small stone fragments (Fig. 30.1). Sofaer and

---

Reproduced by permission from *Astronomy and Ceremony in the Prehistoric Southwest*, ed. J. B. Carlson and W. J. Judge. Papers of the Maxwell Museum of Anthropology, No. 2, 1987.

archaeologist Jay Crotty watched as light shining between Slabs 1 and 2 (Sofaer et al. 1979b:284–285) formed a spot at the top of the large spiral, lengthened to create a vertical dagger shape of light spanning the petroglyph, and then moved downward rapidly to disappear near the bottom of the spiral. The whole event, which has now been well documented with time-lapse photography, takes approximately 18 minutes from start to finish. The effect is striking; observing the time-lapse sequence has convinced most researchers that the large spiral petroglyph was intentionally positioned by its maker to interact in some way with the "sun dagger" phenomenon.

Because of the striking appearance of the dagger of light, the fact that it virtually bisected the spiral petroglyph, and because it was June 29, Sofaer was compelled to believe that the phenomenon was created intentionally to mark the summer solstice specifically—that it was not simply a daily pattern that repeated essentially unchanged throughout the year. As a result of many subsequent observations of the site by Sofaer and others, her initial inspiration that the dagger of light would be found to be centered on the petroglyph around the time of the June solstice has been adequately verified. The petroglyph site serves and probably was created to serve as at least an approximate marker of the summer solstice. The creators were suggested plausibly, but not conclusively, to have been the Anasazi Indians who occupied Chaco Canyon from about AD 400 to 1300 (Reyman 1980; Sofaer et al. 1979b).

If this had been the whole story—that the Fajada site contained a visually pleasing light-and-shadow-on-petroglyph marker of the summer solstice, possibly created by the Chacoan Anasazi—we would have simply yet another example of a Native American archaeoastronomical site involving plays of light and shadow on petroglyphs. Williamson and Young (1979) first presented a discussion of an Anasazi site at Hovenweep National Monument that subsequently (Williamson 1979) was shown to contain a compelling sun-dagger-type summer solstice marking. Other examples include those proposed by Preston and Preston for petroglyph sites in Arizona. Sofaer and her colleagues (Sofaer et al. 1979a, 1979b) propose a far more elaborate set of claims for the Fajada site, however. They claim that the site not only marks the summer solstice with a sun dagger that bisects the large spiral, but also that two daggers would bracket this spiral on either side at the winter solstice and that the second, smaller spiral to the left would be bisected by a short shaft of light on the equinoxes. Furthermore, they claim to demonstrate that the site was a "construct." Unlike all other such petroglyph marker sites that have been proposed, this one is alleged to have been constructed by the Chacoans, who selected and moved the slabs into their near-vertical position and modified the shadow-producing edges (Figs. 30.1–30.3). Finally, they note that "the patterns formed at night by moonlight shining between the slabs are just as clear and noticeable as those formed by the sun" and that "extrapolations from the solar data suggest that a significant marking of the maximum lunar declination may occur" (1979b:287).

In the concluding paragraphs on the "possible significance of the lunar markings," they state that (a) "The Indians of the Southwest knew of the 18.6-year cyclic extremes in lunar declination [the lunar standstills] and incorporated that knowledge

30.1a View of the three-slab site from the south. The three near-vertical sandstone slabs are seen at lower right in the context of their location within 10 m of the summit of Fajada Butte. Photograph by J. Carlson.

30.1b Close-up view of the three-slab site from the south. Slab 1 is to the right (east), Slab 2 is in the center, and Slab 3 is on the left. Note that a portion of a "sun dagger" created by light passing between Slabs 2 and 3 can be seen between Slabs 1 and 2. Photograph by J. Carlson.

30.1c View of the three-slab site looking from the west. Considerable weathering of the exposed surfaces is evident. Photograph by J. Carlson.

30.1d View of the three-slab site looking straight down from the overhanging edge of the mesa summit. Photograph by J. Carlson.

30.1e View of the three-slab site from the east showing the natural lean-to. A "sun dagger" is seen on the cliff face at the approximate location of the large spiral petroglyph. Photograph by J. Carlson.

30.1f View of the three-slab site from the east. Note the sharp fracture plane of the sunrise shadow-casting edge of Slab 1 and its parallel alignment with the inner faces of Slabs 2 and 3. Photograph by J. Carlson.

30.1g View inside the natural lean-to of the three-slab site from the east. Note the crisp, parallel fracture planes of the three slabs and a portion of the sunrise shadow-casting edge of Slab 1. Photograph by J. Carlson.

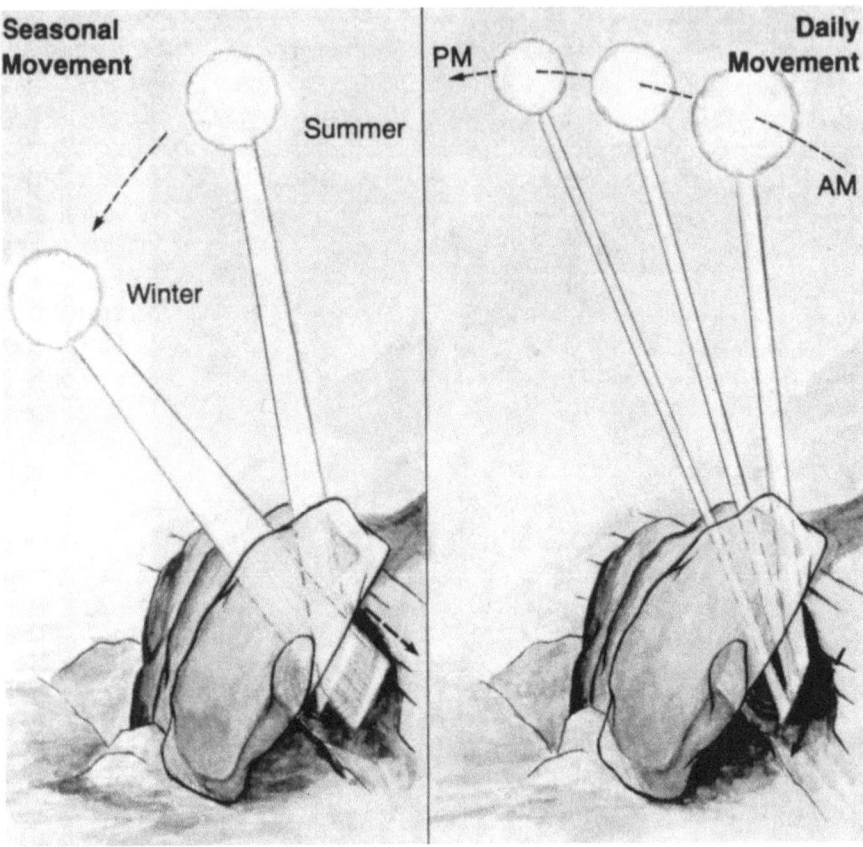

30.2 Diagram prepared by Mary Challinor for the *Science 80* article by Kendrick Frazier (1979:61). "The slab's precise geometry transforms seasonal lowering sun into a rightward shift of light daggers on spirals, and the sun's daily horizontal passage into daggers that move down vertically."

in a major archaeological site [at Casa Grande]" and that (b) "Striking alterations of patterns would indicate the alignment of sun, earth, and moon that makes possible a midwinter or midsummer solstitial or an equinoctial eclipse." This knowledge of the 18.6-year lunar standstill cycle is suggested to be reflected in the 19 half-turns of the large spiral. All of these assertions are based on the flimsiest of evidence and include an apparent confusion of the 29.53-day period of the moon's phases, the subject of the well-known Pueblo interest in the lunation, with the far more subtle 18.6-year cycle of lunistices. To date, there is no convincing evidence that the peoples of native America knew of or recorded the lunistices and their cycle (e.g., see Aveni 1985). As discussed later, Sofaer and Sinclair have abandoned the specific lunar alignments involving the moon near meridian transit and offer a new set of moon*rise* shadow effects that allegedly mark the northern major and minor lunar standstills as well as moon (or sun) rise at zero degrees declination (Sofaer et al. 1982).

John B. Carlson

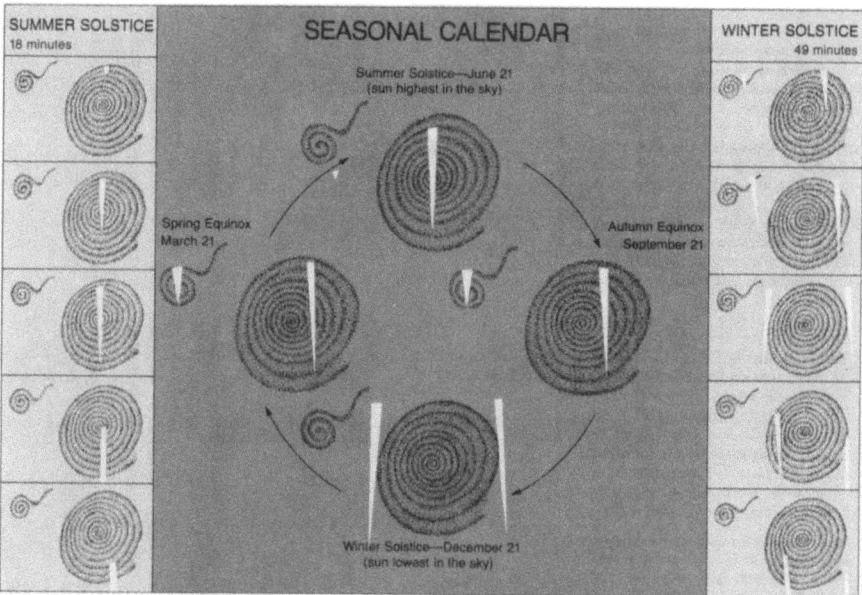

30.3 Schematic diagram of the hypothetical four-part "seasonal calendar" created by daggers of light falling on the two spiral petroglyphs at the three-slab site. Illustration prepared by Mary Challinor for the *Science 80* article by Kendrick Frazier (1979:61); the stylized daggers are quite misleading in that they bear little resemblance to any of the light phenomena that actually appear at the site.

Beyond the initial suggestion that the three-slab site, as it is now called (Sofaer and Sinclair 1985), was a summer solstice marker, most of the proposed astronomical functions of the site are speculative and the subject of considerable scientific controversy. This controversy has been intensified by media coverage that presents the sun dagger and all of its complexity to the public as established scientific fact—"an American Stonehenge" constructed by "Chacoan Einsteins" as it were. Science writer Kendrick Frazier first promulgated the speculative "construct" hypothesis in *Science News* (1978); a year later he created considerable new mythology in an article (Frazier 1979) for *Science 80* with diagrams of a four-part seasonal calendar with neat daggers such as never actually appear at the site (see Figs. 30.2 and 30.3).

The main event of this media coverage was "The Sun Dagger," an hour-long pseudo-documentary produced and directed by Sofaer and Ihde (1982), which was aired nationwide on public television. The story line graphically tells how artist Anna Sofaer, triumphing over scientific disbelievers, has demonstrated that not only does the site mark summer solstice, but that it also indicates the equinoxes and winter solstice, giving a full annual solar calendar. Furthermore, it is called a "construct." The three 1000 kg slabs of sandstone are said to have been moved into place by the Anasazi, who also aligned the stones to work as a *lunar* marking construct to indicate the 18.6-year cycle of major and minor lunar standstills.

701

The final triumph in the story came when Sofaer went to England to present her results before the September 1981 Oxford Archaeoastronomy Symposium, a unique gathering of European and American researchers. To quote from the film, "The conclusion reached by this international group of scholars was that Fajada Butte is presently the only known site in the world where the extreme positions of both the sun and the moon are marked." It is indicative of the film that it characterizes the suggestion that Stonehenge marked both extremes of the sun and moon as "controversial"; the same claim is given the alleged stamp of approval by the Oxford participants for the Sun Dagger. In fact, no such endorsement was ever made.

Following the airing of "The Sun Dagger" there were numerous media excesses, such as the "rock clock" of Fajada Butte being billed as "an astronomical and geometrical marvel" that "ranks with the Pyramids and Stonehenge" (Marshall 1982). All this "hype" greatly obscures reality by creating a durable popular mythology and makes the process of careful scientific evaluation very difficult.

RETHINKING THE SUN DAGGER

Scientific evaluation of the data and of the many multifaceted claims made for the Fajada Butte "three-slab site" in both the professional and popular literature has now begun in earnest. Reyman (1980) was one of the first to question many of the ideas put forward by Sofaer et al. (1979a, 1979b) and Frazier (1979), particularly with regard to their treatment of the relevant archaeological and ethnographic data. Newman et al. (1982) have provided convincing data and analyses demonstrating that the site is a natural rockfall, not a man-made construct. Aveni (1987) provides further evidence casting doubt on Native American knowledge or use of the lunistices and lunar standstill cycle. Carlson (1983, 1984) and Hadingham (1984) reviewed "The Sun Dagger," questioning the methods, conclusions, and strong antiscientific bias of the film.

Zeilik (1985) has performed the most comprehensive reevaluation of the Fajada Butte solar marker. He approached the site on the basis of site integrity; archaeological, geological, and ethnographic context; and usefulness for confirmatory and anticipatory solar observations. He concludes that the site does not function as an accurate solar calendar, where "accurate" is defined in the context of the historical Puebloan culture, and that if used at all intentionally by the Anasazi for its relationship to the sun, the site most likely served as a shrine rather than as an observing station—a conclusion drawn from the characteristics of sun shrines among the historical Pueblos. "The place would be special and sacred, but not calendrical. It would act as a red showing of the sun—rather than an accurate calendar" (Zeilik 1985:S80). Furthermore, he evaluates the claim that the site also marks the northern declinations of the lunar standstill cycle and concludes that

> The case for lunar standstill alignments for the Fajada site fails to have support from ethnographic grounds. It also has no practical value as an adaptation for

survival. Even if the site did mark the northern declinations of the major and minor standstills, it does so crudely and cannot be used to predict the year in which the standstill will occur. As an alternative interpretation, the site can be used at sunrise to provide a planting calendar, to anticipate the summer solstice, and to mark the equinoxes—functions much more in harmony with the ethno-astronomy of the historic Pueblos than observations of the lunar standstill cycle. (Zeilik 1985:S84)

The present essay does not attempt anything like a comprehensive reevaluation of the three-slab site. Instead, I would like to rethink the full scenario of solar and lunar alignments proposed for the three-slab site as outlined by Sofaer et al. (1982) and more recently by Sofaer and Sinclair (Chapter 29). In particular let us consider the three "solar markings at midday" and the three "solar/lunar markings at rising" illustrated in Figure 29.1 (previous chapter) in the context of the site as a natural rockfall, as a slightly modified natural rockfall, and as a substantial, engineered construction. This illustration is a corrected version of a diagram that appeared originally in Sofaer et al. (1982:fig. 4).

THE FACTS AND THE CLAIMS

The specific set of hypotheses that we will take as given is the one outlined by Sofaer et al. (1982) and more recently by Sofaer and Sinclair (Chapter 29). Figures 29.1, 30.2, and 30.3 present this set in simplified diagrammatic form.

The three-slab site would never have been suggested as an archaeological site, man-made or otherwise, had it not been for the two spiral petroglyphs that are pecked into the vertical cliff face behind the slabs. The larger spiral to the right unwinds counterclockwise with approximately nine and one-half turns. It is elliptical and measures 34 by 41 cm. The smaller spiral to the upper left is actually a coiled snake with the head at the center. It has two and one-half turns, measures 9 by 13 cm, and uncoils from head to tail in a clockwise direction.

1. *Solar marking at midday* (+23.6 degrees declination). The vertical dagger formed when light passes between Slabs 1 and 2 (Figs. 30.2 and 30.3) bisects the large spiral at the June solstice. In the proposed model, this determines the position of the center along the vertical axis. The top and bottom edges are fixed by the place where the initial spot of light first appears and roughly by the maximal hilt-to-tip dimensions of the dagger before it disappears. Though quite approximate, during the 18-minute sequence the location of the upper and lower boundaries and hence the center of the large spiral are delimited on the rock face. Because the spiral is elliptical, its width is as yet undetermined.

2. *Solar marking at midday* (−23.6 degrees declination). Two vertical stripes formed by light passing between Slabs 1 and 2 and Slabs 2 and 3 "bracket the large spiral, holding it empty of light" (Sofaer et al. 1982:173). This winter solstice configuration thus operationally fixes the left and right edges of the large spiral. With the constraints imposed by the summer solstice display, the center and the horizontal and vertical boundaries have all now been determined. If

the times of the solstices were already known by the usual horizon calendar methods, a sun watcher could in principle determine these parameters at the three-slab site in one year—given adequate observing conditions.

3. *Solar marking at midday* (0.0 degrees declination). A second dagger formed when light passes between Slabs 2 and 3 bisects the smaller spiral at the site. This dagger's vertical dimensions provide rough upper and lower boundaries for this spiral, thus determining a vertical and horizontal locus for the center. As an elliptical spiral, its width remains undetermined.

4. *Solar/lunar markings at rising* (0.0 degrees declination). At equinox sunrise the outer edge of Slab 1 casts a shadow over the far right groove of the large spiral (Figure 29.1). Of course, a similar shadow will be formed at moonrise when the moon is at zero degrees declination. This alignment is suggested to have been intentionally created. Note that the right edge of this spiral has already been determined by the winter solstice midday marker configuration.

5. *Solar/lunar markings at rising* (+18.4 degrees declination). Around the time of the minor lunar standstill, the shadow cast by moonlight grazing the outer edge of Slab 1 at the northern extreme moonrise covers the left half of the large spiral. This shadow passes roughly through the center of the spiral, and it is further alleged that the shadow is "aligned with a pecked groove [Sofaer et al. 1979b: fig. 7] that runs from the spiral's center to the lower left edge, emphasizing this particular occurrence" (Sofaer et al. 1982:173–175). My own observations of the site have not convinced me that such an intentional groove actually exists. As noted above, the sun at sunrise will also produce such a shadow when it is at +18.4 degrees declination. Note that the center of the large spiral has already been determined by the summer solstice midday marker. Note also that such minor lunar standstills recur in an 18.61-year cycle.

6. *Lunar markings at rising* (+28.7 degrees declination). With calculations and simulation experiments, Sofaer et al. (1979b) show that around the time of major lunar standstill the shadow cast by moonlight grazing the outer edge of Slab 1 at the northern extreme moonrise falls roughly tangential to the far left edge of the large spiral (see Figure 29.1). The shift of the obliquity of the ecliptic for epoch AD 1000 was taken into account in their calculations, as were the effects of lunar parallax, atmospheric refraction, the lunar wobble of 9 arc minutes amplitude, and the local horizon height (Sofaer et al. 1982:176). As with the previously described lunar marking, Sofaer and Sinclair now report that there is a previously unnoticed pecked groove tangential to the large spiral and aligned with the major standstill shadow marking (Chapter 29). From my own inspections I have observed no such groove and reject this as valid data. Note that the dimensions and the left edge of the spiral have already been determined by the winter and summer solstice midday marker configurations. As is the case for the minor standstills, the major lunar standstills recur in an 18.61-year cycle. Sofaer et al. (1982:176) further propose that the nine and one-half turns of the spiral may, as 19 grooves, record this 18.61-year lunar cycle or perhaps "are also a record of knowledge of the 19-year Metonic cycle." Here there are hints of lunar eclipse prediction.

The initial motivations for a search for lunar standstill alignments at such an obviously solar site remain obscure. There do appear, however, to be two main sources of influence. The first is Stonehenge, with its complexity of solar and lunar standstill alignments and possibilities for lunar eclipse prediction, the forerunner of megalithic archaeoastronomical studies. Aveni (1987) has briefly discussed the history of this subject, noting how the paradigm has been uncritically transferred to other cultures and cults. The second influence has been noted by Sofaer et al. (1982:175): "A further stimulus to search for lunar significance within this assembly was provided by [Pueblo Indian anthropologist] Alfonso Ortiz (personal communication) . . . , who suggested that because of the dual roles of sun and moon in Pueblo culture, a site in which the sun was so clearly marked would also include the moon." This may or may not be true. When one looks at the record of known or suggested Pueblo sun-watching sites or sun shrines—be they ancient or contemporary, occurring in natural formations or in Pueblo architecture—one would be hard pressed to find a single example with alignments marking the moon's extremes as well as those of the sun. In any event, the suggestion that historical Pueblo moon watching involves alignments to the lunistices or knowledge of the 18.61-year standstill cycle is misguided. Any Southwestern ethnologist or student of the Pueblo cultures will attest to the symbolic and religious importance of the moon and the calendrically vital knowledge of the monthly lunation cycle, but at present there are no reliable data attesting to an awareness or cultural use of the lunistices and their cycle among the historical or ancient Pueblos.

Failing to discover and verify plausible lunar standstill markings produced near the moon's meridian passage, Sofaer and Sinclair (Chapter 5) sought and subsequently found moon*rise* alignments. The question is, are these alignments intentional or merely fortuitous? Corroboration must wait for further indisputable archaeological examples and solid ethnographic data. For now, we may gain insights by exploring the logic of the claims made for the three-slab site.

The Three-Slab Site: A Natural Rockfall

Despite the claims that the three-slab site was constructed, the best geological and archaeological data now strongly suggest that it is a natural rockfall. A series of facts support this. The three main slabs in their present relative positions form a "sandwich" that fell essentially intact from horizontal strata above and to the left of where the assemblage now lies. This has been documented in an article by two geologists and an archaeologist (Newman et al. 1982). They have shown that such natural configurations are common in Chaco Canyon and that none of the characteristic signs of Chacoan Anasazi masonry or other cultural remains are found at the three-slab site. The slabs are precariously balanced amid their own rubble and earth, with no foundation or buttressing to stabilize them (Fig. 30.1). The outer exposed edges are heavily weathered, whereas the sides and protected inner edges appear to be as crisply fractured as the day they split from the cliff face. This can be seen clearly in Figure 30.1

or in the photograph provided by Sofaer et al. (1979b:cover). With these thoughts in mind, let us explore the ramifications for the program of six hypothetical solar and lunar alignments outlined above.

Obviously, if we accept that the three-slab site is entirely natural and that it functions now virtually as it did at the time of its creation (i.e., with a negligible minimum of erosion and subsidence), then all of the light-and-shadow engineering involved only the specific design and placement of the two spiral petroglyphs behind the slabs. The daggers of light are then purely natural, serendipitous visual phenomena, striking though they may be. If this were the case, the situation would be in accord with all of the comparable Native American rock art sites investigated thus far.

As we have seen, the three solar markings at midday at the solstices and equinoxes uniquely fix the locations and dimensions of the two spirals. For this reason the solar/lunar shadow markings at rising *must* be fortuitous and *not* created intentionally by design. Alternatively, if one were to argue that the *rise* shadow markings were designed, then the midday solstice markings *must* be serendipitous. This latter proposition seems to me to be most unlikely. In any case, if the rockfall was unaltered, it is impossible for both sets of markings to have been created by design. Given the lack of evidence for Pueblo lunistice watching, I reject the intentional creation of lunar markings at rising as an untenable hypothesis for this first scenario.

The Three-Slab Site: A Modified Natural Rockfall

It is well within the realm of possibility that the carvers of the two petroglyphs also slightly modified the natural rockfall and enhanced the serendipitous light phenomena that originally attracted their attention. The inner edges and mutually facing sides of the slabs are unmodified, and the tops and outside edges are so weathered that it is not possible to prove that any modification occurred (Fig. 30.1). A close inspection of the three-slab and other similar sites will show that such daggers of light must necessarily form in vertically aligned sandwiches of rock that have fractured along flat sedimentary planes. In my view, modification of the light-dagger-forming surfaces may have occurred, but probably only by natural weathering. In any case, modification would have only a small effect on the essential visual outcome.

The real challenge relative to the second scenario would have been to make the three lunar markings at rising function in consort with the solstice markings at midday when these latter alignments had already uniquely fixed the location of the large spiral. If the rise alignments are not purely serendipitous, the Chacoan designer would probably have needed to make two types of modifications of the shadow-casting outer edge of Slab 1. First, this edge must have been modified and/or Slab 1 must have been levered in or out, toward or away from the cliff face, to align the shadow edge. Second, the projected distance from the shadow-casting edge to the large spiral would have to have been adjusted by moving Slab 1 toward or away from the spiral to adjust the angles subtended as the moonrise moved from 0.0 to +18.4 to +28.7 degrees declination. Since *any* movement of Slab 1 would disrupt the delicate geometry already established to mark the solstices at midday, only modification of the outer shadow-casting edge of Slab 1 would avoid this problem.

At this point, two relevant observations become important. Even a casual inspection of the shadow-casting edge of Slab 1 shows that it has not been modified by abrasion, grinding, or further fracturing, Figures 30.1e–g and the cover photograph published in Sofaer et al. (1979b) all support this conclusion. The full fracture planes on the insides of all three slabs are in virtually the same condition that they were when the slabs broke away from the point on the cliff face where they match the parent strata. Furthermore, as can also be seen in these photos (which are no substitute for on-site inspection), the inside fracture planes of the three slabs parallel each other rather loosely. Slab 1 rests against the cliff face at roughly the same angle as the other two slabs; if Slab 1 was levered in or out, the others were moved in the same way and to the same degree.

My view of the scenario of minor modifications of a natural rockfall is that the evidence argues strongly against either of the two types of alteration of the moonrise shadow edge that would have been necessary to create these alignments. Hence, they are serendipitous at best and were unintended by and probably unknown to the carvers of the petroglyphs. Thus, I do not accept the alleged lunar significance of the nine and one-half turns of the large spiral.

The Three-Slab Site: A Substantial Engineered Construction

We know from their massive feats of masonry construction in the many ancient pueblos of Chaco Canyon that the Anasazi and later Pueblo peoples were capable of great engineering achievements in stone. For example, thick, one-ton, worked stone disks were placed as foundation supports for the great tree-sized posts that supported the roof beams in the great kivas at Casa Rinconada and Chetro Ketl. The three-slab site shows no evidence of the skilled handiwork of the Anasazi stonemason. Indeed, the 1000 kg slabs are precariously balanced, showing no evidence of foundation or buttressing—Anasazi or otherwise. The subtle, "natural" appearance of the stone was described by mythologist Joseph Campbell in "The Sun Dagger" film as being "somehow very Indian." The thought is good, but in this case the superior hypothesis, following Newman et al. (1982), is that the site looks so natural because it *is* natural, in contrast to the appearance of most other Pueblo sun shrines. If we nonetheless hypothesize that the three-slab site was a construction designed to do everything that has been claimed for it, we must see how this hypothesis fits the facts.

As discussed in detail above, the slabs are found in the same three-layered sandwich configuration that they exhibited when they formed part of the cliff, the facing sides and inner sides are nearly pristine (unweathered and unmodified), and the slabs are precariously balanced against the cliff face with no archaeological evidence for human use save for the two spiral petroglyphs. As also mentioned previously, no one would have suggested that the site was anything other than a natural rockfall had the sun dagger phenomenon not been discovered.

In spite of the seeming freedom to create available to the designers in this third scenario, in fact they would have had a much greater burden. They had somehow to visualize in advance the whole complex creation involving the erection of an existing ensemble of three sandstone slabs, presumably then lying flat below the cliff face amid additional rubble. Whatever they did, their end result would have to maintain this sandwich configuration with the inside edges unmodified because this is the way we find it today. As weathering is evident only on the currently exposed outer surfaces, the implication is clear that the ancient engineers found the slabs largely unweathered; all of the external weathering has to have taken place in the years since the hypothesized human creation of the site. Furthermore, the designer/engineers must have had a very good knowledge of the solar and lunar excursions along the horizon and near meridian transit as viewed from this site, and they must have accurately known the lengths of the annual solar solstice and almost 19-year lunar standstill cycles. To create this unique site today, both in conception and in execution, would be a formidable task for a modern astronomer/engineer, let alone a Pueblo sun priest. Yet the site does exist, and it apparently will "work" in a manner close to the way described by Sofaer et al. (1982).

My view is that, even given the freedom of design implied by this third scenario and the engineering capability for execution of this design, the physical and geometrical constraints as presented make the task almost impossible. I would be the last person to argue against the philosophy that "anything is possible under the sun," and perhaps under the moon, but I reject this third scenario as being far less likely than the second, though it seems at first glance to provide greater creative freedom for the Anasazi designer. At least in the second scenario, the essentials of the design already existed naturally, waiting to be discovered.

CONCLUSIONS

In rethinking interpretations of the three-slab site as comprehensively as possible, I conclude that it is a sun shrine marking the summer solstice and probably the winter solstice and equinoxes. The two spiral petroglyphs were carved behind a natural rockfall by the Chacoan sun watchers to mark these events, taking advantage of the natural patterns of light that have come to be called sun daggers. Any modification of the outer edges of the slabs would have little effect on the essential light phenomena. Crudely executed though they are, the spiral petroglyphs in conjunction with the natural lean-to formed by the slabs create an elegant and striking display of light marking the annual solar events. Previous knowledge of the times of the solstices and, in particular, of the equinoxes would have been necessary for the petroglyph carvers—knowledge certainly derived through the standard Pueblo practices of horizon sun watching. Despite claims to the contrary, no particular precision is manifest in the marking of the solar events at the three-slab site. The design of the site and its inaccessible location can be used to support the argument (Zeilik 1985) that it probably functioned as a sun shrine and not as a sun watcher's station or a "calendrical" site.

Furthermore, there is no compelling evidence for intentional lunar alignments at the three-slab site. Any potential lunar standstill moonrise alignments are serendipitous and were most likely unknown to the Chacoan petroglyph carvers, who were primarily taking advantage of the midday sun dagger phenomena to mark culturally meaningful solar events.

This brings up the question of who carved the two spiral petroglyphs at the site. Sofaer et al. (1979b:289–290) have suggested that "several factors show that the Anasazi inhabitants of Chaco developed the construct between AD 900 and 1300 (the approximate date of Pueblo abandonment of the canyon) and indicate that the specific time was between AD 950 and 1150, the period of greatest population and development in the canyon." This suggestion is not unreasonable, but neither is it supported by any archaeological evidence save the general stylistic analysis of the petroglyphs.

In fact, the only known archaeological remains found on Fajada Butte are two sites attributed to the late occupation of Chaco Canyon by Mesa Verde people—or Chacoans influenced by Mesa Verdean cultural traits—around AD 1220–1300 (Zeilik 1985:S72–73).

> This period is characterized by occupational use of cave sites and the tops of buttes (in addition to more traditional locations), a change in burial patterns, and the appearance of Mesa Verde Black-on-white and St. Johns Polychrome ceramics. Although once interpreted as evidence of a migration of Mesa Verde Anasazi into the Chacoan area, later analyses suggest a change to increased economic interaction with Mesa Verde Anasazi rather than a migration. By 1300, however, Chaco Canyon was abandoned. (Cordell 1984:102)

Though no artifacts have been found at the three-slab site itself, it was originally included as part of one of the two "Mesa Verdean talus-cliff sites" by the National Park Service Chaco Center Inventory surveyors in the summer of 1972 (Zeilik 1985:S72). This proximity does not prove that the three-slab site petroglyphs are from the Mesa Verde phase, but this must be taken as one of the best-supported hypotheses.

We may never know who carved the two spiral petroglyphs behind the sandstone slabs resting against the cliff face atop Fajada Butte. When and why they were created will likely remain the subject of controversy for years to come. What is certain, however, is that, with the mass of speculation and misinformation that has been purveyed to the public, the sun dagger will become a pseudoscientific legend of our time and times to come. Feeding the legend is the persistent confusion of the well-known Pueblo interest in moon watching involving the 29.5-day lunation cycle with the question of observing the lunar standstills and their 18.6-year period. As in the case of England's Stonehenge, we are witnessing an anthropological odyssey, the creation of a modern myth. This process is worthy of anthropological study in its own right. As romance novelist and heroine Joan Wilder discovered in a popular, tongue-in-cheek adventure movie of 1984, there is boundless excitement and adventure to be had in "Romancing the Stone."

REFERENCES

Aveni, A. 1987. Archaeoastronomy in the Southwestern United States: A Neighbor's Eye View. In *Astronomy and Ceremony in the Prehistoric Southwest*, ed. J. B. Carlson and W. J. Judge, pp. 9–24. Papers of the Maxwell Museum of Anthropology, no. 2.

Carlson, John B. 1983. The Selling of Fajada Butte: An Anacalypsis. *Archaeoastronomy* 6:156–160.

Carlson, John B. 1984. Film review of "The Sun Dagger" (produced by Anna Sofaer, directed by Albert Ihde). *Archaeology* 37 (2):78–79.

Cordell, Linda S. 1984. *Prehistory of the Southwest*. New York, Academic Press.

Frazier, Kendrick. 1978. Solstice-Watchers of Chaco. *Science News* 114:148–151.

Frazier, Kendrick. 1979 The Anasazi Sun Dagger. *Science* 80 (1):56–67.

Hadingham, Evan. 1984. Film review of "The Sun Dagger" (produced by Anna Sofaer, directed by Albert Ihde). *American Anthropologist* 86:232–233.

Marshall, Leslie. 1982. Secret of the Sun Dagger. *The Washington Post Magazine,* 18 July, 12–17.

Newman, Evelyn B., Robert K. Mark, and R. Gwinn Vivian. 1982. Anasazi Solar Marker: The Use of a Natural Rockfall. *Science* 217:1036–1038.

Reyman, Jonathan E. 1980. Letters: An Anasazi Solar Marker? *Science* 209:858, 860.

Sofaer, Anna, and Albert Ihde. 1982. "The Sun Dagger." The Solstice Project. Bullfrog Films, distributor.

Sofaer, A., R. M. Sinclair, and L. E. Doggett. 1982. Lunar Markings on Fajada Butte, Chaco Canyon, New Mexico. In *Archaeoastronomy in the New World*, ed. A. F. Aveni, 169–181. Cambridge, Cambridge University Press.

Sofaer, Anna, Volker Zinser, and Rolf Sinclair. 1979a. A Unique Solar Marking Construct of the Ancient Pueblo Indians. In *American Indian Rock Art*, vol. 5, ed. Frank G. Bock, Ken Hedges, Georgia Lee, and Helen Michaelis, 115–125. Papers Presented at the Fifth Annual A.R.A.R.A. Symposium, The Dalles, Oregon, May 27–29, 1978. El Toro, CA, American Rock Art Research Association.

Sofaer, Anna, Volker Zinser, and Rolf Sinclair. 1979b. A Unique Solar Marking Construct. *Science* 206:283–291.

Williamson, Ray A. 1979. Field Report: Hovenweep National Monument. *Archaeoastronomy* 2 (3):11–12.

Williamson, Ray A., and Mary Jane Young. 1979. An Equinox Sun Petroglyph Panel at Hovenweep National Monument. In *American Indian Rock Art*, vol. 5, ed. Frank G. Bock, Ken Hedges, Georgia Lee, and Helen Michaelis, 70–80. Papers Presented at the Fifth Annual A.R.A.R.A. Symposium, The Dalles, Oregon, May 27–29, 1978. El Toro, CA, American Rock Art Research Association.

Zeilik, Michael. 1985. A Reassessment of the Fajada Butte Solar Marker. In Archaeoastronomy Supplement No. 9. *Journal for the History of Astronomy* 16:S69–S85.

CHAPTER THIRTY-ONE

Lunar Standstills at Chimney Rock

J. McKim Malville, Frank W. Eddy, and Carol Ambruster

INTRODUCTION

Whereas prehistoric populations in North America paid attention to the horizon sun at winter or summer solstice, evidence for observations of comparable phenomena involving the moon remains elusive and controversial. The amplitude of the monthly movement of the moon varies from a minimum at minor standstill to a maximum at major standstill with an 18.6-year period. Studies of megalithic sites in Britain dating between 3500 and 1500 BC provide evidence of deliberate orientation to lunar rising and setting points, especially those of maximum southern standstill.[1]

In North America the most thoroughly studied standstill marker is the three-slab site of Fajada Butte in Chaco Canyon.[2] The central slab has recently shifted, causing a change in the shadow patterns of solar apparitions, but the lunar standstill effects may

---

From *Archaeoastronomy* 16 (*JHA* 22) (1991):S43–S50. Reproduced by permission of the authors.

have remained unaltered.[3] In addition, there are indications of standstill observations at Casa Grande[4] and Zodiac Ridge.[5]

Although the lunar aspects of the three-slab site were confirmed during the recent major standstill,[6] questions remain as to the intentionality and significance of the proposed lunar markings. It is puzzling that there is no ethnographic evidence that the historic Pueblo Indians observed or had knowledge of the 18.6-year cycle.[7] The astronomy of the Anasazi may indeed have been more elaborate than that of the historic Pueblo.[8] For cultures relying upon lunar-based calendars, systematic observations of the moon could have readily revealed the monthly standstill and probably also the systematic shift of standstill positions.

We present evidence for observation of lunar standstill within the Chimney Rock Archaeological Area (CRAA) in Southwestern Colorado (Fig. 31.1). The CRAA contains evidence of eight semi-independent, integrated communities of late Pueblo I and Pueblo II age (AD 925–1125).[9] This collection of sites is dominated by the double natural pinnacles of Chimney Rock and by the Chimney Rock Pueblo, 5AA83, which has been identified as a Chaco outlier.[10] The chimneys and the associated high mesa have been recognized by the Taos Pueblo of northern New Mexico to be a shrine dedicated to the Twin War Gods of Pueblo mythology.[11]

Dramatically set on the high mesa, 1200 feet above the water and agricultural lands of the valley floor, the Chimney Rock Pueblo contains two kivas and thirty-five ground floor rooms. A second story may have contained an additional twenty rooms. The core-and-veneer masonry is similar in style to the Great Houses of Chaco Canyon located some 150 straight-line kilometers to the south.[12] Among the proposed outliers of Chaco Canyon, the Chimney Rock Pueblo, 5AA83, is distinguished by being the most isolated (72.5 km to Aztec), the highest (2316 m), the most remote from arable land (2 km), and one of the first to be constructed (AD 1076).[13]

The Chaco-style East Kiva of 5AA83 contains two subfloor ventilator tunnels, one on top of the other, a southern vertical ventilator shaft and a 5-ft-high banquette upon which were built eight beam rests. Other features characteristic of Chaco-style kivas are a western subfloor vault and no evidence of a sipapu. The East and West Kivas were set within a quadrangle of walls, which had been filled by the prehistoric inhabitants to the level of the kiva roofs, forming courts within the building. The East Kiva, in particular, is enclosed in a rectangular dead space which is further enclosed on the south and east by small rooms which were probably filled during the occupancy of the Pueblo. The general layout is similar to tri-wall structures which may have been built to provide ceremonial platforms.[14] In the case of the East Kiva we propose it provided an observing platform for viewing the double chimneys. The L-shaped building encloses the East Court. A bench has been built along the east-facing exterior wall in the East Court, providing another place from which to view the double chimneys.

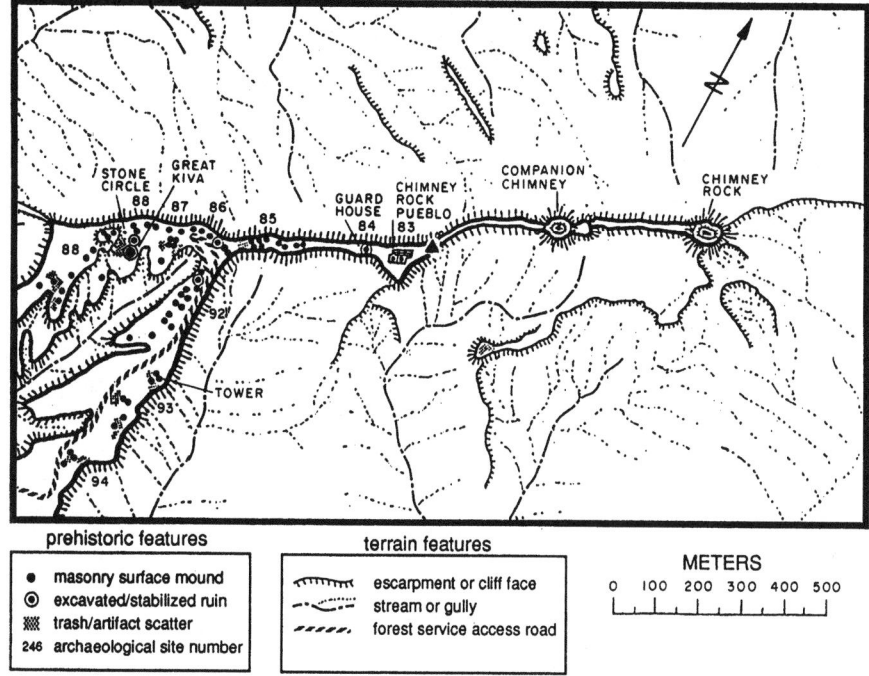

31.1 Map of the high mesa of Chimney Rock; the location from which the lunar standstill was observed in 1988 is marked by a triangle. Map prepared by Dale Lightfoot, based on aerial photography.

OBSERVATIONS

Following his excavation in the Chimney Rock area, one of the authors (FWE) postulated that 5AA83 was built by Chacoan priests not from practical considerations but from a religious motivation to view and live near the double chimneys. In the spring of 1988 we advanced the hypothesis that the Pueblo was constructed, at least in part, in order to use the double chimneys as a foresight for astronomical phenomena. It appeared possible that as viewed from 5AA83, the sun, Venus and/or the moon could rise between the chimneys; Venus was particularly implicated because of its association with the Twin War gods in some Pueblo traditions.

Our survey of the Chimney Rock area during May 1988 established that neither the sun nor Venus reaches sufficiently high northern declination to rise in the gap between the chimneys as viewed from 5AA83. However, we found that the moon should rise in the gap at times of major northern standstill when viewed from the vicinity of 5AA83, as indicated in Figure 31.1.

A watchtower has been constructed within recent years on the high mesa between 5AA83 and the chimneys, thereby blocking the view of the gap between the double chimneys from the Pueblo. Our measurements were made on a prominent rock

LUNAR STANDSTILLS AT CHIMNEY ROCK

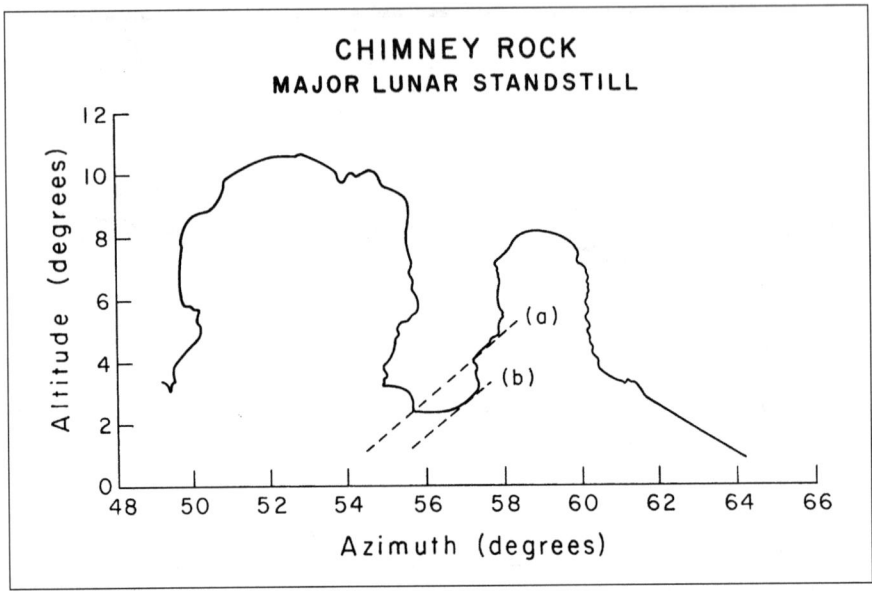

31.2 Chimney Rock as viewed from the high mesa, northeast of 5AA83. The path of the centre of the moon at two declinations is indicated: (a) 28.7°, (b) 27.9°.

platform at the northeast corner of the high mesa, which is along a line connecting the Pueblo with the gap. During the summer of 1988 sufficient time remained to observe moonrise at high declinations, and on 8 August 1988, we were able to confirm our prediction, as shown in Figures 31.2 and 31.3.

LUNAR STANDSTILLS DURING OCCUPATION OF CHIMNEY ROCK

Dates of lunar standstills were determined by calculating the longitude of the ascending node of the lunar orbit using the equation:

$$\Omega = 259.183° - 0.05295°d + 0.002078°T^2 + 0.000002°T^3,$$

where d and T are respectively the days and Julian centuries from J.D. 2415020. The declination of the moon was estimated by the function: $23.56° + 5.14°\cos(2\pi t/18.613)$, where t is the time from the date of zero longitude of the ascending lunar node; the inclination perturbation is not included. The minimum lunar declination visible between the chimneys is 27.9° as seen from ground level near 5AA83, taking into account lunar parallax and refraction. We estimate that within 1.2 years from the date of zero longitude of the ascending node, every lunar month should have had at least one moonrise that cleared the gap between the spires as viewed from 5AA83. During the period from 1050 to 1120, the predicted dates of consecutive moonrises between the chimneys are 1056.0–1058.4, 1074.6–1077.0, 1093.2–1095.6, and 1112.8–1114.2.

31.3 Moonrise on the morning of 8 August 1988. The moon's declination was 28.5°. The bright stars rising ahead of the moon were θ Aur and β Tau.

The availability of tree ring dates from beams excavated by Eddy from the East Kiva and Room 8 of 5AA83 provides us with the opportunity for testing whether observability of lunar standstills may have had an influence upon construction of the Chimney Rock Pueblo. Specimens from 74 original trees were collected from the East Kiva and from the remains of the roof of Room 8 of 5AA83. Of the 47 datable specimens, 41 came from Room 8, and of those 26 were dated at AD 1093. All of the remaining 15 dates, which fall between 1066 and 1092, are questionable because they were obtained from logs in which an unknown number of outer rings had been lost. A majority of the logs (17) that date to 1093 contained continuous outermost rings, which indicate cutting dates; two specimens contained bark and therefore give definite cutting dates. No logs were dated after 1093, indicating that it was not necessary to repair or renovate the roofing for the remainder of the life of the building which was destroyed by fire around 1125.

The six datable logs from the East Kiva yield two dates. One pole taken from the lower (and therefore the earlier) of the two horizontal ventilator tunnels contained a continuous outer ring and dates to AD 1076. A second log from the roof fall with a continuous outer ring dates to 1093.

Eddy[15] has concluded that the initial construction of the Pueblo was associated with the dendro-date of AD 1076. Some seventeen years later, about 1093, the floor of the East Kiva was raised and a second horizontal ventilator tunnel was constructed. The renovated kiva, Room 8, and by implication other portions of the Pueblo, were re-roofed with logs cut in 1093.

Comparing these dendro-dates with those of lunar standstills, we find that both of the two sets of cutting dates coincide with major lunar standstills. No trees with datable outer rings were discovered from the Pueblo with cutting dates other than 1076 and 1093.

LUNAR STANDSTILL AT WINTER SOLSTICE

The most important astronomical event visible from 5AA83 may have been the rising of the full moon between the chimneys at the time of winter solstice. Among the historic Pueblo, observations of the sun and moon established ceremonial calendars. The Zunis attempted the difficult task of organizing their calendar such that winter solstice occurred at or near full moon. Within certain Pueblo traditions, White Shell Woman (the moon) helps to persuade the sun to return north at winter solstice.[16] At Hopi, the sun chief watched the setting sun to establish the date of Soyal, the winter solstice ceremony.

In the case of these two historic Pueblos, the rising moon was carefully watched near winter solstice. The inhabitants of the Chimney Rock Pueblo may have been similarly watchful of the rising moon. Near the winter solstice of AD 1055, or perhaps earlier, residents of the Chimney Rock area may have noted the full moon rising between the chimneys. Using Goldstine[17] and Bretagnon and Simon[18] we find that the full moon of 14 December 1057 occurred at a time when the sun had a declination less than 1' from its value at solstice. A pre-telescopic observer could not have distinguished the date from that of solstice. On such an event a solar calendar and a lunar calendar could have been brought into agreement.

LUNAR STANDSTILL PRECEDING WINTER SOLSTICE

We suggest that the apparitions of the moon during the autumn prior to full moon standstill at winter solstice constituted a major astronomical event for the Chimney Rock population. During that autumn season the percentage of illumination of the moon rising between the chimneys systematically increased each month, culminating in the rising of the full moon at dusk at winter solstice. These six successive moonrises should have provided an unmistakable opportunity for anticipation and may have provided confirmation of the predictive power of the priests as well as the order of the heavens. Of the other proposed instances of lunar standstills observed in either the American Southwest or megalithic Britain, none appears to have the arresting quality of the event at Chimney Rock: the moon, growing in size each month, appears in the gap in a darkened sky during the autumn; soon after rising it is eclipsed by the high pinnacle. In the east court, in the courts above the kivas, and on roof tops people could have gathered to observe moonrise through the foresight of the double spires.

It was not necessary for those responsible for planning and construction to have knowledge of the 18.6-year lunar cycle. Anticipatory observations of moonrise at

Table 31.1 Chronology of Chimney Rock.

	1056.0–1058.4	
Full moon rising between the chimneys:		25 Dec. 1056
Full moon rising between the chimneys:		14 Dec. 1057
Winter solstice:		15.6 Dec. 1057
	1074.6–1077.0	
Full moon rising between the chimneys:		6 Dec. 1074
Full moon rising between the chimneys:		25 Dec. 1075
Solar eclipse:		7 March 1076
Cutting of trees for 5AA83:		July–Aug. 1076
Full moon rising between the chimneys:		13 Dec. 1076
Winter solstice:		14.2 Dec. 1076
	1093.2–1095.6	
Cutting of trees for 5AA83:		July–Aug. 1093
Full moon rising between the chimneys:		5 Dec. 1093
Full moon rising between the chimneys:		24 Dec. 1094
Full moon rising between the chimneys:		14 Dec. 1095
Winter solstice:		15.8 Dec. 1095

the start of each northern major standstill were clearly possible, and an accurate year count would have been unnecessary.

As indicated in Table 31.1, during the latter half of the eleventh century the rise of the full moon between the chimneys recurred at winter solstice at intervals of the Metonic cycle of 19 years while the "window of opportunity" for moonrise between the chimneys recurred at intervals of 18.6 years. Construction of the Pueblo or portions thereof may have been completed before the full moon of 13 December 1076. It is worthy of note that one of the two total solar eclipses in the Southwest during the latter half of the eleventh century occurred on 7 March 1076.[19] Although the eclipse was partial at Chimney Rock, the conjoining of an eclipse with the lunar apparitions between the chimneys may have further intensified interest in astronomical observation. Tree cutting commenced some four months after the eclipse.

Trees for the second episode of construction, perhaps an occasion of ceremonial reconstruction, were cut in the summer of 1093 and the first moonrise that cleared the chimneys occurred in the spring of that year. The full moon could have been seen rising between the chimneys near three winter solstices.

RELATIONSHIP TO THE THREE-SLAB SITE OF FAJADA BUTTE

Knowledge of lunar cycles may have been shared between the outlier at Chimney Rock and the centre at Chaco Canyon. As we have noted, there has been scepticism that the Anasazi actually paid attention to lunar standstills at Fajada Butte.[20] Besides the lack of appropriate ethnographic evidence, such scepticism has been prompted by the difficulty of dating the three-slab site, the absence of any obvious ceremonial or practical relevance of standstill, and the difficulty of anticipation of the event.

At Chimney Rock we encounter a well-dated site at which the major standstill is unmistakable and fully predictable. Since the relationship between the lunar and solar calendars could be re-established on such an occasion, both the Metonic cycle and the standstill cycle may have been ceremonially important and calendrically relevant. (During the 19 years between 1076 and 1095 there were only two full moons, in addition to that at standstill, when the sun was within 1' of winter solstice declination: in 1084 and 1092.) The sight of the major lunar standstill at the Chimney Rock outlier may have been an astronomical phenomenon of such acknowledged magnitude and relevance that it inspired the lunar aspects of the three-slab site at Fajada Butte. The tree cutting dates clearly indicate that people were working near the Chimney Rock Pueblo during the two lunar standstills that occurred during the florescence of Chacoan culture.

ACKNOWLEDGMENTS

This work was supported in part by a grant from the Council for Research and Creative Work of the University of Colorado. We are grateful to Milan Halek of the Department of Civil Engineering of the University of Colorado for loan of electronic distance meters and theodolites. We thank the personnel of the U.S. Forest Service for encouragement and assistance in our field work: Sharon Hatch, Peggy Jacobson, Gary Matlock, and Robert York.

NOTES

1. C.L.N. Ruggles, *Megalithic Astronomy* (Oxford, B.A.R., 1984); R. Norris, "Megalithic Observatories in Britain: Real or Imagined?" in *Records in Stone,* ed. C.L.N. Ruggles (Cambridge, Cambridge University Press, 1988), 262–276.

2. A. Sofaer, R. M. Sinclair, and L. E. Doggett, "Lunar Markings on Fajada Butte, Chaco Canyon, New Mexico," in *Archaeoastronomy in the New World,* ed. A. F. Aveni (Cambridge, Cambridge University Press, 1982), 169–181.

3. A. Sofaer and R. M. Sinclair, "Changes in Solstice Marking at the Three-slab Site, New Mexico, USA," *Archaeoastronomy* (Supplement to *Journal for the History of Astronomy*), 15 (1990):S59–60.

4. J. Evans and H. Hillman, "Documentation of Some Lunar and Solar Events at Casa Grande," in *Archaeoastronomy in the Americas,* ed. R. Williamson (Los Altos, CA, Balena Press, 1979), 133–136.

5. N. Autrey and W. Autrey, "Zodiac Ridge," in ibid., 81–100.

6. R. Sinclair, A. Sofaer, and J. J. McCann Jr., "Marking of Lunar Major Standstill at the Three-slab Site on Fajada Butte," *Bulletin of the American Astronomical Society* 19 (1987): 1043.

7. M. Zeilik, "A Reassessment of the Fajada Butte Solar Marker," *Archaeoastronomy* (Supplement to *Journal for the History of Astronomy*) 9 (1985):S69–85.

8. F. H. Ellis, "A Thousand Years of the Pueblo Sun-moon-star Calendar," *Archaeoastronomy in Precolumbian America,* ed. A. F. Aveni (Austin, University of Texas Press, 1975), 59–87.

9. F. W. Eddy, *Archaeological Investigations at Chimney Rock Mesa: 1970–1972* (Boulder, Colorado Archaeological Society, 1977); L. D. Webster, *An Archaeological Survey of the West Rim of the Piedra River* (Durango, CO, USDA Forest Service, 1983).

10. R. P. Powers, W. B. Gillespie, and S. H. Lekson, *The Outlier Survey: A Regional View of Settlement in the San Juan Basin* (Albuquerque, Division of Cultural Research, National Park Service, U.S. Department of the Interior, 1983).

11. Eddy, *Archaeological Investigations*.

12. S. W. Lekson, *Great Pueblo Architecture of Chaco Canyon* (Albuquerque, National Park Service, U.S. Department of the Interior, 1984).

13. Powers, Gillespie, and Lekson, *Outlier Survey*.

14. R. G. Vivian, *The Hubbard Site and Other Tri-wall Structures in New Mexico and Colorado* (National Park Service Archaeological Research Series, no. 5; Washington, DC, 1959).

15. Eddy, *Archaeological Investigations*.

16. F. H. Ellis and L. Hammack, "The Innersanctum of Feather Cave, a Mogollon Sun and Earth Shrine Linking Mexico and the Southwest," *American Antiquity* 30 (1968):25–44.

17. H. H. Goldstine, *New and Full Moons 1001 BC to AD 1651* (Philadelphia, American Philosophical Society, 1973).

18. P. Bretagnon and J. Simon, *Planetary Programs and Tables from −4000 to +2800* (Richmond, VA, Willmann-Bell, 1987).

19. H. Mucke and J. Meeus, *Canon of Solar Eclipses* (Vienna, Astronomisches Buro, 1983).

20. A. F. Aveni, "Archaeoastronomy in the Southwestern United States: A Neighbor's Eye View," in *Astronomy and Ceremony in the Prehistoric Southwest,* ed. J. B. Carlson and W. James Judge (Albuquerque, Maxwell Museum of Anthropology, 1987), 9–23; J. B. Carlson, "Romancing the Stone, or Moonshine on the Sun Dagger," in ibid., 71–88; M. Zeilik, "A Reassessment of the Fajada Butte Solar Marker"; idem, "The Ethnoastronomy of the Historic Pueblo: Moon Watching," *Archaeoastronomy* (Supplement to *Journal for the History of Astronomy*) 10 (1986):S1–22; A. Sofaer and R. M. Sinclair, "An Appraisal of Michael Zeilik's 'A Reassessment of the Fajada Solar Marker,'" ibid., S59–66; M. Zeilik, "Response," ibid., S66–69.

PART VI

The Present and Future of Cultural Astronomy

In the last part of *Foundations*, I bring together what I consider to be some of the most revealing discussions of basic issues that persist in the development of the still relatively young interdisciplines of archaeoastronomy and ethnoastronomy. Here readers will discern that students of culture (anthropologists) and students of nature (scientists) indeed make strange bedfellows. (I speak from experience: I have had one foot in each of two such academic departments for thirty years at my university!) Knowledge that seems important to the archaeologist or anthropologist may matter less to the trained astronomer (and vice versa). In my view, real progress is achieved only when individuals schooled in diverse disciplines work together, each giving something substantial from these disciplines to the others.

We live in an interdisciplinary age. If we think of the disciplines as floorboards in the house of knowledge, then the interdisciplines are concerned with what happens in the cracks among them. As a teacher for more than forty years, I am well aware of courses and programs of high quality that exist in a number of university curricula—from bio-

PART VI: THE PRESENT AND FUTURE OF CULTURAL ASTRONOMY

physics and neuroscience to several different kinds of environmental studies, cognitive psychology, astrobiology, astrogeophysics, and so forth. Because it came along in advance of the now-accepted trend toward interdisciplinary learning, cultural astronomy (then archaeoastronomy) tended to be viewed with considerable skepticism. And as I mentioned earlier, its affiliation with crackpot ideas did not help its cause.

To begin this final section, I offer an introductory essay to an archaeoastronomy conference by astronomer Clive Ruggles and anthropologist Nicholas Saunders. It is important because it proposed nothing less than a reorientation, replete with renaming, of the interdiscipline of archaeoastronomy, suggesting the term "cultural astronomy," or astronomy *in* culture. For an interesting dialogue on where archaeoastronomy was thought to belong a decade earlier, see the symposium section in Williamson 1981. I follow this with a series of ongoing debates on specific issues. This includes a discussion of whether "cosmograms" played a role in Mesoamerican (especially Maya) city planning; then archaeologist Jim Judge, in a piece that originally appeared in an excellent publication that unfortunately never received wide distribution, weighs in on the situation in cultural astronomy in the U.S. Southwest. This section concludes with a brief dialogue, in the little-known interim publication *Archaeoastronomy and Ethnoastronomy Newsletter*, between me and archaeologist Keith Kintigh. Our two pieces briefly articulate where obstacles to building disciplinary bridges may yet lie. Updates to a number of such dialogues and debates are provided in the section on further reading.

THINGS TO THINK ABOUT

1. Write a short essay on the difference between the following two statements: (1) Maya cosmological principles are reflected in city planning, and (2) Maya cities are cosmograms.
2. Write short essays on what you believe cultural astronomy has contributed to each of the following disciplines: archaeology, anthropology, history, history of astronomy, and the modern science of astronomy.

SUGGESTIONS FOR FURTHER READING

Aveni, A. 1989. "Whither Archaeoastronomy?" In *World Archaeoastronomy*, ed. A. Aveni, 3–12. Cambridge, Cambridge University Press.

Aveni. A. 1987. "Archaeoastronomy in the Southwestern United States: A Neighbor's Eye View." In *Astronomy and Ceremony in the Prehistoric Southwest Papers Maxwell Museum of Anthropology* 2:1–8.

Aveni, A. 2006a. "Evidence and Intentionality: On Method in Archaeoastronomy." In *Viewing the Sky through Past and Present Cultures,* ed. T. Bostwick and B. Bates, 51–70. Phoenix, Pueblo Grande Museum Anthropological Papers.

Aveni, A. 2006b. "Schaefer's Rigid Ethnocentric Criteria." In *Viewing the Sky through Past and Present Cultures,* ed. T. Bostwick and B. Bates, 79–84. Phoenix, Pueblo Grande Museum Anthropological Papers.

Carlson, J., D. Dearborn, and S. McCluskey. 1999. "Astronomy in Culture." *Archaeoastronomy Journal of Astronomy in Culture* 16:3–21.

Schaefer, B. 2006a. "Case Studies of Three of the Most Famous Claimed Archaeoastronomical Alignments in America." In *Viewing the Sky through Past and Present Cultures,* ed. T. Bostwick and B. Bates, 27–56. Pueblo Grande Anthropological Papers No. 15, Phoenix.

Schaefer, B. 2006b. "No Astronomical Alignments at the Caracol." In *Viewing the Sky through Past and Present Cultures,* ed. T. Bostwick and B. Bates, 71–78. Pueblo Grande Anthropological Papers No. 15, Phoenix.

Williamson, R., ed. 1981. *Archaeoastronomy in the Americas.* Los Altos, CA, Ballena, and College Park, Center for Archaeoastronomy.

Zeilik, M. 1985. "A Reassessment of the Fajada Butte Solar Marker." *Archaeoastronomy* 9 (Supplement to *Journal for the History of Astronomy* 16):S69–S85.

Zeilik, M. 1986. "The Fajada Butte Solar Marker: A Reevaluation." *Science* 228:1311–1313.

CHAPTER THIRTY-TWO

The Study of Cultural Astronomy

Clive L.N. Ruggles and Nicholas J. Saunders

THE SKY AS A CULTURAL RESOURCE

What do people see when they look at the night sky? The answer is as much a cultural as an astronomical one. Thus, the study of cultural astronomies is concerned with the diversity of ways in which cultures, both ancient and modern, perceive celestial objects and integrate them into their view of the world. This fact, by definition, illustrates that a society's view of and beliefs about the celestial sphere are inextricably linked to the realm of politics, economics, religion, and ideology. In this sense, cultural astronomy is but part of the wider endeavor of investigating and interpreting human culture.

Anthropologists and archaeologists are concerned with retrieving data on cultural systems, identifying patterns in these data, and attempting to elicit meanings from these patterns. A crucial aspect of this process, and one that has hitherto been

---

From *Astronomies and Cultures*, ed. C.L.N. Ruggles and N. J. Saunders (Niwot: University Press of Colorado, 1993).

largely overlooked, is the consideration of perceived relationships between a society and its environment. How cultures conceive of themselves in relation to what may be called the "natural world" determines their view of and attitude toward all beings and entities that they recognize as coexisting in their phenomenological universes.

In one sense, every culture makes an appraisal of its own environment by selecting certain aspects of its physical surroundings—such as other people, animals, mountains, meteorological and celestial phenomena—and assigning meaning to them and to the relationships that they regard as existing between them. Interaction with different aspects of the "natural world" can be seen as taking place through "channels of interaction" that are culturally defined (Wobst 1990: 327). Thus, cultures can be seen as creating or constructing their own meaningful universes (Kus 1983: 278), without which human behavior would be virtually ungovernable (Hodder 1985: 2).

In order to explore these ideas more fully, to illustrate their relevance to cultural astronomy, and to situate cultural astronomy within the debate about the epistemic limits of inquiry, we will begin by analyzing what we consider to be three distinct but inevitably linked processes that hitherto have often been lumped together with a consequent loss of analytical precision.

Observation

In general terms, the observation of celestial objects can be either casual or deliberate (that is, more or less organized). In many cases, for example, where recurrent phenomena are involved, a certain level of observational competence is required. All those who attain this level of competence can, in one sense, be called astronomers, although they will in general bear little resemblance to astronomers in the modern sense of the word.

Perception

Quite distinct from observation is the process of perception: the process, in other words, of making sense of and attaching meaning to particular observations. Different cultures and individuals may "see" the same objects in the sky, but the significance that they attach to them will be influenced by their classification of the natural world and the various ways in which they interact with it. Thus, when the Bororo of central Brazil lie down at dusk and look at the night sky, they ignore meteors when they are distant and silent. However, when they hear or see one land they react with great alarm and consult the local shaman (Crocker 1985), an act that links astronomy directly to other spheres of cultural activity, such as religious belief and mythology.

Perceptions are guided by both cultural and psychological factors (Neisser 1976). They may also be channeled (though not necessarily constrained) either explicitly or implicitly by a culture's level of socioeconomic and political organization, its ideological imperatives, and the nature of its relations with the environment. In particular, and despite statements to the contrary (for example, Vogt 1990: 52), star lore cannot necessarily be expected to survive cultural attrition any better than any other kind of knowledge. This is because the practice of observing and making sense of celestial phenomena is as much a cultural activity as any other, and consequently it is equally sensitive to the varied processes of culture change.

Different societies discern in the sky culturally appropriate patterns of both appearance and movement. The former may be correlated with latitude. For example, in the western Classical tradition, rooted in the northern hemisphere, patterns of appearance tend to be composed of groups of stars that are regarded as entities in their own right (for example, constellations such as Orion). By contrast, the definition of constellations among traditional cultures of equatorial and southern latitudes may include either groups of stars, the prominent light and dark areas of the Milky Way that are only visible in southern skies such as the Andean llama (Urton 1981a), or a combination of both. However, even where there exists a concordance of shape or form between constellations recognized by different cultures this does not necessarily imply a convergence of meaning.

Perceived patterns of movement are strongly latitude-dependent. Thus, in the tropics, where the characteristic motion of the heavenly bodies is predominantly vertical, cultures tend to be concerned with up-down movement. The Borana of southern Ethiopia, for instance, reckon days by identifying the stars that rise and set on a level with the moon—in other words, the stars at the same right ascension as the moon (Bassi 1988; Ruggles 1987, 1993). A circular motion is easily perceived at these latitudes, in which the heavenly bodies continue round below the horizon when not visible above it.[1] Special importance may be attached to their passage (and particularly that of the sun) through the zenith and nadir, as is widely evidenced in the American tropics, both in Pre-Columbian times and among modern indigenous groups (Aveni 1981a: 40–45; Urton 1981a: 76; Zuidema 1981).

On the other hand, at higher latitudes in the northern hemisphere, where the heavenly motions are predominantly horizontal, there is a widespread image of the sky as a huge bowl rotating overhead with an axis at Polaris and held up by a "world pole" (Eliade 1964: 262). Such world poles have ancient and widespread roots (for example, Martynov 1988: 12–29).

By contrast, in subpolar regions the long winter nights give rise to very different realms of perception, such as fortnightly lunar "days." The Inuit, for instance, have little in the way of star lore because the stars and planets are seldom very visible due to sunlight, moonlight, and bad weather (Pearce 1993).

Use

The final process to consider is the political and ideological "use" of the astronomical knowledge gained through observation and perception. This knowledge, like any other, may be seen as a resource that can be brought to bear in support of the dominant ideology to reaffirm and reinforce the structures of a particular society. A particularly good example is the way in which the so-called priest-astronomers of Pre-

Columbian Mesoamerica used the movements of Venus to regulate their elaborate and often dynastic system of warfare (Kelley 1975; Lounsbury 1982).

The "use" that can be made of celestial phenomena, as with any aspect of the natural world, depends upon the meaning and significance that culture ascribes to its constructed universe. While use-values are established by consensus within a society, in practice it is those who guide or construct the social consensus who have the power to persuade and influence, with the likely outcome of reinforcing their own position and status. For example, in terms of religious activity, considerable social prestige and political power may accrue to those individuals who are regarded as effecting a degree of control or influence over otherwise "manifestly" uncontrollable natural phenomena. For example, an astronomer-priest may hold the solstice ceremony around the time that the sun sets in a particular horizon notch so that he can demonstrate the power of "turning it around." As, in one sense, celestial (and meteorological) phenomena are the most uncontrollable of all, the psychological effect of appearing to control them is, presumably, one of the most impressive displays of "supernatural power" (and its sociocultural correlates) available to elites.

PERCEPTION AND THE NATURAL WORLD

An essential distinction among the stages of observation, perception, and use is that the first is universal whereas the second and third are culture-specific. What one society might perceive as significant might arouse little or no interest in another. For example, the supernova of AD 1054 that resulted in the formation of the Crab nebula was marked as a dire event in China (Ho et al. 1972),[2] but left no known comment in the histories of northern Europe. Similarly, there may be a diversity of views about the sky within the same society, as the following account of the Mursi of Ethiopia illustrates:

> On one occasion a man, who had been wearing a knotted cord round his ankle for several weeks, announced to a group of bystanders that 72 days had elapsed [between] the planting and first harvesting of his sorghum crop. He had been keeping track of this interval by successively knotting a piece of cord for every day that passed. The other members of the group treated this information as a curiosity without relevance to their daily lives, not as a "discovery" to be added to their total stock of knowledge about the world. Their main reaction, in fact, was one of mild surprise that anyone should have taken the trouble to record such a trivial fact, and it was, without doubt, quickly forgotten. (Turton and Ruggles 1978: 592)

What we consider an empirical reality does not necessarily have any relevance for other cultures; even if it does, its meaning is likely to be very different from ours. Thus, significance and meaning in the night sky, as elsewhere, are neither given nor self-evident, but are culturally established by social agreement and use (Saunders 1991: 13). In other words, individuals and cultures negotiate the meaning of their constructed "natural world" and arrange the material and non-material aspects of it accordingly (Hodder 1985: 4–5). In acknowledging that things only mean what people and cultures agree that they mean, concepts of "nature" and the "natural world," together with the resources contained within them, are revealed as cultural appraisals (Sauer 1954: 2–3). What we recognize as "empirical reality" is merely our own cultural appraisal couched in the framework of a "rational," Cartesian, and thus quintessentially Western view of the world.

In attempting to understand how and why cultures assign meanings to objects, the objects themselves may be seen as possessing a set of "invariant properties" (Ingold 1988: 13). Formalized in Gibson's (1979) theory of "affordances," natural or artifactual objects are regarded as affording different uses to different people and cultures at different times. For example, the possible affordances of a standing stone might be as a trackway marker, a boundary stone, a focus of witchcraft or folk superstitions, or an integral part of a megalithic structure. In more recent times, such standing stones may also have afforded local inhabitants a good source of building material. Each of these affordances may, of course, lead to particular cultural attitudes toward the object, which may in turn lead to its modification or destruction, and consequently to bias in preservation. The difficulties that this can produce in attempting to reconstruct the object's original form and meaning is well illustrated by Ucko et al. (1991) in relation to Avebury. Similarly, Stonehenge always possessed the possibility of being used as a computer, but this has only become one of its affordances since about AD 1950, a fact that illustrates the important point that only certain types and levels of society can recognize and make use of a particular affordance.

In different cultural contexts, or merely in different situations, the same affordance may have similar or very different meanings. A jaguar pelt, for example, afforded a use as a prestigious item of clothing both to fashion-conscious women in the 1960s and 1970s and to (male) Amazonian shamans and warriors from time immemorial. However, while no Westerner would regard the wearing of such garments as bestowing supernatural powers on the wearer, those native Amazonians who wore such clothing were often regarded as particularly brave and powerful, and consequently attracted more lovers (Roe 1998: 3). Thus, despite wholly different cultural aesthetics, the connotations of sexual attractiveness afforded by a jaguar's skin were in one sense shared, whereas other connotations remained distinct. Meaning and significance are therefore situation-dependent—even within the same society—and apparent similarities between societies may have quite different explanations.

STRUCTURING THE NATURAL WORLD

In order to understand a culture's perception of its environment we need to consider how it has constructed its "natural world." All cultures seek to structure the world according to classification schemes that codify the meanings attached to objects. This is because, as Lévi-Strauss (1976: 15) noted, any classification is preferable to chaos, and through assigning meaning and order both within and without society a culture creates and sustains its own meaningful universe (Kus 1983: 278).

Indigenous cultural attitudes toward and perceptions of what constitutes the natural world may be very different from ours. As significance, like similarity, is not an inherent quality but is culture-dependent (Douglas 1990: 26), this will be reflected in a culture's classification scheme. Thus, something that seems significant to us might be acknowledged by another culture yet not be visible in their classification schema. On the other hand, their animated natural world may include what, within the framework of Western science, would be classified as inanimate objects, such as stones, mountains, and stars (for example, Bastien 1985; Lévi-Strauss 1976: 184–185), or not recognized as an object or entity at all, such as earth spirits and sky gods.

In this sense, a culture's constructed world is free to include creatures and objects we would term "fantastical"—entities that appear to us as disarticulations and amalgamations of "real" objects, animals, and people, and yet may embody vital cultural qualities. For example, "Old Star," chief protector of the inhabitants of the sacred *He* world of the Barasana of the Colombian Amazon, is at once a short trumpet, Orion, the fierce thunder jaguar, and a human warrior (Hugh-Jones 1979: 138–146). To our understanding, myth and reality are here inextricably mixed. Yet in the historically contingent classifications of other world-views, fantastical beings such as the "fish-man" or "bird-woman" are regarded as every bit as much a meaningful part of the perceptual universe as physically real ones. According to their type of significance such fantastical creations may populate art, mythology, or the night sky; and they may quite possibly be perceived to do so alongside what are to us "natural" objects (for example, Sagittarius and Leo). For non-Western traditional societies, objects or entities we regard as inanimate or fantastical may be seen as "fellow beings" with which the individual or society may enjoy a more or less advantageous relationship (Wax 1968: 235).

Ordering within classification schemes is based in part upon correspondences that are perceived to exist within the boundaries of a society's phenomenological experience. The Barasana, for example, link the "Caterpillar Jaguar" constellation to the appearance of living caterpillars, which form an essential part of their diet. Regarded as the "father of caterpillars," the Caterpillar Jaguar is believed to be responsible for the increase in their numbers as he rises higher and higher in the sky at dusk. To us, the correlation can be seen to be due to the fact that the constellation is in the eastern sky at dusk at the time of year when caterpillars pupate and come down from the trees on which they feed (Hugh-Jones 1982: 191). In Barasana thought there exists a correspondence between two entities that to us are quite distinct: a constellation and its position in the sky and an important part of the seasonal food supply.

Correspondences between objects or entities in the natural world may be perceived by means of characteristics that are regarded as being shared in common. In other words, a correspondence is predicated on the social acceptance of a stated affinity existing between two or more types of "being," "object," or activity (Saunders 1991: 85). Thus, to the Mursi there was a direct association among four stars in Centaurus and the Southern Cross, the river Omo, certain flora, and the successive disappearance of the stars in the morning sky—all of which were in turn correlated with terrestrial events such as the flooding of the river and the flowering of plants (Turton and Ruggles 1978: 590). Similarly, to the people of the Andean village of Misminay, the Vilcanota river is regarded as a reflection of the Milky Way in the sky, and both are conceived as part of an integrated system that serves to circulate water through the terrestrial and celestial spheres (Urton 1981a: 64).

It is clear that the acknowledgment of such correspondences depends upon a definition of meaning that is not inherent but relative, variable, and culture-dependent (Douglas 1990: 26). It follows that the resulting classification is, in part, psychological and may bear very little resemblance to the principles of Linnaean classification that underpin Western (that is, "rational" or "objective") scientific thought. Linnaean procedures have the advantage for our society that categorization is, by definition and design, predictable. Indigenous folk-categories are not necessarily predictable in the Linnaean sense, either in their meaning or their ordering, but they may nonetheless be methodically established in their own terms and exhibit a greater or lesser degree of correspondence with Western scientific categories (Saunders 1991: 16).

Whereas in Western classification membership of a category (taxon) depends, as Darwin (1985: 408–409) noted, on genealogical connections regardless of appearance or emotional reactions by humans, any entity may be a member of a folk taxon if enough society members agree that it should be (Clark 1988: 18–19). This does not mean that the folk taxonomy is unsystematic—merely that it adheres to its own principles rather than Linnaean ones (Saunders 1991: 16). In the twentieth-century Western world the Linnaean system has become the dominant one; but in terms of human societies in general it is nevertheless one system of classification among many.[3]

Classifications need to be expressed; qualities that the process of classification attaches to objects in the natural world must be brought to the fore and displayed for all to see. This is effected, in part at least, by the use of symbols, expressed as verbal or visual metaphors mediating among individuals, culture, and the environment. Thus, for the Aztecs, the jaguar *(ocelotl* in Nahuatl) symbolized qualities such as strength, courage, haughtiness, and pride (Sahagún 1950–1982, book 11: 1), human characteristics assigned to the animal in Aztec classification. In the same way that qualities of human behavior and animal behavior may be linked, so may human concerns and the movements of the heavenly bodies. Thus, to the Aztecs, people born under the calendrical day-sign *Ocelotl* were "destined to be like the jaguar" and demonstrated the qualities associated with it (Durán 1971: 402). In Aztec perception, therefore, there were symbolic associations—that is, correspondences of meaning—among a particular animal, certain individuals, and a constellation.

THE SPECIAL ROLE OF CULTURAL ASTRONOMY

Even though we may fully appreciate that many systems of classification exist in different cultures, we inevitably attempt to understand and to assess other cultural systems from a perspective created by and framed within the "rational" structures of Western scientific thought.

From this point of view, relationships between people and their surroundings fall broadly into three categories: those with other people, those with the terrestrial environment (fauna, flora, landscape, and sea), and those with the celestial environment (that is, astronomical and meteorological phenomena). Of these, only the third deals with that part of the environment that is not susceptible to physical alteration by humans. In other words, the celestial sphere can be regarded as an untouched, untouchable, and hence immutable part of the environment.

The study of astronomical phenomena in this context has a significant advantage over studies of other aspects of people's interaction with their surroundings, because the "raw resource" is directly accessible to us. Modern astronomy and physics can reconstruct important components of the night sky at any place on earth and any time during the last several millennia, to within certain largely determinable margins of error.[4] The results can then be recorded on computer or reconstructed in the planetarium. Thus, subject to uncertainties about the time and place being studied, we can construct directly a part of the celestial environment of a group of people in any remote culture. This ability to access an exact configuration of one aspect of the past is manifestly not available to other kinds of anthropological investigation.

Within this immutable database of the night sky we can pick out, at various levels, features that are common to different cultural contexts: at one level, for instance, the appearance of the same objects such as Polaris or the Pleiades, and at another level, the same general pattern of the motions of the heavenly bodies at similar latitudes—tropical, temperate, or subpolar. The immutability of the sky, the commonality of the features within it, and the fact that we can reconstruct it directly give a number of special advantages to the study of cultural astronomy within the study of cultural systems as a whole.

Cross-Cultural Comparisons

The first advantage concerns cross-cultural comparisons which are particularly valuable at the level of seeing how the same celestial resource is perceived, integrated into different systems of classification and belief, and manipulated to different or similar ideological ends. Especially interesting are comparisons of cultures for whom the skies are the same or very similar, but who might have very different perceptions of the celestial sphere. As a well-known example, the ancient Egyptians saw the sky as populated by gods in the form of constellations: the Greeks did not.

In many cases, however, common features are evident in the classification systems of adjacent societies: cross-cultural comparisons are then most useful. Examples abound at various levels. In northeastern South America, the architectural design of Yekuana roundhouses reflects the cosmology of the adjacent Warao, inasmuch as the cylindrical floor of the world ocean and earth disk of the latter correspond to the ground plan and concentric rooms of the former (Wilbert 1981: 48).

A view prevalent in Pre-Columbian Mesoamerica partitioned the horizontal world into five parts: the center plus four quadrants associated with the cardinal directions (Carrasco 1987: 140–141), the latter being demarcated by the solstice sunrise and sunset directions (Köhler 1982; Broda, this volume). Four-directionality and six-directionality (including the vertical axis, zenith, and nadir) were of great significance. A similar view is evident among the Zuni to the north (Young 1988, 1989: 170).

Similarly, the recognition of radiality as an influential general concept is found in a variety of cultural contexts in Pre-Columbian, historic, and modern South America. In particular, this has led to new insights into the possible ritual, kinship, and socioeconomic significance of the prehistoric Nazca lines (Aveni 1990).

A particularly interesting example is provided by two distinct cultural traditions, one from Andean Peru and the other from the tropical forests of the montaña to the east. In both cases the same configuration of dark cloud features in the Milky Way is recognized as a predator-prey scene—an interesting fact in itself—but whereas in the Andes it is conceived of as a black highland fox chasing a llama, in the montaña it is seen as a lowland black jaguar chasing a deer (Roe 1990).

Comparative studies of cultural astronomy are thus shown to have particular potential to bring into focus the social and political dynamics of cross-cultural interactions.

Cultural Correlates for Astronomy

Another advantage in studying cultural astronomy is that aspects of a culture's astronomy might bear some correlation to the nature and concerns of that society. The process of classification in general informs us of a society's perception of itself. The celestial part of the classification assumes special significance precisely because it is "locked away" in the immutable part of the natural environment. Yet this is a point that has been much undervalued in anthropological work to date.

Cultural correlates for astronomy have long been recognized, though usually at a superficial level. Literacy is a prime example, the supposed dichotomy between "pre-literate" and "literate" astronomical activities being adopted as a primary classification within many conferences and publications, such as the 1974 joint Royal Society–British Academy conference on the place of astronomy in the ancient world (Kendall et al. 1974). Similarly, Aveni (1989b: Ch. 2) has more recently divided examples of early time-reckoning into the oral and written. However, literacy, taken in the narrow sense, is a crude and arbitrary indicator of cultural transition (albeit no more so than the technological boundaries used to denote prehistoric "ages"), and hence limited in value as a correlate for astronomical activity.

Of more interest are examples of symbolic notation in the wider sense, such as writing, numbers, and "notation" encapsulated in art and architecture. Various levels of symbolic notation evidently act as prerequisites for, or at least significant enablers of, certain levels of mathematical and astronomical activity. For example, high-precision lunar observations of the type envisaged by Thom (1971) require extrapolation calculations (ibid.: Ch. 8) performed, if not using written notation, by using an alternative medium such as grids of standing stones encoding the relevant information geometrically (ibid.: Ch. 9).

Crump (1990: 6–10) has identified three types of symbolic number, representing three levels of conceptual evolution, and surveys their uses in different cultural contexts (ibid.). This begins to suggest how studies of orders of relationship between levels of numerical notation in particular, and symbolic notation in general, and levels of astronomical activity could be put on a more solid footing.

Other correlates are also of interest. Thus, for example, Iwaniszewski (1993) explores the relationship between the conceptualization of east and west and the rising and setting of the sun. Such explorations need to be developed much further and within a rigorous anthropological framework. In the longer term, an "anthropology of astronomy" (Platt 1991), adopting an approach analogous to that of Crump, is much needed.

Cultural Responses to Astronomy

A third advantage in studying cultural astronomy is that despite the untouchability of the celestial sphere, astronomical expertise can, as has been noted, yield a degree of apparent control by virtue of the ability to predict certain kinds of phenomena, such as solar and lunar eclipses and heliacal risings and settings. The predictive capacity of astronomy, available to those with even a limited observational expertise, is not only a unique characteristic but also one of the few universal realities in the study of human societies, and illustrates the potential of cultural astronomy to shed light on the diversity of cultural responses to the "natural world."

It is, therefore, through the manipulation of perceptions that form the basis of a structured world-view that objects within the natural world serve as a cultural resource. In practice, this resource serves, politically and ideologically, to reaffirm and reinforce the structures of a particular society by serving as an impressive display and justification of elite power and prestige. For example, the very location and layout of the medieval Hindu city of Vijayanagara symbolized the place of kingship within the sacred and cosmic order.

If an elite has persuaded others that they have powers and influence over the celestial sphere (on the basis of "understanding" patterned movements) then unpredicted or unpredictable events in the sky may undermine their authority and cause social and political instability. Examples abound where celestial events are correlated with the fate of the elite: an unexpected lunar eclipse appears to have caused two Teutonic armies to flee the scene of battle and postponed the Roman occupation of the Carpathian basin (Pásztor 1993); the appearance of Halley's Comet portended disaster for King Harold at the battle of Hastings in 1066; a spectacular meteor observed in Tenochtitlan around 1515 was taken as a bad omen and preceded the Spanish conquest (Sahagún 1950–1982, book 8: Fo. 12r; Köhler 1989). The degree to which people were panicked by unforeseen celestial events reflects the extent to which prestige and status were invested in the "control" and interpretation of celestial events by the elites of various cultures. Thus, our knowledge of the use of the sky as a cultural resource is informed equally from when it breaks down as from its everyday manipulation. This is because astronomy, like myth and religion, is not an end in itself but can serve the ends of political ideology.

THE NATURE OF THE EVIDENCE

How can we, as outsiders, penetrate the natural world of another culture? As we have seen, each culture makes its own sense of the environment, and so classifications of the world may be very different, even for contiguous cultures. Correspondences may be predicated upon empirical reality (for example, rivers flooding when certain stars disappear), or upon something that appears to us as physically impossible and/or esoteric, such as the idea that an individual's personality or future life can be affected and predicted through the astrological interpretation of astronomical phenomena surrounding the person's birth.

How, then, can we attempt to understand and describe another culture's classification? Other cultures' notions of reality and unreality, if they exist at all, will be different from ours, and will certainly be expressed in ways that are difficult (if not impossible) for us to comprehend. This is due, at least in part, to differences of ontological perspective, and between the classificatory structures and terms employed (Hallowell 1969: 50–58). One approach to overcoming this problem is to acknowledge that correspondences in a classificatory scheme are essentially psychological and expressed by metaphorical representation, which takes the form of symbols and their manipulation in ritual activity. One way to penetrate another classification scheme, therefore, is partly through the recognition of metaphor—by, for example, recognizing the display of distinctive combinations of symbols in ritual activity.

Here, once again, the sky assumes special significance because it can only be used symbolically to express correspondences in the culturally constructed "natural world." It is an exclusively metaphorical resource—a never-changing database of potential symbols.

The symbolic use of the sky is not restricted to celestial objects in isolation. There is a correlation between a society's classification of things on the ground and those in the sky. Thus, at Misminay there is a correspondence between the gradual rise of the celestial llama in the sky each morning and the breeding period of terrestrial llamas (Urton 1981a: 187; 1981b). While it may be assumed that the earthbound phenomenon came first, it is arguable that Misminay's inhabitants would have perceived a llama in the sky regardless of the actual distribution of celestial objects. Similarly, symbolic relationships may find expression as, say, a recumbent stone circle oriented upon the southern moon (Ruggles and Burl 1985); a cluster of aligned temple-pyramids at Uxmal (Aveni and Hartung 1982, 1986); the supernatural ordering of the universe reflected in the layout of a kiva (Williamson et al. 1975; Williamson 1982); an Amazonian roundhouse (Wilbert 1981); or a whole scheme of "state-sponsored" sacred geography, as in the Valley of Mexico (Broda, this volume; Tichy 1981), the Inca ceque system (Aveni 1981b; Zuidema 1964), and the Islamic world.

METHODOLOGICAL CONFLICT

In order to recognize an object or a relationship between objects that may have had symbolic significance to another culture, we have to consider not only the different types of available evidence, but also the paradigm-dependent interpretational biases through which we perceive pattern and meaning.

Hitherto, the study of cultural astronomy has brought together three major disciplines: cultural anthropology, archaeology, and ethnohistory. Each of these has its problems and limitations, which, in turn, create problems of integration (Charlton 1981). Thus, for example, the Yanomamo of South America conceive of themselves as the "fierce people," and actively conduct an elaborate and graded system of clubfights, raiding, and warfare. According to Yanomamo myth, it is in their nature to fight because the blood of the moon spilled upon their layer of the cosmos, causing them to become fierce (Chagnon 1977). Their "use" of the moon in this instance is mythological, and could only have been retrieved by cultural anthropology or ethnohistory. It would be invisible to the archaeologist.

Each of the three disciplines also has very different methodologies of explanation. The cultural anthropologist observes a living society and attempts to understand its totality by studying the articulation of its component parts. This "top-down" approach must acknowledge problems such as prior biases, the linguistic competence of the anthropologist, the reliability of informants, and the appropriateness of questions asked and the meaning of answers given. Ethnohistory inherits many of the problems of cultural anthropology but has in addition to cope with the often even more pronounced ideological persuasions and prejudices of those who collected, preserved, and interpreted the historical documents in the first instance.

The problems facing the archaeologist are the most severe of all. Extreme financial, political, and temporal constraints often force the archaeologist to sample a site with particular research questions in mind, and hence to rely heavily on inference. The archaeological approach also suffers from problems related to the limitations of the material record and its interpretation, such as distortion processes, possible bias in sampling strategy, and objectivity in the selection of data. There is, in addition, an ethical problem, as "viewing" by excavation necessarily involves partial or total site destruction.

The problems of recovery and interpretation of archaeological data pertain to the dominant set of beliefs that left a physical pattern retrievable from the material record. We should not forget, however, that these may have coexisted with other systems of belief that are much less likely to have produced recoverable or identifiable physical remains. In addition, a culture may have had more than one system of classification or belief operating side by side. For example, an official ideology may have coexisted with a local folk tradition, as happens in the Western world today with scientific astronomy and astrology.

The clash of methodologies of the three "parent disciplines" creates severe problems for the study of cultural astronomy. This is evident from debates within archaeo- and ethnoastronomy, the twin banners under which most work in cultural astronomy has taken place during the last two or three decades. These differences have created a methodological divide characterized hitherto by Aveni's "green" and "brown" classification (Aveni 1986, 1989a). Emanating respectively from different approaches in Europe (Heggie 1982) and the Americas (Aveni 1982), "green" methodology came to epitomize a prepossession with using statistical rigor to interpret the archaeological record in the absence of other kinds of evidence; "brown" methodology, faced with a rich ethnographic and ethnohistoric record, often felt it could relegate statistical arguments to a secondary supporting role.

While this bipartite division has often hindered the development of a cohesive identity for cultural astronomy, it is nevertheless true that statistical rigor is a prerequisite for estimating the likelihood that observed patterns have real (and quantifiable) cultural significance. Thus, the cautious use of appropriate statistical methods is relevant to any study in cultural astronomy that includes patterns in the material record among its dataset. However, before the use of such methods can even be considered, a more fundamental and underlying problem must be addressed: the need for objective selection of data. While objectivity is itself a problematic concept (see what follows), it is at least clear that negative data—that is, those that do not fit a given theory—should be considered on an equal basis alongside positive data in weighing up the idea, and presented with equal prominence so that others can judge it fairly. While the many critiques within "green archaeoastronomy" in the late 1970s and early 1980s addressed such issues (for example, Moir 1981; Heggie 1981a, 1981b; Ruggles 1981, 1982), similar "brown" critiques (for example, Ruggles and Saunders 1984; Köhler 1991) have been comparatively rare.

It is evident from this discussion that a serious problem in studying cultural astronomy is the lack of a rigorous methodology for *combining* evidence from the three main disciplines, as well as associated subjects such as folklore, mythology, art history, and the history of religion. How should we, for example, assess different types of ethnographic and archaeological data? This is, of course, merely one aspect of a general problem of data integration, a crucial issue that has been thrown into sharp relief by the recent excavations at El Templo Mayor in Mexico City (Boone 1987; Broda et al. 1987; Matos Moctezuma 1988).

Despite these problems, it is precisely cultural astronomy's integration of different but complementary types of evidence that has led to the establishment of a new and unique cross-disciplinary approach followed by those working in the mainstream of the parent disciplines.

TOWARD A METHODOLOGY OF DATA INTEGRATION

The breadth of view of cultural astronomy is its great strength and its main justification for recognition as a discipline in its own right. In order fully to exploit its cross-disciplinary potential, the integration of diverse data must be given a sound methodological underpinning. This would ensure that, among other things, cultural astronomy has something more to hand back to its parent disciplines than interesting—but mostly peripheral—data (Kintigh 1992).

As an illustration of the sort of methodological issues needing to be addressed, we shall consider one aspect of the problem of assimilating different types of cultural data, namely, how to weigh patterns observed in the archaeological record against other kinds of cultural data. First, we consider some of the special difficulties that arise in dealing with the material record.

Dealing with the Material Record

A number of studies, particularly of astronomy in the prehistoric British Isles, have adopted an approach led by a statistical methodology, paying close and often meticulous attention to rigorous criteria for the selection of data in an objective manner (Cooke et al. 1977; Hawkins 1968; Patrick 1993; Ruggles 1984a). Recently, however, there has been an increasing awareness of the limitations of classical statistics in the context of archaeoastronomy (Aveni 1988: 471; Ruggles 1988: 248–249), as indeed more widely within archaeology as a whole (Bell 1986; Orton 1980: 220; Ruggles 1986). There are a number of reasons for this.

First, statistical arguments can only enable us to pick out the most general trends of repeated patterns in the material record, and ignore the rich variety and diversity of symbolism that was almost certainly perceived in the celestial and terrestrial environment by a particular culture. Thus, in the context of "green archaeoastronomy," superficially similar sites may have had complex, differing, and changing functions of which astronomy, if it played a part at all, may have entered in various ways, differing from site to site (Ruggles 1988: 249).

A second and more fundamental problem is that the methodological foundations underlying classical statistics itself are based on the hypothetico-deductive paradigm (Popper 1959). Arguable enough in the context of the physical sciences (for example, Feyerabend 1978), this paradigm is totally inappropriate for studying the material record left by human activity, despite the efforts of the "new archaeology" to place archaeology on "objectively scientific" foundations (Binford 1967: 10, 1968: 18). Whereas in the numerate sciences explanation is prediction, archaeological evidence, because it was produced by humans as part of social processes, is not amenable either to prediction or the formulation of universal laws (*pace* Binford). On a practical level, in the hypothetico-deductive model theories are tested by repeated experiment, and any set of data may be used only once to test a hypothesis and then must be thrown away. In the material record we have only small amounts of data, which we may wish to examine over and again for patterning in the light of new ideas and theories.

Finally, the notion of objective data is itself problematic, and leads to a fundamental paradox. Objectivity is necessary because we work within the framework of rational, "scientific" thought. Thus, for example, the idea that certain horizon notches were used as high-precision lunar indicators is undermined when the full extent of the "beaver-gnawed" horizon at some alleged megalithic lunar observatories is revealed (Burl 1976: 143; Gingerich 1981). Similarly, the idea that three pecked cross-circles in the vicinity of Teotihuacan acted as surveyors' benchmarks (Aveni et al. 1978) is undermined by the presence of at least eleven that are in no position to do so (Ruggles and Saunders 1984). It is clear that unless we can demonstrate that we have not simply selected the data that fit an idea and ignored the rest, any conclusions will tend to reflect our own cultural input rather than providing any insights about the culture being studied.

However, in order to be "objective"—in other words, to formulate selection criteria—we need to make prior decisions about which types of pattern in the material record to consider and which to reject. These decisions will inevitably be made on a prior assessment of which patterns have the greatest potential significance—for example, which types of architectural alignment have the greatest astronomical potential—in order to make the ensuing investigation worthwhile. This is evident, for example, in Ruggles's selection criteria for examining megalithic alignments in western Scotland (Ruggles 1984a: Ch. 3). The eventual results will depend upon how closely the prior ideas reflect the classification of the target culture. For example, a "bad" prior guess at which megalithic alignments to choose might result in other alignments that really did have symbolic significance to the builders being missed in the statistical analysis. In short, any criteria for the "objective" selection of data will themselves reflect a cultural input; where there are little or no independent data considered relevant to the target culture, which could be used to guide the formation of prior decisions, this cultural input is entirely our own.

The recognition of these problems carries the danger, evident in some reactions to the "new archaeology" (Hodder 1985, 1989: 68–71; Shanks and Tilley 1987), of concluding that there is unlikely to be any single, unifying theoretical approach to archaeological knowledge. This implies the abandonment of any attempt to supply a general methodological framework for interpreting material data. Such an extreme view is unproductive, since it implies that all interpretations ultimately reflect little other than our own cultural input, in which case the only sensible course of action is to abandon interpretation altogether.[5]

However, while human behavior cannot be predicted and while some cultural input of our own can never be escaped, objectivity and quantitative support for particular conclusions on the basis of given data are still of paramount importance. To take an example from the archaeoastronomical literature, while the recognition that the Mesoamerican cross-symbol had a variety of possible meanings and uses (Aveni 1988; Iwaniszewski 1993) serves to emphasize the limitations of the statistical approach in examining a single possible use in isolation, it should not lead us to abandon methodological rigor in attempting to assess these possible meanings in their broad cultural context. The key point is that this assessment needs to be achieved within a methodological framework suitable to investigations of cultural data, rather than with one imported from an entirely different area with a totally inappropriate methodological framework.

The Role of Analogical Inference

How, then, should we proceed? It is clear that, working with the material record alone, we can never reconstruct the original meanings of objects retrieved by excavation. However, the limitations of the archaeological record can be partly offset and our interpretive horizons widened by the use of analogy. Despite Gould's (1980) view that analogy is an unwarranted assimilation of the past to the present, it can be properly used as an inductive, probabilistic argument, keeping in mind that only a partial similarity is suggested, never a complete or exact identity (Wylie 1982: 392–393). In addition, and significantly, the formulation of our contemporary views of the past has been irrevocably shaped by the ever-increasing availability of ethnographic sources since the sixteenth century (Orme 1981: 3–14).

Analogical inference is of considerable importance because the multidisciplinary nature of cultural astronomy, like post-positivist archaeology, is increasingly concerned with the epistemic limits of inquiry, in other words, with the status of what can be understood of the past. It is important to note that what can be known of the past is seen to depend on the theoretical and ideological presuppositions that are brought to the inquiry (Wylie 1989: 94). Thus, "facts" about what constitutes the past are themselves neither stable nor self-evident, but are paradigm-dependent and consequently open to continuing revision in the light of empirical and theoretical developments.

Through the use of analogy it may be possible to ascertain certain correlations in the original culture's classification scheme by recognizing the symbolic associations between celestial and terrestrial features that were used to express them. Thus, for example, changes from lunar to solar symbolism in the alignments in burial and ceremonial monuments in prehistoric Britain (Burl 1987) tell us something about changing views of the natural world and their prevalence in different areas at different times. Similar insights may be gained by studying the evolution of architectural alignments through time at ceremonial complexes such as Group E, Uaxactún (Aveni and Hartung 1989).

The central problem is to assess our degree of confidence or belief that an association we observe in the material record represents one of symbolic significance to the culture being studied—one, in other words, which did express a correlation in their classification scheme. This assessment is not made in isolation, but in the context of a class of associations about which we have a degree of confidence prior to examining the data in question. For example, consider the class of associations between the principal axis of a Scottish recumbent stone circle and the path of the moon near major standstill (Ruggles and Burl 1985). A particular *instance* of an association found in the material record, such as the alignment of the Cothiemuir Wood recumbent stone circle in Scotland upon the moon setting at declination −30°, needs to be assessed in the light of our prior degree of belief in the association (strictly, class of associations) between recumbent stone circle orientations and the moon. Thus, the original question might be rephrased as follows. Given a prior degree of belief in a class of associations, how should this degree of belief be modified in the light of new data from the material record? If we could express our prior model of the world (that is, of the class of associations currently under consideration) in terms of probability distributions of suitable parameters, and develop a formalism for how this model should be modified in the light of new data, then we could justifiably have more confidence in our assessment of the situation. This is the Bayesian paradigm (Barnett 1982: Ch. 6), and suggests that Bayesian statistics might provide a suitable formalism for the integration of material data and other cultural data, both in the study of cultural astronomy (Ruggles 1986, 1990: 247–248) and in a wider context (Buck et al. 1991; Buck and Litton 1991).

Data Interpretation and Integration Methodology as the Determination of Priors and Likelihoods

In order to use a Bayesian approach to a given problem of data integration and interpretation, it is necessary for the anthropologist to quantify the prior probability density ("prior"), which expresses his or her beliefs about the unknown parameters before the data are examined, and for the statistician to quantify the likelihood function ("likelihood"), which expresses his or her model of how likely the observed data are to arise, given specific values of the unknown parameters. By combining the two using Bayes's theorem, we arrive at a posterior probability density ("posterior"), which expresses how the anthropologist's beliefs should be updated in the light of the data. What interpretation is placed upon these modified beliefs is entirely the concern of the anthropologist and cannot be prescribed by the statistician.

The Bayesian approach has a number of attractive features in the context of the development of methodologies for the integration of diverse cultural data. Two are paramount. First, data of many different kinds may be considered in the formulation of a prior, and indeed must be if they have contributed to the anthropologist's belief about a particular issue. (Classical statistics, in contrast, lacks any means of taking into account "background" or "corroborating" evidence relating to the hypothesis being tested.) Second, data may legitimately be reused to examine how the same data affect different prior ideas. In addition, knowledge may be accumulated, for the posterior obtained after the consideration of one set of data may subsequently be used as a new prior.

The formulation of suitable quantifications for the prior and likelihood is far from easy, and has been attempted in very few archaeological contexts, exceptions being the megalithic yard (Freeman 1976), paleoethnobotany (Kadane and Hastorf 1988), cluster analysis and chronological seriation (Buck and Litton 1991), and radiocarbon dating (Buck et al. 1991). Such attempts are under way in the context of cultural astronomy. However, a few general observations are of value here.

The anthropologist may have acquired his or her background knowledge of the culture being studied, and hence his or her prior beliefs about the parameter distribution of a model, through data of various kinds, including all those of concern to cultural astronomy. How removed each of these sources is from the primary activities of the culture concerned, and hence which should be given greater prominence or afforded greater weight, must ultimately be the decision of the anthropologist. The Bayesian paradigm represents a framework within which the methodological basis for such decisions could be couched. As far as the material record is concerned, classes of association are best evidenced by repeated trends. This harks back to the old "green" approach, but now diverse background data can be taken into account to permit an assessment of whether alignment trends reinforce or counteract current cultural ideas (Ruggles 1986).

The problem of quantifying background data into parameter distributions in a prior model also bears strongly upon the question of data selection and "objectivity." Data selection criteria formulated in the absence of cultural input, such as those of Ruggles (1984a), amount to the prior selection of data—alignments, horizon features, and so on—considered plausible as astronomical indicators per se. The decision on what is plausible per se is inevitably made in the context of our own cultural perspective. It would, however, be wrong to dismiss such analyses out of hand—that is, to take the view that there is no point in studying astronomical alignments without some independent cultural evidence. Such a view would eliminate not only "green archaeoastronomy" but also most of prehistoric archaeology in general.

We must, however, develop criteria for these situations that are more rigorous in the anthropological sense. For example, selection criteria in the absence of direct cultural input could likely be improved through techniques such as cultural deconstruction or simulation. A recent exercise was undertaken using Earthwatch volunteers, largely ignorant of background work on astronomy in prehistoric Britain, who were asked to give their personal assessments of the relative prominence of horizon features from positions at or near to certain sites on the Scottish island of Mull. Only a small proportion manifested a view at all similar to those propounded by Thom (for example, Thom 1971) and generally followed in studies of prehistoric British astronomy.

Discussion

The Bayesian paradigm is not necessarily the panacea for the integration and interpretation of diverse data, and there are a number of significant problems to be overcome in defining suitable priors and likelihoods. The general approach does, however, seem to be a promising one in a number of respects, as already outlined.

The facility of repeatedly using the same data in the light of new ideas is, we believe, closely linked to recent developments in the theory of analogy. A powerful argument for using an analogical approach is that the assessment of the limits of what is knowable about the past is open to continuing revision, and so different kinds of interpretive inference can be assessed against continually changing background knowledge (Wylie 1989: 108). The distinctive character of Bayesian statistics together with recent thoughts on analogical reasoning combines not only to highlight the inappropriateness of classical statistics but also to undermine the "objectively scientific" foundations of the "new archaeology's" hypothetico-deductive philosophy. Unlike in the physical sciences, cultural "facts" vary in time and space and thus are neither immutable, stable, nor "given." Thus, Bayesian statistics would seem to be particularly well suited not only to addressing the problems confronting cultural astronomy per se but also to a more considered use of analogical reasoning.

Thus, cultural astronomy, with its cross-disciplinary origins and approach and its avowed aim of integrating a diversity of data, follows—at least in part—Hodder's (1985: 9) suggestions for the development of post-processual archaeology, where the challenge is to develop theories that specify integration and allow for both general principles of human activity and the particularity of cultural constructions. As Hodder (ibid.: 7) noted, the problem with processual archaeology was that it subverted the role of the individual to that of the system; it was part of a trend toward determinism and the quest for causal relationships and laws. Cultural astronomers, as anthropologists, cannot make general deterministic laws about human behavior; however, they can perhaps aid the search for general principles by which individuals construct their "natural worlds" within cultural-historical contexts.

CONCLUSION

Any culture must inhabit a meaningful universe, the content and boundaries of which it creates and sustains by classification. Within this phenomenological world of collective and individual experience, cultures attach meanings to objects and use symbols to express those meanings. They also perceive correspondences to exist among people, animals, and the terrestrial and celestial spheres, and use symbols to express those correspondences. Thus, cultural classifications lie at the heart of all meaningful social activity. Albeit mostly unconsciously, they guide patterns of attitude, thought, speech, and activity. In both secular and sacred situations, symbols are used to display the qualities that are considered to exist and to reinforce their meanings.

From ancient Babylonia to China, Mexico, and Peru, from empire and city state to tribe, astronomical information was gathered, recorded, and used by those whose interests lay as much in the spheres of status enforcement and political ideology as in predicting rainy seasons or planning agricultural schedules. By making natural phenomena appear liable to social manipulation in the form of cosmological myth, cleverly aligned architecture, and appropriately timed religious ritual, the elite used the predictive value of astronomical knowledge as an impressive display and justification of their power and prestige (Saunders 1989: 58).

Cultural astronomy affords direct access to a vital part of the cultural resources of any remote culture and is in a unique position to use this information to seek meaning within particular cultural contexts and through cross-cultural comparisons.

Astronomy can provide unequivocal knowledge of what was there to be seen; ethnohistory and the ethnographic record can provide an indication of types of perceptual explanation (and so widen our interpretive horizons); and the archaeological record can provide material evidence of symbolic associations through time.

Cultural astronomy provides a unique opportunity to explore the relationships between human societies and their wider environment. Cross-disciplinary perspectives, access to immutable data, the use of appropriate statistical methods, and the cautious use of analogical reasoning can together provide a unique insight into the diverse ways in which human societies have conceived of the heavens and how they have situated themselves in relation to them.

NOTES

1. We also, however, encounter the perception—for example, in Vedic hymns (Lyle 1991: 93–95, 163)—of a sun which, after completing its daily journey from east to west, rises again into the sky with a dark side turned toward the earth and retraces its journey back to the east during the night. This dark sun, moving invisibly in the opposite direction to the other celestial bodies, tends to be associated with misfortune and death.

2. Claims have also been made that this event is represented in rock art of the US Southwest (Brandt and Williamson 1979); but see Young (1986).

3. Even in the twentieth century we find the Western so-called "rational" tradition coexisting both with various religious classifications and with powerful "folk" classifications, such as the one that lies at the basis of modern astrology.

4. The most reliable components are the positions of "fixed" objects, such as stars, nebulae, and the Milky Way. Also generally reliable are the positional aspects of recurrent phenomena, such as the limits of the solar and lunar motions. Their theoretical positions in the sky can be calculated with very great precision, and their actual appearance, which is subject to uncertainties caused by atmospheric effects such as refraction and extinction, can be modeled and subjected to empirical verification. Temporal aspects of periodic phenomena—such as the position of a planet relative to the stars at a particular time, or the timing of coincidental events such as conjunctions and eclipses, and hence whether they are visible above the horizon at particular places—are subject to cumulative errors as we extrapolate back in time from the present, and estimates become less accurate further into the past. Generally, we cannot accurately reconstruct non-recurrent or very long-period events such as the appearance of novae or comets. For a detailed survey, see Schaefer (1993).

5. For a more recent debate concerning this topic, see Hodder 1991; Mithen 1991; Tilley 1991; for another view, see Binford 1989.

REFERENCES

Aveni, A. F. 1981a. *Skywatchers of Ancient Mexico*. Austin, University of Texas Press.

Aveni, A. F. 1981b. Horizon Astronomy in Incaic Cuzco. In *Archaeoastronomy in the Americas,* ed. R. A. Williamson, 305–318. Los Altos, CA, Ballena Press/Center for Archaeoastronomy.

Aveni, A. F. 1986. Archaeoastronomy: Past, Present, and Future. *Sky and Telescope* 72:456–460.

Aveni, A. F. 1988. The Thom Paradigm in the Americas: The Case of the Cross-circle Designs. In *Records in Stone,* ed. C.L.N. Ruggles, 442–472. Cambridge, Cambridge University Press.

Aveni, A. F. 1989a. Introduction: Whither Archaeoastronomy? In *World Archaeoastronomy,* ed. A. F. Aveni, 3–12. Cambridge, Cambridge University Press.

Aveni, A. F. 1989b. *Empires of Time.* New York, Basic Books.

Aveni, A. F., ed. 1982. *Archaeoastronomy in the New World.* Cambridge, Cambridge University Press.

Aveni, A. F., ed. 1990. *The Lines of Nazca.* Philadelphia, American Philosophical Society.

Aveni, A. F., and H. Hartung. 1982. Precision in the Layout of Maya Architecture. In *Ethnoastronomy and Archaeoastronomy in the American Tropics,* ed. A. F. Aveni and G. Urton, 63–80. New York, Annals of the New York Academy of Sciences 385.

Aveni, A. F., and H. Hartung. 1986. Maya City Planning and the Calendar. *Transactions of the American Philosophical Society* 76:7–87.

Aveni, A. F., and H. Hartung. 1989. Uaxactun, Guatemala, Group E and Similar Assemblages: An Archaeoastronomical Reconsideration. In *World Archaeoastronomy,* ed. A. F. Aveni, 441–461. Cambridge, Cambridge University Press.

Aveni, A. F., H. Hartung, and B. Buckingham. 1978. The Pecked Cross Symbol in Ancient Mesoamerica. *Science* 202:267–279.

Barnett, V. 1982. *Comparative Statistical Inference,* 2nd ed. Chichester, Wiley.

Bassi, M. 1988. On the Borana Calendrical System: A Preliminary Field Report. *Current Anthropology* 29:619–624.

Bastien, J. W. 1985. *Mountain of the Condor.* Prospect Heights, IL, Waveland Press.

Bell, J. A. 1986. On Applying Quantitative and Formal Methods in Theoretical Archaeology. *Science and Archaeology* 28:1–8.

Binford, L. R. 1967. Smudge Pits and Hide Smoking: The Use of Analogy in Archaeological Reasoning. *American Antiquity* 32:1–12.

Binford, L. R. 1968. Archaeological Perspectives. In *New Perspectives in Archaeology,* ed. S. R. Binford and L. R. Binford, 532. Chicago, Aldine.

Binford, L. R. 1989. Science to Seance, or Processual to "Postprocessual" Archaeology. In *Debating Archaeology,* a collection of papers by L. R. Binford, 27–40. London, Academic Press.

Boone, E. H., ed. 1987. *The Aztec Templo Mayor.* Washington, DC, Dumbarton Oaks.

Brandt, J. C., and R. A. Williamson. 1979. The 1054 Supernova and North American Rock Art. *Archaeoastronomy* 1 (supplement to *Journal for the History of Astronomy* 10):S1–S38.

Broda, J., D. Carrasco, and E. Matos Moctezuma, eds. 1987. *The Great Temple of Tenochtitlan: Center and Periphery in the Aztec World.* Berkeley, University of California Press.

Buck, C. E., J. B. Kenworthy, C. D. Litton, and A.F.M. Smith. 1991. Combining Archaeological and Radiocarbon Information: A Bayesian Approach to Calibration. *Antiquity* 65:808–821.

Buck, C. E., and C. D. Litton. 1991. A Computational Bayes Approach to Some Common Archaeological Problems. In *Computer Applications and Quantitative Methods in Archaeology, 1990,* ed. K. Lockyear and S.P.Q. Rahtz, 93–99. Oxford, Ternpus Reparaturn (BAR International Series 565).

Burl, H.A.W. 1976. *The Stone Circles of the British Isles.* New Haven, CT, Yale University Press.

Burl, H.A.W. 1987. *The Stonehenge People.* London, Dent.

Carrasco, D. 1987. Myth, Cosmic Terror and the Templo Mayor. In *The Great Temple of Tenochtitlan: Center and Periphery in the Aztec World,* ed. J. Broda, D. Carrasco, and E. Matos Moctezuma, 124–162. Berkeley, University of California Press.

Chagnon, N. 1977. *Yanomamo, the Fierce People.* New York, Holt, Rinehart, and Winston.

Charlton, T. H. 1981. Archaeology, Ethnohistory and Ethnology: Interpretive Interfaces. In *Advances in Archaeological Method and Theory,* vol. 4, ed. M. B. Schiffer, 129–174. London, Academic Press.

Clark, S.R.L. 1988. Is Humanity a Natural Kind? In *What Is an Animal?* ed. T. Ingold, 17–34. London, Unwin Hyman.

Cooke, J. A., R. W. Few, J. G. Morgan, and C.L.N. Ruggles. 1977. Indicated Declinations at the Callanish Megalithic Sites. *Journal for the History of Astronomy* 8:113–133.

Crocker, J. C. 1985. *Vital Souls: Bororo Cosmology, Natural Symbolism and Shamanism.* Tucson, University of Arizona Press.

Crump, T. 1990. *The Anthropology of Numbers.* Cambridge, Cambridge University Press.

Darwin, C. 1985. *The Origin of Species.* London, Penguin.

Douglas, M. 1990. The Pangolin Revisited: A New Approach to Animal Symbolism. In *Signifying Animals: Human Meaning in the Natural World,* ed. R. G. Willis, 25–36. London, Unwin Hyman.

Durán, D. 1971. *Book of the Gods and Rites and the Ancient Calendar,* trans. D. Heyden and F. Horcasitas. Norman, University of Oklahoma Press.

Eliade, M. 1964. *Shamanism: Archaic Techniques of Ecstasy,* trans. W. R. Trask. Princeton, NJ, Princeton University Press.

Feyerabend, P. 1978. *Against Method.* London, Verso.

Freeman, P. R. 1976. A Bayesian Analysis of the Megalithic Yard. *Journal of the Royal Statistical Society* A139:20–55.

Gibson, J. J. 1979. *The Ecological Approach to Visual Perception.* Boston, Houghton Mifflin.

Gingerich, O. 1981. Comment on "Stone Age Science in Britain?" by A. Ellegård. *Current Anthropology* 22:121–122.

Gould, R. A. 1980. *Living Archaeology.* Cambridge, Cambridge University Press.

Hallowell, A. I. 1969. Ojibwa Ontology, Behavior, and World View. In *Primitive Views of the World,* ed. S. Diamond, 49–82. New York, Columbia University Press.

Hawkins, G. S. 1968. Astro-archaeology. *Vistas in Astronomy* 10:45–88.

Heggie, D. C. 1981a. *Megalithic Science: Ancient Mathematics and Astronomy in Northwest Europe.* London, Thames and Hudson.

Heggie, D. C. 1981b. Highlights and Problems of Megalithic Astronomy. *Archaeoastronomy* 3 (supplement to *Journal for the History of Astronomy* 12):S17–S37.

Heggie, D. C., ed. 1982. *Archaeoastronomy in the Old World.* Cambridge, Cambridge University Press.

Ho Peng-Yoke, F. W. Paar, and P. W. Parsons. 1972. The Chinese Guest Star of AD 1054 and the Crab Nebula. *Vistas in Astronomy* 13:1–13.

Hodder, I. 1985. Postprocessual Archaeology. In *Advances in Archaeological Method and Theory,* vol. 8, ed. M. B. Schiffer, 1–26. London, Academic Press.

Hodder, I. 1989. Post-modernism, Post-structuralism, and Post-processual Archaeology. In *The Meanings of Things,* ed. I. Hodder, 64–78. London, Unwin Hyman.

Hodder, I. 1991. To Interpret Is to Act: The Need for an Interpretive Archaeology. *Scottish Archaeological Review* 8:8–13.

Hugh-Jones, S. 1979. *The Palm and the Pleiades: Initiation and Cosmology in Northwest Amazonia.* Cambridge, Cambridge University Press.

Hugh-Jones, S. 1982. The Pleiades and Scorpius in Barasana Cosmology. In *Ethnoastronomy and Archaeoastronomy in the American Tropics,* ed. A. F. Aveni and G. Urton, 183–201. New York, Annals of the New York Academy of Sciences 385.

Ingold, T. 1988. Introduction. In *What Is an Animal?* ed. T. Ingold, 1–16. London, Unwin Hyman.

Iwaniszewski, S. 1993. Some Social Correlates of Directional Symbolism. In *Archaeoastronomy in the 1990s,* ed. C.L.N. Ruggles, 45–56. Loughborough, Group D Publications.

Kadane, J. B., and C. A. Hastorf. 1988. Bayesian Paleoethnobotany. In *Bayesian Statistics* 3, ed. J. M. Bernardo, M. H. DeGroot, D. V. Lindley, and A.F.M. Smith, 243–260. Oxford, Oxford University Press.

Kelley, D. H. 1975. Maya Astronomical Tables and Inscriptions. In *Archaeoastronomy in Pre-Columbian America,* ed. A. F. Aveni, 57–73. Austin, University of Texas Press.

Kendall, D. G., S. Piggott, D. G. King-Hele, and I.E.S. Edwards, eds. 1974. The Place of Astronomy in the Ancient World. *Philosophical Transactions of the Royal Society of London* A276:1–276.

King, D. A. N.d. The Sacred Geography of Islam. In press.

Kintigh, K. W. 1992. Archaeoastronomy and Archaeology. *Archaeoastronomy and Ethnoastronomy News* (Center for Archaeoastronomy) 5:1, 4.

Köhler, U. 1982. On the Significance of the Aztec Day Sign "Ollin." In *Space and Time in the Cosmovisión of Mesoamerica,* ed. F. Tichy, 111–128. Munich, Wilhelm Fink Verlag (Lateinamerika-Studien 10).

Köhler, U. 1989. Comets and Falling Stars in the Perception of Mesoamerican Indians. In *World Archaeoastronomy,* ed. A. F. Aveni, 289–299. Cambridge, Cambridge University Press.

Köhler, U. 1991. Pitfalls in Archaeoastronomy: With Examples from Mesoamerica. In *Colloquio Internazionale Archeologia e Astronomia,* ed. G. Romano and G. Traversari, 130–136. Rome, Giorgio Bretschneider Editore (Supplementi alla RdA, 9).

Kus, S. M. 1983. The Social Representation of Space: Dimensioning the Cosmological and the Quotidean. In *Archaeological Hammers and Theories,* ed. J. A. Moore and A. S. Keene, 278–300. London, Academic Press.

Lévi-Strauss, C. 1976. *The Savage Mind.* London, Weidenfeld and Nicolson.

Lounsbury, F. 1982. Astronomical Knowledge and Its Uses at Bonampak, Mexico. In *Archaeoastronomy in the New World,* ed. A. F. Aveni, 143–168. Cambridge, Cambridge University Press.

Lyle, E. 1991. *Archaic Cosmos: Polarity, Space and Time.* Edinburgh, Polygon.

MacKie, E. W. 1981. Wise Men in Antiquity? In *Astronomy and Society in Britain during the Period 4000–1500 BC,* ed. C.L.N. Ruggles and A.W.R. Whittle, 111–152. Oxford, British Archaeological Reports (BAR British Series 88).

Martynov, A. I. 1988. The Solar Cult and the Tree of Life. *Arctic Anthropology* 25 (2):12–29.

Matos Moctezuma, E. 1988. *The Great Temple of the Aztecs.* London, Thames and Hudson.

Mithen, S. 1991. Archaeologies of Dissonance and Interpretation: A Comment on Hodder and Tilley. *Scottish Archaeological Review* 8:23–32.

Moir, G. 1981. Some Archaeological and Astronomical Objections to Scientific Astronomy in British Prehistory. In *Astronomy and Society in Britain during the Period 4000–1500 BC,* ed. C.L.N. Ruggles and A.W.R. Whittle, 221–241. Oxford, British Archaeological Reports (BAR British Series 88).

Neisser, U. 1976. *Cognition and Reality: Principles and Implications of Cognitive Psychology.* San Francisco, W. H. Freeman.

Orme, B. 1981. *Anthropology for Archaeologists.* London, Duckworth.

Orton, C. R. 1980. *Mathematics in Archaeology.* London, Collins.

Pásztor, E. 1993. Some Remarks on the Moon Cult of Teutonic Tribes. In *Archaeoastronomy in the 1990s,* ed. C.L.N. Ruggles, 98–106. Loughborough, Group D Publications.

Patrick, J. D. 1993. The Astronomy and Geometry of Irish Passage Grave Cemeteries: A Systematic Approach. In *Archaeoastronomy in the 1990s,* ed. C.L.N. Ruggles, 198–216. Loughborough, Group D Publications.

Pearce, S. M. 1993. Moon Man and Sea Woman: The Cosmology of the Central Inuit. In *Archaeoastronomy in the 1990s,* ed. C.L.N. Ruggles, 59–68. Loughborough, Group D Publications.

Platt, T. 1991. Review of "World archaeoastronomy," ed. A. F. Aveni. *Archaeoastronomy* 16 (supplement to *Journal for the History of Astronomy* 22):S76–S83.

Popper, K. R. 1959. *The Logic of Scientific Discovery.* London, Hutchinson.

Roe, P. G. 1990. It Takes Two to Launch a Dart: Dual Triadic Dualism in Chavín Iconography. Paper presented at the 23rd annual Chacmool conference. Calgary, AB, University of Calgary.

Roe, P. G. 1998. Paragon or Peril? The Jaguar in Amazonian Indian Society. In *Icons of Power: Feline Symbolism in the Americas,* ed. N. J. Saunders, 171–202. London, Routledge.

Ruggles, C.L.N. 1981. A Critical Examination of the Megalithic Lunar Observatories. In *Astronomy and Society in Britain during the Period 4000–1500 BC,* ed. C.L.N. Ruggles and A.W.R. Whittle, 153–209. Oxford, British Archaeological Reports (BAR British Series 88).

Ruggles, C.L.N. 1982. Megalithic Astronomical Sightlines: Current Reassessment and Future Directions. In *Archaeoastronomy in the Old World,* ed. D. C. Heggie, 83–105. Cambridge, Cambridge University Press.

Ruggles, C.L.N. 1984a. *Megalithic Astronomy: A New Archaeological and Statistical Study of 300 Western Scottish Sites.* Oxford, British Archaeological Reports (BAR British Series 123).

Ruggles, C.L.N. 1984b. Megalithic Astronomy: The Last Five Years. *Vistas in Astronomy* 27:231–289.

Ruggles, C.L.N. 1986. You Can't Have One without the Other? I.T. and Bayesian Statistics, and Their Possible Impact within Archaeology. *Science and Archaeology* 28:8–15.

Ruggles, C.L.N. 1987. The Borana Calendar: Some Observations. *Archaeoastronomy* 11 (supplement to *Journal for the History of Astronomy* 18):S35–S53.

Ruggles, C.L.N. 1988. The Stone Alignments of Argyll and Mull: A Perspective on the Statistical Approach in Archaeoastronomy. In *Records in Stone,* ed. C.L.N. Ruggles, 232–250. Cambridge, Cambridge University Press.

Ruggles, C.L.N. 1990. A Statistical Examination of the Radial Line Azimuths at Nazca. In *The Lines of Nazca,* ed. A. F. Aveni, 245–269. Philadelphia, American Philosophical Society.

Ruggles, C.L.N. 1993. Four Approaches to the Borana Calendar. In *Archaeoastronomy in the 1990s,* ed. C.L.N. Ruggles, 117–122. Loughborough, Group D Publications.

Ruggles, C.L.N., and H.A.W. Burl. 1985. A New Study of the Aberdeenshire Recumbent Stone Circles, 2: Interpretation. *Archaeoastronomy* 8 (supplement to *Journal for the History of Astronomy* 16): S25–S60.

Ruggles, C.L.N., and N. J. Saunders. 1984. The Interpretation of the Pecked Cross Symbols at Teotihuacan: A Methodological Note. *Archaeoastronomy* 7 (supplement to *Journal for the History of Astronomy* 15):S101–S110.

Sahagún, B. de. 1950–1982. *Florentine Codex: General History of the Things of New Spain*, ed. and trans. A. O. Anderson and C. E. Dibble (12 books, in 13 parts). Santa Fe, NM, School of American Research; Salt Lake City, University of Utah Press.

Sauer, C. 1954. *Agricultural Origins and Dispersals*. New York, American Geographical Society.

Saunders, N. J. 1989. Review of "World archaeoastronomy," ed. A. F. Aveni. *New Scientist* 123:57–58.

Saunders, N. J. 1991. The Jaguars of Culture: Symbolising Humanity in Pre-Columbian and Amerindian Societies. Ph.D. thesis, Department of Archaeology, Southampton University.

Schaefer, B. E. 1993. Basic Research in Astronomy and Its Applications to Archaeoastronomy. In *Archaeoastronomy in the 1990s,* ed. C.L.N. Ruggles, 155–177. Loughborough, Group D Publications.

Shanks, M., and C. Tilley. 1987. *Re-constructing Archaeology: Theory and Practice.* Cambridge, Cambridge University Press.

Thom, A. 1971. *Megalithic Lunar Observatories*. Oxford, Oxford University Press.

Tichy, F. 1981. Order and Relationship of Space and Time in Mesoamerica: Myth or Reality? In *Mesoamerican Sites and World-views,* ed. E. P. Benson, 217–245. Washington, DC, Dumbarton Oaks.

Tilley, C. 1991. Materialism and an Archaeology of Dissonance. *Scottish Archaeological Review* 8:14–22.

Turton, D. A., and C.L.N. Ruggles. 1978. Agreeing to Disagree: The Measurement of Duration in a Southwestern Ethiopian Community. *Current Anthropology* 19:585–600.

Ucko, P. J., M. Hunter, A. J. Clark, and A. David. 1991. *Avebury Reconsidered*. London, Unwin Hyman.

Urton, G. 1981a. *At the Crossroads of the Earth and Sky: An Andean Cosmology*. Austin, University of Texas Press.

Urton, G. 1981b. Animals and Astronomy in the Quechua Universe. *Proceedings of the American Philosophical Society* 125 (2):110–127.

Vogt, D. E. 1990. An Information Analysis of Great Plains Medicine Wheels. Ph.D. thesis, Simon Fraser University, Vancouver, BC.

Wax, M. 1968. Religion and Magic. In *Introduction to Cultural Anthropology,* ed. J. A. Clifton, 225–242. New York, Houghton Mifflin.

Wilbert, J. 1981. Warao Cosmology and Yekuana Roundhouse Symbolism. *Journal of Latin American Lore* 7 (1):37–72.

Williamson, R. A. 1982. Casa Rinconada: A Twelfth Century Anasazi Kiva. In *Archaeoastronomy in the New World,* ed. A. F. Aveni, 205–219. Cambridge, Cambridge University Press.

Williamson, R. A., H. J. Fisher, A. F. Williamson, and C. Cochran. 1975. The Astronomical Record in Chaco Canyon, New Mexico. In *Astronomy in Pre-Columbian America,* ed. A. F. Aveni, 33–42. Austin, University of Texas Press.

Wobst, H. M. 1990. Afterword: Minitime and Megaspace in the Palaeolithic at 18K and Otherwise. In *The World at 18,000 BP:* vol. 2: *Low Latitudes,* ed. C. Gamble and O. Soffer, 322–335. London, Unwin Hyman.

Wylie, A. 1982. An Analogy by Any Other Name Is Just as Analogical. *Journal of Anthropological Archaeology* 1 (4):382–401.
Wylie, A. 1989. Matters of Fact and Matters of Interest. In *Archaeological Approaches to Cultural Identity,* ed. S. J. Shennan, 94–109. London, Unwin Hyman.
Young, M. J. 1986. The Interrelationship of Rock Art and Astronomical Practice in the American Southwest. *Archaeoastronomy* 10 (supplement to *Journal for the History of Astronomy* 17):S43–S58.
Young, M. J. 1988. Directionality as a Conceptual Model for Zuni Expressive Behavior. In *New Directions in American Archaeoastronomy,* ed. A. F. Aveni, 171–182. Oxford, British Archaeological Reports (BAR International Series 454).
Young, M. J. 1989. The Southwest Connection: Similarities between Western Puebloan and Mesoamerican Cosmology. In *World Archaeoastronomy,* ed. A. F. Aveni, 167–179. Cambridge, Cambridge University Press.
Zuidema, R. T. 1964. *The Ceque System of Cuzco.* Leiden, Brill.
Zuidema, R. T. 1981. Inca Observations of the Solar and Lunar Passages through Zenith and Anti-zenith at Cuzco. In *Archaeoastronomy in the Americas,* ed. R. A. Williamson, 319–342. Los Altos, CA, Ballena Press/Center for Archaeoastronomy.

CHAPTER THIRTY-THREE

Cosmograms and Maya City Planning
Selected Articles

Michael E. Smith, Wendy Ashmore, Jeremy A. Sabloff, and Ivan Šprajc

Can We Read Cosmology in Ancient Maya City Plans?
Comment on Ashmore and Sabloff
Michael E. Smith

In a recent paper, Ashmore and Sabloff argue that the "position and arrangement of ancient Maya buildings and arenas emphatically express statements about cosmology and political order" (Ashmore and Sabloff 2002:201); see also the Spanish version (Ashmore and Sabloff 2000). Given current understandings of Mesoamerican cultures—and of ancient urban societies in general—it is certainly plausible to suggest a role for these two forces in the planning and layout of Maya cities. Personally, I agree with Ashmore and Sabloff that cosmology must have played a role in generating the

---

Reproduced by permission of the Society for American Archaeology from *Latin American Antiquity* 14:2 (2003).

layouts of cities among the Maya and other Mesoamerican societies. Nevertheless, the arguments they present for the influence of cosmology are vague, weak, and unconvincing. What kind of a role did cosmology play? How large a role? Can we reconstruct that role? Instead of presenting rigorous methods for investigating this issue, the authors rely upon assertions and subjective judgments backed not by empirical evidence but by uncritical citations of the works of others who agree with them. In this comment I first explore some of the complexities of studying cosmology from ancient city plans. I then address one component of Ashmore and Sabloff's cosmological model—the hypothesized north-south axis of Classic Maya cities.

COSMOLOGY AND ANCIENT URBAN PLANNING

The planning and layouts of ancient cities have long fascinated archaeologists, architects, and other scholars. Site maps often suggest that some sort of spatial order existed in ancient cities, but scholars have yet to develop systematic approaches to the study of the nature and origin of that order. The influence of cosmology, symbolism, and metaphor on ancient urban plans is an especially difficult topic for archaeologists. Some scholars are of the opinion that such research reveals more about the minds of modern scholars than about the minds of the ancients (e.g., Flannery and Marcus 1993; Kemp 2000; Prem 2000). For this reason, research in this area requires rigorous and explicit methods if it is to have credibility within the archaeological community.

Three urban traditions of the ancient world are particularly notable in comparative perspective for the large role played by cosmology in city planning—China, India, and Cambodia. These urban traditions share several characteristics: the layouts of numerous cities and public buildings within each tradition exhibit close similarities, there are ancient textual descriptions and images of the layout of the cosmos, there are plans and descriptions of the layout of the ideal city, and there are ancient textual sources stating that rulers deliberately followed cosmological models in laying out their capital cities (for China see Chang 1976; Steinhardt 1990; Wheatley 1971; for India see Allchin 1995; Coningham 2000; Spodek and Srinivasan 1993; for Cambodia see Dumarçay and Royère 2001; Higham 2000, 2002; Mannikka 1996).

In an earlier paper Ashmore (1992:173) classifies the Maya with these Asian urban traditions as a culture in which cosmology played a significant role in urban planning. To me, this does not appear to be a close fit. Apart from the existence of a few descriptions of the spatial layout of the cosmos (mostly from Postclassic codices and colonial texts, not Classic-period sources), the traits listed above are lacking for the Maya and other Mesoamerican urban cultures. The similarities among Maya cities are much less striking than the resemblances among Chinese, Cambodian, and perhaps Indian cities. More importantly, there are no surviving texts from anywhere in Mesoamerica that describe ideal cities or the efforts of kings to follow cosmological models in laying out their capitals.

The Aztec capital Tenochtitlan illustrates some of the difficulties involved in identifying the role of cosmology in urban planning. Although there is a large corpus of documentary sources on Aztec history and society (Smith 2003), there are only a few scraps of information describing the nature of urban planning. There is a large body of scholarship on the role of cosmology (usually termed "cosmovision") in the design and meaning of the Templo Mayor of Tenochtitlan (e.g., Broda et al. 1987; Carrasco 1991, 1999; López Luján 1998; Matos Moctezuma 1995), but this research relies almost exclusively on subjective interpretations of Aztec myths and rituals. The Templo Mayor, however, furnishes what may be the only direct evidence for the explicit use of astronomical or cosmological factors in the planning and layout of Mesoamerican urban architecture. A statement in the "Motolinía Insert no. 1," a document published in Friar Motolinía's (1971:51) *Memoriales*, suggests that Motecuhzoma had part of the Templo Mayor torn down and rebuilt so that the sun would rise directly over the temple of Huitzilopochtli on the equinox.[1]

Assuming for the sake of argument that the cosmovision scholars have correctly interpreted the cosmological significance of the Templo Mayor, can their results be extended to the whole city of Tenochtitlan? For some authors this is a straightforward interpretation. The passage of the sun across the sky was one of the most important elements of Aztec cosmology (e.g., Graulich 1997), and it seems natural to interpret the east-west avenues and alignments of Tenochtitlan in terms of the passage of the sun. As Tenochtitlan is one of the few Mesoamerican cities with orthogonal planning, the roles of cosmology and astronomy would seem particularly prominent in its layout (e.g., Broda et al. 1987; Carrasco 1999). But there is an alternative interpretation of the grid layout of the Aztec capital focusing instead on political and historical factors. It is clear from numerous studies that the Mexica rulers drew on images and concepts of the ancient Classic-period metropolis of Teotihuacan to reinforce their imperial legitimacy (Carrasco et al. 2000; López Luján 1989; Smith and Montiel 2001; Umberger 1987, 1996). Teotihuacan was an earlier large city not far from Tenochtitlan whose orthogonal planning would have been obvious to the Mexica. Given our knowledge of Mexica attitudes toward Teotihuacan, it would make sense for the Mexica rulers to imitate Teotihuacan's grid layout in planning their own capital Tenochtitlan, irrespective of any cosmological notions of their own.

Or perhaps the grid layout of the Aztec capital had nothing to do with the passage of the sun or with Teotihuacan, but instead originated for reasons of energetic efficiency. A grid is the most efficient layout for dividing up new land (Carter 1981:151; Stanislawski 1946). Much of the surface of the island city of Tenochtitlan was formed by fill brought from the mainland. The edges of the city were farmed with *chinampas*, agricultural fields with a rectilinear layout (Calnek 1974, 1976). It seems logical to assume that as chinampas were filled in to accommodate the growing population and prosperity of the city, their orthogonal layout would influence or determine the arrangement of lots, buildings, and streets. How can we decide between the cosmological, political, and energetic interpretations of Tenochtitlan's grid? I have no answer to this question and must admit that I have resorted to a hybrid explanation suggesting, that all three factors probably played a role in shaping the layout of Tenochtitlan (Smith 1997, 2003: Chapter 8). I do not find this a particularly satisfying explanation, but I have trouble thinking of methods for evaluating the relative importance of the three factors.

The Aztec case illustrates the difficulty of inferring the ideas and intentions of rulers and builders from the material remains of urban sites, even when there is a corpus of written documentation. Another pertinent example is the layout of Inka Cuzco. Several early chroniclers wrote that the imperial capital had been built in the form of a puma. As discussed by Hyslop (1990:50–51), it is difficult to determine today whether these writers were speaking literally or metaphorically (see also D'Altroy 2002:114–15). Modern scholars are similarly divided, some declaring that the city does indeed resemble a puma (Moseley 2001:85; Rowe 1967:60) and others viewing the model as a metaphor without direct and obvious physical expression (Hyslop 1990:51; Zuidema 1990). Gasparini and Margolies (1980:48) provide three maps with alternative spatial models for Cuzco as a puma, none of which look at all convincing to me.

Gutschow (1993) provides an even more striking example from Bhaktapur, Nepal, a city whose layout is said by its priests to conform closely to an ideal mandala form. The actual city layout, however, looks nothing like the mandala that they sketched for Gutschow (1993:170). These and other examples in which emic spatial models of cities conflict with actual urban layouts should give archaeologists pause; we would never be able to reconstruct the cosmological models behind Cuzco or Bhaktapur in the absence of written documentation. It is very likely that the layouts of Maya cities had symbolic associations known to some or all of their elite residents, but can we recover those meanings today with confidence and rigor?

It is also instructive to consider the inverse situation, in which apparently meaningful spatial patterns may have arisen from random factors unrelated to any cosmological ideas of the builders. Kemp (2000), for example, illustrates a simulation model that generates urban spatial layouts whose implication is that the apparently ordered layout of residential districts at the Egyptian city of Amarna—interpreted by some as evidence for the influence of cosmology on urban planning—may have arisen instead from random factors of urban growth. Similarly, Banning (1996) argues that the seemingly planned and cosmologically significant layouts of a number of ancient Near Eastern towns could have arisen unintentionally through nonlinear growth dynamics. To return to the Aztec example, it is entirely possible that Tenochtitlan's grid layout originated solely for reasons of energetic efficiency, and thus we are wasting time today searching for cosmological or political interpretations of that grid. Could the hypothesized north-south axis of Maya sites be another example of a trait that arose through random or stochastic growth processes? Before that question can be addressed, we need to determine the nature of the north-south axis. Is it an empirical pattern or a symbolic construct?

DO CLASSIC MAYA CITY PLANS HAVE A NORTH-SOUTH AXIS?

The existence of some sort of north-south pattern is a major part of Ashmore and Sabloff's cosmological model. Among Ashmore's statements of her views on city layout and cosmology, she lists "a strongly marked north-south axis" (Ashmore 1989:273, 1992:174) as one of the "five principal components" (1992:174) of the Classic Maya cosmological template. In another article she describes this principle as "emphatic reference to a north-south axis in site organization" (Ashmore 1991:200), and in the recent paper, Ashmore and Sabloff (2002) write of north-south "axial dominance" (p. 203) and of a "pronounced north-south axis" (p. 206). Just what does this mean? The fact that Ashmore's five principles (1989:273, 1992:174) combine empirical spatial patterns of architecture (e.g., architectural groups that form triangles) with symbolic interpretations (e.g., "north stands for the celestial supernatural sphere") confuses the question. Is the north-south axis an empirical phenomenon—something that archaeologists can identify from site maps—or is it a symbolic construct used to interpret site maps?

My own understanding of the north-south principle, based upon reading Ashmore's articles (1989, 1991, 1992), is that it is meant to be an empirical pattern that once identified at a site, can be given symbolic content through reference to external information on Maya cosmology (from sources like the *Popol Vuh*). But this causes problems—for me, at least—because I cannot seem to find the pattern in the site maps published by Ashmore. She says that the cosmological template (which includes the north-south axis) "can be seen most easily at Tikal, and there, most readily in the famous Twin Pyramid groups" (1992:194). To me, the east-west axis in twin pyramid groups seems stronger, or at least more prominent, than the north-south axis. As for the overall plan of Tikal, I do not see any dominant cardinal (or other) axis.

Ashmore also illustrates her model with maps of Cerros, Quirigua, Copan, and Gualjoquito. Perhaps there is some kind of north-south axis at Quirigua, but I fail to see the pattern in the other site maps. I find her discussion of Gualjoquito particularly puzzling. She compares the site to Copan, asserting that both have a north-south linear arrangement of a public plaza, a ballcourt, and an enclosed compound (1989:281, 1992:181). At Gualjoquito this linear pattern runs east-west, however, and Ashmore claims that the pattern is "skewed counterclockwise" (1989:281). At what point does a skewed north-south axis become an east-west axis?

Perhaps others can see north-south axes at these and other Maya sites, but I am biased or incapable of seeing them. I wrote my undergraduate thesis on Teotihuacan and lived in San Juan Teotihuacan my first summer in Mexico. I readily admit that Teotihuacan looms large in my thinking about Mesoamerican cities. In my mind, Teotihuacan is a site with a "strongly marked" or "pronounced" north-south axis. Perhaps this "Teotihuacan bias" blinds me to more subtle spatial patterning at Classic Maya sites. Perhaps I simply do not have the perceptual or cognitive ability to see north-south axes at the Maya sites. Or maybe the emperor has no clothes; maybe there are no clearly discernable north-south axes at these sites. My reading of Ashmore's articles could be in error and the north-south axis a symbolic construct, not an empirical pattern.

If this latter suggestion is correct, it is not at all clear just how one goes about applying the model. Ashmore and Sabloff do not provide enough information on how to select particular structures or groups that can be given a north-south cosmological interpretation. In one article Ashmore (1989:274) suggests that the cosmological template can apply to "the pairing of open, public gathering spaces on the north with enclosed, private (residential/administrative) groups on the south." Does any case of a plaza located somewhere north of an elite residential compound fit the model? Do the features have to be contiguous? Do they have to be the largest plaza and compound at a site? Does this symbolism refer to certain kinds of plazas and compounds but not others?

Several of the site maps published in the recent article (Ashmore and Sabloff 2002:206–207) do seem to have north-south axes. These are not complete site maps, but rather plans of key groups selected from the overall site maps to illustrate a (quite reasonable) suggestion of architectural emulation among sites. The architectural groups depicted at Xunantunich, Naranjo, and Calakmul do appear to share some spatial characteristics, although we are not told exactly what is similar or different about these plans other than "the pronounced north-south axis arguably linked to royal authority and continuity" (2002:206). The Spanish version of the article (Ashmore and Sabloff 2000) includes more complete site plans of Xunantunich and Naranjo, but as in the cases of the sites mentioned above, I have trouble seeing clear north-south axes in these maps. The east-west axis—at Xunantunich at least—seems to me equally prominent. In Figure 33.1, I present my own maps of selected buildings from those site plans that highlight what appear to be pronounced east-west axes. At Xunantunich a constructed causeway runs east-west to join the central plaza on the east side, and what appears to be a cleared east-west avenue runs from Structure A-21 to join the plaza on the west side. At Naranjo, an east-west corridor or axis of major public architecture extends from the tall pyramids on the east through several plazas and buildings, including a large platform and a series of enclosures adjacent to a bajo on the west side.

I freely admit to selecting out only those buildings that fit my preconceived goal of defining east-west axes at Xunantunich and Naranjo, and I claim no cosmological (or other) significance for the plans in Figure 33.1. These are perverse fantasy figures whose sole purpose is to challenge Ashmore and Sabloff to clarify their methods and procedures of analysis. I am sure that Ashmore and Sabloff used more rigorous criteria in creating their Figures 4 and 5 from the same base maps that I used. But what are those criteria? The reader needs to know. And what is the empirical basis for the judgment that the plans of architectural groups at Xunantunich, Naranjo, and Calakmul—or Copan and Gualjoquito—are similar? What would have to be different to conclude that these groups were *not* similar?[2] Just how does one decide that a complicated urban plan has a "pronounced" north-south (or east-west) axis? Are there degrees of adherence to an ideal north-south model? Do Xunantunich and Naranjo—or Tikal and Copan—fit the model closely, or do they only bear a vague resemblance to it? Because the authors fail to present objective methods or criteria for comparison, their interpretations sound highly subjective.

Analyses of the role of cosmology in ancient urban planning do not have to be vague and subjective. Urban planner Kevin Lynch (1981:73–81) discusses ancient

Michael E. Smith, Wendy Ashmore, Jeremy A. Sabloff, and Ivan Šprajc

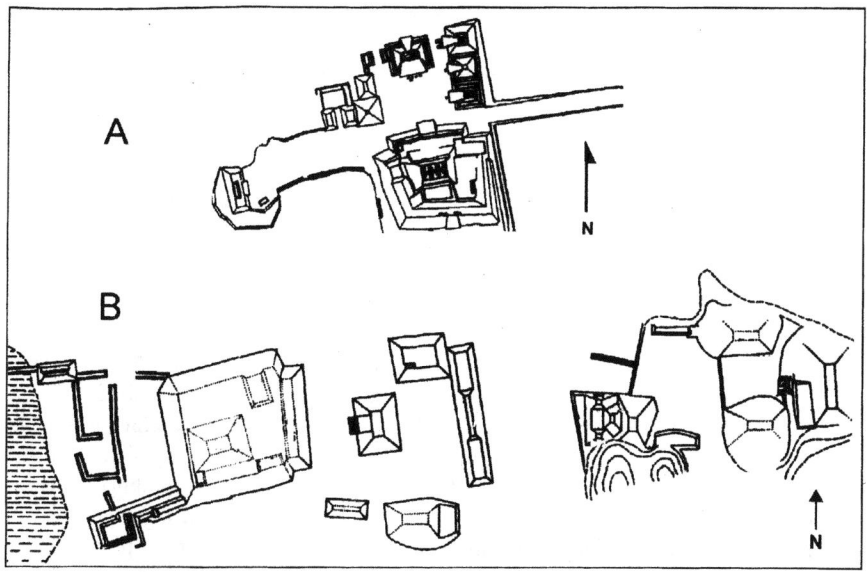

33.1 Plans suggesting possible east-west orientations at Maya cities: A) Xunantunich (modified after Ashmore and Sabloff 2000:Figure 4); B) Naranjo (modified after Ashmore and Sabloff 2000:Figure 5). I make no claims for cosmological or other significance for these plans.

Chinese, Indian, and other patterns of urban planning under the label of an emic "theory of magical correspondences" and identifies a number of cross-cultural architectural expressions of that cosmological model (Table 33.1). Environmental psychologist Amos Rapoport (1993) discusses a similar cosmological model, drawing upon a much-cited passage in Eliade (1959:5–12). Rather than present a single cross-cultural set of architectural manifestations like Lynch, Rapoport applies the conceptual model to a variety of ancient cities and describes its architectural expressions in reference to individual cases (Table 33.1). Similarly, studies of the expression of ancient political ideologies in urban plans by archaeologists (e.g., Blanton 1989; DeMarrais 2001; Prem 2000) present specific architectural and spatial features as possible manifestations of state or elite ideologies.

Compared to such works, Ashmore and Sabloff's cosmological model lacks specificity and rigor, largely because the material expressions of their cosmological and political models are vague or unspecified. In earlier publications on this theme, Ashmore was careful to label her conclusions provisional and exploratory. Now, ten years later, Ashmore and Sabloff (2002:202) again state that their work is "provisional." But until they devise more objective methods with a firmer grounding in empirical data, the argument for cosmological principles in Maya urban planning will remain weak, speculative, and provisional.

Table 33.1 Architectural Expressions of Cosmological Symbolism in Ancient Cities.

Lynch (1981:75–79)	Rapoport (1993:43–52)
axial line of procession	city walls with gates
encircling enclosure with gates	orientation to the cardinal directions
dominance of up versus down	vertical markers at the center
grid layout	open sacred plazas
bilateral symmetry	tombs in key locations

Note: These features do not exhaust either author's lists of architectural expressions.

NOTES

1. The statement in the Motolinía Insert no. 1 (published in Motolinía 1971:50–54) was first noted by Maudslay (1913:175), who stated, "Motolinía says that the festival called Tlacaxipeualistli 'took place when the sun stood in the middle of Huichilobos, which was at the equinox, and because it was a little out of the straight, Montezuma wished to pull it down and set it right'" (quotation from Rowe 1977:229; the Spanish original is in Motolinía 1971:51). This quote, whose importance for the study of Mesoamerican astronomical alignments is obvious, has been much discussed in the literature. Maudslay's interpretation is accepted by Aveni (2001:236–238) and Aveni et al. (1988:290, 294–295), who analyze the nature of the likely observational practices in relation to the architecture of the Templo Mayor. Rowe (1977:229–230), on the other hand, expresses reservations about Maudslay's interpretation. Hanns Prem (personal communication 2002) follows Rowe and suggests that two separate issues may be conflated in the Motolinía Insert: whether Tlacaxipehualiztli fell on the equinox, and the meaning of the obscure statement that the sun was "in the middle of Huichilobos" (which Maudslay and Aveni interpret as meaning that the sun rose behind the Huitzilopochtli temple on the Templo Mayor). Anthony Aveni (personal communication 2002) acknowledges Rowe's and Prem's reservations but prefers his own published interpretation. Although I am hesitant to venture very far into the realm of archaeoastronomy, Aveni's interpretation seems to me the most logical one. I thank Hanns Prem and Anthony Aveni for their opinions and citations on this issue.

2. Another similarity in site plan discussed by Ashmore and Sabloff focuses on Sayil and Labná (2002:207–208). Prem (2000:66) points out disagreement over the degree of similarity of these site plans, however. He contrasts Ashmore and Sabloff's (2000) views with those of Andrews (1975:326), who finds "almost no consistency in their general configurations."

REFERENCES

Allchin, F. Raymond, ed. 1995. *The Archaeology of Early Historic South Asia: The Emergence of Cities and States.* New York, Cambridge University Press.

Andrews, George F. 1975. *Maya Cities: Placemaking and Urbanization.* Norman, University of Oklahoma Press.

Ashmore, Wendy. 1989. Construction and Cosmology: Politics and Ideology in Lowland Maya Settlement Patterns. In *Word and Image in Maya Culture: Explorations in Language, Writing, and Representation,* ed. William F. Hanks and Don S. Rice, 272–286. Salt Lake City, University of Utah Press.

Ashmore, Wendy. 1991. Site-Planning Principles and Concepts of Directionality among the Ancient Maya. *Latin American Antiquity* 2:199–226.

Ashmore, Wendy. 1992. Deciphering Maya Architectural Plans. In *New Theories on the Ancient Maya*, ed. Elin C. Danien and Robert J. Sharer, 173–184. University Museum Monograph, vol. 77. University Museum, University of Pennsylvania, Philadelphia.

Ashmore, Wendy, and Jeremy A. Sabloff. 2000. El orden de espacio en los planes cívicos mayas. In *Arquitectura e ideología de los antiguos mayas: Memoria de la Segunda Mesa Redonda de Palenque*, ed. Silvia Trejo, 15–34. Mexico City, Instituto Nacional de Antropología e Historia.

Ashmore, Wendy, and Jeremy A. Sabloff. 2002. Spatial Orders in Maya Civic Plans. *Latin American Antiquity* 13:201–215.

Aveni, Anthony F. 1992. Moctezuma's Sky: Aztec Astronomy and Ritual. In *Moctezuma's Mexico: Visions of the Aztec World*, ed. Davíd Carrasco and Eduardo Matos Moctezuma, 149–158. Boulder, University Press of Colorado.

Aveni, Anthony F. 2001. *Skywatchers*. 2nd ed. Austin, University of Texas Press.

Aveni, Anthony F., Edward E. Calnek, and Horst Hartung. 1988. Myth, Environment and the Orientation of the Templo Mayor of Tenochtitlan. *American Antiquity* 53:287–309.

Banning, E. B. 1996. Pattern or Chaos? New Ways of Looking at "Town Planning" in the Ancient Near East. In *Debating Complexity: Proceedings of the 26th Annual Chacmool Conference*, ed. Daniel A. Meyer, Peter C. Dawson, and Donald T. Hanna, 510–518. Calgary, Archaeological Association of the University of Calgary.

Blanton, Richard E. 1989. Continuity and Change in Public Architecture: Periods I through V of the Valley of Oaxaca, Mexico. In *Monte Alban's Hinterland, Part 2. Prehispanic Settlement Patterns in Tlacolula, Etla, and Ocotlan, the Valley of Oaxaca, Mexico*, ed. Stephen A. Kowalewski, Gary M. Feinman, Laura Finsten, Richard E. Blanton, and Linda M. Nicholas, 409–447. Memoirs, vol. 23. Museum of Anthropology, University of Michigan, Ann Arbor.

Broda, Johanna, Davíd Carrasco, and Eduardo Matos Moctezuma. 1987. *The Great Temple of Tenochtitlan: Center and Periphery in the Aztec World*. Berkeley, University of California Press.

Calnek, Edward E. 1974. Conjunto urbana y modelo residencial en Tenochtitlan. In *Ensayos sobre el desarollo urbano de México*, ed. Woodrow Borah, 11–65. Mexico City, Secretaría de Educación Pública.

Calnek, Edward E. 1976. The Internal Structure of Tenochtitlan. In *The Valley of Mexico: Studies of Pre-Hispanic Ecology and Society*, ed. Eric R. Wolf, 287–302. Albuquerque, University of New Mexico Press.

Carrasco, Davíd. 1991. To *Change Place: Aztec Ceremonial Landscapes*. Boulder, University Press of Colorado.

Carrasco, Davíd. 1999. *City of Sacrifice: The Aztec Empire and the Role of Violence in Civilization*. Boston, Beacon Press.

Carrasco, Davíd, Lindsay Jones, and Scott Sessions, eds. 2000. *Mesoamerica's Classic Heritage: From Teotihuacan to the Aztecs*. Boulder, University Press of Colorado.

Carter, Harold. 1981. *The Study of Urban Geography*. 3rd ed. London, Edward Arnold.

Chang, Kwang-Chih. 1976. Towns and Cities in Ancient China. In *Ancient Chinese Civilization: Anthropological Perspectives*, ed. Kwang-Chih Chang, 61–71. Cambridge, MA, Harvard University Press.

Coningham, Robin. 2000. Contestatory Urban Texts or Were Cities in South Asia Built as Images? *Cambridge Archaeological Journal* 10:348–354.

D'Altroy, Terence N. 2002. *The Incas*. Oxford, Blackwell.

DeMarrais, Elizabeth. 2001. The Architecture and Organization of Xauxa Settlements. In *Empire and Domestic Economy*, ed. Terence N. D'Altroy and Christine A. Hastorf, 115–153. New York, Plenum.

Dumarçay, Jacques, and Pascal Royère. 2001. *Cambodian Architecture, Eighth to Thirteenth Centuries*. Handbook of Oriental Studies, Section Three, South-East Asia, vol. 12. Leiden, Brill.

Eliade, Mircea. 1959. *Cosmos and History: The Myth of the Eternal Return*. Trans. Williard R. Trask. New York, Harper and Row.

Flannery, Kent V., and Joyce Marcus. 1993. Cognitive Archaeology. *Cambridge Archaeological Journal* 3:260–270.

Gasparini, Graziano, and Luise Margolies. 1980. *Inca Architecture*. Trans. Patricia J. Lyon. Bloomington, Indiana University Press.

Graulich, Michel. 1997. *Myths of Ancient Mexico*. Trans. Bernard R. Ortiz de Montellano and Thelma Ortiz de Montellano. Norman, University of Oklahoma Press.

Gutschow, Niels. 1993. Bhaktapur: Sacred Patterns of a Living Urban Tradition. In *Urban Form and Meaning in South Asia: The Shaping of Cities from Prehistoric to Precolonial Times*, ed. Howard Spodek and Doris Meth Srinivasan, 163–183. Studies in the History of Art, Center for Advanced Study in the Visual Arts, Symposium Papers 15, vol. 31. National Gallery of Art, Washington, DC.

Higham, Charles. 2000. The Symbolism of the Angkorian City. In Viewpoint: Were Cities Built as Images? *Cambridge Archaeological Journal* 10:355–357.

Higham, Charles. 2002. *The Civilization of Angkor*. Berkeley, University of California Press.

Hyslop, John. 1990. *Inka Settlement Planning*. Austin, University of Texas Press.

Kemp, Barry J. 2000. Bricks and Metaphor. In Viewpoint: Were Cities Built as Images? *Cambridge Archaeological Journal* 10:335–346.

López Luján, Leonardo. 1989. *La recuperación mexica del pasado, teotihuacano*. Instituto Nacional de Antropología e Historia, Mexico City.

López Luján, Leonardo. 1998. Re-creating the Cosmos: Seventeen Aztec Dedication Caches. In *The Sowing and the Dawning: Termination, Dedication, and Transformation in the Archaeological and Ethnographic Record of Mesoamerica*, ed. Shirley Boteler Mock, 177–188. Albuquerque, University of New Mexico Press.

Lynch, Kevin. 1981. *A Theory of Good City Form*. Cambridge, MIT Press.

Mannikka, Eleanor. 1996. *Angkor Wat: Time, Space, and Kingship*. Honolulu, University of Hawaii Press.

Matos Moctezuma, Eduardo. 1995. *Life and Death in the Templo Mayor*. Trans. Bernard R. Ortiz de Montellano and Thelma Ortiz de Montellano. Boulder, University Press of Colorado.

Maudslay, Alfred R. 1913. A Note on the Position and Extent of the Great Temple Enclosure of Tenochtitlan and the Position, Structure, and Orientation of the Teocalli of Huitzilopochtli (Abstract). In *Proceedings, 18th International Congress of Americanists* (London, 1912), vol. 1, 173–175. London, Harrison and Sons.

Moseley, Michael E. 2001. *The Incas and Their Ancestors: The Archaeology of Peru*. Rev. ed. New York, Thames and Hudson.

Motolinía, Fray Toribio de Benavente. 1971. *Memoriales, o libro de las cosas de la Nueva España y de los naturales de ella*. Ed. Edmundo O'Gorman. Mexico City, Universidad Nacional Autónoma de México.

Prem, Hanns J. 2000. ¿Detrás de qué esquina se esconde la ideología? In *Arquitectura e ideología de los antiguos mayas: Memoria de la Segunda Mesa Redonda de Palenque*, ed. Silvia Trejo, 55–70. Mexico City, Instituto Nacional de Antropología e Historia.

Rapoport, Amos. 1993. On the Nature of Capitals and Their Physical Expression. In *Capital Cities, Les Capitales: Perspectives Internationales, International Perspectives*, ed. John Taylor, Jean G. Lengellé, and Caroline Andrew, 31–67. Ottawa, Carleton University Press.

Rowe, John H. 1967. What Kind of a Settlement Was Inca Cuzco? *Nawpa Pacha* 5:59–77.

Rowe, John H. 1977. Archaeoastronomy in Mesoamerica and Peru. *Latin American Research Review* 14 (2):227–233.

Smith, Michael E. 1997. City Planning: Aztec City Planning. In *Encyclopaedia of the History of Non-Western Science, Technology, and Medicine,* ed. Helaine Selin, 200–202. Dordrecht, Kluwer Academic.

Smith, Michael E. 2003. *The Aztecs.* 2nd ed. Oxford, Blackwell.

Smith, Michael E., and Lisa Montiel. 2001. The Archaeological Study of Empires and Imperialism in Prehispanic Central Mexico. *Journal of Anthropological Archaeology* 20:245–284.

Spodek, Howard, and Doris Meth Srinivasan, eds. 1993. *Urban Form and Meaning in South Asia: The Shaping of Cities from Prehistoric to Precolonial Times.* Studies in the History of Art, Center for Advanced Study in the Visual Arts, Symposium Papers XV, vol. 31. Washington, DC, National Gallery of Art.

Stanislawski, Dan. 1946. The Origin and Spread of the Grid Pattern Town. *Geographical Review* 36:105–120.

Steinhardt, Nancy S. 1990. *Chinese Imperial City Planning.* Honolulu, University of Hawaii Press.

Umberger, Emily. 1987. Antiques, Revivals, and References to the Past in Aztec Art. *RES: Anthropology and Aesthetics* 13:62–105.

Umberger, Emily. 1996. Art and Imperial Strategy in Tenochtitlan. In *Aztec Imperial Strategies,* ed. Frances F. Berdan et al., 85–106. Washington, DC, Dumbarton Oaks.

Wheatley, Paul. 1971. *The Pivot of the Four Quarters: A Preliminary Enquiry into the Origins and the Character of the Ancient Chinese City.* Chicago, Aldine.

Zuidema, R. Tom. 1990. *Inca Civilization in Cuzco.* Austin, University of Texas Press.

Interpreting Ancient Maya Civic Plans: Reply to Smith

Wendy Ashmore and Jeremy A. Sabloff

We welcome Smith's critique of our recent inferences about spatial orders in Maya civic plans. As we stated near the outset of that article, our intent was programmatic and our conclusions provisional (Ashmore and Sabloff 2002:202). The crux of Smith's critique is that until we "devise more objective methods, with a firmer grounding in empirical data," our conclusions must always remain "weak, speculative, and provisional" (p. 752). To the extent that his conclusion is correct, the programmatic aspects of our article would seem effectively moot. To the contrary, however, we believe that our arguments fit comfortably within the scientific process of generating and testing ideas, whether or not the latter are expressed formally as hypotheses.

Reproduced by permission of the Society for American Archaeology from *Latin American Antiquity* 14:2 (2003).

In the paragraphs that follow, we consider how our line of ideas and hypotheses were generated, how—explicitly—they have been tested, by ourselves and by others, and with what results. Specifically, we focus, as does Smith, on our contention that Maya civic centers include emphatic expression of a north-south axis, and that this axis orientation is based *in part* in Maya cosmology. We cite sources of those inferences and review explicitly the criteria that Smith faults us for having omitted before. We suggest anew that these ideas are worth continued examination, and welcome renewed and expanded collaboration as to methods appropriate for doing so. In the spirit of such collaborative efforts, we close by offering some specific programmatic suggestions for how this collaborative inquiry might be undertaken.

HYPOTHESIS GENERATION

As is often the case in archaeology, our ideas and formal hypotheses were generated by pattern recognition and efforts to understand the patterns perceived. In this instance, the pattern initially recognized paired "palace" buildings and complexes with "temple" buildings and complexes in the civic centers of a specified set of Maya sites, with each member of the pair occupying one end of a north-south axis. Clemency Coggins (1967) identified this pattern at 13 sites in the Guatemalan Petén and adjacent lowlands of Belize, "an area which, with Uaxactun as its center, stretches for about 100 kilometers in every direction" (Coggins 1967:4). Using the best published maps available at that time, she identified this pattern at Baking Pot, Balakbal, Benque Viejo (Xunantunich), Chochkitam, La Honradez, Naachtun, Nakum, Naranjo, Polol, Tikal, Uaxactun, Xultun, and Yaxha.

In her 1967 paper, Coggins ascribed specific cosmological meaning to the spatial arrangements cited, drawing on sixteenth-century sources, especially Bishop Diego de Landa (Coe 1965; Tozzer 1941). In later work, she turned her focus to the spatial orders embodied in quartered circles widespread in Mesoamerican iconography, whose component parts included the pecked circles, crossed axes that divided the circles in four, and an emphasis within the designs on cardinal and intercardinal directions and positions. Drawing again on sixteenth-century sources, as well as colonial and modern Maya dictionaries, decipherment of Classic hieroglyphs, interpretation of Classic iconography, and archaeoastronomical studies of pecked circles (which circles date from at least the second century AD) and of the orientation of their axes, Coggins tied the circles to calendrical completion cycles (especially solar cycles), to civic design, and to expressions combining political and cosmological symbolism (Coggins 1980; compare Aveni 2002:231; Aveni et al. 1978).

These ideas Coggins considered to be manifest in architecture. "E-Groups" known from Uaxactun and other lowland Maya sites exemplified early expressions of annual solar cycles, linked in space to activities led by sovereigns or others in the ruling class. Tikal's twin-pyramid groups (TPG) exemplified the four-part spatial order, with pointed inclusion of the king, through his visage and a likely seat of his participation in public ritual. Coggins interpreted the distinctive set of TPG elements as collectively mapping the vertical path of the sun, rotated 90° to a horizontal plan in which "north" was "up," and "south" was "down" (compare Guillemin 1968). As support for the latter inferences, Coggins offered the preponderance of underworld imagery associated with "south" (Thompson 1960), and Brotherston's reasoned assertion that the Maya words we gloss as "north" and "south" were perhaps better understood as "moments between" east and west, and furthermore, that "betweenness" could be horizontal, vertical, or temporal (Brotherston 1976; Coggins 1980). Gossen (1974) provided support for these inferences in Chamula ethnography, and Bricker (1983) gave subsequent linguistic support for readings of directional glyphs. And in the twin-pyramid groups, as well as Tikal's North Acropolis/Great Plaza/Central Acropolis, the same directional positions place images of rulers (and sometimes their actual remains) metaphorically in the supernatural realm of the heavens, the apogee of the sun's daily path, its (and by extension, the ruler's) position of greatest strength. Also consistent with these inferences is the fact that the southern position in each TPG is occupied by a single-room building with nine doorways, which Coggins identified metaphorically with the Nine Lords of the underworld.

Ashmore encountered Coggins's unpublished 1967 seminar paper in the late 1970s, while seeking to account for similar regularities of architectural location, arrangement, and orientation at Quiriguá, patterns that spatial models then current in anthropology seemed unable to explain (Ashmore 1986, 1989; compare parallel pattern identifications by Hammond 1981; Jones in Jones, Ashmore, and Sharer 1983). In 1983, Ashmore brought together the aforementioned lines of thought to *hypothesize* that particular, specified architectural assemblages at Tikal and other sites of the lowland Maya and at least some of their southern neighbors embodied the kinds of structuring principles and *political and cosmological meanings* for which Coggins and others had laid interpretive foundations (Ashmore 1989, 1992).[1] The specifically Maya bases for inference of beliefs and spatial organization were coupled with theory drawn less from conceptions of cosmologic space than of political space and rhetoric (e.g., Kuper 1972; Rapoport 1982), a point to which we return below.

Certainly, Maya civic centers include complexes of buildings that do not occupy any identifiable north-south axis, and in our article, we acknowledged explicitly the additional existence of prominent east-west axes of spatial distribution. Other buildings are situated along no perceptible civic axis at all. We continue to believe, however, that disposition of prominent construction along a north-south line *does* dominate parts or all of many Maya civic precincts in Classic times. Indeed, demonstrable modification of topography sometimes complements the distribution of formal construction, which we take to support inferences of intentionality and *planning* in such alignments (e.g., the ridge top setting of Xunantunich). The alignment involved need not be strictly cardinal, however. In a related example, orientations recorded in "E-Group"-like complexes other than at Uaxactun do not consistently replicate the astronomical precision of that earliest observatory complex; this fact led Vilma Fialko (1988) to argue that the other groups more likely commemorate acts of observation rather than facilitating actual astronomical observance. Our parallel suggestion is that the *approximate* cardinal positions suffice in many cases to make mutually equivalent positional points.

Of course, Smith rightly decries indiscriminate "skewing" of cardinality to "fit" the model. He cites the case of Gualjoquito, as an example. The lack of ground contours on the published maps (e.g., Ashmore 1989:283, 1992:180) omits graphic recognition of the ridge that abruptly bounds the site to east and northeast, and of the watercourses and gullies bounding it to south and north; to the west, the Ulua river is indicated verbally. In Ashmore's estimate, the site's layout is as close to a north-south arrangement as constricted topography and cumulative construction history allow (e.g., Ashmore 1987). For inferentially comparable reasons, the "eastern" pyramid in Tikal's most spatially expansive political-cosmological expression, Temple VI, is actually southeast of civic center, its placement skewed by the large *bajo* (wetland, swamp) directly east.

The foregoing seems to us quite consistent with other considerations of civic planning, in Mesoamerica and beyond. Besides Fialko's consideration of "aberrant" E-Groups, Aveni (1980:277) cautions more generally against astronomical bases for every construction alignment. And without allusion to specifics of astronomy, cosmology or political expression, Marcus notes that many Mesoamerican cities combine both planned and unplanned aspects, having a clearly planned "inner city," which is the locus for public secular and religious structures, and an unplanned "outer city," which shows haphazard residential growth (Marcus 1983:197). It seems plausible to us that different placement and orientation principles can apply in different zones.

To this point, our research had comprised what we view as aspects of hypothesis generation. The hypothetical specification of elements constituting the pattern under investigation was published most explicitly for the Tikal case in 1989, although elements were designated less clearly for other sites mentioned in the 1989 and 1992 articles. Put succinctly, the basic criteria for recognizing constructions relevant to assessing the pattern and defining (any) axes were arguably civic buildings and spaces that seemed to stand out as prominent amid the surrounding architectural. milieu, especially when these were associated with attention-calling elements, such as carved monuments, or entry into attached causeways.

These are not easily quantifiable characterizations; they are, however, susceptible to critique, response, and ongoing discussion. The maps in our 2002 article highlight what we believe are the most pertinent constructions by eliminating others; earlier articles did no such graphic editing, and therefore provide larger contexts for understanding the highlighted elements. The criteria cited provide first-round designations, subject to modification (as in any sequential hypothesis testing) as examination proceeds. Two critical aspects in selection and modification involve (1) refinements to the map from which elements are available to be selected (e.g., compare early and often incorrect maps with subsequent instrument maps of the same sites) and (2) detailed documentation of construction history, as when one considers Tikal's reconstructed architectural history at the close of The University of Pennsylvania Museum's project and then incorporates the significant enhancements to chronological and other inferences by subsequent Guatemalan projects (e.g., Jones 1991; Laporte and Fialko 1995).

HYPOTHESIS TESTING

From the outset, Ashmore knew well that the data on which the interpretive civic-planning inferences were based had all been recovered for other reasons, not related to the hypotheses here in question. For that reason, she organized and implemented the Copán North Group Project, specifically as a test of the hypothesis that the ancient Maya linked the cardinal position of "north" symbolically with the celestial realm ("up"), and politically with supernatural sanctioning of royal authority. That is, and specifically proposed as a test, if the principles and meanings previously inferred held merit, Copán's Group 8L-10 and 8L-12, a locally imposing complex 1 km north of the Main Group, should yield evidence of royal ancestors and ritual activities, distinct from those of the royal precinct (i.e., the well-studied Main Group), and unique within the surrounding settlement.

With regard to Smith's critique of criteria for selecting the elements for inquiry, we reply that the hypotheses stated in earlier articles, and the criteria noted earlier here, led directly to identification of a particular set of buildings as marking north in the comprehensive map of ancient Copán valley settlement. And in short, the findings were that what others had seen as a fairly undistinguished "elite residential compound" did, in fact, yield hieroglyphic, sculptural and mortuary evidence that seemed to support the original hypothesis (Ashmore 1991). This was *far from* a conclusive demonstration, but a piece of evidence in a larger, open-ended scientific inquiry.

Another test took place, if less formally, at Xunantunich. From the early 1980s on, Norman Hammond had asserted, in conversation, that Ashmore's model "didn't work" in Belize. So in the early 1990s, she undertook to consider at least one Belizean civic center, and examine its form and constructional history. From that, she came to infer that the elements, axes and meanings were present, but that the localization of meanings at north and south ends of the dominant axis were reversed. In other words, this inquiry led to revision of the original hypothesis about specific forms, largely in terms of localization of the posited meanings (Ashmore 1998; Ashmore, in LeCount et al. 2002:43; Ashmore and Leventhal 1993).

Others have embarked on additional tests, or *proposed alternative hypotheses*. We take these as constructive and necessary, whatever the outcome. Brett Houk, for example, examined the form and constructional history of the site of Dos Hombres and found that "the original Late Classic site plan was based on ... a variation of the model proposed by Ashmore ... [and] appears to have been used by colonizing or conquering lineages as a form of expression and proclamation of cultural and political identity" (Dunning et al. 1999; Houk 1996:317). On the other hand, Allan Maca (1999, 2002) proposed a distinctly different model for *civic planning* and *bounding* the community of Copán, focusing on U-shaped groups that define a rhomboid, whose vertices mark the cardinal boundary corners neatly encompassing what others had recognized earlier as a dense urban core (e.g., Webster 1999). For Maca, drawing on ethnographic accounts of comparable practices, Copán's U-shaped groups mark positions on a ritual circuit traversed repeatedly by Copanec Maya, to re-establish their community and their world (compare Coe 1965). And Angela Keller is currently exploring the idea that Xunantunich was planned to emphasize a *cruciform* layout, harking back to some of Coggins's and others' arguments cited earlier (Keller 1995).

To us, the alternatives (to date) do not necessarily invalidate the model outlined earlier. Clearly, Houk's research is most directly supportive, as are some other analyses (e.g., Tourtellot et al. 2002). Maca's ideas, on the other hand, suggest that while Ashmore's inferences about Copán Groups 8L-10 and 8L-12 may relate to concepts of "north," they do not necessarily reflect *civic planning,* in the sense of overall design program for placement of architectural complexes. Keller's research suggests different scales of consideration; Ashmore sees a cruciform plan as complementing or incorporating, not contradicting, the dominance of a north-south axis—for enhancement of which the Xunantunich ridge top seems to have been deliberately shaped, by quarrying. At Dos Hombres, as well, "builders of the site maintained the north-south orientation by intentionally selecting which hills in the bajo would be utilized as natural construction platforms" (Houk 1996:315). Perhaps *royal precinct planning* would be a clearer designation in this and some other instances.

PROGRAMMATIC SUGGESTIONS

Certainly, the ideas we have offered are not demonstrated to be true. Our 2002 article marked another round of hypothesis generation, one that explicitly linked political history and construction sequences with the inferences made earlier. In our view, cosmically sanctified civic plans of antiquity were as much a vehicle for royal promotion and political opinion shaping as were public texts and imagery found within their confines (e.g., Marcus 1992; Schele and Mathews 1998; Steinhardt 1986). Although little grounding in spatial theory was included explicitly in early expressions of our arguments, we have certainly drawn from multiple sources of inspiration and support, ranging from general theoretical works concerning the expressive capacities of architecture and space (e.g., Blanton 1994; Kuper 1972; Leach 1983; Rapoport 1982) to pertinent comparative applications in and well beyond the Maya area (e.g., Fritz et al. 1984; Low 1995; Moore 1992; O'Connor 1989; Preucel 2000). Our thinking is increasingly informed by the importance of both settlement and landscape archaeologies, their intersection, and theoretical developments that link the built environment with ideology and belief (e.g., Ashmore and Knapp 1999; Bender 1993; Bradley 1993; Sabloff and Ashmore 2001). And we believe that the time is right for renewed tests of the hypotheses advanced here.

The central challenge is not whether political or cosmological symbolism might be expressed in architecture and space, but whether and how one can recognize when such symbolic communication has taken place (compare Carl et al. 2000). For example, as Smith indicates, Hanns Prem (2000) expresses this concern at some length, with Prem pointedly citing his dubiousness about our work, in the context of a more general critique of the 1997 Round Table in which Prem and we were participants. As counter examples, Prem cites disagreements over the supposed similarity of specific civic plans in the Puuc region, including the Sayil/Labná pair we discuss (Prem 2000:66). The final reports of Tomás Gallareta and his colleagues on their mapping and excavations at Labná should provide a stronger data set for the testing of our hypothesis (Gallareta N. et al. 1995).

We take these and other critiques seriously, in the spirit of intellectual discourse and debate. Although we may not have a definitively acceptable resolution for disagreements about specific cases, we suspect that Smith's critique pertains in at least the instance just cited, for Sayil and Labná: that those who disagree are taking different elements for comparison, and remain unclear about their selection process among mapped buildings and spaces, and thus remain mutually unclear about resulting comparison sets. We were perhaps overly elliptical in using graphic editing to highlight the relevant elements in our sets, and should have been more explicit about how these elements were selected. We continue to believe that the elements highlighted in our 2002 maps identify what were, in antiquity, the most publicly visible, visited, and acknowledged buildings and spaces of the civic centers under consideration, and that the layout of Labná does resemble that of Sayil closely.

Clearly, we have difficulty with the notion that random accretions of construction would account for the patterns that seem, *on the basis of our stated selection and recognition criteria,* so pervasive and consistent. Such an assertion seems significantly at odds with what linguistic, iconographic, and epigraphic evidence suggests was a highly spatially structured Maya world in a time of highly propagandistic state systems (e.g., Culbert 1991; Freidel et al. 1993; Hanks 1990; Martin and Grube 2000). On the other hand, the possibility of random accretion remains a null hypothesis. We acknowledge that it is difficult to falsify our hypotheses definitively, but contend that there are ways of knowing that may not be fully susceptible to the kind of falsification testing associated with positivist approaches of early processualism (e.g., Preucel 1991). Moreover, by extending case examples, we hope to stimulate new examinations of these data sets that will provide stronger and tighter means of testing our hypotheses.

The programmatic suggestions we offer toward greater resolution of these issues do presume that the questions posed merit inquiry. If so, what we urge begins with wider discussion *and critique* of the models and inferences proposed, and examination of as wide an array of civic centers as possible with these ideas in mind.[3] Initial consideration can take place with published maps. Fundamental for meaningful examination, however, is the combination of high-quality, spatially comprehensive maps *and* detailed information on construction history. Practice-oriented models have potential for finer grained consideration of how the architecturally structured messages were effectively communicated, and possibly how they were received (e.g., Keller 1995; Maca 1999, 2002). Moreover, finely detailed stratigraphic and other chronological analyses of architecture now increasingly join with textual evidence, when available, to define building programs and "planning agendas" of particular sovereigns (e.g., Jones

1991; Sharer et al. 1999). Testing the ideas does not depend on accompanying texts, however. That is accomplished principally through archaeological investigation.

We summarize the foregoing by reiterating our appreciation of Smith's thoughtful critique. Our reply does not provide a topically exhaustive response to the questions he raises, but does seek to continue the dialogue, with him and other colleagues. We agree that methods can be sharpened, and heartily invite his and others' collaboration on doing so. As Lewis Binford (1962), Olivier de Montmollin (1989) and others have asserted, in varying ways, the first issue is to formulate interesting and meaningful questions, and then the challenge is to propose methods appropriate to addressing those questions.

ACKNOWLEDGMENTS

We thank Michael Smith for his thoughtful critique of our work and Suzanne Fish and Maria Dulce Gaspar, co-editors of this journal, for giving us the opportunity to respond to Smith's remarks. We are grateful to Clemency Coggins and Gair Tourtellot for early stimulus in this research, and to them as well as Chelsea Blackmore, Christine Carrelli, George Cowgill, Olivier de Montmollin, Norman Hammond, Brett Houk, Angela Keller, Richard Leventhal, Allan Maca, Saburo Sugiyama, and Thomas Patterson for helpful, ongoing contributions and constructive criticism of this work that is still very much in progress.

NOTES

1. The 1989 paper was originally written in 1983, and the 1992 paper, in 1987.

2. Non-civic architecture, especially domestic compounds of various classes, may evince some of the same spatial principles (Johnston and Gonlin 1998; Robin 2002), but their expression as propaganda is clearer in civic building programs.

REFERENCES

Ashmore, Wendy. 1986. Peten Cosmology in the Maya Southeast: An Analysis of Architecture and Settlement Patterns at Classic Quiriguá. In *The Southeast Maya Periphery,* ed. Patricia A. Urban and Edward M. Schortman, 35–49. Austin, University of Texas Press.

Ashmore, Wendy. 1987. Cobble Crossroads: Gualjoquito Architecture and External Elite Ties. In *Interaction on the Southeast Mesoamerican Periphery: Prehistoric and Historic Honduras and El Salvador,* ed. Eugenia J. Robinson, 28–48. BAR International Series 327. British Archaeological Reports, Oxford.

Ashmore, Wendy. 1989. Construction and Cosmology: Politics and Ideology in Lowland Maya Settlement Patterns. In *Word and Image in Maya Culture: Explorations in Language, Writing, and Representation,* ed. William F. Hanks and Don S. Rice, 272–286. Salt Lake City, University of Utah Press.

Ashmore, Wendy. 1991. Site-Planning Principles and Concepts of Directionality among the Ancient Maya. *Latin American Antiquity* 2:199–226.

Ashmore, Wendy. 1992. Deciphering Maya Site Plans. In *New Theories on the Ancient Maya*, ed. Elin Danien and Robert J. Sharer, 173–184. Museum Monographs, 77. Philadelphia, University of Pennsylvania Museum.

Ashmore, Wendy. 1998. Monumentos Políticos: Sitios, Asentamiento, y Paisaje por Xunantunich, Belice. In *Anatomia de una Civilización: Aproximaciones Interdisciplinarias a la Cultura Maya*, ed. Andres Ciudad Ruiz, Yolanda Fernández Marquinez, José Miguel García Campillo, Ma. Josefa Iglesias Ponce de León, Alfonso Lacadena García-Gallo, and Luis T. Sanz Castro, 161–183. Publ. No. 4. Madrid, Sociedad Española de Estudios Mayas.

Ashmore, Wendy, and A. Bernard Knapp, eds. 1999. *Archaeologies of Landscape: Contemporary Perspectives*. Oxford, Blackwell.

Ashmore, Wendy, and Richard M. Leventhal. 1993. Xunantunich Revisited. Paper presented at the Conference on Belize, University of North Florida, Jacksonville.

Ashmore, Wendy, and Jeremy A. Sabloff. 2002. Spatial Orders in Maya Civic Plans. *Latin American Antiquity* 13:201–216.

Aveni, Anthony F. 1980. *Skywatchers of Ancient Mexico*. Austin, University of Texas Press.

Aveni, Anthony F. 2002. *Skywatchers*. Austin, University of Texas Press.

Aveni, Anthony F., Horst Hartung, and Beth Buckingham. 1978. The Pecked-Cross Symbol in Ancient Mesoamerica. *Science* 202:267–279.

Bender, Barbara, ed. 1993. *Landscape: Politics and Perspectives*. Oxford, Berg.

Binford, Lewis R. 1962. Archaeology as Anthropology. *American Antiquity* 28:217–225.

Blanton, Richard E. 1994. *Houses and Households: A Comparative Study*. New York, Plenum.

Bradley, Richard. 1993. *Altering the Earth: The Origins of Monuments in Britain and Continental Europe*. Monograph Series Number 8. Edinburgh, Society of Antiquaries of Scotland.

Bricker, Victoria. 1983. Directional Glyphs in Maya Inscriptions and Codices. *American Antiquity* 48:347–353.

Brotherston, Gordon. 1976. Mesoamerican Description of Space II: Signs for Direction. *Ibero-Amerikanisches Archiv* N.F. 2 (1):39–42.

Carl, Peter, Barry Kemp, Ray Laurence, Robin Coningham, Charles Higham, and George L. Cowgill. 2000. Viewpoint: Were Cities Built as Images? *Cambridge Archaeological Journal* 10:327–365.

Coe, Michael D. 1965. A Model of Ancient Community Structure in the Maya Lowlands. *Southwestern Journal of Anthropology* 21:97–114.

Coggins, Clemency Chase. 1967. Palaces and the Planning of Ceremonial Centers in the Maya Lowlands. Unpublished manuscript, Tozzer Library, Peabody Museum, Harvard University, Cambridge, MA.

Coggins, Clemency Chase. 1980. The Shape of Time: Some Political Implications of a Four-part Figure. *American Antiquity* 45:727–739.

Culbert, T. Patrick, ed. 1991. *Classic Maya Political History: Hieroglyphic and Archaeological Evidence*. Cambridge, Cambridge University Press.

Dunning, Nicholas P., Vernon Scarborough, Fred Valdez Jr., Sheryl Luzzadder-Beach, Timothy Beach, and John G. Jones. 1999. Temple Mountains, Sacred Lakes, and Fertile Fields: Ancient Maya Landscapes in Northwestern Belize. *Antiquity* 73:650–660.

Fialko, Vilma. 1988. Mundo Perdido, Tikal: Un Ejemplo de Complejos de Conmemoración Astronómica. *Mayab* 4:13–21.

Freidel, David, Linda Schele, and Joy Parker. 1993. *Maya Cosmos: Three Thousand Years on the Shaman's Trail*. New York, William Morrow.

Fritz, John M., George Michell, and M. S. Nagaraja Rao. 1984. *Where Kings and Gods Meet: The Royal Centre at Vijayanagara, India.* Tucson, University of Arizona Press.

Gallareta Negrón, Tomás, Lourdes Toscano Hernández, and Carlos Pérez Alvarez. 1995. Programa de Investigación del Proyecto Labná: Temporada de Campo 1995. Instituto Nacional de Antropología e Historia, Consejo de Arqueología, Yucatán, Mexico.

Gossen, Gary. 1974. *Chamulas in the World of the Sun: Time and Space in a Maya Oral Tradition.* Cambridge, Harvard University Press.

Guillemin, George. 1968. Development and Function of the Tikal Ceremonial Center. *Ethnos* 33:1–35.

Hammond, Norman. 1981. Settlement Patterns in Belize. In *Lowland Maya Settlement Patterns,* ed. Wendy Ashmore, 157–186. Albuquerque, University of New Mexico Press.

Hanks, William F. 1990. *Referential Practice: Language and Lived Space among the Maya.* Chicago, University of Chicago Press.

Houk, Brett A. 1996. The Archaeology of Site Planning: An Example from the Maya Site of Dos Hombres, Belize. Ph.D. dissertation, Department of Anthropology, University of Texas, Austin. Ann Arbor, University Microfilms.

Johnston, Kevin J., and Nancy Gonlin. 1998. What Do Houses Mean? Approaches to the Analysis of Classic Maya Commoner Residences. In *Function and Meaning in Classic Maya Architecture,* ed. Stephen D. Houston, 141–185. Washington, DC, Dumbarton Oaks.

Jones, Christopher. 1991. Cycles of Growth at Tikal. In *Classic Maya Political History,* ed. T. Patrick Culbert, 102–127. Cambridge, Cambridge University Press.

Jones, Christopher, Wendy Ashmore, and Robert J. Sharer. 1983. The Quiriguá Project: The 1977 Season. In *Quiriguá Reports,* vol. 2, ed. Edward M. Schortman and Patricia A. Urban, Paper No. 6, 1–38. Museum Monograph 49. Philadelphia, University of Pennsylvania Museum.

Keller, Angela H. 1995. Testing and Excavation around *Sacbe* II and Group C. In *Xunantunich Archaeological Project: 1997 Field Season,* ed. Richard M. Leventhal, 96–115. Los Angeles, Cotsen Institute of Archaeology.

Kuper, Hilda. 1972. The Language of Sites in the Politics of Space. *American Anthropologist* 74:411–425.

Laporte, Juan Pedro, and Vilma Fialko. 1995. Un Reencuentro con Mundo Perdido, Tikal, Guatemala. *Ancient Mesoamerica* 6:41–94.

Leach, Edmund. 1983. The Gatekeepers of Heaven: Anthropological Aspects of Grandiose Architecture. *Journal of Anthropological Research* 39:243–264.

LeCount, Lisa J., Jason Yaeger, Richard M. Leventhal, and Wendy Ashmore. 2002. Dating the Rise and Fall of Xunantunich, Belize: A Late and Terminal Classic Lowland Maya Regional Center. *Ancient Mesoamerica* 13:41–63.

Low, Setha M. 1995. Indigenous Architecture and the Spanish-American Plaza in Mesoamerica and the Caribbean. *American Anthropologist* 97:748–762.

Maca, Allan L. 1999. The Urban Panorama of a Classic Maya Center: Site Planning at 8th Century Copán. Paper presented at the First Annual Graduate Symposium, University of Pennsylvania, Philadelphia.

Maca, Allan L. 2002. Spatio-temporal Boundaries in Classic Maya Settlement Systems: Copán's Urban Foothills and the Excavations at Group 9J-5. Ph.D. dissertation, Department of Anthropology, Harvard University. Ann Arbor, University Microfilms.

Marcus, Joyce. 1983. On the Nature of the Mesoamerican City. In *Prehistoric Settlement Patterns: Essays in Honor of Gordon R. Willey,* ed. Evon Z. Vogt and Richard M. Leventhal,

195–242. Albuquerque, University of New Mexico Press, and Cambridge MA, Peabody Museum, Harvard University.

Marcus, Joyce. 1992. *Mesoamerican Writing Systems: Propaganda, Myth, and History in Four Ancient Civilizations.* Princeton, NJ, Princeton University Press.

Martin, Simon, and Nikolai Grube. 2000. *Chronicle of the Maya Kings and Queens: Deciphering the Dynasties of the Ancient Maya.* London, Thames and Hudson.

Montmollin, Olivier de. 1989. *The Archaeology of Political Structure: Settlement Analysis in a Classic Maya Polity.* Cambridge, Cambridge University Press.

Moore, Jerry D. 1992. Pattern and Meaning in Prehistoric Peruvian Architecture: The Architecture of Social Control in the Chimú State. *Latin American Antiquity* 3:95–113.

O'Connor, David. 1989. City and Palace in New Kingdom Egypt. *Cahier de Recherches de l'Institut de Papyrologie et d'Egyptologie de Lille: Societes urbaines en Egypte et au Soudan,* 11:73–87.

Prem, Hanns. 2000. ¿Detrás de qué Esquina Se Esconde la Ideología? In *Arquitectura e Ideología de los Antiguos Mayas: Memoria de la Segunda Mesa Redonda de Palenque,* ed. Silvia Trejo, 55–70. Mexico, CONACULTA/Instituto Nacional de Antropología e Historia.

Preucel, Robert W. 1991. The Philosophy of Archaeology. In *Processual and Postprocessual Archaeologies: Multiple Ways of Knowing the Past,* ed. Robert W. Preucel, 17–29. Occasional Paper 10. Carbondale, Center for Archaeological Investigations, Southern Illinois University.

Preucel, Robert W. 2000. Making Pueblo Communities: Architectural Discourse at Kotyiti, New Mexico. In *The Archaeology of Communities: A New World Perspective,* ed. Marcello A. Canuto and Jason Yaejer, 58–77. London, Routledge.

Rapoport, Amos. 1982. *The Meaning of the Built Environment: A Nonverbal Communication Approach.* Beverly Hills, Sage.

Robin, Cynthia. 2002. Outside of Houses: The Practices of Everyday Life at Chan Nòohol, Belize. *Journal of Social Archaeology* 2:245–268.

Sabloff, Jeremy A., and Wendy Ashmore. 2001. An Aspect of Archaeology's Recent Past and Its Relevance in the New Millennium. In *Archaeology at the Millennium: A Sourcebook,* ed. Gary M. Feinman and T. Douglas Price, 11–32. New York, Kluwer Academic/Plenum.

Schele, Linda, and Peter Mathews. 1998. *The Code of Kings: The Language of Seven Sacred Maya Temples and Tombs.* New York, Scribner.

Sharer, Robert J., William L. Fash, David W. Sedat, Loa P. Traxler, and Richard Williamson. 1999. Continuities and Contrasts in Early Classic Architecture of Central Copán. In *Mesoamerican Architecture as a Cultural Symbol,* ed. Jeff Karl Kowalski, 220–249. Oxford, Oxford University Press.

Steinhardt, Nancy Shatzman. 1986. Why Were Chang'an and Beijing So Different? *Journal of the Society of Architectural Historians* 45:339–357.

Thompson, J. Eric S. 1960. *Maya Hieroglyphic Writing: An Introduction.* 2nd ed. Norman, University of Oklahoma Press.

Tourtellot, Gair, Marc Wolf, Scott Smith, Kristen Gardella, and Norman Hammond. 2002. Exploring Heaven on Earth: Testing the Cosmological Model at La Milpa, Belize. *Antiquity* 76:633–634.

Tozzer, Alfred M., trans. and ed. 1941. *Landa's Relación de las Cosas de Yucatan.* Papers of the Peabody Museum of Archaeology and Ethnology, 18. Cambridge, MA, Harvard University.

Webster, David. 1999. The Archaeology of Copán, Honduras. *Journal of Anthropological Research* 7:1–53.

More on Mesoamerican Cosmology and City Plans
Ivan Šprajc

In a comment published in a former issue of this journal, Michael Smith (2003) challenges the cosmological interpretations of Maya urban layouts proposed by Wendy Ashmore and Jeremy Sabloff (2002). He argues that their hypotheses are weak, vague, and unconvincing and emphasizes the need for rigorous methods in this kind of research. In his subsequent essay, Smith extends his criticism to recent applications of the "cosmogram" concept to Maya architectural layouts. It is not my purpose to comment on his critique of the ideas expressed by Ashmore and Sabloff; their own reply (2003) eliminates many doubts and makes their procedures much more explicit than they had been before. Neither will I debate Smith's contribution published in this issue: even if one would prefer to see a better-founded case-by-case discussion, rather than a sweeping rejection, I agree that many recent interpretations in terms of cosmograms seem to be the result more of a kind of fashion trend than of serious research supported by evidence.

Instead, and following Ashmore and Sabloff's (2003:233–234) invitation to continue the dialogue and discussion, I would like to focus on the more general part of Smith's (2003) argument in his earlier article, in which he discusses the difficulties involved in cosmological interpretations of archaeologically recovered urban layouts. Nobody seriously engaged in any scientific endeavor will question his contention that "research in this area requires rigorous and explicit methods if it is to have credibility within the archaeological community" (2003:221–222). However, when he admits that "site maps often suggest that some sort of spatial order existed in ancient cities" but adds that "scholars have yet to develop systematic approaches to the study of the nature and origin of that order" (2003:221), he makes a subjective and unbalanced generalization, neglecting many recent advances and reliable methodological procedures applied in this field of research.

Presenting some cosmological interpretations of particular city plans from different cultures, Smith argues that they are unconvincing and that alternative hypotheses, unrelated to cosmology or worldview, could be substituted. One of the examples he discusses to support his opinion that the cosmological meanings of Mesoamerican urban layouts have not been recovered with confidence and rigor is the Templo Mayor of Tenochtitlan. He affirms that the large body of scholarship on the role of cosmology in the design of this structure "relies almost exclusively on subjective interpretations of Aztec myths and rituals" (2003:222), and he doubts that the results of these studies can be extended to the whole city of Tenochtitlan. The main temple of the Aztec capital is supposed to illustrate "the difficulty of inferring the ideas and inten-

---

Reproduced by permission of the Society for American Archaeology from *Latin American Antiquity* 16:2 (2005).

tions of rulers and builders from the material remains of urban sites, even when there is a corpus of written documentation" (Smith 2003:223). Smith presents some historical data and archaeoastronomical hypotheses about the meaning of the Templo Mayor and suggests that the urban pattern of Tenochtitlan might be an imitation of the Teotihuacan grid, but in doing so he fails to take into account all the relevant evidence and ignores recent advances both in Mesoamerican archaeoastronomy in general and in the understanding of Teotihuacan and the Templo Mayor of Tenochtitlan in particular.

Because the astronomically derived concepts were an important part of ancient cosmologies or worldviews, it is obvious that archaeoastronomy, specialized in the study of diverse manifestations of these concepts, including architectural orientations, has a prominent role in the search for the cosmological templates of the ancient urban plans. Smith's marginal references to some archaeoastronomical works do not reflect the fact that this field of research-in spite of examples of bad scholarship—has made significant progress precisely in the direction he demands: toward the application of rigorous methods and techniques that yield reliable and testable results. To support this statement I will summarize a few archaeoastronomical studies that have contributed to the understanding of Mesoamerican urban planning, with a special emphasis on two cases discussed by Smith: Teotihuacan and the Templo Mayor of Tenochtitlan.

ARCHAEOASTRONOMY AND MESOAMERICAN URBAN LAYOUTS

The shortest way of summarizing the methodological guidelines for any serious archaeoastronomical study of orientations might be the following: to conclude, with a reasonable degree of confidence, that an architectural orientation, or any alignment recognized in the archaeological record or ancient cultural landscape, had an intentionally chosen astronomical target, we need either a statistically significant number of comparable alignments, incorporated in a coherent set of archaeological features (i.e., of the same type and pertaining to the same cultural complex) and referring to the same position (declination) on the celestial sphere; or independent contextual evidence suggesting an astronomical motive for the alignment in question (iconography, written sources, etc.); or both. On the other hand, the meaning of an alignment, or a homogenous set of alignments with the same astronomical referent, can be properly understood only if we manage to find reasons for which the postulated astronomical phenomenon could have been significant to the society that produced the alignment(s). The viability of archaeoastronomical hypotheses is directly proportional to the degree of significance that can be assigned to the astronomical phenomena involved. Such significance is to be sought in the relationship of the astronomical phenomena with specific environmental and cultural facts (e.g., seasonal climatic changes, subsistence strategies, religion, political ideology, etc.; cf. Aveni 2003; Iwaniszewski 1989; Ruggles 1999).

The application of these general methodological principles in Mesoamerican archaeoastronomy can be illustrated by a number of studies, which have led to the recognition of particular concepts involved in pre-hispanic architectural and urban planning. Systematic research carried out during the last few decades has revealed that the orientations in civic and ceremonial architecture exhibit a clearly nonrandom distribution, which indicates that the buildings were mostly oriented on the basis of astronomical considerations, particularly to the Sun's positions on the horizon on certain dates of the tropical year (Aveni 2001; Aveni and Gibbs 1976; Aveni and Hartung 1986; Tichy 1991). Any skeptic wanting to challenge this conclusion should offer an alternative explanation for the widespread orientation groups (the azimuths clustering around certain values occur at a number of sites in different Mesoamerican regions, some of them over long time spans), as well as for the fact that most of the east-west azimuths lie within the angle of annual movement of the Sun along the horizon.[1]

Furthermore, interpretations based on contextual evidence have been proposed concerning both the practical and the symbolic significance of architectural orientations. Aveni and Hartung (1986), for example, have analyzed a number of alignments in Maya architecture and conclude that they allowed the use of observational calendars based on solar zenith passages and other dates separated by multiples of 20 days, that is, basic periods of the Mesoamerican calendrical system; these observational calendars, they (1986:56–57) argue, must have served agricultural needs. The existence of similar observational schemes, composed of calendrically significant and, therefore, easily manageable intervals, is disclosed by a recent study (Šprajc 2001) based on 37 archaeological sites with monumental architecture in central Mexico: the intervals separating the sunrise and sunset dates recorded by the alignments tend to be multiples of 13 and 20 days. Because the dates included in these patterns are found to correspond to sunrises and sunsets both along architectural orientations and above prominent hilltops on the local horizon, it has been argued that important ceremonial structures were not only oriented but also located on astronomical grounds. The correspondence between the most frequently recorded dates and the crucial moments of the cultivation cycle suggests that the reconstructed observational schemes facilitated a proper scheduling of agricultural and associated ritual activities (Šprajc 2001).

Aveni et al. (2003) recently studied alignments involved in a special type of Maya architectural assemblage located in the Petén area and resembling Group E at Uaxactún, Guatemala. Their analysis, based on a statistically significant and typologically homogenous sample of alignments, led them to abandon a previous hypothesis, which interpreted the greater part of these assemblages as nonfunctional imitations of the (astronomically functional) Group E of Uaxactún (Aveni and Hartung 1989). They then conclude that the alignments reflect the use of observational schemes composed of calendrically significant intervals. They also note that the most frequently recorded dates suggest the importance of anticipatory Sun sightings during the dry half of the year leading up to the planting season (Aveni 2003:161–162; Aveni et al. 2003:163).

Even if the observational function of architectural orientations indicates their relationship with practical needs, which is in accordance with what we know about the adaptive value of astronomical knowledge and its consequent importance in archaic civilizations (Aveni and Hartung 1986:56; Iwaniszewski 1989:28–29; Šprajc 1996a:20–22), the alignments cannot be understood in purely utilitarian terms. As the repeatedly occurring directions are most consistently incorporated in the monumental architecture of civic and ceremonial urban cores, entailing considerable effort, they must have had an important place in the worldview and even in the cosmologically substantiated political ideology. This can be understood if we consider that the apparently immutable and perfect order observed in the sky, obviously superior to the one reigning on the earth, must have been the primary source of the deification of heavenly bodies, whose cyclic behavior thus was not viewed as being simply correlated with seasonal transformations in the natural environment but, rather, as provoking them. Assuming that also the proper annual movement of the Sun was, therefore, believed to be responsible for timely occurrences of these changes, the directions to the points of sunrise and sunset on crucial dates of the agricultural cycle must have acquired a sacred dimension. Because the beliefs composing the worldview were incorporated into the political ideology of rulers, who as man-gods pretended to be responsible for the proper functioning of the universe (cf. López Austin 1973), the alignments reproducing significant astronomical directions in civic and ceremonial architecture can be interpreted not only as a sanctified materialization of the union of space and time (whose importance in the Mesoamerican worldview is attested in different sources) but also as a manifestation of attempts of the governing class to legitimate its power by re-creating and perpetuating the cosmic order in the earthly environment (Aveni 2001:148–152, 217–222; Šprajc 1996a:21–22).

The ability to determine specific dates, whose importance was vital for subsistence, and to lay out accurate alignments to the corresponding solar events was obviously not a public domain based on a commonly shared worldview but, rather, part of the esoteric knowledge reserved for the elite. If these phenomena, which in certain architectural configurations produced light and shadow effects that may have been conceived as solar hierophanies, were observed on predicted dates, they sanctioned the ideology of the ruling class, reinforced social cohesion, and thereby contributed to the preservation of the existing political order (Broda 1982:99–105, 1991:462–463, 491; Iwaniszewski 1989:30–31; Šprajc 2001:121–122, 154–155, 411–415).

Although these are rather general conclusions, the studies summarized above, as well as many others, offer quite specific answers about a significant part of the regularities detected in the spatial ordering of Mesoamerican cities. Beyond merely identifying the astronomical phenomena implicated, they attempt to explain the reasons for their importance in terms of what we know about the economy, worldview, and political organization of the societies involved. However, although "perhaps more often than we have yet recognized, the sky provides the cues to spatial order on the terrestrial plane" (Knapp and Ashmore 1999:3), the following example illustrates archaeoastronomers' awareness of the fact that an objective and comprehensive understanding of this order can only be achieved by exploring both its astronomical and other possible foundations and by placing these efforts within a broader context of landscape archaeology (cf. Aveni 2001:217–222; Ruggles 1999:112–124).

Aveni (1991:63) has observed that in a number of cases in Mesoamerica, a prominent mountain is found to the north of a civic or ceremonial center. Furthermore, in central Mexico there are a large number of structures accurately oriented to mountaintops on the local horizon. Though there is no clear preference for the east- or west-lying mountains, the number of buildings aligned to a peak to the north is nearly twice as large as the number of those oriented to a hill to the south (Šprajc 2001:57). Even if the prominent summits on the eastern and western horizon could have served as precise markers of the Sun's positions and thereby facilitated observations, the relationship of architectural orientations with mountains, in general, may be accounted for by the latter's aquatic and fertility symbolism, an important aspect of the Mesoamerican worldview (Broda 1991), whereas the prevalence of the north-lying mountains probably reflects beliefs connecting not only mountains but also the northern part of the universe with water and fertility (Šprajc 1996b:41–43, 58–61). The discovery of this pattern, which reveals that not all of the evidently intentional alignments were based on astronomical motives, adds another element to our understanding of the complex set of rules that dictated architectural and urban planning in Mesoamerica, in which astronomical considerations were intertwined with beliefs about the symbolic meanings of landscape features and sides of the world.

In some cases, if there is a sufficient amount of supportive contextual data, a plausible interpretation can be proposed even for a single orientation. An illustrative example is the Palace of the Governor at Uxmal, Yucatán, Mexico. The plastic decoration of the façade includes nearly 400 Venus glyphs. The correspondence between five synodic periods of Venus and eight years was well known to the Maya; therefore, the fact that the masks of the rain god Chac adorning the façade are arranged in groups of five as well as the occurrence of eight bicephalic serpent bars above the main entrance and of a numeral eight on a Chac mask at the palace's northeast corner also suggest some relationship of the building with Venus. Aveni (1975:183–186; Aveni and Hartung 1986:22–34) long ago related the orientation of this structure to the southernmost rising point of Venus as the morning star, and my own interpretation links the alignment to the great northerly extremes of the evening star (Šprajc 1996a:173–178, 1996b:75–77). My argument is based on a better agreement of the orientation with the evening star extremes, as well as on the fact that Venus glyphs are placed in the cheeks of the rain god masks, probably alluding to the coincidence of these phenomena, always occurring in late April or early May, with the onset of the rainy season. Aveni (2001:286), on the other hand, attributes less importance to precision; apart from the fact that the building faces east, he mentions other data that, he believes, more strongly support an eastward-directed orientation scheme.

Here it is important to stress that the difference between Aveni's and my own interpretations by no means reflects inconsistencies in the methodology applied or a lack of credibility of archaeoastronomical hypotheses. Our disagreement, which concerns only the final details of our proposals and derives from giving different weights and interpretations to particular types of contextual evidence, may eventually be solved by the application of the very same methodology we have been employing.

What we need is more comparative data. If more orientations that can be associated with Venus extremes are detected, it will be possible to find out how closely they match the morning/evening star extreme rising/setting points; by evaluating the degree of precision involved, it should become easier to identify the phenomena targeted in particular cases. For the moment only a few other structures probably referring to Venus extremes are known (e.g., El Circular at Huexotla, in the Valley of Mexico, and the Caracol at Chichén Itzá, Yucatán [Aveni 2001:273–276; Šprajc 1996a:178–184, 1996b:72–85]), constituting too small a sample to allow any reliable conclusion.

TEOTIHUACAN AND THE TEMPLO MAYOR OF TENOCHTITLAN

To support his overall skepticism concerning the reliability of cosmological interpretations of Mesoamerican urban plans, Smith discusses two examples from central Mexico. If properly viewed in the light of the evidence available, however, they constitute perhaps the most illustrative cases that refute his opinion.

Smith (2003:222–223) suggests that the Aztecs, designing the layout of Tenochtitlan, may have simply imitated the orthogonal grid of Teotihuacan, irrespective of any cosmological notions of their own, and also that this pattern may have nothing to do with the passage of the Sun. In view of the arguments he presents, it might be assumed that a grid layout originated for reasons not related to cosmology; however, such motives can by no means account for the *orientations* of the two urban grids.

The two main orientations embedded in the urban layout of Teotihuacan pertain to the so-called 17° family, which is one of the most widespread alignment groups in Mesoamerica (Aveni 2001:234). Numerous hypotheses have been proposed about the meaning of these orientations (Aveni 2001:223–230, 2003:156–158). Partly in agreement with these former proposals is my own interpretation, based on both contextual evidence and a large sample of comparative alignment data (Šprajc 2000a). Because the whole argument, including an exhaustive discussion of previous hypotheses, has been presented elsewhere (Šprajc 2001:107–120, 201–238), I will only summarize the most important conclusions: (a) the two similar but slightly different orientations dominating the Teotihuacan urban grid must have been dictated by those of the Sun Pyramid and the Ciudadela; (b) both orientations were related to the Sun's positions on the horizon on dates separated by calendrically significant intervals and composing a canonical agricultural cycle; and (c) the Pyramid of the Sun was deliberately located on the spot where the perpendicular to the intended east-west alignment pointed to Cerro Gordo to the north and from where sunrises on a pair of significant dates (recorded at several other sites) could be observed over a prominent mountain on the eastern horizon.

Some of these interpretations may be challenged. It is a fact, however, that architectural alignments at a number of other sites from different periods correspond to sunrises and sunsets on the very same dates as those recorded by the east-west axes of the Pyramid of the Sun and the Ciudadela at Teotihuacan, and we can thus conclude with reasonable certainty that these alignments, indeed, referred to the Sun, and also that the target dates had some practical or ritual significance, or both.[2] Therefore, if urban layouts reproduce such alignments, then they can hardly be explained only in terms of "energetic efficiency" or even "random or stochastic growth processes" (Smith 2003:223).

Although the archaeological information about the urban layout of Tenochtitlan is much poorer than in the case of Teotihuacan, we can reach a similar conclusion and even support it with historical data. It has been commonly held that the streets in the historical center of Mexico City follow the prehispanic urban configuration. This is, indeed, very likely if we consider that the orientation of the colonial grid, skewed 7–8° clockwise from cardinal directions, corresponds with the orientation of Phase II of the Templo Mayor. As in Teotihuacan, the main temple of Tenochtitlan must have dictated the orientation of the prehispanic urban layout, which was later adopted by the colonial grid. Although the grid pattern per se may have been a tradition inherited from Teotihuacan, as Smith (2003:222) suggests, we can, again, hardly find an interpretation other than the astronomical one for its orientation, which is different from that prevailing in the Classic period metropolis but belongs to a group common in Postclassic central Mexico (Šprajc 2001:fig. 7).

However, the famous text inserted in Motolinía's work and mentioned by Smith (2003:222), saying that the feast of Tlacaxipehualiztli "fell when the sun was in the middle of Uchilobos, which was the equinox" (Motolinía 1971:51), obviously refers to the temple that was in use at the time of Spanish conquest. When Aveni and Gibbs (1976:513–516) and Aveni et al. (1988) attempted to reconcile this statement with the orientation of the Templo Mayor, suggesting that the equinox sunrises were observed in the notch between the two upper sanctuaries, they assumed that the orientation of Phase II was preserved by subsequent construction stages. This assumption was supported by the north-south alignment azimuths, which remained virtually the same throughout the temple's construction history. It is now clear, however, because of precise orientation measurements in the Templo Mayor precinct, that in its walls running east-west Phase III adopted a different orientation, which was maintained in all the following phases up to the conquest and was incorporated also in many adjacent structures. One of the two sunset dates corresponding to the east-west axis of the temple's late construction stages, including the last one, is 4 April, which in the Julian calendar of the sixteenth century corresponded to 25 March. In 1519, this was the last day of the month of Tlacaxipehualiztli, according to the day-by-day correlation of the Mexica and Julian calendars established by Caso (1967:58, table 4) and supported by different kinds of evidence (Prem 1991; Šprajc 2000c). According to various sources (including Motolinía 1971:45), the main feast of every month was celebrated on its last day (Caso 1967:39, 51; Prem 1991:395). Furthermore, in medieval Europe, 25 March, the Feast of the Annunciation, was commonly identified with the vernal equinox (McCluskey 1993:110–111, 114; Newton 1972:22–27).[3] We can thus conclude that the author of the statement quoted above did not refer to the astronomical equinox (the date of which would have hardly been known to a non-astronomer at that time). Rather, he only made note of the correlation between the day of the Mexica festival, which in the last years before the conquest coincided with the sunset along the axis of the Templo Mayor, and the date of the Christian (Julian) calendar that corresponded to the traditional day of spring equinox (see the whole argument in Šprajc 2000b, 2001:383–410).

Both the text inserted in Motolinía and the drawing of the Templo Mayor in the map of Tenochtitlan attributed to Cortés, where the Sun disk is shown between the twin sanctuaries, have frequently been interpreted as references to the observation of *sunrises*, but the sources are far from explicit. The fact that Marquina, paraphrasing Motolinía, mentions the Sun "in front of Huichilobos" (1960:113) shows clearly that the text is ambiguous and may well refer to *sunsets* in the axis of the building. Additionally, the Templo Mayor faces west, which might be an indication of the special importance of that direction. Nonetheless, and in spite of the prevalence of west-facing temples, it has been argued that most architectural orientations in central Mexico were astronomically functional in both eastern and western directions (Šprajc 2001:69–71); because the observational scheme proposed for the late stages of the Templo Mayor and composed of calendrically significant intervals includes both sunrise and sunset dates (Šprajc 2000b:S22, 2001:399), the scene depicted in the map of Tenochtitlan may represent a general allusion to the relationship between the temple's orientation and the Sun.

Whereas the hypothesis forwarded by Aveni and Gibbs (1976) and Aveni et al. (1988) implies an oblique alignment (i.e., to a celestial target well above the horizon), the azimuth distribution patterns exhibited by Mesoamerican architectural orientations indicate that in most cases these orientations recorded astronomical phenomena on the horizon (Šprajc 2001:25). There is evidence suggesting that orientations similar to that of the late stages of the Templo Mayor of Tenochtitlan (5°36' south of east) were common in the area of Texcoco (Šprajc 2001:322, 324–325, 330). The agreement between the text in Motolinía and one of the two sunset dates corresponding to the archaeologically attested orientation of the late phases of the Templo Mayor is hardly fortuitous and offers probably the most convincing support to the conclusion that this structure, as so many others, was intentionally oriented to the Sun's positions on the horizon.

Moreover, considering that two prominent mountain peaks on the eastern horizon of the Templo Mayor of Tenochtitlan marked sunrises on significant dates (included in observational calendars reconstructed for both orientations), it seems very likely that even the location of this building, just like that of the Sun Pyramid at Teotihuacan (Šprajc 2000a:410–412, 2001:231–238), was determined by astronomical considerations related to the surrounding topography. The idea is supported by independent evidence indicating that the site where the Templo Mayor was erected was, in practical terms, hardly appropriate for construction. Based on the results of their analyses of soil mechanics, Mazari et al. (1989) argue that no natural island had ever existed on the spot and that the temple was built upon a huge artificial platform some 11 m in height and submerged approximately 6 m below the lake surface (Šprajc 2000b:S22, 2001:397–398).

EPILOGUE

The preceding examples show that archaeoastronomical studies can and do formulate explicit conclusions based on coherent data selection and rigorous methodological procedures. Though they obviously do not represent the only approach to the understanding of Mesoamerican architecture and urbanism, they do offer answers to a number of specific questions concerning the nature of the underlying concepts, their significance with respect to the natural environment and cultural context, and their consequent role in worldview and political ideology.

Some time ago Kintigh asserted that "archaeologists see archaeoastronomers as answering questions that, from a social scientific standpoint, no one is asking" (1992:1). Smith's opinion, if shared by a wider community, gives a somewhat different impression: archaeoastronomers may be viewed as giving relevant answers that, within the "mainstream" archaeological audience, no one reads. My foregoing comments represent an attempt to bridge the communication gap between the two, who are in any case closely related brands of scholars, and an invitation to combine our efforts in the pursuit of common anthropological goals.

NOTES

1. In their study on Maya architectural alignments, Aveni and Hartung comment: "The astronomical hypothesis would seem especially worthy of consideration if we find alignments that are confined to a narrow azimuthal range in a sample of buildings spread far apart in space. In this case, there can be no conceivable way of actually laying out the chosen direction other than by the use of astronomical bodies at the horizon as reference objects" (1986:7–8).

2. A particularly illustrative example is the Acropolis of Xochicalco. Just like in the urban grid of Teotihuacan, the buildings of the Xochicalco Acropolis (including the Pyramid of the Feathered Serpents), sufficiently well preserved to allow precise measurements of orientations, incorporate two slightly different east-west alignments, which correspond to the same declinations (sunrise and sunset dates) as the orientations of the Sun Pyramid and the Ciudadela at Teotihuacan (Šprajc 2000a, 2001:201–238, 258–275). The conclusion that the orientations of the 17° family were solar derives precisely from the fact that the target declinations (dates) remained the same for many centuries: had these alignments referred to the rising or setting point of a star, the corresponding declinations would necessarily exhibit a consistent increase/decrease as a function of time, because of precessional shifts in the star's position on the celestial vault.

3. Even if the canonical date of ecclesiastical equinox established in AD 325 by the Council of Nicaea was 21 March, the Roman tradition associating the equinox with 25 March also survived (Newton 1972:22–27).

REFERENCES

Ashmore, Wendy, and Jeremy A. Sabloff. 2002. Spatial Orders in Maya Civic Plans. *Latin American Antiquity* 13:201–215.

Ashmore, Wendy, and Jeremy A. Sabloff. 2003 Interpreting Ancient Maya Civic Plans: Reply to Smith. *Latin American Antiquity* 14:229–236.

Aveni, Anthony F. 1975. Possible Astronomical Orientations in Ancient Mesoamerican Architecture. In *Archaeoastronomy in Pre-Columbian America,* ed. Anthony F. Aveni, 163–190. Austin, University of Texas Press.

Aveni, Anthony F. 1991. Mapping the Ritual Landscape: Debt Payment to Tlaloc during the Month of Atlcahualo. In *To Change Place: Aztec Ceremonial Landscapes,* ed. Davíd Carrasco, 58–73. Niwot, University Press of Colorado.

Aveni, Anthony F. 2001. *Skywatchers: A Revised and Updated Version of Skywatchers of Ancient Mexico.* Austin, University of Texas Press.

Aveni, Anthony F. 2003. Archaeoastronomy in the Ancient Americas. *Journal of Archaeological Research* 11:149–191.

Aveni, Anthony F., E. E. Calnek, and H. Hartung. 1988. Myth, Environment, and the Orientation of the Templo Mayor of Tenochtitlan. *American Antiquity* 53:287–309.

Aveni, Anthony F., Anne S. Dowd, and Benjamin Vining. 2003. Maya Calendar Reform? Evidence from Orientations of Specialized Architectural Assemblages. *Latin American Antiquity* 14:159–178.

Aveni, Anthony F., and Sharon L. Gibbs. 1976. On the Orientation of Precolumbian Buildings in Central Mexico. *American Antiquity* 41:510–517.

Aveni, Anthony F., and H. Hartung. 1986. *Maya City Planning and the Calendar.* Transactions of the American Philosophical Society Vol. 76, Pt. 7. American Philosophical Society, Philadelphia.

Aveni, Anthony F., and H. Hartung. 1989. Uaxactun, Guatemala, Group E and Similar Assemblages: An Archaeoastronomical Reconsideration. In *World Archaeoastronomy,* ed. A. F. Aveni, 441–461. Cambridge, Cambridge University Press.

Broda, Johanna. 1982. Astronomy, Cosmovisión, and Ideology in Pre-Hispanic Mesoamerica. In *Ethnoastronomy and Archaeoastronomy in the American Tropics,* ed. Anthony F. Aveni and Gary Urton, 81–110. Annals of the New York Academy of Sciences, 385. New York.

Broda, Johanna. 1991. Cosmovisión y observación de la naturaleza: El ejemplo del culto de los cerros en Mesoamérica. In *Arqueoastronomía y etnoastronomía en Mesoamérica,* ed. Johanna Broda, Stanislaw Iwaniszewski, and Lucrecia Maupomé, 461–500. Universidad Nacional Autónoma de México, Mexico City.

Caso, Alfonso. 1967. *Los calendarios prehispánicos.* Universidad Nacional Autónoma de México, Mexico City.

Iwaniszewski, Stanislaw. 1989. Exploring Some Anthropological Theoretical Foundations for Archaeoastronomy. In *World Archaeoastronomy,* ed. A. F. Aveni, 27–37. Cambridge, Cambridge University Press.

Kintigh, Keith W. 1992. I Wasn't Going to Say Anything, but Since You Asked: Archaeoastronomy and Archaeology. *Archaeoastronomy and Ethnoastronomy News* 5:1–4.

Knapp, A. Bernard, and Wendy Ashmore. 1999. Archaeological Landscapes: Constructed, Conceptualized, Ideational. In *Archaeologies of Landscape: Contemporary Perspectives,* ed. Wendy Ashmore and A. Bernard Knapp, 1–30. Malden, MA, Blackwell.

López Austin, Alfredo. 1973. *Hombre-dios: Religión y política en el mundo náhuatl.* Mexico City, Universidad Nacional Autónoma de México.

Marquina, Ignacio. 1960. *El Templo Mayor de México.* Mexico City, Instituto Nacional de Antropología e Historia.

Mazari, Marcos, Raúl J. Marsal, and Jesús Alberro. 1989. Los asentamientos del Templo Mayor analizados por la mecánica de suelos. *Estudios de Cultura Náhuatl* 19:145–182.

McCluskey, Stephen C. 1993. Astronomies and Rituals at the Dawn of the Middle Ages. In *Astronomies and Cultures,* ed. Clive L.N. Ruggles and Nicholas J. Saunders, 100–123. Niwot, University Press of Colorado.

Motolinía, Fray Toribio de Benavente. 1971. *Memoriales o libro de las cosas de la Nueva España y de los naturales de ella,* ed. Edmundo O'Gorman. Mexico City, Universidad Nacional Autónoma de México.

Newton, Robert R. 1972. *Medieval Chronicles and the Rotation of the Earth.* Baltimore, Johns Hopkins University Press.

Prem, Hanns J. 1991. Los calendarios prehispánicos y sus correlaciones: Problemas históricos y técnicos. In *Arqueoastronomía y etnoastronomía en Mesoamérica,* ed. Johanna Broda, Stanislaw Iwaniszewski, and Lucrecia Maupomé, 389–411. Mexico City, Universidad Nacional Autónoma de México.

Ruggles, Clive. 1999. *Astronomy in Prehistoric Britain and Ireland.* New Haven, CT, Yale University Press.

Smith, Michael E. 2003. Can We Read Cosmology in Ancient Maya City Plans? Comment on Ashmore and Sabloff. *Latin American Antiquity* 14:221–228.

Šprajc, Ivan. 1996a. *La estrella de Quetzalcóatl: El planeta Venus en Mesoamérica.* Mexico City, Editorial Diana.

Šprajc, Ivan. 1996b. *Venus, lluvia y maíz: Simbolismo y astronomía en la cosmovisión mesoamericana.* Colección Científica 318. Mexico City, Instituto Nacional de Antropología e Historia.

Šprajc, Ivan. 2000a. Astronomical Alignments at Teotihuacan, Mexico. *Latin American Antiquity* 11:403–415.

Šprajc, Ivan. 2000b. Astronomical Alignments at the Templo Mayor of Tenochtitlan, Mexico. *Archaeoastronomy* 25 (supplement to *Journal for the History of Astronomy* 31):S11–S40.

Šprajc, Ivan. 2000c [2001]. Problema de ajustes del año calendárico mesoamericano al año trópico. *Anales de Antropologia* 34:133–160.

Šprajc, Ivan. 2001. *Orientaciones astronómicas en la arquitectura prehispánica del centro de México.* Colección Científica 427. Instituto Nacional de Antropología e Historia, Mexico City.

Tichy, Franz. 1991. *Die geordnete Welt indianischer Völker: Ein Bespiel von Raumordnung und Zeitordnung im vorkolumbischen Mexiko.* Das Mexiko-Projekt der Deutschen Forschungsgemeinschaft 21. Stuttgart, Franz Steiner Verlag.

Did the Maya Build Architectural Cosmograms?
Michael E. Smith

In 2003 I published a comment (Smith 2003) on a report by Wendy Ashmore and Jeremy Sabloff (2002) in which I criticize their interpretations of possible cosmological influences on Maya city planning. At the time of writing (2002), I was unaware of

---

Reproduced by permission of the Society for American Archaeology from *Latin American Antiquity* 16:2 (2005).

an impending explosion of publications on Maya cosmology and city planning the following year. In comparison with the work of Ashmore and Sabloff, most of these studies are more speculative and less grounded in empirical data. Yet, unlike the cautious and judicious language of Ashmore and Sabloff's article and prior publications by Ashmore (e.g., 1989, 1991, 1992), these recent works are phrased in the language of confident, well-supported research conclusions. My purpose here is not to continue to criticize cosmological interpretations of Maya city plans (my views should be clear in the 2003 comment) but, rather, to point out the degree to which poorly supported speculations are being treated like established empirical findings. I find this trend troubling and worthy of public discussion within the scholarly community.

The studies I am concerned with focus on the concept of the "cosmogram." Although this term has been used in Mesoamerican studies for some time now (Freidel and Schele 1988b; Méluzin 1987–1988), I could find no explicit definition of it until 2004.[1] In a glossary to a textbook, Hendon and Joyce offer the following definition: "Cosmogram. A representation of the entire universe through symbolic shorthand or artistic metaphor" (2004:326). This definition seems to depart slightly from customary usage within the field of Mesoamerican studies, where *cosmogram* typically refers to a graphical representation of particular aspects of cosmology (rather than "the entire universe"). The dominant meaning of *cosmogram* prior to the flurry of the "new cosmogram studies" in 2003 focused on depictions of directional cosmology. Most or all ancient Mesoamerican cultures had a four-directional symbolic-spatial cosmology. The cardinal directions—each associated with particular deities, colors, birds, trees, and other symbolic elements—were important components of Mesoamerican mythology, cosmology, and ritual practice (Boone 2000; Brotherston 1976; Carrasco 1999; León-Portilla 1963; López Austin 2001).

A number of Late Postclassic and early colonial sources depict four-part cosmological scenes that have been called cosmograms. A clear discussion of these can be found in Aveni's work in a section labeled "The Union of Time and Space in Mesoamerican Cosmology" (2001:148–152). Four of these cosmograms are illustrated in Figure 33.2. The first two images are complex cosmological scenes from the Maya Codex Madrid (Fig. 33.2a) and the central Mexican (Borgia Group) Codex Fejérváry-Mayer (Fig. 33.2b) that incorporate multiple levels of symbolism about the 260-day ritual calendar and the iconography of the cardinal directions. These scenes have been much analyzed by Mesoamerican iconographers and others (e.g., Aveni 2001:148–152; Boone 2000; Brotherston 1976). The third image (Fig. 33.2c) is a depiction of the Aztec 52-year calendar round in the form of a circle and cross, with the cardinal directions labeled on the four sides. The fourth image, the face of the "Aztec calendar stone," is less often called a cosmogram, but in Townsend's (1979:63–70) interpretation this monument fuses imperial ideology, the calendar, and the four cardinal directions. His description of the central message of the monument is labeled "Time, Space, and the Ascendancy of Tenochtitlan" (1979:63).

Most scholars agree that the four images in Figure 33.2 are pictorial symbols of Maya and Aztec directional cosmology. Each one incorporates time (in the form of one or more calendrical systems), space (the four directions), and a number of additional symbolic and mythological elements. In short, these are cosmograms. Depictions of the vertical elements of Aztec cosmology as shown in the Codex Vaticanus A (Codex Vaticanus 1979:figs. 1–7), analyzed by Quiñones Keber (1995), might also be called cosmograms, as might other spatial-temporal images in Aztec codices and monumental sculptures (Boone 2000; Townsend 1979; Umberger 1998). A quartered circle figure common at Teotihuacan and other Classic period sites may also be a cosmogram (Coggins 1980).

Freidel and Schele (1988b) published the earliest explicit application of the cosmogram concept to Classic Maya society. They identify recurring sets of iconographic elements in sculptures and stelae as representations of the ancient Maya cosmos. Although my lack of iconographic training prevents me from following all of the details of their rich exposition, their use of numerous examples in diverse media from many sites, coupled with an explicit and clear logic of argument, suggests to me that this is a rigorous and convincing analysis (see also Freidel and Schele 1988a; Freidel et al. 1993; Schele and Freidel 1990). The "new cosmogram" studies, in contrast, are based on the untested assumption that Maya directional cosmology was expressed in many or most buildings and cities. But what is the evidence for this?

The most common interpretations of Maya architectural cosmograms focus on the layouts of key architectural compounds and whole cities. In some cases individual buildings or compounds are interpreted as cosmograms, including the Murcielagos group at Dos Pilas (Demarest et al. 2003:142) and the east court of the Acropolis at Copán, which has been labeled "a giant cosmogram" (Fash 1998:250). In other cases, the layouts of entire cities are interpreted as cosmograms (although that phrase is not always used). For example, at Uxmal, "the quadrilateral layout and approximate correspondence of the principal buildings to the cardinal points represents an effort to replicate the well-documented quadripartite organization of the Maya cosmos" (Kowalski and Dunning 1999:280); the same phrase is repeated by Kowalski (2003:215).[2]

Reputed Maya cosmograms are not limited to buildings. At Tikal, for example, four reservoirs "located approximately in the cardinal directions" formed "a water cosmogram of the site" (Scarborough 1998:154–155). Tate labels certain monuments at Yaxchilan as "cosmogram stelae" (1992:101; see also 119, 131–132). *Sacbes* (raised causeways) are also called cosmograms: "Serving as *axis mundi*, *sacbeob* may have represented the Milky Way. . . . [*Sacbeob*] served as cosmograms, or models, of the Maya universe" (Shaw 2001:266). Even the bodies of Maya kings could be cosmograms! "Thus not only the temple centers from which they ruled but also the rulers' bodies themselves constituted living terrestrial cosmograms" (Gossen 1996:295). The word *cosmogram* is evidently so appealing today that some scholars have decided to use it to replace the term *cosmology*: "The sun rising in the east, climbing to the zenith at noon, setting in the west, and passing through the nadir at night, united the tripartite vertical and four-part horizontal divisions of the world into a holistic cosmogram" (Christie 2003:292). Was the cosmos itself viewed as a model of the cosmos, or does the author just mean "cosmology," not "cosmogram"?

Michael E. Smith, Wendy Ashmore, Jeremy A. Sabloff, and Ivan Šprajc

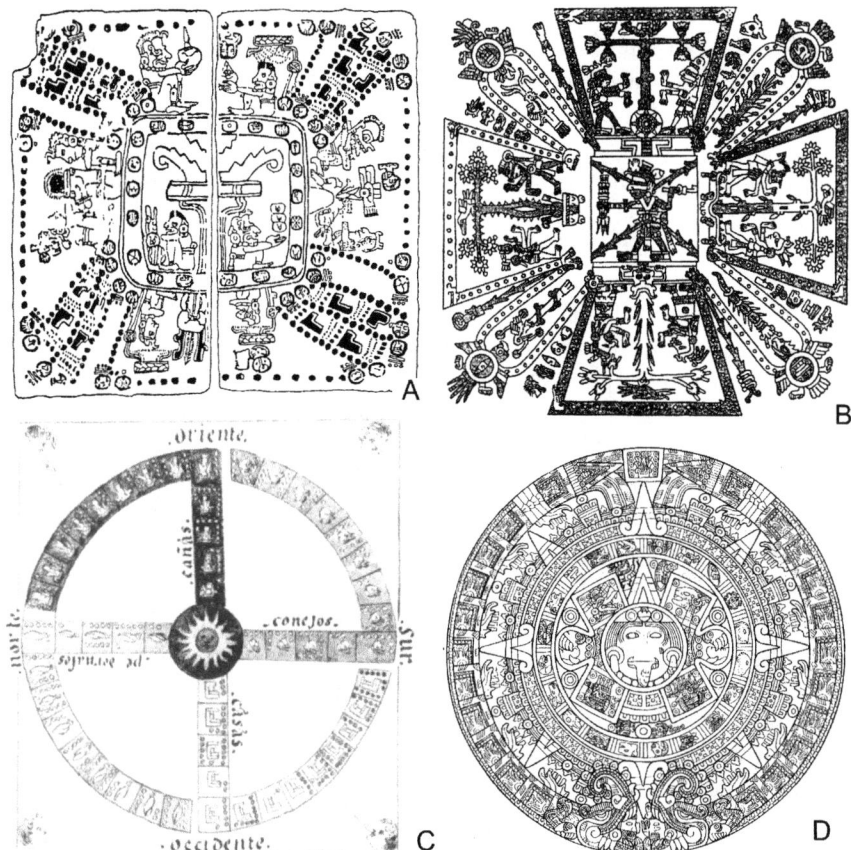

33.2 Conquest-era Mesoamerican cosmograms from Maya (A) and Aztec sources (B–D). A: Cosmological scene from the Codex Madrid (Anders 1967:75–76); the image is from Bricker and Vail 1997:41, after Villacorta C. and Villacorta 1976:374, 376. B: Cosmological scene from the Codex Fejérváry-Mayer, p. 1 (Burland 1971:1); drawing by John M. D. Pohl. C: Calendar wheel from *Book of the Gods and Rites and the Ancient Calendar*, by Fray Diego Durán (1971:plate 35), copyright 1971 by the University of Oklahoma Press. D: The so-called Aztec calendar stone; drawing by Emily Umberger. All images used with permission.

A newly discovered cache at the site of Cival has been interpreted as a cosmogram. (Estrada-Belli et al. 2003). John G. Fox (1996) makes similar cosmological interpretations of caches, although he does not use the term *cosmogram*. For example, a cache with nine obsidian blades demonstrates that ballcourt features symbolized the Maya underworld (Fox 1996:485), and a cache with one shell and one bead in a Copán ballcourt "provides a microcosmic model of the universe, with the bead representing the earth and the shell the cosmic ocean" (Fox 1996:486).

Closely related concepts include the "*axis mundi*" and the world tree: "The Castillo and the Cenote Ch'en Mul formed the *axis mundi* (the primordial mountain-cave) of Mayapán, virtually standing between cosmic planes at the beginning of time" (Pugh 2003:943); also, "the five serpent temples at Mayapan form a quincunx layout, which represents the quadripartite division of the Maya universe" (Pugh 2001:255). And at Xunantunich, the buildings and plazas are interpreted as a world tree (Yaeger 2003:132).

By 2003, usage of the architectural cosmogram concept was rampant in the Maya region, and it had spread to Oaxaca (Joyce 2004; Méluzin 1987–1988), Central America (Graham 2003:291), and even the Andes (Swenson 2003:274). The uncritical acceptance of this concept now appears in popularized accounts: "The ceremonial center was not just the political heart of the kingdom, it was also the sacred center of the polity and was designed as a cosmogram, re-creating the Maya world order" (Foster 2002:229).

Tourtellot et al. take the notion of the cosmogram to a higher spatial level by interpreting the distribution of settlements as a cosmogram: "We argue that the middle-level sites around La Milpa are organized in a concentric and cardinally aligned cosmogram" (2003:95). I find the use of the present tense here significant. In most models, the cosmogram is asserted to be an ancient phenomenon that archaeologists try to identify today in the ruins of ancient Maya cities. By phrasing their cosmogram interpretation in the present, not the past, however, these authors unwittingly suggest the most reasonable interpretation of the phenomenon: Maya architectural cosmograms are modern phenomena, invented by scholars to satisfy their desire to reconstruct ancient cosmology from fragmentary evidence. I am flabbergasted at some of the quotes above for presenting highly speculative interpretations as if they were reasoned and unproblematic conclusions based on empirical evidence.

I find this trend troubling from a methodological viewpoint. These studies contrast with Ashmore's methods. She starts with empirical distributions of buildings and architectural compounds within Maya cities, identifies spatial patterns (e.g., north-south orientations, the placement of ballcourts), and then provides cosmological interpretations for those patterns. My criticism of her work focuses on the subjective and impressionistic nature of her methods, which have proved difficult to replicate or validate. Šprajc provides another example of a rigorous approach to the topic of cosmology and city planning.

The new cosmogram studies, on the other hand, start with the assumption that directional cosmology must have been expressed in architectural settings. They identify a case in which buildings or features seem to have some kind of cardinal orientation or arrangement and then assert confidently that the building/compound/city/reservoir/stelae in question formed a cosmogram. Authors of most of these studies offer little or no iconographic or epigraphic evidence for the presence of a cosmogram or cosmological symbolism in the settings they analyze. They rarely step back to consider the larger issue of whether Mesoamerican cosmograms were ever expressed in architecture and urban planning.[3]

Contributors to a recent special section of the *Cambridge Archaeological Journal* considered the question, "Were cities built as images?" (Carl et al. 2000). The answer is that in some ancient urban traditions, cities and buildings were clearly planned and constructed as cosmograms. Evidence is particularly strong for ancient China, India, and Thailand (see Smith 2003:222). In other urban traditions, such as in Mesoamerica, there is little or no explicit evidence for this practice. The archaeoastronomical research reviewed by Šprajc provides strong empirical support for the astronomical alignments of buildings. The question of whether buildings and cities were viewed as models of the cosmos requires inferences considerably more speculative in scope. I am unaware of any explicit statements in the ethnohistoric or epigraphic sources for direct cosmological influences on Mesoamerican architecture or urbanism.

Given the importance of directional cosmology in ancient Mesoamerica, it seems likely that cosmology may have played a role in architectural symbolism and perhaps even in the design and layout of buildings and cities. But in the absence of the kind of clear and direct evidence available for areas like China and India, scholars need to approach this question cautiously with rigorous and explicit methods. My major criticism of the new cosmogram studies is that few of the authors describe their hypothetical architectural cosmograms using the language of caution and hypothesis; instead, they use the language of confident conclusions. Rather than simply assert that the Maya had architectural cosmograms, however, scholars should undertake empirical research designed to test this notion. Promising directions include the work of Ashmore (1986, 1989, 1991, 1992, 2002; Ashmore and Sabloff 2002) and the numerical data on building alignments assembled by Aveni and Hartung (1987), Šprajc (2000, 2001), and others.

Discussions of ancient Mayan architectural cosmograms appeared at a rate of approximately one publication per year between 1996 and 2002. In 2003, a plethora of such studies appeared (my count of nine works in 2003 does not include unpublished conference papers, Internet postings, and theses). In my view, these confidently phrased speculations are harmful to the discipline of Mesoamerican studies. They set a bad example by suggesting to students and the public that poorly grounded speculation can pass for acceptable scholarship in our field.

EPILOGUE: REPLY TO ŠPRAJC

The underlying motivation for both of my works—the critique of Ashmore and Sabloff (Smith 2003) and the present opinion piece—is to encourage rigorous and explicit methods in the analysis of the relationship between cosmology and urban planning in ancient Mesoamerica. I do not deny the influence of cosmology on ancient architectural practice, but this relationship needs to be demonstrated empirically, not simply assumed. Šprajc suggests that archaeoastronomy provides just the sort of empirical demonstration I am calling for.

I agree with Šprajc, up to a point. I suspect that we may differ in our views of just how far archaeoastronomical data allow us to go in reconstructing patterns of ancient cosmology. Archaeoastronomy does have the ability to identify cosmological influences on ancient building and settlement alignments. Šprajc provides a clear and succinct overview of the kind of rigorous research on this issue conducted by scholars such as Anthony Aveni, Stanislaw Iwaniszewski, Clive Ruggles, and himself. I thank Šprajc for his discussion of the complexities of the topic of astronomical alignments at Tenochtitlan. Although I was aware of his 2001 monograph on central Mexican astronomical orientations, I did not consult it in preparing my articles. This was a scholarly lapse on my part, and I apologize.

I tentatively accept Šprajc's interpretation of the astronomical significance of the layout of Tenochtitlan. I use the word *tentatively* because frankly I do not understand the astronomical details, but the argument seems rigorous and plausible. Nevertheless, it seems to me that these data provide only tenuous support for inferences that go beyond the notion that the buildings and streets of the Aztec capital were aligned with astronomical phenomena. They certainly do not permit the inference that Tenochtitlan was viewed as a model of the cosmos. Yes, there was astronomical influence on the city's layout, and yes, astronomical phenomena were related to various Aztec cosmological beliefs and landscape practices. But in the absence of textual confirmation, the conclusion that Tenochtitlan was a cosmogram requires a leap of faith that exceeds cautious empirical inference.

I second Šprajc's call for greater interaction between archaeologists and archaeoastronomers. Although the situation has improved since discussed over a decade ago by Kintigh (1992) and Aveni (1992) (see Chapter 35), there is still much that can be done. I am certainly among those archaeologists guilty of not paying sufficient attention to archaeoastronomy. The topic of the political uses of astronomical data by elites, touched on in Šprajc's comment, is a promising avenue for joint research, and there are many others. Archaeoastronomical research alone, however, will not permit the identification of architectural cosmograms in ancient Mesoamerica.

ACKNOWLEDGMENTS

I thank Cynthia Heath-Smith, Robert Rosenswig, and Anthony F. Aveni for helpful suggestions on earlier drafts of this article. The comments of three referees were very helpful in clarifying my argument, and I benefited from the opportunity to read Ivan Šprajc's contribution before making final revisions.

NOTES

1. Before I found the definition in Hendon and Joyce's (2004) glossary, all I could find was the definition of *cosmogram* in astrology: "Cosmogram is the cosmobiological term for horoscope. The foundation for casting the cosmogram is the zodiac, through which the Sun moves in one year" (Rauchhaus 1994:147).

2. Kowalski and Dunning (1999) do not use the term *cosmogram,* and two reviewers of this manuscript rightly pointed out that their argument is considerably more rigorous and con-

vincing than many of the other studies I consider in this article. Because this is a brief opinion piece, I do not have space to provide a full discussion of the views of each of the authors I criticize. For a more extensive treatment of the topic of ancient city planning, see Smith 2004.

3. There is a tradition of cosmological interpretations of Mesoamerican urban layouts (see Benson 1981), but the lack of concrete evidence to support those interpretations is striking. To take just one example, Carrasco (1999:43–46) includes a section titled "Architectural Parallelism of Macrocosmos and Microcosmos" in his book on Tenochtitlan, but it includes no evidence for such parallelism apart from his own interpretations. Many accounts (e.g., Carlson 1981) rely on the universalistic models of scholars like Rykwert (1976) and Wheatley (1971) who assert that all ancient cultures had sacred, cosmologically grounded cities and towns. Apart from the anthropological naïveté and empirical inadequacy of such universalistic notions, empirical doubts have recently been cast on Wheatley's analysis of ancient China, his main case study of the cosmological importance of cities (Wiesheu 1997, 1999).

REFERENCES

Anders, Ferdinand. 1967. *Codex Tro-Cortesianus (Codex Madrid): Museo de América Madrid.* Codices selecti phototypice impressi, Vol. 8. Graz, Austria, Akademische Druck- u. Verlagsanstalt.

Ashmore, Wendy. 1986. Peten Cosmology in the Maya Southeast: An Analysis of Architecture and Settlement Patterns at Classic Quirigua. In *The Southeast Maya Periphery,* ed. Patricia A. Urban and Edward M. Schortman, 35–49. Austin, University of Texas Press.

Ashmore, Wendy. 1989. Construction and Cosmology: Politics and Ideology in Lowland Maya Settlement Patterns. In *Word and Image in Maya Culture: Explorations in Language, Writing, and Representation,* ed. William F. Hanks and Don S. Rice, 272–286. Salt Lake City, University of Utah Press.

Ashmore, Wendy. 1991. Site-Planning Principles and Concepts of Directionality among the Ancient Maya. *Latin American Antiquity* 2:199–226.

Ashmore, Wendy. 1992. Deciphering Maya Architectural Plans. In *New Theories on the Ancient Maya,* ed. Elin C. Danien and Robert J. Sharer, 173–184. University Museum Monograph, vol. 77. Philadelphia, University Museum, University of Pennsylvania.

Ashmore, Wendy. 2002. "Decisions and Dispositions": Socializing Spatial Archaeology. *American Anthropologist* 104:1172–1183.

Ashmore, Wendy, and Jeremy A. Sabloff. 2002. Spatial Orders in Maya Civic Plans. *Latin American Antiquity* 13:201–215.

Aveni, Anthony F. 1992. Nobody Asked, but I Couldn't Resist: A Response to Keith Kintigh on Archaeoastronomy and Archaeology. *Archaeoastronomy and Archaeology News* 6:1, 4.

Aveni, Anthony F. 2001. *Skywatchers: A Revised and Updated Version of Skywatchers of Ancient Mexico.* Austin, University of Texas Press.

Aveni, Anthony F., and Horst Hartung. 1987. *Maya City Planning and the Calendar.* American Philosophical Society, Transactions, Vol. 76, Pt. 7. Philadelphia.

Benson, Elizabeth P., ed. 1981. *Mesoamerican Sites and World-Views.* Washington, DC, Dumbarton Oaks.

Boone, Elizabeth H. 2000. Guides for Living: The Divinatory Codices of Mexico. In *In Chalchihuitl in Quetzalli, Precious Greenstone Precious Quetzal Feather: Mesoamerican Studies in Honor of Doris Heyden,* ed. Eloise Quiñones Keber, 69–82. Lancaster, CA, Labyrinthos.

Bricker, Victoria R., and Gabrielle Vail, eds. 1997. *Papers on the Madrid Codex*. Publications, Vol. 64. New Orleans, Middle American Research Institute, Tulane University.

Brotherston, Gordon. 1976. Mesoamerican Description of Space, 11: Signs for Direction. *Ibero-Amerikanisches Archiv* 2:39–62.

Burland, C. A. 1971. *Codex Fejéváry-Mayer: 12014 M. City of Liverpool Museums*. Graz, Austria, Akadem. Druck- u. Verlagsanst.

Carl, Peter, Barry Kemp, Ray Laurence, Robin Coningham, Charles Higham, and George L. Cowgill. 2000. Viewpoint: Were Cities Built as Images? *Cambridge Archaeological Journal* 10:327–365.

Carlson, John B. 1981. A Geomantic Model for the Interpretation of Mesoamerican Sites: An Essay in Cross-Cultural Comparison. In *Mesoamerican Sites and World-Views*, ed. Elizabeth P. Benson, 143–216. Washington, DC, Dumbarton Oaks.

Carrasco, Davíd. 1990. *Religions of Mesoamerica: Cosmovision and Ceremonial Centers*. New York, Harper and Row.

Carrasco, Davíd. 1999. *City of Sacrifice: The Aztec Empire and the Role of Violence in Civilization*. Boston, Beacon Press.

Christie, Jessica Joyce. 2003. The Tripartite Layout of Rooms in Maya Elite Residences. In *Maya Palaces and Elite Residences: An Interdisciplinary Approach*, ed. Jessica Joyce Christie, 291–314. Austin, University of Texas Press.

Codex Vaticanus. 1979. *Codex Vaticanus 3738*. Codices Selecti. Graz, Austria, Akademische Druck- u. Verlagsanstalt.

Coggins, Clemency Chase. 1980. The Shape of Time: Some Political Implications of a Four-Part Figure. *American Antiquity* 45:727–739.

Demarest, Arthur A., Kim Morgan, Claudia Wolley, and Hector L. Escobedo. 2003. The Political Acquisition of Sacred Geography: The Murciélagos Complex at Dos Pilas. In *Maya Palaces and Elite Residences: An Interdisciplinary Approach*, ed. Jessica Joyce Christie, 120–153. Austin, University of Texas Press.

Durán, Fray Diego. 1971. *Book of the Gods and Rites and the Ancient Calendar*. Trans. Fernando Horcasitas and Doris Heyden. Norman, University of Oklahoma Press.

Estrada-Belli, Francisco, Jeremy Bauer, Molly Morgan, and Angel Chavez. 2003. Symbols of Early Maya Kingship at Cival, Petén, Guatemala. *Antiquity Online*. Available at http://antiquity.ac.uk/ProjGall/estrada-belli/ (accessed February 2004).

Fash, William L. 1998. Dynastic Architectural Programs: Intention and Design in Classic Maya Buildings at Copan and Other Sites. In *Function and Meaning in Classic Maya Architecture*, ed. Stephen D. Houston, 223–270. Washington, DC, Dumbarton Oaks.

Foster, Lynn V. 2002. *Handbook to Life in the Ancient Maya World*. New York, Facts on File, Inc.

Fox, John Gerard. 1996. Playing with Power: Ballcourts and Political Ritual in Southern Mesoamerica. *Current Anthropology* 37:483–509.

Freidel, David A., and Linda Schele. 1988a. Kings in the Late Preclassic Lowlands: The Instruments and Places of Ritual Power. *American Anthropologist* 90:547–567.

Freidel, David A., and Linda Schele. 1988b. Symbol and Power: A History of the Lowland Maya Cosmogram. In *Maya Iconography*, ed. Elizabeth P. Benson and Gillette Griffin, 44–93. Princeton, NJ, Princeton University Press.

Freidel, David A., Linda Schele, and Joy Parker. 1993. *Maya Cosmos: Three Thousand Years on the Shaman's Path*. New York, William Morrow.

Gossen, Gary H. 1996. The Religions of Mesoamerica. In *The Legacy of Mesoamerica: History and Culture of a Native American Civilization*, ed. Robert M. Carmack, Janine Gasco, and Gary H. Gossen, 290–320. Englewood Cliffs, NJ, Prentice-Hall.

Graham, Mark Miller. 2003. Creation Imagery in the Goldwork of Costa Rica, Panama, and Colombia. In *Gold and Power in Ancient Costa Rica, Panama, and Colombia,* ed. Jeffrey Quilter and John W. Hoopes, 279–299. Washington, DC, Dumbarton Oaks.

Hendon, Julia A., and Rosemary Joyce. 2004. Glossary. In *Mesoamerican Archaeology: Theory and Practice,* ed. Julia A. Hendon and Rosemary Joyce, 323–331. Oxford, Blackwell.

Joyce, Arthur A. 2004. Sacred Space and Social Relations in the Valley of Oaxaca. In *Mesoamerican Archaeology: Theory and Practice,* ed. Julia A. Hendon and Rosemary Joyce, 192–216. Oxford, Blackwell.

Kintigh, Keith W. 1992. I Wasn't Going to Say Anything, but Since You Asked: Archaeoastronomy and Archaeology. *Archaeoastronomy and Archaeology News* 5:1, 4.

Kowalski, Jeff Karl. 2003. Evidence for the Functions and Meanings of Some Northern Maya Palaces. In *Maya Palaces and Elite Residences: An Interdisciplinary Approach,* ed. Jessica Joyce Christie, 204–252. Austin, University of Texas Press.

Kowalski, Jeff Karl, and Nicholas P. Dunning. 1999. The Architecture of Uxmal: The Symbolics of Statemaking at a Puuc Maya Regional Capital. In *Mesoamerican Architecture as a Cultural Symbol,* ed. Jeff Karl Kowalski, 274–297. New York, Oxford University Press.

León-Portilla, Miguel. 1963. *Aztec Thought and Culture: A Study of the Ancient Náhuatl Mind.* Norman, University of Oklahoma Press.

López Austin, Alfredo. 2001. Cosmovision. In *The Oxford Encyclopedia of Mesoamerican Cultures: The Civilizations of Mexico and Central America,* vol. 1, ed. Davíd Carrasco, 268–274. New York, Oxford University Press.

Méluzin, Sylvia. 1987–1988. Ancient Zapotec Calendrical Cosmogram. *Archaeoastronomy* 10:139–147.

Pugh, Timothy W. 2001. Flood Reptiles, Serpent Temples, and the Quadripartite Universe: The Imago Mundi of Late Postclassic Mayapan. *Ancient Mesoamerica* 12:247–258.

Pugh, Timothy W. 2003. A Cluster and Spatial Analysis of Ceremonial Architecture at Late Postclassic Mayapan. *Journal of Archaeological Science* 30:941–953.

Quiñones Keber, Eloise. 1995. Painting the Nahua Universe: Cosmology and Cosmogony in the Codex Vaticanus A, Part 1. Introduction and Translation. *Latin American Indian Literatures Journal* 11:183–204.

Rauchhaus, Irmgard. 1994. Cosmobiology. In *The Astrology Encyclopedia,* ed. James R. Lewis, 147–155. Detroit, Visible Ink.

Rykwert, Joseph. 1976. *The Idea of a Town: The Anthropology of Urban Form in Rome, Italy, and the Ancient World.* Princeton, NJ, Princeton University Press.

Scarborough, Vernon L. 1998. Ecology and Ritual: Water Management and the Maya. *Latin American Antiquity* 9:135–159.

Schele, Linda, and David Freidel. 1990. *A Forest of Kings: The Untold Story of the Ancient Maya.* New York, William Morrow.

Shaw, Justine M. 2001. Maya *Sacbeob:* Form and Function. *Ancient Mesoamerica* 12:261–272.

Smith, Michael E. 2003. Can We Read Cosmology in Ancient Maya City Plans? Comment on Ashmore and Sabloff. *Latin American Antiquity* 14:221–228.

Smith, Michael E. 2004. Form and Meaning in the Earliest Cities: A New Approach to Ancient Urban Planning. Manuscript on file, Department of Anthropology, University at Albany, State University of New York, Albany.

Šprajc, Ivan. 2000. Astronomical Alignments at Teotihuacan. *Latin American Antiquity* 11:403–415.

Šprajc, Ivan. 2001. *Orientaciones astronómicas en la arquitectura pre-hispánica del centro de México.* Colección Científica, vol. 427. Mexico City, Instituto Nacional de Antropología e Historia.

Swenson, Edward R. 2003. Cities of Violence: Sacrifice, Power and Urbanization in the Andes. *Journal of Social Archaeology* 3:256–296.

Tate, Carolyn E. 1992. *Yaxchilan: The Design of a Maya Ceremonial City.* Austin, University of Texas Press.

Tourtellot, Gair, Gloria Everson, and Norman Hammond. 2003. Suburban Organization: Minor Centers at La Milpa, Belize. In *Perspectives on Ancient Maya Rural Complexity,* ed. Gyles Iannone and Samuel V. Connell, 95–107. Monograph, vol. 49. Cotsen Institute of Archaeology, University of California, Los Angeles.

Townsend, Richard F. 1979. *State and Cosmos in the Art of Tenochtitlan.* Studies in Pre-Columbian Art and Archaeology, vol. 20. Washington, DC, Dumbarton Oaks.

Umberger, Emily. 1998. New Blood from an Old Stone. *Estudios de Cultura Náhuatl* 28:241–256.

Villacorta C., J. Antonio, and Carlos A. Villacorta, eds. 1976. *Códices mayas.* 2nd ed. Guatemala City, Tipografía Nacional.

Wheatley, Paul. 1971. *The Pivot of the Four Quarters: A Preliminary Enquiry into the Origins and the Character of the Ancient Chinese City.* Chicago, Aldine.

Wiesheu, Walburga. 1997. China's First Cities: The Walled Site of Wangcheng-gang in the Central Plain Region of North China. In *Emergence and Change in Early Urban Societies,* ed. Linda Manzanilla, 87–105. New York, Plenum.

Wiesheu, Walburga. 1999. Urban Genesis in China: A Brief Reevaluation of Wheatley's City-as-Temple Thesis. *Wall and Market: Chinese Urban History News.* Available at www.albany.edu/mumford/Center-Act/wall-market.html (accessed January 2004).

Yaeger, Jason. 2003. Untangling the Ties That Bind: The City, the Countryside, and the Nature of Maya Urbanism at Xunantunich, Belize. In *The Social Construction of Ancient Cities,* ed. Monica L. Smith, 121–155. Washington, DC, Smithsonian Institution Press.

CHAPTER THIRTY-FOUR

Archaeology and Astronomy
A View from the Southwest

W. James Judge

Recently I asked a friend of mine, an archaeologist whose opinion I respect very much, what he thought of the subject of archaeoastronomy. He said, "Well, it's one-half archaeology and one-half astronomy, but unfortunately, much of it has turned out to be sort of halfastroarchaeology." This, I am afraid, is a perception of the field shared by a number of archaeologists, a perception that conditions the character of this essay.

I would like to discuss the relationship between archaeology and astronomy, as they are integrated in the emerging field of archaeoastronomy. You must realize that my talk is presented from the perspective of an archaeologist who knows something about the subject of archaeology—at least Southwestern archaeology—but virtually nothing about the subject of astronomy. (I am one of the few people who has trouble finding the Milky Way, let alone the Big Dipper.) Thus, I ask you to please understand and forgive my biases.

Reproduced by permission from *Astronomy and Ceremony in the Prehistoric Southwest*, ed. J. B. Carlson and W. J. Judge. Papers of the Maxwell Museum of Anthropology, No. 2, 1987.

This chapter will offer some thoughts on the distinction between archaeology and archaeoastronomy (as well as between archaeologists and archaeoastronomers), followed by a review of some of the work we have done in Chaco Canyon. I will conclude by offering suggestions about ways in which research in Chaco can help to integrate the two disciplines, illustrating the challenge thereby posed to those who work in the field of archaeoastronomy.

I would like to preface all of this by stating some of the basic assumptions that will underlie this discussion. First, I feel that the field of archaeoastronomy has more ultimate relevance to archaeology than to astronomy, and those should make its home in the camp of the former. Though prehistoric astronomy may be of interest relative to the history of astronomy, I doubt that the key to the origin of the universe will ultimately be found in Pueblo Bonito. Archaeology, on the other hand, can benefit considerably from an understanding of the role of astronomy in prehistoric cultures.

Second, I assume that like it or not, a dichotomy of considerable magnitude exists between archaeology and archaeoastronomy at this time, in this country. Otherwise, why would my archaeological colleague have made the remark he did? My third assumption is that this gap can be, and most certainly should be, bridged—if we work at it. I am convinced that symposia such as the one reported here, which offer opportunities for archaeologists, astronomers, cultural anthropologists, and others with similar interests to get together and talk to each other, are of great value in providing the emerging field of archaeoastronomy with a more fundamentally anthropological foundation from which to develop.

Although I do not know for sure, I would bet that most of those who now consider themselves archaeoastronomers were actually trained in astronomy, rather than in archaeology or, more correctly, in anthropology. In the United States one cannot get a degree in archaeology per se. Archaeology is subsumed within the larger field of anthropology, which is the study of human behavior. This intellectual background, of course, conditions how archaeologists go about their business. My guess is that archaeoastronomers whose degrees are in anthropology are the exception rather than the rule with regard to formal training—Jonathan Reyman, a participant in this symposium, is one of the exceptions.

Dominance of the field of archaeoastronomy by astronomers obviously conditions both the character of the research undertaken and the interpretations derived and does not necessarily foster integration of the discipline with anthropological archaeology. Please understand that I am neither blaming nor belittling the astronomers—in fact, were it not for their interest, the field of archaeoastronomy would have shown very little progress since its inception. It is, however, this distinct intellectual background of archaeoastronomy that I wish to address. Perhaps the best way to begin is to compare astronomy with archaeology.

First, let's consider the public perception of astronomy. When one thinks of astronomy, what comes to mind? Galaxies, black holes, quarks, Carl Sagan, warp factor one, the starship *Enterprise* . . . things like that. Now, think of archaeology and what comes to mind? Dirt, trash, refuse, garbage, potsherds, broken rocks, and generally, dinosaur bones (which aren't even in the domain of the archaeologist). Now, think of archaeoastronomy and what comes to mind? Sun daggers, supernova, calendars, mystery, Carl Sagan, glamour, sun*rise,* the thrill of discovery. Now think archaeology again and what comes to mind? More garbage, more pottery, more rocks, dirt, dust, and sun*burn,* the agony of defeat.

I'm not complaining, you understand. We archaeologists know our place, are content with our lot, and generally enjoy what we do. It's just that we get no respect. I have discussed this elsewhere and have called it the "Rodney Dangerfield" syndrome in archaeology. I am, of course, being facetious in emphasizing our paranoia, but there is a point here, and it has to do with the public perception of what we as archaeologists do for a living and how that relates to the things that are more likely to capture the attention and the interest of the public—things like medicine wheels, stone monoliths, corner windows, kiva niches, and strange petroglyphs—things, incidentally, that are the lifeblood of the archaeoastronomers.

Archaeology deals with, actually archaeology *must* deal with, the mundane, the commonplace, the discarded remnants of human behavior. Therein lies the difference between it and astronomy, and perhaps the key to achieving the integration of the two disciplines. Let me try to explain.

As I noted earlier, archaeologists are trained in anthropology, and thus they are primarily interested in human behavior—specifically, in the processes of change through time in human behavior and, even more specifically, in why such changes take place. I mentioned before that archaeologists deal with trash—we enjoy it and like to work with it. You may be familiar with a project in Tucson, Arizona, in which archaeologists are studying modern human trash—actually going through people's garbage in that fair city to see what they throw away. Dr. Rathje, the project director, has noted that while all archaeologists study garbage, his data were just a little fresher than most. What does this have to do with archaeoastronomy? Please bear with me.

The purpose of the garbage project is to compare what the researchers call "front door" and "back door" behavior in Tucson—in other words, to compare what people say they do with what they actually do. What is the relevance of this to archaeology? Basically, Rathje is trying to get at what we call the "material correlates" of modern human behavior. He wants to understand the relationship between how humans actually behave and what they actually leave in the archaeological record. Understanding this will help archaeologists to comprehend the relationship between what we find as we dig prehistoric trash mounds and other discarded remains and how the people actually behaved in the past. Incidentally, as an interesting aside, Rathje has found that in Tucson people discard between 7 and 14 percent, by weight, of their purchased food. This would come to about $11 billion annually for the nation as a whole, or enough to feed the entire population of Canada for a year.

But I digress. Back to the archaeological correlates of human behavior. This quest to determine what conditions the formation of the archaeological record now constitutes a primary thrust of archaeological research in this country. Theoretical and empirical attempts to verify the inferred function of archaeological manifestations, sometimes called *middle-range theory*, are among the foremost issues confronting archaeologists today. For example, take something as simple as an excavated room in a prehistoric pueblo. How are we to interpret the function of that room? If we call it a habitation room, how are we to know that it indeed functioned in that context? What are the material correlates of domestic living, and how would they differ from the material remains resulting from other possible functions of the room? Is a room without a hearth really a storage room? Or how about a room with heating pits rather than hearths? Was it a living room, a storage room, or both? Sometimes it seems as though we have not progressed very far in interpreting the archaeological record if we cannot even discern something as basic as room function.

I am sure you can appreciate the importance of understanding things as fundamental as this. How, for example, can we reconstruct past population size without being able to determine which rooms served primarily for habitation, which for storage, and for what portion of the year they so served? Thus we see the reasons behind the current emphasis on understanding the material correlates of human behavior.

Let me try to give you an example from the field of archaeoastronomy. Take, for instance, the so-called supernova pictograph located in Chaco Canyon near Peñasco Blanco, a site that is two quite dusty but otherwise very beautiful miles west of Pueblo Bonito. Some astronomers have suggested that this pictograph is an Anasazi record of the Crab Nebula supernova, which was recorded by Chinese astronomers in AD 1054. The time is right, since the Chaco system peaked between AD 1050 and 1130. Some archaeologists, on the other hand, have suggested that this pictograph is nothing more than a prehistoric sunwatcher's station, a phenomenon not uncommon among Pueblo peoples and nothing to get very excited about.

Whom should we believe? Well, for one thing, I agree fully with Zeilik's approach (1987) to the question, which is to try to test the sunwatcher hypothesis. Rather than arguing about it, we should go out and actually observe the sun from the location of the pictograph to see whether the resultant observations make any sense whatsoever, given what we know about solar observations among the modern Pueblos. To my knowledge, Zeilik is the first one to suggest this. I don't know whether he has done it yet, or if so, what he found out. But it is certainly worth a try to determine, in effect, whether this site exhibits the material correlates of sunwatching behavior. That is the anthropological perspective.

Let me take this opportunity to point out something else about the supernova pictograph, something with which many of you might not agree. If it were determined that this site is a sunwatcher's station, that would be interesting, but it would not change our interpretation of the Chaco Phenomenon very much. If, on the other hand, it were determined that this rock art panel is a record of the supernova of 1054, a record of a unique celestial event, what would that mean? Little more, in my opinion. It would be interesting, but it would not change our interpretation of the Chaco Phenomenon very much. My point is this—however the supernova pictograph is interpreted, the result is only of minimal value archaeologically. It simply does not constitute the kind of information that archaeologists rely on to address the questions of cultural process that are of primary interest to them. We already know that the Pueblo Indians watched the sun and were aware of what went on in the heavens. So how would the knowledge that the people of Chaco Canyon recorded a unique event, such as the supernova, help us? It certainly does not contribute much to the explanations of changes in human behavior that the archaeologists of today seek to find. Sad to say, something that the public and at least some archaeoastronomers find so fascinating simply does not contribute to the primary data base of the modern archaeologist. I hope that you are beginning to appreciate the nature of the differences between our two disciplines, as well as the magnitude of effort needed to integrate them.

Basically, these differences hinge on the distinction between the common and the unique, the mundane and the spectacular, the secular and the ceremonial. Astronomers look up, archaeologists look down. By their very nature, the data of the archaeoastronomers are unique and spectacular. The data of the archaeologists are, in contrast, more common and generally much less spectacular. This might be referred to as the "ho-hum" aspect of archaeology. Did you ever go to an archaeological convention and listen to some guy describe the interior dimensions of 34 bell-shaped storage pits? No wonder people yawn when they see us coming.

Again I digress. To return to the two kinds of data bases, the distinction would be less important were it not for the archaeologists' desire to go beyond *description* of past human behavior and to attempt to *explain* it. Explanation is sought in the context of some external factor, such as the physical and/or social environment, and the methodology of explanation frequently lies in the realm of quantitative analysis. Thus, archaeologists search for redundancy and patterns of behavior in order to seek and verify explanations of that behavior. The search includes analysis of covariation within the archaeological record—covariation that can be isolated and verified statistically.

This quest is quite different from that of the archaeoastronomer, who generally deals with the unique, not the redundant, and with the sacred, not the secular, thus making quantitative treatment of the data more difficult and explanation more challenging. To be quite candid, from the archaeological perspective, astronomical interpretations of the archaeological record frequently have been based more on simple correlation than on actual, verifiable covariation. Identification of covariation is difficult to accomplish in the realm of ritual, and this suggests why archaeologists tend to shy away from that realm in seeking explanation of the phenomena with which they deal.

But differences in primary data do *not* mean that archaeoastronomers should not attempt anthropological explanations, or that archaeologists should not consider the realm of ritual. I will deal with the latter in more detail when I discuss Chaco Canyon. With regard to the former, in my opinion astronomers need to go beyond the realm of description of the phenomena they observe, beyond the actual recording of these phenomena (which sometimes becomes an end in itself), and make more attempts to interpret the purpose and function of these phenomena in a larger social context. In brief, they need to explore the explanation of the archaeological sites they find in broader, more anthropological terms in order to contribute significantly to the interpretation of the archaeological record. When astronomers deal more with the cultural content of sites in an area, and when archaeologists deal more with the ritual aspects of the same sites, we will begin to see progress in the maturation of archaeoastronomy as a truly integrated discipline.

I would like to mention our research in Chaco Canyon as an example of how to approach the integration of the two disciplines discussed thus far. By "our," I mean the work that the Chaco Center, or Chaco Project, carried out for 10 years in the San Juan Basin, that 26,000 sq mi expanse of greasewood, saltbush, and strip mines lying to the northwest of Albuquerque. The Chaco Center was a National Park Service research organization located on the campus of the University of New Mexico that carried out archaeological and related research in and around Chaco Canyon between 1972 and 1985.

I mentioned that archaeologists shy away from ritual explanations of the phenomena they observe because such explanations are very hard to verify. For this reason, the ritual explanation is one that is resorted to only when other explanations are exhausted. Put differently, when they can't think of anything else to call it, archaeologists label it "ceremonial." Basically, that is what has happened in our research at Chaco, at least from one perspective.

To summarize our work there, the most parsimonious explanation or *model*, as we call it, of what we see there archaeologically is that the canyon functioned primarily as a ceremonial center, which received periodic pilgrimages from outlying areas, and that a great deal of real "business" involving day-to-day human survival in the San Juan Basin was carried out at those times under this ritual metaphor. The ritual, then, functioned as a means of gathering people together for the exchange of information and material goods. Note that, in our view, the *true* function of the ceremonial system was to regulate the exchange of goods and services throughout the basin.

Although we have arrived at a "ritual explanation" of an archaeological manifestation, it was arrived at neither lightly nor hastily. Only after other possibilities were examined, analyzed, and discarded as empirically invalid did we accept the ceremonial model as the most likely, and then only when the latent economic function could be assumed as an integral component. Thus, this is not a simple "We don't know what it is, thus it must be ceremonial" explanation of something as complex as the Chacoan system.

If this model is true, and right now it is just a hypothesis, then the challenge to archaeologists is to verify it. By the same token, in my opinion, the challenge to the field of archaeoastronomy is to aid in that process. This must be done by analyzing the ceremonial aspects of Chacoan sites in the broader cultural context of economically based alliance and exchange networks, which were basically an ecological adaptation to a marginal environment in the San Juan Basin. Such analysis is much easier to describe than to do, and it will involve close cooperation between archaeoastronomers and archaeologists interested in the Chaco Phenomenon. It will also involve, in my opinion, a change in the research perspective of archaeoastronomy in Chaco: a change from an emphasis on the descriptive aspects of the ceremonial sites to an emphasis on explanation of these sites relative to the larger cultural context.

Not long before this symposium took place my colleague Tom Windes and I were discussing how much we have learned about Chacoan archaeology since we started in 1972. It really is quite fascinating, since at the beginning we had virtually no idea that Chaco was the focus of a system permeating the entire San Juan Basin, no knowledge of the road network, or of the nature and timing of the climatic episodes, or of the importation of 200,000 trees to build the pueblos, no concept of pilgrimage festivals, and so on. We know these things now because we established a series of research objectives and systematically attempted to verify them empirically. The point is that scientific model building progresses through a series of theoretical/empirical trials.

Again, in my view, archaeoastronomy needs to follow roughly the same steps in progressing through its own series of testable models, in Chaco or elsewhere, in order to become more firmly integrated with, and beneficial to, the anthropological discipline from which it draws its primary data.

Going beyond Chaco for a moment, I would like to return to the question of why the public becomes so transfixed by archaeoastronomical discoveries, such as solstice markers. When such a marker is verified by astronomers through observation, the public seems to acquire an almost hypnotic fascination with it, crowding to the site on the appointed day to watch the thing perform (not unlike Old Faithful, actually). When you think about it, these performances are not much of a subject for awe and mystery. Consider, on the other hand, how mysterious it would be if the site did not perform on the solstice.

What puzzles me, I guess, is why laymen are so thrilled by their own personal discoveries that the Indians knew about solstices. Consider for a moment the fact that Native Americans lived very close to nature in this area for at least 12,000 years before the Caucasians arrived. Now, you can't live that close to a phenomenon for that long without observing something about it. Under the same circumstances, even Anglos would have figured out that the sun's position on the horizon varies with the seasons. Thus, the fact that the Anasazi tracked and marked the solstices should not amaze us, or even make us blink twice, yet it seems to do so. My hunch is that what we are witnessing here is a "rediscovery" of the sun's path by modern Anglo populations. Now that we have an energy crisis, and we even get tax breaks for solar improvements, we as a culture have rediscovered the sun's movements and are beginning to track solstices again, as we undoubtedly did in our own European homelands centuries ago. This rediscovery is a function solely of our having been very effectively insulated from our environment for a long time; now, in the process of having to insulate our homes, we are finding out things about the real world we thought we never knew. Thus, in my view, the amazement we display over prehistoric solstice markers is largely a function of our own recently acquired ignorance about the environment.

This brings me to what I consider to be the most important challenge to the field of archaeoastronomy—the need to demonstrate to the public that there is more to all of this than the mechanics and aesthetics of solstice markers and that the achievements of the Anasazi were much more sophisticated than simply tracking the sun's movements along the horizon. We need to examine the behavior behind the markers to see how that behavior was integrated into the larger socioeconomic system that we now postulate as the key to survival in the San Juan Basin in the tenth and eleventh centuries. That socioeconomic system (that is, the Chaco system) was a *truly* significant achievement on the part of the Anasazi, one that may have been a virtually unique adaptation among cultures in similar environments on this continent. The potential contribution of archaeoastronomy to our understanding of that socioreligious system is both considerable and important, and it needs to be addressed systematically by anthropologically oriented astronomers.

Let me hasten to add that I am not the first person to recognize this challenge confronting archaeoastronomy, or even to articulate it. Far more knowledgeable people than I have written of this, two of the most recent being Dr. Anthony Aveni and Dr. Jonathan Reyman, both participants in this symposium. I would like to quote from Dr. Aveni's recent synthesis of archaeoastronomy:

> While we may demand that anthropologists and archaeologists learn astronomy [heaven forbid!, I would add], we have not advocated strongly enough the more difficult maxim that astronomers interested in archaeoastronomical research ought to become anthropologically oriented.... We must become more conscious about the relationships ancient astronomers sought between the heavens and agricultural cycles, meteorology, and environmental affairs in general. The ... course followed by many investigators simply assumes "they were like us." It does not look enough at the needs and behavior patterns of ancient civilizations as they might relate to their astronomical systems. (1981:S6–S7)

Amen, Dr. Aveni, and well said.

Although I began this discussion by emphasizing the disparity between archaeology and astronomy, I will end by stressing the commonality. Working together with a common understanding and a common approach to mutually agreed upon problems, archaeologists and astronomers *can* make significant contributions to the interpretation of past life systems. It will not be an easy task, but then remember that when the going gets rough ... the archaeologists call in the graduate students.

As a keynote lecture, this certainly has been the most low-key aspect of this symposium on astronomy and ceremony in the prehistoric Southwest. Fortunately, the high-key parts, indeed the highlights, are yet to come. Over the next several days we will have the unique opportunity to hear a number of nationally recognized scholars address some of the issues that I have mentioned along with many others that will come to light. The true value of this symposium, of course, lies in the encouragement of communication among astronomers, prehistorians, and the public to facilitate the interplay that Aveni, Reyman, and others have called for as essential to the development and recognition of archaeoastronomy as a valuable discipline in its own right.

NOTE

This chapter is a revised version of the keynote address given by Dr. Judge at the symposium.

REFERENCES

Aveni, Anthony F. 1981. Archaeoastronomy in the Maya Region: A Review of the Past Decade. *Archaeoastronomy* 3 (*JHA* 12):S1–S16.

Zeilik, Michael. 1987. Anticipation in Ceremony: The Readiness Is All. In *Astronomy and Ceremony in the Prehistoric Southwest,* ed. J. B. Carlson and W. J. Judge, 25–41. Albuquerque, Maxwell Museum Press. Papers of the Maxwell Museum of Anthropology, no. 2.

CHAPTER THIRTY-FIVE

I Wasn't Going to Say Anything, but Since You Asked
Archaeoastronomy and Archaeology

Keith W. Kintigh

Let me first propose two hypotheses: (1) archaeoastronomy has had little impact on mainstream archaeology; and (2) archaeoastronomers feel that their work is under-appreciated by archaeologists. For purposes of this essay, I assume that there is an empirical warrant for these hypotheses: that archaeoastronomers do, in fact, feel injured because there has, in fact, been little impact on the larger discipline. My purpose is to provide the perspective of a practicing Southwestern archaeologist on why this situation obtains and the degree to which archaeoastronomy's station within archaeology is commensurate with its contributions.

In this brief essay, I am "shooting from the hip," based on my knowledge and impressions; it is not an argument based on a thorough review of the archaeoastronomical literature. However, it is based on perceptions that are not entirely uninformed and that are shared by other archaeologists. While readers will doubtless wish

Reproduced by permission from *Archaeoastronomy and Ethnoastronomy News* 5:1 (1992):4.

to return fire (perhaps taking careful aim), I hope to provide some constructive insights into what I take to be a chasm (or should I say "void") between the disciplines.

The issue is not so much that archaeologists actively object to archaeoastronomical research as that they ignore it. When there is a response, it seems to be in approximate proportion to the broader scientific visibility of the claims. Presumably, archaeoastronomers would prefer to be accepted, acknowledged, or even argued with, than ignored. In light of the fact that archaeoastronomers bring considerable energy and expertise to their efforts, what accounts for archaeologists' indifference?

I think the principal reason is that archaeologists see archaeoastronomers as answering questions that, from a social scientific standpoint, no one is asking. To put it bluntly, in many cases it doesn't matter much to the progress of anthropology whether a particular archaeoastronomical claim is right or wrong because the information doesn't inform the current interpretive questions. It may be true that a building is lined up within half a degree of true north, but what do I do with that singular fact?

Why is there a divergence between archaeological and archaeoastronomical questions? It is because questions asked by archaeoastronomers generally do not derive from current archaeological debates about theory or regional prehistory. Instead, much archaeoastronomy seems to be a professionalized response to a widespread, seemingly innate, curiosity about the topic. I, too, read *Stonehenge Decoded* when I was in high school and found it fascinating.

The problem is that there are lots of facts that are *neat* but that do not contribute to our body of systematic knowledge. It is neat that a particular architectural configuration would cast a distinctive shadow at the summer solstice in AD 1270, but it is not scientifically *interesting* unless it informs some research question. That the Anasazi probably made this observation is curious, but probably less useful than the determination of the composition of their purple glaze, or the discovery that they managed to find the optimum depth to plant their seed corn. Glaze composition is interesting if, for example, it shows use of a non-local mineral, which bears on the issue of exchange, which is central to our theoretical understandings of the scale and complexity of prehistoric Southwestern social organization.

Like other scientific disciplines, archaeology has a paradigm that, among other things, loosely determines the relevance of research questions. Thus, like astronomers, we spend most of our time pursuing questions that our theoretical ideas suggest are important, not just arbitrarily collecting facts (or for that matter, artifacts) wherever we might find them.

It may be useful to note several parallels between rock art research and archaeoastronomy. There is a widespread non-specialist interest in both; serious and talented people engage in both with the intention of contributing more broadly to archaeology. However, both research domains seem largely self-contained. The practitioners propose and answer their own questions and communicate largely with one another. Finally, for better or worse, both fields are pretty marginal to mainstream archaeology, and neither is entirely happy about it.

Lest anyone suggest we arrogant archaeologists are just desperate to maintain our turf, I note that archaeologists have long-standing and productive relationships with specialists in other fields: geology, chemistry, botany, zoology, and statistics, to name a few. The difference is that, by and large, these specialists aren't out there defining a new field with new questions and demanding that we pay attention; rather, there is joint work with archaeologists toward *shared* goals.

Another reason for the lack of interest shown by archaeologists for archaeoastronomy is some justifiable skepticism concerning research in archaeoastronomy. I am amazed at the way in which otherwise sane and apparently sober physical scientists with respectable academic positions appear to lose all critical ability when utilizing archaeological data, and exhibit stunning naiveté when proposing what are essentially anthropological arguments. Certainly this is not true of all archaeoastronomers; but *there are* guilty parties who are unfortunately oblivious to their offenses. Believe me, the lack of anthropological sophistication (e.g., conspicuous ethnocentrism) is often excruciatingly obvious in archaeoastronomical arguments.

Many physical scientists are only too happy to sneer about the superiority of their fields over the social sciences. Perhaps some feel that so fuzzy an enterprise as archaeology can't be terribly difficult for a *real* scientist. There are reasons why social science (and anthropology, in particular) is inherently difficult, but that is another story.

In any event, you are all aware that some archaeoastronomy isn't really all that good, and some is spectacularly awful. Prudent archaeologists may keep their distance from archaeoastronomy, not wishing to invite more problems than they already have. (Of course, much archaeological research is poorly conceived and executed, but that is beside the point.)

It is not my desire to indict all archaeoastronomy. There has been some fine work that has contributed significantly to archaeology *because* it has been connected to broader intellectual efforts. To the extent that archaeoastronomy remains high-tech, celestial butterfly collecting, then methodological rigor is the principal quality to be desired. To the extent that the goal is contribution to social science, then the research must *also* bear on research issues. In archaeoastronomy, the needed scientific relevance can be approached from a couple of directions.

First, one may study seriously, or work with someone studying, for example, ancient cosmology. Such a study of cosmology may generate anthropological questions that are amenable to investigation by archaeoastronomical methods and whose resolution has real cultural significance. (As an aside, the first thing to recognize in studying cosmology is that anthropologists use the term differently than astronomers. In both fields, cosmology has to do with a philosophical understanding of the origin and structure of the universe. However, the universe of anthropological interest is the universe *defined by a native group,* in its own language. This will often have precious little to do with the Sun and stars and lots to do with things like causality and gender differences.)

Second, if you must start with the demonstration of architectural alignments with celestial events, you need to be able to say what that means about the prehistoric people, and how that relates to current research. Architectural alignments are, after all, artifacts. As with other artifacts, the anthropological interest is *not in the object itself*, it is in what it tells us about culture. The observation of an alignment, however ingenious, is in and of itself not interesting. However, an argument relating control over privileged knowledge of celestial events to the development of hierarchical power relations at some crucial juncture might have substantial theoretical interest (call this the Connecticut Yankee in King Arthur's Court Hypothesis).

In social science, the generation of facts—astronomical observation and identification of alignments—is easy (analogous to excavation, classification and dating). However, it is my suspicion that it will be difficult to make *rigorous and testable* arguments linking archaeoastronomical observations with serious anthropological questions. To continue with the example, this will involve not just proposing the relationship, but demonstrating that the knowledge was esoteric and not public, specifying the societal conditions under which the development of hierarchical power relations would and would not occur, showing that the group in question was in that state, and so on. Note that we are not saddling you with any difficulties that we don't already have; most interesting anthropological arguments are faced with the same problems. I invite or perhaps challenge archaeoastronomers to use their expertise to attack the very much more difficult but infinitely more interesting theoretical and substantive questions.

Nobody Asked, but I Couldn't Resist

A Response to Keith Kintigh on Archaeoastronomy and Archaeology

Anthony F. Aveni

Professor Kintigh's constructive if, to some degree, slightly misinformed essay in the last issue of *A&E News* ought to be read and discussed by all students and practitioners of archaeoastronomy. It raises many important issues on method and practice in our field and it points out some of the difficulties in carrying out serious and meaningful research programs in an interdisciplinary manner. Indeed, Jon Reyman had already made many of these still justifiable complaints two decades ago.

The validity of Kintigh's two hypotheses depends upon which archaeology and what sort of archaeoastronomy we happen to be dealing with. Had he reviewed the literature beyond the Southwest before loading up and firing, Kintigh might have discovered that in Mexico, for example, archaeoastronomers, who used to be tossed out of the ruins like boisterous fans at a football game, are now routinely invited—even cajoled—into participating in joint investigations. Research questions based in the study of culture are jointly formulated and testable hypotheses are communally thought out. (I always said archaeoastronomy belongs in cultural anthropology.)

Reproduced by permission from *Archaeoastronomy and Ethnoastronomy News* 6:1 (1992):4.

Today, archaeoastronomy isn't all alignment hunting, as Kintigh seems to characterize it. Zeilik in the Southwest, Dearborn et al. in Peru and Carlson at Cacaxtla, Mexico, are among more than a handful of contributors to research agendas in archaeology, anthropology, art history, ethnohistory, etc., originally approached via archaeoastronomical inroads. (Archaeoastronomy *does* impact disciplines other than archaeology.) In the Old World, Burl and Ruggles have carried the worn-out debate about whether Stonehenge was a computer into the realm of the study of prehistory. So archaeoastronomy does affect archaeology, even over there.

Like Kintigh, I once was amazed by the phenomenon of the physical scientists who abandon their rigorous, disciplined professional lives when they set off akimbo to practice archaeoastronomy on the weekend. I even remarked about it (in print) at the Maxwell Museum conference several years ago. But now I think I understand why this happens. How can an engineer who has never taken an anthropology course be expected to address questions about whether astronomical knowledge was public or private, much less a part of hierarchical power relations? A better question might be: how can an engineer who has never read a book about the culture, whose alignments he/she measures to the nearest arc minute, even make a meaningful statement about indigenous astronomy? Kintigh is right. The research agendas of physical and social scientists are not the same, and their perspectives are as different as the Sun and Moon.

Archaeologist Jim Judge (Chapter 34) once remarked that a lot of archaeoastronomy is concerned with the Anglo population's rediscovery of how the sky works. Likewise, many amateur Mayanists are enthralled by their revelations about planetary conjunctions acquired with their PCs. That conjunctions and alignments exist in and of themselves may matter an awful lot to many computer buffs and hard scientists, but it carries little force with archaeologists and anthropologists, until one develops some ideas about the impact and uses of such knowledge in the context of culture.

While we and the archaeologists (some of whom are us!) do operate under different paradigms, I have seen not a divergence but rather a slow convergence of research agendas over the years. Today's "product" really is a better one, much more sophisticated and substantial, far more interdisciplinary in scope. A survey I conducted at the September 1990 "Oxford 3" meeting (Aveni 1993) revealed that in recent times there have been more coauthored works, more papers appearing in the disciplinary journals, more and better sessions at the Society for American Archaeology, American Anthropological Association, etc., and the incorporation of more astronomical presentations in sessions that do not bear the title "astronomy." Brighter and better students educated in the ways of archaeoastronomy now receive PhDs and write books in Comparative Religion, Art History, Ethnology, Ethnohistory and the History of Science. We *are* making progress.

We are long past the age of simply measuring and reporting the facts of building alignments or calendrical correlation numbers, handing them over to the culture historians and saying (often with a superior air), "See what you can do with these." I read Kintigh's essay as constructive because he invites those of us interested in the history of astronomies to work further toward shared goals, to think less about natural phenomena for themselves and more about the impact of nature upon people.

But I fear few of our "celestial butterfly collectors" will pay the price exacted by Kintigh. To become true celestial lepidopterists, they will need to escalate their work to address culturally substantive questions. This is hard work and, moreover, it requires a change of intellectual life style many professionals may not be willing to make. Kintigh and his colleagues in all the disciplines that border on archaeoastronomy must, therefore, allow the validity of contributions to archaeoastronomy to be decided by the quality of work that appears in refereed publications, especially those in the standard disciplines. The reputation of our interdiscipline cannot be judged by the all too proliferous reportage of what lines up with what or whether this or that standstill was being observed.

REFERENCE

Aveni, Anthony F. 1993. Archaeoastronomy in the America since Oxford 2. In *Archaeoastronomy in the 1990s,* ed. C.L.N. Ruggles, 15–32. Loughborough, England, Group D.

Index

Page numbers in *italics* indicate illustrations.

Accessions, *591*; Venus associations with, 551, 554(table), 555
Achernar, 411, 427, *431*
Acoma, 651, 656, 665
Acosta, Jose de, 78
Adhara, 470
Adultery, in Kogi culture, 435
Adulthood, Inca rites of passage to, 140
Agriculture, 78, 247, 308, 556; E-group complexes, 356–57, 365; Mayan, 262, 270, 405, 607–15; Nazca lines and, 624, 626, 630, 632; planting season, 67, 116, 143–44, 154–55, 325–26; Quechua, *472,* 475, *476,* 476–77, 490; 260-day calendar and, 235–36; Zuni, 659–60
Aguateca, 574
Ah Cacaw, 336
Aldebaran, 37(n27), 407, 434, 665; Big Horn Medicine Wheel alignment with, 28–33

Alignments, 3, 4–5, 9, 14; Big Horn Medicine Wheel, 24–36
Allauca, 86, 95
Almanacs, in *Dresden Codex,* 541–46, 607–15
Alpha Centauri, 177; Quechua concepts of, 477, 478
α Crucis, and Southern Cross, 475
Altair, 409
Alta Vista, 489
Amarna (Egypt), 754
Amarus, 467–68, *469*
Amazonia, 5, 396, 466, 736; social organizing principles, 427–28, 732
Anacondas, 466, 468, *469*
Anahuarque, 88, 89, 91, 95, 174
Anales de Quauhtitlan, 258
Anasazi, 199, 200, 221, 804; astronomical sites, 223–24; horizon calendars, 208–10; light and

809

INDEX

shadow sites, 210–17, 694; lunar standstills, 711–12; rock art, 201–2; sky watching, 665–66; use of spirals, 688–89. *See also* Chaco Canyon; *various sites by name*
Andes, 5, 72, 267, 396, 491
Animals, 5, 618; celestial, 463–65, 735; Quechua constellations, 457–58, 459–62, 467. *See also by type*
Annunciation, Feast of, 778–79
Antisuyu, 83, 93(table), 123(n1)
Antizenith, *191*, 310; Incan observations of, 117, *119*, 121, 125, 146–55, 156, 161, 178
Apuyauira, 140–41
Archaeoastronomy, 1; history of, 7–9; methodological issues of, 737–43
Archaeology, 37–38(n36); and archaeoastronomy, 793–800, 803–8; methodology, 736, 737, 738
Architecture, 357, 735, 751; archaeoastronomy of, 739, 773–74; calendrical systems and, 363–68; as cosmograms, 782, 784–88; cosmology and, 805–6; Palenque, 306–8, 310–11, 334; ritual cycles, 371–72. *See also* E-group complexes
Arco Punco, 116
Aristotle, 414
Arizona, crescent moon-supernova pictographs in, 618, 636–37, 638
Asterisms: Mayan, *249–50*; Mesoamerica, 244, 245, *249*
Astillejos, 247
Astronomy, and archaeology, 793–800, 803–8
Atarazanas, 258
Atkinson, Richard, 14; "Moonshine on Stonehenge," 3–4
Atlcahualo, 238
Avebury, 729
Avila, Francisco de, 458, 463, 478
Axis mundi, 784, 786; in tropics, 486, 488, 489, 490–91
Ayamarca (Ayarmaca), 87, 147
Ayllus, 631, 632
Aztecs, 73, 232, 254, 261, 553, 659, 731, 784; calendrical system, 236–38; constellations, 72, *251*; day structure, 238–39; founding myth and Templo Mayor, 256–57; mountain cult, 255–56; and Pleiades, 246–47; stars, 245–246, *246;* urban planning and cosmology, 753–54, 772–73, 777–79

Bacabs, 236
Bachué, 436, 438(n16)
Bactun cycles, 511

Badger Clan, 656
Baking Pot, 359, 762
Balakbal, 762
Ballcourts, 785; E-group complexes and, 366–68
Ball games, and Venus dates, 552, 582–83
Barasana, 396, 489, 730
Batammaliba, 396
Baum earthworks, 57
Bayesian statistics, 741, 742–43
Beans, planting and harvesting, 405, 653
Bede, Venerable, 245, 302(n95)
Bejucal, 550
Bellatrix, 431
Belonging, Mesoamerican sense of, 384–86
Benque Viejo (Xunantunich), 762
Beta Centauri, 177, 178, 477, 478
β Crux, 411
Betelgeuse, 29, 178, 248, 431
Bhaktapur (Nepal), 754
Big Dipper, 410, 421, 444, 446, 447, 450, 453, 657, 665
Big Horn Medicine Wheel, 14, 19, 37–38(nn36, 40); authenticity of, 33–35, 37(n35); dating of, 20–21, 32–33; purposes of, 35–36; solstice alignments, 23–32
Binding of the Years (Toxiuhmolpilia), 246
Bird effigy (Newark earthworks), 43, 59(n7)
Black God (Navajo), 246
Blackman Eddy, 359
Blessingway ceremony, 450; goal of, 446–47, 451; *hooghan* and, 447–48
Bloodletting rites, 334, 550, 551
Boas, 466, *468*
Bonampak, 262, 505; commemorative stelae at, 578–80; historical dates and Venus events, 568–69, 570–71, 577; Jupiter and, 593–94; Lintels 2 and 3 at, 583–84; rulership and Venus events, 551, 553, 555, 565; Venus dates, 581–82
Books of Chilam Balam, 242, 387
Borana, 197(n4), 727
Bororo, 396, 489, 726
Bourke, John G., 206, 221
Bow priests, 656–57; sun watching, 659–60, 661
Bóyusú, 468, 469
British Isles, 4, 8
Bureau of Ethnology (Smithsonian), 44
Burials, E-group complexes, 367–68, 371
Burners: Maya ritual, 414, 415–17, 418, 419
Bushmasters, *467*

Caches, in E-group complexes, 371
Cahal Pech, 353, 359

Index

Cahuachi, 632–33
Cairns, 37(n33), 631; at Big Horn Medicine Wheel, 19, 22, 23, 24–32, 34(table)
Cajon Group Ruins, 211, 212(table), 213
Cakchiquel speakers, 400, 401, 405, 406; celestial paths, 402–3; on constellations, 407–10; directions, 401–2; on eclipses, 403–4
Calakmul, 336, 756; E-group complex, 369–70
Calendar priests: Maya, 414, 415–23; Pueblo, 665–66
Calendar rounds (52-year), 182, 246, 590
Calendars, 197(n4), 299(n44), 302(n95), 626; directional cosmology on, 783–84, 785; E-group complexes, 356–57; horizon, 208–10; lunar, 217–21, 512, 514; lunisolar, 239–43; Maya ritual, 519–23; Mesoamerican, 235–36, 250, 783; Templo Mayor as, 73, 259–60, 281–85, 286, 778–79; Tenochtitlan, 255–56
Calendar sticks: Pima and Papago, 649; Puebloan, *219–21*
Calendrical systems, 73, 181, 489, 589; Anasazi, 208–17, *701*; Andean, 72, 87; and E-group complexes, 368–71; Hopi, *206–7*; Incan, 77–86, 91–98, 101–3, 169–70, 171; Maya, 401, 414–23, 505, 510–14, 517–18, 601–2(nn2, 4); Mesoamerican, 235–44, 250, 308, 774, 783; Puebloan, 201, 202, 217–21, 650–654, 664–65; Tenochtitlan, 258–60, 288–89; Zuni, 203–4, 205–6
California, rock art in, *637*, 639–42, 686
Calispuquiu, 140
Callachaca, 104, 165
Canis Majoris, 470
Canis Minor, 408
Canopus, 427
Capac ayllu, 88
Capac Churi, 140
Capac hucha, feast of, 80, 100
Capac Inti raymi, 88
Capac Raymi festival, 140–41, 142, *143*, 159; and Inca origin myth, 145–46
Capella, 39, 307, 427, 434
Captives: E-group complexes, 367–68; raiding for, 552–53
Caracol (Belize), 369, 575
Caracol (Chichén Itza), 356, 777
Cardinal directions, 22; Mayan, 325, 363–64, 763–64, 765, 783; Mesoamerican, 235, 236, 732–33; Navajo, 450, 454(n2); Pueblo cultures, 203, 222–23, 665
Carmenga, 116
Carpathian Basin, 734
Casa de las Monjas (Chichén Itza), 570

Casa Grande, 64, 660, 712
Casa Rinconada, 211, 214, 224, 689
Cassiopeia, 245; Navajo concepts of, 446, 447, 450, 453
Castor and Pollux, 244, 247
Caterpillar Jaguar constellation, 730
Catholic Church, 414
Causeways, at Tenochtitlan, 259
Cave of Life site (Petrified Forest), 216
Caves, 145, 159, 300(n53)
Cayao, 84
Cayocache, 133, 141
Celestial paths, Mayan, 402–3
Cenote (Guatemala), E-group complex, 353, *355*
Centaurus, 411, 730
Ceque systems, 165, 171, 182, 270, 736; astronomical observation, 98–103, 142; and calendar, 94–95; comparison to quipu, 80–82; in Cuzco, 71–72, 78, 79–80, 87, 110–22, 160–61, 178, 264–65, 266–67, 268, 490, 493; December solstices, 142, 144–45; harvest festivals, 163–64; *huacas* associated with, 89–98, 130, *131*; Inkawasi, 174, 175–76; Nazca lines as, 627, 631; structure of, 82–86, 88–89
Ceremonial cycles: Puebloan, 202–7, 650–51, 652–54, 656, 659–62, 664–65
Ceremonialism: Chacoan, 798–99; Maya, 263; Mexico, 201
Cerros (Belize), 364, 755
Chac, 262, 611, 776
Chacaguanacauri, 140
Chac Chel, 611
Chaco Canyon, 208, 222, 665, 690, 794, 798–99; Fajada Butte petroglyphs, 618–19, 669–85, 709; light and shadow sites, 210–15, 685–86; lunar standstills, 683, 700, 701; Peñasco Blanco pictograph, 201–2, *642*, 642–43, 658, 659, 666, 796–97; sun watching at, 650, 658; symbols used in, 688–89; three-slab site, 218, 671–72, *673*, *674*, *695*, *696*, *697*, *698*, *699*, *700*, *701*, 700–4, 711–12
Chakwena, 653
Chalchiuhtlicue (Chalchuihtlicueh), 258, 389
Chama Valley, 208, 666
Chamula, 365, 763
Chan-Bahlum (Serpent-Jaguar), 550, 551, 588; Jupiter and, *590*, *592*, 592–93, *593*, 594–98, 600
Changing Woman (Navajo), 447
Chankillo, 4–5, 72, *182*; dating, 182–83; Thirteen Towers at, *184*, 184–96
Chan Kom, 262
Chanters, Navajo, 440, 442–43, 452

INDEX

Chapultepec, 256
Chayote, 405
Ch'en Mul cenote, 786
Cheyenne, 22
Chiapa de Corzo, 349
Chiapas, 238, 349, 351
Chicchan-Yax combination, 572
Chichén Itza, 64, 306, 325, 356, 368, 372, 570, 686; asterisms recorded at, 249, *250*; equinox sitings, *307*, 315; Venus events at, 552, 777
Chichimecs, 244
Chief Hooghan Songs, 442
Children, Incan sacrifices of, 80, 133, 141
Chillicothe (Ohio), earthworks near, 41
Chilque, 85
Chimalpopoca, 256
Chimney Rock Archaeological Area (CRAA): lunar standstills, 619, 712–17; and three-slab site, 717–18
Chimney Rock Pueblo (5AA83), lunar standstill observations from, 712–18
China, 251; record of Crab Nebula supernova, 635–36, 638, 644, 728
Chinautla, 410
Chinchaysuyu, 83, 84, 111, 115, *116*, 123(n1), 130, 150; *huacas*, 93(table), 104, 113, *114*
Chinchincalla, 103, 106, 111, 133, 142, 160, 165, 166–67
Chinchircuma, 103
Chiquihuite, Cerro, 286
Chochkitam, 762
Chol Maya, 262
Cholula, 286
Chorti Maya, 238, 239, 247
Christianity, 245, 414
Chronicles, use of historical, 168, 169–70, 172; Maya, 509–10. *See also various codices*
Chuj Maya, 407, 411
Chumash, 644
Chuquicancha, 104, 105, 133, 135
Chuquimarca, 104, 114, 115, 133, 134, 140, 156, 164
Cihuacoatl, 389
Cinteotl, 389
Citlalco, 234
Citlalcolotl (Scorpion), *246*, 247
Citlaltlachtli (Star Ballcourt), 248
Citlallicue (Starry Skirt), 234, 248
Cituay, feast of, 82
City/civic planning: Aztec, 772–73, 777–79; cosmology and, 752–58; Maya, 270–71, 751–52, 754–58, 761–68, 780(n1), 782–88; Mesoamerican, 773–77

Ciudadela (Teotihuacan), 267, 353, 777
Cival, 785
Classification, natural world, 731, 740
Coalsack, 473
Coatlinchan, 261
Cobo, Bernabe, 78, 126, 154, 163–64, 172, 463; on *ceque* system, 83–84, 89, 99–100, 101, 102, 111–12, 160, 264
Codex Borbonicus, 388, 391
Codex Fejérváry Mayer, 253–54, 283, *785*
Codex Tellerianus, 391
Codex Tonalamatl of Aubin, 388
Codex Vaticanus A, 234, 784
Codices, 413, 509–10; planetary tables in, 514–16. *See also by name*
Colha, 359
Collaconcho, 265
Collanasayba, *116,* 116–17, 133, 149
Collasuyu, 83, *93,* 99, 104, 123(n1)
Collca, 64, 152
Colombia, 396, 425–26
Colotepec, Cerro, 259
Columns, Muiska solstice observations and, 436
Comalcalco, 359
Comets, 235, 550, 644, 734
Consciousness, Desana, 433
Constellations, 174, 396, 730; Aztec, 72, *251*; cultural contexts of, 727, 730–31; Desana, 431–33; Inca, 129, 152, 178; Maya, 248–50, 407–11, *407, 408, 409, 410, 411,* 569, 570, 572, 584; Mesoamerican, 235, 238, 244–48; Navajo, 440–41, *444, 445,* 444–48, 451–53; Nazca lines as, 618, 626; Quechua, 64, 457–62, 467, 468, 472–74, 475–76, 478–82
Copacabana, 85, 86
Copán, 73, 241, 265, 268, 269(table), 368, 372, 755, 784; calendrical system at, 526, 527–28, 530; city plan, 765–66; historical events and Venus in, 575, 576–77; 260-day calendar, 235–36; Venus observation at, 261–64, 270
Copil, 285
Coricancha, *114,* 122, 129, 139, 142, 148, 160, 165, 269(table); in *ceque* system, 82, 85, 90, 94, 95, *99,* 100, 101, 110, 116–17, 182, 265, 267; as observatory, 105, 111, *112,* 113, 130, 161, 167, 169, 170, 178; solstice observations, 131, 132, 143, 144–45
Cosmograms, Mesoamerican, 782–88
Cosmology, 2, 5, 126, 254, 267, 396, 787, 789(nn2, 3), 805; Amazonian, 426–27, 736; cross-cultural comparison, 732–33; Maya cities, 6, 751–52, 754–58, 762–63; Mesoamerican, *234,* 234–35, 385, 386–92,

783–84; Navajo, 442–43; Puebloan, 222, 223, 224; Quechua, 463–65; Tenochtitlan, 255, 752–54, 772–73; tropical, 487–89
Cothiemuir Wood, 740–41
Course of Santiago (Corrida Santiago; Milky Way), 408
Coyote, and Pleiades, 246
CRAA. *See* Chimney Rock Archaeological Area
Crab Nebula, supernova event, 201–2, 498–99, 500, 618, 635–38, 648, 728, 796
Creation myths: Inca, 145–46; Mesoamerica, 236, 245–46, 248; Navajo, 440–42, 449
Creator God, 239; Bachué, 436
Crescent-shaped mounds, 57
Crescents, in rock art, *636*, 638–39, 640–43, *640*, *641*, *642*, 647, 658
Crosses, 659; in Mesoamerican symbolism, 363–64, 739
Cross Group (Palenque), 310–11, 313, 315–16, 328, 331–32, 334, 342, *343*, 589
Cross of St. James, 248
Crow, and Big Horn Medicine Wheel, 23, 35
Cruz de Corió, 352
Crystals, Inca symbolism, 151
Cuauhtepec, Cerro, 286
Cultural anthropology, 736, 737
Cultural astronomy, methodological issues of, 737–43
Cultural context, 4–5, 725, 732; continuity of, 395–96; knowledge retention and, 728–29; natural world in, 729–31; observation and perception in, 726–27
Cuntisuyu (Kuntisuyu), 84, 89, 90, 91, 102, 111, 123(n1), 142; *huacas* associated with, 93(table); 113; at Inkawasi, 175–76, 179
Cunturpata, 144
Curley, Slim, 449
Cushing, Frank H., 201; on Zuni calendrical system, 203–4, 206; on Zuni sunwatching, 659–60
Cuzco (Cusco), 4, 71–72, 73, 125, 126, 129, 159, 269(table), 466, 754; *ceque* system, 78, 79–86, 89–100, 110–22, 178, 182, 264–67, 268, 270, 490, 491, 493; December solstice in, 141–45, 146; harvest festival in, 163–64; historical chronicles of, 169–70; *huacas* in, 106–7, 127, *131*; June solstice in, 133–34, 156; observatories in, 100–106, 160–61; political systems, 88–89; social and physical structure of, 123(n1), 175; zenith-antizenith observations, 147, 148–51, 157(n2), 161, 195
Cygnus, 245

Dakota, Leonid meteor shower painting, 644
Dark cloud constellations, 5; Quechua, 458, 459–62, 468, 472–74, 476, 477–80, *480*; rainy season and, 462–63
Dawn, Navajo concepts of, 450
Daykeepers, 396, 414, *417*, 418, 419
Day of the Holy Cross, 262
Days: Mayan concepts of, 401, 416; Mesoamerican organization of, 238–39
December solstice, 125, 178, 475, 487; ceremonies of, 140–41; at Chankillo, 188, 192; in Cuzco, 141–45; at Machu Picchu, 145–46
Deities, 238, 385; destinies and dualities of, 389–92; lunar, 239, *240*; Mayan, 510, 511–12; at Palenque, 311, 333, 343; Pueblo, 650, 656, 664–65; solar, 236, *237*, 780(n2)
Delta Orionis, 431, 433
Desana, 437(n2), 475, 489; energy in universe, 429–30; hexagonal organization of, 427–29, 433–34; longhouses, 430–33; origin myths, 426–27
Destinies: stargazing, 392–93; *tonalli*, 386–88, *391*
Directions, 238; Maya concepts of, 401–2; Mesoamerican symbolism, 363–64
Divination, 399; toads used for, 471–72
Dos Hombres, 765, 766
Dos Pilas, 574, 784
Dresden Codex, 239, 265, 388, 413, 422, 484; double tzolkin almanac in, 607–15; eclipse tables in, 241, 419, 421, 520, 524, 526, 527; Mars calculations in, 541–46; planetary tables in, 546–49; Venus tables in, 243, 244, 262, 270, 514–16, 531–41, 551, 565–68, 574–75, 596
Dry seasons, 325, 425, 463, 489; Maya, 401, 402–3, 612
Dualism, duality, 255, 486; Andean/Incan, 121, 147, 161; Mayan, 343, 404; Mesoamerican, 239, 389–92; Puebloan, 222, 223, 650, 654
Dzibilchaltún, 291, 306, *307*, 315, 353, 372

Eagle Pueblo, 662
Earth, 427; and Quechua celestial animals, 463–64
Earth Mother, 223, 491
Earth Surface People (Navajo), 441; and stars, 440, 446, 451, 453
Earthworks, Hopewell, 14–15, 39–49
Eclipses, 251, 426; lunar, 392, 603(n10), 734; Mayan observation and prediction of, 241–42, 403–4, 419, 421, 519–23, 524, 525, 526, 527, 529–30, 613; predicting lunar, 56, 530–31,

INDEX

683; predicting solar, 609–10; and Venus cycle, 536–41
E-group complexes, 5, 73, 196, *348*, 372, 764, 774; calendrical cycles and, 368–71; cosmology of, 762–63; distribution of, 349–55; as observatories, 306, 347, 355–62, 373; ritual cycles and, 362–68
Egypt, 30
Ehecatl, 666
8 Batz' ceremony, 414
819-day cycle, 587, 590, *591,* 599
El Castillo (Chichén Itza), 306, *307,* 325, 786
El Cayo, 556–57
El Mirador, 349, 353, 369
El Mirador, Cerro, 328, 330, 337; equinox observations, 314–16, *316;* solstice observations, 311, *313,* 318
El Venado, 358, 359
Emeralds, Muiska use of, 436
Eototo, 656
Epsilon Orionis, 427, *431, 432,* 433, 434, 435
Equinoxes, 82, 207, 435, 489, 687, 778–79, 780(n3); Chankillo, *191,* 192; dark cloud constellations, 459, 461; Fajada Butte, 670, 675–77, *678, 679,* 708; and E-group architecture, 306, 357–58, 360; Inca observations of, 127–28, 129, 146–47; Mexica observations of, 257, 270, 287–88; Palenque observations of, 314–17, *316, 317;* Teotihuacan plan, 260–61
Ethnoastronomy, 2, 5, 395–96; Mayan, 400–11; Mesoamerican, 232–33
Ethnographic analogy, 200–201, 495, 686–87, 740–41
Ethos, 5, 383–84
Evening Star, 243, 244, 657, 664
Exogamy, Tukanoan social organization, 428–29

Fairground Circle (Newark earthworks), *42, 43, 44,* 46, 55, 59(n10); geometry of, 48, *49,* 60(n21)
Fajada Butte, 208, 687, 689–90; astronomical markings at, 669–70; light and shadow sites at, 214–15; noon and seasonal markings at, *672–81;* rock art at, 6, 210, 618–19; three-slab site at, 214–15, 218, 219, 671–72, *673, 674,* 682–85, 693–94, 695–709, 711–12
Family, Navajo, 449–50
Father Sun, 650
Feast of the Annunciation, 289
Feathered Serpent, 244
Fern Cave, crescent moon-supernova, *637,* 639–41
Fertility, 266, 368, 480; Tukanoan views of, 429–30

Festivals, feasts, 80, 82, 246, 414; first-plowing, 85–86; Incan, 88, 126, 130, 131–40, 147, 163; Pueblo, 203, 217; Tenochtitlan, 258, 259, 286, 287, 288–89, 291, 302(n86), 758(n1), 778–79
Fewkes, Jesse W., 201, 213, 219–20
52-year calendar, 246, 783
Figurines, warrior, *195,* 196
Finca Acapulco, 349
Fire God (Navajo), 245
Fire Serpents, 235
First plowing feast, Cuzco, 85–86
Florentine Codex, 286, 389
Flute ceremony, 218, 654
Flute Society, 656
Formalhaut, 410, *411*
Founding myths, Tenochtitlan, 256, 257, 258, 285–86
Foxes, Quechua and, 478–80, *479, 480*

Garcilaso de la Vega, 122, 126, 129, 159, 163, 172, 458, 466; on Cuzco observatories, 100, 101, 103, 117, 127–28; on equinox observations, 146–47, 154; on festivals, 62, 63, 131–32
Gemini, 408, *431,* 584
Gender: celestial bodies, 236, 239, 464, 651, 656; and eclipse meanings, 403–4
Geoglyphs, 6, 618. *See also* Nasca (Nazca) lines
Geography, sacred, 735–36
Geomancy, E-group complex sitings, 357, 360–61
Geometry, 363(table); Amazonian organizing principles, 427–28; Newark earthworks, 46–49, 59–60(nn20, 21); symbolic notation, 733–34
Glyph Y, 548–49
Glyph Z, 548
Gnomons, in Pueblo sun watching, 216–17
Goddess of the Terrestrial Waters, 389
"God in the Machine" (Hawkes), 4
God K, 320, 338, 559
God L, 262, 308, 340
Gómez Orozco Codex, 389
Gordo, Cerro, 777
Great Acropolis (El Mirador), 369
Great Orion Nebula, 433
Great Spiral (Nazca), *623*
Great Star. *See* Morning Star
Great Triangle (Nazca), *622*
Grolier Codex, 243, 413
Group E, at Uaxactun, 347–49, 364. *See also* E-group complexes
Grus, 410, *411*

814

Gualjoquito, 755, 764
Guamancancha, 140
Guaman Poma de Ayala, Felipe, 78, 94, 126, 147, 163, 172, 266, 267, 475; on *ceque* system, 80–81, 85, 90; on observatories, 106, 110–11, 127, 128; on *quipu* specialists, 97–98
Guanaypata, 105
Guancarcaya (Guancarcalla), 104, 105, 133
Guargua Illapuquiu, *114*
Guarmichaca, 99
Guatemala, 400
Guest Star, 644
Güiro/Wakna, 349

Haile, Berard, 233
Halley's Comet, 550, 734
Halona. *See* Zuni
Hanans: Cuzco, 82, 88, 89, 91, 92–93, 123(n1), 147, 265; Haucaypata, 85, 105, 117, *118*, 121, 266; Inkawasi, 175
Hand designs, Pueblo, 659
Han Dynasty, 251
Hano, 208, 221, 657
Hanp'atu, 470
Harold, King, 734
Harvest season, 163, 204, 405, 654
Hastings, Battle of, 734
Haucaypata, 85, 105, 117, *118*, 121, *132*, 151
Hawkes, Jacquetta, 14; "God in the Machine," 4
Hawkins, Gerald: on Nazca lines, 626–27; *Stonehenge Decoded*, 3–4, 13–14
Healing rituals, Navajo, 440
Heavens, in Mesoamerican universe, 234–35
Heir-designation, 555, 595, 603–4(n11)
He-She God of Maize, 389
Hero Twins, Mayan, 343
Hexagons, 438(n10); Amazonian ordering concepts, 427–35
Hieroglyphic Stairway, Naranjo, 575
Hierophanies, Mayan architecture and, 306–8, 311; Mayan rulership, 308–9
High Bank Works, 15, 48, 57
Highland Maya, 352, 414. *See also various sites*
Hill of the Star, 246
Hipparchus, 30
H'meen, 396
Hok, 593
Holly House, petroglyphs near, 210, 216
Holy People (Navajo), 441, 451
Hooghan, 454(n2); cosmic symbolism of, 396, 440, *445*, 446–48, 453, 454(n5); cultural roles of, 441–43
Hopewell Circle, 14

Hopewell culture, 58(n2); lunar alignments, 50–56; Newark earthworks, 14–15, 39–49, 59–60(nn6, 20, 21); other earthworks, 56–57
Hopi, 37(n27), 181, 210, 222, 224, 643, 649, 657, 665; calendar sticks, 219, 219–21, *221*; calendrical system, 201, 218, 655; ceremonial cycle, 203, 217, 650–51, 653–54, 656; horizon calendar, *206-7*
Hopi-Tewa, 655, 657
Horizon markers, 63, 739, 742; calendars, 181, *206-7*, 206–10, 213; *ceque* system, 98–99, 101, *120*; at Machu Picchu, 64–65; Pueblo use of, 203–4, 205; and Templo Mayor, 276–81, 286
Hotevilla, 651
Hotun-ending dates, at Bonampak, 578–80
House of the Eagles (Templo Mayor), 280
House Blessing Ceremony, 447
Houses: and cosmology, 5, 396; Desana, 430–33; Navajo, 441–43
Hovenweep, light and shadow sites at, 210, 211, 212(table), 213, 216, 685–86, 694
Hovenweep Castle, 211, 213, 216
Hózhó, 451, 453
Huacas: astronomical significance of, 92–98; and *ceque* system, 72, 78, 79, 80, 83–84, 89–92, 99–100, 113–15, 130, *131*, 265; around Cuzco, 127, 265, 266; and December solstice, 140–41, 142, 144; and June solstice, 133–34; zenith-antizenith observations, 149–50, 154
Huanacauri, 91, 140, 141, 145
Huana Picchu, 67, 68
Huanuco Pampa, 270
Huarco, 177
Huari, calendrical textile, 97, 98, 106
Huaro, 85, 87
Huarochiri, 461, 478
Huehuehtlahtolli, 386, 389
Huepango, Cerro, 259
Huexotla, 300(n51), 777
Huey Teocalli, 289
Huichol, astronomy, 233, 247
Huitzilopochtli, 255, 256, 257, 385, 758(n1); and Templo Mayor, 275, 303(n102); Tenochtitlan founding myth, 285–86
Human behavior, and celestial bodies, 425–26
Human beings, relationship with gods, 390–92
Humboldt, Alexander von, 462
Hunahpu, 343
Hurin, Inkawasi, 175
Hurins: at Cuzco, 82, 89, 91, 92–93, 94–95, 123(n1), 147, 265; at Haucaypata, 85, 105, 106, *118*, 120, 121, 131
Hyades, 244, 247

Iconography, Maya, 263–64
Identity, in Mesoamerica, 384–86
Ilhuicatl Mamalhuazocan, 235
Ilhuicatl metzli, 234
Ilhuicatl Tonatiuh, 235
Illness, in Navajo culture, 443, 453(n1)
Impregnation, by Sun, 688
Inca, the, 86, 87
Incas, 73, 254, 490; astronomical methods, 109–10; astronomical observations, 98–106, 110–22, 126–28, 270; calendrical cycle, 67, 77–79; *ceque* system, 71–72, 80–86, 736; constellations, 457–58; *huacas*, 89–98; social organization, 87–89; solstice ceremonies and observations, 131–46; zenith-antizenith observations, 146–55
Incest, Kogi culture, 435
Inkawasi (Incawasi), 270; as observatory, 173–76, 177, 178–79
Interdisciplinary research, 2, 721–22
Inti, 132
Intihuatana Barrio (Pisac), 138–39, *139*
Intihuatana Stone, *150*; observations using, 64–68, 154; as Usno, 152–53, 159
Intimachay, *144*, 145, 146, 156
Inti Raymi festival, 130, 131–32, 140, 142, 151
Iron, meteoritic, 57
Island of the Sun (Lake Titicaca), 175, 196
Isleta, 223
Itzamna, 245
Itztapatl Nanatzcayan, 235
Ixil Maya, 238
Ixtlilxochitl, 258, 260
Izapa, 489
Izcalli, 238

Jalcatec Maya, 238
Janahb Pakal, 309–10, 334, 336, 338, 340
Jemez, 218, 656
Jornada Mogollon, 222
June solstice, 125, 156, 170, 475, 487, 493, 621; at Chankillo, 188, 189(table), *192*, 193(table); in Cuzco, 133–34; at Inkawasi, 174–75, 179; Inti Raymi festival, 131–32; at Machu Picchu, 134–39
Jupiter, 244, 248, 537, 603(n8), 604(n13); and Bonampak rituals, 593–94; and Chan-Bahlum, 310, 336, 589–593, 594–99, 600; Mayan observation of, 310, 333, 406, 505, 516, 517, 518, 546, 602–3(n6); Mayan historical events and, 550, 556–58, 560

Kabah, 359

Kachinas (katcinas), 217, 219, 222, 650, 651, 652, 653, 656
Kaminaljuyu, 352
Kan B'ahlam II, 320, 331, 339(table), 341; history and iconography of, 333–34, 336–38, 340, 342; structures built by, 310–11; transfer of power to, 309–10
Kan-Xul, 593
Katun (360 days) cycles, 511, 512, 513, 518, 520–21, 576; E-group complexes and, 362, 367–70; stelae, 370–71
Keepers of the Rock Crystal, 435
Keres, 651
Kiasi, *220*
Kingship, divine, 308–9
K'inich Ahaw (Ahaw Kin), 333
Kivas, 201, 203, 210, 223, 657, 665, 712, 735
Kogi Indians, 426, 489; ritual cycles, 435–36; zenith observations, 434–35
Kohoutek, Comet, 406
Koshare, 657, 663
Kosok, Paula and Rose, 621–22
Kuaua, 208, 210
Kukulcan, 244, 393
Kuyo Chico, 464
Kuyo Grande, 464

Labná, 767
Lacandon Maya, 245, 247, 248
Lacco, *114*, 115, 122, 134, *135*, 165
Laguna, 219, 651
La Honradez, 762
Lahun Chan, 262
La Libertad, 349
La Malinche, Cerro, 287
Lamanai, 359
Lambayeque, 86
La Milpa, 786
La Muñeca, 359
Landa, Diego de, 238, 239, 242, 762
Las Charcas, 352
Lava Beds National Monument, crescent moon-supernova rock art in, 639–42
La Venta, E-group complexes, 351, *352*
Layutsailunkia, *220*
Leonid meteor shower, 643–44
Light and shadow sites, 687; Anasazi, 210–17, 685–86; Fajada Butte, *672–81*, 693–694, *695–700*
Limapampa, 105, 121
Linguistics, dualism in Mesoamerican, 391–92
Lira, Jorge, 463, 467
Little Dipper (Ursa Minor), 248

Index

Little Miami River, 41
Llamacñawin, 477
Llamas (Yacana), 735; dark cloud constellation of, 458, 461, 463; Quechua and, 476–78
Long Count Calendar, 237, 368, 512, 513–14, 518, 601(n2); Maya lunar calculations and, 240–41
Long Horns, 217
Longhouses, Desana, 430–33
Long Mound (La Venta), 351
Lord and Lady of the Earth, 389
Lord and Lady of the Place of the Dead, 389
Lord of Rain, 389
Lords of the Night, Maya, 238, 511–12
Lords of the Underworld, 763
Los Casorios (constellation), *407*
Lowland Maya, 368, 525; E-group complexes, 347, 349, 350(table), 353
Lunar alignments, 337; Newark earthworks, 50–56, 57–58
Lunar calendar: Inca, 67, 95–96; Maya, 512, 516–17, 550–51; Pueblo, 652–54
Lunar cult, Hopewell, 57
Lunar cycles, 396; Inca calendar, 67, 95–96; Palenque, 587–88; Tukanoan culture, 429–30
Lunar standstills: Chimney Rock, 619, 712–18; Fajada Butte, 683, 700, 701, 702–3, 704, 711–12; Palenque, 313–14; Scottish stone circles, 740–41
Lunations, Mayan calculations of, 240–41

Mach'acuay, 466, 467, 470
Machaquila, 369
Machu Picchu, 15, 61, 125, 126, 159–60, 177; astronomical observation at, 128–29; December solstice observance, 145–46; Intihuatana Stone, 64–68; June solstice at, 134–39, 156; as observatory, 169, 173; as Pachacuti's estate, 165–66; Torreón at, 62–64; zenith-antizenith observations, 151–54, 155
Madrid Codex, 245, 387, 388, 413, 783, *785*
Main Plaza (La Venta), 351
Maize, 263, 365, 368, 405, 653
Maize God, 243
Mama Anahuarque, 88
Mamalhuaztli (Fire Drill), *246,* 247
Mamom, E-group complex at, 369
Manco Capac, 145
Manturcalla (Mantocalla), 114, 115, 133
Maps: Nazca lines as, 626; sky as, 427, 428–29, 430–33, 435, 436–37
Marriage, Amazonian cultures, 434, 435
Mars, 244, 505; Maya observation of, 334, 406, 516, 517, 531, 541–46, *590, 591,* 600, 602–3(n6), 608; Maya historical events, 556–58, 559, 560; Palenque observation of, 336, 337, 595, 598; and Venus, 537, 547
Masewi, 210
Master of Animals (Orion), Tukanoan concept of, 429, 433
Matsakya, 203, *204,* 659, 662, 666; sun watching at, 208, 650
Maya, 5–6, 73, 196, 232, 254, 306, 396, 400, 413, 602(n3), 807; agricultural almanac, 607–15; calendrical system of, 235–37, 238, 239–41, 414–23, 510–14, 519–30, 550–51, 601–2(nn2, 4); on celestial paths, *402,* 402–3, *403*; city planning of, 270–71, 751–52, 754–58, 761–68, 774, 780(n1), 782, 784–88; constellations, 247, *407,* 407–11, *408, 409, 410, 411*; directions, 401–2; eclipse observations by, 241–42, 251, 403–4, 603(n10); historical dates of, 599–600; Jupiter observations, 589, 590, 602–3(n6); katun cycles, 368–72; lunar observations, 239–41, 404–6, 516–18; Mars observations, 541–46; planet cycles, 546–49; possible zodiac, 248–50; royal ceremonies and accessions of, 19–20; rulership and Venus, 551–56; *tonalli,* 386–88; Venus observation by, 242–43, 261–64, 531–41; written records of, 504, 509–10, 549–50. *See also various groups*
Mayapan, 370, 372, 786
Medallion Series (Palenque), 589, 592, 593
Medicine Wheel. *See* Big Horn Medicine Wheel
Medicine Lodge, *22,* 35
Medieval period, 254, 778
Megaliths, 14, 739; alignments with, 4, 8, 740–41
Membilla, 133
Mercury, 244; Maya historical events, 558–59, 577; Mayan observation of, 516, 517, 518, 546
Meridian passage, Pueblo observance of, 687–88
Mesa Verde, 201, 665, 666
Mesoamerica, 2, 4, 5, 37(n27), 72, 181, 182, 201, 233, 273, 399, 491, 686, 739; astronomy in, 231–32; cosmograms, 782–88; cosmology, 234–35, 732–33; lunisolar calendar, 239–42; pre-planting ceremony, 652–53; stars in, 244–48; *tonalli,* 386–88; and U.S. Southwest, 222–23; urban planning/layout, 773–77; Venus in, 242–44; world view, 253–54, 383–85. *See also various cultures; sites*
Meteors, meteorites, 57, 406, 643–44, 726, 734
Mexica. *See* Aztecs
Mexico: calendrical system, 235–36; cosmology, 234–35
Mexico, Valley of, 735

817

INDEX

Mexico City, and Tenochtitlan, 298–99(n33), 778
Mictlan (Place of the Dead), 390
Mictlanteuctli, 389
Mictlancihuatl, 389
Middleton survey, of Hopewell earthworks, 44–45, 46, 57
Milky Way (Course of Santiago), 64, *410*, 426, 430, 450, 627, 643, 665, 744(n4); Andean interest in, 174, 396; as celestial river, 461–62, *463*; dark cloud constellations in, 458, 459–61, 468, 469, 727; Maya views of, 408, 421; Mesoamerican significance, 244, 248; solstices and, 480–81, *481*; Tukanoan concepts of, 429, 436
Milpa crops, 405
Misminay, 122, 142, 143, 154, 461, 735; astronomical lore in, 458–59; on celestial animals, 463–65, 479
Mixcoatl, 248
Mixcoatl-Camaxtli, 244
Mixtecs, cosmology, 389
Moctezuma (Moctezuma Xocoyotzin; Motecuhzoma), 73, 245, 503, 753, 758(n1)
Moieties, Cuzco, 82–83
Molina, Cristobal de, 78, 80, 94, 104, 126, 130, 132, 154, 163, 168, 172
Momostenango, calender ritualists in, 414–23
Monroy, Estrada, Cuzco Plaza Hanan-Haucaypata, *118,* 119
Monte Alban, 308
Months: Andean social organization, 87–88; Incan *ceque* system and, 92–98; Mayan calendar, 239–40; Mexica, 287–89; Pueblo ceremonial cycle, 217–21; Pueblo, 652–54
Moon, 147, 148, 197(n4), *202,* 236, 392, 435, 437(n2), 687, 711, 727, 734, 736; eclipse predictions, 519–23; Fajada Butte observations, 671, 682–85, 704; Incan observation of, 87, 95–96, 126, 129, 130, 490; Mayan observation of, 237, 404–6, 421–22, 516–17, 519–30, 550–51, 608, 613, 614(table), 615–16(n1); Mesoamerican calendar and, 239–42; and Newark earthworks, 50–56, 57–58; Palenque observations of, 313–14, 336, 328–29, 337, 341, *343*; Pueblo calendar and, 201, 203, 217–21, 650; supernova conjunction with, *636*–43; synodic cycles of, 517–18; Tukanoan society and, 428, 430; and Venus cycle, 536–41
Moon dagger, Fajada Butte, 618–19
Moon Goddess, Mesoamerican, 239, *240*
"Moonshine on Stonehenge" (Atkinson), 3–4
Moon watching, Pueblo, 219

Mopan Kekchi, 262
Morning Star, 643, 650, 647, 657; Pueblo concepts of, 658–59, 664, 665, 666; Venus as, 242–43, 244, 262, 566
Mother-child bond, Navajo, 447
Mother Earth, 650, 656, 658, 661, 665
Motolinía, Toribio de, on Templo Mayor, 287–89, 302(n86), 778–79
Mountain cult, Tenochtitlan, 255–26, 285–86
Mountains, 91; symbolic importance of, 301(n61), 776, 777, 779
Mountain Woman (Navajo), 447
Mucho Malo, Cerro, 188, *192*
Mudca, 133
Muiska Indians, 426, 436, 437, 438(n14), 489
Mundo Perdido (Tikal), 369; E-group complex at, 349, *351*, 367
Muñoz Camargo, 258
Murcielagos group (Dos Pilas), 784
Mursi, 728, 730
Muskingum River (Ohio), 41
Muyucmarca, 105

Naachtun, 359, 762
Nadir passage, 478; at Palenque, 324–26, *325, 326, 327,* 342; tropics, 486, 487, 489, 490, 491
Náhookǫs, 446, 447, 450, 453
Nahua speakers, 245, 248; calendrical system, 236–39; cosmology, 389–92; *tonalli,* 386–88; universe, *234,* 234–35
Nahui Ollin, 286
Nakbe, E-group complex at, 349, 353, 369
Nakum, 762
Nambe, 657
Nampallec, 86
Nanahe, *220,* 221
Naranjo, 336, 575, 756, *757,* 762
Natural world, cultural context of, 729–31, 733, 734, 735–36
Navaho Canyon, *636*
Navajo, 233, 396, 643; constellations, *444, 444*–47, *445,* 450–53; creation of stars, 245–46; family, 449–50; *hooghan,* 441–43, 447–48, 454(n5); moral universe, 440–41; sandpaintings, 443–44, 454(n4); social order and behavior, 439–40
Nayuchi, *220,* 221
Nazca culture, 632–33
Nazca (Nasca) lines, 6, 270, 489, 618, 621, *628–29,* 733; interpretations of, 622–27; makers of, 632–33; and water, 630–31
Newark earthworks, 14–15, 39–41, 57, 58(n2); dating of, 41–42, 59(n6); description of, *42,*

818

Index

42–43; lunar alignments of, 50–56; geometry of, 46–49, 59–60(nn20, 21); mapping of, 44–46; restoration of, 43–44, 59(n10)
New Fire ceremony, Zuni, 217, 221
Nezahualpilli, 231–32
Night People, 665
Niman ceremony, 653, 656
Nobility: Incan, 88; Mayan, 19–20, 549; and warfare, 728
Nohmul, 353
Noon: Pueblo observance of, 687–88; three-slab site markings, 703–4
Nopal foundation site, 256
North Group (Copán), 765
Northern Cross, Mesoamerica, 245
North Road, 689
Numerology, 250, 393, 529

Oa'qul basket dance, 654
Oaxaca, 514, 786
Objectivity, 738–39
Observatories, 739; Anasazi, 215–16; Big Horn Medicine Wheel as, 22, 23–36; at Chankillo, 188–96; Coricancha as, 130, 169; Cuzco area, 100–106, 110–13, 117–18, 127, 144; E-group complexes as, 306, 349, 355–62, 373; Inkawasi as, 174–75; at Machu Picchu, 128–29; at Palenque, 311–26; Pueblo structures as, 665–66; Templo Mayor as, 276–92
Observatory Circle (Newark earthworks), *40, 42*, 43, 44, *45,* 57; geometry of, 46, 47–48, *49,* 60(n21); lunar alignments using, *53,* 54, *55*
Observatory Mound (Newark earthworks), 43, 48, 55
Oceania, 396
Ocongate, 464, 465
Ocozocoautla, 349
Octagon earthworks (Newark), 14, *40, 42*, 43, 44, *45*; geometry of, 46–48, 60(n21); lunar alignments and, 51, *53,* 54, *55,* 55–56
Offerings: at Chankillo, 187, *195,* 196; Navajo, 450–51; at Palenque, 330–31, *333*
Ohio, Hopewell earthworks in, 14–15, 39–41, 43
Old Fire God (Aztec), 236, 245
Old Star, 730
Olmecs, 388
Oma (San Jeronimo), 87, 88, 91, 95
Ometeotl, 389
Omeyocan, 235
Omo River, 730–31
Omotourco (Omoto-ynacauri; Mutu), 104, 141, 142, 143
1 Ahau, 531

Oraibi, 656
Origin myths, Desana, 426–27
Orion, 247, 402, 584, 643; Desana views of, *431,* 431–32, *432,* 433; Puebloan concepts of, 650, 657, 664, 665; tropical observation of, 489, 730; Tukanoan concepts of, 429, 438(n12)
Orion's Belt, 238, 247, 405, 427, 429, 435; Desana views of, *431*–32, 433; Maya views of, 407–8; Mesoamerica, 244, 248
Ortiz Mountains, 210
Otolum Group (Palenque), 316
Outside Chief, 656
Oxford International Conferences on Archaeoastronomy, 7–8

Pacaritambo (Pacariqtambo), 140, 146
Pacbitun, 353
Paccari-tampu, 145
Pachacamac, 126, 175
Pachacuti, 126, 128, 129, 145, 151, 165–66
Pachacuti Yamqui, Santacruz, 120, 163
Pachamama, 72, 464
Pachatira (pachatierra), 463–65
Pachawawa, pachakuti, 471
Pahlowahtiwa, *220*
Paint Creek, 41
Pajarito Plateau, 208
Pakal (Pacal), 310, 588, 593, 595
Palace of the Governors (Uxmal), 776
Palenque, 5, 241, *333,* 505, 550, 601(n1), 603(n10); calendrical system at, 526, 527–28, 529, 530; gods of, 512, 587; hieroglyphs and iconography in, 332–40; hierophanies at, 308–9; Jupiter observations at, 589–593, 594–99, 600; nadir passage at, 324–26, *325, 326, 327*; planetary alignments at, *590, 591, 592*; rulers at, 309–10, 551; summer solstice at, 317–21; symbolic architecture, 306–8; Temple of the Sun at, 73, *312,* 314–24, 330–31, 340–41, 342–43, 587, 603–4(nn7, 11); Venus dates and, 576, 581; winter solstice at, 311, *313, 314*
Palenque Triad, 587, 588
Pamurti ceremony, 653
Pan-American Highway, and Nazca lines, 631
Panquetzaliztli, 246
Pantanaya, 142, 160
Pantanayoc, 113
Papre, 85
Paris Codex, *249,* 388, 413, 570
Path of Life, Desana, 433–34
Path of the Sun, Orion's belt as, 431
Patiabu, 662
Pawnee, 396

819

INDEX

Paxcaman, 353
Payao, 84
Peccaries, celestial, 569, 572
Pedernal Peak, and Tsiping Pueblo, *209*
Pekwin, 656; summer solstice ceremony, 660–61; sun-watching, 204–6, 216, 217, 396, 651, 659–60, 663
Peñasco Blanco: horizon watching at, 208, 209; pictograph at, 201–2, 216, *642,* 642–43, *648,* 659, 666
Perception, cultural context of, 726–27
Peru, 6; first plowing feasts in, 85–86; planting season in, 67, 154–55
Petén, 574; E-group structures, 196, 371; Maya city plans in, 762, 774
Petrified Forest National Park, petroglyph sites in, 210, 215–16
Petroglyphs, 6, 261, 427; light and shadow sites, 210–11, 214–16, *672–81;* moon-supernova, *636,* 636–37, *637;* sun dagger, 693–94; Zuni, 496–501
Picho (Piccho), 140–41
Picchu, Cerro, *116,* 117, 120, 122; and ceque system, 266, 267; Pachacuti's estate at, 128, 129; zenith-antizenith observations, 151, 161
Picchu Sucanca, 118
Pictographs: Chaco Canyon, 647, *648,* 796; supernova, 201–2, *636,* 639–40, 796
Piedras Negras: calendrical system, 526, 530; Venus dates at, 551, 552, 553, 580–81, 582, 583
Pilar (Machu Picchu), 66, 67
Pillars (*sucancas; mojones*), 167; in *ceque* system, 89–90, 98–99, 111–13, 266; as observatories, 100–106, 117–18, 127; zenith-antizenith observations, 148–49, 154
Pino, Pedro, 216–17
Pisac, 138–39, *139,* 166, 173, 177
Piscis Austrinus, 410
Plains Indians, 20, 21
Planetary conjunctions: Dresden Codex, 614–15; Palenque, 306, 333–34
Planets: Maya historical events and, 555–60; Mayan records of, 242–44, 306, 310, 337, 406, 514–16, 517, 518, 807. *See also by name*
Planting season, 207; Anasazi prediction of, 209, 213; Maya, 235–36, 262, 405, 612; Palenque, 325–26; in Peru, 67, 116, 143–44, 154–55, 266; Pueblo, 210, 653
Plato, 414
Plaza de Armas: in Cuzco, 118, *119,* 120, 131, *132*
Plaza de Regocijo (Cusipata), 120, 131

Plaza of the Inscriptions (Palenque), 331–32
Plaza of the Seven Temples (Tikal), 367
Plazas, 756; and E-group complexes, 366–68
Pleiades, 64, 138, 238, 643; in Aztec ritual, 246–47; Desana views of, 432, 434; Inca observations of, 129, 152, 165, 169, 170, 178, 267, 493; Inkawasi alignments with, 174–75, 177; Maya observations of, 407, 550; Mesoamerica, 244, 245, *246,* 392; Navajo and, 451, *452;* Palenque sitings, 325, 337; Puebloan concepts of, 650, 657, 664, 665; Tenochtitlan, 491–92; tropical observation of, 486, 489, 490, 491
Poco Uinic, 551
Polaris, 248; Navajo concepts of, *445,* 446, 448
Politico-religious philosophy, 255
Pollux, 427, 434
Polo de Ondegardo, Juan, 126, 145, 163, 457; on animal constellations, 465, 481–82; on *ceque* system, 78, 80, 82, 89, 90, 100, 101, 102
Polol, 762
Pomacucho, 130
Pop, 238
Popol Vuh, 343, 389
Poroypuquio, *116*
Portsmouth (Ohio), 41
Positivism, 158–59, 162
Powamu ceremony, 203, 217, 218, 652–53, 656
Prayers, Navajo, 450–51
Prayer sticks, Zuni, 660, 661
Precision, 166, 168–69, 173
Pre-planting ceremony, 652–53
Priestess of Fecundity, 661
Priests: Aztec, 238–39; Incan, 86, 88, 95; Kogi, 435; Maya calendar, 414, 415–23; Mesoamerican, 727–28; Zuni, 656–57, 661
Primeros Memoriales, 251
Procyon, 408, 427
Propaganda, raiding as, 552–53
Pueblo Bonito, 689, 794, 797; corner openings at, 213–14; horizon watching at, 208, 209; light and shadow sites, 210, *212,* 216–17, 685
Pueblo Indians, 72, 199, 201, 306, 495; calendrical systems, 202–7; deities, 664–65; ethnographic analogy, 686–87; events recorded by, 648–49; lunar calendar, 651–52; lunar observations, 217–21, 712; Mesoamerican influences, 222, 223; months, 652–54; skywatching, 650–51, 655–58, 662–63, 705; solstices, 663–64; symbols used by, 649–50, 658–59, 688–89. *See also* Hopi; Zuni
Puquincancha (Puquinque; Poquen Cancha), 104, 105, *113,* 133, 141, 142, *143,* 145–46, 156, 172

Puuc, 270, 767
Pyramid of the Sun (Teotihuacan), 261, 267, 777

Quadrants, four cosmic, 385
Quadripartite signs, *364*
Quechua, 5, 63, 64; on celestial animals, 463–65; dark cloud constellations, 458–59, 461–63; on foxes, 478–80; and llamas, 476–78; on serpents, 465–70; on tinamous, 473–76; on toads, 470–73
Quest, spiritual, 22, 434
Quetzalcoatl, 222, 236, 244, 248, 262, 389, 390, 393, 656, 666
Quiangalla (Quiancalla), 103, 104, 105, 113–114, *114*, 115, 127, 133–34, 156
Quiche Maya, 248, 400, 404, 405, 406; calendrical ritual among, 414–23; celestial paths, 402–3; directions, 401–2; on stars, 407–11. *See also* Momostenango
Quihuar, 85
Quinoacalla, 140
Quipus, 78; ceque as, 80–82; specialists in, 97–98
Quiquijana, 90
Quirigua, 524, 575, 755, 763
Quisco, *114*
Quispicancha, 132, 151
Quispicanchis, 267
Quito, zenith-equinox observations at, 146–47

Rabbit, 239
Rabinal Maya, 400
Radiality, 733
Radial structures, in E-group complexes, 362–65
Radiocarbon dates, Chankillo, 182, *183*
Raiding, raids, and Venus dates, 552–53, 567, 570
Rainbows, 129, 475, 481, 611; as serpents, 467–68, 469, 470
Rainbow Serpent, 467–68, 469, 470
Rain God, 239
Rainy seasons, 147, 425, 489, 611–12, 653, 776; celestial paths, *402,* 402–3; dark cloud constellations, 461, 462–63, 475; Maya, 262, 401, 405; and serpents, 468–70
Rattlesnakes, 688; Fajada Butte rock art, 673, 675, 676, *677*
Rauaraya, 141
Ravaypampa (Ruauypampa), 113, 142, 160
Rebirth, Aldebaran and, 434
Reciprocity, human-supernatural, 222
Red Temple (Templo Mayor), 280
Regulus, 408–9, 421, 571
Reiche, Maria, 618, 623, 625–26
Religion, as context, 73, 726

Rigel, 29, 31, 32, 248, 431
Rincon (Guatemala), 352
Río Azul, 385
Río Bec, 359
Rio Grande pueblos, 208, 223; ceremonial cycle, 662–65; skywatching, 650, 651; star symbolism, 657–58
Ritual cycles, 370, 490; architecture, 371–72; E-group complexes, 349, 362–68; Kogi, 435–36; Mars in, 542–43; Mayan, 413, 414, 513–14, 519–23, 528–29(table), 547(table); Mesoamerican, 308, 385. *See also* 260-day calendar
Roads, Nazca lines as, 631
Roberts, Frank H. H., Jr., Zuni rock art interpretations, 496–97, 666
Rock art, 222, 644; Anasazi, 210–17; Chaco Canyon, 201–2, 209, 210–11, 213, 214–15; crescents recorded in, 638–39; Fajada Butte, 669–70, 681–82; moon-supernova, *636,* 636–37, *637,* 639–43, 647–48; Puebloan, 649–50; Zuni, 496–501, 662, 666
Rock crystals, 435, 436; Tukanoan world as, 427–29
Roman Catholic Church, 414
Rome, 734
Roof comb, Temple of the Sun, 320, *321, 343*
Rosario-Naranjo, 352
Rulership, 261, 372; divine, 308–9; Mayan, 336, 550, 556–57, 582–83; and Venus associations, 551–56, 567–58, 575–76. *See also by name*

Sacbes, as cosmograms, 784
Sacred directions, Pueblo, 203
Sacrifices, 336; ball games, 582–83; E-group complexes, 367–68; Incan, 80, 133, 141–42, 147, 151; of llamas, 477–78; Mesoamerican, 390–91
Sacsahuaman, 105
Sagittarius, 409, *410. See* Thieves' Dagger/Cross
Sahagún, Bernardino de, 232, 235, 245, 286; on constellations, *246,* 247–48
Saiph, 431
Sandia Mountains, 210
Sandpaintings, Navajo, 440, 443–44, 446–47, 452–53, 453(n1), 454(n4)
Sangre de Cristo Mountains, 210
San Ildefonso, horizon watching, 210, 219
San Isidro, 349
San Isidro II, 352
San José, 359, 367
San Juan Pueblo, 200
San Lazaro, Plaza de, 258

INDEX

San Miguel Mountain: and Intihuatana Stone, 65, 66, 67–68
Santa Ana Pueblo, 210, 655
Santa Cruz Festival (Peru), 68
Santiago (Venus), 406
Santo Domingo, Church of (Cuzco), 111, 129. *See also* Coricancha
Sanu (San Sebastian), 87, 88
Sapawe, 657–58
Sarmiento de Gamboa, 100, 105, 117, 166
Saturn, 604(n13); Maya observation of, 310, 334, 406, 516, 517, 518, *590, 591*, 602–3(n6); Mayan historical events and, 550, 556–58, 560; Palenque, 310, 337, 598
Sausero, 105
Sayatasha, 217
Sayil, 767
Scioto River earthworks, 41
Scorpio, 247, 409–10, 468, 489
Scotland, stone circles, 740–41
Sculptured Stone 1 (Bonampak), 581
Seasons, 401, 486; celestial paths, *402,* 402–3, *403*; Incan observations of, 110, 116; Pueblo ceremonial cycle and, 204, 206–7, 217–21; tropical, 425–26, 489
Seibal: E-group complex at, 367, 369; Venus and, 567, 573, 574, 582
Sekaquaptewa, Abbott, 218
Selden Roll, 389
Serpents, 461, *500*; bicephalic, 263–64; in Milky Way, 430, 459; Quechua concepts of, 465–70
Seven Kids (Siete Cabritos; Pleiades), 407
Sexuality, 435; Tukanoan Indians on, 427, 428, 430, 432
Shabik'eschee, 658
Shalako, 203, 652, 661–62; setting dates for, 217–18
Shamans, 414, 434, 467; Tukanoan, 427, 429, 430, 432, 433, 437(n2)
Shield-Jaguar, 599
Shrines: in Mayan calendrical ritual, 416–23; Pueblo sun, 648, 658, 659, 662–63, 685
Sicllabamba, 116, 117, 133
Sierra Nevada de Santa Maria (Colombia), 426, 435
Siete Cabreras (Big Dipper), 410
Sirius, 30, 37(n33), 248, 434; Big Horn Medicine Wheel alignment with, 29, 31, 32, 34(table); Palenque sitings, 307, 325
65-day cycles, 418–19
Sky: Amazonian concepts of, 427, 428–29, 430–33, 435, 436–37; in cultural context, 725–26, 775; symbolic use of, 735–36

Sky gods, Maya, 510
Skywatching, Pueblo, 655–58
Smoking Squirrel, 336
Snake ceremonies, 688; Hopi, 218, 653–54
Snakes, in Fajada Butte rock art, 673, 675, 676, *677,* 703. *See also* Serpents
Snake society (Hopi), 653–54
Social organization, 268; Cuzco, 82–83, 85–86, 87–88, 110, 123(n1); Navajo, 439–40; Tukanoan, 427–29, 432–33
Sofaer, Anna, 693, 694, 701–2
Solar cycles, 349; eclipse predictions, 609–10, 613; planting season, 154–55
Solstices, 181, 238, 343, 435, 475, 670; Chankillo, 188–92; E-group complexes, 358–59, 360; Inca calendar, 94, 97; Inca feasts related to, 80, 85; Inca observations of, 104, 106–7, 111, 113–15, *117,* 134–39; Milky Way and, *480,* 480–81, *481*; Puebloan focus on, 650, 657, 663–64; tropics, 489–90. *See* December solstice; June solstice; Summer solstice; Winter solstice
Songo, 464, 467
South America, 2, 4, 5, 6, 15
Southern Cross, 238, 244, 462, 470, 475, 730; Maya, 402, 410
Spanish conquest, 734
Spider Woman, 451
Spirals: Fajada Butte, 618, *672, 673, 674, 676, 679, 680,* 682–83, 684, 703–4, 709; Pueblo use of, 688–89; Zuni rock art, *498,* 500–501
Star Chief, 657
Stargazing, 233, 239, 663; *tonalli,* 392–93
Starlore, cultural context of, 726–27
Starry Skirt (Citlallicue), 248
Stars, 37(n27), 307, 427, 464, 727, 744(n4); Big Horn Medicine Wheel alignments with, 28–33; and Desana society, 431–34; Incan observation of, 126, 129, 152; Maya concepts of, 402, *407*–11; in Mesoamerican universe, 234, 236, 244–48; Navajo concepts of, 440–41, 446, 451; Pueblo concepts of, 657–58, 659, 665. *See also* Constellations; *by name*
Stelae, 784; E-group complexes and, 367, 369–70; katun cycles and, 370–71; Seattle, 525, 530; Venus and historical events on, 573, 575, 576, 577, 578–80
Stephen, Alexander M., 219, 650, 654, 655
Stone circles, in Scotland, 740–41
Stonehenge, 1, 3, 56, 729
Stonehenge Decoded (Hawkins), 3–4, 13–14
String figures, Navajo, 451–52
Sucanca, 132, 133, 149, 150, 163, 164, 178

Sucancas: astronomical observations, 98–99, 100–103, 106; in *ceque* system, 89–90, 98–99
Succanca, 132, 141–42
Sucsupanaca, 85
Sulcanca, 132, 151
Sulluullucu, 467
Summer solstice, 210, 216, 306, 431, 436; and Aldebaran, 29–32, 33; at Big Horn Medicine Wheel, 23–28, *28,* 34(table); Chaco Canyon, 209, 214; Fajada Butte, *673, 674,* 675, 684–85, 693–94, 701, 703, 708; measuring, 63–64; Palenque, 308, 317–21, 336, 603–4(n11); Pueblo, 653, 654, 663, 687; Zuni, 660–61
Sun, 147, 181, 341, 404, 425, 437(n2), 447, 744(n1), 775; Incan observations of, 126, 129, 130, 131; and Incan calendar, 87, 96–97, 141; Kogi observations of, 434–35; Maya observations of, 307, *329,* 329–30, 355–62; in Mesoamerican cosmos, 234–35, 236–37; Pueblo calendar, 201, 218, 664–65; Puebloan focus on, 650–51, 655–58, 662–63, 687–88; Pueblo symbols for, 647, 649; and Tukanoan social organization, 428–29
Sun Chiefs, 656, 663
Sun clan, Zuni, 662
Sun dagger (Fajada Butte), 6, 618, 684, 693–94, *697, 701*
Sun Dance Lodge, *22,* 35
Sung Dynasty, record of Crab Nebula supernova, 635–36, 638, 644
Sun Man (Zuni), 662
Sun priests, 203, 208, 306; at Zuni, 204–6, 396
Sun Temple, at Inkawasi, 175
Sunturhuasi, 121, 267
Sun watching: Anasazi, 223–24; Chaco Canyon, 648, 666; Hopi and Zuni, 203–4, 216–17, 659–60; Puebloan, 210, 215, 218, 650, 651, 655–58, 663, 705, 797
Sun Youth, 651, 653
Supernova (1054), 223, 635, 637, 728; in Arizona rock art, 618, *636,* 638; in California rock art, 639–42; in Chaco Canyon rock art, 201–2, *642,* 642–43, *648,* 796; in Zuni rock art, *498,* 498–99, *499,* 643
Surucucú, 467
Susurpuquio, 104
Symbol Bridge site, *641,* 641–42
Symbolic notation, 733–34

Tablet of the Sun (Palenque): iconography of, 314, *341*–42; text of, 333–40
Tachymenis peruviana, 465–66, 470

Tahuantinsuyu, 159
Talking God, 450, 454(n6)
Tamacaz, 248
Tambo Colorado, 270
Tanoan pueblos, calendrical cycle, 662–63
Taos, 712
Taurus, 247, 407, 584, 635–36
Tawantinsuyu, 126, 129
Tayasal, 367
Tehuicocone, Cerro, 286
Telapón, Cerro, 259, 260, 285
Temple 11 (Copán), 575
Temple VI (Tikal), 764
Temple XIV (Palenque), 307, 593
Temple of Huitzilopochtli (Templo Mayor), 257, 276, *277, 290,* 291–92, 303(n102)
Temple of Quetzalcoatl, 287
Temple of the Count (Palenque), 307, 325
Temple of the Cross (Palenque), 308, 309, 310, 311, 330, 334, 337, 588; equinox observations, *314,* 314–16, *316*; moon alignments, 328–29; summer solstice observation, *318,* 319
Temple of the Foliated Cross (Palenque), 307, 310–11, 314, 328, 334, 337, 588, 589
Temple of the Inscriptions (Palenque), 308, *309,* 310, 576, 589, 592
Temple of the Seven Dolls (Dzibilchaltún), 291, 306, *307,* 372
Temple of the Sun (Palenque), 73, 307, 308, *312,* 328, 341, 590, 603–4(nn7, 11); astronomical alignments, 331–32, 335(table), 342–43; construction of, 310–11; equinox at, 314–17; hieroglyphs and iconography in, 332–40; inscriptions at, 587, 588, 595–96; nadir passage at, 324–26, *325, 326, 327*; observations from, 329–30, 336; offerings, 330–31, *333*; roof comb, 320, *321, 343*; summer solstice, 317–21; winter solstice, 311, *313,* 313–14, *314*; zenith passage at, 321–24, *322, 323, 324*
Temple of Tlaloc (Templo Mayor), 276, *277*; observations from, 290–91
Temples, 735; Kogi, 434–35
Temples of the Moon: Machu Picchu, 68; Muiska, 436; Palenque, 314
Temples of the Sun: Copacabana, 85; Inkawasi, 175; Muiska, 436. *See also* Coricancha
Temple 22 (Copán), 269(table); Venus observations, 261–64
Templo Mayor, 73, 255, 269(table), 299(n39), 504, 737, 753; architecture of, 274–76; horizon markers and, 276–81; as observational calendar, 281–85, 286–87, 302(n86), 778–79; orientation of, 256–58, 259–61, 268, 276–81,

823

287–89, 297(n20), 298(nn27, 32), 359, 503, 758(n1), 772–73; possible observations from, 289–92; subsidence and skewing of, 293–95, 296(n7), 299(n47), 300(n48)

Tenochtitlan, 5, 73, 264, 267, 268, 269(table), 270, 280, 298–99(n33), 734, 789(n3); calendrical system of, 259–61; founding myth of, 256, 257, 258, 285–86; layout and orientation of, 752–54, 777–79; mountain cult in, 255–56; observational calendar in, 281–85, 286–87; Pleiades, 491–92; Templo Mayor at, 274–76, 359, 758(n1)

Teotihuacan, 236, 239, 267, 286, 270, 353, 361, 367, 739, 753, 755, 773, 777, 784; equinox, 260–61

Tepetzinco, Cerro (Peñón de los Baños), 256, 260, 285

Territorial conquest, Venus and, 552–53, 575

Teutonic armies, 734

Tewa, 207, 218, 649, 656, 687

Texcoco, Lake, 260, 296(n7)

Textiles, Inca calendrical system, 79, 95, 97, 98, 106

Tezcatlipocas (Smoking Mirrors), 236, 248, 389

Tezozomoc, 245, 247, 256, 260

Thieves' Dagger; Thieves' Cross, Maya views of, *409,* 409–10

Thirteen Birds of the Day, 238

Thirteen Towers (Chankillo), 182, *184*; description of, 185–88; as observatories, 188–96

Thom, Alexander, 4, 14

Thom paradigm, 4, 14, 29

Three Kings, Three Marys (Orion's Belt), 407–8

Three-slab site (Fajada Butte), 210, 214–15, 218, 219, 671–72, *672, 673, 674,* 693, *695–99, 701*; assessment of, 703–5; and Chimney Rock, 717–18; formation of, 705–8; function of, 708–9; lunar markings at, 682–83, 694, 700, 702–3, 711–12; markings and assemblage of, 684–85

Thunder Mountain (To'wayal'lanne), 659, 661

Ticcicocha, *114*

Tihuanaco, 270

Tikal, 336, 577, 600, 784; layout/orientation, 756, 764–65; katun cycles at, 369, 371; Mundo Perdido, 349, *351,* 353, 367, 369; twin-pyramid complex at, 371, 373, 755, 762–763; Uniform System at, 524, 525, 526, 530

Time, 73, 79, 368, 387, 399, 531; architectural representations of, 362–65; cyclical concepts of, 426, 443–44, 454(n3); Maya concepts of, 401, 413–14, 416, 510

Time units, Andean structure of, 87

Tinamous, 459; Quechua and, *473,* 473–76, *476*

Tipón, 151, 165

Titicaca, Lake, 85, 196

Tiwanaku, 489

Tlacaxipehualiztli, feast of, 287, 288, 289, 291, 302(n86), 758(n1), 778

Tlacocomolco, 256

Tlalancaleca, 351–52

Tlaloc, 255, 257, 259, 275, 389

Tlaloc, Cerro, 261; as calendrical marker, 281, 282, 283, 284–85; ruin on, 258–59; Templo Mayor and, 255–56, 400(n48)

Tlalteuctli, 389

Tlamacas, Cerro: as calendrical marker, 281, 282, 283, 284–85; Templo Mayor and, 299(n42), 300(n48)

Tlamacehualiztli, 390–91

Tlazolteotl, *240*

Toads, Quechua concepts of, 470–73, 474

Tocoripuquio, 90

Tonalli, 386–88, 391; and star-gazing, 392–93

Tonalpohualli. See 260-day calendar

Tonatiuh, 236, *237*

Topa Inca, 177

Torquemada, Juan de, 256, 258

Torreón (Machu Picchu), 15, 68, 159, 166, 169, 173, 177; orientation and description of, 62–64, 139–40; solstice observations at, 134–39; zenith-antizenith observations at, 151–52, 155, 156

Tortuguero, warfare, 552, 553

Tower of the Palace, at Palenque, 308

Towers, Anasazi, 665–66

Town Chiefs (Pueblo), 656, 657, 664

Toxiuhmolpilia (Binding of the Years), 246

Tozoztontli, 249

Trance states, drug-induced, 434

Tree-ring dates: Big Horn Medicine Wheel, 20–21, 36(n3); Chimney Rock Pueblo, 715–16

Tres Padres, Pico, 286

Tres Reyes, and planting, 405

Tres Zapotes, 551

Tri-wall structures, 712

Tropics, 396; astronomy in 485–86, 727; *Axis Mundi,* 490–91; cosmologies, 487–89; ethnoastronomy in, 5, 425–26; solstices, 489–90

Tsankawi, 208, 210

Tsiping Pueblo, 208, *209*

T3 Eridani, 427

Tukanoan Indians, 426, 436; energy in universe, 429–30; cosmology of, 427–29. *See also* Desana

Turtle Dance, 200

Turtles, as celestial objects, 569, 570, 572, 584